RICE UNIVERSITY

SEMICENTENNIAL PUBLICATIONS

The Natural
Radiation Environment

EDITORS

John A. S. Adams
Wayne M. Lowder

CONTRIBUTORS

J. A. S. ADAMS
D. B. APPLEBY
H. BARKHAU
ROBERT G. BATES
HAROLD L. BECK
M. L. BORKAR
R. R. BOULENGER
S. S. BRAR
P. R. J. BURCH
R. L. BUSDIECKER
JAMES L. CARTER
MARGARITA CELMA
WILLIAM Y. CHEN
ROBIN D. CHERRY
WILLIAM J. CONDON
T. L. CULLEN
PHILLIP N. DEAN
E. R. DI FERRANTE
R. LOWRY DOBSON
R. T. DREW
J. C. DUGGLEBY
M. EISENBUD
TOM L. ERB
ROBLEY D. EVANS
R. S. FOOTE
GLENN E. FRYER
S. GANAPATHY
M. GEYH
SEYMOUR GOLD
DAVID GOTTFRIED
E. GOURSKI
WERNER N. GRUNE
ROBERT B. GUILLOU
P. F. GUSTAFSON
JOHN E. HAND
KNUT S. HEIER

WALTER HERBST
C. R. HILL
RICHARD B. HOLTZMAN
J. E. HOY
JOHN B. HURSH
R. G. HUTCHINS
CHARLES W. JOHNSTONE
B. KAHN
P. R. KAMATH
CHARLES H. KAPLAN
JACOB KASTNER
M. KAWANO
G. KEGEL
A. A. KHAN
SERGE A. KORFF
HOBART W. KRANER
L. F. LANDON
A. LIBOFF
L. B. LOCKHART, JR.
S. LORCH
ARVIN LOVAAS
WAYNE M. LOWDER
HENRY F. LUCAS, JR.
M. J. MC HUGH
AZIZEH MAHDAVI
L. D. MARINELLI
ERNEST MARSDEN
B. W. MAXWELL
HAROLD A. MAY
W. V. MAYNEORD
JOSEPH H. MEHAFFEY
YASUO MIYAKE
R. M. MOXHAM
S. NAKATANI
SHERMAN K. NEUSCHEL
B. OLDROYD

B. G. OLTMAN
R. V. OSBORNE
JOHN K. OSMOND
R. L. PATTERSON, JR.
H. PETROW
GEORGE PHAIR
T. N. V. PILLAI
CECIL PINKERTON
JAMES A. PITKIN
A. E. PURVIS
PAUL C. RAGLAND
EDUARDO RAMOS
S. R. RAO
KEITH RICHARDSON
JOHN J. W. ROGERS
JOHN E. ROSE
F. X. ROSER
LARRY D. SAMUELS
GERALD L. SCHROEDER
RALPH E. SCHROHENLOHER
ASCHER SEGALL
MORRIS H. SHAMOS
B. SHLEIEN
FRANCIS R. SHONKA
J. SIDEROWITZ
ALAN R. SMITH
F. W. SPIERS
YUKIO SUGIMURA
ALLAN B. TANNER
HIROYUKI TSUBOTA
RICHARD I. WELLER
A. WENSEL
MARVIN H. WILKENING
HAROLD A. WOLLENBERG
D. S. WOODHEAD

The
Natural
Radiation
Environment

PUBLISHED FOR

WILLIAM MARSH RICE UNIVERSITY

BY

THE UNIVERSITY OF CHICAGO PRESS

Library of Congress Catalog Card Number: 64-12256

The University of Chicago Press, Chicago & London
The University of Toronto Press, Toronto 5, Canada

To Our Children

whose future welfare is becoming more and more dependent
upon our understanding stewardship
of the environment

Contents

Introduction

THE AVERAGE LIFE SPAN of little better than thirty years for all human beings who have ever lived is but one measure of the very limited success that man has attained in his struggle to adapt to a predominantly hostile environment. A whole complex of biological, chemical, and physical environmental factors presses in constantly, while others, such as earthquakes and epidemics, are intermittent. The interplay between man and each environmental factor and that between the various factors themselves are fascinating in their great intricacy and delicate balance.

The past few generations of men have achieved a small but rapidly increasing measure of understanding and control of a few of the many environmental factors that significantly affect human viability. By means of this understanding and control, a longer life span has been realized for many. But these same developments have resulted in a rapidly increasing population, and some fear that this expansion itself may become an unbalanced environmental factor, one that can result only in a wrenching readjustment among the various factors until a new dynamic equilibrium has been achieved. This example epitomizes the difficulty and implicit dangers in the manipulation of a limited number of environmental factors to solve one problem without consideration for the balance of other factors, whose alteration might produce new and unforeseen effects.

The manipulation of most environmental factors takes place on such a scale and with such variety and complexity that there is often very little possibility for individual understanding, much less choice. Even when their benefits are most obvious, these manipulations sometimes produce nagging doubts with regard to the possible existence of harmful and unforeseen side effects. It is particularly where such side effects are long-term or cumulative in nature that great difficulty arises in identifying the causative agent. The present controversy over the linkage of lung cancer with cigarette smoking is an example of the difficulty of such problems in all their scientific, economic, political, and propagandistic aspects.

Thus, elementary prudence would appear to dictate that our understanding of the environment must increase at least as rapidly as our ability to manipulate the significant factors that characterize it. The following pages record the proceedings of an international symposium devoted to the subject of ionizing radiation as a factor in man's environment, whether arising from the radioactivity of certain naturally occurring elements in the earth or impinging on the earth from outer space. The symposium is largely addressed to the external natural radiation environment, but by the nature of the subject it is also closely involved with radiation from radioisotopes within the human body and from man-made sources in the environment. The fundamental question of the effects, if any, of these ionizing radiations on man is hardly touched upon directly, although it provides a broad background and often a strong motivation for many of the studies described. The symposium deals with the prior problem of developing basic data on the physical properties and distribution of natural radiation and its sources in the environment, data that are necessary for any attempt at understanding the nature of the long-term interaction between radiation and man.

The desirability of such a symposium on the natural radiation environment occurred to the two editors of these proceedings during their private discussions of some New England radioactivity anomalies that they both had been studying from widely different viewpoints and with quite different instrumentation. It is clear that the increase in both the magnitude and variety of man-made sources of radiation makes it most desirable to obtain a more detailed knowledge of the natural base line of human exposure in order to evaluate realistically the significance of additions to these levels of exposure. Closely related is the increased interest of biological and medical scientists in the more general problem of the effects on man of continuous low-level exposure to ionizing radiation. This last interest has been greatly stimulated by research into the genetic effects of radiation, which has put emphasis on the significance of the cumulative dose to the germ plasm in determining the nature and extent of these effects. In the absence of evidence for a dose or dose-rate threshold for the production of certain genetic mutations, it is not unreasonable to expect that the more penetrating components of the natural radiation environment would be responsible for some fraction of the observed genetic mutation rate in humans, continuously exposed to the low dose rates that appear to prevail in most inhabited parts of the world. Much of the effort now being expended in surveys of natural environmental radiation, particularly in regions of unusually high levels, is directed toward evaluating ionizing radiation as a mutagenic agent by eventually comparing the observable effects in large populations of long-term exposure to different natural radiation levels. The relatively long life spans of humans combined with moral considerations make direct experimentation impossible, and it is the naturally high radiation areas that have been populated for some generations that provide the best opportunity for evaluating the mutagenic effect of increasing the ionizing radiation factor in the environment.

It was obvious to the editors in their private discussions that much of the information relating to the natural radiation environment derives from studies not directed per se toward human environment questions. Both the natural and man-made radioactivities are of great utility as "tracers" in studying a great variety of geologic, meteorologic, oceanographic, and soil-science problems. The devising of ever more sensitive and quantitative techniques for radiation detection and measurement has led to great successes in these sciences. Advanced radiometric equipment has also been developed for the search for economic deposits of uranium, thorium, and potassium, as well as for petroleum and general mineral exploration. Thus, significant data regarding the natural radiation environment are scattered through the literature of a number of disciplines, making it difficult for any one interested investigator to extract and evaluate the information. Under these conditions there exists a clear need for improved communications across geographic and disciplinary boundaries in order to provide some coherence to the vast amount of data potentially available and to bring into clearer focus the "state of the art." With this in mind, the present symposium was organized to provide an open forum for workers from various disciplines and countries to meet and exchange information on techniques, experimental rationale, and recent results. The papers presented necessarily include a wide range of subjects, and some overlap is inevitable. However, they do represent a substantial and representative sampling of what is known and being learned about the natural radiation environment.

The First International Symposium on the Natural Radiation Environment was held at William Marsh Rice University, Houston, Texas, on April 10 through 13, 1963. Approximately two hundred scientists from various parts of the world attended, of whom nearly one-half were authors or co-authors of the papers presented. The first day of the symposium was given over to a field "intercalibration" or "intercomparison" of the various instruments used by different groups to study natural radioactivity. Measurements were made at three locations in and about Houston, and the results are summarized in Appendix 3. Three days of formal non-concurrent sessions followed, during which time fifty-five papers were presented orally and six by title. These proceedings contain the texts of all sixty-one papers, in many cases in revised and amplified form, together with brief summaries of the discussion that the papers provoked. A special effort was made to make the bibliographies as complete and accurate as possible.

The order of papers has been revised somewhat from that of the symposium to provide more coherent groupings, as indicated in the Table of Contents. The reader should note that these groups are not mutually exclusive, since a large number of the papers deal with several aspects of the natural radiation environment. This is particularly true in Part II, where many papers are concerned with a detailed description of instrumentation and methodology as well as a discussion of field measurements and their significance. In these, as in other cases, the assignment of a particular paper to a particular category contains a necessary element of arbitrariness.

The decision was made at an early date to unify the terminology from paper to paper in certain respects. Of particular note is the use of formal isotopic names for all radioisotopes rather than their historical names, although the latter are sometimes given also. Thus, the unmodified terms, "uranium," "thorium," "radium," and "radon," refer to the element and not to any particular isotope of that element. While this usage has several disadvantages, particularly when the uranium and thorium decay chains are being considered in detail, the editors believed that the advantages of uniformity in notation and its easy recognition by scientists from many disciplines made necessary such a step. For the convenience of the reader, the alternate names of the various isotopes in the uranium and thorium series are given in Appendix 4 along with a summary of their significant properties. A list of the abbreviations used throughout the text is given in Appendix 5.

Financial support for the symposium was generously provided by the Division of Special Projects of the U.S. Atomic Energy Commission, by the Division of Radiological Health of the U.S. Public Health Service, and by the William Marsh Rice University. Many participants were supported by funds from their respective institutions. The editors and organizers of the symposium wish to express their thanks and deep appreciation to the above-mentioned organizations, to the students and staff of the Department of Geology of William Marsh Rice University for operating the sessions and editorial services, and, most particularly, to the participants in the symposium for their sustained enthusiasm and co-operation during the sessions and the final preparation of the following contributions.

J. A. S. Adams
W. M. Lowder

GEORGE PHAIR AND DAVID GOTTFRIED

1. The Colorado Front Range, Colorado, U.S.A., as a Uranium and Thorium Province

KNOWLEDGE OF THE DISTRIBUTION of uranium, thorium, and potassium in specific regions of the outer crust is fundamental to an understanding of the earth's heat budget. So long as definitive quantitative information is lacking, the driving forces behind such basic earth processes as mountain-making, magmatism, and metamorphism must remain matters of qualitative, or at best semiquantitative, conjecture.

This paper is, in part, an outgrowth of a conference at Oak Ridge National Laboratory that was aimed at evaluating potential reserves of thorium in the United States. It was concluded that, although other avenues should be explored, the available data indicate that, of all known domestic sources of thorium, only the silicic igneous rocks are present in the tonnages needed to justify any future large-scale industrial conversion to thorium-breeder reactors in power production. The Front Range is of special interest in this connection, including as it does the largest and most thorium-rich areas of true granite in the United States. An important purpose of this report is to present geographic patterns of uranium and thorium distribution as a possible starting point for detailed selected areas studied by others interested in potential "minable" thorium or in variations in background radiation. The petrogenetic implications of those distribution patterns, along with the detailed geology, petrology, and geochemistry, are being studied.

Phair has spent thirteen years in the study of the distribution of uranium and thorium in the igneous rocks (Larsen, Phair, Gottfried, and Smith, 1956), ore deposits (Phair and Shemamoto, 1952; Sims, Phair, and Moench, 1958), and weathering products (Phair and Le-

GEORGE PHAIR and DAVID GOTTFRIED are geologists with the U.S. Geological Survey, Washington, D.C.

Publication authorized by the director, U.S. Geological Survey.

7

vine, 1953) of the Colorado Front Range, with the compilation of useful regional averages for uranium and thorium as one long-range objective. Under test is the extent to which the total radioactivity of rocks now exposed on the surface may be used as an indicator of local "heat pockets" and thus help to explain the pattern of recurrent magmatism in the region.

The procedure has been to reconnoiter, unit by unit, the major rock types that form the building blocks of this mountain mass, an area of some 7,000 square miles, working backward in geological time from the most recent rock units (age 50–60 million years) to the oldest (minimum age, 1,700 million years) as determined by mineral dating using radioactivity techniques. Each of the igneous units studied has been a petrological and chemical research problem in itself. Where possible, the changes in uranium and thorium abundances in space and in time have been examined in the light of what has been learned about the isotopic evolution of the common lead in the genetically related ore deposits (Phair and Mela, 1956).

The point has now been reached at which we can begin to integrate a substantial pool (392 analyzed rocks) of uranium and thorium data and, after taking into account relative rock volumes (outcrop areas), arrive at reasonable regional averages. The averages so obtained indicate a regional enrichment on the order of twofold for each of the radioactive constituents, uranium, thorium, and potassium, when compared with the most recent averages for the continental crust as a whole. Because the Colorado Front Range is the only region in the United States wherein such bedrock enrichment in radioactive constituents can be demonstrated on a crustal scale of many thousands of square miles, it is appropriate to present the results at a symposium devoted to the natural radiation environment. The individual averages for uranium and thorium may change somewhat when the final analytical results are evaluated, but the fact of enrichment seems to be firmly established by the present data.

All the regional enrichment in potassium, much of the regional enrichment in uranium, and part, but only part, of the regional enrichment in thorium stem from the fact that the Front Range includes the largest masses of true granite and closely related siliceous quartz monzonite in the United States. The granites of the Front Range, duly weighted for outcrop area, average 29 p.p.m. thorium—twice that of normal granite—and individual masses studied have outcrop areas of 80–160 square miles in which the thorium content averages four- to fivefold that of normal granite. The thorium enrichment apparently extended backward in time to what are believed to be some of the oldest rocks now exposed in the region, biotite schist layers in the

Idaho Springs formation. Here, the picture becomes clouded by a lack of comparable thorium data on metamorphic rocks elsewhere. Such scanty data as are available (4 samples; Adams *et al.*, 1959) suggest that common schists contain about half as much thorium (av. 7.5 p.p.m.) as those in the Front Range (18.8 p.p.m.).

From these data it appears that rocks of all ages throughout the Front Range are predominantly thorium rich, and the region satisfies the criteria for a *geochemical* thorium province. Within this geochemical province no less than 3 *petrographic* thorium provinces are present, that is to say, 3 separate magma series are unusually enriched in thorium throughout. Finally, that part of the Front Range transected by the silver-gold-lead-zinc mining belt may be considered a *metallogenic* uranium province, meaning by that a region characterized by an unusual concentration of workable uranium deposits. A primary objective of our studies has been to explore the geological and geochemical links that connect these uranium deposits in time and space to uranium-rich and thorium-rich Laramide magmas in the region (Phair, 1952). Although an appreciable tonnage of high-grade uranium has been produced from the veins, the amount becomes negligible in comparison to the vast quantities disseminated as a trace constituent throughout the igneous rocks. Such "minable" uranium can therefore be neglected in any computation of regional or crustal averages in the Front Range.

Approximately half of the 7,000-square-mile area of the Front Range is underlain by silicic igneous rocks made up of distinct types, each relatively uniform over broad areas and having a more or less characteristic range of uranium and thorium contents. In the more enriched series (Silver Plume and Laramide), the average thorium and uranium contents of the separate masses increase toward the central (Fig. 1) area transected by the mineral belt. The average enrichment for all silicic igneous rocks in the region weighted for abundance is about twofold for uranium and threefold for thorium over the average rock of the continental crust. These enriched silicic igneous rocks have been emplaced in 8 large masses that are called batholiths. The separate batholiths underlie areas ranging from 80 to nearly 1,500 square miles. The batholiths are commonly discordant to the structural grain of the older country rocks and have steeply dipping walls; they show no inclination to diminish in size with depth throughout the range of topographic relief over which contact relationships can be observed (about 7,000 feet as a maximum). Extrapolation of these igneous rocks to depths of many miles seems justified. It cannot, of course, be proved that the uranium and thorium enrichment observed in the surface rocks will persist to these very considerable depths. The fact remains, however, that a combination of uplift and erosion has exposed levels deep within the original batholiths, and, throughout these levels of present observation, enrichment is a proved fact. Because there is no reason to postulate abrupt diminishment of uranium and thorium at greater depths, we can begin to think in terms not only of uranium and thorium averages at the surface but of averages for the outer crust in this region.

In addition to an approximate doubling of the separate radiation components attributable to uranium, thorium, and potassium in the bedrock in this region, the contribution of cosmic rays to the total background radiation is also approximately doubled as a consequence of altitude. The entire area of the Front Range is above 5,000 feet in altitude, and 95 per cent of the area is above 7,000 feet. As measured

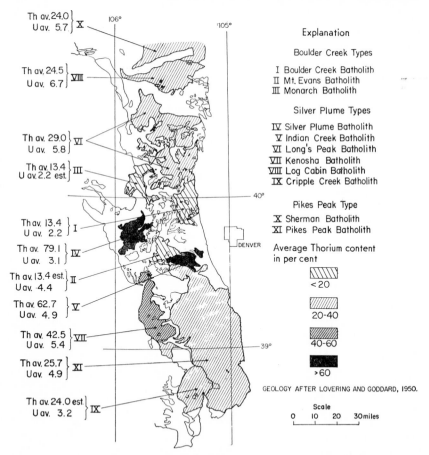

FIG. 1.—The Front Range batholiths, showing average uranium and thorium contents. Shading indicates range of average thorium contents.

by a Geiger-Müller counter (Johnson, 1954), the cosmic-ray intensity increases slowly between 0 and 5,000 feet, is twice that at sea level at about 8,000 feet, and over three times at 14,000 feet. Roughly one-half the area of the Front Range lies between 10,000 and 14,000 feet. The average cosmic-ray activity at the surface throughout the Front Range would be, therefore, approximately twice that encountered near sea level. A station at the 14,000-foot summit of Mount Evans monitors cosmic-ray activity and is a center for co-operative research.

This paper is chiefly concerned with the moderate but widespread regional anomalies that set the Front Range apart as a unit. Unexpectedly high radiation anomalies, purely local in extent, are common in and around the uranium-bearing mines. Under oxidizing conditions the sulfide-rich ores decompose, and sulfuric acid is generated; uranium, once it is oxidized to the hexavalent state, goes rapidly into solution in the acid, leaving the less mobile radium behind. Hot spots on the mine dumps have been noted (Phair and Levine, 1953) to contain 150 times as much radium as would be required to satisfy secular equilibrium with the amount of uranium present. Such radium "hot spots" testify to the quantity of uranium that has entered meteoric waters in and around the mines during the relatively short time since the workings were first opened, a maximum of fifty years. Spring waters having abnormally high radon contents have been reported from Soda Creek near Idaho Springs, Colorado, and from the Jamestown area. No doubt others are present.

AVERAGE URANIUM AND THORIUM CONTENTS OF THE CONTINENTAL CRUST

Fleischer and Chao (1960) have outlined the difficulties inherent in the calculation of the abundances of elements in the lithosphere. By limiting the calculations to the granitic crust, some of these obstacles can be avoided. The granitic crust includes the outer earth layer, characterized mainly by lighter, more alkali-rich rocks, that thins or vanishes entirely over the oceanic basins and thickens over the continents, becoming especially thick in deeply folded mountain regions. True granite is a relatively minor constituent of the granitic crust. Much more abundant are related rocks containing less alkalies and silica than does granite, together with more calcium and magnesium oxides. The terms "uranium enrichment" and "thorium enrichment" used in the preceding section obviously have no meaning unless the average uranium and thorium contents for the continental crust can be estimated within sufficiently narrow limits to give a basis for comparison.

One may follow either of two procedures in estimating averages for the granitic crust:

1. Determine the arithmetic averages of all continental rocks, regardless of bulk composition, that have been analyzed for uranium and thorium. This approach, geochemical in essence, assumes that, so long as the statistical sample is sufficiently large, the number of analyses available for a given rock type will be roughly proportional to its abundance in nature. Unfortunately, rocks that are scarce are likely to be of unusual bulk compositions, therefore of special interest, and as a result tend to be overanalyzed. For similar reasons the available uranium and thorium data are overloaded with analytical results obtained on rocks of exceptionally high radioactivity. From the point of view of computation of simple arithmetic averages, the recent concentrated sampling of the Conway granite and the Front Range intrusives have probably set the state of the art back a good ten years. The question is always which analyses to include and which to reject.

2. Determine the average uranium and thorium contents of each of the major rock types that make up a significant fraction of the continental crust, and, after weighting each rock type according to its areal abundance, compute over-all averages for uranium and thorium. This method, involving as it does the weighting of specific rock types, may be called the petrological approach. The sedimentary rocks are generally insignificant in volume as measured on a crustal scale and can be neglected. The metamorphic rocks are more of a problem but as a rule can be expected to decrease in volume with depth. Moreover, for the purposes of calculation, it is possible to select extensive regions in which both sedimentary and metamorphic rocks, as exposed on the surface, add up to relatively small volumes and can be ignored. In the Front Range, where the volume of metamorphic rocks is large, such rocks have been considered in arriving at regional averages.

This process of elimination leaves the igneous rocks as the main components of the earth's crust. Clarke and Washington (1924) estimated the abundance of igneous rocks in the lithosphere at 95 per cent, and, as Fleischer and Chao (1960) have pointed out, this figure has been used by all succeeding workers in the field with little or no modification. Fortunately, more is known about the distribution of uranium and thorium in the igneous rocks than in all other rocks combined. Data are now available for most of the large intrusive masses in the United States that relate changes in uranium and thorium to changes in bulk chemical composition during magmatic crystallization. (See, e.g., Larsen III and Gottfried, 1960.)

The dependence of uranium and thorium contents upon bulk composition in the rocks of the southern California batholith is illustrated graphically in Figures 2 and 3. These rocks together make up a magma series of the predominant type in the continental crust, the so-called

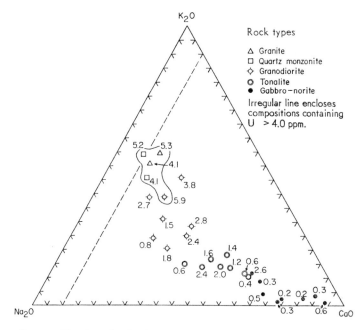

Fig. 2.—Uranium distribution in p.p.m., southern California batholith

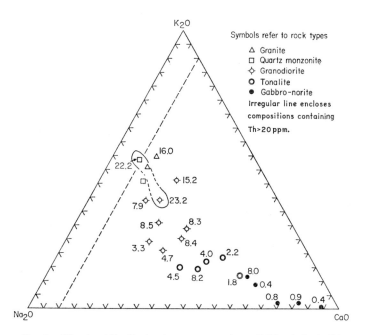

Fig. 3.—Thorium distribution in p.p.m., southern California batholith

calc-alkalic series, in which the early magma enriched in calcium and magnesium oxides give way to later liquid compositions successively depleted in these constituents and correspondingly enriched in sodium, potassium, and silicon oxides. Both uranium and thorium increase along with sodium, potassium, and silicon oxides; the increase in thorium with changing bulk composition is more systematic than is the corresponding increase in uranium. As plotted against various rock-forming constituents alone or in combination, using rectilinear coordinates (the typical variation diagrams of the petrologists), uranium contents are found to fall within a belt marked by increasing scatter toward the uranium-rich end rather than along a smooth curve (Larsen III and Gottfried, 1960). Petrographically, the change in the direction of increasing uranium and thorium content is from dark rocks, the gabbros, through successive intermediate types, the quartz diorites and the granodiorites, to light-colored end members, the granites and closely related silicic quartz monzonites.

The southern California batholith was one of a number of batholiths in the western United States formed during a widespread period of granitic intrusion of about 100 million years' age, the products of which underlie nearly 100,000 square miles in the states of California, Washington, Idaho, and Montana. Moore (1959) calculated the areal abundances of the separate rock types belonging to this intrusive sequence for a total mapped area of 17,374 square miles. Using Moore's data on rock abundances and the data of Larsen III and Gottfried (1960) on the uranium and thorium contents of the constituent rocks, weighted average uranium and thorium contents for the combined intrusions were calculated (Tables 1A, 1B). The results, 2.5 p.p.m. uranium and 11.4 p.p.m. thorium, are in substantial agreement with arithmetic averages on granitic rocks in general obtained by Heier and Rogers (1963), who used empirically determined U/K and Th/K ratios, and Heier and Adams' (in press) figures for the average potassium content of the continental crust. Average thorium in the western batholiths is somewhat higher, and average uranium is somewhat lower, than the corresponding figures calculated by Heier and Rogers (1963). These differences, though small for all present purposes, are opposite in sign and, as a result, introduce a larger difference into the Th/U ratios calculated by the separate methods.

The averages determined in these two ways are those appropriate to siliceous granodiorite, the rock that is thought to represent best the average composition of batholiths in folded mountain areas the world over.

It should be noted that the uranium-rich and thorium-rich end product, true granite, is a relatively scarce rock in batholithic intrusions outside the Front Range. Moore estimates the total area of true granite in the western batholiths to be 7 per cent of the total. In the older geochemical literature it is not uncommon to find the figures representing the uranium and thorium contents of true granite used to represent averages for the continental crust as a whole. These figures are almost certainly high by a factor of 1.5–2.

TABLE 1A

DISTRIBUTION OF URANIUM AND THORIUM IN THE
BATHOLITHS OF THE WESTERN UNITED STATES*

	No. Samples	Uranium Range (p.p.m.)	Uranium Av. (p.p.m.)	Thorium Range (p.p.m.)	Thorium Av. (p.p.m.)
Gabbro, diorite.............	9	0.17–0.82	0.5	0.2– 1.2	0.8
Tonalite...................	17	0.6 –3.3	1.7	1.4–18.5	5.5
Granodiorite...............	38	1.1 –5.2	2.5	3.1–24.0	12.1
Quartz monzonite, granite....	12	2.2 –7.2	4.0	7.6–36.0	18.5

* Combined data for southern California, Sierra Nevada, and Idaho batholiths (Larsen III and Gottfried, 1960).

TABLE 1B

THORIUM AND URANIUM AVERAGES IN WESTERN BATHOLITHS
WEIGHTED FOR ROCK ABUNDANCES*

	Outcrop Area (sq. mi.)	Outcrop Area (per cent)	Per Cent Area×Av. Uranium	Per Cent Area×Av. Thorium
Gabbro, diorite.............	2,186	10.7	0.05	0.86
Tonalite...................	5,846	33.7	0.57	1.85
Granodiorite...............	3,301	19.1	0.48	2.31
Quartz monzonite, granite....	6,016	34.6	1.38	6.40
Av. U (p.p.m.) weighted......................			2.5	
Av. Th (p.p.m.) weighted.............................				11.4
Av Th/Av U=4.6				
Continental crust:				
Heier and Rogers (1963)		U=2.8 p.p.m. Th=10 p.p.m. Th/U=3.6		

* As determined by Moore (1959).

THE COLORADO FRONT RANGE

The location of the Front Range relative to other granitic masses that form the cores of the separate ranges in the southern and middle Rocky Mountains is given in Figure 4. Results of uranium and thorium analyses of 25 rocks collected from Precambrian crystalline areas outside the Front Range are shown in Figure 4 and Table 2. Though the areal sampling was inadequate to provide accurate uranium and thorium averages for the large regions involved, the results strongly suggest that the general levels of uranium and thorium in the very old rocks that make up the granitic cores of the three main ranges in the middle Rocky Mountains are 25–50 per cent lower than the Front Range averages. Only a small part of the difference in average uranium content, about 10 per cent as a maximum, can be attributed to differences in the amount of original uranium lost via radioactive decay resulting from the greater age (2,000–2,500 million years; Bassett and Giletti, 1963) of the rocks in the ranges to the north and west. The batholithic rocks of the Front Range apparently were

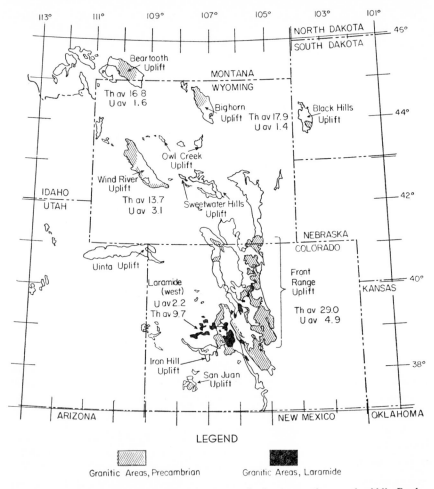

LEGEND

Granitic Areas, Precambrian   Granitic Areas, Laramide

Fig. 4.—Outcrop areas of granitic intrusive rocks in the southern and middle Rocky Mountains. Uranium and thorium averages are for igneous rocks only. All measurements are in p.p.m.

intruded during an interval 1,000–1,500 million years ago, according to mineral ages obtained by a number of investigators, including the present writers. Because the half-life of thorium-232 is much longer than that of uranium-238 and uranium-235, the correction for decay of parent thorium during the intervening 1-billion-year average time interval before the Front Range rocks were intruded is very small, and for all practical purposes the thorium averages for the separate regions can be directly compared.

TABLE 2

URANIUM AND THORIUM IN PRECAMBRIAN GRANITIC
ROCKS OF THE MIDDLE ROCKY MOUNTAINS

	URANIUM CONTENT (p.p.m.)			THORIUM CONTENT (p.p.m.)		
	No. Samples	Range	Av-erage	No. Samples	Range	Av-erage
Beartooth, Montana......	3	1 – 2.3	1.6	3	11.5–21	16.8
Bighorn, Wyoming........	10	0.3– 2.8	1.4	10	1.2–29	17.9
Wind River, Wyoming....	12	0.3–14.0	3.1	11	1.7–35	13.7
Total...............	25			24		

TABLE 3

URANIUM AND THORIUM IN LARAMIDE STOCKS WEST OF THE FRONT RANGE

	URANIUM CONTENT (p.p.m.)			THORIUM CONTENT (p.p.m.)		
	No. Samples	Range	Av-erage	No. Samples	Range	Av-erage
7 Large stocks...........	25	1.1–6.9	2.2	25	4.7–31.0	9.7

Figure 4 and Table 3 show that not only the very old granitic rocks but also the younger (50–60 million years) Laramide stocks decrease in average uranium and thorium contents away from the Front Range. Throughout the central Colorado, Laramide stocks were intruded along a northeast-southwest trending belt 200 miles long; the northeasterly segment of that belt transects the north-south axis of elongation of the Front Range. The data for 25 samples from the Laramide stocks southwest of the Front Range show that these rocks contain on the average about one-third as much uranium and thorium as do the rocks from stocks of similar age in the Front Range proper. Moreover, they contain only about half as much uranium and thorium as do the areally weighted averages for all rocks, regardless of age, in the Front Range.

Phair has obtained voluminous uranium and thorium data on the older granitic country rocks, younger intrusives, and younger volcanic rocks of the Wet Mountains, the offset southerly prolongation of the Front Range. All analyzed rocks came from a single small area, the Wet Mountains thorium district; consequently, for fear of introducing an areal bias, these results have not been included here. Some 20 samples of Precambrian granitic rocks from the San Juan, Black Canyon, and Powderhorn areas in western Colorado have been analyzed for uranium, but not, as yet, for thorium; data on these rocks have been omitted also.

The general geology and ore deposits of the Front Range proper have been summarized by Lovering and Goddard (1950). Lovering and Goddard's maps of the Front Range as a whole, on a scale of 8 miles to the inch, and of the included section of the mineral belt, on a scale of 1 mile to the inch, have been used by the present writers as a basis for calculation of the areal abundances of the major rock types. Mrs. Boos and co-workers (Boos and Boos, 1934, 1957; Boos, 1935; Boos and Aberdeen, 1940) have described the petrology, mineralogy, and structure of the Front Range granitic rocks as a whole and the Silver Plume types in particular. Hutchinson (1960) has described the structure and petrology of the northern end of the Pikes Peak batholith. Lovering and Goddard (1938) presented data on the chemical variation and sequence of intrusion of the Laramide rocks.

Figure 1 gives the locations of the 11 granitic batholiths in the Front Range and their average uranium and thorium contents. A general increase in thorium content of the separate batholiths is apparent inward from the north and south toward the central region transected by the mineral belt and by the coextensive belt of Laramide intrusives. The Laramide stocks are outlined in Figure 1 and are shown unshaded. Within the Laramide belt thorium enrichment in both stocks and dikes reaches a maximum in the area lying between the two Precambrian batholiths showing maximal thorium, the Silver Plume and Indian Creek masses.

Figure 5 gives Na_2O-K_2O-CaO relationships for 164 chemically analyzed igneous rocks from the Front Range together with 30 analyzed Laramide rocks from the region to the west. All but 8 of these rock analyses have been obtained by the writers. Comparison with the more normal calc-alkalic trend (Fig. 2) shows that all rock types in the Front Range, regardless of age, average distinctly higher in potassium oxide than do their counterparts of similar calcium oxide content in the southern California batholith.

In Figure 6 the pattern of thorium enrichment for all Front Range rocks for which precise thorium analyses are available have been indicated by approximate contours on the Na_2O-K_2O-CaO diagram. When compared to the similar plot for the rocks of the southern California batholith (Fig. 3), this diagram brings out two important characteristics of the thorium distribution in the Front Range: (1) the wide range of bulk compositions over which thorium enrichment

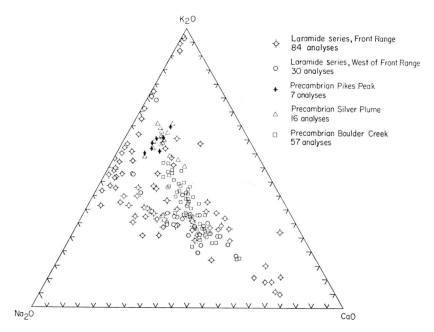

K_2O

◇ Laramide series, Front Range
84 analyses

○ Laramide series, West of Front Range
30 analyses

+ Precambrian Pikes Peak
7 analyses

△ Precambrian Silver Plume
16 analyses

□ Precambrian Boulder Creek
57 analyses

Na_2O

CaO

FIG. 5.—$Na_2O \cdot K_2O \cdot CaO$ analyses of intrusive rocks of central Colorado

occurs and (2) the exceptionally high ($>$ 100 p.p.m.) thorium levels attained by many of the calcium oxide–poor Laramide rocks. Nearly all rocks in the Front Range having more than 50 p.p.m. thorium contain more potassium oxide than sodium oxide, but it does not necessarily follow that the higher the potassium oxide content, the higher the thorium content. In fact, some of the rocks containing the most thorium, calcium oxide–poor dikes in the Central City district, contain almost as much sodium oxide as potassium oxide.

The separate rock types recognized as major units in the Front Range for the purposes of this report are those differentiated on the 8-mile-to-the-inch map compiled by Lovering and Goddard (1950).

The procedure to be followed in presenting the data will be to work backward in geological time from the youngest to the oldest rocks; this approach has the advantage of leading from the simpler to the successively more complex. As a rule, the older a rock, the more complex its geological history, as evident in changes of texture and mineralogy. Listed in order of increasing age: the Laramide rocks are essentially undeformed throughout; the coarse Pikes Peak type is

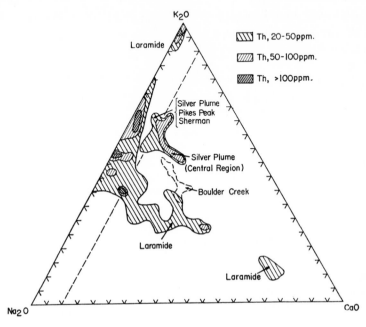

Fig. 6.—Thorium-rich compositions, Front Range intrusive rocks, in p.p.m.

largely massive but is deformed, foliated, and crushed locally about its borders; the Silver Plume type retains primary flow structures over large areas but is cut by belts of crushing; the Boulder Creek type has been foliated throughout by later earth movements, partly recrystallized, and locally strongly crushed; the Idaho Springs formation has been completely metamorphosed and recrystallized. From detailed studies of the relatively old Boulder Creek batholith, we conclude that later crushing and hydrothermal metamorphism tend to reduce, rather than to increase, the original uranium content of a given igneous rock. From this point of view the uranium averages obtained on very old, strongly deformed igneous rocks should be considered to be minimal indicators of "magmatic" uranium.

With the exception of the Silver Plume batholith, the large intrusions lying along the western margin of the Front Range have been studied less than have those to the east. Both uranium and thorium data are lacking for the Mount Evans batholith of Boulder Creek type and for the Cripple Creek batholith of Silver Plume type. The minimal uranium and/or thorium averages for batholiths of similar rock type have been used as estimates of the average for those masses for which data are lacking. These inadequately analyzed igneous intrusions together make up about 6 per cent of the area of the Front Range; hence, the uncertainty with regard to their precise uranium and thorium contents introduces only a very small uncertainty into the computation of the over-all averages for the Front Range.

A greater source of uncertainty stems from the inadequate sampling of the Idaho Springs formation that underlies 40 per cent of the area of the Front Range. The averages as calculated for the very large area rest upon results of 14 uranium analyses and 16 thorium analyses. Reduction of the averages so obtained by 50 per cent under the highly improbable assumption that the averages for this heterogeneous formation as a whole are best approximated by the included uranium-poor and thorium-poor dark layers reduces the regional averages by less than 25 per cent. Therefore, the fact of regional uranium and thorium enrichment seems to be established beyond reasonable doubt.

The uranium and thorium data for each rock type, beginning with the youngest, will be presented in a pair of tables. The first summarizes information on the separate intrusions comprising the particular rock type; the second weights the separate intrusions for areal abundance and sets forth the contribution of each toward the over-all uranium and thorium averages for that rock type. A summary table (Table 10) weights the individual rock types for areal abundance and considers the contribution of each toward the regional uranium and thorium averages for the Front Range as a whole.

THE LARAMIDE INTRUSIVES

Laramide stocks underlie 81 square miles in the Front Range, or 1.2 per cent of the total area. These intrusives were emplaced 50–65 million years ago according to (1) a potassium-argon age on biotite from the Eldora stock obtained by Hart (in Davis *et al.*, 1961) and (2) isotopic uranium-lead ages on pitchblende from veins genetically related to uranium-rich intrusives (Phair, 1952) in the Central City district (Nier *et al.*, 1941; Kulp *et al.*, 1953).

Data on the Laramide intrusives are given in Tables 4A and 4B. Among the most thorium-rich stocks are low-silica types characterized by a lack of quartz. A nepheline-bearing monzonite stock near Empire is notable for a high content of uranium (5.6–18 p.p.m.) and of thorium (28–60 p.p.m.) coupled with a high content of calcium oxide. All other Laramide stocks of similar calcium oxide content have higher silica and one-fourth to one-half as much uranium and thorium.

Among the latest intrusives are calcium oxide–poor dikes that rank among the most radioactive of granitic rocks analyzed to date (Table 4A). Individual dikes averaging as high as 60 p.p.m. uranium and 375 p.p.m. thorium have been found together, with individual samples

TABLE 4A

LARAMIDE INTRUSIVES, FRONT RANGE, URANIUM AND THORIUM ANALYSES

	URANIUM CONTENT (p.p.m.)			THORIUM CONTENT (p.p.m.)		
	No. Samples	Range	Average	No. Samples	Range	Average
CaO-poor dikes.........	15	7.0–139	43.6	15	43.0–400	142
Stocks................	27	0.7– 18	7.6	23	6.8–112	30

TABLE 4B

URANIUM AND THORIUM AVERAGES WEIGHTED FOR AREAL ABUNDANCE

	Area (sq. mi.)	Area (per cent)	Av. Uranium	Per Cent Area×Av. Uranium	Av. Thorium	Per Cent Area×Av. Thorium
CaO-poor dikes......	Neg.	Neg.	....	....	....	....
Stocks..............	81	100	....	7.6	....	30
Av. Th/Av. U = 3.9						

containing as much as 100 p.p.m. uranium and 400 p.p.m. thorium. These highly radioactive dikes are as much as $5\frac{1}{4}$ miles long but are so narrow as to represent negligible total volumes.

PIKES PEAK TYPE

Quartz monzonite and granite of Pikes Peak type formed two batholiths, one the very large Pikes Peak mass, the other the much smaller Sherman intrusion. The Pikes Peak batholith, the largest single intrusion in the Front Range, underlies an area of 1,473 square miles, or 23.6 per cent of the area of the entire region. The smaller Sherman batholith crops out over an area of 166 square miles in Colorado, representing 2.4 per cent of the total area of the Front Range. In addition, it underlies a comparable area in Wyoming not considered in this report. Samples from the Pikes Peak batholith have given isotopic mineral ages, by a variety of methods (Tilton *et al.*, 1957), ranging from 1,000 to 1,100 million years.

Uranium and thorium data on rocks of Pikes Peak type are given in Tables 5A and 5B. Some of the highest thorium results (32–46

TABLE 5A

PIKES PEAK–TYPE INTRUSIVES, URANIUM AND THORIUM ANALYSES

BATHOLITH	URANIUM CONTENT (p.p.m.)			THORIUM CONTENT (p.p.m.)		
	No. Samples	Range	Average	No. Samples	Range	Average
Sherman............	8	3.4– 8.6	5.7	8	16–41	24.0
Pikes Peak..........	30	2.1–30.0	4.9	30	13–46	25.7
Smaller intrusions.....			4.9 est.			25.7 est.
Total...........	38			38		

TABLE 5B

PIKES PEAK–TYPE INTRUSIVES, URANIUM AND THORIUM
AVERAGES WEIGHTED FOR AREAL ABUNDANCE

Batholith	Area (sq. mi.)	Area (per cent)	Av. Uranium	Per Cent Area×Av. Uranium	Av. Thorium	Per Cent Area×Av. Thorium
Sherman..........	166 (Colo.)	10.1	5.7	0.58	24.0	2.42
Pikes Peak........	1,473.3	89.1	4.9	4.36	25.7	22.90
Smaller intrusions...	13.3	0.8	4.9 est.	0.03	25.7 est.	0.21
Total.........	1,652.6	100				

Av. U (p.p.m.) weighted............................5.0 '
Av. Th (p.p.m.) weighted...25.5
Av. Th/Av. U = 5.1

p.p.m.) were obtained on a fine-grained phase. This relatively thorium-rich phase is readily recognized in the field; it crops out over considerable areas, particularly in the region northeast of Lake George and on the summit of Pikes Peak.

SILVER PLUME TYPE

Granodiorite, quartz monzonite, and granite of Silver Plume type formed 6 batholiths ranging in outcrop area from 80 to 643 square miles; such rocks total 1,649 square miles of outcrop over all, or 23.6

per cent of the area of the Front Range as a whole. Isotopic mineral ages indicate an emplacement about 1,250 million years ago for the Silver Plume batholith (Aldrich *et al.*, 1958) and for a smaller intrusion in the Central City district (Phair, unpublished data) correlated with the Silver Plume by geologists mapping that area (P. K. Sims, oral communication).

The thorium and uranium data are shown in Tables 6A and 6B. As a group, the Silver Plume batholiths have the highest thorium contents of all batholiths in the Front Range. The maximum thorium content (av. 79 p.p.m.) is shown by the Silver Plume batholith that underlies 162 square miles, or 2.3 per cent of the entire Front Range.

TABLE 6A

SILVER PLUME–TYPE INTRUSIVES, URANIUM AND THORIUM ANALYSES

BATHOLITH	URANIUM CONTENT (p.p.m.)			THORIUM CONTENT (p.p.m.)		
	No. Samples	Range	Average	No. Samples	Range	Average
Small intrusives....	4	2.0– 7.5	4.3	4	72– 90	80.7
Indian Creek.......	9	3.1– 6.6	4.9	8	38–103	62.7
Silver Plume.......	7	1 0– 7.6	3.1	7	61– 98	79.1
Cripple Creek......	3	2.2– 4.7	3.2	....	....	24.0est.
Kenosha...........	3	2.7– 8.0	5.2	2	16– 69	42.5
Log Cabin.........	11	2.4–11.2	6.7	11	19– 47	24.5
Long's Peak.......	15	0.6–14.3	5.8	13	4– 60	29.0
Total.........	52			45		

TABLE 6B

SILVER PLUME–TYPE INTRUSIVES, WEIGHTED URANIUM AND THORIUM AVERAGES

Batholith	Area (sq. mi.)	Area (per cent)	Av. Uranium (p.p.m.)	Per Cent Area × Av. Uranium	Av. Thorium (p.p.m.)	Per Cent Area × Av. Thorium
Small intrusives.....	10.8	0.7	4.3	0.03	80.7	0.56
Indian Creek........	80.1	4.9	4.9	0.24	62.7	3.07
Silver Plume........	162.2	9.8	3.1	0.30	79.1	7.75
Cripple Creek......	223.2	13.5	3.2	0.43	24.0est.	3.24
Kenosha............	235.6	14.3	5.2	0.77	42.5	6.07
Log Cabin.........	293.6	17.8	6.7	1.19	24.5	4.36
Long's Peak........	643.4	39.0	5.8	2.26	29.0	11.31
Total..........	1,648.9	100				

Av. U (p.p.m.) weighted..............................5.2
Av. Th (p.p.m.) weighted...36.4
Av. Th/Av. U = 7

The largest batholith of Silver Plume type is the Long's Peak mass. Uranium ranges from 1 to 7.6 p.p.m. and thorium from 4 to 60 p.p.m. All the low uranium and thorium results were obtained on a fine-grained phase called the Mount Olympus granite by Boos and Boos (1934). The typical coarser-grained Silver Plume granite from this intrusion ranged in uranium content between 3.2 and 14.3 p.p.m., with an average of 7.9 p.p.m. for 9 samples; thorium for the same samples ranged from 15.9 to 60 p.p.m. and averaged 36 p.p.m.

In addition to the 15 large samples collected from surface outcrops of the Long's Peak batholith, some 42 samples of granite from the same mass, collected underground at depths as much as 4,000 feet below the present surface, were analyzed for uranium. These granites

TABLE 7

URANIUM CONTENTS OF 79 ROCKS COLLECTED
UNDERGROUND IN THE ADAMS TUNNEL*

Rock Type	No. Samples	Av. Uranium Content (p.p.m.)
Granites................	42	5.2
Quartz diorite gneiss.....	9	3.8
Injected schist...........	18	2.4
Injected gneiss..........	8	1.6
Dolerite dike...........	1	0.4
Pegmatite..............	1	4.0

* All samples supplied by Professor Francis Birch, Harvard University.

were part of a suite of 79 samples taken at regular intervals over the 13-mile length of the Adams tunnel and obtained through the courtesy of Professor Francis Birch. The granites collected at these considerable depths did not differ materially in average uranium content (5.2 p.p.m.) from the samples collected at the surface (5.8 p.p.m.) (Table 7).

Insofar as any effect it may have on the regional uranium and thorium averages, the presence in all Silver Plume–type batholiths of considerable barren country rock disposed in masses too small to appear on the 8-mile-to-the-inch geological map is more or less counterbalanced on a regional scale by the presence of unmapped thorium-rich dikes and sills of Silver Plume type in parts of the Boulder Creek batholith and the Idaho Springs formation. Individual dikes cutting the Boulder Creek batholith contain 80–100 p.p.m. thorium, but, though locally abundant, they are insignificant in volume as compared with the total area of the Front Range.

BOULDER CREEK TYPE

Typical Boulder Creek–type intrusives range widely in composition from quartz diorite to siliceous quartz monzonite and should be considered a gradational series rather than a single rock type. Igneous rocks of Boulder Creek type were emplaced in 3 small batholiths ranging in outcrop area from 85 to 127 square miles, and aggregating 405 square miles, or 5.8 per cent of the total area of the Front Range. A suite of rocks from the least deformed part of the Boulder Creek batholith collected by the present writers gave zircon ages by the lead-α method suggesting an original emplacement 1,400–1,500 million years ago (Stern, Phair, Gottfried, unpublished data). Strongly crushed Boulder Creek rocks in the vicinity of younger intrusives gave zircon ages younger than 1,400 million years. Boulder Creek

TABLE 8A

BOULDER CREEK–TYPE INTRUSIVES, URANIUM AND THORIUM ANALYSES

BATHOLITH	URANIUM CONTENT (p.p.m.)			THORIUM CONTENT (p.p.m.)		
	No. Samples	Range	Average	No. Samples	Range	Average
Boulder Creek.......	151	0.5–7.3	2.2	52	1–54	13.4
Monarch..........	....	....	2.2 est.	....	....	13.4 est.
Mount Evans.......	14	1.6–7.0	4.4	....	....	13.4 est.
Smaller intrusions....	....	....	2.2 est.	....	....	13.4 est.
Total..........	165					

TABLE 8B

BOULDER CREEK–TYPE INTRUSIVES, URANIUM AND THORIUM
AVERAGES WEIGHTED FOR AREAL ABUNDANCE

Batholith	Area (sq. mi.)	Area (per cent)	Av. Uranium	Per Cent Area×Av. Uranium	Av. Thorium	Per Cent Area×Av. Thorium
Boulder Creek.......	127.2	31.4	2.2	0.69	13.4	4.2
Monarch..........	104.6	25.8	2.2 est.	0.57	13.4 est.	....
Mount Evans.......	84.8	21.0	4.4	0.92	13.4 est.	....
Smaller intrusions....	88.2	21.8	2.2 est.	0.47	13.4 est.	....
Total..........	404.8	100				

Av. U (p.p.m.) weighted...........................2.6
Av. Th (p.p.m.)...13.4
Av. Th/Av. U=5.1

samples from the contaminated dark borders of the batholith containing numerous inclusions of older country rock gave ages older than 1,500 million years.

The thorium and uranium data are shown in Tables 8A and 8B. The map of the Boulder Creek batholith (Fig. 7), showing the distribution of thorium based on 52 analyzed samples, is of special interest. It demonstrates that an igneous complex of a "normal" low-thorium type may yet include sizable areas (about 16 sq. mi.) containing more than a minimum of 30 p.p.m. thorium. This map also brings out the spatial association of relatively high thorium with the more siliceous rocks of the complex. By far the greater part of the thorium "high" is confined to a part of the batholith containing more than 66 per cent silica.

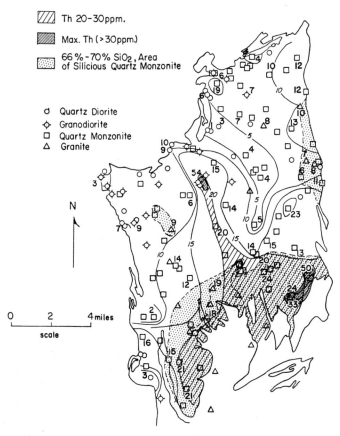

FIG. 7.—Distribution of Th and SiO₂ in the Boulder Creek batholith, Colorado

The Older Country Rocks

THE IDAHO SPRINGS FORMATION

Where sampled, the predominant layers in the Idaho Springs formation consisted of biotite-quartz-feldspar schist with or without garnet and/or sillimanite. Elsewhere the Idaho Springs formation has been injected by granite pegmatites, and mixed rocks have resulted. The relative proportion of biotite schist, gneiss, amphibolite, and injected material varies widely from place to place. Locally, dikes and sills of radioactive Silver Plume granite are abundant within the Idaho Springs section. The Idaho Springs formation has been mapped over an area of 2,801 square miles, comprising 40 per cent of the area of the Front Range. A single lead-α age on zircon from an amphibolitic

TABLE 9

URANIUM AND THORIUM DATA ON TWO CONSTITUENT
ROCKS OF THE IDAHO SPRINGS FORMATION

ROCK	URANIUM (p.p.m.)			THORIUM (p.p.m.)		
	No. Samples	Range	Average	No. Samples	Range	Average
Biotite schist.............	9	2.3–6.6	4.7	11	12.7–26.9	18.8
Biotite, hornblende gneiss, and amphibolite.........	4	1.5–9	4.7	4	5.3–22.5	10.6

layer suggests a minimum age of 1,700 million years (Stern, Phair, Gottfried, unpublished data), but this figure must be considered tentative until confirmed by additional data.

Uranium and thorium data on the Idaho Springs formation as a whole are summarized in Table 10. Data on two of the most important rock types that make up this variable formation are presented in Table 9.

Rocks that are as widespread and at the same time as heterogeneous as these present special problems. Much of the field season of 1963 will be devoted to filling some of the more obvious gaps in the present sampling pattern. Further data on the injected Idaho Springs formation in the vicinity of the Long's Peak batholith will be forthcoming when thorium analyses are completed on the 79 samples from the Adams tunnel previously analyzed for uranium. Similar data will be obtained on the Idaho Springs formation bordering a part of the Silver Plume batholith when uranium and thorium analyses now in progress on powdered splits of 123 drill-core samples from Robert's tunnel are finished. These samples were selected from a much larger suite collected at regular intervals along this 23-mile tunnel by Professor E. E. Wahlstrom, geologist in charge, during its construction. The final analytical results on the 202 samples from the two tunnels will provide a measure of uranium and thorium variation at depth along a composite section through the Front Range totaling 36 miles.

ORTHOGNEISS

Recent field studies have shown that the orthogneiss as indicated on Lovering and Goddard's map (1950) consists of two quite different rock types, which were emplaced at different times and which

TABLE 10

URANIUM AND THORIUM AVERAGES FOR THE FRONT RANGE AS A WHOLE

	AREA AS MAPPED		Av. URANIUM (p.p.m.)	PER CENT AREA ×Av. URANIUM	Av. THORIUM (p.p.m.)	PER CENT AREA ×Av. THORIUM
	Sq. Mi.	Per Cent of Total Area				
Country rocks:						
Idaho Springs........	2,801.2	40.0	4.6	1.8_4	16.3	6.5_2
Undifferentiated......	97.9	1.4	....	....	....	....
Orthogneiss..........	314.7	4.5	1.8	0.0_8	9.0	0.40
Igneous rocks:						
Boulder Creek type...	404.8	5.8	2.7	0.1_6	13.4	0.7_8
Silver Plume type.....	1,648.9	23.6	5.2	1.2_3	36.4	8.5_9
Pikes Peak type.......	1,652.6	23.6	5.0	1.1_8	25.5	6.0_2
Laramide stocks......	81	1.2	7.6	0.0_9	30.0	0.3_6
Av. U (p.p.m.) weighted by area.....................4.5_8						
Av. Th (p.p.m.) weighted by area.....................................22.6_7						
Av. Th/Av. U=4.9						

therefore should be subdivided. The older of the included rock types is the "orthogneiss" as used in this paper. Varying widely in biotite content and commonly banded, it has been called quartz monzonite gneiss by Sims and co-workers in the Central City district (Sims, written communication). A single zircon age determination by the lead-α method suggests a minimum age of emplacement or recrystallization of 1,700 million years. If this figure is close to a true age, such orthogneiss is older than the time of emplacement of the Boulder Creek batholith. Orthogneiss, presumably mainly of this older type, underlies 314 square miles, or 4.5 per cent of the total area of the Front Range. Sparse data on the older orthogneiss are given in Table 10. Contents of both thorium and uranium tend to be low, in fact among the lowest encountered in all rocks in the Front Range.

By contrast, the so-called "orthogneiss" mapped in the vicinity of the Boulder Creek batholith is a relatively uniform, clean, fine-grained granite gneiss that is clearly younger than the Boulder Creek into which it intrudes. This fine-grained granite gneiss, unlike the older ortho-gneiss, contains on the average relatively high thorium (21–62 p.p.m.). It follows that the thorium analyses may provide useful clues to differentiating one from the other when the distribution is not obvious on other grounds. For lack of proved volume, this thorium-rich, fine-grained granite has not been considered in calculating the regional uranium and thorium averages.

Regional Uranium and Thorium Averages

The calculation of the areally weighted uranium and thorium averages for the 7,000-square-mile area of the Front Range is detailed in Table 10. The averages so obtained, 4.6 p.p.m. uranium and 22.7 p.p.m. thorium, represent an enrichment over the averages for the continental crust (2.8 p.p.m. uranium and 10 p.p.m. thorium) of 1.6 and 2.3, respectively. The regional average Th/U ratio is 4.9. In arriving at the figures used in these calculations, conservative estimates were made of the average uranium and/or thorium contents of those few igneous masses for which data were lacking. In each instance the estimates used were the minimal averages for all other intrusive masses of similar rock type.

For those in allied sciences who may be interested in the search for areas of high background radiation, Figure 8 may be of some general interest. The results apply specifically to the Front Range but also in a rough way suggest the types of relationships that may be expected in enriched terrains elsewhere. The uranium and thorium curves plotted as a function of total area were obtained by adding together all areas having uranium and thorium contents equal to or greater than the indicated cutoff, starting with the most enriched areas and proceeding toward the less enriched. Figure 8 shows, for example, that if one is interested in an area having a minimum enrichment in thorium of twofold (20 p.p.m.) over the normal granitic crust, 88 per cent of the area of the Front Range is available for study. If, on the other hand, one seeks an area in which the average thorium enrichment is sixfold or greater (60 p.p.m.), less than 5 per cent of the total area passes the test. This is one more illustration of the rule of diminishing returns as it applies to thorium distribution in nature.

Twice previously the effect of the operation of this rule has been noted: (1) Granite, the normal uranium-rich and thorium-rich end member of common intrusive series, is, for physicochemical reasons, generally much less abundant than are the related igneous rocks. (2)

Given a group of radioactive granites, all of the same age, those having the highest thorium content tend to be distributed in masses of correspondingly small volume.

At the distinctly lower levels of enrichment typical of the uranium distribution in the region, the operation of the rule of diminishing returns is less clean cut. Possibly a part of the difference may lie in the greater leachability of uranium with respect to thorium in natural solutions. Small intrusive masses tend to be more easily deformed than do large masses, thereby providing channels for the entrance of solutions. It follows that, in the same region, a large batholith should, on

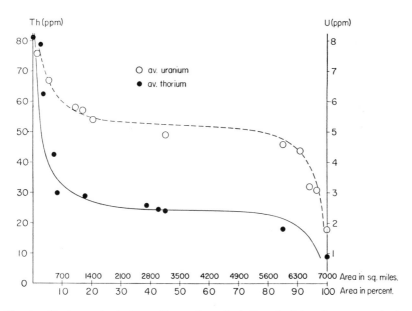

Fig. 8.—Total areas in the Front Range above the indicated minima for average thorium and average uranium contents.

this basis, retain proportionately more of its original uranium content than should a small batholith. The detailed studies of the small Boulder Creek batholith show that, for the same bulk composition, strongly crushed and lineated types are depleted in uranium compared to the more massive types.

Sampling and Analytical Techniques

Time and analytical facilities put limits on the detail of the sampling possible in a region the size of the Front Range. Moreover, the detail of the sampling must vary with the nature of the immediate geological problem. Rarely is it possible or even desirable to sample "on a grid." Instead, the sampling pattern must be adjusted to the outcrops, and specifically to their pattern of distribution, condition of weathering, representativeness, and accessibility. Given sufficient outcrops, samples must be taken where they promise to yield the most information.

Perhaps it is worth re-emphasizing that the data here collected were not obtained as an end in itself but is the accumulated by-product of separate studies on a variety of subjects. To work out the relationships connecting the uranium deposits with the igneous rocks, it was necessary to collect and study approximately 450 samples of Laramide intrusives. Because the emphasis in this work was on "anomalies," less accurate methods of uranium analysis sufficed for the bulk of the samples. This paper summarizes only those data on uranium and thorium obtained by accurate methods on a selected group of 42 of these rocks. Similarly, in the studies of the Boulder Creek batholith, wherein the effects of a number of geological parameters were under test and small differences in uranium and thorium could be important, it was necessary to collect and analyze by accurate methods a large suite of samples; in all, 151 were analyzed for uranium and 52 for thorium. In the studies of the other batholiths in the Front Range, the aim was simply to establish the general level of uranium and thorium contents as one of several clues yielding information on the age and correlation of these granitic masses. "Sampling" of the granitic cores of the Wind River, Bighorn, and Beartooth Ranges was on a different scale and can be considered only a rough sort of preliminary reconnaissance.

The size of the samples collected varied from 3 to 50 pounds, depending upon the grain size and the nature of the analyses anticipated. Samples from the Front Range batholiths averaged 15–25 pounds; four collected especially for isotopic age work averaged 50 pounds. After removal of a representative hand specimen and thin-section chip from these large samples, the bulk of the material was crushed to pass 40 mesh, taking precautions to save the fines. Approximately 1 quart of the granulated material was split from the bulk sample, using a Jones splitter, and fine-ground to a powder for analytical purposes (rapid rock, uranium, thorium, and semiquantitative spectrographic analyses, etc.). The remaining coarser material was screened into 3 sieve fractions to provide material for mineral separations employing magnetic and heavy-liquid methods. A large starting bulk was necessary not only to insure a representative sample of these coarse-grained rocks but also to facilitate final separation of the scarcer accessory minerals in amounts adequate for analysis. The final purification of each mineral fraction was a matter of tedious hand-picking under a binocular microscope. As a result of the analysis of many such mineral separates, it is now possible to draw approximate contours representing not only the distribution of uranium and thorium in the rocks of the Boulder Creek batholith (see Fig. 7 for thorium) but also the distribution of the same elements in several of the constituent accessory minerals as well. Similar but less detailed studies are in progress on the other batholiths.

Wet chemical methods were used for the determination of both thorium and uranium. These highly sensitive methods are adaptable not only to 1-gram samples of igneous rocks but also to the much smaller samples typical of mineral separates. Unlike the γ-ray spectrometer technique, which has proved a powerful tool as applied to igneous rocks, the chemical methods do not depend upon secular

TABLE 11

Sample	No. Analyses	Thorium Range (p.p.m.)	Thorium Mean (p.p.m.)	Std. Deviation (p p.m.)
G-1.......	96	46–55	51.5_6	1.7_6

equilibrium, an important consideration when, as in the Front Range studies, a wide variety of geological materials must be analyzed. The thorium method used is that of Levine and Grimaldi (1957), which employs a mesityl oxide extraction and colorimetric determination using the reagent "thoron." Mesotartaric acid is added to the "thoron" system to suppress any zirconium remaining in solution at this stage. Above a threshold of 2 p.p.m. thorium, the analysis is performed on a 1-gram sample; below 2 p.p.m., 5-gram samples are used and longer decomposition periods are necessary. On the average, 6 thorium analyses can be completed per man-week, and the results are sufficiently reproducible that duplicate runs are not required. A blank of standard granite sample G-1 is run along with each batch of samples. Table 11 shows the results of 96 thorium determinations on G-1 by five Survey chemists over a period of 7 years. Results of 23 thorium determinations of standard sample W-1 used as an additional comparison sample in the low-thorium range are given in Table 12. The method has been cross-checked against the isotope dilution and γ-ray spectrometer methods with generally good results (see Table 13). Additional interlaboratory cross-checks of the Survey's thorium method with γ-ray spectroscopy have been published by Adams, Richardson, and Templeton (1958); others are in process.

Judging by the greater dispersion of the data, whether obtained by wet chemical methods or by the γ-ray spectrometer, uranium is more erratically distributed in the igneous rocks than is thorium. The fluorimetric method described by Grimaldi, May, and Fletcher (1952), adapted for use at lower concentrations, was used for all analyses reported in this paper. By duplicate analyses and, where necessary, additional replicates, the accuracy of the determinations is considerably increased. Each uranium figure reported in this paper is

TABLE 12

Sample	No. Analyses	Thorium Range (p.p.m.)	Thorium Mean (p.p.m.)	Std. Deviation (p.p.m.)
W-1.......	23	2.0–2.7	2.23	0.152

TABLE 13

Sample	Colorimetry, U.S.G.S. (p.p.m.)	γ Ray Spectrometry, Hurley, M.I.T. (p.p.m.)	Isotope Dilution, Doe, U.S.G.S. (p.p.m.)
BD-1.........	13.6	....	13.6
G-1..........	51.6	53.2	....
GP-100.......	72.0	72.0	....
54-G-78.......	97.0	94.0	....
GP-34........	6.7	6.0	....
GP-1.........	18.7	15.7	....

an average of at least two completely independent analyses. In our experience the method yields results reproducible to ±10 per cent in the range above 1 p.p.m.; below this range the uncertainty increases to ±100 per cent at 0.1 p.p.m. Approximately 18 uranium analyses or, allowing for duplicate determinations, 9 rock samples can be run per week. As in the thorium procedure, standards G-1 and/or W-1 are included with each sample batch submitted for uranium analyses. Results obtained on such standards by 6 Survey chemists over a 7-year period are summarized in Table 14. Results of interlaboratory cross-checks on the same standard samples using other techniques are given in Table 15.

Table 16 gives the results of fluorimetric uranium analyses by two chemists of 11 standard rock samples previously analyzed by the

TABLE 14

Sample	No. Analyses	Range (p.p.m.)	Mean (p p.m)	Std. Deviation (p.p.m.)
G-1........	104	2.8–5.0	3.9_6	0.35
W-1........	46	0.3–0.8	0.5_4	0.1_2

TABLE 15

Sample	Fluorimetry, U.S.G.S. (p.p.m.)	Neutron Activation, Hamilton (1950) (p.p.m.)	γ Ray Spectrometry, Hurley, M.I.T.* (p.p.m.)
G-1.......	4 0	3.6	3.8
W-1.......	0.5_4	0.5_3	0.5

* P. M. Hurley, written communication, 1955.

TABLE 16

URANIUM BY RADON METHOD AND BY FLUORIMETRY COMPARED

N.B.S. STD. SAMPLE No.	ROCK TYPE, LOCALITY	URANIUM (p.p.m.) BY RADON METHOD*		URANIUM (p.p.m.) BY FLUORIMETRIC METHOD†	
		Radium (10^{-12} gm/ gm rock)	Uranium (p.p.m.) ($2.77 \times 10^{12} \times$ Ra conc.)	Analyses	Average
4978...	Basalt, Columbia River	0.33 ± 0.03	0.91	0.81, 0.96, 0.84, 1.0	0.90
4985...	Basalt, Deccan Trap, India	0.21 ± 0.04	0.58	0.60, 0.58	0.59
4984...	Diabase, Centerville, Va.	0.18 ± 0.03	0.50	0.46, 0.46	0.46
4982...	Gabbro-diorite, Woburn, Mass.	0.18 ± 0.02	0.50	0.42, 0.46	0.44
4986...	Kimberlite, Kimberly, South Africa	0.59 ± 0.04	1.63	1.8, 2.3	2.0
4979...	Granite, Chelmsford, Mass.	2.96 ± 0.08	8.20	9.6, 11.7, 11.2, 9.4, 10.2, 9.6, 10.4	10.3
4981...	Granite, Graniteville, Mo.	3.3 ± 0.02	9.14	9.0, 9.7, 10.5, 10.9, 11.0	10.2
4983...	Granite, Milford, Mass.	0.23 ± 0.02	0.64	0.96, 0.98	0.97
4976...	Limestone, Carthage, Mo.	0.15 ± 0.03	0.42	0.58, 0.51, 0.30, 0.32, 0.42, 0.42	0.42
4977...	Sandstone (Berea), Cleveland, Ohio	0.24 ± 0.02	0.66	0.78, 0.74	0.76
4980...	Quartzite, Bull Run Mt., Va.	0.06 ± 0.01	0.17	0.33, 0.32	0.32

* Average of determinations made at Massachusetts Institute of Technology, Geophysical Laboratory of the Carnegie Institution, and the National Bureau of Standards.

† Analyses by Caemmerer and Moore, chemists, U.S. Geological Survey.

radon method at 3 laboratories: Massachusetts Institute of Technology, the Geophysical Laboratory of the Carnegie Institution, and the National Bureau of Standards.

ACKNOWLEDGMENTS

Without the painstaking work of a number of Geological Survey chemists who obtained the data, this report could not have been written. We are especially indebted to Lillie Jenkins, Esma Campbell, Roosevelt Moore, Alice Caemmerer, Marjorie Malloy, Marion Schnepfe, Jesse Warr, and John Antweiler. Senior chemists Frank Grimaldi and Irving May provided advice and encouragement throughout. Nelson Hickling supplied the areal measurements used in this report and compiled data. Benjamin McCall assisted in the drafting of illustrations. The writers wish to thank Michael Fleischer and Zell Peterman of the U.S. Geological Survey for careful reviews of the manuscript. The uranium aspects of the studies summarized here were sponsored in large part by the Atomic Energy Commission.

REFERENCES

ADAMS, J. A. S., J. K. OSMOND, and J. J. W. ROGERS. 1959. The geochemistry of thorium and uranium. *In* L. H. AHRENS (ed.), Physics and Chemistry of the Earth, **3**:298–348. London: Pergamon Press.

ADAMS, J. A. S., J. E. RICHARDSON, and C. C. TEMPLETON. 1958. Determination of thorium and uranium in sedimentary rocks by two independent methods. Geochim. & Cosmochim. Acta, **13**:270–79.

ALDRICH, L. T., G. W. WEATHERILL, G. L. DAVIS, and G. R. TILTON. 1958. Radioactive ages of micas from granitic rocks by Rb-Sr and K-A methods. Trans. Am. Geophys. Union, **39**:1124–34.

BASSETT, W. A., and B. J. GILETTI. 1963. Precambrian ages in the Wind River Mountains, Wyoming. Geol. Soc. America Bull., **74**:209–12.

Boos, C. M., and M. F. Boos. 1957. Tectonics of eastern flank and foothills of Front Range, Colorado. Bull. Am. Assoc. Petrol. Geologists, **41**:2603–76.

Boos, M. F. 1935. Some heavy minerals of Front Range granites. J. Geology, **43**:1033–48.

Boos, M. F., and E. ABERDEEN. 1940. Granites of the Front Range, Colorado: The Indian Creek plutons. Geol. Soc. America Bull., **51**:695–730.

Boos, M. F., and C. M. Boos. 1934. Granites of the Front Range–the Long's Peak–St. Vrain batholith. Geol. Soc. America Bull., **45**:303–32.

CLARKE, F. W., and H. S. Washington. 1924. The Composition of the Earth's Crust. U.S. Geol. Survey Prof. Paper 127. Pp. 117.

DAVIS, G. L., G. R. TILTON, B. R. DOE, L. T. ALDRICH, and S. R. HART. 1961. The ages of rocks and minerals. Carnegie Inst. Wash. Yearbook, **60**:190–99.

FLEISCHER, M., and E. C. T. CHAO. 1960. Some problems in the estimation of abundances of elements in the earth's crust. Proc. 21st Internat. Geol. Congr., Sec. 1: Geochemical Cycles, pp. 141–48.

GRIMALDI, F. S., I. MAY, and M. H. FLETCHER. 1952. U.S. Geological Survey Fluorimetric Methods of Uranium Analysis. U.S. Geol. Survey Circ. 199. Pp. 20.

HAMILTON, E. I. 1950. The Uranium Content of the Differentiated Skaergaard Intrusion. Medd Grønland, Vol. **162**, No. 7. Pp. 35.

HEIER, K. S., and J. A. S. ADAMS. 1963. The geochemistry of the alkalis. *In* L. H. AHRENS (ed.), Physics and Chemistry of the Earth. Vol. 5. (In press.)

HEIER, K. S., and J. J. W. ROGERS. 1963. Radiometric determination of thorium, uranium and potassium in basalts and in two magmatic differentiation series. Geochim. & Cosmochim. Acta, **27**:137–54.

HUTCHINSON, R. M. 1960. Structure and petrology of north end of Pikes Peak batholith, Colorado. *In* R. J. WEIMER (ed.), Guide to the Geology of Colorado, pp. 170–80. Denver: Geol. Soc. America, Rocky Mountain Assoc. Geologists, and Colorado Sci. Soc.

JOHNSON, D. H. 1954. Radiometric prospecting and assaying. *In* H. FAUL (ed.), Nuclear Geology, pp. 219–41. New York: John Wiley & Sons.

KULP, J. L., W. S. BROECKER, and W. R. ECKELMANN. 1953. Age determination of uranium minerals by the Pb-210 method. Nucleonics, **11**:19–21.

LARSEN, E. S., III, and D. GOTTFRIED. 1960. Uranium and thorium in selected suites of igneous rocks. Am. J. Sci., **258-A**:151–69.

LARSEN, E. S., JR., G. PHAIR, D. GOTTFRIED, and W. L. SMITH. 1956. Uranium in magmatic differentiation. U.S. Geol. Survey Prof. Paper 300. Pp. 65–74.

LEVINE, H., and F. S. GRIMALDI. 1957. Determination of thorium in the parts per million range in rocks. Geochim. & Cosmochim. Acta, **14**:93–97.

LOVERING, T. S., and E. N. GODDARD. 1938. Laramide igneous sequence and differentiation in the Front Range, Colorado. Geol. Soc. America Bull., **49**:35–68.

———. 1950. Geology and Ore Deposits of the Front Range, Colorado. U.S. Geol. Survey Prof. Paper 223. Pp. 319.

MOORE, J. G. 1959. The quartz diorite boundary line in the western United States. J. Geology, **67**:198–210.

NIER, A. O., R. W. THOMPSON, and B. F. MURPHEY. 1941. The isotopic constitution of lead and the measurement of geological time. III. Phys. Rev., **60**:112–16.

PHAIR, G. 1952. Radioactive Tertiary porphyries in the Central City district, Colorado, and their bearing upon pitchblende deposition. U.S. Geol. Survey Rpt. TEI-247. Pp. 53.

PHAIR, G., and H. LEVINE. 1953. Notes on the differential leaching of uranium, radium, and lead from pitchblende in H_2SO_4 solutions. Econ. Geol., **48**:358–69.

PHAIR, G., and H. MELA, JR. 1956. The isotopic variation of common lead in galena from the Front Range and its geological significance. Am. J. Sci., **254**:420–28.

PHAIR, G., and K. O. SHEMAMOTO. 1952. Hydrothermal uranothorite in fluorite breccias from the Blue Jay mine, Jamestown, Boulder County, Colorado. Am. Mineralogist, **37**:659–66.

SIMS, P. K., G. PHAIR, and R. H. MOENCH. 1958. Geology of the Copper King uranium mine, Larimer County, Colorado. Bull. U.S. Geol. Survey 1032-D. Pp. 171–221.

TILTON, G. R., G. L. DAVIS, G. W. WEATHERILL, and L. T. ALDRICH. 1957. Isotopic ages of zircon from granites and pegmatites. Trans. Am. Geophys. Union, **38**:360–71.

K. A. RICHARDSON

2. Thorium, Uranium, and Potassium in the Conway Granite, New Hampshire, U.S.A.

FOR THE PAST THREE YEARS an investigation into the possibility of recovering thorium from common rocks has been carried out under subcontract from Oak Ridge National Laboratory to Rice University. The investigation has focused on the highly radioactive igneous rocks of New England and, in particular, on the Conway granite of New Hampshire. Whereas the objective of this investigation is not directly related to the study of the radiation environment, the results may be of interest to this symposium. This paper will outline an area of high radioactivity of natural terrestrial origin and indicate the occurrence of radioactive elements in the rocks of the area. This work is an example of the application of some of the techniques that are discussed in the symposium (γ-ray spectrometry, α-particle spectrometry, and α-particle autoradiography) to the study of natural terrestrial radiation. Furthermore, the area underlain by the Conway granite is part of the area surveyed by Lowder et al. (1963) in northern New England.

Average thorium and uranium concentrations in granitic rocks of the upper part of the earth's crust are in the ranges of 10–15 p.p.m. thorium and 3–4 p.p.m. uranium (Adams et al., 1959; Adams, 1962). Thorium and uranium concentrations in basic igneous rocks and many sediments are lower than the concentrations in granitic rocks. In the early part of the present investigation it was found that many of the granites of New England contained more than twice the average thorium concentration of 10–15 p.p.m. (ORNL Progress Report, May 31, 1961; TID-13001). These included the Westerly granite of Rhode Island, some granites in Maine, and, in particular, the Conway granite of New Hampshire. As a result of these findings

K. A. RICHARDSON is research associate in the Department of Geology, Rice University, Houston, Texas.

an intensive study of thorium in the Conway granite has been carried out during the last two years.

Several hundreds of thorium determinations have been made on outcrop surfaces and drill-core samples of the Conway granite. A portable, single-channel γ-ray spectrometer, described by Adams and Fryer (1963), was used for thorium determinations made in the field. Many thorium, uranium, and potassium determinations were made on Conway granite samples with the 256-channel laboratory γ-ray spectrometer described by Adams (1963). Sites of α emitters in the Conway granite were determined autoradiographically, and α-particle pulse-height analysis was applied to measuring thorium/uranium ratios in accessory minerals and uranium-234/uranium-238 ratios in weathered and unweathered samples of the granite.

GEOLOGIC SETTING

A geologic map and description of the bedrock geology of the state of New Hampshire were published by Billings (1956). The White Mountain magma series, a group of related igneous rocks ranging from gabbros to syenites and granites, is one of four major

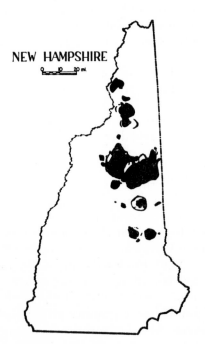

Fig. 1.—Map of the state of New Hampshire showing areas underlain by rocks of the White Mountain magma series.

series of plutonic rocks in New Hampshire. The Conway granite is the youngest intrusive member of the White Mountain series with the exception of some later dike rocks. Age determinations on the Conway granite, summarized by Toulmin (1961), fall between 170 and 190 million years. The White Mountain magma series has been considered Mississippian in age (Billings, 1956), but age determinations indicate a lower Jurassic age according to the geologic time scale of Kulp (1960).

Figure 1 is an outline map of the state of New Hampshire showing the areas underlain by rocks of the White Mountain magma series. The Conway granite, the most extensive unit of the magma series, underlies an area of approximately 450 square miles. The terrain is mountainous, and the permanent population of the region is not dense. Environmental effects of direct radiation from the bedrock are decreased by the fairly continuous blanket of glacial till over the bedrock. The present investigation has been limited to the study of the bedrock; the radioactivity of the glacial till was not measured. However, the Conway granite has certainly contributed material to the blanket of till, and it may also contribute to the radiation environment through migration of radioactive nuclides into the atmosphere and into the ground waters. High concentrations of radium-226 and radon-222 in ground waters of parts of Maine and New Hampshire have been reported by Smith *et al.* (1961).

THORIUM IN THE CONWAY GRANITE

The high radioactivity of the Conway granite has been known for some time. In 1946 Billings and Keevil obtained data by total α counting indicating that rocks of the White Mountain magma series were twice as radioactive as the average of several thousand samples of other North American igneous rocks. Gamma-ray scintillation analyses of 4 samples of Conway granite by Hurley (1956) averaged 51.6 p.p.m. thorium, 11.5 p.p.m. uranium, and 3.8 per cent potassium. Butler (1961) gave analytical data for 12 samples of Conway or biotite granite of the White Mountain magma series, with values ranging from 30 to 77 p.p.m. thorium and from 4.3 to 25.5 p.p.m. uranium. Analyses of 4 Conway granite samples by Rogers and Ragland (1961) average 36 p.p.m. thorium, 9 p.p.m. uranium, and 4.7 per cent potassium.

Figure 2 is a map of the White Mountain batholith (the main intrusive mass of the White Mountain magma series) and the Mad River stock located southwest of the batholith. This map shows the areal extent of the four most widespread units of the magma series. The histograms in Figure 3 illustrate the results of γ-ray thorium determinations made on outcrops of these four major rock units. The approximate mean thorium concentrations are 23 p.p.m. in the syenite, 38 p.p.m. in the porphyritic quartz syenite, 43 p.p.m. in the Mount Osceola granite, and 56 p.p.m. in the Conway granite.

A statistical model has been developed to justify theoretically the lognormal distribution of a trace element (Rogers and Adams, 1963),

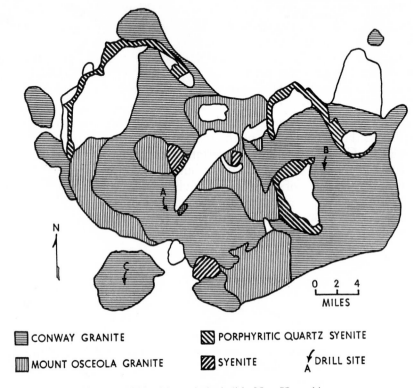

CONWAY GRANITE   PORPHYRITIC QUARTZ SYENITE

MOUNT OSCEOLA GRANITE   SYENITE   DRILL SITE

0 2 4
MILES

Fig. 2.—White Mountain batholith, New Hampshire

and the conclusion has been drawn that thorium is lognormally distributed in the Conway granite (ORNL Progress Report, May 31, 1962; TID-16781). Figure 4 contains cumulative frequency curves of thorium concentrations measured on the surface of the White Mountain batholith and on the drill cores taken from the Conway granite. The drill sites are indicated on Figure 2. Linearity of the cumulative frequency curves for the surface of the White Mountain batholith indicates that the thorium is lognormally distributed over the surface of the batholith; that is, the surface of the batholith is homogeneous

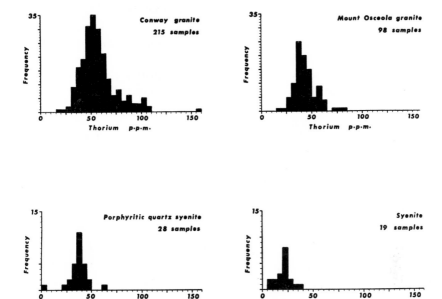

Fig. 3.—Frequency distribution of thorium concentrations in four members of the White Mountain magma series.

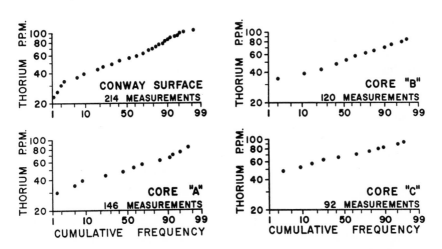

Fig. 4.—Cumulative frequency distributions of thorium in the Conway granite

with respect to thorium. Homogeneity is defined here as the absence of any trend or pattern of variation on a scale commensurate with the size of the batholith. The geometric mean of the measured thorium concentrations in Conway granite outcrops in the White Mountain batholith is 56 ± 6 p.p.m. thorium. The lognormal distribution of thorium concentrations in the drill cores, demonstrated by Figure 4, indicates that thorium concentration in the Conway granite does not change with depth. Consequently, the Conway granite might be considered a low-grade thorium resource containing about three million tons of thorium per hundred feet of depth in the White Mountain batholith (Adams *et al.*, 1962). By including the smaller bodies of Conway granite, such as the Mad River stock, the size of the thorium resource might be increased by 50 per cent.

THORIUM, URANIUM, AND POTASSIUM IN THE CONWAY GRANITE

For the purpose of calibrating the portable γ-ray spectrometer, many outcrop and drill-core samples of Conway granite were analyzed with a multichannel γ-ray spectrometer in the laboratory. Laboratory analyses gave uranium and potassium concentrations in addition to thorium concentrations. The uranium content of the Conway granite is about four times the crustal average of 3–4 p.p.m. Potassium concentrations in the Conway granite are in the range of 3.5–5 per cent. Thorium and uranium concentrations are apparently not related to potassium concentration, but a well-defined linear relationship between thorium and uranium exists in drill-core samples from the White Mountain batholith.

Figure 5a shows the relationship between thorium and uranium in samples from drill cores "A" and "B." The close alignment of these plotted points, and the fact that the samples were taken from two sites separated by almost 15 miles, suggest that this relationship may be representative of all the unaltered Conway granite in the White Mountain batholith. The linear relationship is represented by the equation:

$$0.4(\text{thorium p.p.m.}) - (\text{uranium p.p.m.}) = 5.5.$$

Plotted on Figure 5b are 17 points representing the thorium and uranium concentrations in samples of typical Conway granite taken from outcrop surfaces in the White Mountain batholith. All the points fall below the line representing samples from drill cores "A" and "B." This suggests that outcrop samples have lost uranium as a result of weathering processes. The uranium determinations are based on the measurement of bismuth-214 (1.76 mev.) γ-rays, and the

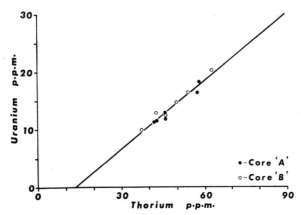

FIG. 5a.—Uranium concentration *vs.* thorium concentration in samples of Conway granite from drill cores "A" and "B."

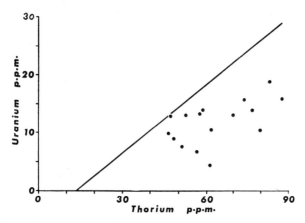

FIG. 5b.—Uranium concentration *vs.* thorium concentration in outcrop samples of Conway granite from the White Mountain batholith.

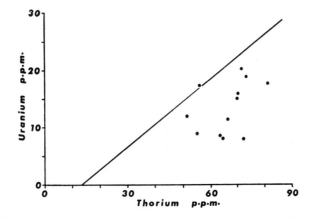

FIG. 5c.—Uranium concentration *vs.* thorium concentration in samples of Conway granite from drill core "C."

assumption that radioactive equilibrium obtains in the uranium-238 decay series. Agreement between γ-ray determinations and chemical uranium analyses made at Oak Ridge National Laboratory indicates that the assumption of equilibrium is generally valid.

Figure 5c is a plot of thorium and uranium values for samples from drill core "C" in the Mad River stock. The failure of samples from core "C" to show a definite relationship between thorium and uranium concentrations may be the result of redistribution of these elements with the late magmatic intrusion of a basic dike. This dike is intersected at a depth of 300 feet in the core, and calculations indicate that the steeply dipping dike is within 30 feet of the drill hole throughout the depth of the hole. The upper few feet of core "C" are extremely weathered, with a resultant loss of uranium and preferential leaching of uranium-234 relative to uranium-238.

RADIOACTIVE DISEQUILIBRIUM IN THE URANIUM-238 DECAY SERIES

Isotopic fractionation of uranium in nature has been reported by various workers in recent years (Chavlov, 1959; Cherdyntsev *et al.*, 1961; Thurber, 1962; Rosholt *et al.*, 1963). Uranium-234/uranium-238 ratios in a few samples of Conway granite have been measured. The procedure that has been used involves separation of uranium from the rock sample (Korkisch *et al.*, 1957, 1960), electrodeposition of uranium on a stainless-steel planchet (Richardson, 1963), and α-particle pulse-height analysis with a silicon semiconductor detector.

The uranium isotopic ratio was measured in samples from outcrops in the northern and central parts of the White Mountain batholith and in samples from drill core "C." A uranium-234 deficiency of 15 per cent was found in the upper 15 feet of drill core "C"; uranium-234 was found to be in equilibrium with uranium-238 in samples from deeper in core "C" and in the few samples taken from outcrops.

SITES OF α EMITTERS IN THE CONWAY GRANITE

The sites of α emitters in the Conway granite have been investigated autoradiographically. Table 1 shows the distribution of α tracks on autoradiographs of thin sections of Conway granite from the White Mountain batholith. The data in Table 1 show the importance of accessory minerals in the Conway granite as sites of thorium and uranium. In order to study the radioactive accessory minerals, grain-mount autoradiographs were made of accessory minerals. The major radioactive accessory minerals were identified by X-ray diffraction as the following: allanite, huttonite (monoclinic $ThSiO_4$), thorite (tetragonal $ThSiO_4$), and zircon. Huttonite and monazite ($CePO_4$) are almost indistinguishable on the basis of physical properties; the observed range of α activity of "huttonite" grains in the Conway granite suggests that extensive isomorphism exists between huttonite and monazite. Variable α activity of zircon and thorite from grain to grain was also noted in the Conway granite.

TABLE 1

DISTRIBUTION OF α EMITTERS IN CONWAY GRANITE
WHITE MOUNTAIN BATHOLITH
(Percentage)

TOTAL No. α TRACKS	SOURCE OF α TRACKS								
	Accessory Minerals	Dispersed	Biotite		Quartz		Feldspar		Grain Boundaries	
			Accessory	Dispersed	Accessory	Dispersed	Accessory	Dispersed	Accessory	Dispersed
27,246......	66	34	42	25	0.2	5	2	3.4	22	0.6

TABLE 2

THORIUM AND URANIUM CONCENTRATIONS IN ACCESSORY
MINERALS OF THE CONWAY GRANITE
(Percentage)

Accessory Mineral	Thorium Average	Concentration Range	Uranium Average	Concentration Range
Allanite............	0.9	0.2– 2	0.06	0.01–0.1
Huttonite..........	21	3 –73	2	0.3 –7
Thorite............	16	2 –60	0.9	1 –4
Zircon.............	0.7	0 – 1	0.2	0 –0.3

Thorium and uranium concentrations in the four major radioactive accessory minerals of the Conway granite are given in Table 2. Using (i) the published values of thorium and uranium concentration in allanite (Smith *et al.*, 1957) of the Conway granite, (ii) thorium/uranium ratios of huttonite, thorite, and zircon calculated from the total α-particle spectra of these minerals, and (iii) the α activities of the four minerals, determined autoradiographically, the values in Table 2 were calculated. The calculations are explained in detail by Richardson (1963).

Finally, on the basis of autoradiographic data and of α-particle and γ-ray spectrometric determinations, the quantitative distribution of thorium and uranium in the Conway granite of the White Mountain batholith has been estimated. Table 3 gives the distribution.

TABLE 3

DISTRIBUTION OF THORIUM AND URANIUM IN
THE CONWAY GRANITE, WHITE MOUN-
TAIN BATHOLITH

(Percentage)

Site	Thorium	Uranium
Allanite....................	7	2
Huttonite...................	28	12
Thorite....................	37	9
Zircon.....................	19	19
Dispersed or unknown phase...	9	58

ACKNOWLEDGMENTS

This work has been done under subcontract No. 1491 from Oak Ridge National Laboratory to Rice University. Funds for the development and construction of the radiometric equipment used in this study were received from the Robert A. Welch Foundation through Grant C-009 to J. A. S. Adams and J. J. W. Rogers and Grant K-054b to J. A. S. Adams.

REFERENCES

ADAMS, J. A. S. 1962. Radioactivity of the lithosphere. *In* H. ISRAËL, and A. KREBS (eds.), Nuclear Radiation in Geophysics, pp. 1–17. Berlin: Springer-Verlag.
———. 1963. Laboratory γ-ray spectrometer for geochemical studies. This symposium.
ADAMS, J. A. S., and G. E. FRYER. 1963. Portable γ-ray spectrometer for field determination of thorium, uranium, and potassium. This symposium.
ADAMS, J. A. S., M.-C. KLINE, K. A. RICHARDSON, and J. J. W. ROGERS. 1962. The Conway granite of New Hampshire as a major low-grade thorium resource. Proc. Nat. Acad. Sci., **48**:1898–1905.
ADAMS, J. A. S., J. K. OSMOND, J. J. W. ROGERS. 1959. The geochemistry of thorium and uranium. *In* L. H. AHRENS (ed.), Physics and Chemistry of the Earth, **3**:298–348. New York: Pergamon Press.
BILLINGS, M. P. 1956. The Geology of New Hampshire. Pt. II. Bedrock Geology, Concord: New Hampshire State Planning and Development Commission. Pp. 200.

BILLINGS, M. P., and N. B. KEEVIL. 1946. Petrography and radioactivity of four Paleozoic magma series in New Hampshire. Geol. Soc. America Bull., **57**:797–828.

BUTLER, A. P., JR. 1961. Ratio of thorium to uranium in some plutonic rocks of the White Mountain plutonic-volcanic series, New Hampshire. U. S. Geol. Survey Prof. Paper 424-B, pp. 67-69.

CHAVLOV, P. I. 1959. The U-234/U-238 ratio in some secondary minerals. Geochemistry, No. 2, pp. 203–10.

CHERDYNTSEV, V. V., D. P. ORLOV, E. A. ISABAEV, and V. I. IVANOV. 1961. Uranium isotopes in nature. II. Isotopic composition of uranium minerals. Geochemistry, No. 10, pp. 927–36.

HURLEY, P. M. 1956. Direct radiometric measurement by gamma-ray scintillation spectrometer. Pt. II. Uranium, thorium, and potassium in common rocks. Geol. Soc. America Bull., **67**:405–11.

KORKISCH, J., P. ANATAL, and F. HECHT. 1960. Adsorption des Urans aus salzsauer alkoholischer Lösung am stark basischen Anionenstauscher Dowex-1. Fresenius' Ztschr. Analyt. Chem., **172**:401–8.

KORKISCH, J., M. R. ZAKY, and F. HECHT. 1957. Schnellbestimmung von Microgrammengen Uran in Mineralen. Mikrochim. Acta (Wien), pp. 485–95.

KULP, J. L. 1961. The geological time scale. Science, **133**:1105–14.

LOWDER, W. M., A. SEGALL, and W. J. CONDON. 1963. Environmental radiation survey in northern New England. This symposium.

OAK RIDGE NATIONAL LABORATORY. 1961. Geologic aspects of thorium recovery from common rocks. Oak Ridge National Laboratory Subcontract No. 1491, Progress Report June 1, 1960 to May 31, 1961. U.S. Atomic Energy Commission, Tech. Inf. Ext. Serv. Pub. TID-13001.

––––. 1962. Geologic aspects of thorium recovery from common rocks. Oak Ridge National Laboratory Subcontract No. 1491, Progress Report June 1, 1961 to May 31, 1962. U. S. Atomic Energy Commission, Tech. Inf. Ext. Serv., Pub. TID-16781.

RICHARDSON, K. A. Radioactivity, Sites of Alpha Emitters, and Radioactive Disequilibrium in the Conway Granite of New Hampshire. Ph. D. thesis, Rice University, Houston, Texas, 1963.

ROGERS, J. J. W., and J. A. S. ADAMS. 1963. Lognormality of thorium concentrations in the Conway granite. Geochim. & Cosmochim. Acta, **27**:775–83.

ROGERS, J. J. W., and P. C. RAGLAND. 1961. Variation of thorium and uranium in selected granitic rocks. Geochim. & Cosmochim. Acta, **25**: 99–109.

ROSHOLT, J. N., W. P. SHIELDS, and E. L. GARNER. 1963. Isotopic fractionation of uranium in sandstone. Science, **139**:224–25.

SMITH, B. M., W. N. GRUNE, F. B. HIGGINS, JR., and J. G. TERRILL, JR. 1961. Natural radioactivity in ground water supplies in Maine and New Hampshire. J. Am. Water Works Assoc., **53**:75–88.

SMITH, W. L., M. L. FRANCK, and A. M. SHERWOOD. 1957. Uranium and thorium in the accessory allanite of igneous rocks. Am. Mineral., **42**: 367–78.

THURBER, D. L. 1962. Anomalous U-234/U-238 in nature. J. Geophys. Res., **67**:4518–20.

TOULMIN, P., III. 1961. Geological significance of lead-alpha and isotopic age determinations of "alkalic" rocks of New England. Geol. Soc. America Bull., **72**:775–80.

J. J. W. ROGERS

3. Statistical Tests of the Homogeneity of the Radioactive Components of Granitic Rocks

THE OBJECT OF THIS PAPER is to discuss the distribution of thorium and uranium in the granites of the United States. The problem will be approached from the point of view of determining various igneous bodies or areas within which the granites can be considered homogeneous with respect to their thorium and uranium contents. Some effort will be made to correlate the populations established in this manner with subdivisions made on other geologic bases, such as age, location, and general lithology. For these purposes it is necessary to define the term "homogeneous" in both a mathematical and a geologic sense, to discuss various methods for the detection and testing of homogeneity, and finally to apply the most appropriate method to the investigation of selected groups of granites.

The analyses used in this paper have either been done at Rice University or have been reported by Gottfried, Larsen, and/or Phair from the laboratories of the U.S. Geological Survey. The reasons for the restriction to these two laboratories are (1) that sizable amounts of data, particularly with regard to thorium contents, are generally not available from other laboratories and (2) that cross-calibration between these two laboratories yields generally the same results on the same samples. It would, of course, be improper to compare analyses from different laboratories unless it is known that the laboratories are obtaining equivalent results. For the purposes of this paper, analyses are not considered unless both thorium and uranium contents are reported. Analyses made at Rice University have been done by γ-ray spectrometry with some chemical checks; analyses by the U.S. Geological Survey have been done by chemical methods.

In this paper the term "granite" applies to any phaneritic quartz-bearing igneous rock in the general range of true granite to granodiorite or quartz diorite.

J. J. W. ROGERS is professor of geology, Rice University, Houston, Texas.

Definition and Detection of Homogeneity

The term "homogeneous" may be defined in a variety of ways. Though a strict mathematical definition is useful for theoretical discussions, a somewhat more vague usage may be necessary in considerations of actual geologic materials.

Mathematical Problems

Mathematically, a body may be described as homogeneous if the value of some measured parameter varies randomly from place to place. Though the term "random" is intuitively clear from a mathematical point of view, it must be defined somewhat differently for different types of actually measured geologic parameters. In particular, as discussed below, the types of frequency distributions exhibited by randomly varying lithologic properties are dependent on the nature of the property. A random variable may be characterized in terms of a succession of random increments about an initial or mean value.

A number of procedures may be used to determine mathematically random variability of a property within a body. One simple method is to use χ-square analysis. A body is divided into a number of regions (e.g., areas of outcrop on the earth's surface), the expected value of the variable in each region is set equal to the mean, and the χ-square for deviations of actual and expected values is calculated. This test, of course, applies only to data that are determined by counting and should not be used for percentages or similar types of data.

More generally applicable methods for the determination of homogeneity are the several techniques of variance analysis. A hypothetical grid or nested set of grids can easily be placed over a surface outcrop, and the variation at different levels of sampling or in different directions can be determined by standard procedures. For this purpose, one-variable-of-classification procedures have been used repeatedly (e.g., see Krumbein and Slack, 1956), and such methods may easily be extended from surface studies to investigation of three-dimensional bodies. Two-variable-of-classification methods have been less commonly used, though one example of the technique is given by Flinn (1959) for surface outcrops. Extension of multiple-variable-of-classification methods to three-dimensional bodies would be extremely cumbersome. Where two or more sets of data exist for the same rock, multivariate methods may be used; an example of this technique in a one-variable-of-classification sense is given by Krumbein and Tukey (1956). In all investigations of this type, lack of significant variation, particularly at the higher levels of sampling, would be considered evidence of homogeneity with respect to the property being measured.

Another approach to the determination of homogeneity would be to investigate whether or not any trends or patterns of areal variation existed in the body or set of bodies being studied. If, for example, a trend-surface analysis (see Whitten, 1959) established a distinct pattern of variability in terms of a small number of parameters (e.g., a cubic equation) and if this trend surface accounted for a large amount of the variability in the original measurements, then the body being studied would clearly not be homogeneous. The various qualitative terms used in the preceding sentence could, of course, be quantified by appropriate statistical techniques, but the final acceptance or rejection of homogeneity would remain a subjective decision.

It is possible to detect homogeneity by a more indirect process than those described above. In particular, in studying the elemental compositions of igneous rocks, the theoretical (random) frequency distributions of the abundances of the different elements can be predicted for homogeneous bodies or sets of bodies (Rogers and Adams, 1963). A trace element (very small mole fraction) is expected to exhibit a logarithmically normal frequency distribution if sampled from a single homogeneous population. Similarly, the distribution of major elements (intermediate values of mole fraction) is expected to be arithmetically normal, and the distribution of an element such as silicon, which presumably has a very large mole fraction, is expected to be skewed in exactly the opposite direction from lognormality. After these theoretical distributions have been established, it is clear that a body or set of bodies may be proved to be non-homogeneous if the frequency distribution of the element under investigation deviates significantly from the theoretical (random) distribution.

It should be noted that random distributions are not identical for all elements; the form of the "random" frequency distribution depends on the average abundance of the element. Furthermore, a body or set of bodies may obviously be homogeneous with respect to one constituent and not with respect to others.

By the foregoing procedure, proof of non-homogeneity is quite definite within the limits of error set by the confidence level chosen in testing the observed frequency distribution against the theoretical one. Conversely, proof of homogeneity is impossible for the same reason that the means of two samples can never be established as exactly equal. From a frequency distribution tested as described above it is possible to establish only one of the following conclusions: (1) at the confidence level chosen the distribution is non-random, and the sample has been chosen from a non-homogeneous population, or (2) the sample shows no evidence of having been drawn from a non-homogeneous population. On the basis of general geologic evidence, a geologist may wish to translate the second conclusion into a statement that the body or set of bodies is apparently homogeneous.

One additional problem must be discussed. Although a non-random frequency distribution cannot be obtained from a homogenous body by random sampling, it is possible that a random distribution could be obtained from a body that exhibits definite areal trends or patterns. A pattern, of course, implies a functional relationship between position and abundance of the element studied, and this functional relationship should destroy any random variability that an elemental abundance might exhibit. It is conceivable, however, that a sampling grid might be laid out on the pattern of variability in such a manner that the resultant frequency distribution of the sample would not be distinguishable from a random one, though such a coincidence seems unlikely.

GEOLOGIC PROBLEMS

The preceding discussion has dealt exclusively with the mathematical aspects of identifying and proving homogeneity in ideal bodies or sets of bodies of rock. In a geologic sense, however, it is doubtful that any actual rock body, such as a granite pluton, is exactly homogeneous in regard to the distribution of any property. Consider, for example, the distribution of an element in a reasonably massive pluton containing a number of widely spaced joints. Along these joints there has probably been leaching or possibly late-stage hydrothermal concentration of the element in question. Both deuteric and weathering processes have undoubtedly affected the surficial materials exposed for sampling. In this case, although the vast bulk of the rock might be homogeneous in regard to the element in question, it would be easily possible to obtain a set of samples near the joint surfaces which could be proved at a high confidence level to come from a population different from that of the main pluton. Mathematically speaking, therefore, the pluton is not exactly homogeneous; geologically, it would be unreasonable to consider it anything but homogeneous.

The basic problem, as exemplified above, is that almost any geologic material has probably been subjected to a long series of processes. Many of these processes, such as weathering along joints, have probably had a small effect, detectable only by special investigations designed for that specific purpose. In order to obtain information about the important processes, therefore, it is necessary to ignore or remove the effects of the minor processes. This decision of relative importance can be made only on the basis of geologic reasoning and not by mathematical methods. A qualitative decision of this type implies that the total amount of data available for some study must be screened in such a manner that some of the data must be ignored in drawing final conclusions. All statistical analyses of the remaining data obviously are partly controlled by this selection procedure.

In the specific case of determining the homogeneity of rock bodies, minor effects can be eliminated by adding a scale factor to the definition of homogeneity. Thus, although inhomogeneities might exist in a pluton on a scale of tens of feet, the pluton might be effectively homogeneous on a scale of miles. A precise definition of this scale factor would be very difficult, but for geologic purposes it is probably sufficient to state the approximate maximum size of inhomogeneity that the investigator expects to be able to detect. For example, a statement might be made to the effect that a pluton is homogeneous if measured in scale units of one mile or more.

For the purpose of detecting homogeneous bodies defined in this manner, an investigation of the frequency distributions obtained for the studied property may be more suitable than other statistical techniques. The reason is that in graphical studies of cumulative frequency distributions, minor deviations from a theoretical curve are immediately obvious. The investigator, then, may ignore these deviations if he has a valid geologic reason for doing so. This subjective choice, which has been shown to be necessary in the paragraphs above, is more difficult to make for data tabulated for other methods of statistical analysis. This graphical procedure may be quantified by suitable statistical techniques if necessary.

One precaution must be taken in determining the frequency distribution for a hypothetical population. For such purposes it is necessary to sample the rock body or bodies studied so as to obtain a reasonably uniform distribution of sample points. For example, if two bodies (A and B) are hypothesized as belonging to one population, it is obviously incorrect to test the hypothesis by constructing a single-frequency distribution of 200 values from A and 10 from B unless the volumes of the two bodies are in approximately the same ratio. In general, individual sample points should represent about equal amounts of rock, but deviations from this ideal situation can obviously be as large as the errors introduced by the qualitative judgment of the investigator in selecting data for analysis.

The following general conclusions may be drawn from the discussions in this section: (1) A rock body or set of bodies is homogeneous with respect to some measured property if the values of that property vary randomly throughout the volume (or area) investigated, with the provision that deviations of the property within volumes (or areas) less than some specified size are not considered. (2) Homogeneity may be tested by comparing the actual frequency distribution of the measured property against the theoretical distribution expected if the property varies randomly.

The following section discusses the application of these concepts to some granites in the United States.

Thorium and Uranium in Granites of the United States

The granites of the United States are subdivided, for the purposes of this discussion, into three groups: (1) Middle to Late Paleozoic granites of New England, (2) Nevadan-Laramide granites of the Cordilleran area, and (3) Precambrian granites of the central and western United States. Some large bodies of granite in the United States are omitted from these categories; they are not discussed simply because the writer does not have available any extensive data concerning their thorium and uranium contents. Comparison of these three groups is given below following a more detailed discussion of some granites of New England.

New England Granites

The most extensive radiometric data on any granite body in the United States are available for the thorium content of the Conway granite of the central White Mountain batholith in New Hampshire. By the use of a portable γ-ray spectrometer (Adams and Fryer, 1963), over 700 thorium determinations were made on surface samples and cores from the main body of the granite. The general nature and economic significance of these results have been discussed by Adams *et al.* (1962).

As reported by Rogers and Adams (1963), the thorium contents of the main central body of the Conway granite are lognormally distributed. Because thorium is a trace constituent, this distribution implies that the granite is homogeneous within the approximately 300 square miles of outcrop area. This implied homogeneity is geologically reasonable in view of the fact that the Conway granite is massive and mineralogically uniform over most of its outcrop area (including several thousand feet of relief) and in cores up to 600 feet into the granite. Unfortunately, the uranium contents have not been determined in enough samples to provide an accurate indication of its distribution. The thorium distribution has been sufficiently discussed in the papers referred to above and will not be considered further here.

An interesting comparison with the Conway granite is provided by the Precambrian Enchanted Rock batholith of central Texas (discussed by Ragland and Adams, in preparation; Hutchinson, 1956). This body is concentrically zoned in its outcrop area of 100 square miles and represents a distinct sequence of magmatic differentiation. In this obviously non-homogeneous body the thorium appears to have been distributed mainly by secondary processes with localization in certain areas. As expected, the thorium contents are not lognormally distributed. The subject is further discussed by Rogers and Adams (1963) and will not be elaborated here.

As discussed earlier in the paper, the concept of homogeneity need not be restricted to individual bodies but may be extended to whole groups of bodies covering broad areas. From this point of view it seems worthwhile to examine both the entire White Mountain magma

TABLE 1

RADIOMETRIC DATA FOR ROCKS OF THE
WHITE MOUNTAIN MAGMA SERIES

Rock Type	No. Samples	Thorium (p.p.m.)*	Uranium (p.p.m.)*	Th/U Ratio†
Conway granite..............	19	48	9.5	5.2
Mt. Osceola granite.........	2	37	10	3.9
Hastingsite granite..........	2	26	7.4	3.6
Syenite and quartz syenite....	9	16	2.9	5.5
Monzonite.................	4	16	3.1	5.7
Granodiorite, etc............	5	18	3.7	4.3
Aplite and miscellaneous granite	3	34	12	2.7

* Thorium and uranium values are cited as parts per million of the metal, assuming radioactive equilibrium in the respective decay series.

† The cited Th/U ratios are the average of the various Th/U ratios of the individual samples, not the ratio of average Th/average U.

series of New Hampshire and also the whole assemblage of Middle and Late Paleozoic granites of New England. In such investigations the sampling problem is, of course, extremely difficult. The idea of taking numbers of samples in proportion to the volume of the rock type that they represent may be virtually impossible to put into practice. The rock types represented in this study and the numbers of samples obtained from each are shown in Tables 1 and 2. The numbers of samples from the White Mountain magma series are in very rough proportion to the abundance of the various rock types, but the general New England sampling is obviously heavily weighted toward certain magma series in New Hampshire. The data used in this compilation do not include the field measurements of the Conway granite and other members of the White Mountain magma series reported by Adams *et al.* (1962); data are assembled from Whitfield *et al.* (1959), Rogers and Ragland (1961), and Adams and Rogers (1961).

The results of investigating the distribution of the thorium and uranium contents of the White Mountain magma series and the whole

TABLE 2

RADIOMETRIC DATA FOR VARIOUS GRANITIC
ROCKS OF NEW ENGLAND

	No. Samples	Thorium (p.p.m.)*	Uranium (p.p.m.)*	Th/U Ratio†
White Mountain magma series.	44	33	6.9	5
New Hampshire series........	25	20	5.6	3.9
Oliverian series..............	17	24	5.9	4.4
Granites from southwestern Maine....................	25	26	6.3	4.2
Granites from Massachusetts..	8	17	4.5	4.1
Granites from Rhode Island...	4	27	6.9	4.9

* Thorium and uranium values are cited as parts per million of the metal, assuming radio-active equilibrium in the respective decay series.

† The cited Th/U ratios are the average of the various Th/U ratios of the individual samples, not the ratio of average Th/average U.

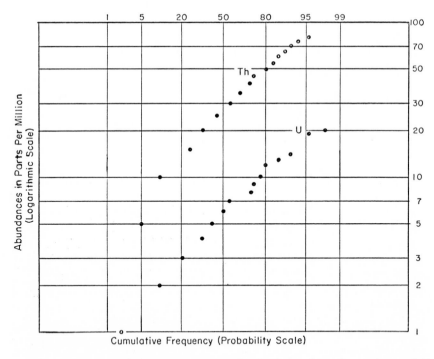

Fig. 1.—Cumulative frequency distributions for thorium and uranium contents of 44 samples of the White Mountain magma series, New Hampshire.

assemblage of New England granites are shown in Figures 1 and 2. Obviously, none of the distributions is lognormal. The non-lognormality is expected for a differentiated sequence of rocks, such as the White Mountain magma series; the process of differentiation is non-random and must produce a non-homogeneous group of rocks. In such a differentiated sequence, an element could be randomly distributed only if its abundance were independent of the differentiation process (e.g., if it were secondarily deposited), and the thorium in the

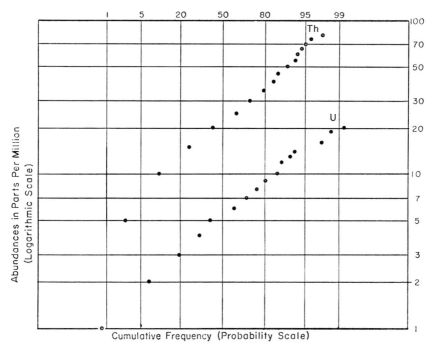

Fɪɢ. 2.—Cumulative frequency distributions for thorium and uranium contents of 123 samples of New England granites.

White Mountain magma series is almost certainly primary (Richardson, 1963). The deviation from lognormality shown in Figure 2 for the whole group of New England granites may simply represent inadequate sampling.

One highly tentative possibility must be discussed in comparing the frequency distributions of thorium with those of uranium for the two groups discussed above. Though the thorium distributions are clearly non-lognormal, those of uranium approach lognormality over considerable portions of the range. It may be that the tendency of uranium toward homogeneity in geologically non-homogeneous rocks is caused by the common secondary redistribution of uranium after the main rock-forming process. The possibility that the highly soluble uranium is less related to rock-forming processes than is thorium has been suggested by Whitfield *et al.* (1959) and may be reinforced by these distributions.

COMPARISON OF GRANITES FROM VARIOUS PARTS
OF THE UNITED STATES

Table 3 shows average thorium and uranium contents and Th/U ratios for different groups of granites among the Nevadan-Laramide and Precambrian suites of the central and western United States.

TABLE 3

RADIOMETRIC DATA FOR VARIOUS GRANITIC ROCKS OF PRECAMBRIAN
AND NEVADAN-LARAMIDE AGE IN THE CENTRAL
AND WESTERN UNITED STATES

	No. Samples	Thorium (p.p.m.)*	Uranium (p.p.m.)*	Th/U Ratio†
Precambrian				
Granites from Oklahoma..............	9	14	3.7	4.2
Granites from Minnesota.............	3	20	3.2	7
Granites from the Front Range, Colorado.	8	27	4.6	5.9
Enchanted Rock batholith, Texas........	81	20	3.5	6
Nevadan-Laramide				
Southern California batholith...........	54	8	2.4	3.8
Granites from Sierra Nevadas, California..	20	18	5.4	3.5
Granites from Washington, Idaho batholith, and Boulder batholith...........	38	10	2.5	4.7

* Thorium and uranium values are cited as parts per million of the metal, assuming radioactive equilibrium in the respective decay series.

† The cited Th/U ratios are the average of the various Th/U ratios of the individual samples, not the ratio of average Th/average U.

Equivalent data for New England granites are given in Table 2. Data for Table 3 have been assembled from Gottfried and Larsen (1959), Larsen (1958), Adams and Rogers (1961), Whitfield *et al.* (1959), and Rogers and Ragland (1961). The reports of Larsen (1958) and Gottfried and Larsen (1959) provide data on some individual rocks but tabulate only the averages of groups of samples for some rock types.

No attempt has been made to construct frequency distributions for any of the data summarized in Table 3. In the first place, the sampling is certainly not fully representative of the "population" of all Nevadan-Laramide or all Precambrian rocks. Second, data are not available on all individual samples.

The following general and highly tentative conclusions may be drawn from the data shown in Tables 2 and 3.

1. The New England granites are obviously rich in thorium and uranium in comparison with the Nevadan-Laramide granites. The difference between these two suites is, of course, reflected in the tendency of the New England granites to be potassic and generally alkaline, whereas the Nevadan-Laramide granites are commonly rich in plagioclase and are most properly classified as quartz monzonites and granodiorites.

2. The Precambrian granites have an unusually high Th/U ratio owing to a comparatively low uranium content. This fact has been noted previously by Whitfield *et al.* (1959). The Precambrian and Nevadan–Laramide granites are obviously different in thorium and uranium contents.

3. The Th/U ratio of the New England granites may be slightly higher than that of the Nevadan-Laramide granites. This hypothesis cannot be quantified by statistical methods owing to the absence of information on some individual samples.

It is obviously unreasonable to determine precise over-all averages for the data in Tables 2 and 3 because of the sporadic nature of the sampling. The number of samples is, however, fairly large, and a number of different types of granites are represented. Consequently, it is possible to estimate the average thorium and uranium contents of "granite." From the data presented in Tables 2 and 3, roughly weighted according to the importance of the various rock types, the average thorium content may be estimated as 15 p.p.m., the average uranium content as 3–4 p.p.m., and the average Th/U ratio as 4–5. These figures should not be taken as the crustal average, for the crust probably does not contain as much true granitic rock as is commonly thought (e.g., note the recent work in shield areas, which shows large amounts of metasedimentary and other rock types in terrains mapped as "granitic shield"). Considering the abundance of rock types other than granite in the continental crust, the average thorium content of the crust may be reduced to approximately 12 p.p.m., the average uranium content to 3 p.p.m., and the average Th/U ratio to 4. These figures are close to those obtained by Adams *et al.* (1959).

ACKNOWLEDGMENTS

This study was made possible both by grant C-009 from the Robert A. Welch Foundation and by subcontract 1491 from the Oak Ridge National Laboratory to John A. S. Adams and John J. W. Rogers.

REFERENCES

ADAMS, J. A. S., and G. FRYER. 1963. Portable γ-ray spectrometer for field determination of thorium, uranium, and potassium. This symposium.

ADAMS, J. A. S., M.-C. KLINE, K. A. RICHARDSON, and J. J. W. ROGERS. 1962. The Conway granite of New Hampshire as a major low-grade thorium resource. Proc. Nat. Acad. Sci., 48:1898–1905.

ADAMS, J. A. S., J. K. OSMOND, and J. J. W. ROGERS. 1959. The geochemistry of thorium and uranium. In L. H. AHRENS (ed.), Physics and Chemistry of the Earth, 3:298–348. London: Pergamon Press.

ADAMS, J. A. S., and J. J. W. ROGERS. 1961. Geologic Aspects of Thorium Recovery from Common Rocks: Annual Report of May 31, 1961, to Oak Ridge National Laboratory (TID-13001).

FLINN, D. 1959. An application of statistical analysis to petrochemical data. Geochim. & Cosmochim. Acta, 17:161–75.

GOTTFRIED, D., and E. S. LARSEN III. 1959. Distribution of uranium and thorium in igneous rocks. In Geologic Investigations of Radioactive Deposits (TEI-752), pp. 70–90.

HUTCHINSON, R. M. 1956. Structure and petrology of Enchanted Rock batholith, Llano and Gillespie counties, Texas. Geol. Soc. America Bull., 67:763–806.

KRUMBEIN, W. C., and H. A. SLACK. 1956. Statistical analysis of low-level radioactivity of Pennsylvanian black fissile shale in Illinois. Geol. Soc. America Bull., 67:739–62.

KRUMBEIN, W. C., and J. W. TUKEY. 1956. Multivariate analysis of mineralogic, lithologic, and chemical composition of rock bodies. Sed. Pet., 26:322–37.

LARSEN, E. S., III. 1958. Thorium in igneous rocks. In Geologic Investigations of Radioactive Deposits (TEI-740), pp. 308–10.

RAGLAND, P. C., and J. A. S. ADAMS. Chemical, radiometric, and mineralogic variation within a zoned granitic batholith. (In preparation.)

RICHARDSON, K. A. 1963. Thorium and uranium in the Conway granite, New Hampshire, U.S.A. This symposium.

ROGERS, J. J. W., and J. A. S. ADAMS. 1963. Lognormality of thorium concentrations in the Conway granite: Geochim. & Cosmochim. Acta, 27:775–83.

ROGERS, J. J. W., and P. C. RAGLAND. 1961. Variation of thorium and uranium in selected granitic rocks. Geochim. & Cosmochim. Acta, 25: pp. 99–109.

WHITFIELD, J. M., J. J. W. ROGERS, and J. A. S. ADAMS. 1959. The relationship between the petrology and the thorium and uranium contents of some granitic rocks. Geochim. & Cosmochim. Acta, 17:248–71.

WHITTEN, E. H. T. 1959. Composition trends in a granite: Modal variation and ghost stratigraphy in part of the Donegal granite, Eire. Geophys. Res., 64:835–48.

K. S. HEIER AND JAMES L. CARTER

4. *Uranium, Thorium, and Potassium Contents in Basic Rocks and Their Bearing on the Nature of the Upper Mantle*

B ASALTIC ROCKS ARE POSSIBLY more important on the surface of the earth than is any other single rock type. The floors of the oceans that cover about 70 per cent of the surface of the earth consist almost entirely of basalt, covered and possibly interlayered with thin pelagic sediments. The Pacific islands are built up almost entirely of basaltic rocks. Similarly, Iceland and other islands along the mid-Atlantic ridge consist almost entirely of basalt. In these areas volcanoes constantly bring basaltic material to the surface. The outpouring of lava is accompanied by large amounts of gases that may bring short-lived radioisotopes directly into the atmosphere.

Current estimates of the composition of the continental crust indicate approximately 20 per cent contribution from basaltic or chemically similar material (Poldervaart, 1955). Though the contribution to the background radiation over the continents is on an average probably more like that from a granodioritic composition (7–10 p.p.m. thorium, 2–3 p.p.m. uranium, 2–3 per cent potassium), large areas of the continents are entirely covered by basaltic material. The plateau basalts are associated with fissure eruptions and are the largest effusions known on earth. Active vulcanism of this type is known from Iceland, but the greatest outpourings are the prehistoric floods that cover vast areas in Brazil, India, Siberia, Oregon, and the Thule province in Greenland.

K. S. HEIER is senior research associate, Department of Geophysics, Australian National University, Canberra, Australia, and JAMES L. CARTER is Robert A. Welch Foundation Fellow in Geochemistry, Rice University, Houston, Texas.

BASALTS

Because of their low concentrations in these rocks, very little data are available on thorium and uranium abundances in basaltic rocks (see Appendix following this article), despite their large areal extension and also their special petrogenic position. Basalt magma is both *parental*, that is, other rocks have been derived from it through fractionation, and *primary*, in that it is not derived from any other magma through a liquid line of descent.

Recently Heier and Rogers (1963) published results of a radiometric examination of thorium and uranium contents in basalts from various provinces. Their results showed that thorium and uranium concentrations vary within nearly 2 orders of magnitude according to basalt type. The alkali olivine basalts had the highest concentrations, and the lowest concentrations were in tholeiitic basalts of orogenic

TABLE 1*

THORIUM, URANIUM, AND POTASSIUM IN MAJOR BASALT TYPES

	Thorium (p.p.m.)*	Uranium (p.p.m.)	Th/U	K (per cent)
Tholeiites of orogenic type......	0.0X–0.32	0.0X–0.26	1.5	0.06–0.36
Tholeiites of non-orogenic type..	0.6 –2.4	0.17 –0.5	2.2–6.8	0.34–1.33
Alkali olivine basalt............	2.0 –8.8	0.44 –1.4	4.3–8.8	0.81–1.57

* These ranges are based mainly on data from Heier and Rogers (1963). Summaries of all available data on thorium and uranium in basic rocks are given in the Appendix and are graphically shown in Figures 5 and 6.

type. Some plateau basalts that have been examined (Palisades, Columbia River, Karroo dolerites) have intermediate concentrations. These latter basalts are tholeiitic but are of a non-orogenic type. Table 1 gives the concentration ranges in the diverse basaltic types.

A close geochemical coherence was found to exist between the elements thorium, uranium, and potassium in basaltic rocks (Figs. 1 and 2). It seems that the potassium concentration, which is easy to determine, is a major guide to the thorium and uranium concentrations in unaltered basalts. The Th/K and U/K ratios do not vary with more than a factor of six, and within the same basalt type the variation is much less. Likewise, the Th/U ratio is remarkably constant in the various rocks examined (Fig. 3). There is a tendency for an increase in all the ratios with fractionation of the rocks. At the same time, the absolute concentrations of all the elements increase. The relative increase is thorium > uranium > potassium.

A striking relationship is found between the Th/U ratio and potassium (Fig. 4). The general chemical properties of thorium and urani-

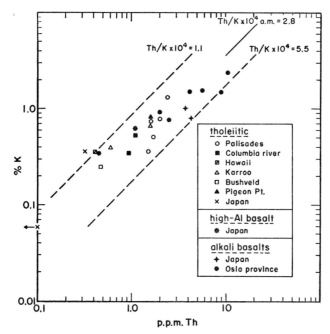

Fig. 1.—Thorium *vs.* potassium in basalts and related rocks. Dashed lines indicate limits of scatter. Solid central line shows the position of the arithmetic mean of the Th/K × 10⁴ ratio. (From Heier and Rogers, 1963.)

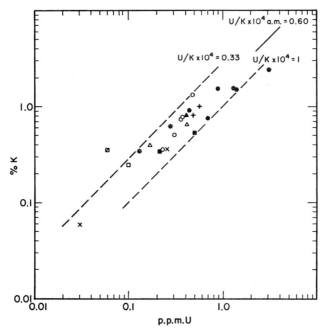

Fig. 2.—Uranium *vs.* potassium in basalts and related rocks. Solid central line shows the position of the arithmetic mean of the U/K × 10⁴ ratio. Symbols as for Figure 1. (From Heier and Rogers, 1963.)

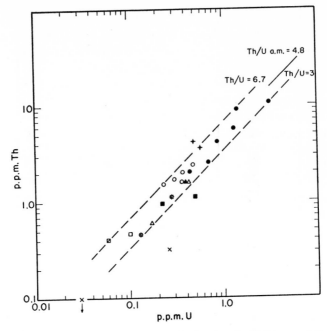

FIG. 3.—Thorium *vs.* uranium in basalts and related rocks. Solid central line shows the position of the arithmetic mean of the Th/U ratio. Symbols as for Figure 1. (From Heier and Rogers, 1963.)

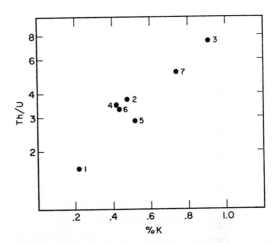

FIG. 4.—Potassium *vs.* Th/U ratio in primary basic magmas. *1*, Japan, tholeiite; *2*, Japan, high-aluminum basalt; *3*, Japan, alkali olivine basalt; *4*, Duluth, layered gabbro; *5*, S. California batholith, gabbro; *6*, Columbia River basalts; *7*, Palisades sill. (From Heier and Rogers, 1963.)

um, on the one hand, and of potassium, on the other, are sufficiently different for one not to expect any relationship between them. The mechanisms for fractionation of thorium and uranium are probably quite different from those which control the distribution of potassium. However, Figure 4 may allow us to draw one very generalized conclusion: the geological processes that control the Th/U ratio (possibly including oxidation) are not independent of the processes that control the major compositional variation of the different rock types as they are exemplified by their potassium contents.

AMPHIBOLITES

It was mentioned that the concentrations of thorium and uranium, as well as the value of the Th/U ratio, were at a minimum in tholeiitic basalts of orogenic type. Recently Billings (1962) determined thorium and uranium concentrations in amphibolites from a Precambrian area in central Texas. It was concluded on the basis of other evidence that the amphibolites were metamorphic equivalents of tholeiitic basalts of orogenic type. This is the commonly accepted origin of amphibolitic layers in old orogens. The history of these rocks, though complex, will be one of basaltic extrusions in an active orogenic area. The initial composition of the lava would be expected to be most similar to that of Japanese tholeiites. Later history would include some weathering (though possibly slight because of relatively rapid burial), interaction with ground water, and subsequent metamorphism, probably also involving hydrous reactions. All these processes will tend to oxidize the rock, and one might expect oxidation of the tetravalent uranium into its hexavalent state with subsequent solution and leaching of the soluble uranyl complex ion. The residual rock would reflect these processes in exhibiting low uranium contents and high Th/U ratios. However, Billings concluded, (1) the uranium contents are quite similar to those in basalts, and (2) the abundance of thorium is consistently less than that of basalts (with the exception of the Japanese tholeiites). He concluded that the low thorium content and low Th/U ratio (average, 0.8) suggest a selective loss of thorium during metamorphism. Later work with refinement of techniques by one of us (J. Carter) seems to indicate that this ratio should be slightly above 1. Such a process is not clearly understood. It may be that partial melting, anatexis, of amphibolites during severe metamorphism (melting beginning as low as 600° C.; Yoder and Tilley, 1962) will lead to this result. This melt, which will have the bulk composition of a plagioclase, would probably mix with the more granitic melt formed simultaneously from the gneissic country rocks and ascend through the geosyncline. Blebs of melt formed at different places might coalesce and crystallize at a higher level as an anatectic granite. If thorium was preferentially concentrated in the palingenic melt, relative to uranium, anatectic granites should be characterized by high Th/U ratios. Heier (1960) described some anatectic granites on a petrographic basis, and these were later shown to have high Th/U ratios (between 6 and 30; Heier, 1962). It is outside the scope of this paper to discuss why thorium should be selectively enriched in melts formed this way. If thorium and uranium coexist in the same minerals, it cannot be deduced from their ionic and atomic properties. However, in most accessory minerals in rocks the Th/U ratio is different from the average crustal ratio of 3–4 (Table 2). The Th/U ratio in the melt may therefore to a large extent depend on the relative melting points of these minerals.

TABLE 2

THORIUM AND URANIUM CONCENTRATIONS AND TH/U RATIOS
IN SOME ACCESSORY MINERALS*

Mineral	Thorium (p.p.m.)	Uranium (p.p.m.)	Th/U
Allanite..........	500– 5,000	30– 700	5 –10
Apatite..........	20– 150	5– 150	1
Epidote..........	50– 500	20– 50	2 – 6
Monazite..........	25,000–200,000	500– 3,000	25 –50
Sphene..........	100– 600	100– 700	1 – 2
Xenotime........	Low	500–35,000	Low (0.01*)
Zircon..........	100– 2,500	300– 3,000	0.2– 1

* From Adams *et al.*, 1959.

It is clear that this is a very important problem, not only for the geochemistry of thorium and uranium, but also for the study of rock-forming processes. Because the chemistry of thorium and uranium in so many respects is closely similar, their fractionation is likely to be petrogenically important.

BASIC INTRUSIVES

Most data indicate that the concentrations of thorium and uranium, and also the Th/U ratio, increase with rock differentiation. Therefore, when basic rocks, instead of being extruded on the surface as basalt lava, are intruded into some deeper part in the crust and allowed to crystallize slowly and differentiate, one would expect the absolute concentrations as well as the Th/U ratio in the basic members to be less than in the parent magma. As the absolute concentrations of thorium and uranium may vary within two orders of magnitude, depending on basalt type, the concentrations of these elements in plutonic basic rocks cannot be safely used as a test of this hypothesis. However, Hamilton (1959*b*), who studied the differentiated Skaergaard intrusion, East Greenland, found 0.2 p.p.m. uranium (0.32 per cent potassium) in the chilled marginal gabbro, which is believed to represent a frozen sample of the original magma. Similar high-uranium concentrations are not found (with one exception) before the ferrogabbro stage is reached in the upper-half portion of the intrusion. Unfortunately, he was not able to determine thorium in this interesting intrusion. Studies of the differentiated series of the Duluth and southern California batholith show the basic rocks of both sequences to have lower Th/U ratios than most basalts (Heier and Rogers, 1963), but they are not as low as the Japanese tholeiites. Adams *et al.* (1959) indicate that Th/U ratios in basic intrusives vary between 3 and 4, and in basic extrusives between 3 and 7.

ECLOGITES

That basaltic magma is primary means that it is not related to a parent magma along any liquid line of descent. Much discussion has arisen about the source of basalt magma. Widely entertained was the concept of a basaltic layer along the continent-mantle boundary. Fermor argued as early as 1913 that such a layer of basaltic composition would exhibit the mineral phases of eclogite. This concept has been supported by a number of workers: Goldschmidt (1922), Holmes (1927), Birch (1952), Lovering (1958), and Kennedy (1959). Yoder and Tilley (1962, p. 501) stated that "for every basalt there is an equivalent eclogite of the same bulk composition." From data presented by them it seems that the Mohorovičić discontinuity under the continents may be the result of the transformation of basalt to eclogite, provided that a layer of this composition exists at this depth. It is equally clear that this transformation cannot define the discontinuity under the oceans. No one has considered the thickness of this layer under the continents. If eclogites and basalts have similar concentrations of the radioactive elements, severe restrictions are imposed on the possible thickness of this layer in order not to give impossible values for the heat flow.

Some information on the concentration of these elements in eclogites was given recently by Heier (1963) and Lovering and Morgan (1963). These data are tabulated in the Appendix. The arithmetic means of the absolute concentrations of thorium and uranium as well as the Th/U ratio are similar in the eclogites from the volcanic and metamorphic environments. The eclogite inclusions in Hawaiian basalt have low concentrations of thorium and uranium, but the Th/U ratio is similar to that in the other eclogites. The three Australian eclogites have consistently low uranium concentrations and high Th/U ratios, possibly because of secondary alteration of the samples. Most alteration and weathering processes will tend to oxidize and leach uranium, resulting in high Th/U ratios.

The average thorium and uranium concentrations and Th/U ratio are most similar to the tholeiitic basalts of orogenic type.

THE UPPER MANTLE

Kuno (1959) assigned different depth levels in the mantle to the origin of the different types of primary basaltic magma. Tholeiites were derived from the mantle above 200 km., high-aluminum basalts from about 200 km., and the alkali olivine basalts from a depth below 200 km. A similar view was argued by Clark (1959) on theoretical grounds. Yoder and Tilley (1962, p. 509), though protesting against the primary nature of the high-aluminum basalt magma, found that their experiments supported Kuno's contention, saying: "A garnet peridotite, for example, would yield nepheline-trending magmas at high pressures, and silica-trending magmas at low pressures. On these grounds one can conclude that the alkali basalt magmas are derived at greater depths than tholeiite magmas." They continued: "The writers believe that this argument constitutes strong support for the concept of a single primary source for all basalt magma." It seems, however, that the differences observed in the thorium and uranium concentrations, and above all in the Th/U ratio, in the various basalt types would indicate a chemical zoning of the upper mantle. The effect of load pressure alone can hardly cause a fractionation of thorium and uranium. Yoder and Tilley's own experiments do not disprove this possibility (1962, p. 509): "That a single primary source can obviously have a relatively wide range of bulk composition. . . ."

More data on meaningful trace elements and element ratios in, for instance, the different types of basalt from Japan may go a long way toward solving the problem of chemical zoning of the mantle (see also Ringwood, 1962*a*, *b*).

CONCLUSIONS

The thorium and uranium data alone cannot define the chemical conditions in the upper mantle. They are consistent, at least for the upper mantle beneath the continents, with a model proposed by Ringwood (1962*a*, *b*). The mantle under the continents immediately below the Mohorovičić discontinuity consists predominantly of dunite and peridotite. This is formed from the more primitive mantle material by a selective loss of lithophile elements that have been concentrated in the crust. This concentration may have taken place through both diffusive processes and fractional melting with ascent of magma. Very little data are available on the concentrations of uranium and thorium in peridotite and dunite (see Appendix). The difference of one order of magnitude between the uranium determinations on the dunite, Twin Sisters, Washington, by isotope dilution and neutron activation methods, respectively, make all these values uncertain. However, the concentrations are small, and we would expect the Th/U ratios to be low. This ultrabasic upper mantle passes downward into the primitive mantle material. (The pyrolite of Ringwood [1962a, b] consists of 4 parts dunite-peridotite to 1 part of basalt, or chondritic material, as assumed by others.) The thorium and uranium concentrations and the value of the Th/U ratio increase downward until the depth of the primitive mantle material is reached (~ 700 km., or the deepest known earthquake focus). At any one depth fractional melting will give a magma with a Th/U ratio characteristic of the parent mantle material.

Yoder and Tilley (1962) give evidence that magma generation takes place at a depth below which basalt is stable, and they argue that the basalt composition arising from the partial melting of the primary source will behave at the depth of the magma generation in the same fashion as eclogite. The magma will erupt as basaltic lava if it ever reaches the surface. If it is trapped below a certain depth, it will crystallize as eclogite. Melting will take place at several places simultaneously, and small quantities of melt generated at each place will converge upward as they move along pressure gradients. It may be significant that the eclogite xenoliths in kimberlite pipes have the form of a "bomb" or "drop."

If large-scale lateral movements of elements do not occur, the upper mantle beneath the oceans cannot have been depleted in oxyphile (lithophile) elements to the same extent as the continental mantle. Apart from yielding material of oceanic basalt composition, the chemical composition of the oceanic mantle would be expected to have changed relatively little, and its composition is more nearly similar to the primary mantle material. It may therefore be significant that the Hawaiian basalts that are tholeiitic in nature are more alkaline than the circumpacific tholeiites. One tholeiitic quartz dolerite intrusion from Hawaii that was examined by Heier and Rogers (1963) had a Th/U ratio of 6.8 (0.41 p.p.m. thorium, 0.06 p.p.m. uranium, 0.36 per cent potassium). However, eclogite inclusions from Salt Lake Crater, Honolulu, have Th/U ratios between 1.8 and 2.4 (see Appendix).

Regardless of the details at the deep-seated origins of basalt and related rocks, basalts represent one of the most widespread of the low-radioactivity rocks.

REFERENCES

ABRAMOVICH, I. I. 1959. Uranium and thorium in the intrusive rocks of central and western Tuva. Geochemistry, No. 4, pp. 442–50.

ADAMS, J. A. S. 1955. The uranium geochemistry of Lassen Volcanic National Park, California. Geochim. & Cosmochim. Acta, **8**:74–85.

ADAMS, J. A. S., J. K. OSMOND, and J. J. W. ROGERS. 1959. The geochemistry of thorium and uranium. *In* Physics and Chemistry of the Earth, **3**:298–348. New York: Pergamon Press.

BILLINGS, G. K. 1962. A geochemical investigation of the Valley Spring gneiss and Packsaddle schist, Llano uplift, Texas. Texas J. Sci., **14**:328-51.

BIRCH, F. 1952. Elasticity and constitution of the earth's interior. J. Geophys. Res., **57**:227–86.

–––. 1954. Heat from radioactivity. *In* H. FAUL (ed.), Nuclear Geology, pp. 148–74. New York: John Wiley & Sons.

CARR, D. R., and J. L. KULP. 1953. Age of a mid-Atlantic ridge basalt boulder. Geol. Soc. America Bull., **64**:253–54.

CARTER, J. L. Unpublished thorium and uranium basalt data. 1962.

CLARK, S. P., JR. 1959. Equations of state and polymorphism at high pressures. *In* P. H. ABELSON (ed.), Researches in Geochemistry, pp. 495–511. New York: John Wiley & Sons.

CLAYTON, D. D. 1963. A calculation of the abundances of uranium and thorium from the primordial Pb-206/Pb-207 ratio. J. Geophys. Res., **68**:3715–21.

DAVIS, G. L. 1947. Radium content of ultramafic igneous rocks. I. Laboratory investigations. Am. Jour. Sci., **245**:677–93.

DAVIS, G. L., and H. H. HESS. 1949. Radium content of ultramafic igneous rocks. II. Geological and chemical implications. Am. Jour. Sci., **247**:856–82.

DAVIS, G. L., G. R. TILTON, B. R. DOE, L. T. ALDRICH, and S. R. HART. 1961. Lead content and radioactivity of dunite and eclogite with application to the problem of the constitution of the mantle. Carnegie Inst. Wash. Year Book 60, pp. 195–97.

EVANS, R. D., and C. GOODMAN. 1941. Radioactivity of rocks. Geol. Soc. America Bull., **52**:459-90.

FERMOR, L. L. 1913. Preliminary note on garnet as a geological barometer and on an infraplutonic zone in the earth's crust. Rec. Geol. Survey India, **43** (Pt. 1). 41–47.

GAST, P. W. 1960. Limitations on the composition of the upper mantle. J. Geophys. Res., **65**:1287–97.

GOLDSCHMIDT, V. M. 1922. Über die Massenverteilung in Erdinneren, verglichen mit der Struktur gewisser Meteoriten. Naturwissenschaften, **42**:918–20.

HAMAGUCHI, H., G. W. REED, and A. TURKEVICH. 1957. Uranium and barium in stone meteorites. Geochim. & Cosmochim. Acta, **12**:337–47.

HAMILTON, E. 1959a. Studies in the distribution of uranium in the Skaergaard intrusion and other igneous and metamorphic rock series. Medd. Dansk Geol. Føren., Vol. **14.**

——. 1959b. The uranium content of the differentiated Skaergaard intrusion, together with the distribution of the alpha particle radioactivity in the various rocks and minerals as recorded by nuclear emulsion studies. Meddel. om Grønland, Vol. **162**, No. 7.

HEIER, K. S. 1960. Petrology and geochemistry of high-grade metamorphic and igneous rocks on Langøy, northern Norway. Norg. Geol. Undersøkelse, No. 207, pp. 1–246.

——. 1962. Spectrometric uranium and thorium determinations on some high-grade metamorphic rocks on Langøy, northern Norway. Norsk Geol. Tidsskr., **42**:143–56.

——. 1963. Uranium, thorium and potassium in eclogitic rocks. Geochim. & Cosmochin. Acta, **27**:849–60.

HEIER, K. S. and J. J. W. ROGERS. 1963. Radiometric determination of thorium, uranium, and potassium in basalts and in two magmatic differentiation series. Geochim. & Cosmochim. Acta, **27**:137–54.

HOLMES, A. 1927. Some problems of physical geology in the earth's thermal history. Geol. Mag., **64**:263–78.

HURLEY, P. M. 1956. Direct radiometric measurements by gamma-ray scintillation spectrometer. II. Uranium, thorium, and potassium in common rocks. Geol. Soc. America Bull., **67**:405–12.

HURLEY, P. M., and C. GOODMAN. 1941. Helium retention in common rock minerals. Geol. Soc. America Bull., **52**:545–60.

KEEVIL, N. B. 1938. Thorium-uranium ratios of rocks and their relation to lead ore genesis. Econ. Geology, **33**:685–96.

——. 1943a. The distribution of helium and radioactivity in rocks. Am. J. Sci., **241**:277–306.

——. 1943b. Helium indexes for several minerals and rocks. *Ibid.*, pp. 680–93.

——.1944. Thorium-uranium ratios in rocks and minerals. *Ibid.*, **242**: 309–21.

KEEVIL, N. B., A. W. JOLLIFFE, and E. S. LARSEN. 1942. The distribution of helium and radioactivity in rocks. Am. J. Sci., **240**:831–46.

KENNEDY, G. C. 1959. The origin of continents, mountain ranges, and ocean basins. Am. Scientist, **47**:491–504.

KUNO, H. 1959. Origin of Cenozoic petrographic provinces of Japan and surrounding areas. Bull. Volcanologique Ser. (2), **20**:37–76.

LARSEN, E. S., JR. 1954. Distribution of uranium in igneous complexes: Geological investigations of radioactive deposits. Semiannual Progress

Report, June 1 to Nov. 30, 1954. U. S. Atomic Energy Commission, TEI-490, pp. 255–61.

LARSEN, E. S., JR., G. PHAIR, D. GOTTFRIED, and W. L. SMITH. 1955. Uranium in magmatic differentiation. U. S. Geol. Survey Prof. Paper 300, pp. 65–74.

LOVERING, J. F. 1958. The nature of the Mohorovičić discontinuity. Trans. Am. Geophys. Union, **39**:947–55.

LOVERING, J. F., and J. W. MORGAN. 1963. Uranium and thorium abundances in possible upper mantle materials. Nature, **197**:138–40.

MARSHALL, R. R. 1960. The amounts and isotopic composition of lead in eclogite from the Münchberg gneiss massiv (Fichtelgebirge). Rpt. Internat. Geol. Cong. XXI, Norden, Part XIII, pp. 404–17.

PATTERSON, C., G. TILTON, and M. INGHRAM. 1955. Age of the earth. Science, **121**:69–75.

PICCIOTTO, E. E. 1950. Distribution de la radioactivité dans les roches éruptives. Société Belge de Géologie de Paléontologie et d'Hydrologie Bull., **59**:170–98.

POLDERVAART, A. 1955. Chemistry of the earth's crust. Geol. Soc. America Spec. Paper, **62**:119–44.

POLYAKOV, A. I., and M. P. VOLYNETS. 1961. Distribution of thorium in a series of ultrabasic (alkalic) rocks of the Kola Peninsula. Geochemistry, No. 5, pp. 446–54.

RICE UNIVERSITY, GEOCHEMISTRY CLASS. Unpublished thorium and uranium basalt data. Courtesy Professor J. A. S. Adams, 1961.

RINGWOOD, A. E. 1962a. A model for the upper mantle. I. J. Geophys. Res., **67**:170–98.

———. 1962b. A model for the upper mantle, II. *Ibid.*, pp. 4473–77.

SMYSLOV, A. A. 1958. Radioactive elements in igneous rocks of northern Kazakhstan. Geochemistry, No. 3, pp. 248–58.

TILTON, G. R., G. W. WETHERILL, L. T. ALDRICH, G. L. DAVIS, and P. M. JEFFERY. 1956. Geochemistry of uranium and lead. Carnegie Inst. Wash. Year Book, **55**, pp. 99–100.

TUROVSKII, S. D. 1957. Geochemistry of uranium and thorium: Distribution of radioactive elements in intrusive rocks of northern Kirgizia. Geochemistry, No. 2, pp. 199–215.

VINOGRADOV, A. P. 1961. The origin of the matter of the earth's crust, I. Geochemistry, No. 1, pp. 3–29.

WOLLENBERG, H. A. Personal communication, 1963.

YODER, H. S., JR., and C. E. TILLEY. 1962. Origin of basalt magmas: An experimental study of natural synthetic rock systems. J. Petrology, **3**:342–532.

APPENDIX

THORIUM AND URANIUM CONTENTS IN BASIC AND ULTRABASIC ROCKS

(See original papers for complete description of samples)

Description	Reference*	Locality	Method of Analysis	Thorium (p.p.m.)	Uranium (p.p.m.)	Th/U
PLUTONIC ROCKS						
Basic rocks (Gabbroids):						
Gabbro	24 (1954)	Boulder batholith, Mont.	Fluorimetry		1.0	
1055 Layered gabbro	16 (1963)	Duluth batholith, Mont.	γ Radiometry	0.96 ±15%	0.37 ±13%	2.6
1956 Layered gabbro	16 (1963)	Duluth batholith, Mont.	γ Radiometry	0.77 ±20%	0.14 ±50%	5.1
1057 Layered gabbro	16 (1963)	Duluth batholith, Mont.	γ Radiometry	0.74 ±20%	0.27 ±20%	2.7
Hornblende gabbro	24 (1954)	Sierra Nevada, Bishop, Calif.	Fluorimetry (2)†		0.42	
Sc-19-66 hbl. pyr. gabbro	16 (1963)	San Marcos gabbro, southern Calif. batholith	γ Radiometry	0.94 ±10%	0.31 ±15%	3
Sc-17-65 hbl. gabbro	16 (1963)	San Marcos gabbro	γ Radiometry	1.5 ±12%	0.50 ±14%	3
Sc-15-63 qtz. hbl. gabbro	16 (1963)	San Marcos gabbro	γ Radiometry	3.1 ±6%	1.1 ±7%	2.8
822 Anorthositic gabbro	16 (1963)	Duluth, Mont.	γ Radiometry	8.7 ±4%	2.5 ±5%	3.5
34 Anorthositic gabbro	16 (1963)	Duluth, Mont.	γ Radiometry	8.8 ±4%	2.8 ±5%	3.1
807 Anorthositic gabbro	16 (1963)	Duluth, Mont.	γ Radiometry	6.1 ±5%	1.9 ±6%	3.2
Gabbro (basic border)	20 (1943a)	Quarry near St. George, New Brunswick	Streaming, direct fusion	1.7 ±0.3	0.51 ±0.03	3.3
Sudbury gabbro	18 (1941)	Graham Township, Ont.	Direct fusion		0.27	
Gabbro, chilled marginal	14 (1959)	Skaergaard intr., E. Greenland	Neutron activation		0.2	
1063 Oslo essexite	16 (1963)	Oslo Province, Norway	γ Radiometry	10.4 ±3%	3.1 ±3%	3.4
Gabbro	1 (1959)	Central and western Tuva, U.S.S.R.	Luminescence (10)		<1.0	
			Radiochemical (3)	3	<0.9	> 3
Gabbro	32 (1958)	West part of northern Kazakhstan, U.S.S.R.	Luminescence (3)		0.4	
Gabbro, small bodies	32 (1958)	East part of northern Kazakhstan, U.S.S.R.	Luminescence (2)		0.6	

* References: 1, Abramovich (1959); 2, Adams (1955); 3, Adams, Osmond, and Rogers (1959); 4, Birch (1954); 5, Carr and Kulp (1953); 6, Carter (1962); 7, Clayton (1963); 8, Davis (1947); 9, Davis and Hess (1949); 10, Davis, Tilton, Doe, Aldrich, and Hart (1961); 11, Evans and Goodman (1941); 12, Gast (1960); 13, Hamaguchi, Reed, and Turkevich (1957); 14, Hamilton (1959b); 15, Heier (1963); 16, Heier and Rogers (1963); 17, Hurley (1956); 18, Hurley and Goodman (1941); 19, Keevil (1938); 20, Keevil (1943a); 21, Keevil (1944); 22, Keevil (1943b); 23, Keevil, Jolliffe, and Larsen (1942); 24, Larsen (1954); 25, Larsen, Phair, Gottfried, and Smith (1955); 26, Lovering and Morgan (1963); 27, Marshall (1960); 28, Patterson, Tilton, and Inghram (1955); 29, Picciotto (1950); 30, Polyakov and Volynets (1961); 31, Rice University Geochemistry Class (1961); 32, Smyslov (1958); 33, Tilton, Wetherill, Aldrich, Davis, and Jeffery (1956); 34, Turovskii (1957); 35, Vinogradov (1961); and 36, Wollenberg (1963).

† Number of samples.

APPENDIX—*Continued*

Description	Locality	Reference*	Method of Analysis	Thorium (p.p.m.)	Uranium (p.p.m.)	Th/U
PLUTONIC ROCKS—*Continued*						
Basic rocks-(Gabbroids)—Continued						
Gabbro........	Shimbone Range, Suak	34 (1957)	Radiochemical	5	0.5	10
Gabbroids....	Ala-Tau, Terek, U.S.S.R.	34 (1957)	Radiochemical (7)	4	0.6	6.6
28B Blueberry Mt. gabbro-diorite (NBS)	Crush stone quarry, Woburn, Mass.	11 (1941)	Total α act., direct fusion	1.6 ±0.3	0.45±0.03	3.6
NBS gabbro-diorite....	Woburn, Mass.	8 (1947)	Refined vacuum fusion	....	0.51±0.03	.....
Gabbro-diorite.......	Central and western Tuva, U.S.S.R.	1 (1959)	Luminescence (33)	....	1.5	.....
Gabbro-diorite.......	Oktorkoi Range, Kirghizian Range, N. Kirghizia, U.S.S.R.	34 (1957)	Radiochemical (2)	9	0.7	12.8
Gabbro-diorite.......	Arsy R., eastern part of Kirghizian Range, U.S.S.R.	34 (1957)	Radiochemical (4)	0.2	.....	.....
1066 Norite.....	Bushveld, S. Africa	16 (1963)	γ Radiometry	0.47 ±22%	0.10±50%	4.7
1068 Olivine dolerite..	Karroo, S. Africa	16 (1963)	γ Radiometry	0.60±20%	0.17±20%	3.5
20B2 Sudbury norite...	Outcrop, Creighton, Ont.	11 (1941)	Total α act., direct fusion	4.3 ±0.8	1.5 ±0.4	2.9
21B2 Sudbury norite..	40' level, Creighton, Ont.	11 (1941)	Total α act., direct fusion	2.1 ±0.4	0.7 ±0.14	3
19B Worthington norite..	Worthington Station, Ont.	11 (1941)	Total α act., direct fusion	15 ±3	2.2 ±0.3	6.8
22B Stillwater norite..	Quad Creek, Mont.	22 (1944)	Streaming, direct fusion	2.13 ±0.3	0.47±0.06	4.5
Stillwater norite...	Quad Creek, Mont.	11 (1941)	Total α act., direct fusion	0.96±0.15	0.37±0.03	2.6
Norite........	Egersund, southern Norway	29 (1950)	Autoradiography	....	0.4	.....
Arithmetic mean of "Gabbroids"	....	.....	....	3.84(24)‡	0.84(34)	4.3(23)
Other "plutonic" rocks:						
824 Diabase....	Pigeon Pt., Minn.	15 (1963)	γ Radiometry	1.6 ±12%	0.41±15%	3.9
Diabase......	Brotherton No. 2, Gogebic Range, Mich.	22 (1944)	Streaming, direct fusion	1.72±0.2	0.4 ±0.02	4.3
Diabase......	Horne mine, Noranda, Que.	22 (1944)	Streaming, direct fusion	2.43±0.23	1.38±0.10	1.8
14B Diabase....	Horne mine, Noranda, Que.	11 (1941)	Total α act., direct fusion	1.8 ±0.4	1.3 ±0.3	1.4
Diabase......	Yellowknife, NWT, Canada	23 (1942)	Total α act., direct fusion	0.63	0.16	3.9
Diabase......	Yellowknife, NWT, Canada	22 (1944)	Streaming, direct fusion	0.45±0.04	0.18±0.02	2.5
15B2 Olivine diabase..	20' level, Creighton mine, Ont.	11 (1941)	Total α act., direct fusion	2.4 ±0.3	0.97±0.2	2.4
Olivine diabase....	Worthington, Ont.	22 (1944)	Streaming, direct fusion	15 ±3	2.2 ±0.3	6.8

‡ Number of values averaged.

APPENDIX—*Continued*

Description	Locality	Reference*	Method of Analysis	Thorium (p.p.m.)	Uranium (p.p.m.)	Th/U
Other "plutonic" rocks—Continued						
15B3 Olivine diabase.	McKim twn., Sudbury, Ont.	18 (1941)	Direct fusion	....	0.74	....
Diabase porphyrites.	Central and western Tuva, U.S.S.R.	1 (1959)	Luminescence (20)	....	1.8	....
		....	Radiochemical (2)	6	1.5	4
Diabase and diabase porphyrites.	West part of northern Kazakhstan, U.S.S.R.	32 (1958)	Luminescence (17)	....	1.1 ±0.3	....
		....	Radiochemical (1)	3	1.1	2.7
Diabase and diabase porphyrites.	East part of northern Kazakhstan, U.S.S.R.	32 (1958)	Luminescence (18)	....	0.9 ±0.2	....
Trap rock.	Long Lake, Wyo.	22 (1944)	Streaming, direct fusion	5.15±0.51	1.29±0.06	4
Trap rock.	2' from margin of 100' dike, Long Lake, Wyo.	21 (1943b)	Streaming, direct fusion	4.08	1.36	3
24B1 Long Lake trap.	Long Lake, Wyo.	11 (1941)	Total α act., direct fusion	2.1 ±0.4	1.3 ±0.2	1.6
37 Basalt (trap rock).	Gogebic Range, Wis.	19 (1938)	Streaming, direct fusion	1.72	0.40	4.3
39B Trap (gogebic sill).	Gogebic Range, Mich.	11 (1941)	Total α act., direct fusion	3.2 ±0.3	0.91±0.17	3.5
40B Gogebic sill.	Wakefield, Mich.	11 (1941)	Total α act., direct fusion	15 ±3	1.3 ±0.2	11.5
41B Gogebic trap.	Gogebic Range, Mich.	11 (1941)	Direct fusion	....	0.28±0.06	....
42B Trap.	Gogebic Range, Mich.	11 (1941)	Direct fusion	....	0.11±0.03	....
43B Trap.	Gogebic Range, Mich.	11 (1941)	Direct fusion	....	0.45±0.09	....
25B Beartooth trap.	Beartooth, Mont.	11 (1941)	Total α act., direct fusion	0.81±0.15	0.60±0.08	1.4
9B1 Watchung trap.	Summit, N.J.	11 (1941)	Direct fusion	....	0.51±0.03	....
9B2 Watchung trap.	Summit, N.J.	11 (1941)	Direct fusion	....	0.45±0.03	....
23B1 Sudbury trap.	Copper Cliff, Ont.	11 (1941)	Total α act., direct fusion	0 ±0.2	2.5 ±0.3	....
23B2 Sudbury trap.	Copper Cliff, Ont.	11 (1941)	Direct fusion	....	1.96±0.09	....
Oliv.-ne.-mel. basalt (trap)						
Vert. col. joint. WSW. wall.	Rock quarry, Knippa, Tex.	31 (1961)	γ Radiometry	7.28±3%	2.23±2%	3.3
Non-vert. col. joint. SSE. wall.	Rock quarry, Knippa, Tex.	31 (1961)	γ Radiometry	5.28±4%	1.66±3%	3.2
Massive plug, S. wall.	Rock quarry, Knippa, Tex.	31 (1961)	γ Radiometry	5.84±4%	1.69±3%	3.5
Upper "basalt."	Rock quarry, Knippa, Tex.	6 (1962)	γ Radiometry	5.96±0.02	1.60±0.04	3.7
Recent workings, stock-pile average.	Rock quarry, Knippa, Tex.	31 (1961)	γ Radiometry	5.77±0.14	1.81±0.03	3.2
Stock-pile average.	Rock quarry, Knippa, Tex.	6 (1962)	γ Radiometry	6.71±0.47	2.05±0.03	3.3
Arithmetic mean of other "plutonic" rocks.	....	....		3.88(19)§	1.02(29)	3.7(18)
Arithmetic mean of all basic "plutonic" rocks.	....	....		3.86(43)§	0.93(63)	4.0(41)

§ Using the stock-pile average as the mean value for the rock quarry, Knippa, Tex.

Description	Locality	Reference*	Method of Analysis	Thorium (p.p.m.)	Uranium (p.p.m.)	Th/U
Ultrabasic rocks:						
Dunite (minus spinel)	Twin Sisters dunite, Wash.	33 (1956)	Isotope dilution	0.05	0.016	3.1
Dunite	Twin Sisters dunite, Wash.	13 (1957)	Neutron activation		0.0011±8	
Dunite No. 1	Yushihkou, China	35 (1961)	Neutron activation		0.0094±8	
Dunite No. 2	Nanshan, China	35 (1961)	Neutron activation		0.0037±4	
Dunite No. 3	Nanshan, China	35 (1961)	Neutron activation		0.0030±3	
Olivine nodule	Hualalai	33 (1956)	Isotope dilution		0.005	
Olivine bomb	Gila, Ariz.	10 (1961)	Improved isotope dilution		0.0025±15	
Olivine nodule (diopsidic)	San Bernardino, Calif.	33 (1956)	Isotope dilution		0.007	
Olivine nodule	Rock quarry, Knippa, Tex.	6 (1962)	γ Radiometry	< 0.03 ±100%	0.013 ±60%	< 3
30B Dunite	Webster, N.C.	11 (1962)	Total α act., direct fusion	0 ±0.03	0.03 ±0.01	
P-140 Dunite	Balsam Gap, N.C.	9 (1949)	Refined vacuum fusion		0.019	
P-186-7 Dunite	Balsam Gap, N.C.	9 (1949)	Refined vacuum fusion		0.009	
P-357 Dunite	Dun Mt, New Zealand	9 (1949)	Refined vacuum fusion		0.006	
P-145 Dunite	Twin Sisters, Wash.	9 (1949)	Refined vacuum fusion		0.023	
P-391 Dunite	Addie, N.C.	9 (1949)	Refined vacuum fusion		0.020	
P-369 Websterite	Webster, N.C.	9 (1949)	Refined vacuum fusion		0.008	
P-403 Dunite	Onverwacht pipe, Bushvelt, Transvaal, S. Africa	9 (1949)	Refined vacuum fusion		0.007	
P-400 Dunite	Mooihoek pipe, Bushvelt, Transvaal, S. Africa	9 (1949)	Refined vacuum fusion		0.017	
P-401 Dunite	Driekop pipe, Bushvelt, Transvaal, S. Africa	9 (1949)	Refined vacuum fusion		0.028	
P-364 Dunite	Stillwater Complex, Mont.	9 (1949)	Refined vacuum fusion		0.043	
P-275 Partially serpentinized dunite.	Margarita I, Venezuela	9 (1949)	Refined vacuum fusion		0.017	
P-307 Part. serp. dunite	Belvidere Mt., Vt.	9 (1949)	Refined vacuum fusion		0.011	
P-320 Part. serp. dunite	King mine, 855' level, Thetford, Que.	9 (1949)	Refined vacuum fusion		0.011	
P-318 Part. serp. dunite	Belvidere Mt., Vt.	9 (1949)	Refined vacuum fusion		0.017	
P-272 Serpentinite	Tagil, Urals, U.S.SR.	9 (1949)	Refined vacuum fusion		0.048	
P-324 Serpentinite	Thetford, Que.	9 (1949)	Refined vacuum fusion		0.031	
P-270 Serpentinite	Saranovsk, Urals, U.S.S.R.	9 (1949)	Refined vacuum fusion		0.057	
P-271 Serpentinite	Thetford, Que.	9 (1949)	Refined vacuum fusion		0.037	
P-141 Serpentinite	Geiger's quarry, Octararo Cr., Pa.	9 (1949)	Refined vacuum fusion		0.045	
P-363 Serpentinite	Newfoundland	9 (1949)	Refined vacuum fusion		0.037	

Description	Locality	Reference*	Method of Analysis	Thorium (p.p.m.)	Uranium (p.p.m.)	Th/U
Ultrabasic rocks—Continued						
P-273 Serpentinite	Tagils, Urals, U.S.S.R.	9 (1949)	Refined vacuum fusion	......	0.057	......
P-325 Serpentinite	Thetford, Que.	9 (1949)	Refined vacuum fusion	......	0.054	......
P-263 Serpentinite	Camaguey, Cuba	9 (1949)	Refined vacuum fusion	......	0.048	......
2050 Ore-bearing olivinite	Lesnaya, Varaka, U.S.S.R.	30 (1961)	Colorimetric	0.5	......	......
2043 Coarse-gr. olivinite	Lesnaya, Varaka, U.S.S.R.	30 (1961)	Colorimetric	0.8	......	......
24K Peridotite	Monche-Tundra, U.S.S.R.	30 (1961)	Colorimetric	0.3	......	......
31B Kimberlite	Kimberly, S. Africa	11 (1941)	Direct fusion	......	1.53 ±0.17	......
NBS Kimberlite	Kimberly, S. Africa	9 (1949)	Refined vacuum fusion	......	1.79 ±0.03	......
2813 Pyroxenite	Monche-Tundra, U.S.S.R.	30 (1961)	Colorimetric	0.2	......	......
2068 Fine-gr. pyroxenite	Afrikanda, U.S.S.R.	30 (1961)	Colorimetric	1.3	......	......
2071 Feldspathic pyrox.	Afrikanda, U.S.S.R.	30 (1961)	Colorimetric	6	......	......
2036 Pyroxenite	Ozernaya, Varaka, U.S.S.R.	30 (1961)	Colorimetric	6	......	......
Pyroxenites, serpentinites	Central and western Tuva, U.S.S.R.	1 (1959)	Luminescence (13)	<3	<1	3
			Radiochemical (2)	......	0.9	
1064 Pyroxenite	Oslo Province, Norway	16 (1963)	γ Radiometry	10.4 ±3%	3.1 ±3%	3.4
Biotite apatite pyroxenite	Bearpaw, Mont.	25 (1955)	Fluorimetry	......	1.13	......
825 Anorthosite	Duluth, Mont.	16 (1963)	γ Radiometry	7.1 ±3%	2.1 ±5%	3.4
Anorthosite	Egersund, southern Norway	29 (1950)	Autoradiography	......	0.3	......
Chondrite		3 (1959)	......	0.05	0.01	5
Solar system abundance		7 (1963)	......	0.184 ±74	0.048 ±19	3.8
ECLOGITES						
Metamorphic environments:						
1096	Volda, Möre, Norway	15 (1963)	γ Radiometry	n.d.	n.d.	......
1071	Almklovdalen, Norway	15 (1963)	γ Radiometry	n.d.	n.d.	......
1072	Hareidland, Norway	15 (1963)	γ Radiometry	0.25 ±25%	0.24 ±10%	1
1073	Hareidland, Norway	15 (1963)	γ Radiometry	0.33 ±15%	0.17 ±13%	1.9
1078	Wüstüben, Münchberg	15 (1963)	γ Radiometry	1.7 ±10%	0.66 ±10%	2.6
R391	Siberbach-Hof, Münchberg gneiss massid, Germany	26 (1963)	Neutron activation	0.6	0.24	2.5
X	Münchberg, Germany	27 (1960)	Deduced from lead isotope	0.41	0.10	4.1
1079	Mittenbachgraben, Austria	15 (1963)	γ Radiometry	0.58 ±17%	0.19 ±20%	3.1
R371	Hof Bavarra, Germany	26 (1963)	Neutron activation	0.28	0.059	4.8
R372 Olivine eclogite	River Tessin, southern Switzerland	26 (1963)	Neutron activation	0.015	0.025	0.6
R375 Bronzite pyropite	Sittanpundi Complex, Madras, India	26 (1963)	Neutron activation	0.018	0.018	1

APPENDIX—Continued

Description	Locality	Reference*	Method of Analysis	Thorium (p.p.m.)	Uranium (p.p.m.)	Th/U
ECLOGITES—Continued Metamorphic environments—Cont.						
R388............................	Junction School, Healsburg Quad., Calif.	26 (1963)	Neutron activation	0.43	0.2	2.2
Arithmetic mean of metamorphic eclogites‖	....	....	....	0.46(9)	0.20(9)	2.2(9)
Volcanic environments "pipe" eclogites:						
1074..........................	Robert Victors mine, S. Africa	15 (1963)	γ Radiometry	0.33±30%	0.45±8%	0.7
1075..........................	Robert Victors mine, S. Africa	15 (1963)	γ Radiometry	0.29±25%	0.15±18%	1.9
1076..........................	Robert Victors mine, S. Africa	15 (1963)	γ Radiometry	0.44±15%	0.80±4%	0.6
1077..........................	Robert Victors mine, S. Africa	15 (1963)	γ Radiometry	1.25±10%	0.51±9%	2.5
1094..........................	Robert Victors mine, S. Africa	15 (1963)	γ Radiometry	0.73±12%	0.35±10%	2.1
1092..........................	Robert Victors mine, S. Africa	15 (1963)	γ Radiometry	0.90±10%	0.43±8%	2.1
1095..........................	Robert Victors mine, S. Africa	15 (1963)	γ Radiometry	0.42±15%	0.20±12%	2.1
R269..........................	Robert Victors mine, Orange Free State, S. Africa	26 (1963)	Neutron activation	0.24	0.041	5.8
1093..........................	Jägersfontein mine, S. Africa	15 (1963)	γ Radiometry	0.17±50%	0.07±30%	2.4
1090..........................	Bultfontein, S. Africa	15 (1963)	γ Radiometry	0.31±18%	0.07±25%	4.4
1091..........................	Dodoma, Tanganyika	15 (1963)	γ Radiometry	0.45±20%	0.17±20%	2.6
1097 Garnet peridotite.....	Kimberley, S. Africa	15 (1963)	γ Radiometry	0.26±25%	0.10±30%	2.6
1098 Garnet peridotite.....	Kimberley, S. Africa	15 (1963)	γ Radiometry	0.44±15%	0.25±10%	1.8
R11...........................	Delegate, N.S.W., Australia	26 (1963)	Neutron activation	0.18	0.052	3.6
R117 Hornblende eclogite..	Delegate, N.S.W., Australia	26 (1963)	Neutron activation	0.29	0.073	3.9
R394 Olivine eclogite.......	Delegate, N.S.W., Australia	26 (1963)	Neutron activation	0.15	0.043	3.5
Arithmetic mean of "pipe" eclogites#	....	....	....	0.38(14)	0.20(14)	2.8(14)
Eclogite inclusions in basalt lava:						
1087..........................	Salt Lake, Honolulu	15 (1963)	γ Spectrometry	0.09±50%	0.05±50%	1.8
R419 Olivine eclogite.......	Salt Lake cone, Oahu, Hawaii	26 (1963)	Neutron activation	0.10	0.042	2.4
Eclogite......................	Tuff at Salt Lake Crater, Hawaii	10 (1961)	Improved isotope dilution	....	0.041	....
Arithmetic mean of eclogite inclusions in basalt lava......	....	....	....	0.10(2)	0.044(3)	2.1(2)

‖ Excluding No. X, which is derived on indirect evidence.

Excluding Nos. 1074 and 1077 because of high K contents and probable contamination.

Description	Locality	Reference*	Method of Analysis	Thorium (p.p.m.)	Uranium (p.p.m.)	Th/U
VOLCANIC ROCKS						
Plateau "basalts":						
16B1 Palisades diabase	Staten Island, N.Y.	18 (1941)	Total α act., direct fusion	2.14	0.68	3.1
16B1 Palisades diabase	Staten Island, N.Y.	18 (1941)	Total α act., direct fusion	2.2 ±0.3	0.54±0.05	4
16B2 Palisades diabase	N. Bergen, N.J.	18 (1941)	Total α act., direct fusion	2.37	0.53	4.5
16B3 Palisades diabase	Guttenberg, N.J.	18 (1941)	Total α act., direct fusion	2.34	0.45	5.2
16B4 Palisades diabase	Edgewater, N.J.	18 (1941)	Total α act., direct fusion	2.08	0.40	5.2
16B5 Palisades diabase	Kingston, N.J.	18 (1941)	Total α act., direct fusion	2.60	0.48	5.4
17B Palisades diabase (NBS)	Somerset Co., N.J.	11 (1941)	Direct fusion	....	0.54±0.06	....
NBS Diabase	Somerset Co., N.J.	8 (1947)	Refined vacuum fusion	....	0.58±0.01	....
18B Palisades diabase	Centerville, Va.	11 (1941)	Total α act., direct fusion	1.5 ±0.15	0.51±0.04	2.9
Palisades diabase	Staten Island, N.Y.	22 (1944)	Streaming, direct fusion	2.45±0.10	0.48±0.06	5.1
Palisades diabase	New Jersey	17 (1956)	γ Spectrometry	1.6	0.7	2.3
828 Pyr. diabase	New Jersey	16 (1963)	γ Spectrometry	2.4 ±7%	0.48±14%	4.8
817 Qtz. pyr. diabase	New Jersey	16 (1963)	γ Spectrometry	1.7 ±9%	0.31±14%	5.7
804 Pyr. diabase	New Jersey	16 (1963)	γ Spectrometry	1.6 ±10%	0.36±14%	4.4
918 Pyr. diabase	New Jersey	16 (1963)	γ Spectrometry	1.5 ±12%	0.23±25%	6.5
826 Pyr. diabase	New Jersey	16 (1963)	γ Spectrometry	2.0 ±9%	0.37±17%	5.4
5B Basalt (NBS)	Columbia River basalt	11 (1941)	Total α act., direct fusion	1.9 ±0.4	0.88±0.06	2.2
NBS Basalt	Columbia River basalt	8 (1947)	Refined vacuum fusion	....	1.08±0.06	....
Shoshone basalt	Snake River, plains of southern Idaho	28 (1955)	Isotope dilution	2.02	0.65	3.3
1064 Pyr. basalt	Columbia River basalt	16 (1963)	γ Spectrometry	1.1 ±16%	0.50±12%	2.2
1065 Basalt	Columbia River basalt	16 (1963)	γ Spectrometry	0.94±17%	0.22±25%	4.3
Fine-grained basalt (Stayton basalt)	Approx. 1 mi. E. of confluence of M. & S. Santiam Rivers	36 (1963)	γ Spectrometry	2.28±10%	0.56±10%	4.1
NBS Deccan trap	India	17 (1956)	γ Spectrometry	2.8	0.7	4.0
NBS Deccan trap	India	8 (1947)	Refined vacuum fusion	....	0.60±0.03	....
8B Deccan trap	Bombay, India	11 (1941)	Direct fusion	....	0.54±0.09	....
Karroo dolerite	South Africa	16 (1963)	γ Spectrometry	1.6 ±7%	0.42±10%	3.8
Arithmetic mean of plateau "basalts"				1.96(21)	0.53(26)	4.2(21)

APPENDIX—*Continued*

Description	Locality	Reference*	Method of Analysis	Thorium (p.p.m.)	Uranium (p.p.m.)	Th/U
VOLCANIC ROCKS—*Continued*						
Island basalts:						
11B Hawaiian basalt	Hawaiian Islands	11 (1941)	Direct fusion	....	0.43±0.09	......
Basalt	Samoa	4 (1954)	Fluorimetry (7)	....	0.96	......
Basalt	Hawaii (surface)	4 (1954)	Fluorimetry (10)	....	0.82	......
Basalt	Hawaii (drill core)	4 (1954)	Fluorimetry (8)	....	1.45	......
Basalt (minus spinel)	Hualalai	33 (1956)	Isotope dilution	1.6	0.50	3.2
Basalt	Hualalai	13 (1957)	Neutron activation	....	0.46	......
1086 Tholeiitic qtz. dol. intrusion	Hawaii	16 (1963)	γ Spectrometry	0.41±25%	0.06±50%	6.8
1084 Alk. ol. basalt	Japan	16 (1963)	γ Spectrometry	4.2 ±6%	0.48±20%	8.8
1085 Alk. ol. basalt	Japan	16 (1963)	γ Spectrometry	3.6 ±6%	0.57±13%	6.3
1082 High-Al basalt	Japan	16 (1963)	γ Spectrometry	1.1 ±14%	0.28±20%	3.9
1083 High-Al basalt	Japan	16 (1963)	γ Spectrometry	0.45±25%	0.13±30%	3.5
1080 Tholeiite	Japan	16 (1963)	γ Spectrometry	0.05	0.03	1.7
1081 Tholeiite	Japan	16 (1963)	γ Spectrometry	0.32±25%	0.26±12%	1.5
Fine-gr. gray basalt boulder	Mid-Atlantic ridge	5 (1953)	Direct fusion		0.085	
Arithmetic mean of island basalts				1.47(8)	0.46(14)	4.5(8)
Other "basalts":						
Basalt	N. Attleboro, Mass.	22 (1944)	Streaming, direct fusion	3.22±0.20	0.78±0.08	4.1
Basalt	Oldwick, N.J.	22 (1944)	Streaming, direct fusion	2.26±0.18	0.65±0.05	3.5
6B Oldwick basalt	Oldwick, N.J.	11 (1941)	Total α act., direct fusion	2.3 ±0.6	0.60±0.11	3.8
Basalt	Overhanging cliff, Yellowstone Park, Wyo.	22 (1944)	Streaming, direct fusion	4.30±0.24	1.18±0.08	3.6
3B Crescent Hill basalt	¼ mi. W. of Geode Creek, Yellowstone Park, Wyo.	11 (1941)	Total α act., direct fusion	7.8 ±1.5	1.1 ±0.2	7.1
7B Basalt	¾ mi. E. of Geode Creek, Yellowstone Park, Wyo.	11 (1941)	Total α act., direct fusion	10 ± 2	0.57±0.11	17.5
1B Porphyritic basalt	Crescent Hill, Yellowstone Park, Wyo.	11 (1941)	Total α act., direct fusion	8.5 ±1.5	2.3 ±0.5	3.7
12B Basalt	Gardiner, Yellowstone Park, Mont.	11 (1941)	Total α act., direct fusion	2.9 ±0.5	0.4 ±0.08	7.1
10B Basalt	Overhanging Cliff, Yellowstone Park, Wyo.	11 (1941)	Total α act., direct fusion	5.9 ±1	0.48±0.09	12.3

APPENDIX—Continued

Description	Locality	Reference*	Method of Analysis	Thorium (p.p.m.)	Uranium (p.p.m.)	Th/U
VOLCANIC ROCKS—*Continued*						
Other basalts—*Continued*						
13B Oxbow Creek basalt....	Oxbow Creek, Yellowstone Park, Wyo.	11 (1941)	Direct fusion	....	0.37±0.06	....
Basalt...	Modoc lavas, Calif.	24 (1954)	Fluorimetry (2)	....	0.88	....
Basalt...	Northern Calif.	25 (1955)	Fluorimetry (2)	....	4.1	....
Basalt...	Northern Calif.	25 (1955)	Fluorimetry (4)	....	0.44	....
Vesicular basalt...	Northern Calif.	25 (1955)	Fluorimetry	....	3.9	....
28 Qtz. basalt...	Old post-cinder flow, Lassen Volcanic Natl. Park, Calif.	2 (1955)	Fluorimetry	....	2.5	....
24 Vesicular basalt...	Cinder cone, Lassen Vol. Natl. Park, Calif.	2 (1955)	Fluorimetry	....	1.7	....
22 Qtz. basalt...	Source of 1851 flow, cinder cone, Lassen Vol. Natl. Park, Calif.	2 (1955)	Fluorimetry	....	1.6	....
31 Eastern basalt...	Butte Lake, Calif.	2 (1955)	Fluorimetry	....	0.86	....
32 Pahoehoe la a...	Subway Cave, outside Lassen Park, Calif.	2 (1955)	Fluorimetry	....	0.36	....
Olivine basalt...	Dish Hill, Ludlow, Calif.	6 (1962)	γ Spectrometry	6.58±1%	1.54±1%	4.3
Basalt...	Boulder batholith, Mont.	25 (1955)	Fluorimetry	....	0.73	....
Basalt...	Hinsdale formation, Calif.	25 (1955)	Fluorimetry	....	1	....
Basalt...	Hill #7073, Williams, Ariz.	6 (1962)	γ Spectrometry	3.25±3%	1.21±2%	2.7
Basalt, non-ves. flow	Hill #7348, Williams, Ariz.	6 (1962)	γ Spectrometry	4.99±2%	1.26±2%	4
Basalt, ejecta block	Kilbourne Hole, N.M.	6 (1962)	γ Spectrometry	3.43±3%	0.93±2%	3.7
Basalt...	Kilbourne Hole, N.M.	6 (1962)	γ Spectrometry	4.52±2%	1.44±2%	3.1
Vesicular basalt...	Peridote Cove, Ariz.	6 (1962)	γ Spectrometry	4.91±2%	1.68±2%	2.9
Solid basalt...	Santiam Summit, Ore.	36 (1963)	γ Spectrometry	0.96±10%	0.32±10%	3.0
Feldspathic basalt...	Santiam Summit, Ore.	36 (1963)	γ Spectrometry	0.79±10%	0.57±10%	1.4
Columnar basalt...	Tombstone Pass, Ore.	36 (1963)	γ Spectrometry	0.96±10%	0.41±10%	2.4
Basalt...	Approx. 2 mi. W. of Cascadia, Ore.	36 (1963)	γ Spectrometry	3.63±10%	0.98±10%	3.7
Basalt...	Santiam Jct., Ore.	36 (1963)	γ Spectrometry	2.00±10%	0.85±10%	2.4

Description	Locality	Reference*	Method of Analysis	Thorium (p.p.m.)	Uranium (p.p.m.)	Th/U
VOLCANIC ROCKS—*Continued*						
Other basalts—Continued						
Basalt..........	Near Laidlaw Butte, Ore.	36 (1963)	γ Spectrometry	2.37 ±10%	1.87 ±10%	1.3
Solid basalt.........	Suttle Lake, Ore.	36 (1963)	γ Spectrometry	0.70 ±10%	0.70 ±10%	1.0
Post-ore lamprophyre.........	Eustis mine, Que.	22 (1944)	Streaming, direct fusion	2.93 ±0.11	1.33 ±0.13	2.2
Pre-ore lamphrophyre.........	Eustis mine, Que.	22 (1944)	Streaming, direct fusion	1.61 ±0.16	0.65 ±0.07	2.5
113 Basalt..........	Eustis mine, Que.	19 (1938)	Streaming, direct fusion	1.61	0.62	2.6
NB2 Basalt..........	St. George, N.B.	19 (1938)	Streaming, direct fusion	1.70	0.49	3.5
Pillow lava..........	Huron Claim, Manitoba	22 (1944)	Streaming, direct fusion	0.08 ±0.02	0.02 ±0.00	4.0
Monchiquite dike.........	Riasg Buidhe, Is. of Colonsay, Scotland	22 (1944)	Streaming, direct fusion	4.40 ±0.20	2.04 ±0.07	2.2
26B Monchiquite dike........	Riasg Buidhe, Is. of Colonsay, Scotland	11 (1941)	Total α act., direct fusion	6.2 ±1.1	1.9 ±0.4	3.3
27B Monchiquite dike........	Kilchattan, Is. of Colonsay, Scotland	11 (1941)	Total α act., direct fusion	3.3 ±0.6	0.91 ±0.09	3.6
Monchiquite dike.........	Kilchattan, Is. of Colonsay, Scotland	22 (1944)	Streaming, direct fusion	3.42 ±0.27	1.15 ±0.08	3.0
1058 Basalt (Bl).........	Oslo province, Norway	16 (1963)	γ Spectrometry	8.8 ±4%	1.4 ±7%	6.3
1059 Basalt (Bl).........	Oslo province, Norway	16 (1963)	γ Spectrometry	4.1 ±4%	0.88 ±7%	4.7
1061 Basalt.........	Oslo province, Norway	16 (1963)	γ Spectrometry	2.0 ±12%	0.44 ±18%	4.5
1062 Lab. porphyr. basalt.........	Oslo province, Norway	16 (1963)	γ Spectrometry	5.6 ±5%	1.3 ±7%	4.3
4B Basalt.........	Giants Causeway, Ireland	11 (1941)	Total α act., direct fusion	6.2 ±1.1	1.9 ±0.4	3.3
2B Vesicular basalt.........	Gravenoire volcano, Clemont, France	11 (1941)	Direct fusion		1.82 ±0.11	
R Basalt.........	Ropruchei, U.S.S.R.	19 (1938)	Streaming, direct fusion	2.04	0.77	2.7
Arithmetic mean of other "basalts"				3.79(37)	1.16(50)	4.1(37)
Composite basalt II...		12 (1960)	Fluorimetry (250)		1.4 ±0.3	
Arithmetic mean of all basic volcanic rocks........				2.93(66)	0.87(90)	4.2(66)

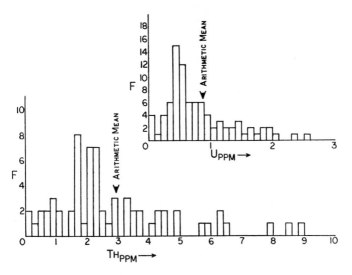

FIG. 5.—Histogram of thorium and uranium in basic plutonic rocks

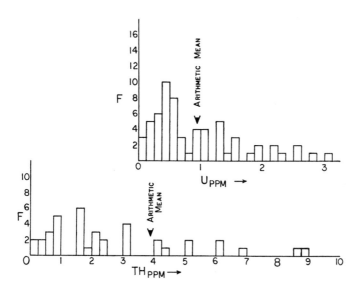

FIG. 6.—Histogram of thorium and uranium in basic volcanic rocks

AZIZEH MAHDAVI

5. The Thorium, Uranium, and Potassium Contents of Atlantic and Gulf Coast Beach Sands

T HE PRESENT WORK concentrated on determining and interpreting the range of thorium, uranium, and potassium concentrations in beach sands. Beach sands were selected for study because (1) little was known about their thorium and uranium content in a descriptive way useful in environmental radiation studies and the empirical interpretation of subsurface and aerial radiometric anomalies, (2) there was the possibility of maximizing the volume of investigation while achieving constant geometry with the field pulse-height analyzer (spectrometer) then under development and calibration, and (3) a number of hypotheses or models of both fundamental and economic geological importance might be tested with the data obtained. Among these models or hypotheses were (a) that the quartz and feldspar in well-mixed mature beach sands should represent a homogenized and representative sample of all crustal environments of mineral formation and thus should have the crustal Th/U ratio of about 3.8; (b) that the thorium and uranium in dense, resistate minerals—for example, zircon and monazite—should be useful as natural radioactive tracers through the processes by which clastic particles are brought to beaches from various sources and moved along and across the beaches by air and water currents; (c) that, once the field instrumentation was developed and the frequency versus concentration data for thorium and uranium obtained, the field instrumentation might prove useful in the exploration for and evaluation of placer deposits of zircon, monazite, xenotime, and other radioactive minerals; and (d) that, even without locating economic deposits of radioactive minerals, the field instrumenta-

Dr. AZIZEH MAHDAVI, formerly at Rice University, Houston, Texas, is at present geologist with the Iranian Geological Survey, Teheran, Iran.

tion might be useful in locating and evaluating placer deposits of non-radioactive minerals—for example, magnetite and ilmenite—that, because they are also dense and resistant, are naturally associated with or "spiked" with minor amounts of radioactive minerals.

PREVIOUS STUDIES

Most of the quantitative data on the thorium and uranium concentrations in common sedimentary rocks have been published in the last 10–15 years. In their review article covering the literature through 1957, Adams *et al.* (1959) cite 316 samples of sedimentary rocks or sedimentary rock aggregates in which both thorium and uranium have been determined; of these, only 18 are sandstones or sands, and 16 of these 18 thorium and uranium determinations are from the study of Murray and Adams (1958). A few additional data of semiquantitative nature are available from total α-activity/chemical uranium determinations (Adams and Weaver, 1958). The intensive search for uranium deposits in the last decade, as well as the development of rapid and reliable chemical analytical procedures for uranium, produced additional data for uranium alone in those rare sandstones with ore or near-ore concentrations of uranium. In short, however, there are few data in the literature on which to base an estimate of frequency versus concentration of thorium and uranium in sands and sandstones.

Estimates of the thorium and uranium concentrations in sandstones have been made on the basis of theoretical geochemical considerations (Adams and Weaver, 1958; Adams *et al.*, 1959). During weathering, thorium and uranium are separated to the extent that tetravalent uranium is oxidized to the hexavalent state and leached and transported in solution as the soluble uranyl ion or its soluble complexes. By comparison, thorium, having only a tetravalent state, remains in primary or secondary resistate minerals (Pliler and Adams, 1962*b*), which concentrate in residual soils (Adams and Richardson, 1960) or are transported as clastic particles. Unoxidized tetravalent uranium in resistates, for example, zircon ($ZrSiO_4$), remains with thorium. In principle, it should be possible to calculate material balances in which the thorium and uranium in primary igneous rocks is balanced against that found in sedimentary rocks that result from the weathering of the igneous rocks. Adams and Weaver (1958) made such a calculation, assuming certain proportions of the major igneous and sedimentary rock types and assuming that the observed average thorium and uranium concentrations in all rock types except sandstones were representative. By this method they estimated 24 ± 7 p.p.m. thorium, 4.1 ± 1.5 p.p.m. uranium, and a Th/U ratio of about 6, with a minimum value of 3 and a maximum value of 11.5 for sandstones. Adams

et al. (1959) repeated the calculation, including the data of Degenhardt (1957) on the geochemically similar zirconium for comparison.
The theoretical basis for such material balance calculations has been
vigorously disputed by many, for example, Poldervaart (1955) and
Pettijohn (1957), and the data for thorium are far less satisfactory than
are those for uranium and zirconium. The material balance or geochemical equation estimates are interesting in that for thorium they
require either (1) that the average for shale is too low or (2) that sandstones do indeed contain something like an average of 24 p.p.m. thorium
and that the high concentrations of thorium in resistate minerals like
monazite found in some of the beach sands of Brazil and India are
common enough to shift the mean concentration far above the mode
of 2 or 3 p.p.m. This mode is found in the quartz and feldspar that
carry most of the thorium in most sandstones. The thorium in question amounts to 10–15 per cent of the total thorium in sedimentary
rocks.

Another theoretical consideration that has been advanced (Adams
et al., 1959) quite independently of the material balance calculations
is the hypothesis that the Zr/Th ratio in all major sedimentary rock
types is close to 13. This hypothesis would require an average thorium
content in sandstones of about 17 p.p.m. As Adams *et al.* (1959, p.
336) conclude:

A major difficulty in accepting such high average thorium values for
sandstones is the qualitative experience that highly radioactive sandstones
are apparently very limited in extent on the earth's surface and apparently
unknown or unrecognized in the subsurface (see Figure 1 and the above
discussion on sandstones). Only more data can finally resolve the several
difficulties indicated in the balances in Table 11.

In addition to the very large, but poorly known, deposits of thorium resistate minerals in the Brazilian and Indian sands, some placer
deposits containing thorium minerals have been reported recently in
the Nile Delta of Egypt (Higazy and Naguib, 1958). Gindy (1961)
determined the thorium and uranium in the monazite, zircon, and
other resistate grains from these deposits, using chemical, radiometric,
and autoradiographic methods. Several square kilometers of delta sand
covered in their investigation contain over 60 p.p.m. thorium, and an
over-all average of 20 p.p.m. does not look completely unreasonable
for the Nile Delta. Higazy and Naguib (1958) concluded that there
were (1) a proportionality between total radioactivity and the
amount of thorium-bearing monazite, (2) some separation of the
various dense resistate minerals, but (3) little separation of monazite
and the equally dense magnetic resistate minerals. The Nile placer

sands described were in deposits a few meters wide and some centi-meters thick, but much elongated. The magnetic and radioactive anomalies caused by these placer deposits could be detected by air-borne instruments only when they were flown 50 feet above the ground. The Egyptian work suggests that most aerial radiometric surveying might easily miss narrow bands of placer sands.

Most aerial surveying is done several hundred feet above flat ter-rain; Guillou (1963) gives 500 feet as the nominal survey altitude. At 500 feet Guillou estimates that air-borne radioactive gases and dust contribute less than 10 per cent of the γ radiation detected; the cosmic-ray contribution is corrected instrumentally by subtracting the read-ings obtained at 2,000 feet from those obtained at 500 feet. Guillou (1963) reported on an aerial survey of the Galveston, Texas, area from a height of 500 feet; 4,400 traverse miles at 1-mile spacings were flown over an area of about 100 square miles. This study was part of the ARMS (Aerial Radiological Measuring Survey) program to de-termine broad levels of environmental radiation. The Galveston study is of particular interest in the present study because it covers the Galveston beaches investigated by Murray and Adams (1958) and in the present work also (see Appendix 3, this volume). It should be noted that surveys from this altitude do not detect small areas of high radioactivity and that the primary interest was in environmental radi-ation and not in detailed aerial geology. The ARMS survey, for ex-ample, failed to detect over 1,800 short tons of zircon sand stored at the Texas City tin smelter; this zircon in two piles (70 $\times$ 30 feet and 10 feet thick; 30 $\times$ 35 feet and 5 feet thick) contains over 700 p.p.m. thorium and nearly 300 p.p.m. uranium. The stockpiles were stored under a thin corrugated-iron roof. The ARMS survey of Galveston Island also indicated that the easternmost tip of the island and adja-cent Bolivar Peninsula had slightly higher total radioactivity than the main part of the island.

Aerial radiometric surveys of the Atlantic Coast beaches from Flor-ida to North Carolina were made by the U.S. Geological Survey in co-operation with the Atomic Energy Commission in the period May–November, 1953. Parts of the western coast of Florida were also surveyed. In general these surveys indicated higher than normal radio-activity in many of these beaches, although the total γ-ray-flux meas-urements provide no information on the relative contributions of the thorium series, the uranium series, and potassium-40. In summary, aerial surveys to date do not provide much indication of the average thorium content of beach sands because (1) much surveying is done at too great an altitude to detect thin, elongated bodies of placer sands, such as those reported from the Nile Delta; and (2) to date, only total radiation has been measured and spectrometric instrumentation is only beginning to come into use.

Another major source of qualitative information about the total γ-ray flux from rocks lies in the large number of γ-ray logs made of petroleum exploration boreholes (Johnstone, 1963). These logs are used for correlation of key beds with anomalous radioactivity and for estimates of gross lithology (shales and, particularly, bentonites being more radioactive than are most sands and carbonate rocks). Adams and Weaver (1958) introduced the term "geochemical facies" in connection with the environmental interpretation of thorium and uranium contents of sediments and indicated how subsurface logs might be used to interpret the source of sediments and their environment of deposition. Rabe (1957) suggested a correlation between the total γ-ray flux and the permeability of the Muddy and Dakota sands of the Denver-Julesburg Basin, the more permeable sands having less clay and other cementing materials and hence lower radioactivity.

SAMPLING

The sampling sites were located on the Gulf and Atlantic Coast beaches (see Figs. 1–5). The samples from the Gulf Coast were taken

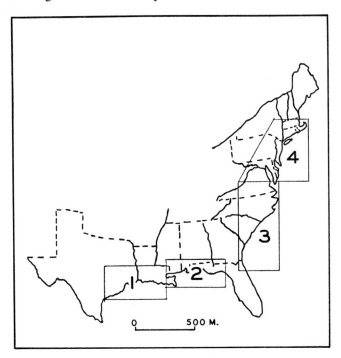

FIG. 1.—Index map of surveyed areas

from the Galveston beach in Texas; from beaches in Louisiana, Mississippi, and Alabama; and from the northwestern Florida coast as far east as the Appalachicola River delta. Along the Atlantic coast, samples were taken from beaches of northeastern Florida, Georgia, South Carolina, North Carolina, Virginia, New Jersey, and Cape Cod in

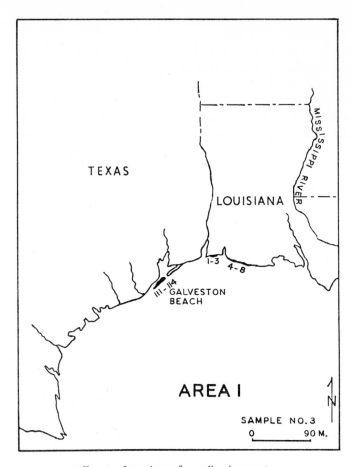

Fig. 2.—Locations of sampling in area 1

Massachusetts. The sites were chosen on the basis of their accessibility and geologic significance rather than on the basis of some regularly spaced interval. At each site two or three points in a line perpendicular to the shore line were selected for measurements with the field γ-ray spectrometer and for the collecting of samples for subsequent laboratory studies. In general, each site had a sample from (1) between the berm and low tide line, (2) the back beach, and (3) the dune inshore from the back beach. All three of these components were not always present or distinguishable at all sites because of the constructive or destructive nature of the beach or because of other morphological features or processes controlling the beach profile. In particular, recent storms had completely swept out the beach profile at Cape Hatteras, North Carolina, and sufficient time had not elapsed for the

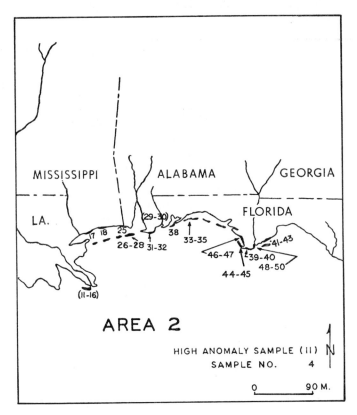

Fig. 3.—Locations of sampling in area 2

development of a new profile. Erosional or destructive-type beaches were also found along the shore line of Louisiana west of the Mississippi River and on the northern edge of Jekyll Island, Georgia.

Other geologic factors affecting the selection of sampling sites were the fact that the Gulf Coast area is a lowland grading into a subsiding basin, and various rivers are the major sources of sediment supplied to the beaches. In Texas, the major sources were the Trinity, Brazos, and Colorado Rivers; in Louisiana, the major source was the Mississippi River; in Alabama, the major source was the Mobile River; in

Florida, the major source was the Appalachicola River. These streams, together with numerous smaller ones, carry eroded sedimentary and crystalline materials from the highlands and distribute them to the beaches. Reworking of this material and older shore deposits by wave action results in the build-up of the modern beaches. Most of these beaches on the Gulf Coast occur as sand bars and longshore islands 1–30 miles long and ½–1 mile wide which are separated from the mainland by lagoons.

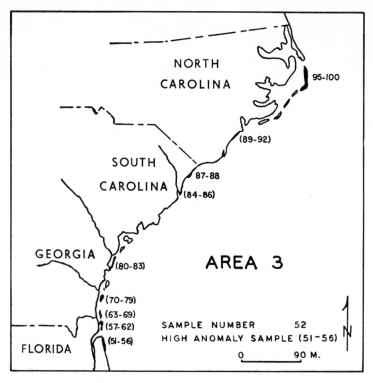

FIG. 4.—Locations of sampling in area 3

FIELD PROCEDURES

Thorium, uranium, and potassium were determined *in situ* by means of a portable γ-ray spectrometer designed and described by Adams and Fryer (1963). To achieve constant geometry at each station, a cylindrical hole 8.5 cm. in diameter and 18 cm. deep was dug so that it just fit the detector unit of the instrument. After temperature equilibration and standardization against a cesium-137 γ-ray source, the count rate was measured at the following γ-ray levels: (1) the 2.62 mev. from thallium-208 in the thorium-232 series; (2)

the 1.76 mev. from bismuth-214 in the uranium-238 series; and (3) the 1.46 mev. from potassium-40. The counting time at each level did not exceed 20 minutes. The data are given in tables below. After each measurement, approximately 1 kilogram of sand was taken and placed in a plastic bag for later laboratory study.

It should be noted that the field instrument integrates the γ radiation coming from a large volume of sand. Experiments with zircon sand, which has a higher density than ordinary beach sand, showed that 2.62-mev. γ photons from the thallium-208 daughter of thorium-232 can be detected with the field instrument through more than 25 cm. of zircon sand. A conservative estimate of the volume of beach sand measured with the field detector buried 18 cm. below the sand surface

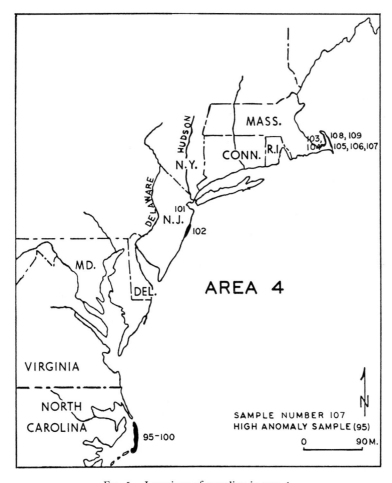

Fig. 5.—Locations of sampling in area 4

is at least 80 liters, which would have a dry weight of 140–160 kg. This large volume of investigation is desirable from the point of view of obtaining the most representative measurement at a particular point on a beach. However, this large volume is difficult to sample for comparative measurements in the laboratory. The usual laboratory sample of 1–2 kg., of which only some 350–400 gm. are actually measured in the laboratory, is almost 100 times smaller than the system measured by the field instrument.

The hundred-fold difference between the field and laboratory samples also created problems in the absolute calibration of the field instrument. One major problem is the fact that beaches are commonly layered, with shell fragments and heavy minerals concentrated at discrete levels. Thus, the field system measured cannot be considered homogeneous. Another problem arises from the fact that the thorium system measured is larger than the uranium system, which in turn is larger than the potassium system, because the γ-ray energies used for maximum volume of investigation decrease in energy in that order. The thorium calibration is easiest and most free of assumptions and complications. This calibration, carried out using an artificial, homogenized zircon sand source over 50 cm. in diameter and 50 cm. thick (essentially infinite to within counting statistics) and of known thorium content, indicated that the field instrument used showed 12 c.p.m. per p.p.m. thorium for a homogenous source in secular radioactive equilibrium. This constant was confirmed to well within 10 per cent by comparison of field and laboratory measurements.

In order to convert c.p.m. in the 1.76-mev. channel to p.p.m. uranium, it is necessary to assume secular radioactive equilibrium in both the uranium and the thorium series and no difference in the concentration and distribution of the γ-ray emitters in the field thorium system, the field uranium system, and the much smaller system of the sample taken for laboratory studies. Making these assumptions, the conversion takes the following general form:

Uranium in p.p.m. = a (c.p.m. at 1.76 mev.) − b (c.p.m. at 2.62 mev.).

The constants a and b are, of course, characteristic of the particular instrument used and the method of taking measurements, which in this case was with the detector buried in the sand. Using all the data (Tables 1 and 2) comparing field and laboratory measurements, a was found to equal 0.018 ± 0.004, and b to equal 0.27.

The conversion of c.p.m. in the 1.46-mev. channel to p.p.m. potassium takes the general form:

Potassium p.p.m. = A (c.p.m. at 1.46 mev.) − B (c.p.m at 1.76 mev.)
$$- C \text{ (c.p.m. at 2.62 mev.)},$$

TABLE 1

RESULTS OF LABORATORY MEASUREMENTS FOR THORIUM,
URANIUM, POTASSIUM, AND TH/U RATIO

Sample No.	Station No.	Thorium (p.p.m.)	Uranium (p.p.m.)	Potassium (per cent)	Th/U Ratio
3	1, C	1.72	0.466	0.377	3.69
4	2, A	4.45	1.01	1.83	4.4
5	2, B	1.83	0.38	0.9	4.8
10	Clay	6.78	2.44	1.55	2.78
11	3, A	3.48	1.13	1.18	3.41
12	3, B	4.88	1.28	1.12	3.82
17	4, A	1.03	0.4	0.019	2.57
18	4, B	0.3	0.16	0.007	1.94
20	5, B	1.83	0.84	0.05	2.17
21	6, A	0.58	0.32	0.008	1.81
22	6, B	0.58	0.24	0.01	2.44
24	7, A	1.08	0.42	0.037	2.57
26	8, A	0.125	0.13	0.008	0.97
27	8, B	1.73	1.26	0.46	1.37
28	8, C	0.28	0.23	0.01	1.22
29	Mobile Bay	6.7	2.47	0.02	2.71
30	" "	10.12	4.79	0.007	2.11
31	9, A	0.19	0.1	0.03	1.9
32	9, B	0.34	0.28	0.001	1.21
33	11, A	0.19	0.19	0.07	1
34	11, B	0.9	0.74	0.01	1.22
35	Santa Rosa	0.25	0.18	0.008	1.4
36	12, A	0.07	0.17	0.005	0.43
38	13, A	0.27	0.6	0.003	0.44
39	14, A	0.499	0.21	0.11	2.31
44	17, A	0.397	0.244	0.015	1.63
45	17, B	5.7	2.03	0.019	2.8
51	19, A	1.19	0.69	0.26	1.72
52	19, B	20.6	5.24	8.74	3.92
53	19, C	2.66	0.93	0.29	2.86
54	20, B	14.33	3.92	1.25	3.65
55	21, A	7.13	2.81	0.091	2.54
56	21, B	8.11	3.49	0.078	2.32
57	22, A	8.9	2.93	0.124	3.02
58	22, B	21.5	6.8	0.12	3.16
59	22, C	8.1	3	0.19	2.7
60	22, D	5.26	2.35	0.14	2.25
61	22, E	2.45	0.998	0.08	2.45
62	22, G	10.37	3.43	0.12	3.02
63	22, F	3.09	1.1	0.37	2.81
64	S. J. River	2.35	0.85	0.27	2.76
65	23, A	3.36	1.28	0.25	2.62
66	23, B	3.51	1.48	0.11	2.37
67	23, C	6.53	2.36	0.14	2.77
68	23, D	5.71	2.09	0.132	2.73
69	23, E	2.04	0.7	0.44	2.91
70	24, A	0.91	0.74	0.22	1.23
71	24, B	1.31	0.9	0.57	1.45
72	24, C	2.15	1.25	0.21	1.71
73	Jekyll offshore	5.26	2.35	0.53	2.23
74	25, A	2.24	1	0.24	2.24
75	25, B	1.82	0.78	0.25	2.33
76	26, A	30.87	9.18	0.156	3.37
77	26, A'	51.27	14.14	0.11	3.62
78	26, B	31.24	8.67	0.17	3.6
79	Ridge sand	14.26	4.64	0.184	3.07
80	27, A	3.64	1.77	0.27	2.06

TABLE 1—*Continued*

Sample No.	Station No.	Thorium (p.p.m.)	Uranium (p.p.m.)	Potassium (per cent)	Th/U Ratio
81	27, B	126.76	31.24	0.129	4.06
82	27, C	8.72	2.56	0.47	3.41
83	Savannah	2.43	1.06	0.24	2.29
84	28, A	8.18	6.08	0.19	1.35
85	28, B	11.56	8	0.31	1.44
86	28, C	3.2	3.55	0.47	0.9
87	29, A	1.88	0.65	0.36	2.89
88	29, B	2.79	1.39	0.31	2
89	30, A	2.02	2.2	0.209	0.918
90	30, B	8.46	9.46	0.08	0.894
91	30, C	4.13	4.69	0.15	8.33
92	W.b.s.	1.32	1.44	0.25	0.92
93	Va. pit.	0.438	0.27	0.43	1.63
94	Old Ridge Sand	3.23	0.88	1	3.69
95	31, A	1.23	0.54	0.29	2.28
96	31, B	1.59	0.7	0.23	2.27
97	31, C	3.11	1.16	0.18	2.68
98	31, C'	43.25	15.21	0.036	2.77
99	32, A	4.96	2.06	0.219	2.41
100	32, B	2.22	0.72	0.49	3.06
101	Glauconite	3.45	7.95	6.08	0.434
102	34	0.26	0.09	0.07	2.67
103	35, A	1.75	0.53	0.66	3.26
104	35, B	1.92	0.54	0.43	3.55
105	36, A	1.56	0.41	0.33	3.80
106	36, B	2.09	0.55	0.03	3.8
107	36, C	1.65	0.41	0.32	3.98
108	37, A	1.68	1.03	0.23	1.62
109	37, B	2.94	1.72	0.42	1.71
110	G.B.S.	1.68	0.6	0.97	2.81
111	1, A	1.69	0.56	0.87	3.01
112	1, B	3.27	1.22	0.73	2.68
113	2, A	1.51	0.48	0.91	3.14
114	2, B	3.3	1.31	0.74	2.52
115	Shell	1.26	0.12	0.025	10

where values of A equal to 0.003 or 0.004, B equal to 0.78, and C equal to 0.5 were found on the basis of a comparison of field and laboratory measurements. A glauconitic sand with 6.1 per cent potassium was particularly useful in this calibration. Although the calculaboratory measurements. A glauconitic sand with 6.1 per cent potastions, it should be noted that in many cases the count rate from the fixed potassium-40 spike in all natural potassium is large relative to the count rates in the C and B terms.

It is estimated that, with the assumptions and constants cited above, thorium, uranium, and potassium can be determined in the field (Table 2) to within ±20 per cent at all levels of concentrations and to within ±10 per cent at the levels of concentration most frequently encountered in this study. This estimation of accuracy rests on the following observations and interpretations.

TABLE 2

RESULTS OF THORIUM, URANIUM, POTASSIUM, AND TH/U RATIOS OBTAINED FROM FIELD DATA

Sample No.	Thorium (p.p.m.)	Uranium (p.p.m.)	Potassium (per cent)	Th/U Ratio
1........	5	....	....	....
2........	6	1.53	1.4	3.9
3........	5	1.53	1.3	3.9
4........	4.8	5.76	2.4	0.8
5........	5	2.28	1.09	2.2
11........	4	1.3	1.02	3.07
12........	3.3	0.95	1.24	3.5
17........	1	0.34	0.06	2.9
18........	1	0.32	0.04	3.1
19........	0.75	0.25	0.03	3
20........	2	0.9	0.09	2.2
21........	1	0.32	0.02	3.1
22........	0.83	0.39	0.04	2.1
24........	1.33	0.4	0.036	3.3
26........	0.58	0.16	0.04	3.6
27........	2.58	1.3	0.054	1.98
31........	0.58	0.2	0.03	2.9
32........	0.75	0.3	0.01	2.5
33........	1	0.18	0.01	2.77
34........	1	0.48	0.045	2.08
36........	0.58	0.2	0.03	2.9
37........	0.58	0.2	0.03	2.9
38........	0.58	0.2	0.033	2.9
39........	0.66	0.3	0.13	2.2
40........	0.75	0.32	0.03	2.34
41........	0.58	0.25	0.03	2.32
43........	0.66	0.23	0.042	2.87
44........	0.58	0.2	0.048	2.9
46........	3.3	1	0.045	3.3
48........	0.83	0.32	0.03	2.6
51........	2.16	0.7	0.224	3.08
52........	7.66	2.09	0.282	3.66
53........	5.2	1.55	0.003	3.35
54........	11	3.6	0.216	3.05
55........	5.2	2.7	0.216	1.92
56........	14.8	5.2	0.165	2.85
57........	11.5	4.37	0.204	2.86
58........	20	6.26	0.222	3.19
59........	6.2	2.7	0.141	2.3
60........	5.8	3.35	0.12	1.73
61........	3	0.97	0.183	3.09
62........	4.7	1.38	0.261	3.44
65........	3	0.86	0.291	3.48
66........	3.3	1.2	0.237	2.75
67........	3.2	1.96	0.189	2.65
68........	6.4	2.18	0.171	2.93
69........	3.7	1.17	0.249	3.16
70........	1.9	0.95	0.228	1.57
71........	1.2	0.54	0.345	2.22
72........	1.8	0.85	0.288	2.15
74........	2.5	0.9	0.252	2.78
75........	3.3	0.99	0.315	3.36
76........	29.16	7.6	0.303	3.83
78........	24.6	6.57	0.228	3.74
80........	12.08	3.76	0.285	3.21
81........	100	26.57	1.2	3.76
82........	4.5	1.4	0.534	3.2

TABLE 2—*Continued*

Sample No.	Thorium (p.p.m.)	Uranium (p.p.m.)	Potassium (per cent)	Th/U Ratio
84.......	9.16	4.68	0.507	1.95
85.......	25.25	9.99	0.456	2.53
86.......	4.4	2.45	0.351	1.8
87.......	3.3	0.95	0.375	3.5
88.......	8.33	2.03	0.42	4.1
89.......	1.75	1.5	0.231	1.16
90.......	8.33	6.77	0.444	1.2
91.......	5.8	4.46	0.282	1.3
95.......	3	0.74	0.216	4.05
96.......	2.66	0.86	0.31	3.09
97.......	7.5	2.14	....	3.51
99.......	6.67	2.2	0.336	3.18
100.......	2.5	0.93	0.354	2.69
102.......	0.75	0.23	0.101	3.2
103.......	4.75	0.85	1.35	3.26
104.......	3	0.73	0.425	4.1
105.......	2	0.68	0.363	2.94
106.......	3	0.95	0.375	3.15
107.......	2	0.57	0.357	3.5
111.......	2.2	0.63	0.5	3.49
112.......	3.2	0.99	0.46	3.23
113.......	2	0.65	0.81	3.07
114.......	3	1.06	0.72	3

TABLE 3

RESULTS OF THE AVERAGES OF THORIUM, URANIUM, AND POTASSIUM IN DIFFERENT SUITES MEASURED IN THE LABORATORY AND FIELD

AREA	LABORATORY					FIELD				
	No. Samples	Thorium (p.p.m.)	Uranium (p.p.m.)	Potassium (per cent)	Th/U Ratio	No. Samples	Thorium (p.p.m.)	Uranium (p.p.m.)	Potassium (per cent)	Th/U Ratio
1........	10	2.78	0.8	1.17	3.3	10	3.97	1.66	1.09	2.39
2........	19	0.86	0.41	0.046	2.1	22	0.98	0.38	0.042	2.57
3........	41	11.27	3.97	0.27	2.82	35	10.43	3.55	0.29	2.93
4........	13	2.07	0.8	0.3	2.59	11	3.44	0.99	0.41	3.47
1 and 2...	29	1.49	0.59	0.43	2.53	32	1.9	0.78	0.37	2.44
2 and 3...	54	9.05	3.21	0.28	2.82	46	8.76	2.94	0.32	2.97
Total.	83	6.42	2.97	0.33	2.82	78	5.87	2.08	0.34	2.8

Table 3 gives a comparison of the average thorium, uranium, potassium and Th/U ratio values for several regional groupings of data from both the field and the laboratory measurements. A comparison of the ranges of individual measurements by each method can be seen in the histograms in Figures 6–10. The over-all averages agree to well within the limits cited, though many individual stations vary much more widely. This variation is interpreted as a measure of the difficulty of obtaining a small laboratory sample that is representative of the hundred-fold larger field system; however, if enough small sample measurements are compared with field measurements, the deviations will average out to within counting statistics. The agreement on the over-all and group averages of individual Th/U ratios is particularly noteworthy.

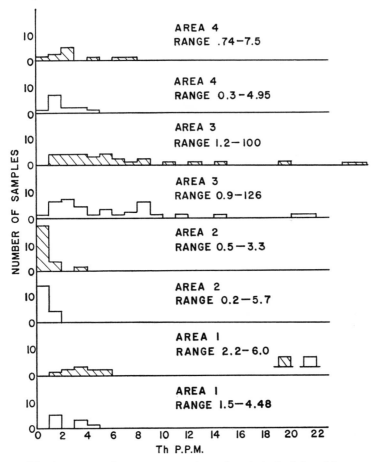

Fig. 6.—Thorium p.p.m. of areas 1, 2, 3, and 4 and total obtained from laboratory and field measurements.

The assumption of secular radioactive equilibrium has not been experimentally proved because of the sampling and experimental difficulties, but the work of Murray and Adams (1958) and the general agreement with the independent and spectrochemical potassium values of Hsu (1960) support the assumption. If the thorium and uranium are largely contained in the quartz and other resistates (e.g.,

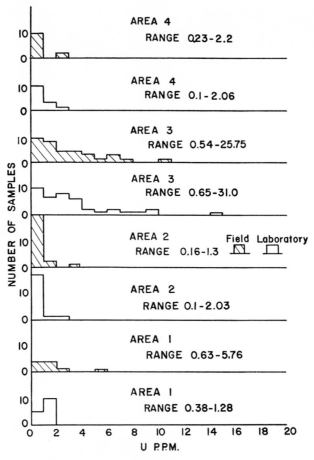

Fig. 7.—Uranium concentrations for areas 1, 2, 3, and 4 from laboratory and field data

zircon and monazite), it is qualitatively unlikely that they have been separated from their radioactive daughters to an important extent. Conversely, any non-resistant or fine-grained mineral whose thorium and uranium are likely to be out of secular radioactive equilibrium is also likely to be quickly removed by the action of the waves on the beaches. Carbonate shells formed in the last hundred thousand years are known to be out of equilibrium in the sense of not having all the radioactive daughters that could be supported by the uranium present. To the extent that shells and shell debris are present on the beach, there will be more uranium present than that indicated by analysis of the γ radiation from the beach.

To date, efforts to determine directly the cosmic-ray contribution to the 2.62-, 1.76-, and 1.46-mev. channels with the field instrument

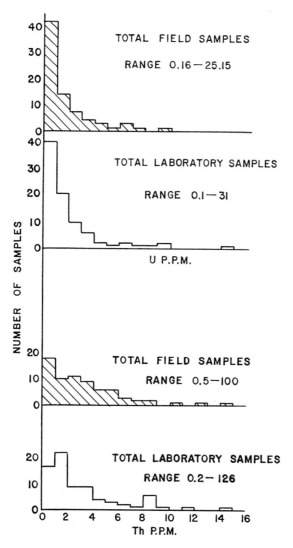

Fig. 8.—Uranium and thorium concentrations for total samples obtained from field and laboratory data.

have failed. By extrapolation of field versus laboratory concentration plots it has been roughly estimated that the cosmic-ray background is of the order of 0.1 p.p.m. thorium and perhaps slightly more uranium. The correction for cosmic-ray background is probably of some importance with the lowest concentrations of thorium and uranium (see Table 3).

In the absence of homogenized, artificial sources of different concentrations of thorium, uranium, and potassium with the thorium and uranium series in secular radioactive equilibrium, the calibration of the field instrument can be refined only by statistical averaging of a still larger number of field and laboratory intercomparisons. For many

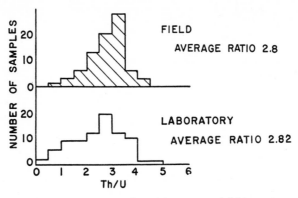

FIG. 9.—Th/U ratios obtained from laboratory and field measurements

geochemical problems, including the exploration for and evaluation of placer deposits, the present instrument and calibration are quite adequate (see comparison of field and laboratory averages in Table 3).

LABORATORY PROCEDURES

The first objective of the laboratory phase of the research was to determine the thorium, uranium, and potassium in the samples in order to calibrate the field instrument and obtain an independent determination of the concentrations of these elements. The beach-sand samples were dried and passed through a coarse sieve to remove large shell fragments and organic debris. After homogenization, an 8-cm. diameter, 3-cm.-high sample can (8 oz. volume) was filled to the top with sand (usually 350–400 gm. of sand), sealed, and the thorium, uranium, and potassium were determined by γ-ray spectrometry. The same energy channels were used for both the field and the laboratory γ-ray spectrometry. The laboratory instrument used a multichannel analyzer and, relative to the field instrument, was much freer of drift due to temperature and shifts in the battery voltage output. The background due to inherent radioactivity in the instruments and shields and to cosmic radiation was about equally important for the small laboratory sample and for the large field system. The laboratory determinations are considered to be accurate to well within ±10 per cent for all but the lowest concentrations. A complete description of the laboratory instrument and its calibration and accuracy is given by Adams (1963).

A second objective of the laboratory research was to determine the sites of the thorium, uranium, and potassium. Depending upon the

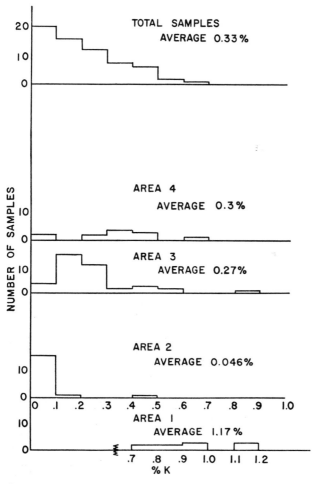

Fig. 10.—Potassium percentage for areas 1, 2, 3, and 4 and totals obtained by laboratory measurements.

megascopic, apparent concentration of dark minerals, 400–700 gm. of original sample were washed with water to remove some of the organic debris. Hydrochloric acid was then added to dissolve any carbonate material, particularly shell debris, present. This treatment probably dissolved any detrital apatite present. The sample was then washed and treated with hydrogen peroxide to oxidize any remaining organic matter and was then washed again, with an effort being made to remove any fine clay material present. After drying, the sample was agitated by bubbling air in bromoform contained in separatory

TABLE 4

HEAVY MINERAL PER CENT IN SOME SAMPLES

Sample No.	Heavy Fraction (per cent)	Sample No.	Heavy Fraction (per cent)
3	0.79	66	1.35
4	0.89	67	2.56
5	0.67	68	2.06
11	3.5	69	1.15
12	3.2	70	1.09
17	0.34	71	1
20	1.2	76	7.8
24	0.3	78	8
27	3.73	80	2.71
29	4.1	81	37.17
30	19.5	82	2.65
45	2.63	84	6.96
51	0.52	85	9.3
52	4.3	86	2.44
54	3.2	95	2.96
55	3.15	96	4.32
56	7.25	97	4.36
57	3.79	105	1.6
58	7.24	106	3.37
59	3.34	107	0.9
65	0.99		

funnels. That portion of the sample that sank in pure bromoform (density of 2.89) was washed, dried, and weighed, as was the portion that floated. The relative percentage of the heavy fraction of the samples studied is given in Table 4.

An attempt was made to determine the concentration of thorium, uranium, and potassium in each of the two bromoform fractions. The successful results are given in Table 5. It should be noted that the heavy fraction weighed much less than the routine 350–400-gm. sample taken for analysis, and thus the accuracy is not ±10 per cent in every case; similarly, the light fraction was very low in thorium and uranium in many cases, and the estimated accuracy in the 0.2–0.5-p.p.m. thorium range is only ±20 per cent.

Thorium and uranium are not evenly distributed throughout the heavy mineral fraction, which is largely ilmenite, tourmaline, kyanite, and other dense, resistate minerals that have very little thorium or uranium. Rather rare grains of monazite carry an important amount of the α activity from the thorium and uranium series, as was shown by a few autoradiographs. Monazite was found to be much more α radioactive than the zircon in the autoradiographic studies. Very rare grains of xenotime and related minerals may be important sites of uranium and thorium, but none was positively identified.

TABLE 5

RESULTS OF RADIOMETRIC ANALYSES OF HEAVY AND LIGHT FRACTIONS

Sample Nos.	HEAVY MINERAL CONCENTRATE			LIGHT PORTION			
	Thorium (p.p.m.)	Uranium (p.p.m.)	Th/U Ratio	Thorium (p.p.m.)	Uranium (p.p.m.)	Potassium (per cent)	Th/U Ratio
11......	77.3	22.6	3.42	....	....	....	....
27......	68.77	63.21	1.09	0.07	0.20	−0.001	0.8
29......	130.75	44.94	2.91	....	....	....	....
30......	66.28	34.99	1.89	0.32	0.26	−0.004	1.23
56......	256.93	80.87	3.18	1.07	0.34	0.083	3.14
57......	350.4	88.07	3.98	....	....	....	....
58......	377.06	92.51	4.07	0.49	0.24	0.122	2.04
59......	316.7	78.68	4.02	....	....	....	....
67......	210.95	60.55	3.48	....	....	....	....
76......	272.08	79.39	3.42	0.61	0.165	0.134	3.69
78......	350.1	88.92	3.94	0.83	0.92	0.48	0.9
80......	75.05	24.72	3.03	....	....	....	....
81......	306.45	77.83	3.94	....	....	....	....
82......	282.03	73.35	3.84	0.73	0.3	0.04	2.45
84......	129.6	41.4	3.13	1.55	3.5	0.27	0.44
85......	193.2	56.55	3.41	1.36	3.47	0.027	0.39
86......	124.38	39.96	3.11	0.68	0.25	0.19	2.72
95......	38.18	19.39	1.97	....	....	....	....
96......	49.45	22.82	2.18	....	....	....	....
97......	42.33	16.83	2.51	0.49	0.14	0.198	3.55
106......	28.91	13.56	2.13	1.37	0.47	0.29	2.9

DISCUSSION

On the basis of the 81 field stations and 91 laboratory samples for which thorium, uranium, and potassium concentrations were determined and on the basis of general geologic considerations, particularly provenance and beach type, the data can be discussed in terms of the following four geographical suites.

Suite I includes those Gulf Coast beaches to the west of the Mississippi River in Louisiana and on Galveston Island, Texas (Fig. 2). The average of 10 laboratory samples is 2.78 p.p.m. thorium, 0.84 p.p.m. uranium, and 1.17 per cent potassium, with a Th/U ratio of

3.3. Most of the samples have a thorium concentration in the 1–2 p.p.m. range. The 10 field stations gave a higher average thorium concentration of 3.97 p.p.m., and this higher value is probably due to the high concentration of clay in the less well-developed Louisiana beaches and the difficulty of obtaining a representative proportion of clay in the laboratory sample. The high potassium concentration in this suite is assignable in part to clay but is mostly due to the presence of potassium feldspar. In general, these potassium determinations confirm the observations of Hsu (1960). At Galveston it was found that the radioactivity of the back beach was almost twice that of the front beach, although the potassium content remained constant. The Galveston beach had a Th/U ratio near that of the crustal average of 3.8.

Suite II includes the Gulf Coast beaches to the east of the Mississippi River. These beaches are characterized by having the lowest radioactivity found in this study. The average for 19 laboratory samples was 0.86 p.p.m. thorium, 0.46 p.p.m. uranium, and 0.046 per cent potassium, with a Th/U ratio of 2.1. The averages for the 22 field stations were in the same general range (see Table 3). Most of the thorium values were in the 0.2–0.5 p.p.m. range (see Tables 1 and 2). In a few localities (Dauphin Island, Alabama, and Cape St. Blas, Florida) the dune contained more heavy minerals and had a higher radioactivity than the front beach. Most of the beaches studied in this suite contained very pure quartz sand, and, although the major rivers of the area (the Appalachicola and Mobile Rivers) supply considerable amounts of heavy minerals, this was not reflected in the beaches. Samples of the sand from Mobile Bay were found to contain 6–10 p.p.m. thorium, well above the average for the beaches nearby. It appears that the only contribution from the rivers that can be recognized is the dune sand of Dauphin Island (see Fig. 3), where the heavy mineral assemblage is similar to that of the rivers. The low potassium content of these high-purity sands indicates a high degree of maturity, but the low Th/U ratio suggests that the sands do not represent a homogeneous sample of crustal quartz, which would be expected to have a Th/U ratio of about 3.8.

Suite III includes the Atlantic Coast beach sands from northeastern Florida to North Carolina, exclusive of Cape Hatteras (Fig. 4). The average of 41 laboratory samples is 11.27 p.p.m. thorium, 3.97 p.p.m. uranium, and 0.27 per cent potassium, with a Th/U ratio of 2.82. The averages for 35 field stations are in excellent agreement (see Table 3) for this large suite that contains almost half the data of this entire study. Thorium occurs most frequently in concentrations of 1–2 p.p.m. and 8–9 p.p.m. (see Tables 1 and 2), although some of the samples contained 100 p.p.m. in this area of commercial beach placers.

The agreement between field and laboratory results is remarkable, considering the difficulty of obtaining representative samples from these beaches, where the radioactive zircon, monazite, and other dense, resistate minerals, such as ilmenite, tourmaline, and rutile, are clearly concentrated in thin, dark layers a few millimeters to a few centimeters thick and of most irregular size and shape.

Suite IV includes the beaches of Cape Hatteras in North Carolina, the beaches of New Jersey, and those of Cape Cod in Massachusetts (Fig. 5). The average of 13 laboratory measurements is 2.07 p.p.m. thorium, 0.8 p.p.m. uranium, and 0.3 per cent potassium, with a Th/U ratio of 2.6. The averages from 11 field stations are in general agreement, considering the small number of determinations in both the laboratory and the field (see Table 5). Recent storms had destroyed the beaches, except at Cape Cod, to such an extent that it was not possible to take measurements across the beach profile. The Cape Cod beach gave a Th/U ratio of 3.7, which is very close to the crustal average, and this beach may be a representative sampling of the earth's crust. In view of the fact that these sands are derived in large part from the reworking of glacial till, it is not surprising that their Th/U ratio is so close to the crustal average. The grains in this suite are coarser than in the other suites, and the composition also is indicative of a lesser degree of maturity. It was noted that the sands in the high cliff and adjacent beaches of the source area were more radioactive and had a lower Th/U ratio than the beaches to the north or south of the source area. Apparently the reworking or dilution of the source material results in a lower total radioactivity and a higher Th/U ratio.

In general, there seemed to be systematic changes from the shore to the dune across all beach profiles. Figure 11 summarizes all the data from the profiles. These changes in the concentrations of thorium and uranium and the Th/U ratio must be due in large measure to the relative proportions of clay and radioactive resistate minerals, particularly where more than 1–2 p.p.m. thorium and 0.3–0.6 p.p.m. uranium are found. These latter concentrations are the ones found most frequently in pure beach sands and the light bromoform fractions (see Tables 1, 2, and 5) and may be considered as the average thorium and uranium content of quartz and feldspar. Potassium feldspar is, of course, the main carrier of potassium, with secondary contributions from clay and biotite.

The data from this study represent a large increment in the available information on the thorium, uranium, and potassium concentrations in beach sands. The data from the various beach profiles and suites are very suggestive of the way in which these natural radioactive tracers can be used to trace the movement of mineral grains across and along the beaches. Only the Galveston and Cape Cod beaches had Th/U ratios near the crustal average of 3.8, and it appears that only few beach sands will meet this criterion of being representative and homogenized samples of crustal quartz with or without representative and homogenized samples of crustal feldspar and accessory minerals. Quartz and feldspar carry most of the thorium and uranium found in most of the beaches studied, and the average concentration of thorium so carried is 1–2 p.p.m., with 0.3–0.6

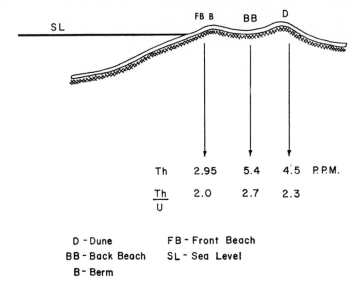

FIG. 11.—Schematic profile of a beach and variations of thorium and Th/U ratios

p.p.m. uranium. From the point of view of geochemical exploration, these last values may be considered the threshold values above which one may begin to see anomalies that may be indicative of economic placer concentrations of zircon, monazite, or related radioactive minerals. Uneconomic concentrations of these minerals may still serve as tracers or pathfinders for economic placer concentrations of non-radioactive ilmenite, magnetite, or similar dense, resistate minerals. The present study also demonstrated that the present field γ-ray spectrometer can be accurately calibrated and efficiently used at and above these threshold values.

The problem of the average thorium and uranium concentrations in beach sands in particular and sands and sandstones in general remains unresolved (see the discussion above of previous work) because of the heterogeneous distribution of monazite, xenotime, zircon, and other thorium- and uranium-bearing minerals in sands. The average concentrations are of considerable geochemical interest in connection with studies of lead isotopes and the distribution of thorium and uranium. The problem is also of potential economic interest because it involves the maximum amount of placer minerals that might be expected to occur in sands. Figure 12 is a plot of thorium concentrations versus frequency for beach sands and is essentially the same as that given by Murray and Adams (1958); the lower mode of 1–2 p.p.m.

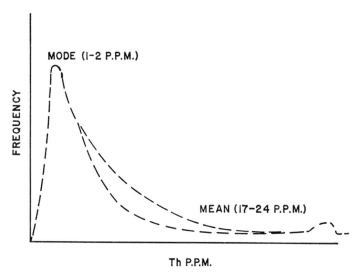

Fig. 12.—Schematic frequency curve for thorium in sand

thorium may be considered as better established as a result of this study. The second mode at high, essentially placer concentrations of thorium is hypothetical. The theoretically calculated means of 17 and 24 p.p.m. are shown (see the discussion above of previous work). If one assumes that sands, including beach sands, do indeed have an average thorium content of 17 p.p.m., then it is possible to relate thorium concentration to the ratio of placer to normal sand as in Figure 13. Figure 13 would qualitatively fit the data from the Nile Delta discussed above but would suggest that there is far more thorium and other metals in dense, resistate minerals in beach sands than has been previously thought. Probably the greater fraction of these minerals is in low concentrations in thin and irregularly distributed placers, some of which may not have great color contrast with the dominant quartz and feldspar. Whether one rejects both the theoretical arguments indicating 17–24 p.p.m. thorium in sands or accepts them with the conclusion that 10–15 per cent of the thorium in sedimentary rocks is in placer and subplacer concentrations of dense, resistate minerals, there is a clear need for more data.

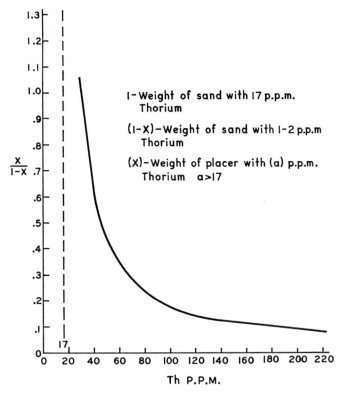

Fig. 13.—Thorium concentration *vs.* placer (high thorium)/sand (low thorium) (theoretical calculation).

Conclusions

The major conclusions of this study are as follows:

1. Thorium, uranium, and potassium in beach sands can be determined in the field by γ-ray spectrometry with an accuracy of better than ±10 per cent in most cases and on a volume of sand 100 times larger and hence much more representative than most laboratory samples.

2. Thorium above 1–2 p.p.m. and uranium above 0.3–0.6 p.p.m. in

beach sands is contained in dense, resistate minerals, such as monazite, zircon, and xenotime.

3. Most beaches have a Th/U ratio of 2.5–3, and only the Galveston Island, Texas, and the Cape Cod, Massachusetts, beaches had a Th/U ratio near the 3.8 crustal average.

4. The concentrations of thorium, uranium, and potassium and the Th/U ratio can be related to provenance and beach processes in several cases and can probably be used as natural radioactive tracers and process indicators in many geologic situations, both modern and ancient.

5. The mode or most frequent concentrations found in beach sands are 1–2 p.p.m. thorium and 0.3–0.6 p.p.m. uranium, but the mean concentrations are experimentally unknown, being greatly affected by irregular occurrence of dense, resistate minerals, such as monazite and zircon.

6. Theoretical estimates of the mean thorium and uranium contents of sands defined geochemically are much higher than are those found experimentally to date, and either these estimates are incorrect or 10–15 per cent of the thorium in such defined sedimentary rocks occurs in placer and subplacer sand deposits.

ACKNOWLEDGMENTS

The research was supported in its entirety by the Robert A. Welch Foundation through Grant C-009 to Professor John A. S. Adams and Professor John J. W. Rogers. Professor Rogers took part in the field measurements and sampling in the summer of 1962, and Professor Adams directed the Ph.D. thesis of which this research is a part.

REFERENCES

ADAMS, J. A. S. 1963. Laboratory γ-ray spectrometer for geochemical studies. This symposium.

ADAMS, J. A. S., and G. E. FRYER. 1963. Portable γ-ray spectrometer for field determination of thorium, uranium, and potassium. This symposium.

ADAMS, J. A. S., J. K. OSMOND, and J. J. W. ROGERS. 1959. The geochemistry of thorium and uranium: *In* L. H. AHRENS (ed.), Physics and Chemistry of the Earth, **3**:298–348. London: Pergamon Press.

ADAMS, J. A. S., and K. A. RICHARDSON. 1960. Thorium, uranium and zirconium concentration in bauxite. Econ. Geol., **55**:1060–63.

ADAMS, J. A. S., and C. E. WEAVER. 1958. Thorium to uranium ratio as indicators of sedimentary processes: An example of the concept of geochemical facies. Bull. Am. Assoc. Petrol. Geologists, **42**:387–430.

DEGENHARDT, H. 1957. Untersuchungen zur geochemischen Verteilung des Zirkoniums in der Lithosphere. Geochim. & Cosmochim. Acta, **2**:279–309.

GINDY, A. R. 1961. Radioactivity in monazite, zircon and radioactive black grains in black sands of Rosetta, Egypt. Econ. Geol., **56**:436–41.

GUILLOU, R. B. 1963. The aerial radiological measuring survey, ARMS Program. This symposium.

HIGAZY, R. A., and A. G. NAGUIB. 1958. A study of the Egyptian monazite-bearing black sands. Proc. 2d United Nations Internat. Conf. on Peaceful Uses of Atomic Energy, **2**:658–62.

HSU, K. J. 1960. Texture and mineralogy of the recent sands of the Gulf Coast. J. Sed. Petrology, **30**:380–403.

JOHNSTONE, C. W. 1963. Detection of natural γ radiation in petroleum exploration boreholes. This symposium.

MURRAY, E. G., and J. A. S. ADAMS. 1958. Thorium, uranium and potassium in some sandstones. Geochim. & Cosmochim. Acta, **13**:260–69.

PETTIJOHN, F. J. 1957. Sedimentary Rocks. New York: Harper & Bros. Pp. 526.

PLILER, R., and J. A. S. ADAMS. 1962*a*. The distribution of thorium, uranium, and potassium in the Mancos shale. Geochim. & Cosmochim. Acta, **26**:1115–35.

———. 1962*b*. The distribution of thorium and uranium in a Pennsylvanian weathering profile. *Ibid.*, pp. 1137–46.

POLDERVAART, A. 1955. Chemistry of the earth's crust. Geol. Soc. America Spec. Paper 62, pp. 119–44.

RABE, C. L. 1957. A relation between gamma-radiation and permeability, Denver-Julesburg Basin. J. Petrol. Tech., Technical Note 403, pp. 65–67.

C. W. JOHNSTONE

6. *Detection of Natural γ Radiation in Petroleum Exploration Boreholes*

FOR MANY YEARS, holes have been drilled thousands of feet deep in the earth to locate and produce oil and natural gas.

Attempts began in 1935 in the U.S.A., and at about the same time in the U.S.S.R., to measure the intensity of natural γ rays in these boreholes. The first measurements (Howell and Frosch, 1939) yielded results that were of sufficient interest to justify the design of more reliable and rugged devices, which used either ionization chambers or Geiger-Müller counters. By 1940, the value of γ-ray logging was recognized, and it had been introduced as a new commercial service (Westby and Scherbatskoy, 1940) to supplement the already established electrical logging services.

By way of explanation, the term "log" means graphical recording *versus* depth of the response of an instrument to some property of the formations through which the hole is drilled. "Logging" is the operation that is performed to record a log.

Well logging is generally not done by the oil companies but is performed for them by one of the service companies, who design, build, and operate their own instruments and equipment. There are many types of these instruments, called "logging tools" or "sondes," which are designed to operate on the end of a long cable. The more important logging services in use include measurements of electrical resistivity, spontaneous potentials, temperature, sonic travel time, formation density, and hydrogen content, in addition to the recording of natural γ-ray intensity (Segesman, Soloway, and Watson, 1962).

Both the oil companies and service companies are continually striving to find and develop logging techniques that will yield new

C. W. JOHNSTONE is head, Nuclear Physics Section, Schlumberger Well Surveying Corporation, Houston, Texas.

or better information to help identify and evaluate underground petroleum reservoirs.

Gamma-ray logging depends upon the presence and distribution of radioactive elements in the earth's crust, namely potassium-40 and members of the uranium and thorium series. Among sedimentary formations below the surface, shales are usually high in γ radioactivity, sands intermediate, and dense carbonates and anhydrites low, although exceptions to this general rule are encountered. The ability to distinguish shales from non-shales, particularly under conditions in which other types of logging tools are useless, makes the γ-ray log very useful for lithology determination. An electrical logging technique (spontaneous potential) is also used to identify shales, but this method cannot be used in oil-base muds, very saline muds, or "cased" holes.* Gamma rays are easily detected through steel casing and cement, and the type of fluid in the hole, whether it be liquid or gas, makes little difference.

Scope of γ-Ray Logging

During the years from 1955 to 1962, the number of new wells drilled in the United States averaged about 50,000 per year (Mott and Ediger, 1959; *Oil and Gas Journal*, 1963). The average depth of these wells was about 4,000 feet, although depths over 10,000 feet are not uncommon and a few wells have exceeded 20,000 feet. Each year about 35,000 γ-ray logs were run. With an average depth of 4,000 feet, this means that during each year in the United States alone a total of about 140,000,000 feet of hole is logged for natural γ radiation. Although the number of wells drilled per year has tended to decrease, the ratio of γ-ray logs to wells drilled has shown a compensating increase.

About the same ratio of γ-ray logs has applied to wells drilled outside the United States, but no recent figures are available.

Instrumentation

Figure 1 illustrates how a logging operation is performed and is largely self-explanatory. It is common practice to protect the instruments for logging against a maximum hydrostatic pressure of 20,000 p.s.i. by incasing them in steel "housings." By the "size" of the "tool," one generally means the outer diameter of the pressure housing. The length of a logging tool is usually not of prime importance. To be

---

* Before petroleum is produced from a well, it is necessary to install steel pipe or "casing" and cement it to the wall of the hole to prevent vertical passage of fluids outside the casing. Perforations are then made through the casing and cement at the correct depth to permit the petroleum to enter the casing and to flow or be pumped to the surface.

able to enter the various sizes of holes, casing, and tubings, logging tools are generally designed to fit in standard pressure housings with outer diameters of $3\frac{5}{8}$, $3\frac{3}{8}$, $2\frac{5}{8}$, or $1\frac{11}{16}$ inches.

Temperature has been a major problem for detectors and circuit components. Present-day temperature requirements call for downhole operation up to at least 300° F., although a rating of 350° F. or even 400° F. is preferable because such high temperatures are occasionally encountered. Another problem, peculiar to radioactivity logging, and particularly to γ-ray logging, has been to achieve a sufficiently high counting rate to minimize statistical fluctuations. The low sensitivity of Geiger-Müller detectors resulted in rather severe compromises

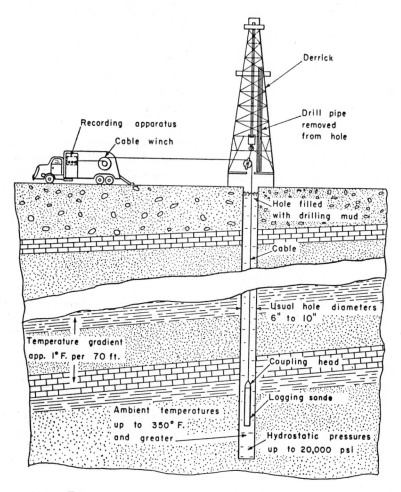

Fig. 1.—Schematic drawing of a well-logging operation

(Kokesh, 1951), taking into consideration detector length, bed definition, statistical variations, and logging speed. The longer the counter was made (to improve the counting rate), the poorer the bed definition became.

With the introduction of scintillation detectors using sodium iodide crystals in the early 1950's, counting rates could be improved by a factor of 2 or 3, using scintillation crystals only 8 inches long compared to an active length of about 40 inches in the Geiger-Müller detectors. Because they provide much better bed definition, scintillation detectors have now almost entirely replaced Geiger-Müller detectors for γ-ray logging.

However, although Geiger-Müller counters had been designed to operate continuously at high temperatures, scintillation counters employing conventional photomultipliers are ordinarily limited to a maximum temperature of about 170° F. It was necessary, therefore, to overcome this temperature limitation. The problem has been solved in two different ways: (1) thermal insulation (with or without a method of cooling) of conventional photomultipliers or (2) development of high-temperature photomultipliers.

Inside a dewar flask made of stainless steel a conventional photomultiplier can operate for several hours at well temperatures in excess of 300° F. without employing any method of cooling that removes heat from inside the dewar. Plate I shows a disassembled scintillation detector for a $3\frac{3}{8}$- or $3\frac{5}{8}$-inch tool, including the crystal, photomultiplier, heat sink, and cork, all of which fit in the dewar flask.

High-temperature photomultipliers (Causse, 1960), developed especially for well logging, operate successfully without insulation at sustained temperatures up to 300° F. Detectors using these photomultipliers are particularly useful in the smaller-diameter logging tools, in which there is little or no space available for a dewar flask. Plate II shows a high temperature photomultiplier, developed in the research laboratories of Schlumberger Well Surveying Corporation. For very high temperature operation, one can, of course, use a high-temperature photomultiplier in a dewar.

Scintillation crystals in hermetically sealed assemblies withstand temperature cycling to 300° F., with little or no deterioration in optical properties.

Shock mounting is very important so that the crystal and photomultiplier can withstand the normal treatment during logging as well as transportation to and from the well site over rough roads. A good detector should survive a drop of 6 inches onto a hard surface without damage.

The development of rugged, stable components for missiles and

PLATE I

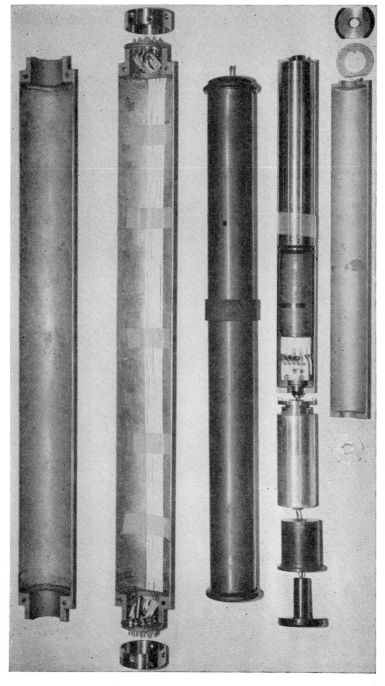

Disassembled γ-ray detector using a dewar flask and conventional photomultiplier

PLATE II

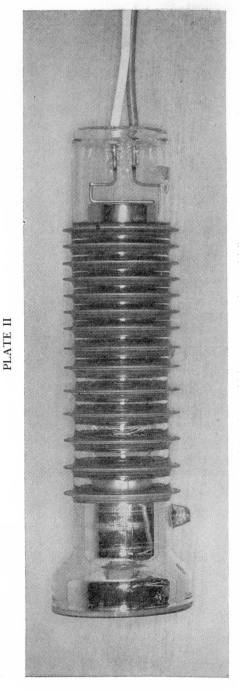

Schlumberger high-temperature photomultiplier

satellites has been a boon to the designers of well-logging instruments. Small resistors, capacitors, and vacuum tubes suitable for high-temperature operation have improved greatly during recent years, as have semiconductor devices.

A scintillation γ-ray logging tool normally requires a reasonably stable (±2 per cent) high-voltage supply and a pulse amplifier, trigger circuit, and output stage to couple the pulses to the logging cable and transmit them to the surface. Although multiconductor cables are sometimes used, the general practice is to design γ-ray equipment that can operate on a single conductor coaxial cable. Primary 60-cycle a.c. power is supplied over the cable, and, in some combination tools, both positive and negative pulses in addition to low-frequency signals from a casing-collar* locator are returned to the surface on the same cable conductor, to be separated by means of filters. Conversion of pulse rate to a.c. current, to operate a recorder, is performed at the surface by a conventional counting rate meter circuit. Plate III is a photograph of a 3⅜-inch O.D. γ-ray tool, showing the pressure housing, monocable head, detector, and electronic circuitry in the form of "potted" modules.

Sufficient gain is employed in the detector and amplifier to make the pulse discriminator respond to essentially every γ ray that has sufficient energy to penetrate the steel pressure housing and cause scintillation to occur in the crystal. Because there is an energy gap between the largest thermal noise pulses and the smallest significant γ-ray pulses, it is possible to operate the detector in a "plateau" region, in which small changes in high voltage or amplifier gain cause practically no change in counting rate. Increasing temperature increases the thermal noise from the photocathode, while reducing the photomultiplier gain and scintillator light output, thus shortening the detector plateau. Since all photomultipliers of a given type are by no means identical, individual detectors have to be carefully heat tested before approval for field operation.

γ-Ray Logging Variables

The counting rate of a scintillation detector during logging ranges from about 50–100 c.p.s. in relatively "clean" formations to 250–350 c.p.s. in typical shales. Thus, with a 3-second time constant, the standard deviation of the instantaneous response due to random occurrence of pulses varies from 2 to 6 per cent. Typical logging speed is 1,800 ft/hr, or one foot every 2 seconds. The "lag" in response is considered to be equivalent to one time constant, or, with

* Sections of steel casing are joined with "collars." These collars can be detected with magnetic devices and serve as depth reference points.

PLATE III

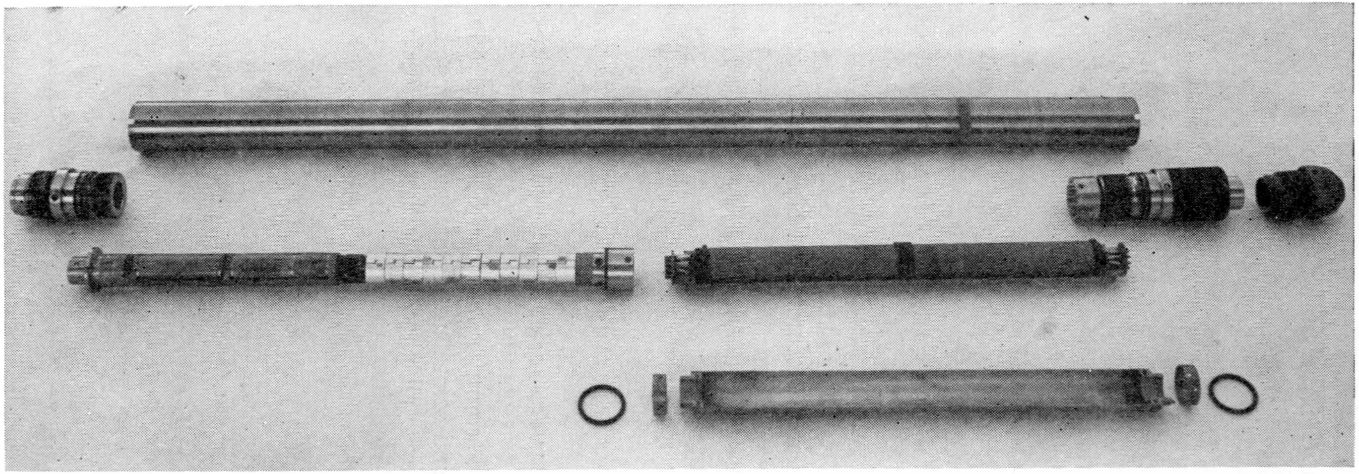

Electronic cartridge and γ-ray detector, with steel pressure housing and connecting heads

the assumptions made above, 1.5 feet in terms of depth. When passing a sharp boundary between formations it would take nearly 4 feet before 90 per cent of the full incremental response is achieved. A considerable improvement in lag and bed definition can be achieved by logging more slowly. Even at very low speed, however, the length (8 inches) of the detector crystal and its sensitivity to γ rays which originate both above and below the detector would cause a sharp boundary to be "smeared" over a distance of 1–1.5 feet.

USES OF γ-RAY LOGS

Quantitative interpretation of γ-ray logs has not been exploited to any great extent (Mott and Ediger, 1959), although interest is increasing in this direction. A quite extensive laboratory investigation of core samples was first reported by Russell (1944). This work was done to relate Geiger counter readings to the average total radioactivity present in common types of sedimentary rock. Independent work was also done with Geiger counters to determine the effect on counting rate of hole size, mud density, casing and cement thickness, and position of the detector in the hole. Appropriate correction charts were published in 1953 (Blanchard and Dewan). These show that variations in hole size and fluid density can change the reading as much as a factor of 2 (comparing an empty hole to a 12-inch-diameter hole containing 16 lb/gal mud). The presence of casing and cement can, under rather extreme conditions (1 inch of steel and 3 inches of cement), attenuate the reading by more than a factor of 5, compared to the uncased hole reading.

The most extensive applications of γ-ray logging do not require quantitative interpretation. Instead, attention is mainly directed to the depths at which bed boundaries occur, the vertical extent of particular beds, and the relative "shaliness" of one bed compared to others in the same well. Gamma-ray logs may be used for the mapping of subsurface strata, which is called "correlation." In this application, logs from different wells in a given region are compared in order to note variations in depth and thickness of particular formations. The ability of the γ-ray tool to "see through" steel casing makes it very useful for "depth control." Here, after the casing has been cemented in the hole, a particular formation can be identified and its depth closely related to adjacent casing collars.

In order to make the best use of the time required to log a well, it is common practice to combine two and sometimes three services in a single logging tool. Thus, the γ-ray log is seldom run by itself. Combination tools include: (1) γ-ray–neutron (hydrogen-content), (2) γ-ray–sonic (acoustic velocity), (3) γ-ray–conduction (resistivity), and (4) γ-ray–laterolog (resistivity). The formation properties measured by the companion services are indicated in parentheses. Explanations of these services are beyond the scope of this paper but will be found in Segesman, Soloway, and Watson (1962), and others.

Figure 2 is a short section of log showing both γ-ray response (*left*) and sonic- or acoustic-wave travel time in μ sec/ft (*right*). For simplicity, only the 50-foot depth lines are shown, omitting the 2- and 10-foot lines. Gamma-ray intensity increases toward the right. A

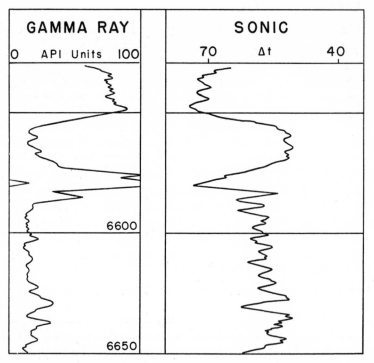

Fig. 2.—Short section of recorded "log" of γ-ray intensity, combined with a recording of acoustic-wave travel time.

rather intense peak of γ-ray intensity occurs in the thin shale at about 6,580 feet. Note that as the deflection passes full scale, a "back-up" trace appears at the left margin, so the peak reading is not lost. A thicker, less radioactive shale is encountered (logging upward) just below 6,550 feet. Also of interest is the almost mirror-image relationship of the γ ray to the sonic curve, which is reflecting changes in lithology. The regions of low γ-ray activity contain dolomite, limestone, and anhydrite. Most of the variations on the γ-ray log (except for the smallest "wiggles") are not statistical but also represent changes in lithology.

CALIBRATION

For many years well-logging service companies had different systems of calibration and different units of measurement of γ-ray intensity. For example, deflections on the log were scaled in such units as inches, counts/second, microroentgens/hour, radiation units, and micrograms of radium equivalent per metric ton. As a result, it was difficult to make a quantitative comparison of the logs of one company with those of another.

To remedy this situation, the American Petroleum Institute, in 1959, established a γ-ray calibration pit (see American Petroleum Institute, 1959) at the University of Houston. The pit is 4 feet in diameter and contains three zones, each 8 feet thick, with $5\frac{1}{2}$-inch O.D. steel casing running down the center. The top and bottom zones are made of low-radioactivity concrete. The center zone contains a radioactive concrete equal in radiation to about twice that of an average shale* and was made by adding uranium, thorium, and potassium to the concrete in the same relative proportions normally found in shales. The difference in response of any γ-ray logging tool when moved from a low-activity zone to the high-activity zone in this pit is defined as 200 API units. Thus the calibration of γ-ray logging equipment has been standardized, and a common unit of scaling is now in use by all logging companies.

It may be of interest that the quantity 200 API units was found to correspond to 13.5 micrograms of radium equivalent per metric ton for the logging tool of one company, and 12.6 microroentgens per hour in the units previously used by another company.

Calibration of a logging tool at the well is done with a test source mounted in a fixture that holds it a given distance from the detector. The increase in counting rate, above background, when the test source fixture is attached, is known by API calibration to represent a certain number of API units. The recorder sensitivity can then be adjusted to provide the desired scale for logging.

SPECTRAL LOGGING

The use of scintillation detectors in commercial logging of natural γ rays has been limited to the recording of the gross intensity of the γ rays. However, the capability for providing spectral information is, of course, inherent in this type of detector.

An investigation of the spectra of natural γ rays in boreholes was made in 1956 by Brannon and Osoba, using a pulse-height analyzer on

* Average shale was assumed to contain 6 p.p.m. uranium, 12 p.p.m. thorium, and 2 per cent potassium.

the surface to sort the pulses from a linear downhole detection system. To accumulate enough counts to determine a spectrum, it was necessary to hold the detector stationary and accumulate counts for 15 minutes.

Other investigations and results obtained from measurement of downhole spectra have been reported (Adams and Weaver, 1958). Relative intensities of γ rays from uranium, thorium, and potassium have been determined, as well as Th/U ratios, both from core samples and from experimental downhole measurements. Quite a wide variation in the relative abundance of these elements has been found. If more information of this nature could be obtained, it should be of significant assistance in geologic interpretation.

From an instrumentation standpoint, it would be feasible to add pulse-height discrimination circuits to γ-ray logging tools so that, during routine logging operations, separate traces could be recorded showing the counting rates in each of two or more energy bands selected to contain spectral peaks representative of the elements of interest. However, aside from the added requirements of good energy resolution, gain stability, and accurate calibration, the principal limitation to this technique is the very low counting rate that can be obtained.

Contributing to the gross counting rate of a scintillation detector, operated on a "plateau" for conventional logging, are a relatively large number of counts from scattered γ rays at the low end of the energy spectrum. By comparison, the counting rate in a rather narrow energy band at the upper end of the spectrum would definitely be too low to provide useful information at conventional logging speeds. Because of this, there has been little development of spectral techniques for commercial natural γ-ray logging. However, spectral techniques have come into use in logging the γ rays from neutron capture (Dewan, Stone, and Morris, 1961) in tools that contain neutron sources of adequate strength to provide useful counting rates.

Conclusion

In an introductory paper of this nature it is not possible to provide detailed information on all aspects of the subject. The intent has been to acquaint the reader with the purpose and scope of γ-ray logging, and to provide a brief description of the instrumentation required for the borehole environment. Because of the fundamental and unchanging nature of the natural γ radiation from subsurface formations, γ-ray logging has been and should continue to be one of the very important well-logging services.

ACKNOWLEDGMENTS

The author wishes to thank the many people at Schlumberger Well Surveying Corporation who have helped in the preparation of this paper.

REFERENCES

ADAMS, J. A. S., and C. E. WEAVER. 1958. Thorium-to-uranium ratios as indications of sedimentary processes: Example of concept of geochemical facies. Bull. Am. Assoc. Petrol. Geologists, **42**:387–430.

AMERICAN PETROLEUM INSTITUTE. 1959. Recommended Practice for Standard Calibration and Form for Nuclear Logs. Am. Petrol. Inst. Rpt. Rp-33, 2d ed. Pp. 12.

BLANCHARD, A., and J. T. DEWAN. 1953. The calibration of gamma ray logs. Petrol. Engineer, **25**:B-76, B-78-80.

BRANNON, H. R., and J. S. OSOBA. 1956. Spectral gamma-ray logging. J. Petrol. Technol., **8**:30–35.

CAUSSE, J. P. 1960. New rugged high-temperature photomultipliers. IRE Trans. Nuc. Sci., NS, **7**:66–71.

DEWAN, J. T., O. L. STONE, and R. L. MORRIS. 1961. Chlorine logging in cased holes. J. Petrol. Technol., **13**:531–37.

Fewer wells, but footage up 3.4 per cent. 1963. Oil & Gas J., **61**:152–53.

HOWELL, L. G., and A. FROSCH. 1939. Gamma-ray well-logging. Geophysics, **4**:106–14.

KOKESH, F. P. Gamma-ray logging. 1951. Oil & Gas J., **50**:284–300.

Most finds still in 5000–7500-foot class. 1963. Oil & Gas J., **61**:157–61.

MOTT, W. E. and N. M. EDIGER. 1959. Nuclear well logging in petroleum exploration and production. Proc. 5th World Petrol. Cong., Sec. X, pp. 195–206.

RUSSELL, W. L. 1944. The total gamma ray activity of sedimentary rocks as indicated by Geiger counter determinations. Geophysics, **9**: 180–216.

SCHLUMBERGER WELL SURVEYING CORPORATION. 1958. Introduction to Schlumberger well logging. Schlumberger Doc. No. 8.

SEGESMAN, F., S. SOLOWAY, and M. WATSON. 1962. Well logging: The exploration of subsurface geology. Proc. IRE, **50**:2227–43.

WESTBY, G. H., and S. A. SCHERBATSKOY. 1940. Well logging by radioactivity. Oil & Gas J. **38**:62–64.

PAUL C. RAGLAND

7. *Autoradiographic Investigations of Naturally Occurring Materials*

T HE DISTRIBUTION of the naturally occurring α emitters in the thorium and uranium series within rocks and minerals has long been of interest to a wide range of scientific disciplines. The distribution of thorium and uranium in naturally occurring materials has proved to be important in such researches as (1) estimates of the level of natural background radiation to which man is exposed; (2) long- and short-range supplies of naturally radioactive materials as energy sources; (3) nature of the earth's interior, as partially based upon heat-flow measurements, and concentrations of radioactive constituents in meteorites and ultrabasic rocks; (4) sources of energy in the geochemical cycle; (5) absolute age determinations, as based upon the uranium, thorium, and actinium disintegration series; and (6) absolute abundance studies utilizing thorium and uranium as physicochemical indicators in a wide variety of natural environments.

In recent years a large quantity of data has been amassed concerning the total α activity, as well as absolute concentrations of thorium and uranium, in rocks and minerals. Much less quantitative information is available, however, concerning the *site* of the radioactive constituents within the commonly occurring rocks. Strutt (1906) first noted that felsic igneous rocks are more radioactive than are mafic rocks. Larsen and Keevil (1942) observed that uranium and thorium are concentrated in igneous rocks in the minor accessory minerals, such as zircon, monazite, apatite, allanite, and sphene. Subsequent studies have substantiated these observations, for example, Keevil *et al.* (1944), Picciotto (1950), Coppens (1951), and Brown and Silver (1956), along with many of the papers that will be subsequently cited. Comparatively few data are available, however, concerning the relative contribu-

PAUL C. RAGLAND is assistant professor of geology, University of North Carolina, Chapel Hill, North Carolina.

tions of the minor accessory minerals, the major rock-forming minerals (the essential minerals), and secondary material and alteration products. The purpose of this paper is to synthesize the available information obtained by the autoradiographic method, to contribute new information on radioactive sites in the common rock types, and to propose some possible applications for the autoradiographic technique.

The Autoradiographic Method

One of the most promising current techniques applicable to studies of the distribution of α emitters is α autoradiography, the observation of the image of α-particle trajectories as recorded in α-sensitive nuclear emulsions. Nuclear emulsions commonly contain a larger amount of silver halide and are much finer grained than the ordinary photographic emulsions. A nuclear track plate (consisting of a layer of nuclear emulsion on a standard glass microscope slide) may be placed in contact with a thin section or a polished section of a rock or mineral, exposed for sufficient time to record a statistically significant number of tracks, removed from the section, and developed (Poole and Bremner, 1949; Picciotto, 1949). The α tracks are then observed within the emulsion under the microscope and correlated with the site on the section from which they were emitted. The location of features on the track plate with respect to the section may be difficult after development, particularly if no precautions are taken beforehand to locate precisely the track plate on the section. In some cases this problem may be solved by pouring and subsequently developing a liquid nuclear emulsion directly on the section, thus facilitating direct identification of the radioactive site (Ford, 1951; Guilbert and Adams, 1955). A stripping emulsion technique has been described by Pelc (1947) and Stieff and Stern (1952). Small rock fragments or mineral grains may be impregnated directly into the emulsion, and liquid emulsion poured over the grains. The grain-mount autoradiograph is then treated as the polished or thin-section autoradiograph.

Beta-sensitive emulsions have been utilized in the identification of potassium-rich minerals (Hée and Jarovoy, 1953). Because of their relatively long range and poor resolution, the autoradiographic investigation of β particles has not proved as useful for studies of the distribution of thorium and uranium as has α autoradiography. Gamma-sensitive emulsions are also available but have not been utilized in studies of rocks.

A general survey of the autoradiographic method has been given by Bowie (1954), including a discussion of the preparation and development of α autoradiographs and a review of the literature.

Yagoda's (1949) pioneering work laid the foundation for many of the quantitative studies that have been undertaken in recent years. The α-track population emitted from a polished section or thin section as recorded in a nuclear emulsion may be closely estimated by (Yagoda, 1949)

$$T_\alpha = \psi(25.73 \; U + 7.80 \; Th),$$

where T_α is the α activity in tracks per cm.^{2} per sec., U and Th are the concentration of uranium and thorium, respectively, in grams per gram, and ψ is the permeability of the mineral. By assuming a constant Th/U ratio, one may calculate a semiquantitative value for the absolute concentrations of the elements. Most of the published data have been based upon a Th/U ratio of 3. This ratio has been demonstrated to be acceptable for the major rock-forming minerals (Hamilton, 1960a); however, it varies quite widely in the bulk rock and the accessory minerals (Whitfield *et al.*, 1959; Rogers and Ragland, 1961; Adams *et al.*, 1959).

Two methods have been proposed to estimate the Th/U ratio by the α autoradiographic method. Both methods facilitate the calculation of absolute concentrations of thorium and uranium in minute grains under the microscope. The first method involves the observation of "five-prong stars" in the nuclear emulsion. The relative half-lives of the α emitters in the uranium and thorium disintegration series are such that, in a 72–720-hour period, as many as five tracks may be emitted from a single thorium nucleus, whereas only three may be emitted from a uranium nucleus. The theoretical considerations involved in this technique have been discussed by Senftle *et al.* (1954). The procedure was utilized to measure thorium isotopes in sea water (Koczy *et al.*, 1957) and in deep-sea sediments (Picciotto and Wilgain, 1954). Some of the problems involved in this method include difficulty in achieving ideal geometry and, in the case of very weakly radioactive minerals, lack of a significantly large population for a statistically valid estimate of the ratio. In theory, assuming radioactive equilibrium, the ratio of five-prong stars to single tracks and/or three-prong stars is directly proportional to the Th/U ratio.

The second method is based upon the observation that the trajectory of the α particle emitted from polonium-214 (mean range in dry air at S.T.P. of 6.91 cm.), the most energetic of the α emitters in the uranium series, is significantly smaller than that from polonium-212 (mean range in dry air at S.T.P. of 8.57 cm.), the most energetic in the thorium series (Bowie, 1954). Curie (1946) has expressed this relationship in the following equation:

$$N_2/N_1 = 0.8 + 3.3 \; C_U/C_{Th},$$

where N_1 is the number of α tracks of length longer than 7 cm. in air, N_2 is the number of tracks shorter than 7 cm. and longer than 5.8 cm. in air, and C_U and C_{Th} are the concentrations in grams per gram of uranium and thorium, respectively. Hayase (1956) has utilized this method in the calculation of the "Th-U tendency," a semiquantitative estimate of the Th/U ratio, in various minerals. Theoretical considerations of α-track-length distributions have been treated by Curie (1946), Poole and Matthews (1951), and Von Buttlar and Houtermans (1951). Problems are involved in this technique similar to those in the observation of five-prong stars. Primarily owing to the tedious procedures involved in both methods, relatively few data are available.

Deutsch et al. (1956a, b, 1957) have assessed the α activity in meteorites by the autoradiographic method, and references are cited above to the measurement of thorium isotopes in sea water and deep-sea sediments. Some information on sedimentary rocks and metamorphic rocks has been cited by Hayase and Tsutsumi (1958) and Coppens (1950). The preponderance of the quantitative studies utilizing this technique, however, has been concerned with the site of the α emitters in igneous rocks and their constituent minerals. Hayase (1954, 1955, 1956, 1957, 1958, 1959) has determined the α activity and estimated the Th/U ratio in a large number of felsic rocks and their constituent minerals from Japan. Picciotto (1950, 1952) has measured the α activity and estimated the uranium content by assuming a Th/U ratio of 3 in several intrusive igneous rocks. Similar studies on a wide variety of igneous-rock types have been made by Hamilton (1959, 1960a, b), Merlin et al. (1957), Deutsch and Longinelli (1958), Longinelli (1959), Hée et al. (1954), and Spears (1961).

Picciotto (1949) has noted that the lower detection limit of α activity as determined by this method corresponds to a uranium concentration of 0.5–10 p.p.m. with a precision of 15 per cent of the value. The lower limit of detection of the α activity in meteorites as reported by Deutsch et al. (1956) corresponds to a concentration of 0.05 p.p.m. uranium. The detection of such low activities requires an exposure time on the order of several months. Errors involved in the quantitative assessment of the concentrations of the α emitters by this method may be caused by (1) difficulty in determining the total number of tracks, owing to such problems as fading of the latent image of the tracks (Yagoda and Kaplan, 1947) and poor resolution, (2) radioactive non-equilibrium, (3) contamination, (4) heterogeneity within the rock, (5) errors in the determination of the Th/U ratio, and (6) variations in the geometry of the system.

Distribution of α Emitters within the Constituent Minerals

Essential minerals in igneous rocks (i.e., major rock-forming minerals, normally in excess of 95 per cent) vary only slightly in α activity and commonly contain comparatively minute quantities of α emitters. Hamilton (1960a) has discussed the α activity and uranium content (based upon a Th/U ratio of 3) of the essential minerals in considerable detail. Data on the α activity of feldspars as determined by α autoradiography are presented in Table 1. In the majority of the papers, potassium feldspar was not distinguished from plagioclase in those rocks which probably contain both minerals; however, there is apparently no real difference in α activity between the two types of feldspar.

Data in the literature are normally reported as the α activity of the essential mineral and of the essential mineral plus inclusions, secondary alteration products, and crack fillings. Values reported in Table 1 and plotted on Figure 1 are those of the individual mineral less inclusions and secondary material. Because several papers fail to distinguish between the α activity of the mineral plus associated material and that of the pure mineral, their data were not incorporated into Figure 1 (for example, Hayase, 1957).

The upper limit of α activity in feldspars from mafic rocks as shown in Table 1 and Figure 1 is approximately 6×10^{-5} α per cm.^{2} per sec. Assuming a Th/U ratio of 3,[*] this value corresponds to an approximate uranium content of 0.3 p.p.m. The α activities of feldspars from felsic and intermediate rocks, however, are considerably more variable, ranging from less than 1 to 25×10^{-5} α per cm.^{2} per sec. The median value for all feldspars is 1.30×10^{-5}, corresponding to an approximate uranium content of 0.05 p.p.m.

Data for pyroxenes compare closely to those for feldspars (Table 2 and Fig. 1). The median value for the pyroxenes is 0.96×10^{-5}, corresponding to an approximate uranium content of 0.04 p.p.m. Pyroxenes from igneous rocks of intermediate composition are apparently slightly more radioactive than are those from mafic rocks.

The paucity of information available on the α activity of pure quartz and biotite renders a median and range of α activity for these two minerals of little significance. Values cited in Tables 3 and 4 and plotted on Figure 1 suggest that quartz is slightly more radioactive than are feldspar and pyroxene and that biotite is considerably more radioactive. Hayase (1955) applied the autoradiographic method to a

---

[*] A Th/U ratio of 3 will be assumed for subsequent references to the uranium content in the essential minerals.

TABLE 1

ALPHA ACTIVITIES OF FELDSPARS

Rock Type	Type Feldspar*	α's/cm²/ sec×10⁻⁵	Reference
Granite, Lac Blanc...............	U	25	Picciotto (1950)
Granite, Elbe....................	U	9.7	" "
Granite, Kasai...................	Or.	9	Picciotto (1952)
Granite, Shap...................	Pl.	6.5	Spears (1961)
Anorthosite, Egersand............	Pl.	6.4	Picciotto (1950)
Trachyandesitic lava..............	Pl.	5.9	Longinelli (1959)
Trachyandesitic lava..............	Or.	5.5	" "
Granite, Shap...................	Or.	4.4	Spears (1961)
Granodiorite, Adamello...........	U	3.7	Merlin et al. (1957)
Granodiorite, Adamello...........	U	3.7	" "
Granophyre, Skye................	U	2.6	Hamilton (1960a)
Ferrogabbro, Skaergaard..........	Pl.	2.4	" "
Ferrogabbro, Skaergaard..........	Pl.	2.3	" "
Tholeiitic basalt, Greenland.......	Pl.	2.3	" "
Rhyolitic ignimbrite..............	Or.	2.1	Longinelli (1959)
Quartzlatitic ignimbrite...........	Pl.	1.9	" "
Tholeiitic basalt, Giants Causeway..	Pl.	1.9	Hamilton (1960a)
Rhyolitic ignimbrite..............	Or.	1.7	Longinelli (1959)
Microgranite, Skye...............	U	1.7	Hamilton (1960a)
Rhyolitic ignimbrite..............	Or.	1.6	Longinelli (1959)
Granite, G-1....................	U	1.6	Hamilton (1960a)
Granodiorite, Adamello...........	U	1.6	Merlin et al. (1957)
Acid granophyre, Skaergaard.......	U	1.5	Hamilton (1960a)
Granophyre, Skye................	U	1.5	" "
Syenite, Biella..................	Pl.	1.3	Deutsch and Longinelli (1958)
"Feldspar rock," Skaergaard.......	Pl.	1.3	Hamilton (1960a)
Olivine basalt, Greenland..........	Pl.	1.2	" "
Granodiorite, Adamello...........	U	1.2	Merlin et al. (1957)
Ferrogabbro, Skaergaard..........	Pl.	1.1	Hamilton (1960a)
Gabbro, Skaergaard..............	Pl.	1.1	" "
Ferrogabbro, Skaergaard..........	Pl.	1	Hamilton (1960a)
Gabbro picrite, Skaergaard........	Pl.	1	" "
Gabbro, Skaergaard..............	Pl.	1	" "
Gabbro, Skaergaard..............	Pl.	1	" "
Microgranite, Skye...............	U	1	" "
Syenite, Biella..................	Or.	1	Deutsch and Longinelli (1958)
Olivine basalt, Iceland............	Pl.	0.87	Hamilton (1960a)
Dolerite, Whin Sill...............	Pl.	0.84	" "
Gabbro, Skaergaard..............	Pl.	0.8	" "
Granite, G-1....................	U	0.65	" "
Olivine basalt, Greenland..........	Pl.	0.65	" "
Basic granophyre, Skaergaard......	U	0.6	" "
Basic granophyre, Skaergaard......	U	0.6	" "
Transgr. granophyre, Skaergaard....	U	0.6	" "
Ferrogabbro, Skaergaard..........	Pl.	0.5	" "
Granophyre, Skye................	U	0.43	" "
Diabase, W-1....................	Pl.	0.24	" "
Granophyre, Loch Ba felsite........	U	0.2	Hamilton (1960b)
Diabase, W-1....................	Pl.	0.19	Hamilton (1960a)
Granite, G-1....................	U	0.06	" "

*Pl. = reported or assumed to be plagioclase; Or. = reported as potassium feldspar; U = undifferentiated.

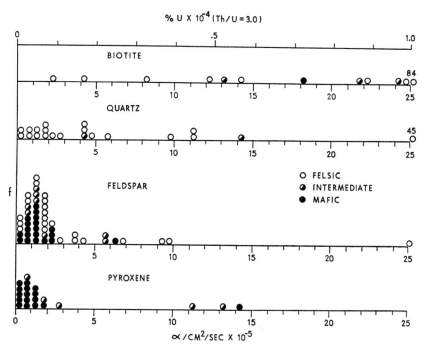

Fig. 1.—Histograms of α activities of biotite, quartz, feldspar, and pyroxene

TABLE 2

ALPHA ACTIVITIES OF PYROXENES

Rock Type	α's/cm²/sec×10⁻⁵	Reference
Anorthosite, Egersund...........	14	Picciotto (1950)
Syenite, Biella..................	13	Deutsch and Longinelli (1958)
Trachyandesitic lava............	11	Longinelli (1959)
Basic granophyre, Skaergaard.....	2.7	Hamilton (1960a)
Ferrogabbro, Skaergaard.........	1.9	" "
Transgr. granophyre, Skaergaard..	1.5	" "
Ferrogabbro, Skaergaard.........	1.2	" "
Gabbro, Skaergaard.............	1.2	" "
Ferrogabbro, Skaergaard.........	1.2	" "
Diabase, W-1....................	1.1	" "
Ferrogabbro, Skaergaard.........	0.96	" "
Olivine basalt, Iceland...........	0.85	" "
"Feldspar rock," Skaergaard......	0.74	" "
Gabbro, Skaergaard.............	0.74	" "
Tholeiitic basalt, Giants Causeway.	0.74	" "
Basic granophyre, Skaergaard.....	0.55	" "
Olivine basalt, Greenland........	0.43	" "
Olivine basalt, Greenland........	0.37	" "
Ferrogabbro, Skaergaard.........	0.27	" "
Gabbro, Skaergaard.............	0.25	" "
Diabase, W-1....................	0	" "

study of the α activity of biotite from the Tanakamiyama granite (Japan) and found biotite to be much more radioactive (as much as 3 orders of magnitude) than the other essential minerals. He did not distinguish, however, between the α activity of the mineral and that of the mineral plus inclusions; therefore, his data were not included in Table 4. Guilbert (1954), Adams *et al.* (1959), and Larsen and Phair (1954) also noted the relatively high concentrations of the α

TABLE 3

ALPHA ACTIVITIES OF QUARTZ

Rock Type	α's/cm²/ sec×10⁻⁵	Reference
Granite, Elbe........................	45	Picciotto (1950)
Basic granophyre, Skaergaard.....	14	Hamilton (1960*a*)
Acid granophyre, Skaergaard......	11	" "
Granite, Kasai.....................	11	Picciotto (1952)
Granophyre, Loch Ba felsite......	9.5	Hamilton (1960*b*)
Transgr. granophyre, Skaergaard..	5.5	Hamilton (1960*a*)
Granite, Shap.....................	4.5	Spears (1961)
Rhyolitic ignimbrite.............	4.4	Longinelli (1959)
Syenite, Biella...................	4.3	Deutsch and Longinelli (1958)
Granodiorite, Adamello..........	4.1	Merlin *et al.* (1957)
Microgranite, Skye...............	2.5	Hamilton (1960*a*)
Granodiorite, Adamello..........	2.4	Merlin *et al.* (1957)
Granodiorite, Adamello..........	1.9	" "
Granophyre, Skye................	1.8	Hamilton (1960*a*)
Granite, Lac Blanc..............	1.6	Picciotto (1950)
Granite, G-1....................	1.4	Hamilton (1960*a*)
Microgranite, Skye.............	1.2	" "
Granophyre, Skye...............	0.9	" "
Granite, G-1....................	0.61	" "
Granite, G-1....................	0.11	" "
Granophyre, Skye...............	0	" "

TABLE 4

ALPHA ACTIVITIES OF BIOTITES

Rock Type	α's/cm²/ sec×10⁻⁵	Reference
Granite, Elbe....................	84	Picciotto (1950)
Rhyolitic ignimbrite............	25	Longinelli (1959)
Quartzlatitic ignimbrite........	24	" "
Rhyolitic ignimbrite............	22	" "
Syenite, Biella.................	22	Deutsch and Longinelli (1958)
Anorthosite, Egersund..........	18	Picciotto (1950)
Granite, Shap..................	14	Spears (1961)
Quartzlatitic ignimbrite........	13	Longinelli (1959)
Extrusive......................	12	" "
Granodiorite, Adamello........	8.2	Merlin *et al.* (1957)
Granodiorite, Adamello........	4.3	" "
Granodiorite, Adamello........	2	" "

emitters in biotite. Guilbert observed, however, that quartz is commonly the essential mineral most devoid of α emitters. The lack of data prohibited the investigation of the concentrations of α emitters in other essential minerals, such as olivine and amphibole; but thorium and uranium seem to be concentrated in these minerals in approximately the same quantities as in other essential minerals.

As previously noted, many workers have observed the concentration of the α emitters in the minor accessory minerals. The information presented herewith substantiates, and to a certain degree quantifies, these observations. More information is available on the α activity as determined by α autoradiography of allanite, $(Ca,Ce,Th)_2(Al,Fe,-Mg)_3Si_3O_{12}(OH)$; sphene, $CaTiSiO_5$; and zircon, $ZrSiO_4$, than on apatite, $Ca_5(Cl,OH,F)(PO_4)_3$; monazite, $(Ce,Y,La,Th)(PO_4)$; xenotime, YPO_4; and other common accessory minerals. Yagoda (1946a) has classified over one hundred uranium- and thorium-bearing minerals on the basis of their relative α activities, but the majority of these minerals do not commonly occur as accessory minerals in igneous rocks.

Figure 2 is a log histogram of the α activity in sphene, allanite, and zircon in igneous rocks as determined by α autoradiography. Data for allanite are from Hayase (1954); the allanites are reportedly from granitic rocks. The α activities of zircon were determined by the writer from a wide variety of igneous and metamorphic rocks. More data from this laboratory on zircon α activity will appear in a subsequent paper. Data for α activities of sphenes are from Hayase (1959) and recalculated from Hée *et al.* (1954). Hayase measured the α activity of sphenes from a large suite of granitic rocks from Japan, whereas Hée *et al.* studied the distribution of the α emitters in the Quincy granite from Massachusetts. With the exception of one value at 700 on the sphene log histogram (Fig. 2), all the values above 10^3 are from the Quincy granite. Whether this discrepancy is real or experimental is not known.

Figure 2 suggests that allanites are the most radioactive, sphenes intermediate, and zircons least. Values for allanite and zircon fall within the range 0–4 α per cm^2/sec proposed by Yagoda (1946a). Xenotime and monazite are probably more radioactive than is allanite, as noted by Yagoda (1946a), Adams *et al.* (1959), and Larsen and Phair (1954). Yagoda cites a range of 5–16 α per cm^2/sec for xenotime and monazite. Another common accessory mineral, apatite, is generally less radioactive than zircon (Adams *et al.*, 1959). A comparison of Figure 1 with Figure 2 reveals that these common accessory minerals are from 1 to 5 orders of magnitude more radioactive than most essential minerals.

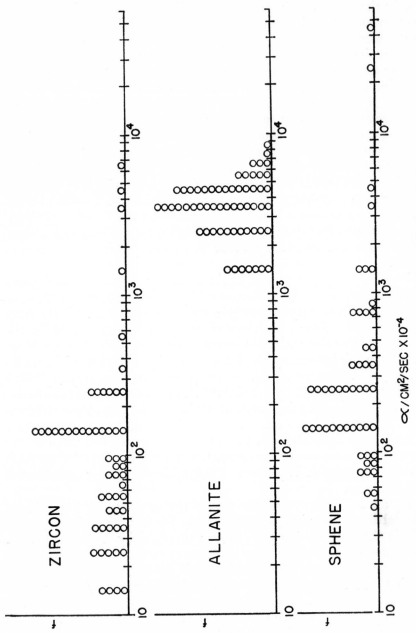

Fig. 2.—Log histograms of α activities of zircon, allanite, and sphene

This marked difference in α activity between the common accessory minerals and the essential minerals leads to some interesting observations concerning the distribution of radioactive constituents within various igneous rock types. These relationships may be seen in Figures 3 and 4, plots of the α activity of two essential minerals (along the ordinate) against the α activity of the minerals plus inclusions and secondary material (along the abscissa). The contribution to the α activity from secondary material within the essential mineral is generally negligible. These plots indicate that there is a much larger contribution to the total α activity from the included accessories in the late than in the early differentiates. Because the relative volumes of inclusions in these different rock types are not determined, the two trends may be caused by a larger volume of accessory minerals or by accessory minerals with higher radioactivity in the later differentiates. Figure 4 suggests that at least the latter explanation is true; probably both are true. The volume of the included accessories generally constitutes less than 5 per cent of the total volume. The proximity of the trends for the early differentiates to the unit ratio line indicates that either highly radioactive inclusions are almost absent from these rocks or their inclusions are comparatively non-radioactive.

The geochemistry and mineralogy of thorium and uranium have been discussed in detail by Adams *et al.* (1959). In general, because of their ionic radii and high electronegativities and valences (ionic radii of tetravalent thorium and uranium ions are 1.10 kX and 1.05 kX, respectively), they do not form isomorphic series with the major rock-forming elements and are concentrated in the residual solution during fractionation. Thus they exist in comparatively minute quantities in the essential minerals and are concentrated in the felsic rocks. Adams *et al.* (1959) have suggested that thorium and uranium may be incorporated within the essential minerals by (1) entrapment in lattice imperfections, (2) entrapment in liquid inclusions, (3) deposition along fractures, or (4) adsorption on crystal surfaces. They also may be incorporated by substitution in the microscopic or submicroscopic crystalline inclusions, by adsorption or intracrystalline surfaces (for example, between the tetrahedral sheets in the micas), and possibly by isomorphic substitution for calcium. Uranium and thorium are known to substitute for calcium in allanite, sphene, and apatite; however, Hayase (1957) has noted that calcic plagioclases are comparatively depleted in uranium and thorium. This discrepancy may be explained by the fact that calcic plagioclase crystallizes relatively early; therefore, at this early stage of magmatic crystallization, uranium and thorium may not be present in sufficiently high concentrations to substitute for calcium to a significant degree.

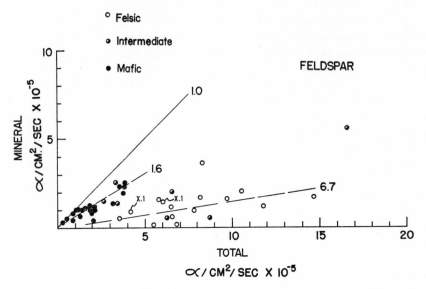

Fig. 3.—Alpha activity of pure feldspar (along the ordinate) *vs.* that of feldspar plus inclusions and secondary material (along the abscissa).

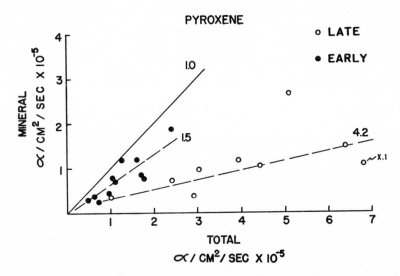

Fig. 4.—Alpha activity of pure pyroxene (along the ordinate) *vs.* that of pyroxene plus inclusions and secondary material (along the abscissa).

There is some evidence, however, that isomorphic substitution of uranium and thorium within feldspar and pyroxene lattices does take place. Figure 5 is a plot of the partitioning of uranium and thorium between pyroxene and feldspar, as evidenced by the α activities of the coexisting minerals in various rock types. Point *1* represents coexisting minerals from a syenite; points *2* and *3*, from granophyres. These rocks are atypical in that they are more felsic than are the majority of the other pyroxene-bearing rocks, which may account for their values' being off the trend. The general positive correlation suggests

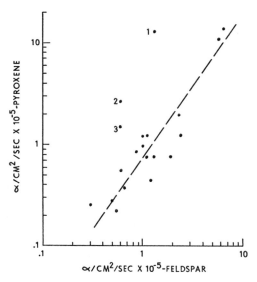

Fɪɢ. 5.—Alpha activity of pyroxene *vs.* that of coexisting feldspar in predominantly mafic rocks. Point *1* represents a syenite; points *2* and *3*, granophyres.

that (1) thorium and uranium did substitute in the crystal lattices of pyroxene and feldspar, probably for calcium, and (2) the minerals crystallized in equilibrium with respect to thorium and uranium. The site of the α emitters within the minerals represented by points *1, 2,* and *3* (Fig. 5) may be submicroscopic inclusions, or the minerals may not have crystallized in equilibrium with respect to uranium and thorium. Figure 6 is a plot of the partitioning of uranium and thorium between feldspar and quartz. The lack of correlation is as expected, inasmuch as there is no suitable cation in the quartz lattice for which uranium and thorium can substitute.

The chemical properties of uranium and thorium enable them to form isomorphic series with zirconium, hafnium, and the rare earths. Thus uranium and thorium are concentrated in zircon, monazite, and xenotime and apparently substitute for calcium in apatite, sphene, and allanite. The mineralogy of thorium and its relation to uranium, hafnium, zirconium, and cerium has been discussed by Frondel (1956). Hurley and Fairbairn (1957) have reported the abundances of uranium and thorium and have discussed their partitioning in various accessory minerals. In addition to the common highly radioactive accessory minerals, Deutsch and Picciotto (1956) identified uraninite, UO_2, in a

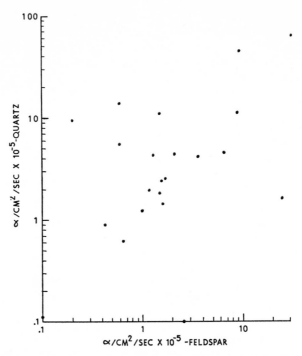

Fig. 6.—Alpha activity of quartz *vs.* that of coexisting feldspar

granite from the Italian Alps and Merlin *et al.* (1957) tentatively identified uraninite and torbernite, $Cu(UO_2)_2(PO_4)_2 \cdot nH_2O$, in a granodiorite from Italy by α autoradiography.

Adams *et al.* (1959, p. 314) have compiled the available information on concentrations of uranium and thorium in both essential and accessory minerals. The majority of analyses were by spectrochemical or radiochemical methods rather than by autoradiography. The ranges of values for concentrations of uranium and thorium in essential minerals as reported by Adams *et al.* are significantly higher than those determined by the autoradiographic method. For example, Adams *et al.* report the range for uranium as 0.2–5 p.p.m., whereas, assuming a

Th/U ratio of 3, the range of uranium values by α autoradiography is 1.0–0.01 p.p.m. The spectrochemical and radiochemical analyses as performed on "pure" mineral separates may yield high values because there is no adequate method for separating an essential mineral from its highly radioactive microscopic or submicroscopic inclusions. The values for the pure essential minerals as determined by α autoradiography may be closer to the true values.

APPLICATIONS OF α AUTORADIOGRAPHY

Advantages of α autoradiography over other radiochemical methods are (1) α activity and, in some cases, absolute concentrations of thorium and uranium in the constituent minerals within a rock may be determined without mineral separations or, at worst, with an impure separation of only a few hundred grains; (2) the method is simple and inexpensive; and (3) relative α activities of individual grains may be determined within a rock or mineral suite. This third advantage may have some important ramifications.

A problem facing geochronologists is obtaining a zircon separate that is a closed chemical system and that represents a homogeneous population of zircons. The first requisite necessitates the absence of uranium, thorium, or their unsupported daughters in secondary crack fillings or coatings associated with the zircon grains. The second necessitates the absence of detrital or relict grains; that is, all grains should be of the same age and genetically related, and the separate should be pure zircon. Marble (1937) noted the possible utilization of α autoradiography in geochronology. Adams and Ragland (1961) suggested that α autoradiographs may be used to screen samples for age determinations, inasmuch as the identification of more than one suite of zircons or of morphologically similar minerals other than zircon may be accomplished by zircon α autoradiography. Figure 7 is a plot of the average number of tracks per grain *versus* the grain size of zircon separates from two samples, a bentonite (volcanic ash) from Tennessee, and a beach sand from Florida. Zircons from the bentonite presumably crystallized within a short range of time and from a single source, whereas zircons from the beach sand most probably were formed at different times from a wide variety of sources. The linear correlation between grain size and α activity suggests a single source for zircons from the bentonite, while the lack of correlation suggests a multiple source for the beach sand. Thus, if radioactive equilibrium is achieved and the suite represents a closed chemical system, the zircon separate from the bentonite should yield a meaningful date, whereas the separate from the beach sand should not. The example of the beach sand and bentonite is a simplification, for these deposits generally may be distinguished on the basis of other criteria; however, α autoradiography may be of use when dealing with separates from "igneous" rocks of questionable origin. These studies, for example, should yield meaningful results concerning the problem of delineating between paragneisses (originally of sedimentary origin) and orthogneisses (originally of igneous origin) or between magmatic and metamorphic granites. A zircon suite from sedimentary material before metamorphism should theoretically be derived from a multiple source, whereas a suite from igneous material should be from a single source. These studies should be made in conjunction with routine petrographic investigations. Eckelmann and Kulp (1959), for example,

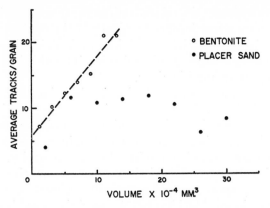

Fig. 7.—Average number of α tracks per grain *vs.* grain size of zircon separates from a bentonite and a placer sand.

concluded that the Cranberry and Henderson "granites" of North Carolina were of sedimentary origin on the basis of morphology of their zircon suites.

The absence of a closed chemical system may also present a problem in geochronology. Yagoda (1946b) noted the presence of radio-colloids, which he considered highly radioactive aggregates of unsupported $RaSO_4$ decaying to $PbSO_4$, deposited on mineral surfaces. Radium is separated from uranium by preferential precipitation of the more insoluble radium from solution as the sulfate. Picciotto (1950) and Brown and Silver (1956) noted the concentration of uranium and thorium along grain boundaries as a coating. The contribution of secondary sources and accessory minerals to the α activity of various rocks is shown in Figure 8. Note the greater contribution to the total α activity from essential minerals in the mafic rocks. The contribution to the total α activity from secondary materials is probably a function of weathering, the greater contribution being in the more weathered rocks. The presence of these secondary materials within a zircon separate may lead to an open chemical system, from which an isotopic analysis may result in a discordant age. The identification of radiocolloids or highly radioactive crack fillings and coatings would be ideally suited to the autoradiographic method.

Distribution of α emitters between silicate minerals and non-silicates coexisting in rocks is of considerable importance in leaching

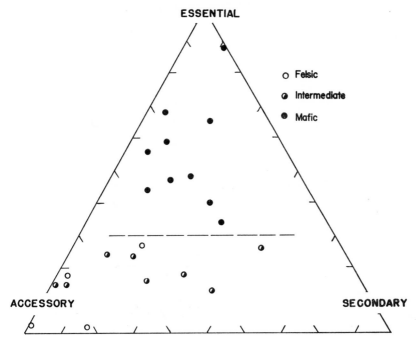

Fig. 8.—Ternary diagram showing the distribution of α activity between secondary, essential, and accessory minerals in some igneous rocks.

processes and the contribution to the internal radiation environment. In general, recovery of uranium and thorium by natural or artificial leaching processes can be facilitated if these elements are concentrated within oxides or phosphates rather than silicates. Much of the highly radioactive opaque coatings and crack fillings observed under the microscope is probably hematite or limonite ($Fe_2O_3 \cdot nH_2O$), and α emitters incorporated within this material are readily leachable. Thorium and uranium would also be leached more easily from phosphate accessory minerals (monazite, apatite, xenotime) than from silicate accessory minerals (zircon, sphene, allanite). Leaching studies by

Hurley (1950), Brown *et al.* (1953), and Brown and Silver (1956) have demonstrated that a large percentage of total uranium and thorium in intrusive acidic rocks is readily leachable. Figure 8 plots percentage of potentially leachable α emitters in various rocks, as present in the secondary material and part of the accessory minerals. Alpha autoradiography is ideally suited for the estimation of α activity attributable to the readily leachable materials within a rock. Recovery of thorium from relatively low-grade (less than 100 p.p.m.) ore has recently become of interest in the search for long-range supplies of thorium for atomic energy (Adams *et al.*, 1962).

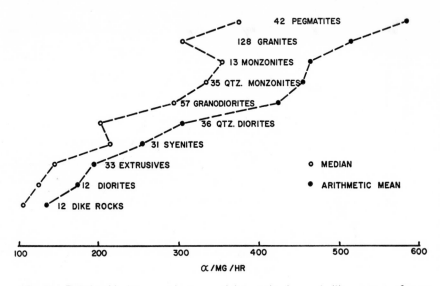

Fig. 9.—Relationship between zircon α activity, grain size, and silica content of some igneous rocks.

As demonstrated above, there is ample evidence that α activity of constituent minerals varies with rock type. Figure 9 demonstrates the relationship between zircon α activity, silica content, and grain size of various silicate rocks (data in conjunction with lead-α age determinations rather than autoradiography; Jaffe *et al.*, 1959). The diagram indicates that zircons from finer-grained and/or relatively mafic rocks are less radioactive than are those from coarser-grained and/or relatively acidic rocks. This relationship is also suggested by Figure 10, *A*, which demonstrates the higher α activity of zircons from the Lausitz granite from Germany as compared to those from a volcanic ash from Tennessee. Because of the faster cooling rate of relatively fine-grained extrusive rocks, thorium and uranium may be more evenly dispersed throughout their constituent minerals, thus accounting for zircons of lower average α activity in extrusive rocks (Rogers and Adams, 1957). The higher α activity of zircons from silica-rich rocks may be explained by the concentration of thorium and uranium in the residual magma during fractionation (Adams *et al.*, 1959).

Not only is the α activity of minerals within dissimilar rock types quite variable, but it is also variable within similar rock types. Thus the α activity of a mineral separate from a granite in an igneous rock series

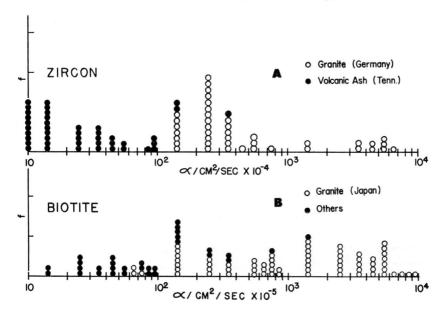

Fig. 10.—Log histograms of (*A*) the α activity of zircons from the Lausitz granite, Germany, and a volcanic ash (bentonite) from Tennessee and (*B*) the α activity of biotites from the Tanakamiyama granite, Japan (Hayase, 1955), and from granitic rocks in other localities.

may be significantly different from that of other rocks within the series, as well as from granites in a different series. Biotites from the Tanakamiyama granite contain significantly greater amounts of radioactive constituents than do those from other granites (Fig. 10, *B*; data on biotites from Tanakamiyama granite from Hayase, 1955). Data on other Japanese granites published by Hayase agree quite well with those published by the European laboratories, suggesting that the difference is real. Thus certain rock types, and, in some cases, certain rock bodies, may be characterized by mineral suites of locally unique radioactivity. These relationships may prove useful in the identification of source areas of detrital grains in a heavy mineral suite from a sedimentary deposit. For example, bimodal distribution of α activities from detrital zircon grains in a sedimentary deposit may result from intrusive and extrusive igneous rocks in the drainage basin.

In conclusion, the potential of the autoradiographic method has not yet been fully realized.

ACKNOWLEDGMENTS

The writer is indebted to Drs. J. R. Butler and C. Everett Brett for their critical review of the manuscript. Professor John A. S. Adams kindly furnished reprints, which aided immeasurably in the review of the literature. Mr. Jay Zimmerman aided in compiling the data. The autoradiographic studies of zircons were conducted at Rice University and sponsored by the Robert A. Welch Foundation grant C-009 to John A. S. Adams and John J. W. Rogers.

REFERENCES

ADAMS, J. A. S., M.-C. KLINE, K. A. RICHARDSON, and J. J. W. ROGERS. 1962. The Conway granites of New Hampshire as a major low-grade thorium resource. Proc. Nat. Acad. Sci., **48**:1898–1905.

ADAMS, J. A. S., J. K. OSMOND, and J. J. W. ROGERS. 1959. The geochemistry of thorium and uranium. *In* Physics and Chemistry of the Earth, **3**:298–348. London: Pergamon Press.

ADAMS, J. A. S., and P. C. RAGLAND. 1961. Autoradiographic and radiometric screening of samples for absolute age determinations. J. Geophys. Res., **66**:2509.

BOWIE, S. H. U. 1954. Nuclear emulsion techniques. *In* H. FAUL (ed.), Nuclear Geology, pp. 148–74. New York: John Wiley & Sons.

BROWN, H., W. BLAKE, A. A. CHODOS, R. KOWALKOWSKI, C. R. McKINNEY, G. J. NEUERBURG, L. T. SILVER, and A. UCHIYAMA. 1953. Leaching studies in interstitial materials in igneous rocks. Geol. Soc. America Bull., **64**:1400.

BROWN, H., and L. T. SILVER. 1956. The possibilities of obtaining long-range supplies of uranium, thorium, and other substances from igneous rocks. U.S. Geol. Survey Prof. Paper 300, pp. 91–95.

BUTTLAR, H. VON, and F. G. HOUTERMANS. 1951. Photographische Messung des U- und Th-Gehaltes nach der Auflagemethode. Geochim. & Cosmochim. Acta, **2**:43–61.

COPPENS, R. 1950. Sur l'emploi de l'émulsion photographique pour la détermination de la radioactivité des roches par l'examen des trajectories des rayons alpha. J. Phys. & Rad., **11**:21–32.

———. 1951. Étude de radioactivité de quelques par émulsion photographique. Bull. Soc. Fran. Minéral, **73**:217–21.

CURIE, I. 1946. Sur la possibilité d'étudier l'activité des roches par l'observation des trajectories des rayons alpha dans l'émulsion photographique. J. Phys. & Rad. (8) **7**:313–19.

DEUTSCH, S., F. G. HOUTERMANS, and E. PICCIOTTO. 1956*a*. Étude de la radioactivité de météorites métalliques par la méthode photographique. Geochim. & Cosmochim. Acta, **10**:166–84.

———. 1956*b*. Radioactivity of iron meteorites by the photographic method. Nature, **177**:885–86.

DEUTSCH, S., and A. LONGINELLI. 1958. Distribuzione della radioattività nella sienite de Biella. Studi e Ricerche della Div. Geomin., **1**:1–16.

DEUTSCH, S., and E. PICCIOTTO. 1956. Présence d'uranite dans les minéraux accessoires du granite de Baveno. Experientia, **12**:333–36.

DEUTSCH, S., E. PICCIOTTO, and F. G. HOUTERMANS. 1957. Alpha radioactivity of iron meteorites (second letter). Geochim. & Cosmochim. Acta, **14**:173–74.

ECKELMANN, F. D., and J. L. KULP. 1959. Sedimentary origin of the Cranberry and Henderson "granites" in North Carolina. Am. J. Sci., **254**:316–24.

FORD, I. H. 1951. Radioactivity of rocks: Improvement in the photographic technique. Nature, **167**:273–74.

FRONDEL, C. 1956. Mineralogy of thorium. U.S. Geol. Survey Prof. Paper 300, pp. 567–79.

GUILBERT, J. M. The autoradiographic determination of alpha-activity distribution in Wisconsin granites and related rocks. M.S. thesis, Univ. Wisconsin, 1954.

GUILBERT, J. M., and J. A. S. ADAMS. 1955. Alpha-particle autoradiography with liquid emulsion. Nucleonics, **13**:43.

HAMILTON, E. 1959. The uranium content of the differentiated Skaergaard intrusion. Meddel. om Grønland, **162**:1–35.

———. 1960*a*. The distribution of radioactivity in the major rock forming minerals. *Ibid.*, pp. 1–41.

———. 1960*b*. The distribution of radioactivity in some fine-grained igneous rocks. Geol. Mag., **47**:255–60.

HAYASE, I. 1954. The radioactivity of rocks and minerals studied with nuclear emulsion. II. Thorium content of granitic allanites. Mem. Coll. Sci., Univ. Kyoto (B), **21**:171–82.

———. 1955. The radioactivity of rocks and minerals studied with nuclear emulsion. III. Radioactivity of biotite of the Tanakamiyama granite, Shiga Pref., Japan. *Ibid.*, **22**:178–84.

———. 1956. The radioactivity of rocks and minerals studied with nuclear emulsion. IV. Thorium and uranium ratio measurement of the minute radioactive minerals by the photographic method. *Ibid.*, **23**:265–74.

———. 1957. The radioactivity of rocks and minerals studied with nuclear emulsion. V. Radioactive behavior of granites. *Ibid.*, **24**:121–54.

———. 1958. The radioactivity of rocks and minerals studied with nuclear emulsion. VI. Radioactivity of some Japanese liparites. *Ibid.*, **25**:81–87.

———. 1959. The radioactivity of rocks and minerals studied with nuclear emulsion. VII. Radioactivity of granitic sphene. *Ibid.*, pp. 205–13.

HAYASE, I., and T. TSUTSUMI. 1958. On the emanating power of powdered rocks. Mem. Coll. Sci., Univ. Tokyo (B), **24**:315–23.

HÉE, A., R. P. DERVILLE, and M. JAROVOY. 1954. Determination of the radioactivity of the Quincy granite by the photographic method. Am. J. Sci., **252**:736–44.

HÉE, A., and M. JAROVOY. 1953. Autoradiographie des rayons beta du potassium. Ann. Géophys., **9**:153–58.

HURLEY, P. M. 1950. Distribution of radioactivity in granites and possible relation to helium age measurements. Geol. Soc. America Bull., **61**:1–8.

HURLEY, P. M., and H. W. FAIRBAIRN. 1957. Abundance and distribution of uranium and thorium in zircon, sphene, apatite, epidote, and monazite in granitic rocks. Trans. Am. Geophys. Union, **38**:939–44.

JAFFE, H. W., D. GOTTFRIED, C. L. WARING, and H. W. WORTHING. 1959. Lead-alpha age determinations of accessory minerals of igneous rocks (1953–1957). U.S. Geol. Survey Bull. 1097, pp. 65–148.

KEEVIL, N. B., E. S. LARSEN, JR., and F. J. WANK. 1944. The distribution of helium and radioactivity in rocks. VI. The Ayer granite migmatite at Chelmsford, Mass. Am. J. Sci., **242**:345–53.

KOCZY, F. F., E. PICCIOTTO, G. POULAERT, and S. WILGAIN. 1957. Mesure des isotopes du thorium dans l'eau de mer. Geochim. & Cosmochim. Acta, **11**:103–29.

LARSEN, E. S., JR., and N. B. KEEVIL. 1942. The distribution of helium and radioactivity in rocks. III. Radioactivity and petrology of some California intrusives. Am. J. Sci., **240**:204–15.

LARSEN, E. S., JR., and G. PHAIR. 1954. This distribution of uranium and thorium in igneous rocks. *In* H. FAUL (ed.), Nuclear Geology, pp. 75–88. New York, John Wiley & Sons.

LONGINELLI, A. 1959. Distribuzione della radioattivita in alcune sezioni di rocce della serie effusiva del Trentino-Alto Adige. Atti Soc. Toscana di Sci. Nat. (A), **66**:1–13.

MARBLE, J. P. 1937. The analysis of allanite for age determination. Nat. Res. Council (U.S.), Rpt. Comm. Geol. Time, pp. 65–77.

MERLIN, O. H., E. PICCIOTTO, and S. WILGAIN. 1957. Étude photographique de la distribution de la radioactivité dans la granodiorite de l'Adamello. Geochim. & Cosmochim. Acta, **11**:171–88.

PELC, S. R. 1947. Autoradiograph technique. Nature, **160**:749–50.

PICCIOTTO, E. 1949. L'étude de la radioactivité des roches par la méthode photographique. Bull. Soc. Belge Géol. Paléontol. & Hydrol., **58**:75–90.

———. 1950. Distribution de la radioactivité dans les roches éruptives. *Ibid.*, **59**:170–98.

———. 1952. Distribution de la radioactivité dans les roches éruptives. *Ibid.*, **61**:215–22.

PICCIOTTO, E., and S. WILGAIN. 1954. Thorium determination in deep-sea sediments. Nature, **173**:632–33.

POOLE, J. H. J., and J. W. BREMNER. 1949. Distribution of the radioactive elements in rocks by the photographic method. Nature, **163**:130–31.

POOLE, J. H. J., and C. M. MATTHEWS. 1951. The theory of the use of alpha ray ranges in nuclear emulsions for the determination of the radioactive contents of materials. Sci. Proc. Roy. Dublin Soc., **25:** 305–16.

ROGERS, J. J. W., and J. A. S. ADAMS. 1957. Autoradiography of volcanic rocks of Mount Lassen. Science, **125:**1150.

ROGERS, J. J. W., and P. C. RAGLAND. 1961. Variation of thorium and uranium in selected granitic rocks. Geochim. & Cosmochim. Acta, **25:** 99–109.

SENFTLE, F. E., T. A. FARLEY, and L. R. STIEFF. 1954. A theoretical study of alpha star populations in loaded nuclear emulsions. Geochim. & Cosmochim. Acta, **6:**197–207.

SPEARS, D. A. 1961. The distribution of alpha radioactivity in a specimen of Shap granite. Geol. Mag., **48:**483–87.

STIEFF, L. R., and T. W. STERN. 1952. Preparation of nuclear-track plates and stripping films for the study of radioactive minerals. Am. Mineralogist, **37:**184–96.

STRUTT, R. J. 1906. On the distribution of radium in the earth's crust, and on the earth's internal heat. Proc. Roy. Soc. London A, **77:**472–88.

WHITFIELD, J. M., J. J. W. ROGERS, and J. A. S. ADAMS. 1959. The relationship between the petrology and the thorium and uranium contents of some granitic rocks. Geochim. & Cosmochim. Acta, **17:**248–71.

YAGODA, H. 1946*a*. The localization of uranium and thorium minerals in polished section. I. The alpha ray emission pattern. Am. Mineralogist, **31:**87–124.

———. 1946*b*. Radiocolloid aggregates in uranium minerals. *Ibid.*, pp. 462–70.

———. 1949. Radioactive Measurements with Nuclear Emulsions. Pp. 356. New York: John Wiley & Sons.

YAGODA, H., and N. KAPLAN. 1947. Fading of latent alpha ray image in emulsions. Phys. Rev., **71:**910–11.

J. K. OSMOND

8. The Distribution of the Heavy Radio-elements in the Rocks and Waters of Florida

O NE OF THE ANOMALOUS ASPECTS of the study of the distribution of uranium and thorium in the rocks of the outer crust has been the observation that sedimentary rocks seem to contain less of the heavy radioelements than do the igneous rocks from which they must ultimately have been derived. The extent of this sediment deficit may be as much as 30 per cent for uranium and 50 per cent for thorium, as indicated in Table 1.

The radioelements are not held to any large extent in sea water, and so the explanation for the imbalance cannot be due to the same process that accounts for a similar imbalance in the case of sodium; nor are the deep-ocean sediments thought to be enriched in uranium and thorium to the extent that they might represent a siphoning-off of the radio-elements, in a fashion analogous to the concentration of manganese, cobalt, and other elements in deep-ocean nodules. The answer to the anomaly in the geochemical balances of uranium and thorium must be found in some defect in our knowledge of the distribution of these elements in sedimentary rocks. More particularly, there must be sedimentary rock types that are either more enriched in uranium and thorium or more widespread than has been realized. Of the sediments known to be enriched in uranium, black shales and phosphorites are among the most common, while thorium is known to be especially abundant in residual deposits and in certain heavy-mineral–bearing sands.

Both phosphatic beds and heavy-mineral sands occur as important sedimentary deposits in Florida. The present report describes the be-

J. K. OSMOND is assistant professor, Department of Geology, Florida State University, Tallahassee, Florida.

ginnings of a program at Florida State University of investigation into the distribution of the heavy radioelements in the sediments and waters of Florida—a program aimed at achieving a better understanding of the processes of weathering, transportation, and deposition of these elements. An important aspect of the study is the attempt to sort out and identify the specific isotopes of uranium, thorium, and their longer-lived radioactive daughter elements. One result of the study will be a better definition of the states of equilibrium between uranium and thorium series elements in portions of the transportation and sedimentation cycle. A second result may be a partial resolution of the anomaly of the sedimentary-igneous geochemical balances for these elements.

TABLE 1

GEOCHEMICAL BALANCE FOR THORIUM AND URANIUM

Rock Type	Relative Abundance*	Thorium Concentration† (p.p.m.)	Relative Abundance × Thorium Concentration (p.p.m.)	Uranium Concentration (p.p m.)	Relative Abundance × Uranium Concentration (p.p.m.)
Shales........	46	12	5.5	3.7	1.7
Sandstones....	32	1.8	0.6	0.5	0.2
Limestones....	22	1.7	0.4	2.2	0.5
		Av. Th in sedimentary rocks, 6.5 p.p.m. Av. Th in igneous rocks, 13.5 p.p.m.		Av. U in sedimentary rocks, 2.4 p.p.m. Av. U in igneous rocks, 3.5 p.p.m.	

* Rock abundance estimates are from Pettijohn (1957).
† Uranium and thorium concentration figures are from Adams, Osmond, and Rogers (1959).

The program to date has been concerned primarily with the development of techniques for measuring the specific radioisotopes at low concentration levels. The procedure used is that of a pulse-height analysis, to identify the radioisotopes by the energy of their emitted a particles. The advantages of the method are numerous. Most of the naturally occurring heavy radioisotopes decay by this mode of activity; only actinium-227, radium-228, and lead-210, of the eleven radioisotopes having half-lives greater than one year, do not. The a emission of an isotope is often monoenergetic and usually distinctly different from the energy levels of other isotopes. Alpha-particle energies can be measured under suitable conditions with a resolution of a few per cent, which is usually sufficient to distinguish two isotopes in the same sample. The a count rate of each isotope of a radioactive series is generally the same as the count rate of the parent isotope if the series is at equilibrium, and the departure from equilibrium is generally a simple function of the variations in count rate. Analysis by a pulse height lends itself to isotope dilution techniques, which eliminate many of the problems of chemical yield and counting geometry. Finally, low background conditions are readily obtained.

The principal disadvantages of the method are that chemical extractions and elemental separations are necessary, low-mass counting samples must be prepared, and low counting rates require long counting times.

The steps in the procedure are as follows: (1) About 10 grams of sample are dissolved, and the resulting solution is split into 2 aliquots. To one of the aliquots an isotopic "spike" is added, whose nature depends on the isotopes to be analyzed. Uranium-234, thorium-232 and -230, and radium-226 and -224 have been used. Artificial isotopes, such as uranium-233 or uranium-232, have not been used, but such spikes might make the splitting of the sample unnecessary. (2) The radioelements are removed from the presence of interfering ions by solvent extraction with thenoyl trifluoroacetone, ethyl acetate, or mesityl oxide. (3) Uranium, thorium, and radium fractions of the sample are obtained by anion exchange procedures (Choppin and Sikkeland, 1959). Actinium and protactinium fractions can also be obtained if desired. (4) Counting planchets of each fraction are prepared either by solvent extraction and evaporation or by electrodeposition. (5) Alpha detection and pulse-height analysis are accomplished by means of a cesium iodide scintillation crystal and phototube connected to a 3-channel analyzer. Since elemental separations have been accomplished, one seldom is interested in measuring more than three energy levels simultaneously. (6) Comparison of the spiked and unspiked samples yields isotopic ratios from which the original isotopic composition and the procedural yield can be calculated.

The foregoing procedure may be applied to water analysis once the original sample has been evaporated to a small volume.

Results to date have been encouraging, though the full potential of the method has not yet been realized. The principal problems have been met in trying to predict the proper extraction procedures to eliminate interfering elements and in preparing high-yield, low-mass counting planchets.

The few samples that have been analyzed by this general procedure include the Florida rocks and waters listed in Tables 2 and 4. These data vary in reliability depending on the stage of development of the procedures at the time they were run. The samples were chosen primarily to test the method rather than to be truly representative of the strata and waters of Florida. Nevertheless, the results offer some interesting clues to the distribution of the heavy radioelements in this region.

The limestones analyzed (Table 2) include samples from six Tertiary formations of fairly wide extent. Excluded from this category are phosphatic, dolomitic, or aragonitic limestones, which are listed separately. The uranium and thorium values are typical of limestones elsewhere.

The "corals" category includes aragonitic carbonates—both the Key Largo limestone and the Miami oölite. These have uranium contents much higher than those of normal limestones. The important determining factor is probably the "room" in the crystal lattice of aragonite rather than some function of the organic activity involved in their formation.

TABLE 2

THE DISTRIBUTION OF THORIUM AND URANIUM IN THE
SEDIMENTARY ROCKS OF FLORIDA

Rock Type	No. Samples	Uranium p.p.m. Average	Uranium p.p.m. Range	Thorium p.p.m. Average	Thorium p.p.m. Range
Limestone.......	10	2	0.5– 6	1.5	0.2– 4
Coral..........	8	10	0.5– 18	0.5	0.2– 4
Dolomite.......	4	3.5	2 – 8	6	3 – 11
Attapulgite.....	10	5	3 – 10	15	12 – 18
Montmorillonite.	5	6	4 – 8	19	14 – 24
Kaolinite.......	3	2.5	1.5– 3	12	6 – 19
Phosphorite.....	2	250	135 –350	12	4 – 20
Sand..........	3	1.2	0.8– 1.8	3	2 – 6
"Black" sand....	5	50	15 –150	300	250 –1,000

The dolomites analyzed were also higher in uranium than average carbonates, probably due to the presence of either clay or phosphate in the samples. Three of the four samples are from the Hawthorn formation, in which both clay and phosphate are abundant. The high value for thorium, for a carbonate, suggests that clay in the samples carries the radioelements.

Attapulgite, montmorillonite, and kaolinite are clay samples from the Hawthorn formation of northern Florida. All are fairly high in uranium and thorium and could have some phosphate present. Illite is a common clay constituent of Florida rocks, particularly in association with other clay minerals, but no independent analyses of illite are as yet available.

Only two phosphorites have been run, but the results agree well with extensive data in the literature (McKelvey, 1956; Cathcart, 1956; Altschuler *et al.*, 1956). Phosphorite is the most hospitable host for uranium in Florida rocks. Although it has been reported that the "aluminum phosphate zone" is especially high in uranium content in the phosphate mining district (Altschuler *et al.*, 1956), elsewhere in the Hawthorn formation it has been observed that wavellite nodules, $Al_3(OH)_3(PO_4)_2 \cdot 5H_2O$, are less radioactive than are francolite nodules, $Ca_5F(PO_4,CO_3)_3$.

Clean Recent beach sands and red Pliocene sands of the Citronelle formation are no more radioactive than are average sands elsewhere. However, certain Pleistocene terrace sands are enriched in heavy minerals, called "black sands" because of their high ilmenite content, and these carry enough monazite and zircon to be radioactive. The monazite (Ce, La, Y, Th)PO_4 is enriched primarily in thorium—up to 8 per cent, according to Twenhofel and Buck (1956)—while zircon is

TABLE 3

AVERAGE URANIUM AND THORIUM CALCULATION FOR FLORIDA ROCKS

Rock Type	Uranium p.p.m. Average	Thorium p.p.m. Average	Rock Abundance	Uranium p.p.m. Contribution	Thorium p.p.m. Contribution
Limestone	2	1.5	0.3	0.6	0.5
Coral	10	0.5	0.04	0.4	0.0
Dolomite	3.5	6	0.08	0.3	0.5
Attapulgite	5	15	0.04	0.2	0.6
Montmorillonite	6	19	0.06	0.4	1.1
Kaolinite	2.5	12	0.04	0.1	0.5
Phosphorite	250	12	0.02	5	0.2
Sand	1.2	3	0.4	0.5	1.2
"Black" sand	50	300	0.02	1	6
				8.5 Av. U p.p.m.	10 6 Av. Th p.p.m.

enriched primarily in uranium. The samples analyzed, and listed in Table 2, are probably unusually high in "heavies" and are thus somewhat more radioactive than are many black sands.

Table 3 summarizes the uranium and thorium averages and attempts to calculate the average concentrations of these elements in Florida rocks by estimating the near-surface abundances of rock types. It is seen that the calculation is extremely sensitive to the estimated abundances of phosphorite beds and black sands. While the exact "average" uranium and thorium values calculated have little meaning, the significance of the radioactive rock terms (phosphorite and black sands) to the geochemical balance mentioned earlier is obvious. Both phosphatic beds and black-sand beds are thought to be concentrated in portions of the subsurface of Florida (Reynolds, 1962; Tanner *et al.*, 1961), and, to the extent that this may be true, it suggests that the "lost" uranium and thorium of the geochemical balances may be found in just such deposits.

The natural waters of Florida have been reported by Fix (1956) and by Scott and Barker (1958) to be below average in uranium and radium, with no other radioelements reported. Table 4 shows preliminary results of radiometric analysis of several Florida waters. The surface waters have uranium and radium values not unlike those reported by Scott and Barker. The well waters are distinctive in that the presence of thorium is detected in all samples, and radium-224 in two samples. It is not certain whether the thorium detected is thorium-232 or thorium-230.

TABLE 4

DISTRIBUTION OF HEAVY RADIOELEMENTS IN SELECTED SURFACE
AND UNDERGROUND WATERS OF FLORIDA

Sample	Uranium p.p.b.*	Radium-226 pc/l†	Thorium p.p.b.	Radium-224 pc/l
Gulf of Mexico..........	0.4	0.1	n.d.	n.d.
Suwannee River.........	n.d.	0.2	n.d.	n.d.
Wakulla spring..........	0.5	n.d.	n.d.	n.d.
Leon well...............	n.d.	0.2	0.1	n.d.
Martin well (T-82°)......	n.d.	0.3	0.5	n.d.
Martin well (T-85°)......	1.5	2	0.1	2
Martin well (T-91°)......	4	3.3	2	1

* Parts per billion.   † Picocuries per liter.

These are not typical well-water samples, it should be noted, since three of the four are distinctly thermal. The warmest well sample, at 91° F., contains detectable amounts of all the radioelements listed. The significance of this fact is as yet unknown; however, the thermal wells in Martin County have been drilled through the phosphatic Hawthorn formation. This might explain the high uranium and radium values. The detection of thorium in these samples is an unexpected development, which suggests that a more comprehensive study of the heavy radioelements in the ground waters of Florida may lead to new concepts in the mobility of these elements.

ACKNOWLEDGMENTS

This research has been supported by a grant from the American Chemical Society Petroleum Research Fund.

REFERENCES

ADAMS, J. A. S., J. K. OSMOND, and J. J. W. ROGERS. 1959. The geochemistry of thorium and uranium. *In* Physics and Chemistry of the Earth, **3**:298–348. London: Pergamon Press.

ALTSCHULER, Z. S., E. B. JAFFE, and F. CUTTITTA. 1956. The aluminum phosphate zone of the Bone Valley formation and its uranium deposits. Proc. Internat. Conf. on Peaceful Uses of Atomic Energy, **6**:507–13.

CATHCART, J. B. 1956. Distribution and occurrence of uranium in the calcium phosphate zone of the Land Pebble phosphate district of Florida. Proc. Internat. Conf. on Peaceful Uses of Atomic Energy, **6**: 514–19.

CHOPPIN, G. R., and T. SIKKELAND. 1959. Scheme for the separation of the elements francium through uranium: The radiochemistry of thorium. Univ. Calif. Res. Lab. Pub. 8703. Pp. 85–86.

FIX, P. F. 1956. Geochemical prospecting for uranium by sampling ground and surface waters. Proc. Internat. Conf. on Peaceful Uses of Atomic Energy, **6**:788–91.

McKELVEY, V. E. 1956. Uranium in phosphate rock. Proc. Internat. Conf. on Peaceful Uses of Atomic Energy, **6**:499–502.

PETTIJOHN, F. J. 1957. Sedimentary Rocks, p. 10. New York: Harper & Bros.

REYNOLDS, W. F. The Lithostratigraphy and Clay Mineralogy of the Tampa-Hawthorn Sequence of Peninsular Florida. M.S. thesis, Florida State Univ., 1962.

SCOTT, R. C., and F. B. BARKER. 1958. Radium and uranium in ground water of the United States. Proc. 2d United Nations Conf. on Peaceful Uses of Atomic Energy, **2**:153–57.

TANNER, W. F., A. MULLINS, and J. D. BATES. 1961. Possible masked heavy mineral deposit, Florida panhandle. Econ. Geol., **56**:1079–87.

TWENHOFEL, W. S., and K. L. BUCK. 1956. The geology of thorium deposits in the United States. Proc. Internat. Conf. on Peaceful Uses of Atomic Energy, **6**:562–67.

ALLAN B. TANNER

9. Radon Migration in the Ground: A Review

Radon isotopes are practically inert and have the properties of gases under conditions of geologic interest. During their brief lives, their atoms are capable of moving from the sites of their generation. How far they move before their radioactive disintegration has been a topic—usually implicit—of many hundreds of reports of radon in the atmosphere, soil gas, natural waters, petroleum, and natural gas. The purpose of this review is to point out the factors that determine the distance that radon isotopes may migrate in the ground, to discuss the geophysical and geochemical implications of radon migration, and to provide an extensive, though not exhaustive, bibliography.

MIGRATION PROCESSES

The migration of radon isotopes may be conveniently considered in several steps: (1) the radioactive formation and recoil of the newly formed ion from its precursor, a radium isotope, (2) the diffusion of the neutral atom through the interior of a mineral grain, and (3) the diffusion and transport of the neutral atom through permeable rock and soil. For the first two steps, the great differences in half-life among the isotopes (radon-222, 3.8 days; radon-220, 52 seconds; radon-219, 3.9 seconds) are rather unimportant. But the subsequent migrations of radon-220 and radon-219 in the ground are severely restricted in comparison with that of radon-222.

MIGRATION OF RADON ISOTOPES INTO ROCK PORES

When a radium atom of mass 226, 224, or 223 disintegrates, it yields an α particle and a radon atom of mass 222, 220, or 219, respectively. The atom is stripped of its outer electrons and expends its

ALLAN B. TANNER is geophysicist, U.S. Geological Survey, Washington, D.C.

Publication authorized by the Director, U.S. Geological Survey.

kinetic energy of recoil, about 1×10^5 electron volts, along a track that is roughly 3×10^{-6} cm. long in minerals of normal rock density and from about 6×10^{-3} to 9×10^{-3} cm. long in air, depending on the isotope (Zimens, 1943). The neutral atom may then diffuse until it disintegrates or escapes from a mineral grain. The fraction of the atoms formed that escape is termed the *emanating power* of the mineral for the radon isotope in question. (In the Soviet literature it is usually termed the *coefficient of emanation*, and "emanating power" is usually equivalent to the concentration of radium in a mineral or rock multiplied by its coefficient of emanation.)

In the early 1920's Otto Hahn developed his "emanation method" of studying the internal surface and grain size of fine artificial powders. Hahn's use of the method (1936), and the development of the theory by Flügge and Zimens (1939) and by Zimens (1943), have not been aimed at understanding the process of emanation from minerals and rocks, but they are most useful in doing so. The following discussion draws heavily upon Zimens' concluding paper on the subject (1943).

According to the theory of the emanation method, an isolated, isotropic, spherical grain containing a uniform distribution of a radium isotope is large enough to contain virtually all the recoiling ions unless the grain is of less-than-micron size (as small as a very fine clay particle). If a number of such very small particles, each capable of emanating a substantial fraction of radon isotopes, are compacted as in the usual geologic situation, almost all the atoms escaping from grains in which they originated bury themselves in other grains, and the fraction terminating their recoil paths in the interstices is again negligible.

In a rock whose smaller pores and fractures are filled with water, however, a recoil ion escaping from one grain into a pore encounters a far more dense absorber of its energy. The recoil range in water should be about 1×10^{-5} cm.—or about three times greater than that in glass and two orders of magnitude less than that in air. Recoil ions entering pores of such size or larger are likely to end their recoil in the pores, where diffusion may proceed at a rate many orders of magnitude greater than in an ionic solid. The effect of moisture in a rock, then, should be to increase its emanating power, as is often observed (Kirikov, Bogoslovskaya, and Gorshkov, 1932; Hahn, 1936; Starik and Melikova, 1957). In rock having grains much larger than 10^{-5} cm. and little internal surface, the effect should be negligible.

If the radon isotope completes its recoil at a point within a relatively intact crystal, its chance of escape thereafter must be extremely small. Many studies have recently been made of the similar problem of leakage of radiogenic argon from minerals. Amirkhanoff *et al.* (1961) report diffusion constants at 300° K., ranging from 10^{-27} to 10^{-65} cm²/sec for several common rock-forming minerals, which implies negligible leakage of argon from grains larger than 10 microns even during geologically long periods. The same authors infer that the diffusion constants obtained are characteristic of the minerals rather than of the diffusing gases (helium and argon, in their investigations), so the constants obtained should be valid also for the radon isotopes. Nicolaysen (1957) has reported the somewhat greater, but still very small, values of 10^{-21}–10^{-23} cm²/sec for the diffusion of lead in zircon and monazite. The brief lives of the radon isotopes further limit their escape.

Kosov and Cherdyntsev (1955) have given the most direct evidence that diffusion of the radon isotopes from minerals is slight; they found that most minerals have nearly equal emanating powers for both radon-222 and radon-219. Each isotope may be considered as emanating from a surface layer of a grain, the layer thickness being roughly equal to the diffusion length of the isotope. Because the diffusion lengths of radon-222 and radon-219 differ by the square root of the ratio of their half-lives, a factor of 291, the volume of the surface layer must be two orders of magnitude greater for radon-222 than for radon-219. This difference should lead to a large difference in emanating power unless (1) the actinium series precursors of radon-219 are systematically separated from the uranium series precursors of radon-222 in such a way that the radon-219 source is distributed nearer the surface, just compensating for the greater diffusion of radon-222—an unlikely explanation—or (2) the contributions of both radon-222 and radon-219 by diffusion from the grain are small compared with the contributions by recoil into pores.

Flügge and Zimens, in their 1939 work, discounted the possibility that a recoil ion escaping from one grain and burying itself in another could diffuse back through the crater formed during its recoil; the cooling of the melted material, they pointed out, was too fast. Zimens later (1943) revised his view, saying that diffusion might progress at a more rapid rate in the resolidified crater than in the surrounding crystal because of the disorder resulting from the recoil. Emanating power may be augmented by this process. For the same reason as that given in the preceding paragraph, however, the effect should be small in rocks composed of sound crystals 10^{-5} cm. or larger in size. In the extremely fine size range, where the recoil range exceeds the size of intact grains, the reasoning is not applicable.

In approximate agreement with theory, the emanating power is quite low for artificially prepared salts in which radium is isomorphous with the principal cation. Hahn (1936, p. 222) cites F. Strassmann's determinations of the emanating power of prepared mixed crystals of barium and radium chlorides, nitrates, and sulfates. The emanating power of such preparations was somewhat dependent on grain size, not more than a few tenths of 1 per cent for grains of 200-micron average diameter and as much as a few per cent for much smaller crystals. Extensive study of the emanating power of natural minerals has been reported by Starik and Melikova (1957). Determinations of the emanating power or radon leakage of more-than-usual numbers of mineral or rock samples are given by Kirikov, Tverskoy, and Grammakov (1932), Breger (1955), Giletti and Kulp (1955), Kosov and Cherdyntsev (1955), and Tokarev and Shcherbakov (1956, Table 10). In contrast with the less than 1 per cent emanating power expected for aggregates of all but fine grain size, most of the measurements have yielded emanating powers of the order of tens of per cent. The conditions of uniform distribution of the radium isotope or structural soundness of the host mineral are evidently not often satisfied.

In many samples the radium isotope probably is not uniformly distributed because it is contained in films or crusts on the surfaces of the rock pores or because it is rejected from host crystals. Secondary deposition of radium or of substances containing its precursors obviously must be a common cause of non-uniform radium distribution. The rejection mechanism appears much less likely. In a recent study of the behavior of recoil atoms in ionic solids, Anderson *et al.* (1963, pp. 19–20) observed the emanation of radon-220 from thoria indexed with thorium-228 (radiothorium). The decay product intermediate between thorium-228 and radon-220, that is, radium-224, was inferred to be rejected from the interior of the thoria grains only above 1300° C. Using Nicolaysen's (1957) largest value, 10^{-21} cm²/sec, for the diffusion coefficient of lead in zircon and monazite, one obtains a mean diffusion length of 10^{-5} cm. (0.1 micron) for radium-226 at room temperature. The other shorter-lived radium isotopes would have correspondingly shorter diffusion lengths.

From the discussion given above, we may conclude that any appreciable emanations of radon-222, radon-220, or radon-219 atoms come from radium isotopes distributed in secondary crusts or films or in the shallow surface layers (approximately as deep as the recoil range) of intact crystals of the host minerals. Radon isotopes in the deeper regions of the crystals are unavailable without the development of large internal surface, such as may result from chemical corrosion, weathering, or intensive fracturing on a microscopic scale. Starik and Melikova (1957) and Hayase and Tsutsumi (1958) have noted that fresh samples, lacking evidence of alteration, tend to have low emanating power; altered or weathered samples tend to have high emanating power. Uranium minerals that have high emanating power are usually secondary minerals. The principal mechanism by which the radon isotopes enter the pores, capillaries, or microfractures is radioactive recoil into liquid-containing spaces or diffusion from solid material appreciably smaller than the diffusion length of the most short-lived isotope observed.

MIGRATION OF RADON ISOTOPES IN A POROUS MEDIUM

Many variables are significant in the migration of radon isotopes in the ground: the decay rate of the isotope, the diffusion constant for the isotope in the pore-filling fluid, the porosity of the ground, the velocity of the fluid, the composition of the fluid, and, if the fluid has more than one phase, the temperature-dependent distribution of the radon isotope among the phases.

For practical purposes the underground migration of radon isotopes may be considered for media having several different sets of characteristics: (1) rock or soil unsaturated by liquids and free of open fractures or channels, (2) fractured rock or disturbed soil, unsaturated by liquids, and (3) saturated rock or soil.

Two different mechanisms of migration should be distinguished: diffusion, where the radon isotope moves with respect to the fluid filling the pores of the medium; and transport ("convection" in the Soviet literature), where the fluid itself moves through the porous medium and carries the radon isotope along with it. Either or both mechanisms may be important in a given place.

DIFFUSION OF RADON ISOTOPES IN A POROUS MEDIUM

The tendency of radon isotopes to diffuse in the ground (assuming it to be a porous, permeable medium) may be measured by the constant of diffusion of radon in the fluid filling the pores, provided that corrections are made for the interference of the solid matrix and for the presence of more than one fluid phase in the pores. These corrections may be incorporated into the diffusion coefficient itself, making it a function of porosity and moisture content in addition to the composition and temperature of the fluid phase (usually air) in which the diffusion constant is greatest. Table 1 presents a summary of coefficients determined experimentally for the diffusion of radon isotopes in various continuous and porous media. A lengthy list of diffusion coefficients for geologic materials of low moisture content may be found in the Soviet manual for radiometric prospecting (Alekseyev, 1957, Table 50). Principal references on the effects of moisture are Grammakov (1936) and Baranov and Novitskaya

TABLE 1

Summary of Diffusion Coefficients (D) Reported for Radon Isotopes in Various Media

Diffusing Isotope	Fluid	Medium	D(cm²/sec)	Conditions	Authority
Rn, Tn, An	Air	Continuous	1.0×10^{-1}	Various	Various, quoted in Zimens, 1943
Rn	"	"	1.20×10^{-1}	15° C., 76 cm. Hg	Hirst and Harrison, 1939
"	H₂	"	4.76×10^{-1}	"	"
"	He	"	3.51×10^{-1}	"	"
"	Ne	"	2.17×10^{-1}	"	"
"	Ar	"	9.2×10^{-2}	"	"
"	Alcohol	"	2.69×10^{-5}	18° C.	Róna, 1917, and Ramstedt, 1919, quoted in Zimens, 1943
"	Toluene	"	2.67×10^{-5}	"	"
"	Benzene	"	2.36×10^{-5}	"	"
"	Water	"	1.13×10^{-5}	"	"
"	Air, 4% moisture	Building sand 1.40 gm/cm³, 39% porosity	5.4×10^{-2}	Not stated	Bulashevich and Khayritdinov, 1959
Tn	Air, no moisture	Fine sand (mostly quartz)	6.8×10^{-2}	Not stated	Grammakov, 1936
"	Air, 8.1% moisture	"	5.0×10^{-2}	"	"
"	Air, 15.2% moisture	"	1.0×10^{-2}	"	"
"	Air, 17% moisture	"	5.0×10^{-3}	"	"
Rn	Air	Eluvial-detrital deposits of granodiorite	4.5×10^{-2}	Mean effective value in natural occurrence	Popretinskiy, 1961
"	"	Diluvium of metamorphic rocks	1.8×10^{-2}	"	"
"	"	Eluvial-detrital deposits of granite	1.5×10^{-2}	"	"
"	"	Loams	8.0×10^{-3}	"	"
"	"	Varved clays	7.0×10^{-3}	"	"
"	Air, 37.2% moisture	Mud, 1.57 gm/cm³	5.7×10^{-6}	19°–20° C.	Baranov and Novitskaya, 1949
"	Air, 85.5% moisture	Mud, 1.02 gm/cm³	2.2×10^{-6}	19°–20° C.	"
"	Solid	Barium nitrate	8.0×10^{-20}	Not stated (room temperature)	Strassmann, quoted by Flügge and Zimens, 1939

(1949); the latter also investigated the diffusion rate as a function of temperature. Zimens (1943) has discussed the use of emanating-power measurements to obtain the diffusion coefficients for solids. Many other studies listed in the bibliography have involved either the laboratory determination of the diffusion coefficient or its calculation from field measurements. Caution must be exercised in accepting the published values, however. Bulashevich and Khayritdinov (1959) have shown experimentally that calculations of diffusion of radon isotopes in porous media should use a concentration parameter that does not include the solid matrix and should use a diffusion coefficient that has been corrected for porosity; the boundary conditions imposed for particular solutions of the diffusion equation are otherwise likely to be incorrect.

Several generalizations appear reasonable from the literature:

1. Diffusion coefficients for the several radon isotopes in identical media are practically equal.

2. The diffusion coefficient is not sensitive to the pore diameter in the usual range of geologic materials, which have pores much larger than the mean free path of the diffusing isotope (of the order of 10^{-6} cm. in gas under normal conditions).

3. The diffusion coefficient is less than the diffusion constant in the presence of a solid matrix. According to Penman (1940a, b, as quoted by Baver, 1956, pp. 213–14), the ratio of the diffusion coefficient for a vapor in a porous medium to the diffusion constant in air is proportional to the pore space, the proportionality constant being 0.66. Within the considerable uncertainty of experiments, Penman's relation appears to give fair values for radon diffusion in dry soils.

4. Increasing moisture in a porous medium causes a reduction of diffusion coefficient greater than can be accounted for by the pore space occupied by water. Grammakov's data, quoted in Table 1, show a reduction of diffusion coefficient by more than an order of magnitude with the addition of 17 per cent moisture to a fine sand. The cited data of Baranov and Novitskaya (1949) for (finer-grained) muds of greater moisture content suggest that the diffusion constant for water, reduced to compensate for the solid matrix, is appropriate for porous media of moderate or high water content.

MIGRATION IN UNSATURATED, UNDISTURBED ROCK AND SOIL

Of great interest in environmental radioactivity studies and geophysical prospecting is the migration of radon isotopes in undisturbed soil and porous rock near the earth's surface, in which situation it is generally assumed that the diffusion mechanism predominates over the transport mechanism. The assumption is based on two observations: (1) The rate of exhalation of radon at the interface between ground and atmosphere corresponds to that calculated from diffusion theory, using experimentally obtained diffusion coefficients (Alekseyev, 1957, p. 433; Wilkening and Hand, 1960). (2) The concentration of radon in ground air is distributed with depth approximately according to a diffusion model (Bulashevich, 1946). Although the observations justify the use of diffusion theory for the solution of many problems of underground radon distribution, they do not justify the conclusion that transport phenomena are negligible under the given conditions. Grammakov (1961*b*) has pointed out that the steady-state solution of the one-dimensional diffusion equation may take transport into account merely by substituting an "effective" diffusion coefficient that is determined by the steady-state convection velocity, in addition to the factors inherent in diffusion. The range of diffusion coefficients observed under various field and laboratory conditions is sufficiently large that the influence of transport may easily be obscured by uncertainty of the diffusion coefficient.

Transport effects have in fact been noted in investigations both of radon concentration versus depth and of the radon exhalation rate. Meteorologic factors have a pronounced effect on the transport of radon isotopes in the uppermost ground layers. Insofar as the upward movement of radon isotopes is complementary with the (downward) aeration of soil, Baver's review (1956, pp. 209–22) of the influence of meteorologic factors on gas exchange is useful in gaining an idea of the relative importance of the factors. In order of decreasing contribution to normal soil aeration, they are rainfall, variation of barometric pressure, variation of soil and atmospheric temperature, and wind; collectively, these factors contribute to less than 10 per cent of normal soil aeration. It should not be expected that the transport and diffusion effects be in the same proportion for radon-isotope migration as for soil aeration, however; the rate of disintegration of a radon isotope and the rate of consumption of the gases serving as indices of soil aeration are generally different, and the factors affecting transport are more transitory than are those affecting diffusion.

In the bibliography are listed many papers whose titles indicate the subject of radon in soil gas or soil air; nearly all these papers discuss the effects of meteorological variables. Owing to non-ideal conditions in the various test areas and to the understandable difficulty of isolating the effects of different variables, the results of these investigations do not present a very coherent picture; the variable inferred to dominate in one investigation is found to be of little significance in another.

The three factors most commonly observed to have pronounced

effects on radon concentration and exhalation are rainfall (Wright and Smith, 1915, and many others), freezing, and snow cover (Bender, 1934; Zupancic, 1934; and many others). Both transport and diffusion are affected. With heavy rainfall, the soil gases near the surface tend to be displaced upward, carrying radon with them and increasing the exhalation rate temporarily. Thereafter, the reduced diffusion coefficient and reduced permeability (Alvarez, 1949) of the wet ground restrict migration by both mechanisms. Exhalation of both radon-222 and radon-220 (and presumably of radon-219) is markedly reduced; the concentration of the radon isotopes in the ground is commensurately increased. In order to account fully for the radon not exhaled after heavy rainfall, it is necessary to consider the distribution of radon between the soil gas and the soil water. Especially when the water temperature is near the freezing point and the distribution ratio is about 0.5 (in water) to 1 (in gas) (Kofler, 1913), a substantial part of the radon should be in the water phase.

The effect of change of atmospheric pressure, to raise or depress the gaseous volume within the ground without substantially altering the shape of the curve of radon distribution, is implied or noted in the works of Bogoslovskaya *et al.* (1932), Hatuda (1953), Shevchenko (1958), and Kraner *et al.* (1963, and written communication). In addition, Zeilinger's data (1935, p. 293) show a tendency for slight-to-moderate increases in the exhalation rate with falling barometric pressure; the effect has been observed also by Rosen (1957) and in the cited work of Kraner *et al.*

The literature contains solutions to various problems of radioactive gas movement through porous media. A problem of wide application is the movement of radon isotopes upward through soil from a subsurface source of infinite lateral extent. In the simplest and most-often-solved problem, a radioactive gas of decay constant λ, in concentration C at a distance z from a plane source in which the concentration is C_0, diffuses in the z direction through an infinite, porous half-space of diffusion coefficient D:

$$C = C_0 \exp(-\sqrt{\lambda/D}\,z). \qquad (1)$$

Hundred-fold diminution of concentration will therefore take place at a distance $z = 4.6\sqrt{D/\lambda}$, where $\sqrt{D/\lambda}$ is the diffusion length. Substituting reasonable values of 5×10^{-2} cm²/sec for the diffusion coefficient in dry sand and either $\lambda_{Rn} = 2.1 \times 10^{-6}$/sec or $\lambda_{Tn} = 1.3 \times 10^{-2}$/sec, we find that the diffusion lengths of radon-222 and radon-220 are 160 cm. and 2 cm., respectively, and that hundred-fold diminution will take place over distances of 730 cm. and 9 cm., respectively.

In 1936 Grammakov published solutions to several problems of diffusion in the presence of a steady component of transport in the direction of diffusion. With a transport component v added to the conditions and notation given above, his one-dimensional solution is

$$C = C_0 \exp\left[\left(\frac{v}{2D} - \sqrt{\frac{v^2}{4D^2} + \frac{\lambda}{D}}\right)z\right]. \tag{2}$$

The two solutions are similar in form; under a given set of conditions, the change in concentration is an exponentially decreasing function of z. Unless the diffusion coefficient is accurately known for the particular column of material through which migration is taking place, one cannot distinguish between the diffusion and the transport components with only one isotope. According to data of R. D. Evans, H. W. Kraner, and G. L. Schroeder (written communication, 1962) on the displacement of the vertical profile of radon concentration in the upper few meters, average vertical transport velocities of roughly 2×10^{-4} cm/sec and 6×10^{-5} cm/sec occurred in weathered tuff during two 2-day periods of barometric pressure change. Such transport velocities significantly increase the effective diffusion distance. Bogoslovskaya et al. (1932) calculated the rate of movement of radon in sand above an artificial source at 7×10^{-4} cm/sec; transport was evidently the dominant mechanism.

Although equations (1) and (2) are strictly valid for unrealistically idealized situations, they do show the essentially exponential and acute attenuation of the flux of a radon isotope in steady-state one-dimensional movement. More realistic solutions, accounting for radioactive overburden (Grammakov et al., 1958) and differences in porosity between the diffusion media (Bulashevich and Khayritdinov, 1959), are expressed in the hyperbolic functions. Okabe (1956a) includes corrections for absorption of radon by soil water and for a soil temperature gradient.

Other idealized problems have been solved for steady-state diffusion. Koenigsberger (1928) and Bulashevich (1946, 1947) derive expressions for the concentration of radon-222 or radon-220 at the surface in alluvium covering a linear source. Khaykovich (1961) considers the distribution of radon migrating through an ore layer into a cylindrical borehole. Grammakov (1961a) gives solutions for several problems in spherical and cylindrical geometry. Solutions for various source distributions have been summarized in the Soviet prospecting manual (Alekseyev, 1957, sec. 49).

Published solutions of problems of transient or non-stationary conditions are few. Kikkawa (1954) and Khaykovich and Khalfin (1961) have considered the effect of sealing the ground surface (as by rain or freezing) on the distribution of radon in the ground. Sakakura *et al.* (1959) have derived expressions for transport (neglecting diffusion) of radioactive gases into wells in fields of natural gas; both steady-state and transient conditions are considered. Grammakov (1961*a*) has solved the transient problem for movement from a sphere into a concentric outer sphere and for movement from infinite source media into inactive spherical and cylindrical regions. The author is not aware of published solutions for one-dimensional diffusion of radon from a source layer through a radioactive overburden to the atmosphere in the presence of a periodically varying transport component, induced by changing atmospheric pressure. Such a solution would be of considerable interest in studies of near-surface atmospheric radioactivity and in prospecting for uranium.

MIGRATION IN UNSATURATED, FRACTURED ROCK OR DISTURBED SOIL

Suppose that unsaturated rock or soil has been disturbed to the extent that open fractures or channels exist in the ground; the mathematical analysis reviewed in the section above is not then directly applicable. In the discrete blocks or zones through which no fractures or channels pass, the diffusion coefficient and transport velocity may be considerably smaller than in the fractures and channels themselves. In a manner analogous with the movement of water in dendritic drainage patterns, radon isotopes should tend to migrate through the undisturbed zones and into fractures and channels mainly by diffusion; thenceforth transport should play a progressively greater role. (In response to changes in atmospheric pressure, the flow of air into or out of the ground should be greater along the paths of least resistance, the fractures and channels, than through the undisturbed zones.) Fracture systems or channels may also permit migration of radon isotopes from localized and "point" sources with less dispersion than would take place with diffusion in all directions from the source.

Some experimental verification of this speculation is available. Okabe (1956*b*) found that the atmospheric concentration of radon decay products increased significantly for several days following local earth shocks. Kraner *et al.* (1963 and written communication, 1962) measured concentrations of radon near the surface in welded tuff, weathered tuff, and alluvium before and after underground nuclear explosions. The results were consistent with a model of more rapid upward movement of radon following disturbance of the ground; macrofissures were sometimes observed. In unconsolidated ground the fissure systems apparently sealed themselves in about a week's time.

In a study of radon behavior in drill holes penetrating uranium ore (Tanner, 1958, 1959), the writer found a distinct relation between the total amount of radon in a drill hole and change of atmospheric pressure. Within the drill holes, the transport mechanism predominated. Transport was also important within the ore layers, since the observed flux of radon into a drill hole was greater than could be accounted for by diffusion alone.

Radon is usually present in moderate or high concentration in the air of poorly ventilated uranium mines (see, e.g., Holaday *et al.*, 1957). Kapitanov and Serdyukova (1962) have treated analytically the migration of radon into and in mine workings but have neglected the transport component of movement from the rock into the workings. An analysis including a periodically varying transport component seems desirable.

If fractures or channels are present in ground in the vicinity of a source of a radon isotope, no particular limit can be placed on the migration distance of the isotope because the controlling factor, the transport velocity, is determined by the local conditions. For the same reason, the overburden will not generally be homogeneous or isotropic; migration will be unusually great only along the high-speed paths and only in the directions traversed by those paths. The chance of detecting a radon anomaly at a distance greater than would be found for homogeneous, isotropic overburden will depend on the chance that a sample is taken at a fracture or channel. An anomaly once detected may be displaced in any direction, and by an indeterminate distance, from the sampling site. Thus the very conditions that extend the migration range of a radon isotope tend to reduce the reliability and usefulness of prospecting methods based on such migration.

MIGRATION IN SATURATED ROCK AND SOIL

The diffusion coefficients of radon isotopes in water-saturated porous media are extremely small. If the diffusion coefficient for water is used as an upper limit, the one-dimensional diffusion equation indicates that radon-222 will undergo hundred-fold diminution in the short maximum distance of 10.7 cm. Radon-220 will travel a distance smaller by a factor of $\sqrt{\lambda_{Tn}/\lambda_{Rn}} = 80$, and radon-219 will travel a distance shorter by a factor of $\sqrt{\lambda_{An}/\lambda_{Rn}} = 291$. The data of Baranov and Novitskaya (1949, Table 5) show that diffusion coefficients in nearly saturated muds are about an order of magnitude smaller than in water and that the effect of temperature is to double the diffusion coefficient with each rise of $10°-20°$ C. in the range of practical interest. Diffusion coefficients in liquids other than water are somewhat greater, but not enough to result in significant movement of radon isotopes by diffusion.

If radon isotopes in liquid-saturated rock or soil are to migrate distances significant in environmental radioactivity studies or in prospecting, the transport mechanism must predominate. Because the flow of underground liquids varies over a great range and depends on local conditions, only a few useful generalizations can be made.

Where water flows through permeable, more-or-less homogeneous material, two cases of particular interest: (1) gravitational flow in one dimension and (2) flow radially inward to a cylindrical well. These are limiting cases describing the practical problem of radon transport to a well. Case (1) describes an idealized situation where so little water is being withdrawn by a well that the pattern of ground-water flow is not disturbed, and case (2) describes the idealized situation where the flow due to withdrawal is large compared with the gravitational flow.

Let us define the mean migration distance of a radon isotope as the distance traveled by the transporting liquid during one mean life $(1/\lambda)$ of the isotope. For flow according to case (1), the mean migration distance is simply proportional to the rate of flow of underground water. According to Todd (1959, p. 53), such flow may be more than 30 m/day but normally ranges from 5 ft/yr (1.5 m/yr) to 5 ft/day (1.5 m/day). The mean migration distances of radon-222 (mean life, 5.5 days), radon-220 (75 sec.), and radon-219 (5.7 sec.) are evidently quite short at normal ground-water flow rates.

For flow according to case (2), we may make the additional observation that the migration distance is proportional to only the square root of the flow rate from the well, because the flow is radial. In another paper in this symposium (Tanner, 1963), I have discussed the analysis of radon migration distance at a specific well, where the mean migration distance was 5.5 m. at an observed discharge rate. When the same well was open to unrestrained flow, greater by a factor of 30 than that of the observation period, the mean migration distance was about 30 m. It seems likely that the mean migration distance is similarly short or moderate for the majority of water wells, such as might be tested in environmental health studies or in prospecting for uranium.

Little can be said about the migration distance of radon isotopes in distinctly anisotropic and fractured media because of its heavy dependence on local conditions. Tracers have shown that ground water may flow as fast as a hundred meters or so per day (A. E. Peckham, oral communication, 1963), and in such a situation the practical problem is to determine the proportion of water flowing at such high rates and its proximity to radon sources. These questions obviously require careful study of the hydrology and geology of a given area.

Few estimates have been made of the migration distances of radon isotopes at specific locations. Trombe and Henry la Blanchetais (1947) inferred that radon-222-bearing water in a subterranean river of Saint-Paul, Haute-Garonne, was derived from great depth. Kimura (1949) studied the equilibrium relations among radon-222, radon-220, their precursors, and their decay products and concluded that the radon isotopes had not traveled far. Okabe (1956*a*) concluded that radon in hot spring waters was derived from sources near the points of issuance rather than from the sources of the waters themselves.

TABLE 2

SUMMARY OF MIGRATION DISTANCES FOR RADON-222

STEADY-STATE CONDITIONS	APPROXIMATE MIGRATION DISTANCE (METERS)		
	Maximum in 5.5 Days $(C/C_0 = 1/e)$	Maximum in 25 Days $(C/C_0 = 1/100)$	"Normal" Mean $(C/C_0 = 1/e)$
One-dimensional diffusion+transport; dry soil.................................	5?	?	1–2
One-dimensional diffusion; dry soil.......	2	10	1
One-dimensional diffusion; moist soil.....	0.5	2	0.1
One-dimensional diffusion; wet soil.......	0.04	0.2	0.02
One-dimensional diffusion; saturated ground.............................	0.02	0.1	0.01
One-dimensional transport by ground water	500	2,300	0.01–2
Radial transport by ground water........	....	....	1–10

SUMMARY OF MIGRATION DISTANCES

Table 2 is a summary of migration distances for radon in one-dimensional and cylindrical flow, corresponding respectively to vertical flow from horizontal sources and to flow into a water well. The figures have been obtained by insertion of reasonable values of diffusion coefficient and transport velocity into equations given above; some choices are obviously qualitative only.

Other estimates are available in the literature. For the maximum distance at which radon in soil gas would signal a buried deposit of radium-bearing material in the one-dimensional situation, Alekseyev (1957, sec. 51) estimates about 6 meters, Budde (1958*b*) estimates several meters, Grammakov *et al.* (1958) estimate 13 meters, and Márton and Stegena (1962) estimate 5–10 meters (over petroleum accumulations). Sharkov (1959) states that radon in ground water is removed in tens of meters and, more rarely, in the first hundred meters. Iwasaki *et al.* (1956) calculate from the radon-220 to radon-222 ratio observed in a single gas sample from a volcanic fumarole that the transit time from source to sampling was only ten minutes.

The brief lives of radon-220 and radon-219 so limit their travel that transport in unsaturated ground and diffusion in saturated ground are negligible. Under optimum conditions the mean migration distances for dry soil should be about 2 cm. for radon-220 and 6 mm. for radon-219. Except for the infrequent instances of very high speed flow, the mean migration distances in ground water should be of the same order of magnitude.

ENVIRONMENTAL RADIOACTIVITY

By their migration into the upper soil layers, into the atmosphere, and into natural water supplies, radon isotopes and their decay products contribute a substantial part of man's radiation environment. The means by which these isotopes migrate has been discussed above. Let us now consider what their migration characteristics imply about the radiation environment.

In dry or moist soil there is depletion of the radon isotopes in the uppermost layer owing to their exhalation into the atmosphere. The layer of twofold or greater depletion of the fraction emanated into the soil pores is of the order of a meter for radon-222, a centimeter for radon-220, and a few millimeters for radon-219. The γ-emitting short-lived decay products, bismuth-214 (RaC), bismuth-212 (ThC), thallium-208 (ThC''), lead-211 (AcB), and others, are consequently depleted from the uppermost half-meter of the soil, which is virtually the entire ground source of γ radiation. Because the depletion thicknesses for radon-220 and radon-219 are minor compared with this source thickness, their migration should not cause appreciable loss of γ-ray sources in the soil. On the other hand, a soil having high emanating power for radon may lose a major fraction of its radon-222 and γ-emitting bismuth-214 through migration into the atmosphere. Retention of the radon-222 and its decay products in the atmosphere near the ground should cause an increase in the γ field because of the lesser absorption of γ rays by air. If wind and convection disperse the radon and decay products, the γ field should be less than that above soil in which no radon migration takes place. Variations in the γ field from soil sources as a result of radon migration into the atmosphere have been treated analytically by Timofeyev (1959) and by Khaykovich and Khalfin (1961). Appreciable precipitation, freezing, or snow cover will tend to seal the ground, to cause build-up of the radon concentration in the important uppermost layer, to increase the γ-ray source strength in the ground, and to deplete the γ-ray source strength in the atmosphere. Some reduction of the γ field at the surface may take place with heavy precipitation because some radon will be washed down to deeper soil layers and the water will increase the effective γ-ray absorption coefficient in the ground.

In both saturated and unsaturated ground the amount of radon-222, radon-220, or radon-219 free to migrate is dependent on the emanating power of the source material. In natural occurrences the emanating power may be from nil to nearly 100 per cent, although it normally ranges between much narrower limits (Tokarev and Shcherbakov, 1956, Table 19). The γ radioactivity due to radon-222, radon-220, radon-219, and their decay products in rocks and soils issues from decay products that have *not migrated* from the "sample." Because there is no general relation between the source strength and emanating power or extent of migration, *there is no necessary correlation between the γ radiation field and the amount of radon, thoron, or actinon migrating into the atmosphere or into natural waters.* In environmental health studies, there is no basis for equating the health hazard due to ingested radon and its decay products with the surface intensity of γ rays from radon decay products unless such a relation has been demonstrated in the area of interest.

Geophysical Applications

For half a century the inherent mobility of radon has stimulated interest in its use for detecting small, near-surface tectonic features and buried radioactive deposits. Ambronn (1928) has summarized the early results, which generally indicated marked increases of α radioactivity and slight increases of γ radiation over faults and structural lines. Koenigsberger (1928) provided a mathematical basis for using the emanation method to determine the thickness of alluvium above such structures. Lane and Bennett (1934) considered their measurements of radon in well waters as indicative of a known fault. Further positive results have been reported by Israël and Becker (1935*a*), Ochiai (1951), Lauterbach (1953), Hradil (1955), Herzog (1956), Teuscher and Budde (1957), Lecoq *et al.* (1958), Budde (1958*b*), Löser (1959), Lorenz *et al.* (1961), and Coppens (1962). Israël(-Köhler) and Becker (1936) did not find the method reliable enough for quantitative treatment, and others have noted the interfering effects of radioactive heavy minerals (Clark and Botset, 1932) and the lack of correlation between faults and radioactive highs in some areas (Botset and Weaver, 1932; Howell, 1934).

Radon and γ-ray measurements may be useful in connection with prospecting for petroleum deposits. Bogoyavlenskiy (1927) attributed enhanced γ radiation above the Maikop oil field to highly penetrating radiation from concentrations of radon in the oil pools and radium below them—an explanation that is now untenable. Stothart (1948) reported that well-to-well changes of radon concentration in petroleum may be used to guide "wildcat" prospecting in certain areas. Lundberg (1952) introduced the concept of a minimum and an encircling maximum of radioactivity at the surface above a petroleum deposit. He attributed the phenomenon to the deflection by the petroleum deposit of upward-migrating solutions bearing radium. Gregory (1956) re-evaluated published aeroradiometric anomalies over oil fields, found that the anomalies could be correlated with areal geology, distribution of soils, and surface and ground waters, and concluded that any correlation with deeply buried oil pools was fortuitous. Langford (1962) attributed the phenomenon to transport of uranium-series decay products by rising methane. Bisir (1958) reported anomalies in Romania according to Lundberg's concept and obtained more positive results with the emanation method than with γ-ray surveys. According to Alekseyev (1959), rising hydrocarbons displace various elements, including the radioactive ones, from adsorption sites on soil grains above a petroleum deposit; the displaced ions then migrate to soil bordering the affected area and are readsorbed. Although the emanation method is a sensitive way of measuring the effects of this phenomenon, if it exists, the principle involved is migration not of radon but of its precursors.

The emanation method of prospecting for uranium is based on migration of radon and its precursors. The method appears to be used widely in the Soviet Union (Grammakov *et al.*, 1958), but only for special applications in most other countries. Běhounek (1927) observed greatly enhanced radioactivity in spring waters, soil and mine air, and the atmosphere in the vicinity of the uranium mines at St. Joachimsthal, Bohemia, and suggested that therein lay a prospecting method. Kirikov, Tverskoy, and Grammakov (1932) and Bogoslovskaya *et al.* (1932) made extensive investigations into the suitability of a prospecting method based on the concentration of radon in soil air near the surface. Developments and applications of the method are given in many subsequent papers, including those by Baranov and Gracheva (1933), Grammakov (1934), Gangloff *et al.* (1958), Grammakov *et al.* (1958), Sarcia (1958), Wennervirta and Kauranen (1960), Peacock and Williamson (1961), and Simič (1961). Baranov (1956, 1957) and Alekseyev (1957) have published manuals on uranium prospecting that give much attention to the emanation method. Gamma-ray surveys, although less specific and less sensitive than the emanation method, may also detect anomalies caused by radon migration (Ridland, 1945; Herzog, 1956). Radon migration into drill holes near uranium ore may be extensive enough to permit greater spacing of holes without decreasing the detectability of ore bodies (Tanner, 1958).

The success of many of the geophysical investigations cited suggests that substantial radon migration takes place. Few of the investigations, however, have permitted discrimination between migration of radon and that of its precursors. Migration of the several most mobile members of the uranium series, uranium-238, uranium-234, radium-226, and radon-222 is probably responsible for many anomalies, particularly those which imply movement through distances of many meters.

Radon migration may even be minor compared with migration of its precursors in some places. In a survey (Tanner, 1957) at a uranium prospect in Karnes County, Texas, the writer found a halo of enhanced radon concentration in soil gas at all sampling points within about 30 m. of a uranium ore body. Without further knowledge the halo might have been ascribed to radon migration. However, the vertical distribution of radon in the ground was inconsistent with upward radon migration of more than a few centimeters, and the horizontal gradient of radon concentration was inadequate to account for the anomaly. Displacement of the anomaly in the downhill direction, correspondence between the γ-ray intensity and radon concentrations in holes near each other, and uranium-series disequilibrium at the prospect (J. N. Rosholt, Jr., written communication) indicated that the anomaly resulted from migration of the precursors of radon.

Uranium series disequilibrium prevails in and about many uranium ores (Rosholt, 1959; Granger, 1963). Ground-water movement is common in fault zones such as are likely to be revealed by the emanation method. Anomalies arising from the migration of radon only should probably be regarded as exceptional.

ACKNOWLEDGMENTS

I wish to thank H. B. Evans for providing access to his literature search of *Chemical Abstracts* through 1955. S. T. Vesselowsky's unpublished translations of Gracheva (1938) and Baranov and Novitskaya (1949) and A. J. Shneiderov's unpublished translation of Grammakov (1936) were valuable in preparation of the review.

REFERENCES AND BIBLIOGRAPHY

In addition to references cited, the following bibliography contains many others pertinent to the migration of radioactive gases, to diffusion in porous media, and to the emanation methods. The early literature and many papers giving concentrations of radon isotopes in various media have been omitted. Some papers that essentially duplicate others by the same authors are also missing.

Titles are given in the language of the paper unless otherwise indicated.

ALEKSEYEV, F. A. 1959. Radiometricheskiy metody poiskov nefti i gaza (o prirode radiometricheskikh i radiogeokhimicheskikh anomaliy v rayone neftyanykh i gazovykh mestorozhdeniy) ("Radiometric method of oil and gas exploration: Nature of radiometric and radiogeochemical anomalies in the region of oil and gas fields"). Yadernaya Geofizika ("Nuclear Geophysics"), pp. 3–26. Moscow: Gostoptekhizdat.

ALEKSEYEV, V. V. (ed.). 1957. Radiometricheskiye metody poiskov i razvedki uranovykh rud ("Radiometric Methods in the Prospecting and Exploration of Uranium Ores"). SSSR Ministerstvo Geologii i Okhrany Nedr. Moscow: Gosgeoltekhizdat. Pp. 610. (Eng. trans., USAEC Rpt. AEC-tr-3738, Books 1 and 2, 1959.)

ALIVERTI, G., and G. LOVERA. 1949*a*. Sulla esalazione del radon dal suolo ("On the emission of radon from the soil"). Ann. Geofisica, **2**:137–41.

———. 1949*b*. Su la influenza di alcuni elementi meteorologici su la diffusione del radon nell'aria tellurica ("Influence of certain meteorological factors on the diffusion of radon into soil air"). *Ibid.*, pp. 92–102.

ALVAREZ, M. 1949. Porosidad y permeabilidad en la relacion con la inyeccion de gas en un campo petrolero ("Porosity and permeability in relation to gas injection in an oil field"). Petroleos Mexicanos, No. 72, pp. 100–111.

AMBRONN, R. 1926. Methoden der angewandten Geophysik ("Methods of Applied Geophysics"). Dresden and Leipzig: Verlag von Theodor Steinkopff. Pp. 258.

———. 1928. Elements of Geophysics, as Applied to Explorations for Minerals, Oil and Gas. New York: McGraw-Hill Book Co. Pp. 372.

AMIRKHANOFF, K. I., S. B. BRANDT, and E. N. BARTNITSKY. 1961. Radiogenic argon in minerals and its migration. Ann. New York Acad. Sci., **91** (art. 2): 235–75.

ANDERSON, J. S., D. J. M. BEVAN, and J. P. BURDEN, 1963. The behavior of recoil atoms in ionic solids. Proc. Roy. Soc. London A, **272**:15–32.

ARNDT, R. H., and P. K. KURODA. 1953. Radioactivity of rivers and lakes in parts of Garland and Hot Spring Counties, Arkansas. Econ. Geol., **48**:551–67.

BARANOV, V. I. 1956. Radiometriya ("Radiometry"). 2d ed. Rev. Moscow: Akad. Nauk SSSR Izdatel'stvo. Pp. 343. (Eng. trans., USAEC Rpt. AEC-tr-4432, 1961.) Pp. 381.

——— (ed.). 1957. Spravochnik po radiometrii dlya geofizikov i geologiv ("Handbook of radiometry for geophysicists and geologists"). Moscow: Gosgeoltekhizdat. Pp. 199.

BARANOV, V. I., and YE. G. GRACHEVA. 1933. K teorii emanatsionnoy razvedki ("On the theory of emanation prospecting"). Gosudarstvennyy Radiyevyy Inst. Trudy, No. 2, pp. 61–67.

BARANOV, V. I., and YE. G. GRACHEVA. 1937. K metodike izucheniya pronitsayemosti gornykh porod dlya radioaktivnykh emanatsiy ("On a method of study of the permeability of rocks to radioactive emanations"). Gosudarstvennyy Radiyevyy Inst. Trudy, No. 3, pp. 117–22.

BARANOV, V. I., and A. P. NOVITSKAYA. 1949. Diffuziya radona v prirodnykh gryazyakh ("Diffusion of radon in natural muds"). Akad. Nauk SSSR Biogeochim. Lab. Trudy, No. 9, pp. 161–71.

BARANOV, V. I., and YU. A. VACHNADZE. 1960. Correlation of natural radioactive emanations in the air in relation to geologic conditions in the example of areas of certain crystalline and sedimentary rocks. (In Russian.) Akad. Nauk Gruzin. SSR Inst. Geofiziki Trudy, 19:151–58.

BARKER, F. B., and R. C. SCOTT. 1958. Uranium and radium in the ground water of the Llano Estacado, Texas and New Mexico. Trans. Am. Geophys. Union, 39:459–66.

BAVER, L. D. 1956. Soil Physics. 3d ed. New York: John Wiley & Sons. Pp. 489.

BECKER, A., and K. H. STEHBERGER. 1929. Über die Adsorption der Radiumemanation ("On the adsorption of radium emanation"). Ann. d. Phys., 1:529–55.

BĚHOUNEK, F. 1927. Über die Verhältnisse der Radioaktivität in Uranpecherzbergbaurevier von St. Joachimsthal in Böhmen. Radioaktivität der Quellen, Boden- und Grubenluft und der Atmosphäre ("On the proportions of radioactivity in the uranium mining district of St. Joachimsthal in Bohemia. Radioactivity of springs, ground and mine air, and the atmosphere"). Phys. Ztschr., 28:333–42.

BELIN, R. E. 1959. Radon in the New Zealand geothermal regions. Geochim. & Cosmochim. Acta, 16:181–91.

BELLUIGI, A. 1942. Determinazione della potenza della coperture e profondità di alimentatori geologici metaniferi con misure gasometriche ("Determination of the depth and stratum thickness of geological accumulations of methane by means of gasometric measurements"). Metano 4:13.

BENDER, H. 1934. Über den Gehalt der Bodenluft an Radiumemanation ("On the content of radium emanation in ground air). Gerlands Beitr. Geophys., 41:401–15.

BISIR, D. P. 1958. Discussion of some results of experiments in radioactive prospecting for petroleum and natural gas in Romania. Proc. 2d United Nations Internat. Conf. on Peaceful Uses of Atomic Energy, 2:837–39.

BOGOSLOVSKAYA, T. N., A. G. GRAMMAKOV, A. P. KIRIKOV, and P. N. TVERSKOY. 1932. Otchet k rabote Kavgolovskoy opytno-metodicheskoy partii v 1931 g. ("Report of the work of the Kavgolovo experimental-methodical party in the year 1931"). Vses. Geol.-Razved. Ob"yedineniye Izv., 51:1283–93.

BOGOYAVLENSKIY, L. N. 1927. Radiometricheskaya razvedka nefti ("Radiometric prospecting for petroleum"). (Leningrad) Inst. Priklad.

Geofiziki Izv., No. 3, pp. 113–23. (Eng. trans. by A. SELETZKY, U.S. Bur. Mines Inf. Circ. 6072, pp. 13–18, 1928.)

BOTSET, H. G., and P. WEAVER. 1932. Radon content of soil gas. Physics, **2**:376–85.

BRANDT, S. B. 1962. Po povodu diskussi s E. K. Gerlingom ("In connection with the discussion by E. K. Gerling"). Geokhimiya, No. 12, pp. 1118–22.

BREGER, I. A. 1955. Radioactive equilibrium in ancient marine sediments. Geochim. & Cosmochim. Acta, **8**:63–73.

BUDDE, E. 1958a. Bestimmung des Beweglichkeits-koeffizienten der Radiumemanation in Lockergesteinen ("Determination of the diffusion coefficient of radon in unconsolidated rocks"). Ztschr. Geophys., **24**: 96–105.

———. 1958b. Radon measurements as a geophysical method. Geophys. Prosp., **6**:25–34.

BULASHEVICH, YU. P. 1946. K teorii interpretatsii radioaktivnykh anomaliy ("On the theory of interpretation of radioactive anomalies"). Akad. Nauk SSSR Izv., ser. geog. geofiz., **10**:469–81.

———. 1947. Diffuziya emanatsii v pochve s uchetom konvektsii ("Diffusion of emanation in the ground with a correction for convection"). *Ibid.*, **11**:93–96.

BULASHEVICH, YU. P., and R. K. KHAYRITDINOV. 1959. K teorii diffuzii emanatsii v poristykh sredakh ("On the theory of diffusion of emanation in porous media"). Akad. Nauk SSSR Izv., ser. geofiz., No. 12, pp. 1787–92.

BUMSTEAD, H. A., and L. P. WHEELER. 1904. On the properties of a radioactive gas found in the soil and water near New Haven. Am. J. Sci., **17** (ser. 4): 97–111.

CHKHENKELI, SH. M. 1953. Interpretatsiya nekotorykh vidov emanatsionnykh anomaliy ("Interpretation of certain types of emanation anomalies"). Akad. Nauk Gruzin, SSR Inst. Geofiziki Trudy, **12**: 73–81.

CLARK, R. W., and H. G. BOTSET. 1932. Correlation between radon and heavy mineral content of soils. Bull. Am. Assoc. Petrol. Geologists, **16**: 1349–56.

COPPENS, R. 1962. Radioactivité et tectonique ("Radioactivity and structure"). C.R. 86e Congrès National des Sociétés Savantes, Montpellier, 1961, Sec. des Sci., pp. 425–28.

CORRY, A. V. 1929. Radioactive Atmospherical Method of Measurement for Geophysical Prospecting. Am. Inst. Mining & Metall. Eng. Tech. Pub. No. 200. Pp. 4.

CULLEN, T. L. 1946. On the exhalation of radon from the earth. Terr. Magnet. & Atmos. Elec., **51**:37–44.

ERIKSSON, E. 1962. Radioactivity in hydrology. *In* H. ISRAËL and A. KREBS (eds.), Nuclear Radiation in Geophysics, pp. 47–60. New York: Academic Press.

EVE, A. S., and D. A KEYS. 1929. Radioactive and other methods. *In* Applied Geophysics in the Search for Minerals, pp. 210–18. Cambridge: Cambridge Univ. Press.

FAUL, H. 1954. Helium, argon, and radon. *In* H. FAUL (ed.), Nuclear Geology, pp. 133–43. New York: John Wiley & Sons.

FLÜGGE, S., and K. E. ZIMENS. 1939. Die Bestimmung von Korngrössen und Diffusionskonstanten aus dem Emaniervermögen. Die Theorie der Emaniermethode ("The determination of grain size and diffusion constant by the emanating power. The theory of the emanation method"). Ztschr. phys. Chemie, **B42**:179–220.

GANGLOFF, A. M., C. R. COLLIN, A. GRIMBERT, and H. SANSELME. 1958. Application of the geophysical and geochemical methods to the search for uranium. Proc. 2d United Nations Internat. Conf. on Peaceful Uses of Atomic Energy, **2**:140–47.

GARRELS, R. M., R. M. DREYER, and A. L. HOWLAND. 1949. Diffusion of ions through intergranular spaces in water-saturated rocks. Geol. Soc. America Bull., **60**:1809–28.

GERLING, E. K., and I. M. MOROZOVA. 1962. Opredeleniye spektra znacheniy energii aktivatsii vydeleniya argona i geliya iz mineralov ("Determination of the spectrum of values of activation energy for the liberation of argon and helium from minerals"). Geokhimiya, No. 12, pp. 1108–18.

GILETTI, B. J., and J. L. KULP. 1955. Radon leakage from radioactive minerals. Am. Mineralogist, **40**:481–96.

GOCKEL, A. 1903. Über die Emanation der Bodenluft ("On the emanation of soil air"). Phys. Ztschr., **4**:604–5.

GORSHKOV, G. V. 1933. K voprosu o vliyanii narusheniy v nanose i ego sloistosti na skorost' rasprostraneniya v nem emanatsii radiya ("On the effect of disturbances and stratification on the speed of radium emanation migration in alluvium"). Zhur. Geofiziki, **3**:292–98.

GORSHKOV, G. V., A. G. GRAMMAKOV, I. YE. STARIK, and V. A. SHPAK. 1940. Nekotoryye dannyye o fizicheskikh osnovakh radioaktivnykh geofizicheskikh metodov razvedki ("Some data on the physical foundations of radioactive geophysical methods of exploration"). 17th Internat. Geol. Cong., Moscow, 1937, Repts., **4**:495–97.

GRACHEVA, YE. G. 1938. Vliyaniye struktury i poristosti porod na diffuziyu radioaktivnykh emanatsiy ("Influence of structure and porosity of rocks on the diffusion of radioactive emanations"). Gosudarstvennyy Radiyevyy Inst. Trudy, **4**:228–33.

GRAMMAKOV, A. G. 1934. Emmanatsionnyy (radonovyy) metod poiskov, issledovaniya i razvedki radioaktivnykh ob"yektov ("Emanation [radon] method of search, investigation, and prospecting for radioactive objects"). Tsentr. Nauchno-Issled. Geol.-Razved. Inst. Trudy, No. 7. Pp. 48.

———. 1936. O vliyanii nekotorykh faktorov na rasprostraneniye radioaktivnykh emanatsiy v prirodnykh usloviyakh ("On the influence of some factors in the spreading of radioactive emanations under natural conditions"). Zhur. Geofiziki, **6**:123–48.

———. 1937. Povedenie radona v pochvennykh kapillyarakh vblizi dnevnoy poverkhnosti ("The behavior of radon in soil capillaries near the surface of the earth"). Tsentr. Nauchno-Issled. Geol.-Razved. Inst. Materialy, Geofizika, No. 4, pp. 32–40.

———. 1961*a*. K teorii emanatsionnogo metoda dlya ob"yektov lokal'nogo tipa ("On the theory of the emanation method for objects of a local kind"). *In* Voprosy Rudnoy Geofiziki ("Problems of Mining Geophysics"). Moscow, Vses. Nauchno-Issled. Inst. Razved. Geofiziki, No. 2, pp. 135–48.

———. 1961*b*. Nekotoryye voprosy teorii emanatsionnogo metoda ("Some problems of the theory of the emanation method"). *In* Voprosy Rudnoy Geofiziki ("Problems of Mining Geophysics"). Moscow, Vses. Nauchno-Issled. Inst. Razved. Geofiziki, No. 2, pp. 86-93.

GRAMMAKOV, A. G., N. V. KVASHNEVSKAYA, A. I. NIKONOV, M. M. SOKOLOV, N. N. SOCHEVANOV, S. A. SUPPE, and G. P. TAFEYEV. 1958. Some theoretical and methodical problems of radiometric prospecting and survey. Proc. 2d United Nations Internat. Conf. on Peaceful Uses of Atomic Energy, **2**:732–43.

GRAMMAKOV, A. G., and N. M. LYATKOVSKAYA. 1935. O diffuzii radio-aktivnykh emanatsiy v gornykh porodakh ("On the diffusion of radioactive emanations in rocks"). Zhur Geofiziki, **5**:290–306.

GRAMMAKOV, A. G., and I. F. POPRETINSKIY. 1957. Raspredeleniye radona v rykhlykh otlozheniyakh pri nalichii oreolov rasseyaniya radiya ("Dis-tribution of radon in porous deposits in the presence of aureoles of dispersed radium"). Akad. Nauk SSSR Izv., ser. geofiz., No. 6, pp. 789–93. (Eng. trans. by R. E. DAISLEY, pp. 113–19. New York: Perga-mon Press.)

GRANGER, H. C. 1963. Radium migration and its effect on the apparent age of uranium deposits at Ambrosia Lake, New Mexico. U.S. Geol. Survey Prof. Paper 475-B, pp. B60–B63.

GREGORY, A. F. 1956. Analysis of radioactive sources in aeroradiometric surveys over oil fields. Bull. Am. Assoc. Petr. Geologists, **40**:2457–74.

GREGORY, J. N., and S. MOORBATH. 1951*a*. The diffusion of thoron in solids. Part I. Investigations on hydrated and anhydrous alumina at elevated temperatures by means of the Hahn emanation technique. Faraday Soc. Trans., **47**:844–59.

———. 1951*b*. The diffusion of thoron in solids. Part II. The emanating power of barium salts of the fatty acids. *Ibid.*, pp. 1064–72.

HAHN, O. 1936. Applied Radiochemistry. Ithaca, N.Y.: Cornell Univ. Press. Pp. 278.

HARRIS, S. J. 1954. Radon levels in mines in New York State. Archives Ind. Hyg. Occupational Med., **10**:54–60.

HATAYE, I. 1962. Radiochemical studies on radioactive mineral springs. III. Leaching experiment of radium from basalts and sinter-deposit.

Shizuoka Univ. Liberal Arts and Sci. Fac. Rpts., Nat. Sci. Sec., **3**: 135–43.

HATUDA, Z. 1953. Radon content and its change in soil air near the ground surface. Kyoto Univ. Coll. Sci. Mem., Ser. B., **20**:285–307.

HAYASE, I., and T. TSUTSUMI. 1958. On the emanating power of powdered rocks. Kyoto Univ. Coll. Sci. Mem., Ser. B., **24**:319–23.

HERZOG, G. 1956. Geophysical prospecting by use of radioactivity surveying. Mines Mag. (Colorado), **66**:25–28.

HILPERT, L. S., and C. M. BUNKER. 1957. Effects of radon in drill holes on gamma-ray logs. Econ. Geol., **52**:438–55.

HINAULT, J. 1961. Prospection des minerals radioactifs en Guyane Française ("Prospecting for radioactive minerals in French Guiana" [with English Abstract]). Proc. 5th Inter-Guiana Geol. Conf., Georgetown, British Guiana, 1959, pp. 239–46.

HIRST, W., and G. E. HARRISON. 1939. The diffusion of radon gas mixtures. Proc. Roy. Soc. London A, **169**:573–86.

HOLADAY, D. A., D. E. RUSHING, R. D. COLEMAN, P. F. WOOLRICH, H. L. KUSNETZ, and W. F. BALE. 1957. Control of Radon and Daughters in Uranium Mines and Calculations on Biologic Effects. U.S. Public Health Service Pub. 494. Pp. 81.

HOOGTEIJLING, P. J., G. J. SIZOO, and J. L. YNTEMA. 1948. Measurements on the radon content of groundwater. Physica, **14**:73–80.

HOWELL, L. G. 1934. Radioactivity of soil gases. Bull. Am. Assoc. Petrol. Geologists, **18**:63–68.

HRADIL, G. 1955. Zur Messung des Emanationsgehaltes der Bodenluft über Strukturlinien ("On the measurement of the emanation content of soil air over structural lines"). Berg. u. Hüttenmänn. Monatsh., **100**: 145–47.

ISRAËL, H. 1934. Emanation in Boden- und Freiluft ("[Radium] Emanation in ground and atmospheric air"). Ztschr. Geophys., **10**:347–56.

———. 1959. Der Diffusionskoeffizient des Radons in der Bodenluft (Bemerkungen zur Arbeit von E. BUDDE, Bestimmung des Beweglichkeits-Koeffizienten der Radium-Emanation in Lockergesteinen) ("The diffusion coefficient of radon in soil air [Remarks on E. BUDDE, Determination of the coefficient of diffusion of radium emanation in porous rocks]"). *Ibid.*, **25**:104–8.

———. 1961. Der Diffusions-Koeffizient des Radons in Bodenluft ("The diffusion coefficient of radon in soil air"). *Ibid.*, **27**:13–17.

———. 1962. Die natürliche und künstliche Radioaktivität der Atmosphäre ("The natural and artificial radioactivity of the atmosphere"). *In* H. ISRAËL and A. KREBS (eds.), Nuclear Radiation in Geophysics, pp. 76–96. New York: Academic Press.

ISRAËL, H., and F. BECKER. 1935*a*. Die Bodenemanation in der Umgebung der Bad Nauheimer Quellenspalte ("Ground emanation in the environment of the fissure of the Bad Nauheim spring"). Gerlands Beitr. Geophys., **44**:40–55.

——. 1935*b*. Emanationsgehalt der Bodenluft und Untergrundtektonik ("Emanation content of soil air and underground structure"). Natur-wiss., **23**:818.

ISRAËL, H., S. BJÖRNSSON, and S. STILLER. 1962. Emanometrische Messungen von Radon und Thoron in Bodenluft ("Emanometric measurements of radon and thoron in soil air"). Ann. Geofisica, **15**:115–26.

ISRAËL-KÖHLER, H., and F. BECKER. 1936. Die Emanationsverhältnisse in der Bodenluft ("The proportion of emanation in ground air"). Gerlands Beitr. Geophys. **48**:13–58.

IWASAKI, I., T. KATSURA, H. SHIMOJIMA, and M. KAMADA. 1956. Radioactivity of volcanic gases in Japan. Bull. Volcanol., **18** (ser. 2): 103–23.

JAKI, S. L., and V. F. HESS. 1958. A study of the distribution of radon, thoron, and their decay products above and below the ground. Jour. Geophys. Res., **63**:373–90.

JURAIN, G. 1962. Contribution à la connaissance géochimique des familles de l'uranium-radium et du thorium dans les Vosges méridionales: Application de certains résultats en prospection des gisements d'uranium ("Contribution to the Geochemical Knowledge of the Families of Uranium-Radium and of Thorium in the Southern Vosges: Application of Some Results to Prospecting of Deposits of Uranium"). Sci. de la Terre Mém. 1. Pp. 349.

KAPITANOV, YU. T., and A. S. SERDYUKOVA. 1962. K raschetu kolichestva vozdukha, neobkhodimogo dlya provetrivaniya gornykh vryabotok uranovykh rudnikov ("On the calculation of the amount of air necessary for ventilation of uranium ore mining operations"). Vyssh. Ucheb. Zavedeniy Izv. Geologiya i Razvedka, No. 6, pp. 112–20.

KHAYKOVICH, I. M. 1961. Raspredeleniye radona v rudnom plaste peresechenom tsilindricheskoy skvazhinoy ("The distribution of radon in an ore layer cut by a cylindrical borehole"). *In* Voprosy Rudnoy Geofiziki ("Problems of Mining Geophysics"). Moscow, Vses. Nauchno-Issled. Inst. Razved. Geofiziki, No. 2, pp. 94–101.

KHAYKOVICH, I. M., and L. A. KHALFIN. 1961. Izmeneniye intensivnosti gamma-izlucheniya radioaktivnogo poluprostranstva pri zakrytii yego gazonepronitsayemoy poverkhnostyu ("The change in gamma-ray intensity of a radioactive half-space due to its sealing by a surface impervious to gas"). *In* Voprosy Rudnoy Geofiziki ("Problems of Mining Geophysics"). Moscow, Vses. Nauchno-Issled. Inst. Razved. Geofiziki, No. 2, pp. 131–34.

KIKKAWA, KYOZO. 1954. Study on radioactive springs. Japanese J. Geophys., **1**:1–25.

KIMURA, K. 1949. Geochemical studies on the radioactive springs in Japan. Proc. 7th Pacific Sci. Assoc. Cong., **2**:485–99.

KIRIKOV, A. P., T. N. BOGOSLOVSKAYA, and G. V. GORSHKOV. 1932. Emaniruyushchaya sposobnost' rud i gornykh porod Tabosharskogo urano-radiyevogo mestorozhdeniya ("Emanating power of ores and rocks of the Taboshar uranium-radium deposit"). Vses. Geol.-Razved. Ob"yedineniye Izv., **51**:1293–99.

KIRIKOV, A. P., P. N. TVERSKOY, and A. G. GRAMMAKOV. 1932. K voprosu ob emanatsionnom metode poiskov radioaktivnykh o"yektov ("On the question of the emanation method of searching for radioactive deposits"). Vses. Geol.-Razved. Ob"yedineniye Izv., **51**:1269–82.

KOENIGSBERGER, J. 1928. Mächtigkeitsbestimmung von Deckschichten über Spalten durch Radioaktivitätsmessungen ("Determination of overburden thickness above faults by radioactivity measurements"). Ztschr. Geophys., **4**:76–83.

KOFLER, M. 1913. Löslichkeit der Ra-Emanation in wässerigen Salzlösungen ("Solubility of radium emanation in aqueous salt solutions"). Akad. Wiss. Wien, Math.-Naturw. Kl., Sitzungsber., Part IIa, **122**:1473–79.

KOSMATH, W. 1934. Die Exhalation der Radiumemanation aus dem Erdboden und ihre Abhängigkeit von den meteorologischen Faktoren. Part II. ("The exhalation of radium emanation from the ground and its dependence upon meteorological factors.") Gerlands Beitr. Geophys. **43**:258–79.

KOSOV, N. D., and V. V. CHERDYNTSEV. 1955. Emanirovaniye mineralov i opredeleniye absolyutnogo geologicheskogo vozrasta ("Emanation from minerals and the determination of absolute geologic age"). Akad. Nauk SSSR Komiss. opredel. absolyut. vozrasta geol. formatsiy Byull., vypusk 1, pp. 22–28.

KOVACH, E. M. 1944. An experimental study of the radon-content of soil-gas. Trans. Am. Geophys. Union, Part IV, **25**:563–71.

———. 1945. Meteorological influence upon the radon-content of soil-gas. *Ibid.*, Part II, **26**:241–48.

———. 1946. Diurnal variations of the radon-content of soil-gas. Terr. Magnet. & Atmos. Elec., **51**:45–46.

KRANER, H. W., G. L. SCHROEDER, and R. D. EVANS. 1963. Measurements of the effects of atmospheric variables on radon-222 flux and soil-gas concentrations. This symposium.

LANE, A. C., and W. R. BENNETT. 1934. Location of a fault by radioactivity. Beitr. angew. Geophys., **4**:353–57.

LANGFORD, G. T. 1962. Radiation surveys aid oil search. World Oil, **154**:114–19.

LAUTERBACH, R. 1953. Zur Frage tektonischer Untersuchungen mit Hilfe emanometrischer Messungen ("On the question of structural investigations by means of emanometric measurements"). Karl Marx Univ. (Leipzig) wiss. Ztschr. math-naturwiss. Reihe, **3**:291–92.

LECOQ, J. J., G. BIGOTTE, J. HINAULT, and J. R. LECONTE. 1958. Prospecting for uranium and thorium minerals in the desert countries and in the equatorial forest regions of the French Union. Proc. 2d United Nations Internat. Conf. on Peaceful Uses of Atomic Energy, **2**:744–86.

LINDNER, R., and H. MATZKE. 1960. The diffusion of radon in oxides after recoil doping. (In German.) Ztschr. Naturforsch., **15a**:1082–86.

LÖSER, G. 1959. Radioaktive Bodenluftmessungen als Beitrag zur Klärung tektonischer Probleme am Südwestrand des Thüringer Waldes ("Radioactive soil air measurements as a contribution to clarification of structural problems on the southwest border of the Thüringer Wald"). Geophys. & Geol. No. 1, pp. 97–103.

LORENZ, P. J., O. C. RODENBERG, L. G. SHALDE, A. C. ANTES, and W. D. HESS. 1961. Background radioactivity in the Decorah fault region. Proc. Iowa Acad. Sci., **68**:397–403.

LUNDBERG, H. 1952. Airborne radioactivity surveys. Oil and Gas J., **50**: 165–66.

MÁRTON, P., and L. STEGENA. 1962. On the basic principles of geophysical radioactive methods. Geofis. Pura e Appl., **53**:55–64.

MOSES, H., H. F. LUCAS, JR., and G. A. ZERBE. 1962. The effect of meteorological variables upon radon concentration three feet above the ground. Argonne Nat. Lab., Radiological Phys. Div., Semiannual Rpt. ANL-6474, pp. 58–72.

MOSES, H., A. F. STEHNEY, and H. F. LUCAS, JR. 1960. The effect of meteorological variables upon the vertical and temporal distributions of atmospheric radon. J. Geophys. Res., **65**:1223–38.

NICOLAYSEN, L. O. 1957. Solid diffusion in radioactive minerals and the measurement of absolute age. Geochim. & Cosmochim. Acta, **11**:41–59.

NORINDER, H., A. METNIEKS, and R. SIKSNA. 1953. Radon content of the air in the soil at Uppsala. Arkiv Geofys., **11**:571–79.

NUSSBAUM, E. 1962. Diffusion of Radon and Tritium through Semi-permeable Materials. Upland, Indiana, Taylor Univ. Technical Progess Rpt. TID-15160. Pp. 27.

OCHIAI, T. 1951. Radioactive exploration on the faults. (In Japanese with English summary.) Geophys. Explor., **4**:78–83.

OKABE, S. 1956a. On some relations between the hot spring and radio-activity. Kyoto Univ. Coll. Sci. Mem., Ser. A, **28**:39–71.

———. 1956b. Time variation of the atmospheric radon-content near the ground surface with relation to some geophysical phenomena. *Ibid.*, pp. 99–115.

PEACOCK, J. D., and R. WILLIAMSON. 1961. Radon determination as a prospecting technique. Trans. Inst. Mining & Metall. (London), **71** (Part II): 75–85, and discussions, Part V, pp. 271–77 (1962), and Part VIII, pp. 497–98 (1962).

PENMAN, H. L. 1940a. Gas and vapour movements in the soil. I. The diffusion of vapours through porous solids. J. Agr. Sci., **30**:437–62.

———. 1940b. Gas and vapour movements in the soil. II. The diffusion of carbon dioxide through porous solids. *Ibid.*, pp. 570–81.

POPRETINSKIY, I. F. 1961. Opredeleniye koeffitsiyentov diffuzii radona i emaniruyushchey sposobnosti gornykh porod po krivym emanatsionnogo zondirovaniya ("Determination of the coefficients of diffusion of radon and of the emanating power of rocks from emanation sounding curves"). *In* Voprosy Rudnoy Geofiziki ("Problems of Mining Geophysics"), Moscow, Vses. Nauchno-Issled. Inst. Razved. Geofiziki, No. 2, pp. 105–14.

PRADEL, J. 1956. La prospection de l'uranium par le radon ("Prospecting for uranium by radon"). Comm. Énergie Atomique Rpt. 588. Pp. 8.

RAMSTEDT, E. 1919. Diffusion der Radiumemanation in Wasser ("Diffusion of radium emanation in water"). Chem. Zentralbl., **90**:993.

RIDLAND, G. C. 1945. Use of the Geiger-Müller counter in the search for pitchblende-bearing veins at Great Bear Lake, Canada. Trans. Am. Inst. Mining & Metall. Eng., **164**:117–24.

ROGERS, A. S. 1958. Physical behavior and geologic control of radon in mountain streams. U.S. Geol. Survey Bull. 1052-E, pp. 187–211.

RÓNA, E. 1917. Diffusiongrösse und Atomdurchmesser der Radiumemanation ("Magnitude of diffusion and atomic dimensions of radium emanation"). Ztschr. phys. Chemie., **92**:213–18.

ROSEN, R. 1957. Note on some observations of radon and thoron exhalation from the ground. New Zealand J. Sci. Tech., **38** (Sec. B): 644–54.

ROSHOLT, J. N., JR. 1959. Natural radioactive disequilibrium of the uranium series. U.S. Geol. Survey Bull. 1084-A, pp. 1–30.

SAKAKURA, A. Y., C. LINDBERG, and H. FAUL. 1959. Equation of continuity in geology with applications to the transport of radioactive gas. U.S. Geol. Survey Bull. 1052-I, pp. 287–305.

SARCIA, J. A. 1958. The uraniferous province of northern Limousin and its three principal deposits. Proc. 2d United Nations Internat. Conf. on Peaceful Uses of Atomic Energy, **2**:578–91.

SATTERLY, J. 1912. A study of the radium emanation contained in the air of various soils. Proc. Cambridge Phil. Soc., **16**:336–55.

SCHMELING, P., and F. FELIX. 1961. Experimental Methods and Equipment for Diffusion Measurements of Radioactive Rare-Gases in Solids (Rare-Gas Diffusion in Solids 7). Berlin, Hahn-Meitner Inst. für Kernforschung Rpt. HMI-B-19. Pp. 42.

SHARKOV, YU. V. 1959. Poiski mestorozhdeniy atomnogo syr'ya ("Exploration for deposits of atomic raw materials"). Priroda, **48**:13–21.

SHEVCHENKO, N. F. 1958. K voprosu ob izmenenii soderzhaniya emanatsii v pochvennom vozdukhe ("On the problem of variation of the emanation content of soil air"). Sredneaziatskiy Univ. Trudy aspirantov. No. 5, pp. 69–74.

SIMIČ, V. 1961. Ispitivanje kontinuiteta i uranonosnosti tektoniskih zona primenom metoda specifičnog električnog otpora i radioaktivne emanacije ("Investigation of the continuity and uranium possibilities of structural zones by the methods of specific electrical resistivity and radioactive emanations." With English summary). (Belgrade) Zavod Nukl. Sirovine Sek Ist. Nukl. i Dr. Mineral Sirovina Radovi, **1**:97–104.

SMYTH, L. B. 1912. On the supply of radium emanation from the soil to the atmosphere. Philos. Mag., **24**:632–37.

STARIK, I. YE. 1943. Forma nakhozhdeniya i usloviya pervichnoy migratsii radioelementov ("Form of occurrence and conditions of primary migration of radioelements"). Uspekhi Khimii, **12**:287–307.

STARIK, I. YE., and O. S. MELIKOVA. 1941. Migration of ionium under natural conditions. (In English.) Akad. Nauk SSSR Doklady, **31**: 911–13.

————. 1957. Emanating power of minerals. Radiyevyy Institut imeni V. G. Khlopina Trudy, **5**:184–202. (Eng. trans. U.S. Atomic Energy Comm. Rpt. AEC-tr-4498, pp. 206–26.)

STCHEPOTJEVA, E. S. 1944. On the conditions under which natural waters become enriched with radium and its isotopes. (In English.) Akad. Nauk SSSR Doklady, **43**:306–9.

STERRETT, E. 1944. Radon emanations outline formations. Oil Weekly, **115**:29–32.

STOTHART, R. A. 1948. Tracing wildcat trends with radon emanations. World Oil, **127**:78–79.

TANNER, A. B. 1957. Physical behavior of radon. *In* U.S. Geol. Survey, Geologic Investigations of Radioactive Deposits, Semiannual Progress Report, June 1 to November 30, 1957. U.S. Geol. Survey TEI-700, pp. 243–46.

————. 1958. Increasing the efficiency of exploration drilling for uranium by measurement of radon in drill holes. Proc. 2d United Nations Internat. Conf. on Peaceful Uses of Atomic Energy, **3**:42–45.

————. 1959. Meteorological influence on radon concentration in drill-holes. Mining Eng., **11**:706–8.

————. 1963. Physical and chemical controls on distribution of radium and radon in ground water near Great Salt Lake. This symposium.

TEUSCHER, E. O., and E. BUDDE. 1957. Emanationsmessungen im Nab-burger Fluss-spätrevier. I. Geologischer Teil: Neuere Untersuchungen auf Bayerischen Fluss-spätvorkommen. II. Geophysikalischer Teil: Rn-Messungen auf Fluss-spätgangen ("Emanation measurements in the Nabburger fluorspar district. I. Geological part: New investigations on the Bavarian fluorspar occurrences. II. Geophysical part: Radon measurements on fluorspar veins"). Geol. Bavarica, No. 35. Pp. 59.

TIMOFEYEV, A. N. 1959. K teorii gamma-razvedki ("On the theory of gamma prospecting"). Akad. Nauk SSSR Izv., ser. geofiz., No. 12, pp. 1873–75.

TODD, D. K. 1959. Ground Water Hydrology. New York: John Wiley & Sons. Pp. 336.

TOKAREV, A. N., and A. V. SHCHERBAKOV. 1956. Radiohydrogeology. Moscow: Gosgeoltekhizdat. (Eng. trans., U.S. Atomic Energy Comm. Rpt. AEC-tr-4100 [1960]. Pp. 346.)

TROMBE, F., and C. HENRY LA BLANCHETAIS. 1947. Ionisation: Sur la présence de radiations pénétrants telluriques dans la rivière souterraine de Saint-Paul, Haute-Garonne ("Ionization: On the presence of a penetrating telluric radiation in the subterranean river of Saint-Paul, Haute-Garonne"). C. R. Acad. Sci. (Paris), **224**:207–9.

VINCENZ, S. A. 1959. Some observations of gamma radiation emitted by a mineral spring in Jamaica. Geophys. Prosp., **7**:422–34.

VINOGRADOV, A. P. 1962. K diskussii o mekhanizme vydeleniya radio-gennykh gazov ("To the discussion on the mechanism of liberation of radiogenic gases"). Geokhimiya, No. 12, p. 1108.

VOGT, W. 1935. Radiologische Untersuchungen in Radiumbad Brambach ("Radiological investigations in the radium springs of Brambach"). Ztschr. Geophys., 11:29–35.

WENNERVIRTA, H., and P. KAURANEN. 1960. Radon measurement in uranium prospecting. (Finland) Comm. Géol. Bull. 188, pp. 23–40.

WILKENING, M. H., and J. E. HAND. 1960. Radon flux at the earth-air interface. J. Geophys. Res., 65:3367–70.

WILLIAMS, W. J., and P. J. LORENZ. 1957. Detecting subsurface faults by radioactive measurements. World Oil, 144:126–28.

WRIGHT, J. R., and O. F. SMITH. 1915. The variations with meteorological conditions of the amount of radium emanation in the atmosphere, in the soil gas, and in the air exhaled from the surface of the ground, at Manila. Phys. Rev., 5 (ser. 2): 459–82.

ZEILINGER, P. R. 1935. Ueber die Nachlieferung von Radiumemanation aus dem Erdboden ("On the release of radium emanation from the ground"). Terr. Magnet. & Atmos. Elec., 40:281–94.

ZIMENS, K. E. 1943. Surface area determinations and diffusion measurements by means of radioactive inert gases. Part III. The process of radioactive emission from disperse systems. Conclusions pertaining to the evaluation of EP (emanating power) measurements and the interpretation of results. (In German.) Ztschr. phys. Chemie. 192:1–55. (Eng. trans., East Orange, N.J., Assoc. Tech. Services, No. 43G7G [1955]. Pp. 38.)

ZUPANCIC, P. R. 1934. Messungen der Exhalation von Radiumemanation aus dem Erdboden ("Measurements of the exhalation of radium emanation from the ground"). Terr. Magnet. & Atmos. Elec., 39:33–46.

HOBART W. KRANER, GERALD L. SCHROEDER,
AND ROBLEY D. EVANS

10. *Measurements of the Effects of Atmospheric Variables on Radon-222 Flux and Soil-Gas Concentrations*

THE EFFECTS of meteorological variables on radon-222 flux and radon-222 concentrations (3.82-day half-life) in the soil gas were studied at two sampling sites as a prerequisite for a program of radon-222 measurements to determine the feasibility of using the effect of underground nuclear detonations on this natural radiation as a possible on-site inspection tool (Evans *et al.*, 1962). The experience gained from this study has provided a background for understanding the mechanisms that affect underground radon-222 concentrations and radon-222 flux at the earth-air interface and that may also be applicable to the control of radon-222 exhalation and concentration in deep mines.

SAMPLING LOCATIONS

Extensive measurements of radon-222 concentration in soil gas and of radon-222 flux at the earth-air interface were conducted at two sites having extremely different soil and climatological conditions: at an unworked orchard in Lincoln, Massachusetts, and on the Rainier Mesa at the U.S. Atomic Energy Commission's Nevada Test Site (NTS). The Lincoln site provided a soil of glacial debris covered by a mat of weeds. The soil was always moist, and in the spring the water table reached to 16 feet below ground level. Soil-gas sampling at depths of 3, 5, and 8 feet from permanent installations was conducted daily from March to September, 1961, and thereafter less regularly.

HOBART W. KRANER is DSR staff member, GERALD L. SCHROEDER is research assistant in geology, and ROBLEY D. EVANS is professor of physics, Department of Physics, Massachusetts Institute of Technology, Cambridge, Massachusetts.

Work supported in part by AEC Contract AT(30-1)-952 and by AFTAC Contract AF49(638)-1110.

Intermittent sampling to depths of 16 feet, as well as in shallow depths (6–15 inches), was also carried out. Flux measurements were made at this site after June, 1962. The Rainier Mesa site, area U12e.04 ("Orchid" site), consisted of 90 inches of weathered tuff overlying a welded-tuff base. The soil was normally dry except following precipitation, which occurs predominantly during the winter. Ground cover

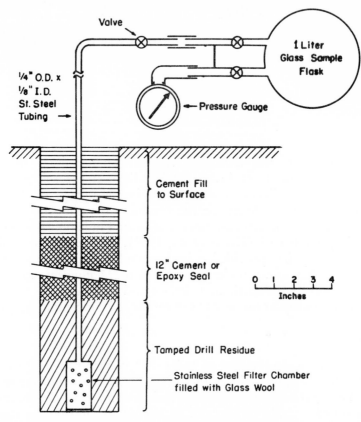

Fɪɢ. 1.—Permanent installation of soil-gas sampling tube showing sample collection flask attached.

was light. Soil gas at depths ranging from 20 to 96 inches was sampled three times per week from January through September, 1961. At other areas in the NTS, brief periods of intensive measurements of radon-222 concentration in soil gas were conducted from September, 1961, through May, 1962. Soil types encountered in these areas were granite, welded tuff, and relatively homogeneous alluvium. Flux measurements were carried out concurrently with all soil-gas sampling from January to August, 1962.

Radon-222 Concentration on Soil-Gas Samples

SAMPLING PROCEDURE

Soil-gas samples were drawn from depths into an evacuated 1-liter glass sample flask, using a ⅛-inch-inner-diameter stainless-steel tube sealed into the ground. Figure 1 illustrates the complete sampling system. Holes were drilled by several types of mechanical drilling rigs, but they were all about 4–5 inches in diameter. After the tube was inserted, 6 inches of drilling residue were repacked about the

PLATE I

Flux sampling container installed on Yucca Flat, NTS

lower end of the tube, followed by a foot of either cement or non-brittle epoxy, and then cement to the surface. The cement and epoxy used for sealants were checked by γ-ray counting in the Radioactivity Center COBAFAC (Controlled Background Facility) to assure low intrinsic radium-226 content and were further tested for low radon-222 adsorptivity. Withdrawal of the sample generally took about 20 seconds, and the attached pressure gauge assured completion of the sampling. The filter chamber on the bottom of the tube was filled with glass wool to prevent dirt particles from being drawn up into the sampling flask. The absolute volume of each flask was measured and used in computations, since it was found that the volume of commercially available 1-liter flasks varied by as much as 4 per cent.

For concentration samples from temporary locations and at depths less than 24 inches, $\frac{1}{8}$-inch-inner-diameter, thick-walled steel pipes were driven into the ground by hand, using no sealant. Shallow-depth samples were withdrawn into 100-ml. flasks to reduce the volume of soil from which gas is drawn and to define more precisely the depth at which the sample is taken.

SAMPLE MEASUREMENTS

One-liter soil-gas samples were analyzed for radon-222 in a lucite-walled cylindrical chamber of our own design, coated internally with zinc sulfide (silver activated), shown in Plate II. These chambers are 12-cm. inside diameter, are 4 cm. high, and have a volume of 450 cc. They are viewed by two 5-inch photomultipliers, one at each cylinder face, with outputs summed and counted by conventional apparatus, using an integral discriminator. The radon-222 sample was introduced into the previously evacuated chamber through a copper-tubing vacuum manifold that connects with the chamber by means of a vacuum fitting in the cylinder wall. This same inlet to the chamber was also used to introduce a rod on which was cemented an α source for daily comparative system evaluation.

Silver-painted spiderweb configurations on the outside cylindrical faces of the chambers were connected to aluminum-foil shields around the perimeter of the chamber and to the aluminum-foil electrostatic shield around the photomultiplier (PM) tubes; all electrostatic shields were grounded. Mu-metal shields were also used on the PM tubes. Positive high voltage with the PM photocathode very nearly at ground was used on all tubes. To maintain precision for the large number of samples analyzed, sample manipulation was simplified. The sample was introduced into the chamber from the 1-liter glass collection flask by volume sharing with the flask and the evacuated chamber. The chamber was then returned to atmospheric pressure with oil-pumped nitrogen. The portion of the sample remaining in the flask was saved for duplicate analyses. The use of the external electrical grounds and operation at atmospheric pressure (which reduces the counting efficiency) considerably reduced the possibility that electrostatic fields would develop within the chamber, which would cause uneven daughter-product distribution and, therefore, variable counting efficiency. Variations of the relative counting efficiencies of the radon-222 daughter products will, if present, produce changes in the

PLATE II

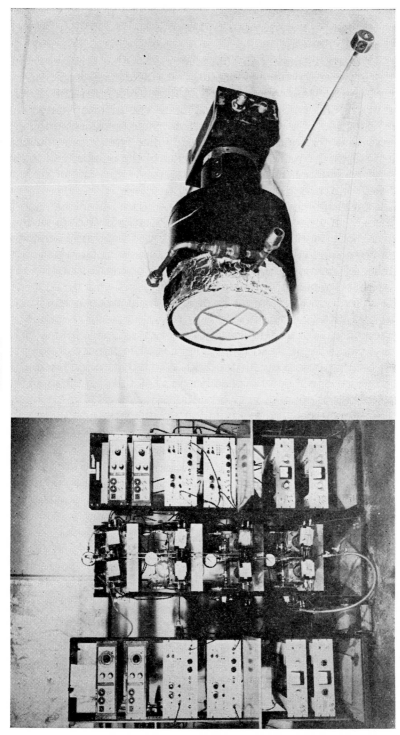

Radon-222 scintillation counter and electronics used to measure radon-222 concentration in soil gas

shape of the daughter-activity build-up curve after sample introduction into the chamber. Build-up curves are recorded during the calibrations of the chambers and are compared with a previous standard curve to assure reproducibility. The relative counting efficiency for radon-222 (Rn), polonium-218 (RaA), and polonium-214 (RaC') α rays was found from an empirical fit to the build-up curve to be 1.00:2.11:2.14. This is qualitatively reasonable, considering the α-ray ranges and deposition sites (Lucas, 1957). Backgrounds of new chambers are about 1 c.p.m.; however, because of the number and strength of samples counted, the backgrounds during two years of operation rose to 20 c.p.m.

Absolute calibration is achieved by simultaneously introducing into each chamber a portion of a 2-liter sample of radon-222 in air drawn from a National Bureau of Standards (NBS) nanocurie radium-226 standard in solution. The absolute efficiency varies from chamber to chamber by no more than 10 per cent and, for a representative chamber with a sample in transient equilibrium, is found to be 0.720 ± 0.004 counts/α ray, where the error is the standard error from 28 calibrations over a 2-year period.

Daily checks for system gain shifts, etc., were made by monitoring the α-ray activity of a lead foil, having a surface activity of polonium-210 (RaF) supported by 20-year lead-210 (RaD) within the foil. The foil was introduced into the chamber on a rod through the filling hole. The source was counted in two positions, facing each PM individually. Day-to-day stability was found in this manner to be within 5 per cent. Fluctuations greater than this value were remedied by routine maintenance. For PM stability, the PM high voltage remained on during insertion and extraction of this source (room lights were lowered).

Radon-222 Flux Measurements

SAMPLING PROCEDURE

Radon-222 flux was measured by determining the rate of build-up of radon-222 in a container, 22 inches in diameter and 33 inches high, which was sealed over the ground surface. Installation of this apparatus is shown in Plate I. A stainless-steel rim on which the container was to be mounted was driven into the ground to a depth of 5 inches. This rim was used to achieve a good seal with the soil surface and to define accurately the sampling-surface area. A polyethylene container was fitted over the rim and sealed to it by black electrical tape and by a screw-tightened, inch-wide steel band. Two vacuum-valve inlets provided access to the container; one was used for a vent to the atmosphere, and the other for withdrawal of the radon-222 sample. The vent valve was opened before the container was placed around the rim and remained open during the entire sample collection. This vent was found to be necessary to maintain gross-pressure equilibrium with the free atmosphere. Experiments using two microbarovariographs, one within the polyethylene flux container and one outside the container, showed that micro-oscillations of barometric pressure are not significantly attenuated by the flexible walls of the container. In areas of intense insolation (e.g., Nevada Test Site), the container was covered with an aluminum reflector to reduce heating of the inclosed air. Internal temperatures were regularly monitored, and an equilibrium rise above outside temperature of less than 6° C. was normally attained within 1 hour after installation. One-liter air samples were withdrawn from the container at 45-minute intervals for 3 hours after installation. Since the container volume was approximately 220 liters, the slow withdrawal of 1 liter is believed to have produced little perturbation on the flux. The sample concentration was found to be independent of the height along the container at which the sample was withdrawn. If the residence time for the collected radon-222 in the container is short compared with the radon-222 half-life, and if the radon-222 concentration in the container is at all sampling times small compared with the radon-222 concentration in the soil gas at the base of the rim (5 inches), the rate of radon-222 concentration build-up in the container will be constant and directly proportional to the flux. A linear build-up of radon-222 concentration was indeed observed until the concentration within the container reached approximately 10 per cent of the soil-gas radon-222 concentration at the 5-inch depth. The flux determinations were made from this linear portion of the build-up curve with an estimated precision of 10 per cent.

SAMPLE MEASUREMENT

Activities encountered in the flux samples and in the 100-ml. soil-gas radon-222-concentration samples taken from shallow depths were considerably lower than the 1-liter radon-222 soil-gas samples. To analyze these weak samples, low-level ion chambers were used; the apparatus is shown in Plate III. The ion chambers are slow, positive ion chambers, used routinely in this laboratory to count breath samples for radon-222. The entire sample is transferred into the 3-liter chamber by flushing the sampling flask with aged, oil-pumped nitrogen. Chamber backgrounds are about 10–20 c.p.h. Periodic washings with versene and nitric acid help maintain a continuously low background. Absolute calibrations are carried out, using 110-picocurie radon-222 samples drawn from NBS standard solutions containing nanocuries of

PLATE III

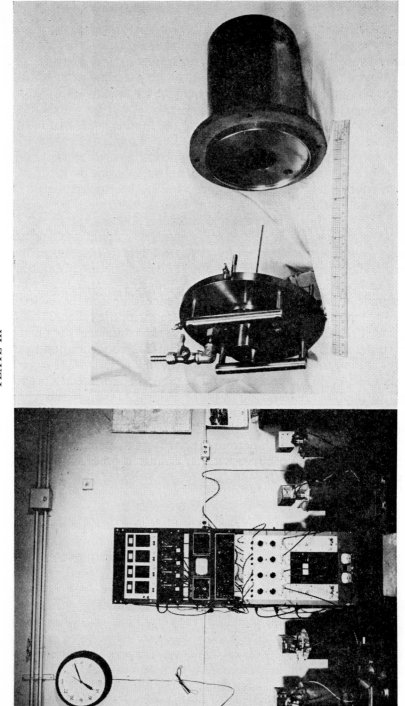

Low-level ion chambers and electronics used to measure low-activity radon-222 samples from flux container and shallow depths

radium-226, one and one-half hours after total radon-222 de-emanation of the standard. A switch pulser, which introduces pulses at the
input of the preamplifier, is used for daily calibration of the electronic
gain and discriminator level. Samples having intermediate activities
(30–60 pc. radon-222/l) were measured on both systems, and agreement was found to be within 5 per cent.

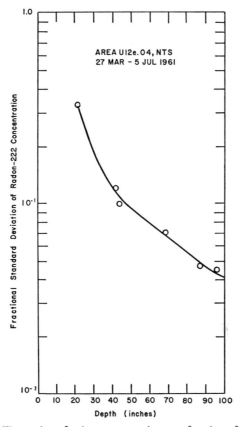

Fig. 2.—Fluctuation of radon concentration as a function of depth

SOIL POROSITY

Soil porosities were measured for surface layers only. No correction for the variation of soil porosity as a function of depth has been
applied to the radon-222 concentration data.

RESULTS

GROSS FLUCTUATIONS AS A FUNCTION OF DEPTH

It is generally accepted that atmospheric variables produce variations in the radon-222 concentration in soil gas predominantly in the

shallow surface layers (Israël and Krebs, 1962; Kovach, 1944, 1945, 1946). Figure 2 displays the gross fluctuation of the concentration data in weathered tuff on the Rainier Mesa at the NTS from March 27 to July 5, 1961, as a function of depth. These data represent at least 40 samples for each depth. It can be seen that the observed fractional standard deviation (F.S.D.) at 96 inches was less than 5 per cent. At

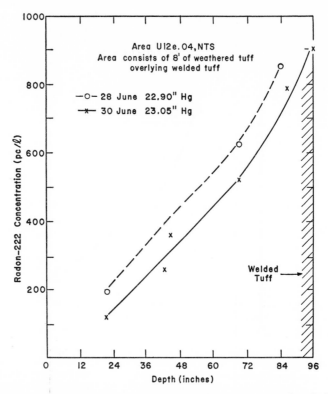

Fig. 3.—Effect of rising barometer on underground radon-222 concentrations during period of low winds, June 28–30, 1961.

the Lincoln site in the period from June to September, 1961, during which relatively moderate precipitation occurred, the F.S.D. at 96-inch depth was only 8 per cent, in spite of more extreme meteorological variations.

BAROMETRIC PRESSURE EFFECTS

Representative effects of barometric pressure changes on the radon-222 concentration as a function of depth are shown in Figures 3 and 4. The figures compare radon-222 concentrations in the weathered tuff at the NTS on alternate days during good weather and

periods of low wind, before and after barometric pressure changes. It
is observed that barometric pressure change is effective to rather great
depths and appears to shift the entire concentration gradient. Using
a piston analogy of the atmosphere, one may interpret this effect as
a short-range displacement of soil gas, moving under the influence of a
pressure differential between the atmosphere and soil gas at depth. As

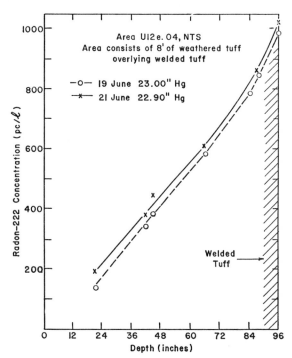

Fig. 4.—Effect of falling barometer on underground radon-222 concentrations during period
of low winds, June 19–21, 1961.

the barometer rises, radon-222-free air flows into the surface soil
layers, displacing and forcing to greater depth the soil gas initially in
these surface layers. This displacement is continued progressively to
greater depths. The opposite effect would occur during a falling
barometer. If a new barometric pressure equilibrium is established for
several days, the radon-222 concentration gradient may be expected to
return to its former equilibrium value by diffusion, altered only
slightly by a small change in the diffusion coefficient through its functional dependence on pressure. Statistically, the barometric pressure
effect is clearly seen in Figure 5, which plots barometric pressure
change versus per cent change in radon-222 concentration between
consecutive samples at 69 inches in weathered tuff at the NTS.

Flux measurements have not yet been made specifically to confirm experimentally the behavior of flux during periods of barometric pressure change. The flux at the surface would be expected to increase during a falling barometer, owing to both the "piston withdrawal" of the surface soil-gas layer and the increased radon-222 gradient in the surface layers. A rising barometer, by lowering the surface radon-222 gradient and introducing radon-222-free air into the soil, would decrease the flux, mainly by its effect on lowering the radon-222 concentration gradient.

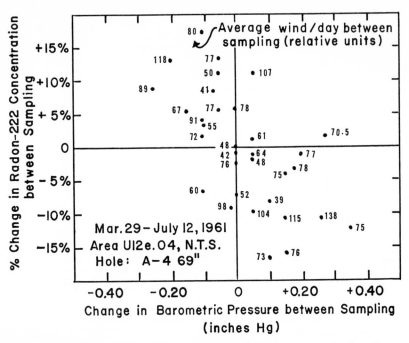

Fig. 5.—Effect of barometric pressure changes on radon-222 concentration in soil gas at 69-inch depth in weathered tuff, NTS.

PRECIPITATION EFFECTS

Figures 6 and 7 display effects of precipitation on soil-gas radon-222 concentrations at several depths, which are in qualitative agreement with the observations of previous workers (Kovach, 1944, 1945, 1946; Wright and Smith, 1915; Smyth, 1912; Kosmath, 1933). These increases are interpreted as a "capping effect," in which the moisture significantly reduces the "vertical porosity" of the surface layers of the soil. It is seen that the immediate and major effects on the concentration are at the shallowest depths. The effect with increasing depth (see Fig. 6, hole *13*) appears to be somewhat delayed as well as being reduced in magnitude. The data of Figures 6 and 7 specifically show the precipitation effect; however, the phenomenon in other cases is not so apparent, and in fact no precipitation effect has been observed when the soil is dry prior to the precipitation. The inhomogeneity of glacial debris and its effect on radon-222 concentration can be noted from Figure 6, wherein hole *21* at 133 inches consistently shows radon-222 concentrations lower than hole *13* at 94 inches. The great rise in concentration noted in Figure 6, following 4 inches of rain

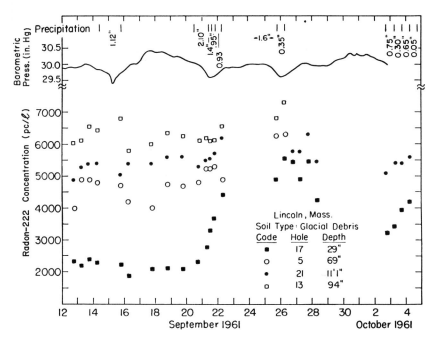

Fig. 6.—Effect of precipitation on radon-222 concentration in soil gas, Lincoln, Mass.

resulting from a hurricane that started on September 20, is an exceptional case comparable with the striking effect observed by Kovach (1945) during January, 1945, when an ice cap allowed the radon-222 concentration at 25-cm. and 75-cm. depths to reach nearly the value at 150 cm. Values of flux at the Lincoln site in December, 1962, taken when the ground was frozen to a depth of 6 inches or greater, but not capped with ice, were on the average about 0.70×10^{-16} curies radon-222/cm² · sec, which is about 40 per cent less than the flux observed during stable summer conditions (Fig. 10). Well-frozen ground thus appears only to reduce the flux.

WIND AND ATMOSPHERIC INSTABILITY

Figure 8 displays the statistical aspect of the effect of wind on the radon-222 concentration at the NTS in weathered tuff for 42-inch and 44-inch depths. The data include effects from other variables as well, but a definite relationship is seen in which the wind causes a depletion of radon-222 concentration at these depths. This trend is evident to a lesser extent at a depth of 70 inches and is absent in the

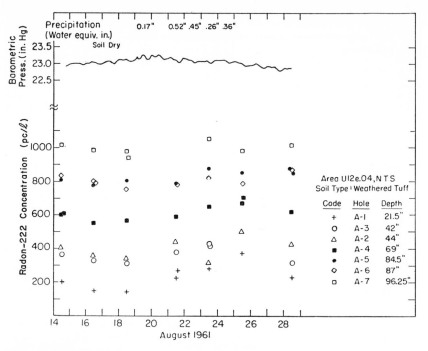

Fig. 7.—Effect of precipitation on radon-222 concentration in soil gas in weathered tuff, NTS.

data at 90-inch depth. Figure 9 represents a series of flux measurements made at the NTS in a virgin area on the alluvium flats, a desert region, to determine the constancy of flux over specific local areas. The measurements made during high winds in clear, hot weather on August 8 and 9, 1962, show higher fluxes than were observed during light breezes on August 12 and 28, 1962. This would be expected if surface radon-222 concentrations were being depleted by the wind.

Figure 10 shows the radon-222 flux measured at the Lincoln site for several days in June and July, 1962, as a function of shallow-depth radon-222 gradient, for both thermally stable (*S*) and unstable

(*U*) atmospheres.* Atmospheric thermal instability is evidenced by a temperature lapse rate of greater than 1° C/100 meters of ascent, the adiabatic lapse rate in the lower region of the atmosphere. Such instability increases convective overturning of the atmosphere through mixing of the warm, light surface air with the colder, heavy air from above. It is observed that the thermally stable points represent to some extent a family with the flux increasing almost linearly with increasing shallow radon-222 concentration gradient, as would be expected. The

* An observation during stable conditions on June 28, having a flux of 0.9 × 10⁻¹⁶ c/cm² · sec for a concentration gradient of 9 × 10⁻¹⁵ c/cm³ · cm, has been excluded from this figure because of a suspected error in the gradient determination.

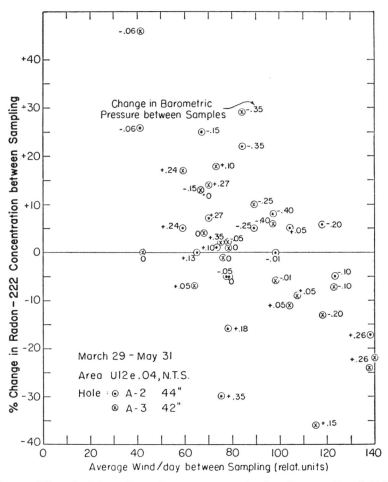

Fig. 8.—Effect of wind speed on radon-222 concentration in soil gas at 42- and 44-inch depths in weathered tuff, NTS.

two flux measurements during distinctly unstable conditions represent a trend to higher fluxes for a given concentration gradient. It should be noted that on July 10 at 1400 hours air motion was present; however, the atmosphere was observed to be thermally stable, and the measurement groups with the other stable points. A similar study of flux was made in alluvium at the NTS during normal summer desert weather; the results are shown in Figure 11. Measurements on August

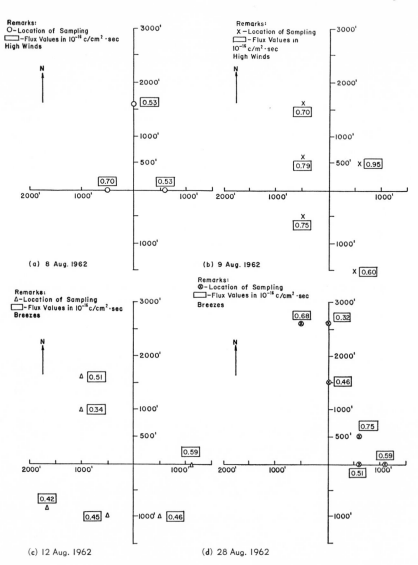

Fig. 9.—Radon-222 flux, area U9ad, NTS

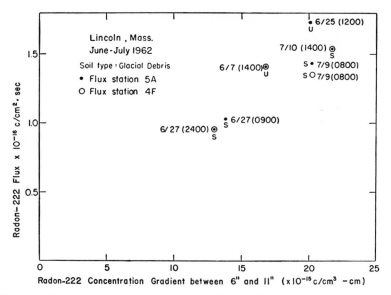

FIG. 10.—Radon-222 flux at the earth-air interface as a function of radon-222 concentration gradient between 6 and 11 inches. Points are shown for thermally stable (S) and unstable (U) atmospheres.

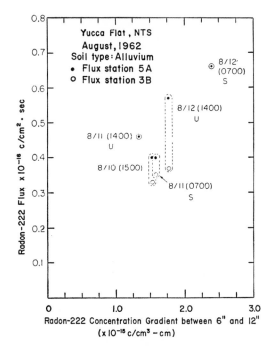

FIG. 11.—Radon-222 flux at the earth-air interface as a function of radon-222 concentration gradient between 6 and 12 inches. Points are shown for thermally stable (S) and unstable (U) atmospheres.

11 and 12 at 1400 hours were made under typical thermally unstable daytime conditions, having both large atmospheric thermal instability and wind. On August 10 at 1500 hours a less pronounced thermal instability was observed with no wind. The flux on this windless day was consequently lower, comparable with the flux for the more stable conditions of August 11 at 0700 at nearly the same concentration gradient. Both flux stations were generally in excellent agreement with each other, and no explanation is advanced for the large discrepancy between stations on August 12 at 1400 hours. The high flux value for August 12 at 1400 hours and the values for August 11 at 1400 hours are, however, distinctly indicative of the effect of unstable atmospheric conditions accompanied by wind on the radon-222 flux. The mechanism by which wind and atmospheric thermal instability deplete the surface layers of radon-222 is believed to be a limited convective motion induced in the shallow layers of soil gas by micro-oscillations in barometric pressure associated with wind. These oscillations give rise to a "turbulent pumping" that dilutes and partially exchanges shallow layer soil gas with radon-222-free air from above the ground surface. Wind has been shown to be a definite cause of micro-oscillations (Clark, 1950); a thermally unstable atmosphere, which forces overturning of air and hence increases turbulence, will very likely increase barometric micro-oscillations.

To date, radon-222 measurements have been made while monitoring wind speed and atmospheric instability but not while studying their direct combined effect on surface air turbulence and micropressure oscillations. We plan in the future to correlate radon-222 changes directly with a measurement of micro-oscillations in the local barometric pressure.

It is probable that, in comparison with the "d.c." pressure differential caused by barometric pressure variations that affect radon-222 concentrations to large depths, the "a.c." pressure fluctuations from turbulent pumping will affect the radon-222 concentration only in the shallow layers of the soil.

The application of the classical Bernoulli effect to radon-222 depletion in the surface soil layers does not appear to be credible for two reasons. First, the wind velocity at the surface must decrease to zero, and therefore any Bernoulli pressure gradient must also decrease to zero. During the hurricane condition at the Lincoln site in September, 1961, wind speeds at $2\frac{1}{2}$ feet above the ground never exceeded 3 m.p.h. Second, from elementary considerations of the resistance to flow of gas in soil compared with the resistance to flow in the free atmosphere, any Bernoulli-caused pressure gradient existing several feet above the ground surface would more likely cause lateral horizontal air movement above ground rather than vertical gas removal from the soil interstices.

COMPARISON OF EXPERIMENTAL DATA WITH DIFFUSION
THEORY PREDICTIONS

A quantity of concentration data, taken at three depths extending to 113 inches in alluvium in Yucca Flat, is available, which may be compared with the predictions of one-dimensional diffusion theory. These data are used because the alluvium is probably the most homogeneous soil type to great depths encountered in this study. The application of the diffusion theory assumes a constant, homogeneous radium-226 concentration in the soil, constant radon-222-emanating power of the soil particles, and a soil porosity that does not vary with depth. The laws of diffusion will be briefly presented. These results will be found to be slightly different from those of other authors (Israël and Krebs, 1962; Wilkening and Hand, 1960).

The first law may be related to a porous medium

$$J = - D \frac{\partial C}{\partial x},$$ (1)

where $J = $ Rn flux in units of c/cm²-sec across a spatial area or total section of the medium, $D = $ the bulk diffusion coefficient for Rn through the volume of the porous medium, $C = $ concentration of Rn in the interstitial gas (curies/liter), and $x = $ linear dimension of the total volume or bulk of the medium under consideration.

The second law follows from equation (1) by a conservation principle with the added terms due to Rn decay and emanation from the solids of the medium.

$$\frac{\partial C}{\partial t} = \frac{D}{S} \frac{\partial^2 C}{\partial x^2} - C\lambda + \beta,$$ (2)

where $S = $ porosity of the medium defined as the ratio of void volume to total volume, $\lambda = $ the Rn decay constant, and $\beta = $ a constant, related to the emanating power of the medium into the interstitial volume. Notice that the porosity appears in the term with $\partial^2 C / \partial x^2$ owing to the fact that the change in the number of Rn atoms in the interstitial volume is caused by flow across a spatial and not interstitial plane as defined by J in equation (1).

Equation (2) may be solved in the steady state using the boundary conditions that $C \to 0$ at $x = 0$ and $C = C_0$ at $x = \infty$.

$$C = C_0 (1 - e^{-\sqrt{[\lambda/(D/S)]}x}) .$$ (3)

The flux J as a function of depth is therefore

$$J = - DC_0 \sqrt{\frac{\lambda}{D/S}} \, e^{-\sqrt{[\lambda/(D/S)]x}} \, . \tag{4}$$

Curve B of Figure 12 is the "best fit" to the average concentration values at 34, 59, and 113 inches and yields empirical values for D/S of 0.036 cm²/sec and for C_0 of 580 pc/l. The surface flux J_0 may also be

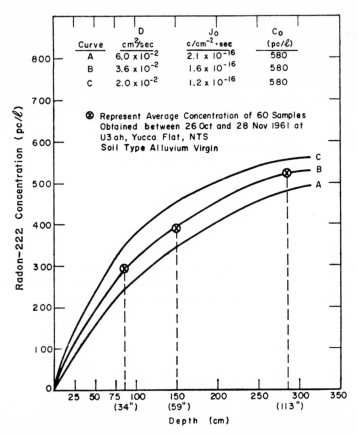

Fig. 12.—Diffusion theory predictions of radon-222 concentrations in soil gas as a function of depth compared with observed concentration values at three depths.

predicted from these data to be $S \times 1.6 \times 10^{-16}$ c/cm² • sec. Curves A and C are included to illustrate the use of different diffusion coefficients, while retaining the best-fit C_0. The use of other than the "best-fit" diffusion coefficient and a C_0 determined by normalization at the 59-inch depth is illustrated in Figure 13.

It might first be noted from Figure 13 that the determination of D

from the concentration gradient is not extremely definitive. For example, the $D/S = 0.02$ cm²/sec curve might well fit the data at 113 inches if a slightly lesser porosity were used at 113 inches than at 59 inches in the solution of the equation for concentration with depth. Owing to the possibility of soil porosity and emanating power variations with depth, an independent measurement of C_0 will not add substantially to the credibility of D as determined from these fits.

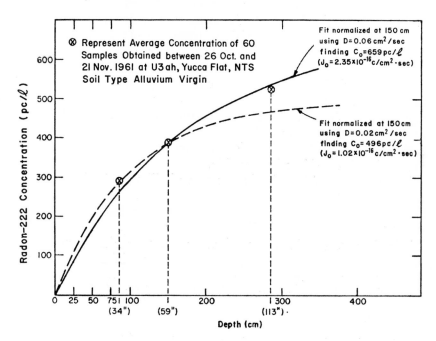

Fig. 13.—Diffusion theory predictions of radon-222 concentrations in soil gas as a function of depth, using selected diffusion coefficients and normalizing to the observed concentration at 59 inches.

The best-fit value for D/S of 0.036 cm²/sec from these concentration data for depth gives a value of 0.009 cm²/sec for D when the measured Yucca Flat porosity of 25 per cent is applied. The data from Figure 11, relating the measured shallow depth Rn concentration gradient with the measured Rn flux at the earth surface, yield by equation (1) a diffusion coefficient of 0.027 cm²/sec. The relatively large diffusion coefficient obtained from the shallow layer concentration gradient and surface flux may contain a convective contribution operating in addition to diffusion.

According to Penman (1940), the diffusion coefficient for carbon bisulfide and acetone vapor through a porous medium is related to the product of the porosity, S, and the diffusion coefficient in air, D_0, by the following equation:

$$D = 0.66 \, SD_0 . \tag{5}$$

The porosity, S, is defined as the ratio of void volume to gross volume. If it is assumed that this same proportionality constant is applicable for radon-222, which is a much smaller molecule than carbon bisulfide or acetone, one may predict D using measured porosity and the tabulated value of D_0. An average of values for D_0 at standard temperature and pressure (S.T.P.) from the literature (Jost, 1960; National Research Council, 1926) may be taken as 0.10 cm^2/sec, which, when corrected for the barometric pressure and soil temperature at the NTS, gives $D_0 = 0.11$ cm^2/sec. Using the porosity of 25 per cent as measured in shallow layers of alluvium, one predicts, with equation (5), $D = 0.018$ cm^2/sec. To the extent of the validity of the assumptions, the agreement between values of D derived from the diffusion theory and from the effect of a porous medium on the free-air value of D_0 is considered reasonable.

The value of J_0/S obtained from curve B, Figure 12, yields a surface flux of 0.40×10^{-16} c/cm$^2 \cdot$ sec. From Figure 9, c, and 9, d, the observed flux in Yucca Flat on days of light wind averaged about 0.51×16^{-16} c/cm$^2 \cdot$ sec. In Figure 9, a, and 9, b, data for days of high winds, the observed flux averaged 0.70 c/cm$^2 \cdot$ sec, illustrating the convective component of flux augmenting the diffusion component strongly suggested by the agreement between diffusion-theory predicted flux and observed flux during days of light wind.

SUMMARY

Data have been obtained on the effects of barometric pressure changes, precipitation, wind, and thermal atmospheric instability in soil gas and on the radon-222 flux at the earth-air interface. Simple, relatively large scintillation chambers have been developed to measure reliably the radon-222 concentration in 1-liter soil-gas samples having activities greater than 30 picocuries per liter. Lower activities, found in samples from very shallow depths and in samples taken from a large container sealed over the surface of the earth, which are used to measure radon-222 flux, are counted in low-background ion chambers.

Extensive studies have been made in undisturbed areas both in the glacial debris at Lincoln, Massachusetts, and in weathered tuff and alluvium at the USAEC Nevada Test Site (NTS).

Barometric pressure changes were found to affect the radon-222 concentration to depths greater than 96 inches in weathered tuff. It appears that the entire radon-222 concentration gradient is shifted by the barometric pressure changes and that the effect may be physically interpreted as a direct displacement of concentrations from adjacent depths, as the atmosphere above, acting as a piston or diaphragm, either forces (rising barometer) radon-222-free air from the atmosphere into the soil or pulls (falling barometer) radon-222-rich soil gas from the soil into the atmosphere.

Precipitation causes a "capping effect," in which the surface layers become clogged with moisture, reducing their vertical porosity for radon-222 transport. Radon-222 concentrations predominantly in shallow depths (<36 inches for each soil type studied) are observed to increase and remain high while the soil is extremely moist. In some cases, if precipitation is severe, the shallow-depth radon-222 concentrations may reach levels found normally at depths as much as three times deeper. Flux measurements in winter at the Lincoln site indicate that the flux from ground frozen to at least 6 inches is reduced by about 40 per cent from the average summertime value.

A definite trend indicating that high wind speeds produce a depletion of radon-222 concentrations in soil gas down to 44 inches has been observed in weathered tuff at the NTS. A somewhat higher flux during periods of high winds has also been noted.

Thermal atmospheric instability, producing vertical air motion above the ground, appears to increase the flux by convection. For a given shallow-depth radon-222 concentration, on which diffusion-based flux depends, higher fluxes were observed at both sampling sites during unstable conditions than were observed for thermally stable conditions. It is believed that wind and thermal atmospheric instability cause the same physical mechanism to act on the soil gas and consequently on the contained radon-222. Nearly all air motions near the surface of the earth are turbulent, under which condition micro-oscillations in pressure must exist and can cause "turbulent pumping" of the soil gas.

The solution of simple, one-dimensional diffusion theory yielding the radon-222 concentration as a function of depth has been fitted to average concentration values at three depths in alluvium at the NTS. A reasonable, although not definitive, value of the diffusion coefficient for radon-222 in soil at depth of 0.009 cm²/sec is thereby obtained. This is a porosity-dependent value. An empirical diffusion coefficient of 0.027 cm²/sec was found as the proportionality constant between observed flux and shallow-depth radon-222 concentration gradient for the same soil type. This discrepancy may indicate convective processes operating in the shallow layers of the soil in addition to diffusion. Furthermore, the diffusion-theory prediction of surface flux is in good agreement with the average observed values of flux in alluvium on days of light wind. The average observed flux on days of high wind was nearly twice the diffusion-theory prediction, again illustrating the effect of convection.

ACKNOWLEDGMENTS

In acknowledgment of their help on this project, the authors wish to thank Anthony R. Lewis, who was active in all phases of the work, Marie M. Costello, Alvin Warwas, A. T. Keane, and Joseph E. Annis for much of the data processing, and J. P. Morris and Marcel Semo, who maintained and contributed to the electronics. The authors are grateful to Professors H. G. Houghton and J. W. Winchester for very helpful discussions on particular aspects of the problem. Mr. J. B. Bulkley graciously consented to the use of his orchard and home in Lincoln, Massachusetts, for the local radon-222 sampling site and was an active administrator during much of the project. Professor John W. Irvine, Jr., helpfully supplied the lead-210 foils used for daily calibration sources for the scintillation chambers. Professor D. B. Keily, of the Massachusetts Institute of Technology, and Mrs. E. Iliff, of the Geophysical Research Group at the Air Force Cambridge Research Center, kindly made available some of the meteorological instrumentation used in this study.

Thanks are also extended to the AFTAC Element at NTS, in particular to Majors Adams, Galiney, Mendenhol, and Antonio and Captain Woodford for their close co-operation, which greatly facilitated the NTS measurements. We wish to acknowledge the assistance and co-operation of Texas Instruments, Inc., particularly Mr. Dallas Russell, in drilling sampling holes at NTS. Miss Mary-Margaret Shanahan and Mr. Edward D. Dillon were invaluable in administration and in ministering to all details of the work.

REFERENCES

CLARK, R. D. M. 1950. Atmospheric micro-oscillations. J. Meteorol., 7: 70–75.

EVANS, R. D., H. W. KRANER, and G. L. SCHROEDER. Edgerton, Germeshausen and Grier, Inc., Report B-2516, December 6, 1962.

ISRAËL, H., and A. KREBS. 1962. Nuclear Radiation in Geophysics, pp. 76–86. New York: Academic Press.

JOST, W. 1960. Diffusion in Solids, Liquids, Gases. New York: Academic Press.

KOSMATH, W. 1933. Die Exhalation der Radiumemanation aus dem Erdboden und ihre Abhängigkeit von den meteorologischen Faktoren. Gerlands Beitr. Geophys., 40:226–37.

KOVACH, E. M. 1944. An experimental study of the radon-content of soil-gas. Trans. Am. Geophys. Union, 25:563–71.

———. 1945. Meteorological influences upon the radon-content of soil-gas. Ibid., 26:241–48.

———. 1946. Diurnal variations of the radon-content of soil-gas. Terr. Magnet. & Atmos. Elec., **51**:45–55.

LUCAS, H. F. 1957. Improved low-level alpha-scintillation counter for radon. Rev. Sci. Instr., **28**:680–83.

NATIONAL RESEARCH COUNCIL. 1926. International Critical Tables of Numerical Data in Physics, Chemistry, and Technology, **1**:376. New York: McGraw-Hill Book Co.

PENMAN, H. L. 1940. Gas and vapour movements in the soil. J. Agr. Sci., **30**:437–62.

SMYTH, L. B. 1912. On the supply of radium emanation from the soil to the atmosphere. Phil. Mag., **24** (Ser. 6): 632–37.

WILKENING, M. H., and J. E. HAND. 1960. Radon flux at the earth-air interface. J. Geophys. Res. **65**:3367–70.

WRIGHT, J. R., and O. F. SMITH. 1915. The variation with meteorological conditions of the amount of radium emanation in the atmosphere, in the soil gas, and in the air exhaled from the surface of the ground at Manila. Phys. Rev., **5**:459–82.

Discussion, Chapters 1–10

(Numbers in parentheses refer to chapter authored by discussant.)

In response to a question from Miyake (11) concerning the accuracy of the radiometric analyses of the natural radioisotopes in minerals, rocks, and sand, Adams (30, 34) noted that the accuracy of his γ-spectrometric analyses of salts and beach sands was ± 10 per cent or better. Osmond (8) indicated that his uranium-238 and uranium-234 determinations could be carried out to ± 5–10 per cent at the present time, and other laboratories claim ± 1–2 per cent. He thought that this kind of accuracy was much less likely with thorium or radium. Wollenberg (32) remarked on the low accuracies obtained for 7 spectrometric measurements on ultramafic rocks, where uranium and thorium contents are of the order of 0.01 p.p.m. An accuracy of ± 100 per cent would not be unexpected for such low concentrations.

Hill (24, 28) asked Ragland (7) for an estimate of the absolute sensitivity of his autoradiograph technique. The latter replied that the method of α track counting was only semiquantitative, since the assumption of a thorium-to-uranium ratio of 3.0 was necessary. Measuring track lengths is tedious but provides a minimum detection limit of 0.1 p.p.m. uranium. At that concentration, the precision is no better than ± 100 per cent. Adams (30, 34) mentioned the technique used by Picciotto on meteorites using 6-month exposures, which lowered the detection limits down to a few hundredths of a p.p.m.

L. Bagaret (Texas Christian University, U.S.A.) asked whether any correlation studies had been made of the variations in areal concentrations of radioisotopes with heat-flow balance. Adams (30, 34) indicated that Burch (Harvard) had determined the geothermal gradient in the core holes drilled in the Conway granite of New Hampshire in an attempt to ascertain whether the high activity of the granite had any effect on the heat gradient. Phair (1) commented on the fact that Burch had done an exhaustive heat-flow study in the Colorado Front Range and that the newer data on the distribution of radioactivity in this region would eventually provide useful information for such a correlation study.

YASUO MIYAKE, YUKIO SUGIMURA, AND
HIROYUKI TSUBOTA

11. Content of Uranium, Radium, and Thorium in River Waters in Japan

IT IS OF INTEREST to know the concentration of uranium, radium, and thorium in river waters from the viewpoint of the geochemical cycles of these natural radioelements. It is also important to evaluate the effects on man and other living things of the nuclear radiation emitted by these elements through fresh water. Up to now there have been few measurements of the content of these elements in Japanese rivers, except for a preliminary study on radium carried out by Iwasaki (1953).

The present authors wish to present a brief report on the results of determination of the contents of uranium, radium, and thorium in fresh waters collected in ten representative rivers of Japan.

RESULTS OF DETERMINATION

The names of the rivers investigated are Ishikari (Hokkaido Island), Mogami, Kitakami, Shinano, Tone, Kiso, Yodo, Asahi (Honshu Island), Yoshino (Shikoku Island), and Chikugo (Kyushu Island). The locations of the rivers and the water-sampling sites are shown in Figure 1. Sampling of water was done during the period from the end of July to early August, 1961. Sampling sites were selected so as to avoid industrial and urban pollution as well as inflowing of sea water.

Each 100-liter water sample was filtered with filter paper Toyo Roshi No. 4 (Whatman No. 50 eq.) and passed through a 20-cm.-long cation exchange column (Dowex 50W-x8, 50–100 mesh, Na+—

YASUO MIYAKE is director and YUKIO SUGIMURA is Ph.D. at the Geochemical Laboratory, Meteorological Research Institute, Mabashi, Suginami, Tokyo, Japan; HIROYUKI TSUBOTA is research scientist at the Division of Chemistry, National Institute of Radiological Sciences, Chiba, Japan.

form, 500 ml.) in order to retain uranium, radium, and thorium in the resin.

Uranium, radium, and thorium trapped in the resin were eluted first with a 5 per cent solution of oxalic acid, followed by a 3N solution of hydrochloric acid. Uranium and thorium were determined in the former solution, radium in the latter.

The oxalic acid in the effluent was decomposed by digesting it with the mixture of nitric and perchloric acid. Uranium and thorium were determined by an absorption spectrophotometric method with di-benzoyl methane (Tatsumoto and Goldberg, 1959; Sugimura and Sugimura, 1962) and neo-thoron (Ishibashi and Higashi, 1954), after

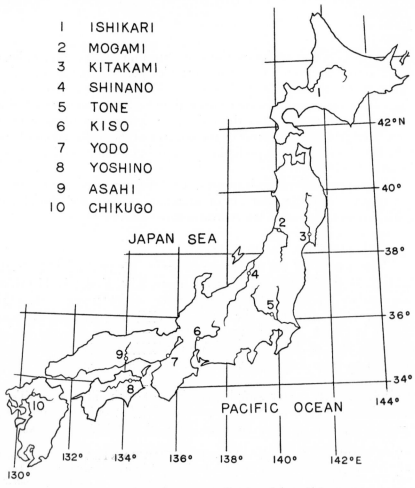

1	ISHIKARI
2	MOGAMI
3	KITAKAMI
4	SHINANO
5	TONE
6	KISO
7	YODO
8	YOSHINO
9	ASAHI
10	CHIKUGO

FIG. 1.—Location and water-sampling site of rivers in Japan

isolating each radioelement with solvent extraction and ion exchange (Korkisch and Tera, 1961). Radium was determined by an α-ray spectrometric method (Sugimura and Tsubota, 1963).

The results of the analyses are given in Table 1. Weighted averages of concentration were 0.57×10^{-6} gm/l for uranium, 8.2×10^{-14} gm/l for radium, and 2.7×10^{-8} gm/l for thorium.

As to the mean values for uranium and radium in river waters in the continents, 1×10^{-6} gm/l and 7×10^{-14} gm/l, respectively, were given by previous researchers (Koczy, 1954; Judson and Osmond, 1955; Fix, 1956; Rona, Gilpatrick, and Jeffrey, 1956; Koczy, Tomic, and Hecht, 1957). Koczy (1954) gave the estimated value of 2×10^{-8} gm/l for thorium; however, no direct measurement has been made on

TABLE 1

CONTENTS OF URANIUM, RADIUM, AND THORIUM
IN RIVER WATERS IN JAPAN

Name of River	Uranium (10^{-6} gm/l)	Radium (10^{-14} gm/l)	Thorium (10^{-8} gm/l)
1. Ishikari.........	0.41	13.7	0.87
2. Mogami.........	0.38	3.7	1.2
3. Kitakami........	0.65	7.8	2.7
4. Shinano........	0.58	4.4	3.7
5. Tone...........	0.62	13.3	2.9
6. Kiso............	0.58	8.2	2.3
7. Yodo...........	0.34	4.1	4.8
8. Yoshino........	0.77	11.9	3.9
9. Asahi..........	0.67	4.4	4.8
10. Chikugo........	1.23	8.4	3.1
Weighted mean..	0.57	8.2	2.7

thorium content in river waters. So far as uranium and radium are concerned, their contents in Japanese river waters are not greatly different from those in continental rivers.

From the analytical results, the total amounts of annual discharge of uranium, radium, and thorium through Japanese rivers are estimated to be, respectively, 319 tons, 46 gm., and 15 tons, assuming that the total rate of runoff of river water is 5.6×10^{14} liters per year (Water Utilization Res. Inst., 1962).

It is to be noticed that, despite the abundance ratio of thorium and uranium in land materials of about 3 (Adams *et al.*, 1959), that of river waters of Japan is only 0.05 (see Table 2), which suggests the fractionation of uranium and thorium during weathering processes. It is also interesting that radium content in river waters in Japan is equivalent to only 34 per cent of the uranium-supported radium in the equilibrium state (see Table 3). In regard to the geographical distribution of uranium and thorium in river water in Japan, it is statistically clear that the contents of these radioelements in the total dissolved matter are much greater in the western part of Japan (U, 10.6 p.p.m.; Th, 0.6 p.p.m.; Ra, 1.2×10^{-12} gm/gm) than that of the eastern part (U, 5.7 p.p.m.; Th, 0.23 p.p.m.; Ra, 0.8×10^{-12} gm/gm), as shown in Table 2, which is in good accordance with the fact that the eastern part of Japan is mostly covered by volcanic eruptives, whereas in the western part granitic material prevails.

TABLE 2

CONTENTS OF URANIUM, RADIUM, AND THORIUM IN DISSOLVED
MATTER IN JAPANESE RIVERS

Name of River	Salinity (mg/l)	Uranium (p.p.m.)		Radium (10^{-12} gm/gm)		Thorium (p.p.m.)	
1. Ishikari........	107	3.8		1.3		0.081	
2. Mogami.......	83	4.6		0.45		0.15	
3. Kitakami......	89	7.3	5.7 (av.)	0.88	0.82 (av.)	0.31	0.23 (av.)
4. Shinano.......	95	6.1		0.46		0.39	
5. Tone..........	127	4.9		1.1		0.23	
6. Kiso..........	55	10.6		1.5		0.42	
7. Yodo..........	59	5.8		0.69		0.81	
8. Yoshino.......	47	16.4	10.6 (av.)	2.5	1.2 (av.)	0.83	0.60 (av.)
9. Asahi..........	70	9.6		0.63		0.68	
10. Chikugo.......	117	10.5		0.72		0.27	
Weighted mean...	87	6.6		0.96		0.31	
Annual discharge.	4.8×10^7 tons	319 tons		46 gm.		15 tons	

TABLE 3

RATIO OF TH/U AND RA/RAU* IN
JAPANESE RIVER WATERS

Name of River	Th/U	Ra/RaU
1. Ishikari......	0.021	0.95†
2. Mogami.....	0.032	0.29
3. Kitakami....	0.042	0.34
4. Shinano.....	0.064	0.22
5. Tone........	0.047	0.62
6. Kiso........	0.040	0.41
7. Yodo........	0.141	0.35
8. Yoshino.....	0.051	0.44
9. Asahi........	0.071	0.19
10. Chikugo.....	0.025	0.20
Mean.......	0.05 ± 0.03	0.34 ± 0.10

* RaU represents the equilibrium amount of radium to uranium.

† This value is omitted in calculating averaged value because the pH of water in Ishikari River is low (6.6), which suggests that the retention factor of radium may be different from those in other rivers of water with pH from 7.0 to 7.4.

GEOCHEMICAL CONSIDERATIONS

As mentioned above, the differences between the relative abundances of uranium, radium, and thorium in river waters and those for the land materials suggest the fractionation of uranium, radium, and thorium during the weathering process. With regard to uranium, the greater part of this element may exist in water as the uranyl-carbonate complex anion $UO_2(CO_3)^{-2}$ or $UO_2(CO_3)^{-4}$ in oxidizing environments. And these complex anions may have little tendency to be fixed again in the land materials, owing to the smaller capacity of the anionic exchange of the latter.

According to Adams *et al.* (1959), the average contents of uranium in igneous and sedimentary rocks are, respectively, 3 p.p.m. and 1 p.p.m. It is known that the surfaces of the Japanese islands are covered by igneous and sedimentary rocks with the approximate ratio of 3:2. The uranium content of the soil will be about 2 p.p.m. on an average. This figure is in close agreement with the results of γ-ray measurements of radium in the soil of Japan performed by Watanabe (1961), who gave a mean value for the radium content of about 0.8×10^{-12} gm/gm of soil. This value of radium is equivalent to about 2 p.p.m. of uranium when radioactive equilibrium is established, although there is some doubt whether the two elements are present in the equilibrium state in soil.

Since the total annual rate of discharge of uranium from Japan is estimated to be 319 tons, and the estimated total amount of dissolved matter in the effluent from the land of Japan is 4.8×10^7 tons per annum (Miyake *et al.*, 1962), the averaged concentration of uranium in the dissolved matter will be 6.6 p.p.m., which is about three times that in the soil. This relative enrichment of uranium in river waters suggests a greater tendency of elution of this element from the land material, or a smaller retention from the solution relative to the other chemical elements.

On the other hand, if we assume the average concentration of uranium of 2 p.p.m. in the land material, the corresponding total amount of the land material to 319 tons of uranium will be about 1.6×10^8 tons. However, the total amount of effluent of dissolved material from Japan is 4.8×10^7 tons in a year, which is only 30 per cent of the total amount of the land substance mentioned above that is considered to have once undergone weathering.

In the process of erosion, a number of chemical constituents of the land materials are dispersed or dissolved, while some will remain in the siliceous or other resistate detritus. In addition, among chemical elements that are once dispersed or dissolved in water, some of them are removed again from the aqueous solution and refixed in the land substances through adhesion, adsorption, coagulation, ionic exchange, precipitation, chemical combination, and the like. If we define the "retention factor" as the fraction of chemical substances retained or refixed in the land material during the process of weathering and effluence of surface and ground water, the over-all "retention factor" of the total dissolved matter will be 0.7, which is obtained by the data for the total annual rate of discharge (4.8×10^7 ton/yr) and the calculated total amount of weathered substances (1.6×10^8 ton/yr) on the basis of uranium concentration in water and land materials.

It is to be noted that this value of 0.7 gives the minimum for the over-all retention factor, since it is assumed that the retention factor of uranium is negligible because of its formation of anionic complex compounds.

With respect to thorium in the soil in Japan, it is estimated to be about 6 p.p.m. of thorium, assuming a ratio of 3:2 for the igneous and sedimentary rocks that cover the surface of Japan. With γ-ray measurement, Watanabe (1961) gave values for thorium content in Japanese soil ranging from 2.8 to 20 p.p.m. and a weighted mean of about 8 p.p.m. When the value of 6–8 p.p.m. is taken for the average content of thorium in the land material, the corresponding amount of thorium in 1.6×10^8 tons of terrestrial material may be 1,000–1,300 tons, of which only 15 tons are discharged through rivers. Therefore, the retention factor for thorium will be somewhere between 0.98 and 0.99.

A similar estimation of the retention factor can also be done for radium. The average concentration of radium in Japanese soil is approximately 1×10^{-12} gm/gm, and the amount of radium that is leached out from the land material is estimated to be 160 gm. per year. In reality, however, the amount of annual discharge of radium by river waters is only 46 gm., as given in Table 2. Therefore, the retention factor will be 0.7 for radium, which is approximately equal to the mean value of retention of other chemical constituents. In other words, in river waters neither enrichment nor depletion of radium occurs in relation to the average content in the land material.

In conclusion, it may be said that the depletion of radium in river water, relative to uranium, and the remarkable deficiency in thorium, may be explained in terms of the difference in "retention factors" among uranium, thorium, and radium in the process of weathering and effluence of water into rivers.

REFERENCES

ADAMS, J. A. S., J. K. OSMOND, and J. J. W. ROGERS. The geochemistry of thorium and uranium. *In* Physics and Chemistry of the Earth, **3**:298–348. London: Pergamon Press.

FIX, P. F. 1956. Hydrochemical exploration for uranium. U.S. Geol. Survey Prof. Paper 300, pp. 667–71.

ISHIBASHI, M., and S. HIGASHI. 1954. Method of thorium determination. Jap. Analyst, **3**:213–16. (In Japanese.)

IWASAKI, I. 1953. Chikyukagaku Gaisetsu ("Introduction to Geochemistry"). Tokyo: Dai Nippon Tosho Pub. Co. Pp. 296.

JUDSON, S., and J. K. OSMOND. 1955. Radioactivity in ground and surface water. Am. J. Sci., **253**:104–16.

KOCZY, F. F. 1954. Geochemical balance in the hydrosphere. *In* H. FAUL (ed.), Nuclear Geology. New York: John Wiley & Sons. Pp. 414.

KOCZY, F. F., E. TOMIC, and F. HECHT. 1957. Zur Geochemie des Urans im Ostseebecken. Geochim. & Cosmochim. Acta, **11**:86–102.

KORKISCH, J., and F. TERA. 1961. Separation of thorium by anion exchange. Anal. Chem., **33**:1264–66.

MIYAKE, Y., *et al*. Chemical studies of river waters in Japan. (To be published.)

RONA, E., L. O. GILPATRICK, and L. M. JEFFREY. 1956. Uranium determination in sea water. Trans. Am. Geophys. Union, **37**:697–701.

SUGIMURA, Y., and T. SUGIMURA. 1962. Uranium in recent Japanese sediments. Nature, **194**:568–69.

SUGIMURA, Y., and H. TSUBOTA. 1963. A new method for the chemical determination of radium in sea water. J. Marine Res., **21**:74–80.

TATSUMOTO, M., and E. D. GOLDBERG. 1959. The marine geochemistry of uranium. Geochim. & Cosmochim. Acta, **17**:201–8.

WATANABE, H. 1961. A new method of measurement of absorption dose rate from terrestrial background radiation. J. Radiation Res., **2**:61–67.

WATER UTILIZATION RESEARCH INSTITUTE. 1962. Applied Hydrology. Tokyo: Chizin-Shokan Pub. Co. (In Japanese.)

RICHARD B. HOLTZMAN

12. Lead-210 (RaD) and Polonium-210 (RaF) in Potable Waters in Illinois

A STUDY OF THE LEAD-210 and polonium-210 concentrations in potable well and surface waters in Illinois was undertaken for the purposes of elucidating their nature and, in particular, of estimating the contribution of potable water to the lead-210 and polonium-210 content of the human body. These nuclides are of particular interest because the lead-210 (in radioactive equilibrium with its decay product, polonium-210) contributes a sizable fraction of the total background-radiation dose to the human skeleton (Holtzman, 1960). Stehney and Lucas (1956) and Lucas (1961) showed a strong correlation between the radium-226 content of the human body and that in potable water; therefore, by analogy, water, along with food and air, appears to be a likely source of lead-210. Although this nuclide is a decay product of radium-226, the two nuclides are not necessarily associated; their chemistries are different—there exists in the intermediate decay chain an easily translocated inert gas, radon-222, and the lead-210 is long lived (21.4 years).

Lead-210 and its daughters have been found in many parts of the human environment, the atmosphere (Lockhart et al., 1958; Burton and Stewart, 1960), plants and animals (Hill, 1960, 1962); Black 1961; Di Ferrante, 1961), and in the human body itself (Holtzman, 1960; Hill, 1962; Black, 1961; Hursh, 1960). Kuroda and Yokoyama (1948) found polonium-210, a member of the lead-210 decay chain, in hot springs in Japan. Both these nuclides have been found in rain water, lead-210 in concentrations of about 2.5 pc/l (Lockhart, Baus, Patterson, and Blifford, 1958; Burton and Stewart, 1960). Preliminary results from this laboratory have shown the existence of lead-210 in

RICHARD B. HOLTZMAN is associate chemist, Radiological Physics Division, Argonne National Laboratory, Argonne, Illinois.

Work performed under the auspices of the U.S. Atomic Energy Commission.

well waters (Holtzman, 1960), and, more recently, Rama and Goldberg (1961) have found appreciable amounts of this nuclide in ocean and fresh water. Some measurements are given in Table 1.

The lead-210 and polonium-210 originate from various sources. In surface and shallow-well waters these nuclides may originate from rain and the leaching of surrounding material and suspended solids. In deeper-well waters they may originate from leaching of the surroundings and the decay of soluble radium-226 and its daughter, radon-222. Knowledge of these concentrations may help in understanding the nature of these waters.

TABLE 1

LEAD-210 CONTENT OF WATERS (PREVIOUS RESULTS)

Type of Water	Activity (pc/l)	Reference
Rain......................	2.5 0.36–3.2 2.4	Lockhart *et al.* (1958) Rama and Goldberg (1961) Burton and Stewart (1960)
Ocean surface...............	0.054	Rama and Goldberg (1961)
Colorado River water.........	0.13–6.7	Rama and Goldberg (1961)
Tap water, La Jolla, Calif.....	0.054	Rama and Goldberg (1961)

EXPERIMENTAL

a) Sampling procedure. Samples from towns in Illinois were obtained mostly at the pumping station of the water system except for a few samples from water taps in private homes. Some samples were collected 1–3 years prior to measurement and stored in 1-gallon glass bottles. The others were collected within 2 weeks of measurement and stored in 32-ounce polyethylene bottles. Ten ml. of concentrated nitric acid was added to each bottle.

b) Analysis of lead-210 and polonium-210. The lead-210 was determined from the amount of its daughter, polonium-210, which was plated onto a silver disk and counted. The procedure, modified from that of Black (1961), was as follows:

1. A sample of suitable size (up to about 2 l.) was evaporated to 200 ml.

2. Twenty ml. of 72 per cent perchloric acid was added. In the presence of organic matter, 10 ml. of concentrated nitric acid was also added to remove the easily oxidized material.

3. In order to remove nitrates that interfere with the final plating procedure, the sample was heated to perchloric acid fumes. It was then cooled, 10 ml. of concentrated hydrochloric acid was added, and the solution again fumed. This step was repeated 3 more times.

4. The sample was diluted to about 200 ml., and concentrated hydrochloric acid was added dropwise to give a concentration of 0.5N hydrochloric (pH 0.3). Most precipitates present at this time could be dissolved by the addition of a few drops of hydrofluoric acid and heat.

5. The sample was added to a plating cell that consisted of an inverted 8-ounce Pyrex baby bottle with the bottom removed. A 1½-inch silver disk was placed against the mouth of the bottle. The cap, with a neoprene gasket the size of the disk, was then tightened to form a watertight seal.

6. About 100 mg. of ascorbic acid was added to prevent interference by ferric ion with the plating of the polonium-210.

7. The sample, kept at 90°–100° C. in a water bath, was stirred for 7 hours.

8. The disk was then removed, washed with a few ml. of 1N hydrochloric acid and water, and dried on a water bath. It was counted in a low-background (0.010–0.030 c.p.m.) Nuclear Measurements Corporation internal α proportional counter with an over-all counting efficiency (geometry, self-absorption, backscatter, etc.) of about 51 per cent. The yield of the chemical procedure was 92 per cent.

c) *Analysis of radium-226.* Radium-226 was determined by the radon-emanation technique of Lucas (1961). The sample was placed in a glass flask, the radon removed by flushing with radon-free air, and the flask sealed by stopcocks. After 3 or more days of storage the radon was collected in a charcoal trap at dry-ice temperature. The charcoal was then heated to 500° C., and the radon flushed with helium into a zinc sulfide–phosphor-coated chamber. The α particle scintillations from the radon and radon-daughter decays were detected by a photomultiplier tube and counted to determine the radon content of the bottle. The amount of radium-226 was then calculated from these data.

d) *Radon-222 retention of the containers.* In order to correct the measured lead-210 concentrations for build-up from radium-226 and its daughters during storage, a measure of the radon-222 retention was necessary because of the permeability of the polyethylene bottles and plastic caps on the glass bottles to radon. In order to measure the radon in the bottle, the cap of the bottle was removed and the mouth quickly sealed by a rubber stopper through which extended a bubbler tube and outlet. Radon-free air was bubbled through the sample in the bottle, and the radon was collected on a charcoal trap at dry-ice temperatures and treated as described above. Tests indicated that losses during the 10 seconds necessary to insert the bubbler tube amounted to less than 2 per cent.

The fractional retention during storage is principally a function of the container, that is, a 1-gallon glass bottle sealed by a plastic cap with a paper insert or an 8- or 32-ounce polyethylene bottle. Actual water samples of known volume and radium-226 concentration were stored in the bottles for longer than 1 month (radon-222 at radioactive equilibrium) before the radon content was determined. The fractional radon retention, shown in Table 2, ranges from 25 to 80 per cent in our bottles. This may be compared with the 81–89 per cent retention in $4\frac{1}{2}$-l. polyethylene bottles computed from the half-life data of Abbatt *et al.* (1960), in which half-lives of 3–3.3 days were observed rather than the accepted 3.823-day value. The large standard deviations of the results shown in Table 2 are probably due to variations in construction and to the tightness of the bottle caps. The precise value

TABLE 2

RETENTION OF RADON-222 IN
STORAGE CONTAINERS

Container	Radon-222 Retention (per cent)
8-oz. polyethylene bottle..............	25± 8
32-oz. polyethylene bottle.............	67± 3
1-gal. glass bottle (plastic cap)........	80±23
$4\frac{1}{2}$-l. polyethylene bottle..............	81–89*

* Estimated from the data in Scholl (1917).

of the retention is not critical, however, in that most of the samples stored for long periods of time had low radium-226 content and those samples of high radium-226 content were usually stored less than 2 weeks, so only small errors were produced in the lead-210 content.

The activity of the lead-210 at the time of collection, A_2^0, as derived from the Bateman equation is

$$A_2^0 = \left\{ A_3 - K A_1^0 \lambda_2 \lambda_3 \left[\frac{1}{(\lambda_2 - \lambda_1)(\lambda_3 - \lambda_1)} + \frac{e^{-\lambda_2 T}}{(\lambda_1 - \lambda_2)(\lambda_3 - \lambda_2)} \right. \right.$$

$$\left. \left. + \frac{e^{-\lambda_3 T}}{(\lambda_1 - \lambda_3)(\lambda_2 - \lambda_3)} \right] \right\} \left\{ \frac{\lambda_3 \lambda_2}{\lambda_3 (e^{-\lambda_2 T} - e^{-\lambda_3 T})} \right\},$$

where the subscripts 1, 2, and 3 refer to radium-226, lead-210, and polonium-210, respectively; A_i^0 is the activity of the *i*th nuclide ($i = 1, 2,$ or 3) at the time of collection $t = 0$; A_i is the activity at the time of measurement T; and λ_i is the radioactive decay constant of nuclide *i*. K is the fractional retention of radon in the storage container. Polonium-210 originally present was determined in samples analyzed shortly after acquisition. After several months polonium-210 is redetermined, and the lead-210 is calculated from the polonium-210 growth.

Results and Discussion

Lead-210 content in surface-water samples from 18 communities was compared to the radium-226 values of Lucas and Ilcewicz (1958), as shown in Table 3. These waters had been treated by either filtration or settling. In addition, samples of untreated water from 4 of the communities were also analyzed (Nos. 19–22, Table 3).

TABLE 3

Radium-226 and Lead-210 Content of Illinois
Surface-Water Supplies

Town*	Radium-226 (pc/l)	Lead-210 (pc/l)
1. Pana (T)	0.06 ±0.03	0.000±0.007
2. Chicago	0.03	0.003±0.005
3. Canton	0.11 ±0.02	0.005±0.006
4. Mount Vernon	0.08 ±0.02	0.007±0.008
5. Macomb	0.16 ±0.02	0.009±0.010
6. Newton (T)	0.10 ±0.02	0.010±0.006
7. Nashville	0.09 ±0.02	0.015±0.007
8. Kankakee	0.10 ±0.02	0.015±0.008
9. West Frankfort	0.06 ±0.02	0.020±0.008
10. Carlinville	0.05 ±0.02	0.023±0.006
11. Grafton	0.21 ±0.03	0.024±0.008
12. Bloomington	0.12 ±0.02	0.025±0.011
13. Carbondale	0.06 ±0.017	0.027±0.006
14. Springfield	0.03 ±0.01	0.028±0.006
15. Eureka (T)	0.01 ±0.01	0.030±0.006
16. East St. Louis	0.06 ±0.02	0.031±0.008
17. Virginia	0.10 ±0.02	0.035±0.018
18. Effingham (T)	0.07 ±0.02	0.040±0.010
19. Pana (R)	0.06 ±0.03	0.085±0.017
20. Eureka (R)	1.32 ±0.05	0.101±0.017
21. Effingham (R)	0.14 ±0.04	0.117±0.013
22. Newton (R)	0.20 ±0.04	0.206±0.017
Averages: All	0.14 ±0.27	0.039±0.043
(T)	0.084±0.048	0.019±0.012
(R)	0.43 ±0.60	0.127±0.054

* Letters in parentheses: (T) treated water; (R) untreated water (raw).

The average radium-226 concentrations in these waters was 0.14 ± 0.27 pc/l in all waters, 0.084 ± 0.048 pc/l in treated waters, and 0.43 ± 0.60 pc/l in untreated waters. The average lead-210 concentrations were somewhat lower: 0.039 ± 0.076 pc/l in all waters, 0.019 ± 0.012 pc/l in the treated waters, and 0.127 ± 0.054 pc/l in the untreated waters (Pana, Eureka, Effingham, and Newton, Ill.). The treated waters from these latter towns have an average concentration of 0.020 ± 0.018 pc/l, indicating an average removal of 85 per cent of the lead-210 by treatment. However, the removal was highly variable, ranging from 40 to 90 per cent for the individual samples.

The low lead-210 concentrations, one-hundredth to one-tenth that in rain water, indicate that the waters are very old with respect to lead-210 decay—that is, the average water precipitated 70 years prior to measurement—or that removal by biological and chemical activity is quite significant. The importance of the latter mechanism was shown by Rama and Goldberg (1961) from measurements of waters along the Colorado River in which the lead-210 depletion rate was much greater than could be accounted for by radioactive decay alone.

TABLE 4

RADIUM-226, LEAD-210, AND POLONIUM-210 CONTENT OF UNTREATED
ILLINOIS WELL-WATER SUPPLIES

Town, Well	Radium-226 (pc/l)	Lead-210 (pc/l)	Polonium-210 (pc/l)	Well Depth (ft.)	Aqui-fer*
1. Toulon 2	3.32 ± 0.06	0.000 ± 0.008	-0.010 ± 0.011	780	L
2. De Kalb 1	3.16 ± 0.06	0.009 ± 0.012	0.000 ± 0.007	1331	DS
3. Prophetstown 1	0.57 ± 0.03	0.014 ± 0.009	0.056 ± 0.011	235	U
4. Granville 1	7.92	0.014 ± 0.016	0.069 ± 0.023	1792	DS
5. Cuba	22.2 ± 0.3	0.014 ± 0.016	0.006 ± 0.007	1600	DS
6. Aledo 3	1.80 ± 0.10	0.014 ± 0.010	0.066 ± 0.012	1172	DS
7. Rockford 4	3.65 ± 0.08	0.017 ± 0.016	0.007 ± 0.008	1400	DS
8. Granville 2	8.20 ± 0.16	0.018 ± 0.016	-0.007 ± 0.009	1792	DS
9. Lockport 2	10.3 ± 0.4	0.022 ± 0.019	-0.004 ± 0.009	1555	DS
10 Lockport 3	9.34 ± 0.20	0.022 ± 0.019	-0.001 ± 0.009	1579	DS
11. Ottawa 8	6.60 ± 0.20	0.022 ± 0.012	0.016 ± 0.010	1180	DS
12 De Kalb 7	3.16 ± 0.06	0.030 ± 0.01616	-0.002 ± 0.009	1330	DS
13. Shannon 1, 2	12.4 ± 0.22	0.035 ± 0.016	0.005 ± 0.007	700	DS
14. La Grange 3	0.95 ± 0.02	0.035 ± 0.011	-0.001 ± 0.005	475	L
15. Toulon 1	22.6 ± 0.4	0.036 ± 0.029	-0.004 ± 0.007	1448	DS
16. Joliet (Jasper)	9.41 ± 0.27	0.049 ± 0.016	-0.003 ± 0.006	1565	DS
17. Rockford 14	0.08 ± 0.02	0.051 ± 0.012	0.011 ± 0.009	235	U
18. La Grange 1	0.95 ± 0.02	0.052 ± 0.016	. . . .	277	L
19. Keithsburg	0.28 ± 0.02	0.062 ± 0.015	. . . .	30	U
20. Hennepin	0.19 ± 0.02	0.066 ± 0.017	. . . .	800	L
21. Peoria 12	0.02 ± 0.05	0.080 ± 0.038	0.022 ± 0.013	140	U
22. Ottawa 7	4.38 ± 0.10	0.081 ± 0.025	0.000 ± 0.010	1185	DS
23. Peoria 4	0.02 ± 0.05	0.081 ± 0.016	0.008 ± 0.006	50	U
24A. La Salle 3, 4, 5	0.45 ± 0.07	0.138 ± 0.019	0.066 ± 0.012	50–60	U
24B. La Salle 3, 4, 5	0.55 ± 0.02	0.21 ± 0.03	. . . .	50–60	U
25. Peoria 7	0.47 ± 0.04	0.15 ± 0.02	. . . .	91	U

* Aquifer: L, limestone; DS, deep sandstone; U, unconsolidated.

The radium-226, lead-210, and polonium-210 concentrations, well depth, and type of aquifer (limestone, deep sandstone, or unconsolidated) in each of 28 untreated well waters are given in Table 4 in order of increasing lead-210 concentration. The average values are given in Table 5. The radium-226 in solution, which is a likely source of the lead-210, ranges in concentration from 0.02 to 22.6 pc/l with an average of 5.1 ± 3.9 pc/l. This may be compared to the lead-210 concentrations, which average 0.051 ± 0.042 pc/l and range from 0 to 0.21 pc/l, which is from a fraction of a per cent up to 40 per cent of the radium-226 concentrations. The low lead-210–radium-226 ratio indicates an effective loss of the lead-210, due possibly to the presence of lead precipitants, hydrogen sulfide, sulfate ion and biological activity, and the highly porous aquifer. The loss of lead-210 is even more striking when it is considered that the concentration of its precursor, radon-222, is about 100 pc/l and ranges upward to 3×10^5 pc/l (Smith, Grune, Higgins, and Terrill, 1961). In New Hampshire and Maine the average values are 2,000 and 5,000 pc/l, respectively (Smith, Grune, Higgins, and Terrill, 1961). In Illinois waters, however, the concentrations are usually less than 1,000 pc/l, as shown by Scholl (1917) and by our measurements ranging from 30 to 300 pc/l. Since the radon is probably in a steady state prior to removal of the water from the well, radon loss is not a likely cause of

TABLE 5

AVERAGE LEAD-210, POLONIUM-210, AND RADIUM-226 IN WELL WATERS

Source	Lead-210 (pc/l)	Polonium-210 (pc/l)	Radium-226 (pc/l)	Well Depth (ft.)
1. Over-all...............	0.051±0.042	0.016±0.030	5.1 ±3.9	879±572
2. Wells, 0–300 ft...........	0.083±0.050	0.033±0.041	0.35±0.33	130± 95
3. Wells, 300–1,792 ft.	0.028±0.021	0.011±0.022	7.6 ±6.6	1,276±390
4. Wells with RaF>0.010....	....	0.044±0.031	....	....

the low lead-210 concentrations, and thus these low lead-210 values indicate that the residence time of this nuclide in the water is short, since decay of 1,000 pc/l of the 3.8-day radon-222 would contribute 0.08 pc/l of lead-210 per day and the observed average concentration is only 0.051 pc/l.

The lead-210 concentration, with a large scatter, appears to be inversely correlated with the radium-226 concentration; that is, the lead-210 decreases with increasing radium-226, as shown in Figure 1. More significantly, Figure 2 shows that the lead-210 concentration decreases with increasing well depth; the concentration appears to be greater in the waters originating in shallow wells, which, as shown in Table 4, consist mostly of unconsolidated formations. These formations either remove the nuclide less effectively than do consolidated ones or contribute more of the nuclide itself to the water.

There appears, in well waters, to be no correlation between lead-210 and mineral content as represented by sulfate, calcium, and fluoride concentrations reported by the Illinois State Water Survey (1961).

The polonium-210 concentrations were not determined in the surface waters because at the time of measurement most of the original material had been lost through radioactive decay. In well waters no correlation was apparent between the concentrations of polonium-210 and those of radium-226 and lead-210 or between the polonium-210 and well depth or aquifer (Table 4). The average polonium-210 concentration of 0.016 ± 0.030 pc/l was lower than that of the lead-210, and the wells appear to be classifiable into two categories, those

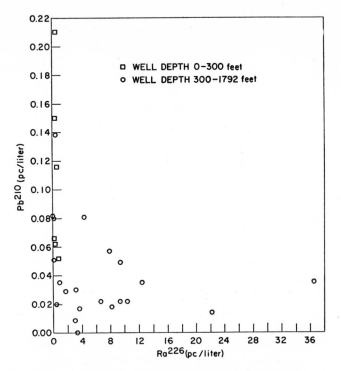

Fig. 1.—Lead-210 *vs.* radium-226 content of well waters

containing less than and those more than 0.01 pc/l of polonium-210. The average concentration in the latter group was 0.044 ± 0.031 pc/l. The low average polonium-210 concentration indicates, first, that this nuclide, relative to lead-210, is not selectively leached from the aquifer and, second, that either it is selectively precipitated or the concentration of its precursor, lead-210, averaged over time is less than the measured values.

Because of interest in the contribution of water to the total lead-210 content of the human body and because most samples were obtained at the pump, it is desirable to estimate similarity of the lead-210 content of this water to that of the water consumed. Surface water is probably identical at both locations, since the radon-222 concentration is usually low and precipitation might also be expected to be small. Well waters, as sampled here, despite the relatively high radon content, are also probably very similar to that at the tap because these waters were stored before analysis for periods of time comparable to transit time through the distribution system. In general, the low concentrations indicate that no gross changes are likely to occur during storage or distribution.

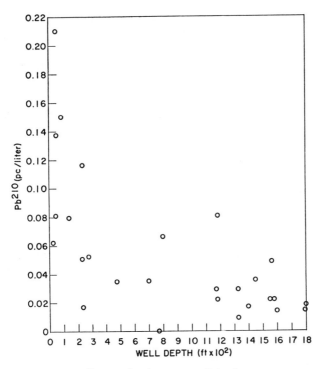

Fig. 2.—Lead-210 *vs.* well depth

The effect of lead-210 concentration in the water on the skeletal content of populations consuming this water was estimated in a previous calculation of lead and lead-210 metabolism in the human body (Holtzman, 1960). The exponential model of isotope metabolism of the ICRP report (1960) was used. It is assumed that the effective half-life of lead-210 in the skeleton is 3,350 days (Holtzman, 1962), consumption of water is 1.5 l/day, the concentration of lead-210 is 0.2 pc/l (the highest value of the series in this paper), the skeletal mass is 2,800 gm. of ash (Holtzman, 1962), and the fraction of the ingested lead entering the skeleton is 0.08. The contribution of lead-210 to the skeleton is therefore about 0.017 pc/gm ash. If a lead-210 concentration in water of 0.051 pc/l (average of the well waters) is assumed, the contribution is 0.004 pc/gm of skeletal ash. Since the average skeletal concentration of subjects investigated in this laboratory was 0.15 pc/gm ash, the maximum water content and the average water content seem to contribute 11 and 3 per cent, respectively. Average surface water (0.019 pc lead-210/l) would contribute about 1 per cent only of the total body content.

In conclusion, except for cistern water, which is usually unpotable anyway (American Water Works Assoc., 1950), the sampling of waters appears to be representative of the water consumed by most of the population of Illinois and probably of the United States. We conclude, therefore, that potable water does not constitute an important source of lead-210 in the human skeleton. Study of lead-210 and polonium-210 concentrations, however, may be useful, along with other naturally occurring nuclides such as tritium and carbon-14, in the determination of the origins and ages of water sources and of the history of radon-222 in the water. In addition, as suggested by Rama and Goldberg (1961), it may be useful as a guide to the study of the chemical and biological processes modifying the composition of these waters.

Acknowledgments

Special thanks are due Mr. H. F. Lucas for the use of some of his samples and for his many helpful suggestions. Thanks are also due Mr. L. D. Marinelli for his helpful discussions and Messrs. F. Markun and T. Kinsella and Mrs. J. Kann for technical assistance.

References

ABBATT, J. D., J. R. A. LAKEY, and D. J. MATHIAS. 1960. Natural radioactivity in West Devon water supplies. Lancet, **2**:1272–74.

AMERICAN WATER WORKS ASSOCIATION. 1950. Water Quality and Treatment: A Manual, 2d ed. New York: American Water Works Assoc.

BLACK, S. C. 1961. Low-level polonium and radiolead analysis. Health Physics, **7**:87–91.

BURTON, W. M., and N. G. STEWART. 1960. Use of long-lived natural radioactivity as an atmospheric tracer. Nature, **186**:584–89.

DI FERRANTE, E. R. 1961. Further studies on natural radioelements in bovine bones and teeth. Argonne Nat. Lab., Radiological Phys. Div., Semiannual Rpt. ANL-6398, pp. 71–76.

HANSON, R. Additions to public ground-water supplies. 1961. Ill. State Water Survey, Suppl. to Bull. 40.

HILL, C. R. 1960. Lead-210 and polonium-210 in grass. Nature, **187**: 211–12.

——. 1962. Identification of alpha emitters in normal biological materials. Health Physics, **8**:17–25.

HOLTZMAN, R. B. 1960. Some determinations of the RaD and RaF concentrations in human bone. Argonne Nat. Lab., Radiological Phys. Div., Semiannual Rpt. ANL-6199, pp. 94–118.

——. 1961. Critique on the half-lives of lead and RaD in the human body. Argonne Nat. Lab., Radiological Phys. Div., Semiannual Rpt. ANL-6297, pp. 67–80.

——. 1962. Desirability of expressing concentrations of mineral-seeking constituents of bone as a function of ash weight. Health Physics, **8**:315–19.

HURSH, J. B. 1960. Natural lead-210 content of man. Science, **132**:1666–67.

INTERNATIONAL COMMISSION ON RADIOLOGICAL PROTECTIONS. 1960. Report of Committee II on Permissible Dose for Internal Radiation (1959). Health Physics, **3**:1–233.

KURODA, K., and Y. YOKOYAMA. 1948. On the equilibrium of the radioactive elements in the hydrosphere. II. Bull. Chem. Soc. Japan, **21**:58–63.

LOCKHART, L. B., JR., R. A. BAUS, R. L. PATTERSON, JR., and I. H. BLIFFORD, JR. 1958. Some Measurements of the Radioactivity of the Air during 1957. Washington, D.C., Naval Research Lab. Rpt. NRL-5208. Pp. 19.

LUCAS, H. F., JR. 1961. Correlation of the natural radioactivity of the human body to that of its environment: Uptake and retention of Ra-226 from food and water. Argonne Nat. Lab., Radiological Phys. Div., Semiannual Rpt. ANL-6297, pp. 55–66.

LUCAS, H. F., JR., and F. H. ILCEWICZ. 1958. Natural radium-226 content of Illinois water supplies. J. Am. Water Works Assoc. **50**:1523–32.

Public ground-water supplies in Illinois. Ill. State Water Survey Bull. 40.

RAMA, M. K., and E. D. GOLDBERG. 1961. Lead-210 in natural waters. Science, **134**:98-99.

SCHOLL, C. 1917. Radioactivity of Illinois waters. Univ. Illinois Bull., Water Survey Series, No. 14, pp. 114–39.

SMITH, B. M., W. N. GRUNE, F. B. HIGGINS, JR., and J. G. TERRILL, JR. 1961. Natural radioactivity in ground water supplies in Maine and New Hampshire. J. Am. Water Works Assoc., **53**:75–88.

STEHNEY, A. F., and H. F. LUCAS, JR. 1956. Studies on the radium content of humans arising from the natural radium of their environment. Proc. Internat. Conf. on Peaceful Uses of Atomic Energy, **11**:49–54.

LARRY D. SAMUELS

13. A Study of Environmental Exposure to Radium in Drinking Water

R ADIUM IN SMALL AMOUNTS is almost ubiquitous in the world. It is found in surface soil, in surface and ocean water, and in underground strata.

Arising from these strata, it has been measured in water obtained from wells in several parts of the world. Among these studies in the United States are the measurements of radium-226 in New England by Grune and his associates (Smith *et al.*, 1961) and in the Midwest by Stehney, Lucas, and co-workers at Argonne National Laboratory (Lucas and Ilcewicz, 1958). Hursh (1957) has also measured the radium-226 content in many water supplies. The Argonne discovery of unusually high radium-226 levels was made almost by accident in the testing of municipal water supplies of the towns surrounding Argonne National Laboratory. The testing has been extended throughout the Midwest until now the involved area has been mapped out rather clearly.

STUDY DESIGN

The population exposed to radium-226 is now the subject of intensive investigation by the U.S. Public Health Service, Division of Radiological Health. The Midwest Environmental Health Study has attacked the problem from several aspects:

1. Delineation of exposed group and investigation of hydrological variables influencing exposure.
2. Measurement of uptake of radium-226 in persons consuming involved waters.

LARRY D. SAMUELS is project officer, Midwest Environmental Health Study, Research Branch, Division of Radiological Health, Public Health Service, Department of Health, Education, and Welfare, Iowa City, Iowa.

3. Analysis of factors affecting uptake of radium-226 into the exposed population.
4. Search for effects consequent to exposure to radium-226 at the measured environmental levels.

The last aspect is, of course, of great public health significance. At present our efforts are directed to a thorough investigation of bone cancer in the study area, as well as an investigation of deaths from other diseases, in an effort to ascertain whether excess deaths exist in towns with high radium-226. Since this symposium is not directly concerned with the effects studies, I shall not elaborate further on this aspect of the study.

GEOLOGY AND GEOGRAPHY OF STUDY AREA

The involved geological formations are those deep sandstone aquifers known as the Galesville, Mt. Simon, and St. Peter formations, prominent in Illinois and Wisconsin, and the Jordan formation, of importance across the state of Iowa (see Fig. 1). The Galesville, Mt. Simon, and Jordan formations date from the Cambrian period, while the St. Peter formation dates to the Ordovician system, all about 500 million years old (Iowa State Geologic Survey, Iowa City, Iowa, 1962). The exact origin of the radium found in these aquifers is uncertain. The suggestion has been made that it percolates from shale overlying the aquifers, but this is not confirmed (Brown, personal communication).

Geographically (Fig. 2), the area involved covers 300,000 square miles in four states, from central Wisconsin and southern Minnesota southward to central Illinois and southern Iowa, including the northern extremes of Missouri. An additional area may be of importance in southwestern Missouri.

Especially at the southern limits of the mapped area, the involved aquifers are deep, about 2,000 feet below the surface, so that they are penetrated only by municipal and some industrial wells, since aquifers adequate for private needs are generally available at much shallower depths. Some towns utilize a combination of deep wells and shallow wells or surface water for their water supplies (Brown, personal communication).

POPULATION EXPOSED TO HIGH-RADIUM WATER

The population exposed to the elevated-radium water supplies numbers about 1 million. This figure includes the people exposed to some elevation above the level of 3 picocuries (pc.) per liter set as the lower limit of an "elevated radium level." This figure was selected because 3 pc/l is the suggested limit of radium-226 content of drink-

SYSTEM	SERIES	GROUP	FORMATION	DESCRIPTION	THICKNESS	(Million Years)
Ordovician	Cincinnatian		Maquoketa	dolomite and shale	300'	
	Mohawkian		Galena	dolomite and chert	320'	
			Decorah	limestone and shale	70'	
			Platteville	limestone, shale and sandstone		
	Chazyan		ST. PETER	sandstone	50'–230'	
	Beekmantown		Prairie du Chien	sandy and cherty dolomite and sandstone	290'	500
Cambrian	St. Croixan	Trempealeau	Madison[a]			
			JORDAN	sandstone	185'	
			Lodi[a]			
			St. Lawrence	dolomite		
		Dresbach	Franconia	glauconitic sandstone, siltstone, shale	160'	
			GALESVILLE	sandstone		
			Eau Claire	sandstone and shale, dolomite	550'	
			MT. SIMON	sandstone		600?
Precambrian				sediments (sandstones), igneous, and metamorphic rocks		

[a] recognized only in extreme northeast Iowa

Adapted from Iowa Geological Survey Mapping

Fig. 1.—Stratigraphic mapping of high radium-226 aquifers. Aquifers under study are shown in boldface

ing water (Public Health Service, 1962). Within this population there are about 50,000 people exposed to over 10 picocuries of radium-226 per liter. In contrast, exposure to radium-226 from foodstuffs alone has been estimated as averaging 1 picocurie per day. This estimate is borne out by measurement of the whole-body radium-226 burden of individuals without known exposure to high radium-226 water supplies. Since the majority of radium-226 consumed in the study towns is from water consumption, the slight variation due to dietary variation has been ignored.

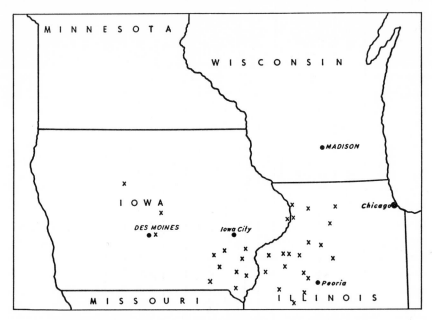

Fig. 2.—Map of radium-226 water study area; x = study towns

A control population, exposed to radium-226 from foodstuffs only, has been selected from the same area as that of the high radium-226 towns, except that the control towns have surface or shallow-well water supplies containing very small quantities of radium-226.

Factors Influencing Exposure and Uptake: Hydrologic Variables

The absolute level of exposure of the population to radium-226 is influenced by several factors, among the most important of which is water treatment. Further study is needed to determine the extent to which various water treatments modify radium-226 levels. Limited studies carried out in Iowa have involved the investigation of one system, the Zeolite softening process. In several Iowa towns this system is used at the municipal level. Morris and Klinsky (1962) have found that the Zeolite softening process removes up to 99 per cent of the radium-226 contained in raw water. In the tested towns, the usual process of "blend back," mixing raw and softened water, left a net reduction of about 70 per cent in radium-226 content of finished water.

Other factors of obvious importance in influencing the level of exposure to radium-226 are the admixtures of other, low radium-226 supply sources, home water-softening systems, private shallow wells providing drinking water, and variations in the radium-226 level of the raw water.

Pumping rate may sharply influence radium-226 levels observed in raw water. One city where this has been observed has had raw water tested at levels of radium-226 ranging from 3 to 80 pc/l. Local commingling of high and low radium-226 aquifers is facilitated by numerous uncased or poorly cased wells, many abandoned, which are scattered throughout the study area (Brown, personal communication).

Stability of water supply in the study towns, especially in Illinois, has been sought by comparing recent and past chemical assays of the water supplies. Where there has been little or no change in the chemical profile of a water supply, it has been accepted as reasonably stable in character (Larson, personal communication). These estimates are necessary, of course, because radium-226 measurements in water supplies are a relatively recent innovation.

With one exception, all Illinois towns chosen for study have stable water supplies, as determined by comparative chemical studies. In this way we have minimized the effect of hydrological variables on the exposure of the study population. Such chemical studies were not possible in choosing all the Iowa study towns, with the results apparent in Table 3 of varied tooth to water radium ratios.

FACTORS INFLUENCING EXPOSURE AND UPTAKE:
OTHER VARIABLES

A human variable of considerable importance is the great variation in individual water consumption. In the youthful school population from whom teeth have been collected, the most frequent estimate of water consumption is under 4 glasses per day. Only a few children estimate consumption as over 6 glasses per day.

The *in utero* deposition, of course, depends on maternal water consumption, a variable difficult to quantitate with confidence. Exposure in early infancy, of great importance to the radium-226 burden acquired by the deciduous teeth, is influenced by the mode of feeding, with respect to whether the child is breast or bottle fed, whether liquid or powdered formula is used, and whether any special water is used to make the formula. The importance of these factors is under study, but no data are yet available. Stehney and Lucas (1955a) have shown that retention of radium-226 ingested in water is greater than that of radium-226 in food.

The metabolic activity of hard tissue is of importance in determining calcium turnover, which in turn affords an opportunity for radium-226 atoms to be inserted into the bone and tooth matrix. Varied activity of calcium turnover causes fluctuation in radium-226 uptake and excretion, which also adds to the difficulty of metabolic studies of radium-226 at environmental levels of exposure.

TABLE 1

EFFECT OF FLUORIDE ON RADIUM-226 UPTAKE

Radium-226 Content of Water (pc/l)	Fluoride Content of Water (p.p.m.)	Ratio Body Burden (pc.)* to Water Content (pc/l)
12.0 (av.)	0–0.9	27.9
8.2 (av.)	1–2.9	43.9
11.4 (av.)	3+	34.2
Average		35.3

* Body burden computed for standard men, with 3,000 gm. skeletal ash.

In vitro evidence by Posner has suggested that fluoride influences the uptake of strontium into hard tissue. Posner (personal communication) has further suggested that radium-226 should be similarly affected by fluoride. Since fluoride is found at various levels in the high radium-226 water supplies being studied, this relationship is being investigated (see Table 1).

RADIUM BODY-BURDEN MEASUREMENT:
REVIEW OF METHODS

Measurement of radium body burdens was begun over thirty years ago in persons exposed to radium-226 occupationally and therapeutically. The levels studied at this time were very much higher than the levels of body burden under consideration here and lent themselves to the rather crude methods of study available (Barker and Schlundt, 1930; Evans, 1933).

With the advent of modern instrumentation and methodology, many of this same group of people have been extensively studied by workers at the Massachusetts Institute of Technology, Argonne National Laboratory, and the New Jersey Department of Health. Whole-body counting of γ emission from radium-226 and its daughters has afforded a reasonably accurate measure of body burdens above 1,000 pc. (Evans, personal communication). *In vitro* measurement of

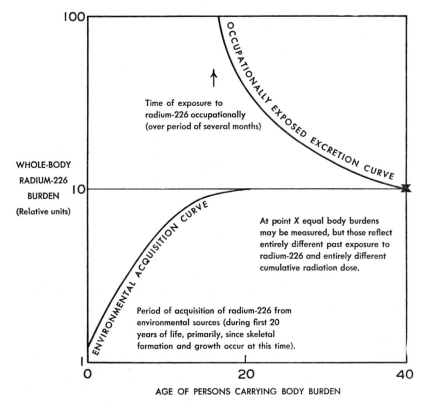

Fig. 3.—Comparison of occupational and environmental exposure to radium-226 with respect to observed body burdens.

radium-226 body burden has been successfully accomplished by analysis of bones removed from persons bearing high radium-226 burdens. This has been possible in persons exposed to radium-226 occupationally but is not readily adaptable to a large population study such as was needed in the Midwest. In fact, environmental exposure results in a different body-radiation exposure situation (Fig. 3).

Measurements of radium-226 body burdens of individuals exposed only environmentally have been carried out by several workers

(Hursh and Gates, 1950; Palmer and Queen, 1958; Walton *et al.*, 1959). Whole-body radium-226 burdens determined from ashed whole-body samples range from 48 to 150 picocuries. Measurement of radium-226 body burden by this method has not been done with persons known to be exposed to elevated radium-226 drinking water, although Stehney and Lucas (1955*b*, 1956) have studied it by other methods, and their well-known studies of Lockport boys and Illinois convicts have provided the estimate that the radium-226 body burden following long-term ingestion is about equal to the radium-226 contained in 40–50 liters of water.

Other workers have suggested that radium-226 body burdens may be measured by the analysis of breath radon-222, urinary or fecal excretion of radium-226, or measurement of radium-226 in hair or

TABLE 2

COMPARISON RESULTS OF TOOTH ANALYSIS

Investigator	Method of Measure	Exposure History	Radium-226 Content of Tooth (pc/gm ash)
Lucas (1960)	Radon-222 emanation	Water 4.6 pc/l	0.103
Radford (1962)	α-prop. count	Food only	0.016
Samuels (this study)	Radon-222 emanation	{Food only	0.013
		{Water 5.3 pc/l	0.11

nails. Breath radon-222 has not yet been evaluated in our study. Analysis of urinary or fecal excretion of radium-226 is fraught both with the problems of very low-level analysis, where sample differences are masked by errors in measurements, and with sampling problems. Since ingested radium-226 is largely eliminated from the bowel without being absorbed, most of the fecal radium-226 does not originate from the internal body burden. For this reason, fecal radium-226 content can be of value in determining ingestion, and as such it has been used by Stehney and Lucas (1955*a*, *b*). Urinary excretion of radium-226 averages only about 10 per cent of fecal excretion, so urinary radium-226 is even smaller in amount (Evans, personal communication).

A recent report of the strontium analysis of hair indicates a wide variability of strontium content of hair compared with actual hard-tissue body burden and points up the large error involved in measuring a hard-tissue body burden by analysis of an uncalcified tissue (Hopkins *et al.*, 1963). Strontium injected into rats in known amounts was recovered from shaved hair in about the same amounts in animals with a hundred-fold difference in body burden.

Evans and co-workers (Evans, 1960) at the Massachusetts Institute of Technology have advanced the suggestion that teeth afford a valid measurement of radium-226 body burden. Workers at Argonne National Laboratory have also measured teeth (Lucas, 1960). This has been pursued by Radford (personal communication) at Harvard, who has determined that teeth and bones contain about equal amounts of radium-226, in terms of either calcium or ashed weight. Radford's specimens were collected from New England residents, whose exposure to radium-226 had presumably been entirely environmental.

Radium Body-Burden Measurement: Methods Used in This Study

Tooth measurement has been used in our study as the principal means of studying body burden. In addition, selected bones and matched bone–tooth samples are to be analyzed. To minimize the effect of population mobility and to facilitate collection, deciduous teeth from school-age children have been collected from selected study towns throughout the involved area of the Midwest.

After teeth are obtained from the contributing children, the children are rewarded with a button proclaiming "I gave a tooth to science," in return for which a detailed history of lifetime residence and water-consumption habits is obtained by questionnaire. Teeth are grouped into samples of 1 or 2 grams, containing from 5 to 20 teeth, on the basis of matching exposure histories as listed on the questionnaires. Some samples were grouped as to type of tooth, some according to amount of water consumed, and others by residence alone.

All tooth analyses to date have been carried out at the Radiological Physics Division, Argonne National Laboratory, by Mr. Henry Lucas (1957), using the radon-222 emanation technique developed in his laboratory.

Secondary teeth have been collected, in limited quantity, from dentists practicing in the towns under study. The analysis of these teeth has not yet begun because of several problems. Very seldom is an adult donor a lifelong resident of a given study town. In most cases the residence in childhood, at the time of formation of the adult teeth, was not even within the study area. As a further complication, most of the adult teeth collected contain extensive caries, which have been shown by workers in St. Louis to influence the adult uptake of strontium and may also influence uptake and retention of radium-226 (Bird, personal communication).

Analysis of the few bones that it has been possible to obtain is in progress. From the reports of Radford (personal communication), a significant correlation between bones and teeth is expected. Lucas (1960) has already reported the analyses of a small series of bones taken from residents of the study area.

Results of Tooth Analysis

From the 35 towns being studied we have collected about 5,000 deciduous teeth to this date. About 1,000 of these fulfill the criterion of lifelong residence in the study town and are suitable for analysis. Preliminary results of analysis of several hundred of these teeth, grouped into samples of 5–15 teeth, representing 12 study towns, with 1–3 pooled samples per town, are shown in Table 3.

TABLE 3

RESULTS OF ANALYSIS OF DECIDUOUS TEETH

Town	Radium-226 (pc/l)	Type of Teeth	Gm. Ash Sample	pc/gm Ash (±S.E.)
Iowa				
Eldon..................	41.0	Mixed	1.01	0.013±0.011
Fairfield..............	11.1	Molar	2.36	0.023±0.006
Fairfield..............	11.1	Incisor	1.4	0.023±0.008
Morning Sun..........	8.5	Molar	1.94	0.070±0.011
Morning Sun..........	8.5	Incisor	0.4	0.087±0.030
Mount Pleasant.......	11.1	Molar	2.73	0.09 ±0.01
Wellman..............	1.7	Mixed	1.24	0.036±0.013
Wellman..............	1.7	Molar	1.01	0.02 ±0.01
Illinois				
Alpha-Woodhull........	4.2– 6.4	Incisor	0.77	0.12 ±0.02
Alpha-Woodhull........	4.2– 6.4	Cuspid	2.13	0.10 ±0.01
Depue................	9.4	Incisor	0.75	0.11 ±0.03
Depue................	9.4	Cuspid	1.02	0.11 ±0.02
Depue................	9.4	Molar	2.79	0.12 ±0.01
Glasford.............	11.6	Mixed	0.76	0.180±0.028
Knoxville.............	6.3–10.3	Incisor	0.63	0.13 ±0.03
Shannon.............	12.4–17.1	Molar	2.15	0.11 ±0.01
Viola................	8.1	Mixed	1.49	0.11 ±0.01

Table 1 shows the preliminary data relating radium-226 and fluoride. It appears that while fluoride may play a significant role in the uptake of heavy elements, such as strontium or radium-226, into hard tissue already formed, fluoride has a less important role in the exclusion of these heavy ions at the time of first laying-down of the crystal lattice.

Value of Tooth Analysis for Radium-226: Specific Examples

Some examples of problems of radium-226 body-burden measurement from deciduous teeth are the following:

I. Eldon, population 1,386, had its water supply from low radium-226 shallow supplies until 1960, when Jordan sandstone wells were drilled by both the city and the school. Analysis of deciduous teeth, formed from 1951 to 1955, exposed only to low radium-226 water until 1960 and to high radium-226 water averaging 40 pc/l until exfoliation in 1962, show 0.013 pc/gm ash, no elevation above levels reported from radium-226 in food alone, suggesting that radium-226 burden is acquired at the time of formation of hard tissue. Presumably, growing bones of Eldon children are now incorporating radium-226, so in this town tooth measurement provides an erroneously low estimate of radium-226 body burden.

II. Fairfield, population 8,054, has had a Jordan sandstone supply since 1957. Prior to that time, low radium-226 shallow wells were used. During this 5-year exposure, there has been a slight increase in tooth radium-226 content, to 0.023 pc/gm ash. The increase is only about one-tenth of the radium-226 burden expected if the radium-226 exposure had existed for the lifetime of the teeth, 6–11 years. Very soon, however, incisor teeth formed in 1957 will be shed, and they should show a significant elevation in radium-226 content. In this instance the radium-226 content would be lower in molar teeth shed at the same time as the high radium-226 incisors because the molar teeth were formed before the high radium-226 exposure began.

III. Mount Pleasant, population 7,339, has multiple wells, only some of which produce water high in radium-226 content. Integrated sampling of uptake, as measured by tooth assay, suggests that over half the water consumed is that from the measured high radium-226 wells, since the observed radium-226 content of teeth is 0.09 pc/gm ash. Measurement of the long-term contribution of high radium-226 wells to the town's water supply has not henceforth been possible.

VALUE OF TOOTH ANALYSIS FOR RADIUM-226:
GENERAL IMPRESSIONS

A. Radium-226 burden from environmental sources is mainly acquired at the time of the laying-down of hard tissue.

B. Exposure to elevated radium-226 water after the teeth are formed has a relatively slight effect on uptake of radium-226 in deciduous teeth.

C. Measurement of radium-226 contained in teeth affords a good index to the radium-226 content of the water supply of the town of which the donors of the teeth were resident during infancy.

D. There is no significant difference in radium-226 uptake and retention by type of teeth, comparing molar, cuspid, and incisor groups.

E. Uptake and retention of radium-226 from foodstuffs alone, where water contains very little radium-226, totals about 0.01–0.015

pc/gm tooth ash. This uptake and retention reflects a combination of maternal uptake and placental transfer and subsequent uptake and retention in infancy.

F. Fluoride does not have an apparent effect on the initial uptake and retention of radium-226 in teeth.

G. Tooth analysis has shown that average calculated body burdens in residents of Midwest towns who are consuming water containing elevated levels of radium-226 are equal to the radium-226 contained in 35 liters of water, or about 35 times the amount of radium-226 ingested per day. Potential body burdens in the study towns would thus reach a calculated 1,400 pc. While this is a hundred-fold less than a radium body burden known to produce any effects, it is sufficient to permit further study.

Summary

I have presented a brief outline of the organization and operation of an environmental health survey of a population exposed to increased amounts of naturally occurring radium-226. Included in the presentation are representative results of tooth analysis and some tentative conclusions suggested by the results of these analyses. It should be emphasized that no firm conclusions should be drawn until the study has proceeded to its conclusion and all data are assembled for study.

Acknowledgments

The Midwest Environmental Health Study herein described is a cooperative venture of the Research Branch, Division of Radiological Health, Public Health Service, and the Atomic Energy Commission.

I am especially indebted to Messrs. Henry Lucas, John Rose, and L. D. Marinelli, of the Radiological Physics Division, Argonne National Laboratory, for performing radium-226 assays on tooth and bone specimens.

I am also indebted to Mr. T. E. Larson, of the Illinois State Water Survey Division, for surveying Illinois water supplies; to Mr. R. L. Morris, of the Iowa State Hygienic Laboratory, for information on Iowa Water Supplies; to Mr. C. N. Brown, of the I.S.G.S. Iowa City office, for helpful discussions of the geology of the region; and to Miss Mildred Jones for assistance in obtaining tooth samples.

References

BARKER, H. H., and H. SCHLUNDT. 1930. The detection, estimation and elimination of radium in living persons. Am. J. Roentgenol. & Radium Therapy, **24**:418–23.

BIRD, J. T. Personal communication, 1962.

BROWN, C. N. Personal communication, 1963.

EVANS, R. D. 1933. Radium poisoning: A review of present knowledge. Am. J. Pub. Health, **23**:1017–23.

――. 1960. Reliability of teeth as indicators of total body radium burdens. Massachusetts Inst. Tech. Ann. Prog. Rpt. (May), pp. 5–12.

――. Personal communication, 1962.

HOPKINS. B. J., L. W. TUTTLE, W. J. PORIES, and W. H. STRAIN. 1963. Strontium-90 in hair. Science, **139**:1064–65.

HURSH, J. B. 1957. Natural occurrence of radium-226 in human subjects, in water and in food. Brit. J. Radiol. Suppl., **7**:45–53.

HURSH, J. B., and A. A. GATES. 1950. Body radium content of individuals with no known occupational exposures. Nucleonics, **7**:46–59.

IOWA STATE GEOLOGIC SURVEY. 1962. Stratigraphic column map. Iowa City, Iowa.

LARSON, T. E. Personal communication, 1962.

LUCAS, H. F. 1957. Improved low-level alpha-scintillation counter for radon. Rev. Sci. Instr., **28**:680–83.

――. 1960. Correlation of the natural radioactivity of the human body to that of its environment: Uptake and retention of Ra-226 from food and water. Argonne Nat. Lab., Radiological Phys. Div. Semi-annual Rpt. ANL-6297, pp. 55–66.

――. Personal communication, 1963.

LUCAS, H. F., JR., and F. H. ILCEWICZ. 1958. Natural radium-226 content of Illinois water supplies. J. Am. Water Works Assoc., **50**:1523–32.

MORRIS, R. L., and J. KLINSKY. 1962. Radiological Aspects of Water Supplies. Iowa City: State Hygienic Lab., Univ. Iowa.

PALMER, R. F., and F. B. QUEEN. 1958. Normal abundance of radium in cadavers from the Pacific Northwest. Am. J. Roentgenol. Radium Therapy Nuc. Med., **79**:521–29.

POSNER, A. S. Personal communication, 1962.

PUBLIC HEALTH SERVICE. 1962. Drinking water standards. Public Health Service, Pub. No. 956, par. 6.21. Washington, D.C.: Government Printing Office.

RADFORD, E. P., JR. Personal communication, 1963.

SMITH, B. M., W. N. GRUNE, F. B. HIGGINS, JR., and J. G. TERRILL, JR. 1961. Natural radioactivity in ground water supplies in Maine and New Hampshire. J. Am. Water Works Assoc., **53**:75–88.

STEHNEY, A. F. 1955. Radium and thorium X in some potable waters. Acta Radiol., **43**:43–51.

STEHNEY, A. F., and H. F. LUCAS. 1955a. Retention of environmental radium by young adult males. Argonne Nat. Lab., Radiol. Phys. Div. Semiannual Rpt. ANL-5456, pp. 95–98.

――. 1955b. Intake and retention of radium at natural levels. Argonne Nat. Lab., Radiol. Phys. Div. Semiannual Rpt. ANL-5378, pp. 8–15.

――. 1956. Studies of the radium content of humans arising from the natural radium of their environment. Proc. Internat. Conf. on Peaceful Uses of Atomic Energy, **11**:49–54.

WALTON, A., R. KOLOGRIVOV, and J. L. KULP. 1959. The concentration and distribution of radium in the normal human skeleton. Health Physics, **1**:409–16.

ALLAN B. TANNER

14. Physical and Chemical Controls on Distribution of Radium-226 and Radon-222 in Ground Water near Great Salt Lake, Utah

O F THE MANY REPORTS of radon-222 and radium-226 concentrations in ground water, most have added to the catalog of known measurements without providing enough information about the source to make more quantitative our knowledge of the control of these nuclides in nature. Notable exceptions are the radon-222 and radium-226 studies of Hoogteijling *et al.* (1948), Rogers (1954, 1955*a*, *b*), Miholić (1958), and Mazor (1962) and the studies of radium-226 distribution with respect to geologic formation or terrain by Scott and Barker (1958, 1962) and Lucas and Ilcewicz (1958). Investigations relating radioactive waters to geologic structure have been made by Lane and Bennett (1934), Rogers (1955*a*, *b*), Miholić (1958), and Bondar (1962); in addition, numerous studies have been made of radioactive waters in and near oil fields. With the exception of general treatments of the hydrogeochemistry of radium-226 and radon-222 in the Soviet literature (Stchepotjeva, 1944; Tokarev and Shcherbakov, 1956; Germanov *et al.*, 1958; Sharkov, 1959), data on the conditions affecting their control in ground water appear infrequently. It is the purpose of the present study to provide chemical data in addition to radioactivity data in order that the behavior of radium-226 and radon-222 in the given environment may be better understood. Various mechanisms of radium-226 and radon-222 control in ground water are suggested, and many are rejected for the conditions prevailing in this environment.

ALLAN B. TANNER is geophysicist, Branch of Theoretical Geophysics, U.S. Geological Survey, Washington, D.C.

Publication authorized by the director, U.S. Geological Survey.

Geologic and Hydrologic Setting

Two groups of irrigation or stock wells were studied. One group is located about 1½ km. west of Woods Cross, Davis County, Utah, and about 8 km. southeast of Great Salt Lake (Fig. 1). The geography, geology, and hydrology of the district that includes the Woods Cross wells have been described by Thomas and Nelson (1948). The other group is located about 3 km. south of Burmester, Tooele County, and about 5 km. west of the southern tip of Great Salt Lake; the

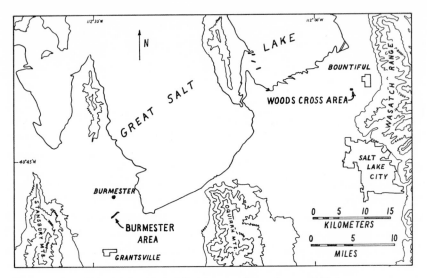

Fig. 1.—Locations of study areas in north-central Utah

area has similarly been described by Thomas (1946). Both groups are in basins of the Basin and Range province and produce water from artesian aquifers in unconsolidated sediments of Quaternary age. The flow of underground water corresponds generally in direction to the surface drainage pattern.

Ground water in the Woods Cross area moves from the Wasatch Range westward toward Great Salt Lake at an average rate that is probably between 1 foot (0.3 m.) per day, estimated by Thomas and Nelson (1948, p. 184) for an area toward the mountains, and several inches (*ca.* 0.1 m.) per day, estimated for the lakeward areas. The hydraulic gradient is about 4 m. per km. in the area of study. All wells included in the group issue water from either a shallow or an intermediate artesian aquifer. Because the aquifers had approximately coincident piezometric surfaces in the study area in 1946, Thomas and Nelson (1948, p. 164, Plate 2, and Fig. 17) infer that

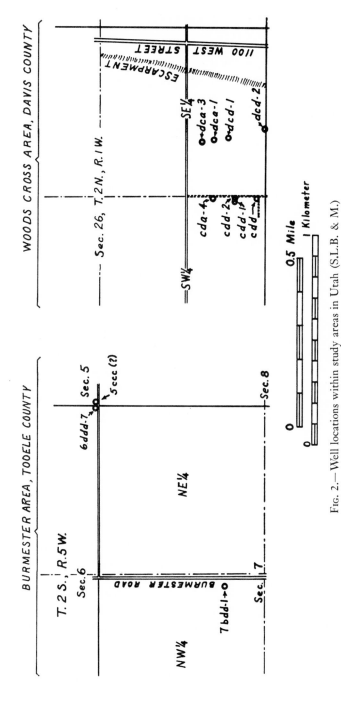

Fig. 2.— Well locations within study areas in Utah (S.L.B. & M.)

they are interconnected. Faulting in the underlying consolidated sediments has resulted in a zone of sufficient distortion of the unconsolidated sediments that an escarpment is present about $\frac{1}{2}$ km. to the east of the group of wells studied (Fig. 2). The quality of water is apparently affected. Several seeps or springs exist along the escarpment, and water temperatures of wells west of the escarpment are anomalously high (Thomas and Nelson, 1948, p. 117). The wells are in a discharge area; that is, a slight component of upward motion of ground water has been inferred by Thomas and Nelson (1948, Fig. 20).

The unconsolidated sediments are thick in both areas: in the Woods Cross area, more than 210 m.; in the Burmester area, at least 180 m.

Ground water in the Burmester area moves from the Stansbury Mountains northeastward toward Great Salt Lake at a rate that has not been estimated. The hydraulic gradient, estimated by Thomas (1946, p. 200) at about 6 feet per mile (1.1 m. per km.), is comparable with the 4 m. per km. gradient of the Woods Cross area. The Burmester wells also are in a discharge area, where ground water seeping to the surface evaporates and leaves a visible frost of salts in many places. Although a subsurface fault has been inferred for the area by Johnson (1958) on the basis of gravity data, no hydrologic effects have been noted, and the inferred fault does not correlate with a distinct pattern of water quality observed by J. S. Gates (Nolan, 1962, p. A46; Gates, written communication). Gates (1962) has observed marked effects of subsurface faulting on ground-water movement and quality in an area about 15 km. to the southeast.

METHOD OF INVESTIGATION

Most of the constituents of interest in the waters studied have a tendency to escape into or to be absorbed from the atmosphere. The special sampling precautions and expeditious analyses needed made it desirable to take a mobile laboratory to the well sites and, except as indicated, to perform the analyses on the spot.

Radon-222 was measured by extracting a water sample into a deemanation flask, boiling the dissolved gases into an evacuated ionization chamber using an argon carrier gas, and measuring the ionization current with a vibrating-reed electrometer and recording potentiometer during radioactive equilibrium among radon-222 and its short-lived decay products. After storage of the sample for a week or more, radon-222 produced by the radium-226 in the sample was measured by the same method.

Certain things that might affect the behavior of radium-226 or radon-222 were also measured: pH, redox potential (Eh), dissolved oxygen, bicarbonate, dissolved iron, water temperature, and barometric pressure.

The pH measurement was performed on an unaerated sample within several minutes of the time of collection, using glass and saturated calomel electrodes and a vibrating-reed electrometer calibrated with suitable buffers. After the pH determination the sample was titrated with a standard dilute sulfuric acid solution to pH 4.5, yielding the approximate concentration of bicarbonate (Rainwater and Thatcher, 1960, pp. 94–95).

Redox potential was measured with platinum thimble and saturated calomel electrodes in a flow chamber of the same general characteristics as that described by Back and Barnes (1961). The electrical system differed from theirs in that a continuous measurement was made with the vibrating-reed electrometer, rather than discrete measurements with a pH meter. When it was verified that the redox potential measurements were superfluous, they were discontinued.

Samples for the dissolved oxygen determination were collected in biochemical-oxygen-demand bottles. Within 20 minutes, and usually much sooner, they were fixed, and within an hour they were analyzed by the modified Winkler method (Rainwater and Thatcher, 1960, pp. 29, 233–36).

Water for dissolved iron determination was collected through plastic tubing and pipetted, with minimum exposure to the atmosphere into a polyethylene bottle containing 8 ml. of 30 per cent hydrochloric acid per 200-ml. sample. All samples except one were apparently clear in the pipette and were not filtered; a second sample of the one slightly turbid water was taken with a filter interposed between the source and the delivery tube to prevent aeration of the sample during the filtration process. The iron analyses were made by a colorimetric technique in the U.S. Geological Survey's laboratory in Salt Lake City.

All water temperatures were measured with the same precision-grade mercurial thermometer, graduated in $\frac{1}{5}°$ F. increments.

Barometric pressure was recorded on a microbarograph located at the base of operations in Salt Lake City, 23 km. from the Woods Cross area and 50 km. from the Burmester area.

RESULTS

The waters available for analysis in the areas studied, the respective results of analysis, and additional quality-of-water data, where available, are listed in Table 1 and shown in part in Figure 3. In Figure 3, wells are presented in profile along traverses passing through the well sites (Fig. 2). The Burmester traverse is nearly straight, trending

TABLE 1

ANALYSES OF ARTESIAN WELL WATERS IN UTAH

Well Designation	Reported Depth (m.)	Temperature (°C.)	Est. Flow (m³/day)	pH	Eh (volt)	Radon-222 (10⁻¹⁰ c/l)	Radium-226 (10⁻¹³ gm/l)	Fe* (p.p.m.)	O₂† (p.p.m.)	Ca (p.p.m.)	HCO₃ (p.p.m.)	Sulfide (p.p.m.)	SO₄ (p.p.m.)	Cl (p.p.m.)
BURMESTER AREA, TOOELE COUNTY														
(C-2-5)5ccc(?)	110(?)	16.2	3	7.8	......	5	20	0.06	<0.1	......	66	......	......	3620§
(C-2-5)6ddd-7	110	16.3	3	7.7	......	6	2	0.04	<0.1	......	85	......	8.6‡	1640§
(C-2-5)7bdd-1	92	14	3	8.1	......	5	≤ 1	0.02	6.5	30§	176§	......	13§	47§
WOODS CROSS AREA, BOUNTIFUL DISTRICT, DAVIS COUNTY														
(B-2-1)26cda-4	87	18.3	160	7.6	+0.45?	9	≤ 2	0.01	1.1	......	188	......	......	1169§
(B-2-1)26cdd	113	18.3	27	7.5	+0.28	18	6	0.17	<0.1	181§	133§	<0.2	37§	
(B-2-1)26cdd-1	34	17	110	7.9	+0.3-	4	≤ 1	0	4.6	......	259	......	......	
(B-2-1)26cdd-2	105	17	160	7.6	+0.5	5	2	0	5.3	......	254	......	......	
(B-2-1)26dca-1	37–52	14.4	220	......	+0.52	9	1	0.04	5.8	......	......	......	......	
(B-2-1)26dca-3	58–64	17.2	110	8.3	......	8	≤ 1	0	4.8	15§	223§	......	12§	51§
(B-2-1)26dcd-1	61–85	14.7	27	7.5	......	7	≤ 1	0.02	7.1	45§	276§	......	35§	62§
(B-2-1)26dcd-2	92	16.4	160	7.7	......	8	......	0.01	4.8	......	239	......	......	

* Dissolved or in suspension fine enough to pass through filter. Analysts, W. K. Hall and L. E. Anderson.

† Dissolved.

‡ Analyst, W. K. Hall.

§ Average value of analyses by U.S. Geological Survey.

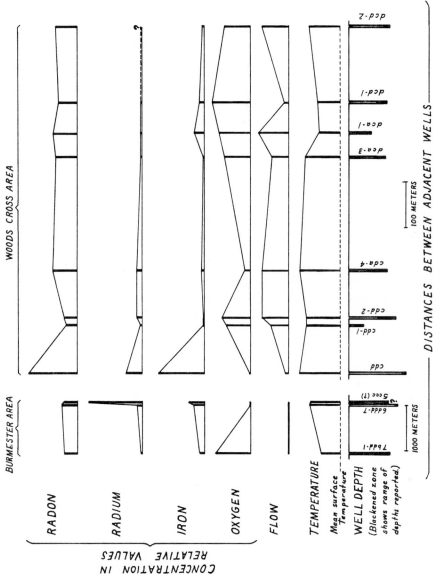

Fig. 3.—Values of selected parameters for well waters studied

northeast; the traverse in the Woods Cross area is horseshoe-shaped, concave toward the south.

In addition to the 11 wells within the study areas, 10 wells near the Burmester area were sampled for dissolved oxygen analysis. Chloride analyses are available for 11 of the 21 wells. The several samples containing more than 500 parts per million (p.p.m.) chloride contained less than 0.1 p.p.m. of dissolved oxygen. Samples containing less than 500 p.p.m. of chloride generally contained oxygen in concentrations that are appreciable fractions of the oxygen concentration at saturation (9.5 p.p.m.) under the mean surface conditions of the locality. From the quality-of-water data and the field relations it appears that the several waters having both intermediate chloride and appreciable oxygen concentrations are mixtures of low-oxygen, high-chloride water and high-oxygen, low-chloride water. One sample was anomalously low in both constituents.

Measured values of redox potential (Eh) were consistent with a plot of Eh versus dissolved oxygen shown by Germanov *et al.* (1958, Fig. 3): the presence of more than about 0.1 p.p.m. of oxygen will cause the Eh to exceed about $+0.3$ volt.

To ascertain the availability of sulfur in the (-2) state for removing oxygen in one of the low-oxygen waters, sulfide tests were made by the iodometric method (Rainwater and Thatcher, 1960, pp. 30, 287–88). One test, completed within about 15 minutes of the time of sampling, indicated a sulfide concentration of less than 0.2 p.p.m. Another test, in which the sample was fixed immediately but titrated two days later, indicated less than 0.1 p.p.m.

Bicarbonate and chloride are the two most prevalent anions in the waters of both areas, as may be seen from inspection of the quality-of-water data in Connor and Mitchell (1958, pp. 60–61 for sec. 26; p. 151 for secs. 5, 6, and 7). Bicarbonate is characteristically lower in concentration in waters containing more than several hundred parts per million chloride than in those containing less chloride. Sulfate, the third most prevalent anion, does not appear to enter into any such complementary relationship: in the Woods Cross area a high sulfate content is taken as evidence of recharge from the Upper Bonneville Canal, about $2\frac{1}{2}$ km. to the east (Thomas and Nelson, 1948, pp. 139–40); in the Burmester area the sulfate content is rather uniformly low.

The two water samples containing the greatest concentrations of iron, 0.17 and 0.06 p.p.m., were low in oxygen. Of the two samples containing 0.04 p.p.m. iron, one was low in oxygen and one was high. The latter sample was filtered within the source through three thicknesses of Whatman No. 1 filter paper; however, there is no assurance that any of the samples was free of iron in colloidal suspension. All seven samples containing 0.02 p.p.m. iron or less contained more than 1 p.p.m. oxygen.

The radium-226 analyses were difficult because of the low radioactivities involved. For samples from wells (B-2-1)26cda-4, cdd, and cdd-2 it was particularly difficult to establish absolute values, but the chart record of ionization current unquestionably shows a greater radium-226 activity in (B-2-1)26cdd than in the other two. In both areas, the greater radium-226 concentrations are associated with high-chloride, low-oxygen waters that tend to have relatively more dissolved (or colloidal) iron and relatively less bicarbonate.

As has been noted in many places, the radioactivity of radon-222 in the waters of both areas exceeds the radioactivity of its immediate precursor, radium-226, by 2–4 orders of magnitude. The range of radon-222 concentration in the Woods Cross area is about fourfold, the greatest concentration being found in the water with the highest concentrations of radium-226, iron, and chloride. Among the three wells analyzed for radioactivity in the Burmester area there is no such variation in radon-222 concentration, although the contrast in concentration is comparable for radium-226, chloride, and—to a lesser degree—iron.

Field conditions were such that it was not feasible to standardize the flow of the wells before sampling or even to measure the flow of some of them. The possible influence of flow on radon-222 concentration is discussed below.

Water Temperatures

Separate graphs (not shown) were made of water temperature against reported well depths in each area. Excluding points that are clearly anomalous, profiles of temperature *versus* depth are obtained that extrapolate to about 11.7° C. at zero depth. This value is assumed to be the mean surface temperature and is the base value for the temperature graph in Figure 3. In the absence of thermal anomalies, the temperature and well-depth bars for each well should be in proportion. That they are not indicates uncertainties in the well depths, influx of water at points above the well bottoms, thermal anomalies, vertical movement of ground water, or variation in the thermal conductivity of the sediments.

That uncertainties in well depths exist in both areas is indicated by soundings in two wells, and conflicting reports of depths are on file for several others. Almost all the wells are at least 35 years old; the casings may be corroded enough to admit sediment that would tend to partly fill the borehole.

The wells of the Woods Cross area are situated within 0.7 km. of an escarpment that is inferred to be the surface expression of bedrock faulting (Thomas and Nelson, 1948, pp. 115–16). Several springs and seeps rise there that issue water at 11.1°–11.7° C., which, according to Thomas and Nelson, corresponds to the temperature of water from wells 100–140 feet (31–43 m.) deep. Because extrapolation of the water temperatures in the wells studied yielded a similar value of about 11.7° C. at the surface, however, it is questionable whether the nearby springs and seeps are thermally anomalous. About 6 km. to the south, the escarpment leads to the surface exposure of the Warm Springs fault; Becks Hot Springs in this vicinity do yield water in the anomalous temperature range 52.2°–56.6° C. (Thomas and Nelson, 1948, p. 116).

J. S. Gates (Nolan, 1962, p. A46) has noted that a well-defined tongue of water of relatively low mineral content extends from the southwest corner of the valley in which the Burmester area of study is located toward Great Salt Lake. The tip of this tongue is defined by wells (C-2-5)6ddd-7 and 5ccc(?), which have waters high in chloride, and by well (C-2-5)7bdd-1, which has water of more than an order of magnitude lower chloride concentration. Of eleven wells in the vicinity (C-2-5)6ddd-7 is the only one producing water warmer (about 1° C.) than would be expected from the thermal profile established from the other wells. (Well [C-2-5]5ccc[?] is of unknown depth.) As (C-2-5)6ddd-7 is one of the low-oxygen, high-chloride wells, the question arises whether it and its similar companion well, (C-2-5)5ccc(?), may not indicate a thermal disturbance such as that caused by an influx of water from a subsurface fault. A well, (C-2-5)-5dcd-4, of similarly low-oxygen water with intermediate chloride concentration (232 p.p.m.) is located 1.1 km. to the east and is not thermally anomalous. Because the thermal anomaly in the area is rather insignificant, because no surface evidence of faulting has been observed, and because no evidence of ground-water barriers, such as a discontinuity in the piezometric surface, has been noted, it is assumed that the inferred subsurface fault does not cause differences in water quality.

Upward movement of ground water from the artesian aquifers is probably taking place in both areas. Thomas and Nelson (1948, pp. 184–85) mention the correlation of the level of ground water very near the surface with fluctuations in the artesian water levels in the area including the Woods Cross wells. The Burmester wells also are located in a discharge area, where ground water seeping to the surface evaporates and leaves a visible frost of salts in many places.

Variations in thermal conductivity are to be expected in sediments of such heterogeneity as are found in the two areas. The variations may not be great, however, because the sediments may be assumed to be saturated with water and not to include the very coarse material characteristic of the areas closer to the mountain fronts. The fair regularity of the temperature-depth profile in the Burmester area suggests that the variations in thermal conductivity are averaged out in the volume from which water is gathered for each well.

SOURCES OF RADON-222

Less than 0.5 per cent of the radon-222 in any of the water samples could have been derived from the radium-226 accompanying it in the sample. It is therefore necessary to look backward along the migration path for an adequate supply of the radium-226 parent. Because of the short half-life of radon-222, 3.8 days, the presence of radon-222 at a substantial distance from its parent requires either that the radon-222 move rapidly or that its initial concentration be quite large. Neither condition is likely to prevail in the study areas.

In the saturated sediments the diffusion coefficient of radon-222 can be expected not to exceed its diffusion coefficient in water, which is reported by Baranov and Novitskaya (1949, Table 4) to be 1.15×10^{-5} cm^2/sec at 18° C. One-dimensional diffusion at this rate results in a hundred-fold diminution of concentration over a diffusion distance of only about 0.1 m. If there is to be significant migration of radon-222 in this environment, transport, rather than diffusion, must be the dominant mechanism. But transport by water moving in the generalized flow patterns cited earlier, that is, nearly horizontally at a rate of about 0.3 m. per day, or less, would take radon-222 only a short distance in the 25 days required for hundred-fold diminution by radioactive decay.

Because of differences in permeability from one place to another in the heterogeneous aquifer materials, the rate of water movement at a given place may deviate by a large factor from the estimated average rate. It is conceivable that paths of high permeability exist along which water moves at rates much greater than average and along which radon-222 is transported quickly enough that much of it is undecayed after the passage. Direct evidence for or against the existence of such high-speed channels is lacking. It is geologically reasonable that old stream channels, approximately in line with the direction of ground-water movement, exist in the unconsolidated sediments. Weighing against the possibility that such channels serve as conduits for the high-speed transport of radon-222 are several considerations:

1. From well to well in both areas there are substantial differences

in the locations and thicknesses of aquifer materials. These differences are illustrated by Figure 3 of Thomas (1946) and by Figures 4 and 5 of Thomas and Nelson (1948). It is likely that continuously high permeability would not be maintained over a path of such length (about 100 m.). Even a short section of low permeability in an otherwise highly permeable channel would retard water movement long enough to permit substantial decay of radon-222.

2. Unrestrained flow of water from a well causes marked decreases of pressure in other wells of the Woods Cross area in the vicinity of the flowing well. Thomas and Nelson (1948, pp. 178–79) describe the effects of water flow from two closely spaced wells in the center of the section containing the Woods Cross area. During a 75.4-hour flowing period, water pressures decreased in wells located in a wide fan as far as 1,600 m. from the flowing wells. The effects were similar for wells both in the direction of ground-water movement and transverse to it, suggesting that the aquifer is not greatly anisotropic, as would be expected if the ground-water movement were along channels of large permeability and flow.

3. If water issuing from a fault zone in bedrock is the source of anomalous amounts of radon-222 moving along high-capacity channels, water containing anomalously large concentrations of radon-222 must be derived solely from the bedrock sources, or the radon-222 concentration in the bedrock sources must be sufficiently great to offset dilution by ground water moving in the normal cycle. Neither condition is likely: in comparison with the water of Becks Hot Springs, which should typify water from bedrock sources, the waters anomalously high in radon-222 concentration are much more dilute and are chemically different; however, they contain radon-222 in about the same concentrations as does the hot spring water.

4. Contributions of water from bedrock sources could be expected to vary little with the small changes in pressure accompanying changes in the discharge rate of a given well. If a large part of the radon-222 were derived from bedrock sources, its component of the radon-222 concentration should vary inversely as the discharge rate. As noted by Rogers (1954), the radon-222 concentration is insensitive to the discharge rate.

Vertical transport of the anomalous amounts of radon-222 from bedrock sources is improbable. Rogers (1954) measured the radon-222 concentration of waters issuing from Becks Hot Springs and obtained a minimum value of 13.5×10^{-10} c/l. The radon-222 anomaly measured between wells (B-2-1)26cdd and cdd-1 was about the same value, 14×10^{-10} c/l. If we overestimate the radon-222 concentration at a bedrock source by an order of magnitude, the transport from bedrock to the well—without dilution—must be accomplished in the corresponding interval of about 13 days or less. Assuming that the minimum depth to bedrock is 210 m., the speed of upward transport of radon-222 from the bedrock source to the bottom of well (B-2-1)26cdd at 122 m. would have to be at least 6.7 m/day. In the absence of a profound thermal anomaly or other evidence of rapid vertical movement of the ground water, it may be dismissed as hydrologically unreasonable.

It follows that the radon-222 source is in the vicinity of each well. Moreover, the relative uniformity of radon-222 concentrations in the waters of this study and the correspondence between radon-222 concentration (averaged over areas of square kilometers) and the probable sources of the sediments noted by Rogers (1954) suggest that the radium-226 source of the radon-222 is a dispersed one rather than a concentrated one.

To make a rough estimate of the distance traveled by radon-222 from its source to the well, let us calculate the volume and radius of a cylinder of height equal to the combined thicknesses of aquifer materials in a given well and volume corresponding to the amount of water discharged by the well during one mean life of radon-222 at the observed steady-state discharge rate. Well (B-2-1)26cdd, the only one for which a driller's log is available and the one showing the greatest radon-222 anomaly, will be used for the calculation. The log of this well shows a total of 48 ft. (15 m.) of gravel only in the perforated section. For months prior to the taking of the radon-222 sample it had been discharging at a rate of about 27 m.^{3} per day. With 10 per cent effective porosity in the aquifer, a geometrical volume of 1,400 m.^{3} will hold the amount of water discharged during one mean life (5.5 days) of radon-222. For a cylinder of 15-m. height this volume corresponds to a radius of 5.5 m. This radius is subject to the errors of the several crude approximations made, of course, but the approximations have been chosen so as to cause the calculated radius to be too large. Because it is proportional to the square roots of the discharge rate and of the reciprocal of the porosity, the error of the calculated radius is smaller than the corresponding errors in discharge rate and porosity. An unknown quantity implicit in this calculation is the uniformity of water production from the aquifer materials. If most of the water is entering from a relatively thin aquifer, the calculated radius is too small, but the error is again proportional only to the square root of the error in the thickness estimate. Thus, most of the radon-222 measured in the water probably comes from radium-226 that is quite close to the well bore and is immobile enough that it is not present in the water sample.

PHYSICAL CONTROLS ON RADON-222 DISTRIBUTION

Within a small volume of the source sediments, the radon-222 concentration in the interstitial water is dependent not only on the source density (mass of radium-226 per unit mass of saturated sediment) but also on the ease with which radon-222 escapes from the sites of disintegration of radium-226 atoms into the interstices of the sediments, and on their porosity. These latter factors are physical controls on the radon-222 distribution and should be independent of the source density because of the extreme dispersion of the radium-226 on an absolute basis (from 10^{-13} to 10^{-11} grams of radium-226 per gram of sediment).

The ease with which radon-222 can get into the interstitial water depends on the material in or on which radium-226 is immobilized. The range of a radon-222 ion recoiling from a disintegrating radium-226 atom and the diffusion rate of the radon-222 atom thereafter are so small in an intact crystal (Flügge and Zimens, 1939) that the contribution of radon-222 to interstitial water from the interiors of sedimentary grains is likely to be small. The contribution from exterior surfaces and microfissures may be appreciable if weathering or chemical corrosion has formed coatings or crusts, in which radon-222 may diffuse much more rapidly than in a crystal (Starik and Melikova, 1957; Hayase and Tsutsumi, 1958). Films of hydrous or gelatinous iron oxide, such as might be deposited on the grain surfaces, will also yield a large fraction of the radon-222 emanating from radium-226 trapped in them (Hahn, 1936, pp. 200–218). In addition, radon-222 should be readily available from radium-226 adsorbed on clay particles or on the exterior surfaces of clay lenses.

One would intuitively expect that a physical disturbance of the sediments, such as that accompanying faulting, would expose a greater emanating surface and therefore would tend to cause a radon-222 anomaly in the zone of disturbance. The sources of major radon-222 contribution described above, however, are those already exposed to interstitial water. Although direct experimental evidence is lacking, it seems likely that corrosion films on grains in near contact would emanate little more radon-222 if exposed by crushing than they do already. Experiments by Starik and Melikova (1957, Table 13) and by F. J. Davis, of the Oak Ridge National Laboratory (written communication), have revealed that the amount of radon-222 made available to passing fluids may be the same or *less* from rock *after* it is crushed than before. The emanation from materials that tend to be deformed plastically, such as the clay lenses, should not change their exposed surface areas by the large factors required to affect the emanation from them.

Associated with some faults are higher water temperatures, as has already been discussed. Although we may expect an increase in the diffusion rate with temperature, the effect for the small thermal anomalies noted (less than 1 per cent of the absolute temperature) should be negligible. Starik and Melikova (1957, Table 14) measured the temperature dependency of emanation of radon-222 from uranium ore; at 20° C. the increase was approximately 1 per cent/°C.

As a result of the foregoing considerations, one should not expect that physical disturbances as such would cause marked changes in the availability of radon-222 from radium-226 present in the sediments.

For a given amount of emanation of radon-222 into the interstices of the sediments, the radon-222 concentration will be greatest in sediments of the lowest porosity. Substantial porosity differences may be found among unconsolidated sediments. Spicer (1942, pp. 19–20) lists sands having porosities ranging from 30.2 to 63.2 per cent and gravels having porosities from 20.2 to 37.7 per cent. The porosities listed by Manger (1962, p. 9) show a similar range. Within a given type of sediment the porosity range is considerably less. Among heterogeneous sediments the porosity range is also smaller. In the study areas it is unlikely that porosity differences greater than twofold exist among the zones from which radon-222 is gathered by the different wells.

The effect of faulting on porosity, either concurrently or after the deposition of the unconsolidated sediments, is difficult to assess. From Rogers' work (1955a) it is known that radon-222 concentrations in waters from wells near the surface trace of the Warm Springs fault tend to be higher than they are elsewhere. If a porosity difference were responsible for the anomalous radon-222 concentrations, it would imply that the sediments near the fault plane were of relatively low porosity—perhaps a poorly sorted gravel mixed with sand and clay. Lithologic logs for wells along a traverse about 1 km. north of the Woods Cross group of wells are shown by Thomas and Nelson (1948, Fig. 5). Much of the aquifer material in each well, including those away from the fault, is gravel and probably is of similar low porosity. Twofold or greater differences in average porosity of the aquifer materials between the fault zone and the areas on either side seem improbable under these conditions.

Tokarev and Shcherbakov (1956, pp. 44–45) point out that radon-222 is a rare gas and is not adsorbed on rocks. It is assumed that differences in adsorption of radon-222 from well to well are negligible.

Although there is necessarily a relation between the discharge rate of a well and the radon-222 concentration in the water flowing from it, the radon-222 distribution and flow data in Figure 3 do not support the idea that the discharge rate was an important factor in the range represented (2–160 m.³ per day). Rogers (1954) saw no apparent relation between discharge rate and radon-222 concentration in his studies of the well waters of the Bountiful district, in which the Woods Cross area is located.

As a result of the foregoing considerations I do not believe that any one of the possible physical controls is responsible for the pattern of radon-222 distribution in ground water of the areas studied. The physical controls could, of course, combine fortuitously to yield the observed pattern, but no basis for their systematic reinforcement is seen. If this interpretation is correct, the radon-222 distribution must be determined by the distribution of its radium-226 parent in the sediments.

CONTROLS ON RADIUM-226 DISTRIBUTION

It is inferred that the radon-222 concentrations in the well waters are roughly proportional to the concentrations of immobilized radium-226 in the gathering zones of the respective wells. Virtually all the radium-226 in these zones is immobile, as indicated by the large factor by which the radon-222 concentrations exceed the radium-226 concentrations in the water samples. The lack of correspondence between the observed radium-226 concentrations in water and the inferred radium-226 concentrations in the ground implies the presence of mechanisms tending to cause this uneven distribution. What mechanism controls the small fraction of the radium-226 that is present in the samples?

The thorium-230 (ionium) parent of radium-226 is an isotope of one of the least mobile trace elements. Radium-226 in the water is consequently likely to be derived not from thorium-230 in the water but by extraction from the sediments or by introduction from radium-226 sources outside the areas studied.

RADIUM EXTRACTION MECHANISMS

Radium-226 may be extracted from the sediments if the water is of low enough pH to dissolve the alkaline earth carbonates, if chelating agents are available to remove the alkaline earth or iron cations from precipitates in which radium-226 is trapped, or if other ions are present in sufficient concentration to displace radium-226 and its captors. Because the waters are of pH 7.5 or more, solution of the carbonates is unlikely. Nor are chelating agents likely to be available, because calcium ion is present in appreciable concentrations in all water samples in which it has been determined.

Ions are available for displacement of the alkaline earth cations and iron. The same waters previously noted as containing the greatest concentrations of radium-226 and iron and the least concentrations of oxygen are those containing high concentrations of chloride. Tokarev and Shcherbakov (1956, Table 27), Hataye (1962), and Mazor (1962, p. 772) have noted the tendency of ground waters high in chloride to be enriched in radium-226 or to have greater power to leach radium-226 from rock. The importance of water that is high in chloride is its complement of positive ions, which tend to compete for adsorption sites with radium-226 and the other alkaline earth ions, and its tendency to be concentrated enough to lower the activity coefficients of the constituents, thereby reducing their tendency to precipitate.

TABLE 2

CALCULATED ACTIVITY COEFFICIENTS OF IONS AND pH'S OF WELL WATERS
FOR WHICH COMPLETE ANALYSES ARE AVAILABLE

WELL DESIGNATION	C_L (p.p.m.)	ACTIVITY COEFFICIENT			CALCULATED pH	MEASURED pH
		Fe^{+3}	Ca^{+2}	HCO_3^-		
BURMESTER AREA, TOOELE COUNTY, UTAH						
(C-2-5)7bdd-1......	47	0.51	0.73	0.92	7.8	8.1
WOODS CROSS AREA, BOUNTIFUL DISTRICT, DAVIS COUNTY, UTAH						
(B-2-1)26dca-3.....	46	0.52	0.74	0.93	8.0	8.3
(B-2-1)26dcd-1.....	62	0.45	0.69	0.91	7.5	7.5
(B-2-1)26cdd.......	985	0.21	0.46	0.80	7.5	7.5

An index of the tendency of water to promote desorption of the alkaline earth ions in favor of sodium ion is the sodium adsorption ratio (SAR) (Hem, 1959, pp. 148–49, 247–50). The SAR's were calculated for waters from 9 wells in the vicinity of the Burmester area and 27 wells in the vicinity of the Woods Cross area. The mean SAR's were 1.0 and 2.6 for waters containing less than 200 p.p.m. chloride in the respective areas, and 8.3 and 6.5 for waters containing more than 200 p.p.m. chloride. High-chloride waters should therefore have a greater ability to extract adsorbed radium-226 and other alkaline earth ions.

Activity coefficients of selected ions in 4 water samples were calculated using the graphs given by Hem (1961a) and are given in Table 2. The activity coefficients for calcium and bicarbonate ions are evidently lower by a factor of 1.5 in the high-chloride water of (B-2-1)26cdd than in the low-chloride waters of the other three wells. Other high-chloride waters in the two areas may also be expected to have reduced activity coefficients. The solubilities of the alkaline earth compounds in the high-chloride waters are thus effectively increased, and there is a greater tendency to retain them in solution.

The two water samples highest in radium-226 concentration are also high-chloride waters (Table 1); of the next four highest in radium-226, two are high in chloride and two are presumably low. The remaining five are low in radium-226 and low or presumably low in chloride.

INTRODUCTION OF RADIUM-226 FROM SOURCES OUTSIDE THE STUDY AREAS

Four directions of movement may be suggested in connection with introducing radium-226 from outside sources: (1) diffusion from postulated radium-226-rich saline waters on the lakeward sides of the study areas, (2) percolation of radium-226-bearing water from the surface downward into the aquifers, (3) migration with ground water from the mountain-front sides of the areas, and (4) migration with water influent from bedrock faults.

Direction (1) opposes the direction of ground-water movement. Although radium-226 is relatively mobile in saline water, it is doubtful that its diffusion rate would exceed the rate of ground-water movement. The concentration gradient should be low. Movement in direction (1) is thus dubious.

Except for local radioactive sinters in the outwash areas of radioactive hot springs, none of which are within about 5 km. of either study area, radium-226-rich surface deposits in these areas of valley fill would be quite unexpected. Because it would be in the zone of oxidation, radium-226 at the surface would tend to be immobile, according to Germanov *et al.* (1958). Downward percolation or diffusion would oppose the upward component of ground-water movement inferred for these discharge areas. Furthermore, the greater radium-226 concentrations are found in waters from the deeper wells. Movement in direction (2) is thus unlikely.

Movement of radium-226 in direction (3) is unlikely because water derived by recharge along the mountain fronts characteristically contains calcium and bicarbonate as its major ions and is rich in oxygen. In such water, radium-226 is relatively immobile, as shown in this study and as generalized by Tokarev and Shcherbakov (1956).

Movement of radium-226 with water from the Warm Springs fault toward the Woods Cross area in direction (4) is plausible. Some of the ground water west of the Warm Springs fault is inferred by Thomas and Nelson (1948, p. 139) to be derived from the fault itself. It may also be expected to carry substantial concentrations of radium-226 with it, because the water from the Warm Springs fault leaves a typical radium-226-bearing sinter.

REMOVAL OF RADIUM-226 FROM GROUND WATER

Of the many processes that might remove radium-226 from water, one is its precipitation as radium sulfate. Using the solubility product $[Ra^{+2}]\,[SO_4^{-2}] = 4.25 \times 10^{-11}$ (mole/l)^2 at 20° C., given by Nikitin and Tolmatscheff (1933), we may calculate that the concentration of radium-226 permitted by the mass-action law is 9.2×10^{-4} gm/l divided by the sulfate concentration in p.p.m. Sulfate concentrations of 400 p.p.m., which are high for both areas, would still permit radium-226 concentrations of about 2×10^{-6} gm/l, about 6 orders of magnitude greater than those observed.

It is well known that radium-226 ions are efficiently removed from solution by coprecipitation with barium as the sulfate (Gordon and Rowley, 1957). The solubility product of barium sulfate is so low that barium ion should not be present in the waters of the study areas in concentrations greater than 1 p.p.m. Let us assume that the sulfate concentration in the water provides an inverse index to barium concentration. (The assumption should be valid in the direction of decreasing barium and increasing sulfate concentrations.) Now we may consider the possibility of radium-226 coprecipitation as the high-sulfate water of the Becks Hot Springs type (about 800 p.p.m. sulfate), issuing from the subsurface fault, mixes with ground water containing much less sulfate and correspondingly more barium. Most of the coprecipitation of radium-226 would occur at the locus of first contact of the presumed barium-containing water with the radium-226 and sulfate-bearing water from the fault; thenceforth, although new supplies of barium might become available for continued precipitation of barium sulfate, the radium-226 would already have been removed. As a result, the radium-226 should be concentrated in the deeper sediments. A complicating factor is that barium sulfate precipitates emanate virtually none of the radon-222 produced within them (Hahn, 1936, pp. 222–29). Thus the radon-222 distribution cannot be used to test the importance of this process in the present problem. For the same reason, this process cannot account for the radon-222 distribution in the water.

Coprecipitation of radium-226 with calcium sulfate may be disregarded because the solubility product for calcium sulfate exceeds the concentration product of its ions by 1–2 orders of magnitude in the study areas. The efficiency of removal of radium-226 by coprecipitation with calcium sulfate is also questionable; according to Tokarev and Shcherbakov (1956, p. 50), radium-226 does not coprecipitate with gypsum ($CaSO_4 \cdot 2H_2O$) during its formation.

A common mechanism of radium-226 immobilization is coprecipitation of radium-226 with calcium carbonate sinter as water bearing calcium and bicarbonate evolves carbon dioxide to the atmosphere. (See, e.g., the discussion of the spa of Višegrad by Miholić, 1958, pp. 226–27.) Coprecipitation of radium-226 with calcium carbonate is also possible under other conditions leading to calcium carbonate saturation. The accurate pH measurements of this study permit comparison with pH's calculated from the activities of calcium and bicarbonate ion in four samples; the results appear in Table 2. Following the reasoning of Hem (1961a, pp. C7–C15), the equal calculated and measured pH's of the last two samples suggest that they are saturated with respect to calcium carbonate; the excess measured pH's of the first two suggest that they are supersaturated. Coprecipitation of radium-226 with calcium carbonate is therefore possible in all the waters for which saturation data are available.

Among the correlations observed in this study were those of Eh and concentrations of oxygen, iron, and radium-226 in the water. The correlation between Eh and oxygen concentrations is to be expected, as noted earlier. In the range of pH, Eh, and bicarbonate concentration prevailing in the waters studied, ferrous ion is unstable and ferric ion tends to precipitate as a hydroxide; the total solubility of iron under these conditions is less than 0.01 p.p.m., according to Hem (1961b, Fig. 3). In most of the water samples, however, iron was present in greater concentrations. Because the sampling procedure did not remove iron in very fine or colloidal suspension, not visible in the pipette used in the sample collection, it is probable that the excess iron was suspended, rather than dissolved, in the samples containing appreciable oxygen. In the samples having vanishingly small oxygen concentrations,[*] the excess iron may be indicative of lack of equilibrium, caused by the unavailability of oxygen for the oxidation of ferrous ions to the ferric state. It was observed that ferric precipitates were forming at the points of effluence of the waters low in oxygen, particularly at well (B-2-1)26cdd in the Woods Cross area. One would expect similar precipitation of ferric ion with mixing of the low-oxygen and oxygen-rich waters. With such mixing, radium-226 tends

---

[*] In the range of oxygen concentration less than 0.1 p.p.m., the method used is likely to give results that are too high; the analytical result is therefore more indicative of an upper limit of concentration.

to be coprecipitated to the extent that ferric hydroxide precipitation has been used as a "scavenging" method for removing radium-226 from solution. (See, e.g., Fry, 1962.) Thus, this process should be important in zones of present or past mixing of iron-bearing, low-oxygen water with water that contains appreciable amounts of oxygen. If the interface between the two types of water stays in the same position for periods comparable with the half-life of radium-226 $(1.6 \times 10^3$ years), a well-defined anomaly in the *radium-226* concentration in the *sediments* and in the *radon-222* concentration in the *water* should appear. Secular movement of the interface, such as must have occurred during the recession of the shore line of Lake Bonneville, should result in an anomaly correspondingly diffuse.

Conclusions

Many arguments based on indirect evidence have been used to develop the conclusions of this study. The more important assumptions are as follows.

1. Porosity differences among the wells on the scale of gathering zones of 10^3–10^4 m.^{3} are small (a few per cent or less).

2. On the same scale, radon-222 is available for concentration in the interstitial water in proportion to the radium-226 concentration in the sediments: they have uniform average emanating power.

3. Channels of high permeability, if they exist in the unconsolidated sediments, are shorter than the well-to-well spacing, are of small capacity compared with the well discharge rates, or are not directly connected with sources of radon-222 much more concentrated than the water of Becks Hot Springs.

From the arguments I derive the following conclusions.

1. Radon-222 concentration in the water is mainly dependent on radium-226 concentration in the sediments in the immediate vicinity of each well.

2. Much of the radium-226 in the sediments is derived locally from thorium-230 (ionium) in the sediments. Less than one-thousandth of this radium-226 is mobile. Water containing large concentrations of cations and having little tendency to reprecipitate alkaline earth compounds or ferric hydroxide may increase the mobile fraction to the order of $\frac{1}{100}$.

3. Additional radium-226 may flow in with water from subsurface fault zones or from sources "upstream," adding to the mobile fraction. The mobile fraction tends to be reduced simultaneously by coprecipitation with calcium carbonate or by coprecipitation with ferric hydroxide. Coprecipitation with ferric hydroxide is likely at interfaces

between iron-bearing, low-oxygen water of the lakeward or of the fault zones and oxygen-rich water derived from recharge "upstream."

4. If radon-222 anomalies in ground water of the study areas are related to subsurface faults, the relation exists through the sequence of radium-226 migration and immobilization upon change of chemical environment rather than by migration of radon-222 itself. A relation between radon-222 anomalies and subsurface faults is not conclusively shown, however, for the immediate areas studied.

ACKNOWLEDGMENTS

I wish to thank J. S. Gates for calling my attention to Burmester area, Ivan Barnes for valuable discussion of iron stability in ground water, and W. K. Hall for assistance in some of the field work.

REFERENCES

BACK, W., and I. BARNES. 1961. Equipment for field measurement of electrochemical potentials. U.S. Geol. Survey Prof. Paper 424-C, pp. C366–68.

BARANOV, V. I., and A. P. NOVITSKAYA. 1949. Diffuziya radona v prirodnykh gryazyakh ("Diffusion of radon in natural muds"). Akad. Nauk SSSR, Biogeokhimicheskaya Laboratoriya Trudy, No. 9, pp. 161–71.

BONDAR, A. G. 1962. New data on radon waters of Byelaya Tserkov district. (In Ukrainian.) Akad. Nauk Ukrain. RSR Dopovidi, No. 6, pp. 789–92.

CONNOR, J. G., and C. G. MITCHELL. 1958. A compilation of chemical quality data for ground and surface waters in Utah. Utah State Engineer Tech. Pub. 10.

FLÜGGE, S., and K. E. ZIMENS. 1939. Die Bestimmung von Korngrössen und Diffusionskonstanten aus dem Emaniervermögen. (Die Theorie der Emaniermethode.) Ztschr. phys. Chemie, B42:179–220.

FRY, L. M. 1962. Radium and fission product radioactivity in thermal waters. Nature, 195:375–76.

GATES, J. S. 1962. Geohydrologic evidence of a buried fault in the Erda area, Tooele Valley, Utah. U.S. Geol. Survey Prof. Paper 450-D, pp. D78–D80.

GERMANOV, A. I., S. G. BATULIN, G. A. VOLKOV, A. K. LISITSIN, and V. S. SEREBRENNIKOV. 1958. Some regularities of uranium distribution in underground waters. Proc. 2d United Nations Internat. Conf. on Peaceful Uses of Atomic Energy, 2:161–77.

GORDON, L., and K. ROWLEY. 1957. Coprecipitation of radium with barium sulfate. Anal. Chem., 29:34–37.

HAHN, O. 1936. Applied Radiochemistry. Ithaca: Cornell University Press.

HATAYE, I. 1962. Radiochemical studies on radioactive mineral springs. III. Leaching experiment of radium from basalts and sinter-deposit. Shizuoka Univ. Liberal Arts and Sci. Fac. Rpts., Nat. Sci. Sec., **3** (No. 3): 135–43.

HAYASE, I., and T. TSUTSUMI. 1958. On the emanating power of powdered rocks. Kyoto Univ. Coll. Sci. Mem., **24** (Ser. B): 319–23.

HEM, J. D. 1959. Study and Interpretation of the Chemical Characteristics of Natural Water. U.S. Geol. Survey Water-Supply Paper 1473. Pp. 269.

———. 1961*a*. Calculation and use of ion activity. U.S. Geol. Survey Water-Supply Paper 1535-C, pp. C1–17.

———. 1961*b*. Stability field diagrams as aids in iron chemistry studies. J. Am. Water Works Assoc., **53**:211–28.

HOOGTEIJLING, P. J., G. J. SIZOO, and J. L. YNTEMA. 1948. Measurements on the radon content of groundwater. Physica, **14**:73–80.

JOHNSON, W. W. Regional Gravity Survey of Part of Tooele County, Utah. M.S. thesis, Univ. Utah, 1958.

LANE, A. C., and W. R. BENNETT. 1934. Location of a fault by radioactivity. Beitr. angew. Geophysik, **4**:353–57.

LUCAS, H. F., JR., and F. H. ILCEWICZ. 1958. Natural radium-226 content of Illinois water supplies. J. Am. Water Works Assoc., **50**: 1523–32.

MANGER, G. E. 1962. Porosity and bulk density, dry and saturated, of sedimentary rocks. U.S. Geol. Survey Rpt. TEI-820. Pp. 10.

MAZOR, E. 1962. Radon and radium content of some Israeli water sources and a hypothesis on underground reservoirs of brines, oils and gases in the Rift Valley. Geochim. & Cosmochim. Acta, **26**:765-86.

MIHOLIĆ, S. 1958. Radioactive waters from sediments. Geochim. & Cosmochim. Acta, **14**:223–33.

NIKITIN, B., and P. TOLMATSCHEFF. 1933. Ein Beitrag zur Gültigkeit des Massenwirkungsgesetzes. II. Quantitative Bestimmung der Löslichkeit des Radiumsulfats in Natriumsulfatlösungen und in Wasser. Ztschr. phys. Chemie, **A167**:260–72.

NOLAN, T. B. 1962. Geological Survey research 1962: Synopsis of geologic, hydrologic, and topographic results. U.S. Geol. Survey Prof. Paper 450-A, pp. A1–A257.

RAINWATER, F. H., and L. L. THATCHER. 1960. Methods for Collection and Analysis of Water Samples. U.S. Geol. Survey Water-Supply Paper 1454. Pp. 301.

ROGERS, A. S. 1954. Physical behavior of radon. *In* U. S. Geological Survey, Geologic Investigations of Radioactive Deposits: Semiannual Progress Report, June 1 to November 30, 1954. U.S. Geol. Survey Rpt. TEI-490, pp. 294–96.

———. 1955*a*. Physical behavior of radon. *In* U.S. Geological Survey, Geologic Investigations of Radioactive Deposits: Semiannual Progress Report, December 1, 1954 to May 31, 1955. U.S. Geol. Survey Rpt. TEI-540, pp. 270–71.

Rogers, A. S. 1955*b*. Physical behavior of radon. *In* U.S. Geological Survey, Geologic Investigations of Radioactive Deposits: Semiannual Progress Report, June 1 to November 30, 1955. U.S. Geol. Survey Rpt. TEI-590, pp. 337–43.

Scott, R. C., and F. B. Barker. 1958. Radium and uranium in ground water of the United States. Proc. 2d United Nations Internat. Conf. on Peaceful Uses of Atomic Energy, **2**:153–57.

———. 1962. Data on Uranium and Radium in Ground Water in the United States, 1954 to 1957. U.S. Geol. Survey Prof. Paper 426. Pp. 115.

Sharkov, Yu. V. 1959. Poiski mestorozhdeniy atomnogo syr'ya ("Exploration for deposits of atomic raw materials"). Priroda, **48**:13–21.

Spicer, H. C. 1942. Porosity and bulk density, dry and saturated, of sedimentary deposits, table 2-6. *In* F. Birch, J. F. Schairer, and H. C. Spicer (eds.), Handbook of Physical Constants. Geol. Soc. America Spec. Paper 36, pp. 17–26.

Starik, I. Ye, and O. S. Melikova. 1957. Emanating power of minerals. Radiyevyy Institut imeni V. G. Khlopina Trudy, **5**:184–202. (In English trans.: U.S. Atomic Energy Comm. Rpt., AEC-tr-4498, pp. 206–26. 1961.)

Stchepotjeva, E. S. 1944. On the conditions under which natural waters become enriched with radium and its isotopes. Akad. Nauk SSSR Doklady, **43**:306–9.

Thomas, H. E. 1946. Ground water in Tooele Valley, Tooele County, Utah. Utah State Engineer Tech. Pub. 4. *In* Utah State Engineer 25th Bienn. Rpt., pp. 91–238.

Thomas, H. E., and W. B. Nelson. 1948. Ground water in the East Shore area, Utah, Pt. 1. Bountiful district, Davis County. Utah State Engineer Tech. Pub. 5. *In* Utah State Engineer 26th Bienn. Rpt., pp. 59–206.

Tokarev, A. N., and A. V. Shcherbakov. 1956. Radiohydrogeology ("Radiogidrogeologiya"). Moscow: Gosgeoltekhizdat. (Eng. trans., U.S. Atomic Energy Comm. Rpt., AEC-tr-4100 [1960]. Pp. 346.)

(Numbers in parentheses refer to chapter authored by discussant.)

In response to a question by Tanner (14), Samuels (13) stated that unpublished data on a few water wells indicate a relationship between the radium concentration in the water and the discharge and draw-down rates of the well. This relationship is being studied further, together with the influence of communicating aquifers. Abandoned, uncased wells provided artificial communication between aquifers in some cases. In certain areas of the midwestern U.S.A. the relationship between low-radium water and draw-down is most marked.

In response to a question by Lucas (17), Samuels (13) confirmed that one of the principal areas discussed in his paper is the Toulon area but that similar situations also exist in Iowa.

Grune (40) reported that in two wells his group studied in Maine there was a distinct relationship between the pumping of the well and the concentration of radium-226 and radon-222. In one case an industrial well that was shut in on Saturday afternoon had 50 per cent more activity when pumping was resumed at 8 A.M. on Monday; after 4 hours there was only 20 per cent additional activity, and after 6 hours the activity returned to a stable level that was maintained until the well was shut in the following Saturday. The same phenomenon was noted in a similar well after a resting period of 24–48 hours.

In response to a question from Grune (40), Tanner (14) commented that he had not had time to discuss the source of the radon-222 in ground waters or the distance that radon-222 could travel before it decayed. Flow to the well might bring radon-222 from a distance of a few meters, depending upon the production rate. Transport by linear diffusion limits migration of radon-222 to about 10 cm. In the wells he studied, the gathering volume for the radon-222 was limited to a cylinder of radium between 1 and 30 meters around the well, with the most likely value being near 10 meters.

Chen (57) noted that the area of the midwestern U.S.A. discussed by Samuels (13) had a population of over 1 million and asked how many of these people were taking in unusually high amounts of radium-226 in their drinking water. Samuels (13) replied that he assumed that the radium-226 intake in food was about one pc/day on the average but that estimates varied and some hearty eaters might take in two pc/day. The population studied by Samuels (13) was exposed to 3–40 pc/liter of drinking water. Assuming the ingestion of one liter of drinking water per day, 1,300 persons in this population are taking in 40 pc/day; there are over 50,000 consuming over 10 pc/day, or ten times the daily intake estimated for average food. It should be noted that the population under study is highly mobile and, of children in the age group 5–10 years, only some 20 per cent were lifelong residents of some of the towns studied.

Tanner (14) commented upon the reciprocal relationship between lead-210 and radium-226 in Holtzman's paper (12) and suggested that this reciprocal relationship may arise in deeper waters with more reductants such as sulfide ion. Under these reducing conditions, the radium-226 would be more mobile than would the lead-210. Holtzman (12) replied that this mechanism had been considered but that unfortunately the wells with the highest concentrations are in unconsolidated formations that may tend to be more oxidizing.

L. B. LOCKHART, JR., AND R. L. PATTERSON, JR.

15. Techniques Employed at the U.S. Naval Research Laboratory for Evaluating Air-borne Radioactivity

THE UNITED STATES NAVAL RESEARCH LABORATORY has had an active interest in the subject of atmospheric radioactivity since 1948, when it initiated a program on the detection of nuclear explosions at long range through the collection and identification of air-borne fission products. Though the methods originated were primarily for that purpose, they also served to give information on some of the natural radioactive components of the atmosphere.

Sampling procedures have involved both rain-collection and air-filtration techniques (Lockhart et al., 1959). In the former, collecting surfaces of 1,000–10,000 square feet of aluminum were employed; volumes of rain water collected ranged up to 2,250 gallons (8,500 liters). In the latter, filtration devices ranged from positive displacement blowers capable of giving about 30 cubic feet per minute (c.f.m.) air flow against a pressure drop of 15-cm. mercury across the filter to centrifugal or turbine-type blowers having capacities up to 2,000 c.f.m. For reasons of convenience, reliability, and simplicity, air filtration has been preferred to the rain-water collections. However, for geophysical purposes both types of collections are of interest, since the information obtained is quite different, though obviously interrelated in a rather complex manner.

LEAD-210 IN RAIN WATER

A brief mention is made here of rain collections because this was our first venture into the realm of atmospheric radioactivity. The

L. B. LOCKHART, JR., and R. L. PATTERSON, JR., are with the U.S. Naval Research Laboratory, Washington, D.C. The former is head of the Physical Chemistry Branch, Chemistry Division, and the latter is head of the Radiochemistry Section, Physical Chemistry Branch, Chemistry Division.

problem of separating isotopes of interest from the large volumes of rain water was solved by coprecipitation with aluminum hydroxide or ferric hydroxide of those isotopes forming hydroxides of low solubility. Among those collected with regularity was lead-210 (RaD), a long-lived descendant of radon-222 (King *et al.*, 1956). It was identified by measurement of its daughter, bismuth-210 (RaE), having a 5-day half-life and a β emission of 1.17-mev. energy, following a radiochemical separation and purification procedure. The long processing time precluded the use of this method for measuring the activity of short-lived natural radioactive isotopes in rain water.

Lead-214 and Lead-212 in Air

One of the early developments of this air-monitoring program was the utilization of changes in the apparent rate of decay of radon-220 daughter products collected on air filters to detect the presence of fission products in the air mass sampled (Blifford *et al.*, 1956). This procedure was of interest because it did not necessitate complete removal of the 10.6-hour lead-212 through radioactive decay before evaluation of the fission product content of the sample.

A high capacity, positive-displacement blower (Leiman Model 29-6) draws air from near ground level down a vertical stack 10–15 feet in length and through an efficient filter pad $2\frac{1}{2}$ inches in diameter for a period of 24 hours at a nominal rate of 30 c.f.m. (Pl. I). The flow rate is monitored by means of a vacuum gauge attached to the intake manifold between the filter and the blower; the actual flow is calculated for each collection from the absolute pressure behind the filter and the known rotation rate and capacity of the pump.

The filter generally employed has been a cellulose-asbestos filter (Army Chemical Corps Type 5) having a reasonably good retentivity for small aerosol particles and a relatively low resistance to air flow. (A 3-h.p. motor is used to power the blower, since rather rapid buildup of resistance occurs with increasing dust loading.) Retentivities of 75 per cent for natural radioactive aerosols and 100 per cent for fission products have been assumed, based on some experimental evaluations.

At the end of the 24-hour collection period, the sample is immediately placed in a β-counting assembly employing a halogen-filled end-window Geiger-Müller tube (window thickness, 5.6 mg/cm^2), and an initial 10-minute count is made. Thereafter, cumulative hourly counts are taken for a period of 16 hours, using automatic count recorders. An analysis of the decay curve permits calculation of the contribution of radon-222 daughters (lead-214, bismuth-214), radon-

PLATE I

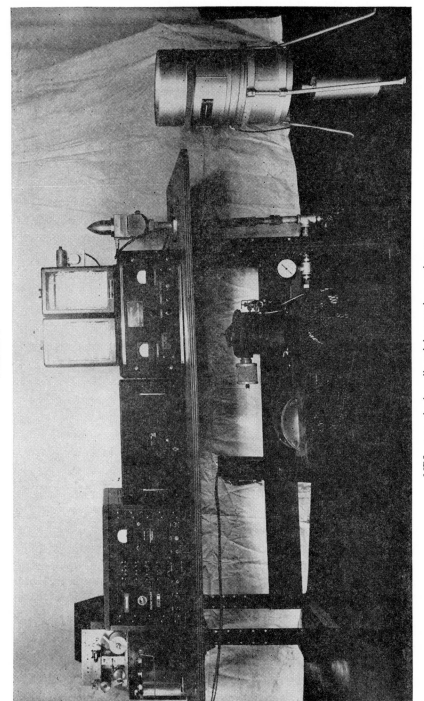

NRL atmospheric radioactivity monitor equipment

220 daughters (lead-212, bismuth-212), and gross fission products to the total activity on the filter.

Daily measurements are made of the counter background and of the counter efficiency toward protoactinium-234 (UX₂) (2.3-mev. maximum β energy) from uranium oxide mounted in a geometry similar to that of the filter and shielded with a 5-mil. thickness of aluminum to prevent the counting of α and low-energy β particles.

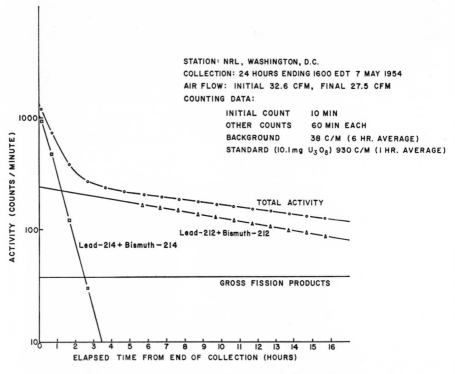

Fig. 1.—Radioactive decay of a typical air-filter collection

A disintegration rate of 620 c/m/mg of uranium oxide is used in estimating the counter efficiency. In calculating the disintegration rate of lead-214 plus bismuth-214, lead-212 plus bismuth-212, and gross fission products from the counting rates, correction factors of 1.05, 1.30, and 1.05, respectively, are employed to correct for the differences in β energy between the activity in question and the standard; no corrections for self-absorption of β energy by the filter are applied, since the collected dust is essentially all on the filter surface.

A typical decay curve is shown in Figure 1. The rapid initial decay is due to the radon-222-daughter activity (lead-214 and bismuth-214 with half-lives of 26.8 and 19.7 minutes, respectively) collected on the filter, while the slower rate of decay exhibited after the first few hours is due to radon-220 daughters (lead-212 and bismuth-212, half-lives 10.6 hours and 60.5 minutes, respectively). Any fission product mixture collected would generally have an effective half-life much greater than 10.6 hours and can be treated as a constant over this time interval. By comparing the initial counting rate with the counting rates existing during the intervals 5–6 hours and 15–16 hours following the end of the collection, it is possible to assign values to the

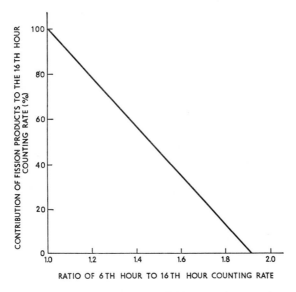

Fɪɢ. 2.—Determination of the contribution of fission products to the average counting rate during hour 16 after collection.

concentrations of radon-222 daughters, radon-220 daughters, and gross fission products in the air mass sampled (Lockhart *et al.*, 1958). This method has been employed to obtain useful information on the major radioactive components of the lower atmosphere at a number of sites throughout the world. A survey of this information is presented elsewhere in this symposium (Lockhart, 1963).

The contribution of fission products can be calculated from the counting rates at two different times following decay of the short-lived radon-222-daughter activity; in Figure 2, the relative fission-product activity during hour 16 is plotted as a function of the ratio of the counting rate during hour 6 to that during hour 16.

The concentration of lead-214 or lead-212 in the air can be calculated from the expression

$$\text{Concentration (pc/m}^3) = \frac{kCE\lambda \, e^{\lambda t_1}}{V\gamma(1 - e^{-\lambda t_2})},$$

where $k = 0.45$ picocuries per count per minute,

$C =$ average count rate (counts/min) at time t_1,

$E =$ correction for filter efficiency (about 75 per cent) and counting efficiency (about 15 per cent),

$\lambda =$ decay constant (min.$^{-1}$),

$t_1 =$ elapsed time (minutes) between end of collection and mid-point of counting interval,

$t_2 =$ length of collection (minutes),

$V =$ air-flow rate (m^3/min),

$\gamma =$ activity of decay series relative to parent isotope (2.3 for lead-214 plus bismuth-214 and 2.73 for lead-212 plus bismuth-212 and includes contribution of conversion electrons).

This equation should be modified to take into account the change in the ratio of lead-214 to bismuth-214 during the collection and also to take into account the extent of lead-214/bismuth-214 equilibrium in the air. Lack of equilibrium between these two would be indicative of a lack of equilibrium between radon-222 and its short-lived daughters. This ratio is not calculable from the information obtained from this air-monitoring system; however, recent information obtained by Hosler (1963) from short-term measurements of lead-214/bismuth-214 ratios in the air near the surface indicates that a lack of equilibrium does exist under some conditions. Fortunately, this lack of equilibrium in the air is offset by growth of bismuth-214 relative to lead-214 on the filter and tends to reduce some of the error inherent in the equation given above. These considerations are not so important in the case of lead-212 and bismuth-212 because of the great disparity in their half-lives. Furthermore, lead-212 concentrations cannot be equated to radon-220 concentrations, since the latter is a gas of short half-life, but are related to the rate of radon-220 exhalation and diffusion.

One interesting sidelight to this method of measurement involved the continuous recording of activity during the collection by use of a Geiger-Müller counter suspended above the filter with its output fed to a rate meter and pen recorder. Only rarely was it possible to identify by this technique the influx of fission debris into an area, because of the large and highly variable background of radon daughter activity. However, a strong diurnal variation in this short-lived activity (lead-214 plus bismuth-214) was observed at most sites. A typical example is shown in Figure 3.

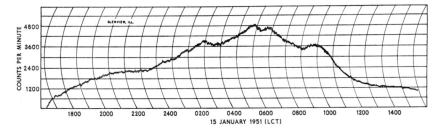

Fɪɢ. 3.—Typical recording of radioactivity decay rate during the collection

PLATE II

Simple air-filtration equipment employed in the 80th meridian air-sampling network

LEAD-210 IN AIR

Considerable data on the lead-210 (RaD) content of the air at ground level have been obtained from air-filter samples collected along the 80th meridian (W.) as part of a co-operative program administered by the U.S. Naval Research Laboratory. Sampling consisted of the operation of a simple positive-displacement blower at each site with return of the samples to Washington, D.C., for gross β assay and radiochemical analysis. The sampling unit shown in Plate II draws air

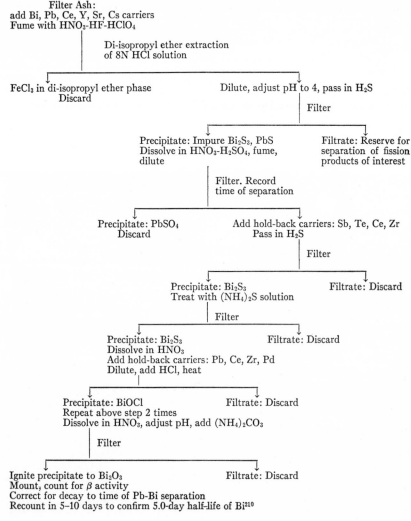

FIG. 4.—Diagram of the procedure for the radiochemical separation and purification of lead-210 (bismuth-210) from air-filter samples.

at the rate of about 30 c.f.m. through an 8-inch circle of cellulose-asbestos filter paper of high efficiency. Measured penetration of this filter by the short-lived radon-222 daughters, lead-214 and bismuth-214, which are attached to extremely small aerosol particles (estimated by Wilkening [1952] to be about 0.01–0.001 μ in diameter), has been found to be essentially zero. On arrival at NRL each filter was ashed, compressed to a pellet, and assayed for gross β activity, most of which was associated with fission products. Later, samples from a given site were combined on a monthly or bimonthly basis and subjected to radiochemical analysis (Baus *et al.*, 1957). A diagram of the radiochemical scheme is shown in Figure 4. The results of these analyses are reported in a separate paper (Patterson and Lockhart, 1963).

RETENTION OF AEROSOLS BY FILTERS

Part of the current program under way at NRL has involved a study of the effectiveness toward radioactive aerosols of a number of filter media used in air-monitoring systems. In principal this has in-

TABLE 1

RETENTIVITY OF SOME FILTER MEDIA TOWARD RADIOACTIVE
AEROSOLS IN THE GROUND-LEVEL AIR*

FILTER TYPE	DESIGNATION	NATURAL RADIOACTIVE AEROSOLS (RAB+C)		FISSION PRODUCTS	
		Velocity through Filter (ft/min)†	Retentivity (per cent)	Velocity through Filter (ft/min)	Retentivity (per cent)
Cellulose............	⌠IPC-1478	⌠121 ⌡730	7 24	97 736	45 82
	⌊Whatman 41	⌠117 ⌡446	69 94	110 366	99 100
Cellulose-asbestos.....	⌠Type 5	⌠121 ⌡662	66 86	121 363	96 100
	⌊Type 6	⌠114 ⌡458	100 100	114 599	100 100
Glass fiber..........	⌠Hurlburt 934 AH ⟨MSA 1106B ⌊Whatman GF/A	113 418	100 100	115 485	100 100 ......
Polystyrene.........	Microsorban	⌠117 ⌡530	95 96	113 349	100 100
Membrane..........	Polypore AM-1	⌠117 ⌡480	85 96	257	100 ....

* Radioactivity on filter compared with that retained by a back-up filter (Type 6).
† Average of initial and final linear velocities (1 ft/min = 0.508 cm/sec).

volved comparing the radioactivity (β) of the aerosol retained on a filter with that of the particles that penetrate the filter and are retained by a so-called ultimate filter. This has been done at several linear velocities of air through the filter and with two different radioactive aerosols, the short-lived radon-222 daughters (lead-214 plus bismuth-214) and the gross fission products present in the atmosphere.

TABLE 2

DETERMINATION OF RELATIVE AEROSOL SIZES
BY THEIR RETENTION ON FILTERS OF
DIFFERENT CHARACTERISTICS
(Percentages)

Filter pack: IPC-1478 (top, initial)
Type 5G (middle)*
Type 6 (bottom, final)

Air flow: 280 ft/min (approx.)

FILTER	NATURAL ACTIVITY (RaB+C) RETAINED COLLECTED 10/19/62	FISSION-PRODUCT ACTIVITY RETAINED	
		Collected 11/9–13/62	Collected 11/23–26/62
IPC-1478.........	7.0	65.9	72.6
Type 5G.........	53.9	28.5	24.6
Type 6..........	39.1	5.6	2.7
Total........	100.0	100.0	99.9
		Collected 1/18–23/63 Duplicate Runs	
IPC-1478.........		51.3	51.5
Type 5G.........		44.9	44.4
Type 6..........		3.8	4.1
Total........		100.0	100.0

* A cellulose-glass fiber filter similar in behavior to Type 5 cellulose-asbestos paper.

Some examples of filter retentivity toward these aerosols are given in Table 1. It is evident that linear velocity through the filter is an important factor in filter retentivity; furthermore, the aerosol particles with which the natural radioactivity (lead-214 and bismuth-214) is associated are evidently considerably smaller than those containing fission products. This latter factor is pointed up even more strongly by collections on packs of three filters of widely different retentivities (Table 2). These differences offer a means of studying the particle-size distribution by such filter-pack collections and also offer a means of studying the isotopic distribution (fractionation) of fission products in the various size fractions.

REFERENCES

BAUS, R. A., P. R. GUSTAFSON, R. L. PATTERSON, JR., and A. W. SAUNDERS, JR. 1957. Procedure for the Sequential Radiochemical Analysis of Strontium, Yttrium, Cesium, Cerium and Bismuth in Air-Filter Collections. Project NR-571-003, Washington, D.C., Naval Research Lab. NRL Memo 758. Pp. 24.

BLIFFORD, I. H., JR., H. FRIEDMAN, L. B. LOCKHART, JR., and R. A. BAUS. 1956. Geographical and time distribution of radioactivity in the air. J. Atmos. & Terr. Phys., 9:1–17.

HOSLER, C. R. U.S. Weather Bureau. Personal communication, 1963.

KING, P., L. B. LOCKHART, JR., R. A. BAUS, R. L. PATTERSON, JR., H. FRIEDMAN, and I. H. BLIFFORD, JR. 1956. The collection of long-lived natural radioactive products from the atmosphere. Nucleonics, 14:78, 80–83.

LOCKHART, L. B., JR. 1963. Radioactivity of the radon and thoron series in the air at ground level. This symposium.

LOCKHART, L. B., JR., R. A. BAUS, P. KING, and I. H. BLIFFORD, JR. 1959. Atmospheric radioactivity studies at the U.S. Naval Research Laboratory. J. Chem. Educ., 36:291–95.

LOCKHART, L. B., JR., R. A. BAUS, R. L. PATTERSON, JR., and I. H. BLIFFORD, JR. 1958. Some measurements of the radioactivity of the air during 1957. Washington, D.C., Naval Research Lab. Rpt. NRL-5208. Pp. 19.

PATTERSON, R. L., JR., and L. B. LOCKHART, JR. 1963. Geographical distribution of lead-210 (RaD) in the ground-level air. This symposium.

WILKENING, M. H. 1952. Natural radioactivity as a tracer in the sorting of aerosols according to mobility. Rev. Sci. Instr., 23:13–16.

M. KAWANO AND S. NAKATANI

16. Some Properties of Natural Radioactivity in the Atmosphere

Many authors (Israël, 1951; Reiter, 1957; Wilkening, 1959; Jacobi et al., 1959; El-Nadi and Omar, 1960) have studied the natural radioactivity in the atmosphere; in particular, the time variations of the concentration of the radioactive substances have been studied in many places in the world. Although some investigators (Hess, 1953; Cotton, 1955; Kawano and Nakatani, 1958) have used an ionization chamber for measuring the concentrations of the radioactive substances in the atmosphere, many researchers used a filter paper and an electrostatic precipitator for concentrating the daughter products of radon-222 and radon-220 (thoron) in the atmosphere. As is well known, a filter and an electrostatic precipitator are not sufficient for collecting perfectly the daughter products of radon-222 and radon-220 contained in the air passed through these filters.

In order to find the difference between the results of measurements of the concentrations of the radioactive dust collected by using an electrostatic precipitator and those of the radioactive substances measured by an ionization chamber, simultaneous measurements of these two elements were carried out in Tokyo for 12 months.

In this paper we shall discuss the results of these simultaneous measurements, the constancy of the collecting efficiency of an electrostatic precipitator, the counting efficiency of a Geiger-Müller counter used for detecting the radioactivity trapped by a precipitator, and the influence of the presence of radon-220 (thoron) and its daughter products on the results of measurements of the concentrations of radon-222 and its daughter products by using an ionization chamber.

M. KAWANO is professor of applied nuclear physics, Department of Nuclear Engineering, Nagoya University, Nagoya, Japan; S. NAKATANI is research physicist, Division of Radiology, Electrotechnical Laboratory, Tanashi, Kitatamagun, Tokyo, Japan.

INSTRUMENTS

INSTRUMENTS USED FOR MEASURING CONCENTRATION OF RADIOACTIVE SUBSTANCES IN THE ATMOSPHERE

Two ionization chambers were used for measuring the concentration of the radioactive substances in the atmosphere. One of them measured the ionization current by α, β, and γ radiations, and the other

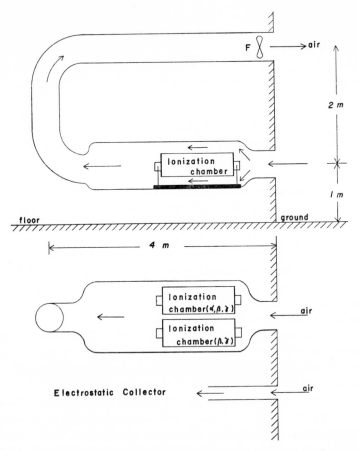

Fig. 1.—Instrument for measuring the concentration of radioactive substances

one measured only the current produced by β and γ radiations in the atmosphere. The wall of the former one was Mylar sheet, 7 microns thick (8.5×10^{-1} mg/cm^2), and that of the latter one was aluminum sheet, 40 microns thick (11 mg/cm^2). Since the lengths and diameters of these chambers were 90 cm. and 28 cm., respectively, their volumes were 56 liters. Although the α particles can penetrate the wall of the former chamber, they cannot penetrate the wall of the latter. In the case of measuring the natural radioactivity, the ionization by β and γ radiations is small compared with that by α, β, and γ radiations. Therefore, the ionization current measured by the aluminum chamber may be considered to be the background value of the Mylar chamber, and the difference between the ionization current of the former one and that of the latter one to be proportional only to α radiation in the atmosphere. Figure 1 shows the instrumentation used for measuring the concentration of the radioactive substances. The air from outdoors was drawn through the cylinder by a motor-driven fan, F. A is the air intake, and B is the air outflow. The chambers were airtight, and the air within them remained there until the Mylar and aluminum sheets required replacing. The walls were kept at a constant potential ($+630$ volts). This potential was sufficient to obtain the saturation current in the ionization chambers. The inner electrode of each chamber consisted of a brass rod about 5 mm. in diameter and was supported through the insulator, teflon, by the wall. A guard ring was connected to the ground, and the leakage current across the teflon insulating the inner electrode was thus minimized. A potential difference between the inner electrode and the wall was sufficient for obtaining the saturated ionization current even in the case of α radiation.

A vibrating-reed electrometer connected to the inner electrode by a fine shielded wire measured the ionization current, which seemed to be almost proportional to the concentration of the α emitters in the air. The ionization current measured by using a vibrating-reed electrometer was recorded on a self-recording microammeter.

INSTRUMENTS USED FOR MEASURING CONCENTRATION OF
RADIOACTIVE DUST IN THE ATMOSPHERE

The monitor for the natural radioactive dust was designed by Wilkening (1952a), and its principle was applied to a measuring instrument by the present authors (Kawano and Nakatani, 1958) and Yano (1961). The monitor used for the present work was an improved model. The aluminum tape maintained at the ground potential was stopped, and the collection of the radioactive dust in the air was continued for 3 hours in each measurement. Immediately after the end of the collection, the tape was moved by a motor, and the accumulated radioactivity on the tape was brought under the end window of a Geiger-Müller counter. The decay curve of the β radioactivity accumulated on the tape was recorded by using a self-recording ammeter until the next tape-moving. By this method, the concentrations of β emitters contained in radioactive dust, accumulated on the aluminum tape for 3 hours, were measured, and the ratio of the radon-220 daughter products to the radon-222 daughter products was continuously estimated. In the cases of these measurements, the discharge current between the corona points and the aluminum tape was 0.1 mamp., and the flow-rate was 20 l/min (see Fig. 2).

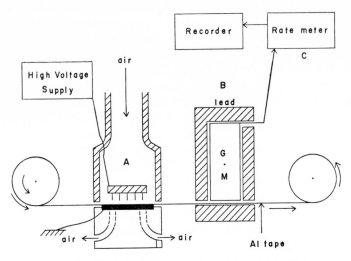

Fɪɢ. 2.—Instrument for measuring the concentration of radioactive dust

CALIBRATION FOR CONCENTRATION OF RADIOACTIVE SUBSTANCES MEASURED BY IONIZATION CHAMBER

The ionization chambers measured the ionization current caused by α, β, and γ radiations and β and γ radiations in the atmosphere, respectively. The difference between the ionization current measured by the Mylar chamber and that by the aluminum chamber may be almost proportional to the concentration of α emitters in the atmosphere. If the radon-222 daughter products are kept in radioactive equilibrium with radon-222, the concentrations of α emitters in the series of radon-222 and its daughter products may be proportional to those of the total radioactive substances. In order to calibrate the ionization current measured by the ionization chamber described above, the standard ionization chamber is used for obtaining the relation between the ionization current and the concentration of radon-222, which is kept in the equilibrium condition with its daughter products. Figure 3 shows the standard ionization chamber used for the calibration (Kawano and Nakatani, 1959). The ionization current measured by this standard chamber was calibrated by using radon-222

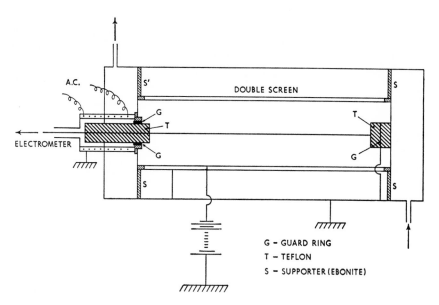

DOUBLE SCREEN

A.C.

ELECTROMETER

G – GUARD RING
T – TEFLON
S – SUPPORTER (EBONITE)

Fig. 3.—Screen ionization chamber

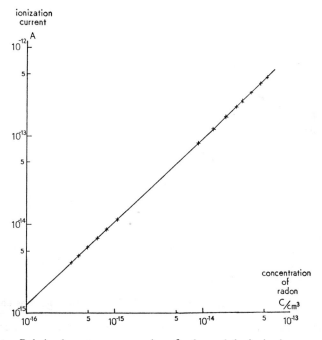

ionization
current

concentration
of
radon
C/cm^3

Fig. 4.—Relation between concentration of radon and the ionization current

gas emanated from a standard radium-226 solution. The activated charcoal trap, consisting of 200 gm. of coconut charcoal in granular form, was confined in a glass tube and was used for obtaining the radon-222-free air needed for deciding the background value of the ionization current.* Figure 4 shows the relation between the concentration of radon-222 inside the ionization chamber and the ionization current.† Simultaneous measurements of the concentrations of the radioactive substances were carried out using the standard ionization chamber and the ionization chamber shown in Figure 1. Figure 5 shows an example of the record of the ionization current measured

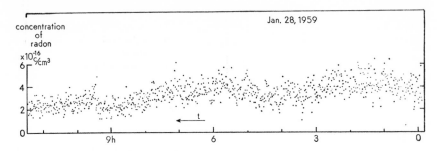

Fig. 5.—Example of record of ionization current with the standard ionization chamber

by the standard ionization chamber. As this current gave the concentration of the radioactive substances, the current measured by the ionization chamber shown in Figure 1 was calibrated by comparing them with the current simultaneously measured by the standard chamber.

COLLECTING EFFICIENCY OF ELECTROSTATIC PRECIPITATOR FOR RADIOACTIVE DUST

As is well known, the collecting efficiencies of a filter paper and an electrostatic precipitator are not perfect. As shown by Schumann (1956), the size of radioactive dust contained in the fallout seems to be considerably larger than that of the naturally occurring radioactive dust. The main part of the latter one seems to be concentrated below about 0.5 micron in diameter (Wilkening, 1952b, Kawano and Nakatani, 1961). Therefore, the complete collection of this kind of dust seems to be very difficult.

* The trapping efficiency of the charcoal trap for radon-222 gas was checked by an induced method and a low-background counter. The result of measurement showed that the charcoal trap mentioned above could trap radon-222 gas almost completely.

† The ionization current was measured at 3 hours after introducing radon-222 gas into the ionization chamber.

Figure 6 shows the instruments used for measuring the collecting efficiency of the electrostatic precipitator. Since the concentrations of radon-222 and its daughter products in the atmosphere are too small to measure the collecting efficiency for lead-214 and bismuth-214 (RaB and RaC), radon-222 and its daughter products exhausted from a standard radium-226 solution were used for the sample.* The radium-226 content in the standard radium solution (radium chloride) was 1×10^{-4} c. (N.B.S.). This radium solution was distilled with a solution of radium chloride (0.01 N.). The ionization chamber was designed to measure the ionization current produced by β radiation, and the wall was an aluminum sheet 40 microns thick (11 mg/cm²).

The inner electrode of brass was supported through a teflon insulator and a guard ring by the wall holder and was connected to a vibrating-reed electrometer. This ionization chamber was put inside the iron box, which was coaxial with the ionization chamber. The diameters of the ionization chamber and the iron box were 10 cm. and 20 cm., respectively. The air containing radon-222 and its daughter products was introduced into the gap between the wall of the chamber and the iron box.

The relation between the ionization current and the concentration of lead-214 and bismuth-214 was determined (Fig. 7).

The counting values of the Geiger-Müller counter for the radioactivity trapped on the aluminum tape were calibrated as follows: A solution of strontium-90–yttrium-90 (6.8 $\times$ 10⁻⁹ c.) was distributed in the same area as the corona discharge covered on the tape inside the electrostatic precipitator used here and was evaporated. Therefore, strontium-90–yttrium-90 remained on the surface. The β energies of strontium-90–yttrium-90 are quite close to those of lead-214 + bismuth-214.† According to the results of measurements, the counting efficiency of the Geiger-Müller counter for the aluminum tape was determined to be 28.6 per cent.

The collecting efficiency of the electrostatic precipitator was measured as follows: Radon-222 gas emanated from the radium-226 solution was introduced into an initially evacuated large-volume bell jar. Radon-222 gas introduced into the bell jar decayed gradually, and the concentration of its daughter products increased.

Three hours after the introduction of the air containing radon-222, radon-222 reaches secular radioactive equilibrium with its daughter products. The air containing radon-222 and its daughter products was

---

* Alpha emitters cannot be used for measuring the collecting efficiency because radon-222 gas is itself an α emitter.

† Strontium-90, 0.61 mev.; lead-214 (RaB), 0.59 mev.; yttrium-90, 2.26 mev.; bismuth-214 (RaC), 1.65 mev.

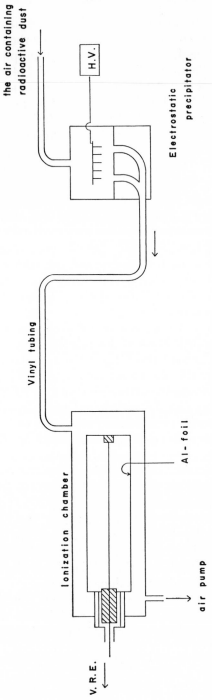

Fig. 6.—Instruments used for measuring collecting efficiency

then introduced into the ionization chamber through the electrostatic precipitator.

Beta radioactivity trapped by the electrostatic precipitator was measured by the Geiger-Müller counter, and β radioactivity passed through the electrostatic precipitator was measured by the ionization chamber. The collecting efficiency of the electrostatic precipitator used for the present work was determined 12 times, and the maximum and minimum values were 18 per cent and 6 per cent, respectively. The mean value was 11 per cent.

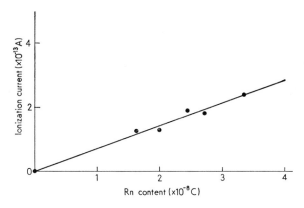

Fig. 7.—Relation between ionization current and radon content introduced into ionization chamber.

Siksna (1949), Okabe and Kitamoto (1957), and Kawano and Nakatani (1961) used autoradiography for studying the properties of the natural radioactive dust. The fourth stage of the cascade impactor used for the present work was designed to trap dust of diameters between 0.3 and 0.5 microns. In a previous paper, the present authors observed that the α tracks carried by dust larger than 0.5 microns seemed to be much fewer than those carried by dust smaller than 0.5 microns. Figure 8 shows the autoradiograph of α emitters collected by a cascade impactor (fourth stage) after passing through the electrostatic precipitator; α emitters passing through no filter were trapped by the cascade impactor (fourth stage). As shown in these photographs, the greater part of the radioactive dust of diameter 0.3–0.5 microns seemed to be trapped by the electrostatic precipitator.

Still, the collecting efficiency of the electrostatic precipitator is only about 10 per cent. This result seems to show that the main part of the naturally occurring radioactive dust is concentrated below 0.3 microns, and to collect them by an electrostatic precipitator is very difficult.

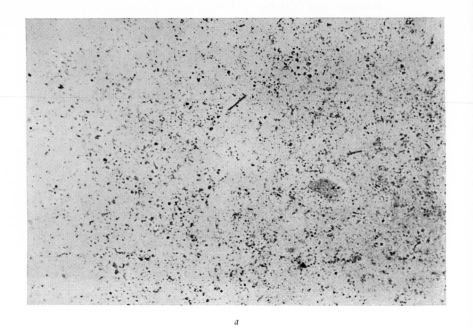

a

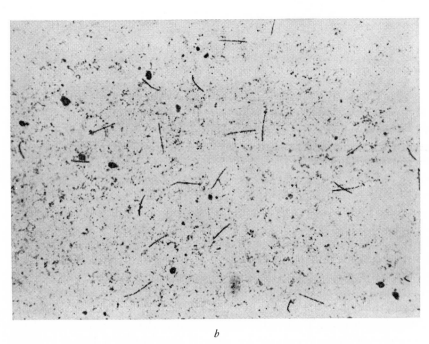

b

Fig. 8.—*a*, Alpha autoradiograph of dust collected by cascade impactor (4th stage). *b*, Alpha autoradiograph of dust collected by electrostatic precipitator.

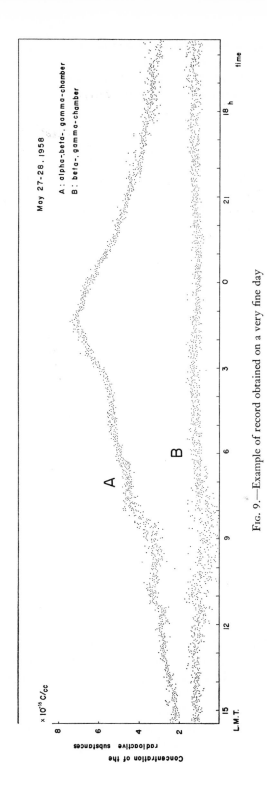

Fig. 9.—Example of record obtained on a very fine day

DIURNAL VARIATIONS OF CONCENTRATIONS OF
RADIOACTIVE SUBSTANCES AND RADIOACTIVE
DUST IN THE ATMOSPHERE

The routine measurements of the concentrations of the radio-active substances and the radioactive dust were carried out in Tokyo for 12 months. The concentration of the radioactive substances changed with time even on the quiet days, and it changed remarkably during the period of rainy, cloudy, and other abnormal weather conditions. In this paper, the values obtained during the periods of abnormal weather conditions were excluded from the calculations of the hourly mean values.

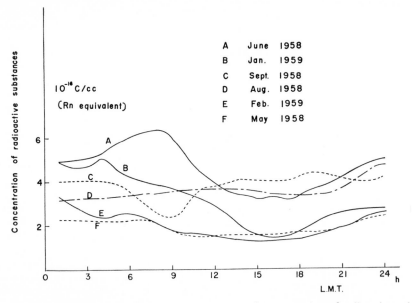

Fig. 10.—Monthly mean diurnal variation curves of concentration of radioactive substances in the atmosphere.

Figure 9 shows an example of the record of the concentration of the radioactive substances as measured by the ionization chamber on a very quiet day. Although the diurnal variation curves vary considerably among one another even on clear days, this example is taken to be typical of a clear day. Figure 10 shows the mean diurnal variation curves over several months. As shown in this figure, the maximum values occur in the early morning, and the minimum in the afternoon, throughout the year. Figure 11 shows an example of the records of the concentration of radioactive dust collected by the electrostatic

Fig. 11.—Example of the record of the concentration of radioactive dust collected with the electrostatic precipitator

precipitator. Figure 12 shows the mean diurnal variation curves over several months. The mode of time variation curves seems to be almost the same as that of the concentration of the radioactive substances shown in Figure 10. Figure 13 shows the diurnal variation curves of the lead-214 and bismuth-214 activities obtained on February 6, 1959. The correlation between the precipitator and ionization chamber data seems to be very good. The values of both elements are high in

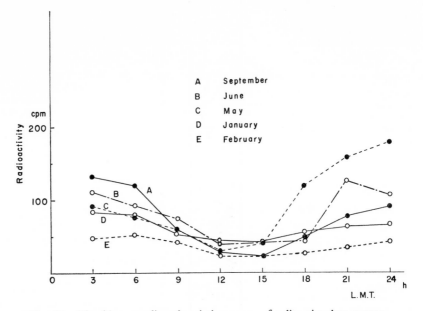

Fig. 12.—Monthly mean diurnal variation curves of radioactive dust content

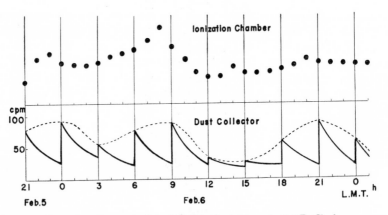

Fig. 13.—Example of record of lead-214(RaB) + bismuth-214(RaC) decay curve on Feb. 6, 1959.

the early morning and suddenly decrease with the sunrise. They decrease gradually until midday, and the minimum values appeared from 14:00 to 16:00. These minimum levels persisted for several hours. The diurnal variations of both elements seem to be almost reverse to the diurnal variation of the eddy diffusion (Kawano, 1957). Comparing the two variation curves obtained in each month in detail, it is noticed that the variation curves of the concentration of the radioactive dust seem to be considerably different from those of the radioactive substances.

INFLUENCE OF PRESENCE OF RADON-220 (THORON) AND ITS
DAUGHTER PRODUCTS ON RESULTS OF MEASUREMENTS
OF RADON-222 CONCENTRATION BY USING THE
IONIZATION CHAMBER

The direct measurement of the concentration of radon-220 (thoron) gas in the atmosphere seems to be difficult because of its very short (54.5 sec.) half-life. Hess (1953) and Jaki and Hess (1958) measured the ratio of the concentration of polonium-212 (ThC') to that of radon-222 + polonium-218 + polonium-214 with an ionization chamber.

The ratio of the concentration of lead-212 + bismuth-212 to that of lead-214 + bismuth-214 using the collecting methods described has been obtained by several authors (Jacobi *et al.*, 1959; El-Nadi and Omar, 1960; Wilkening, 1952a). As previously described here, the decay curves of radioactivity collected on the aluminum tape in the electrostatic precipitator were continuously recorded. Figure 13 shows an example of the records of these decay curves. In general, the activity of radon-222 daughter products is negligible 3 hours after the end of collection, and the activity that remains after 3 hours or more is due to radon-220 (thoron) daughter products. Because the half-decay period of lead-212 (ThB) is 10.6 hours and is very much longer than that of lead-214 or bismuth-214, the initial value of (lead-212 + bismuth-212) may be estimated by extrapolating the decay curve to zero time. When there is radioactive equilibrium between radon-222, radon-220, and their daughter products, respectively, the concentrations of radon-222 and radon-220 may be estimated from the radioactivity collected on a filter.

The time variation of the radioactivity collected on a filter may be expressed by the following equations:

$$\frac{dn_G}{dt} = p - \lambda_G n_G , \tag{1}$$

$$\frac{dN_A}{dt} = \eta q \frac{p}{\lambda_A} - \lambda_A N_A , \tag{2}$$

$$\frac{dN_B}{dt} = \eta q \frac{p}{\lambda_B} + \lambda_A N_A - \lambda_B N_B, \tag{3}$$

$$\frac{dN_C}{dt} = \eta q \frac{p}{\lambda_C} + \lambda_B N_B - \lambda_C N_C, \tag{4}$$

where n_G = the number of atoms of the radioactive gas in the unit volume of the air, p = supply rate of atoms of radioactive gas into the referred unit volume, η = collecting efficiency, and q = flow rate (cc/s). N_A, N_B, and N_C = the number of atoms of daughter products A, B, and C, respectively, collected on a filter at the time t. Initial conditions at $t = 0$, $N_A = N_B = N_C = 0$.

The solutions of the equations are as follows:

$$N_A = \frac{\eta q p}{\lambda_A{}^2}(1 - e^{-\lambda_A t})$$

$$N_B = \frac{\eta q p}{\lambda_B}\left(\frac{1}{\lambda_A} + \frac{1}{\lambda_B}\right)(1 - e^{-\lambda_B t}) - \frac{\eta q p}{\lambda_A(\lambda_B - \lambda_A)}(e^{-\lambda_A t} - e^{-\lambda_B t}),$$

$$N_C = \frac{\lambda_C(\lambda_A + \lambda_B) + \lambda_A \lambda_B}{\lambda_A \lambda_B \lambda_C{}^2}\eta q p(1 - e^{-\lambda_C t})$$

$$-\frac{\lambda_B \eta q p}{\lambda_A(\lambda_B - \lambda_A)(\lambda_C - \lambda_A)}(e^{-\lambda_A t} - e^{-\lambda_C t})$$

$$+\frac{\lambda_A \eta q p}{\lambda_B(\lambda_C - \lambda_B)(\lambda_B - \lambda_A)}(e^{-\lambda_B t} - e^{-\lambda_C t}).$$

In these formulas, the time, t, refers to the duration of collection. In order to distinguish the duration of collection from the time of decay, t is exchanged with T in the formulas described above.

$$N_A{}^T = \frac{\eta q p}{\lambda_A{}^2}(1 - e^{-\lambda_A T}),$$

$$N_B{}^T = \frac{\eta q p}{\lambda_B}\left(\frac{1}{\lambda_A} + \frac{1}{\lambda_B}\right)(1 - e^{-\lambda_B T}) - \frac{\eta q p}{\lambda_A(\lambda_B - \lambda_A)}(e^{-\lambda_A T} - e^{-\lambda_B T}),$$

$$N_C{}^T = \frac{\lambda_C \lambda_A + \lambda_B \lambda_C + \lambda_A \lambda_B}{\lambda_A \lambda_B \lambda_C{}^2}\eta q p(1 - e^{-\lambda_C T})$$

$$-\frac{\lambda_B \eta q p}{\lambda_A(\lambda_B - \lambda_A)(\lambda_C - \lambda_A)}(e^{-\lambda_A T} - e^{-\lambda_C T})$$

$$+\frac{\lambda_A \eta q p}{\lambda_B(\lambda_C - \lambda_B)(\lambda_B - \lambda_A)}(e^{-\lambda_B T} - e^{-\lambda_C T}).$$

The decay of each product after the end of collection is given by the following equations:

$$\frac{dN_A}{dt} = -\lambda_A N_A,$$

$$\frac{dN_B}{dt} = \lambda_A N_A - \lambda_B N_B,$$

$$\frac{dN_C}{dt} = \lambda_B N_B - \lambda_C N_C.$$

Initial conditions:

$$\text{at } t = 0, \quad N_A = N_A{}^T, \quad N_B = N_B{}^T, \quad N_C = N_C{}^T.$$

The solutions of these equations are as follows:

$$N_A = N_A{}^T e^{-\lambda_A t},$$

$$N_B = \frac{\lambda_A N_A{}^T}{\lambda_B - \lambda_A}(e^{-\lambda_A t} - e^{-\lambda_B t}) + N_B{}^T e^{-\lambda_B t},$$

$$N_C = \frac{\lambda_A \lambda_B N_A{}^T}{(\lambda_B - \lambda_A)(\lambda_C - \lambda_A)}(e^{-\lambda_A t} - e^{-\lambda_C t})$$

$$- \frac{\lambda_A \lambda_B N_A{}^T}{(\lambda_B - \lambda_A)(\lambda_C - \lambda_B)}(e^{-\lambda_B t} - e^{-\lambda_C t})$$

$$+ \frac{\lambda_B N_B{}^T}{\lambda_C - \lambda_B}(e^{-\lambda_B t} - e^{-\lambda_C t}) + N_C{}^T e^{-\lambda_C t}.$$

Therefore, the α and β radioactivities trapped on a filter may be given as follows.

α activity:

$$\frac{I_\alpha}{\eta q p} = \lambda_A N_A + \lambda_C N_C$$

$$= \frac{1}{\lambda_A}(1 - e^{-\lambda_A T})e^{-\lambda_A t} + \frac{\lambda_B \lambda_C}{\lambda_A(\lambda_B - \lambda_A)(\lambda_C - \lambda_A)}$$

$$\times (1 - e^{-\lambda_A T})(e^{-\lambda_A t} - e^{-\lambda_C t})$$

$$- \frac{\lambda_C \lambda_A}{\lambda_B(\lambda_C - \lambda_B)(\lambda_B - \lambda_A)}(1 - e^{-\lambda_B T})(e^{-\lambda_B t} - e^{-\lambda_C t})$$

$$+ \left\{ \left(\frac{1}{\lambda_A} + \frac{1}{\lambda_B} + \frac{1}{\lambda_C} \right)(1 - e^{-\lambda_C T}) - \frac{\lambda_B \lambda_C}{\lambda_A(\lambda_B - \lambda_A)(\lambda_C - \lambda_A)} \right.$$

$$\times (e^{-\lambda_A T} - e^{-\lambda_C T})$$

$$\left. + \frac{\lambda_C \lambda_A}{\lambda_B(\lambda_C - \lambda_B)(\lambda_B - \lambda_A)}(e^{-\lambda_B T} - e^{-\lambda_C T}) \right\} e^{-\lambda_C t}.$$

β activity:

$$\frac{I_\beta}{nqp} = \lambda_B N_B + \lambda_C N_C = \frac{\lambda_B}{\lambda_A(\lambda_B - \lambda_A)}(1 - e^{-\lambda_A T})(e^{-\lambda_A t} - e^{-\lambda_B t})$$

$$+ \left\{ \left(\frac{1}{\lambda_B} + \frac{1}{\lambda_A}\right)(1 - e^{-\lambda_B T}) - \frac{\lambda_B}{\lambda_A(\lambda_B - \lambda_A)}(e^{-\lambda_A T} - e^{-\lambda_B T}) \right\} e^{-\lambda_B t}$$

$$+ \frac{\lambda_B \lambda_C}{\lambda_A(\lambda_B - \lambda_A)(\lambda_C - \lambda_A)}(1 - e^{-\lambda_A T})(e^{-\lambda_A t} - e^{-\lambda_C t})$$

$$- \frac{\lambda_C \lambda_A}{\lambda_B(\lambda_C - \lambda_B)(\lambda_B - \lambda_A)}(1 - e^{-\lambda_B T})(e^{-\lambda_B t} - e^{-\lambda_C t})$$

$$+ \left\{ \left(\frac{1}{\lambda_A} + \frac{1}{\lambda_B} + \frac{1}{\lambda_C}\right) \cdot (1 - e^{-\lambda_C T}) - \frac{\lambda_B \lambda_C}{\lambda_A(\lambda_B - \lambda_A)(\lambda_C - \lambda_A)} \right.$$

$$\times (e^{-\lambda_A T} - e^{-\lambda_C T})$$

$$\left. + \frac{\lambda_C \lambda_A}{\lambda_B(\lambda_C - \lambda_B)(\lambda_B - \lambda_A)}(e^{-\lambda_B T} - e^{-\lambda_C T}) \right\} e^{-\lambda_C t} = g(t).$$

In these formulas, I_a and I_β express the measured values of a and β radioactivity in disintegration per second. The theoretical curves calculated by the formulas given above are compared with those determined by a activity measurements (zinc sulfide scintillation counter) and β activity measurements (Geiger-Müller counter). Figure 14 shows an example of the decay curves of radioactivity on the Al tape.

Assuming that $dn_G/dt = 0$ in equation (1), the following relation is given between the number of atoms of radon-222 or radon-220 in the unit volume and the supplying rate of atoms of these gases into the unit volume. In this case, the supplying rate is the same as the concentration of the radioactive gas in the unit of disintegration per sec. cm.³. Therefore, the concentrations of radon-222 or radon-220 may be estimated by the following relations:

$$p = \frac{I_a}{nqf(t)} = \frac{I_\beta}{nq\,g(t)} \qquad (\text{dps/cm}^3);$$

$$Q = \frac{p}{3.7 \times 10^{10}} \quad (\text{c/cm}^3).$$

Corresponding to the values shown in Figure 14, the concentrations of radon-222 and radon-220 are 4×10^{-16} c/cm³ and 3×10^{-17} c/cm³, respectively. Figure 15 shows the diurnal variation curves of the value

$$\frac{(\text{lead-214} + \text{bismuth-214})}{\text{lead-212}}$$

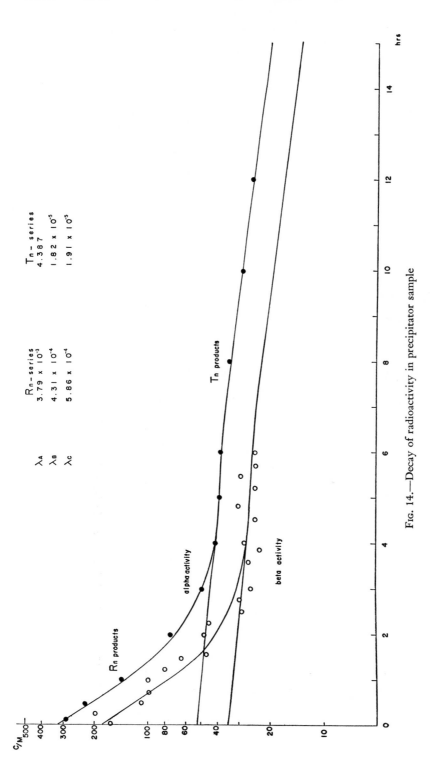

Fig. 14.—Decay of radioactivity in precipitator sample

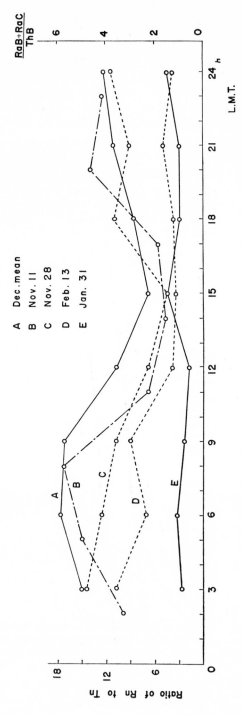

Fig. 15.—Diurnal variation curve of lead-214(RaB) + bismuth-214(RaC)/lead-212 (ThB) and the value of radon-222/radon-220

obtained from the decay curves of radioactivity on the electrostatic precipitator. The ratio of the concentration of radon-222 to that of radon-220 is estimated from the relation described above for each value shown in Figure 14.

CONCLUSION

The concentration of the radioactive dust collected by using an electrostatic precipitator is not always proportional to that of the radioactive substances measured by an ionization chamber.

The main cause of this observation seems to be that the collecting efficiency of an electrostatic precipitator is not constant.

The values of radon-222/radon-220 were estimated from the results of analysis of β-decay curves and were large (12–16) at night and small (4–7) in the daytime.

Taking the results of estimations described above into consideration, the influence of the presence of radon-220 and its daughter products on the results of measuring the concentrations of radon-222 and its daughter products, by using the ionization chamber, was discussed. Since the energies of the α particles from radon-220 (6.280 mev.), polonium-216 (6.774 mev.), and polonium-212 (8.776 mev.) are larger than those from radon-222 (5.484 mev.), polonium-218 (5.998 mev.), and polonium-214 (7.68 mev.), the contributions of radon-220 series to the results of measurements by the ionization chambers seem to be almost 10–20 per cent.

Assuming the radioactive equilibrium between radon-222 or radon-220 and their respective daughter products, the results of continuous measurements of the natural radioactivity are discussed in this paper.

Although several authors (Kawano and Nakatani, 1958; Jacobi, 1959; Yano, 1961) have discussed the radioactive equilibrium in the atmosphere, they used the collecting method for measuring the concentration of the radioactive substances. As shown recently by the present author (Kawano, in press), the collecting method is not adequate for studying this problem.

REFERENCES

COTTON, E. S. 1955. Diurnal variations in natural atmospheric radioactivity. J. Atmos. & Terr. Phys., **7**:90–98.

EL-NADI, A. F., and H. M. OMAR. 1960. Radon and thoron content in atmospheric air at Giza, Egypt. Geofisica Pura e Applicata, **45**:261–66.

HESS, V. F. 1953. Radon, thoron, and their decay products in the atmosphere. J. Atmos. Terr. Phys., **3**:172–77.

ISRAËL, H. 1951. Radioactivity of the atmosphere. *In* T. F. MALONE (ed.), Compendium of Meteorology, pp. 155–61. Boston: Meteorological Soc.

JACOBI, W., A. SCHRAUB, K. AURAND, and H. MUTH. 1959. Über das Verhalten der Zerfallsprodukte der Radons in der Atmosphäre. Beitr. z. Phys. Atmos., **31**:244–57.

JAKI, S. L., and V. F. HESS. 1958. A study of the distribution of radon, thoron, and their decay products above and below the ground. J. Geophys. Res., **63**:373–90.

KAWANO, M. 1957. The coefficient of eddy diffusivity estimated from the atmospheric electricity. J. Met. Soc. Japan, **35**:339–44.

KAWANO, M. (In press.)

KAWANO, M., and S. NAKATANI. 1958. The results of routine observations of the ionization and the natural radioactive dust concentration in the atmosphere in Tokyo. J. Met. Soc. Japan, **36**:135–45.

———. 1959. The absolute measurement of the concentration of the radioactive substances in the atmosphere in Tokyo. J. Geomag. Geoelec., **10**:56–63.

———. 1961. Size distribution of naturally occurring radioactive dust measured by a cascade impactor and autoradiography. Geofisica Pura e Applicata, **50**:243–48.

OKABE, S., and A. KITAMOTO. 1957. Measurement of the naturally occurring alpha-radioactivity by use of nuclear plate. J. Appl. Phys. Japan, **26**:247.

REITER, R. 1957. Schwankungen der Konzentration und des Verhältnisses der Radon- und Thoronabkömmlinge in der Luft nach Messungen in den Nordalpen. Ztschr., Naturforsch., **12a**:720–31.

SCHUMANN, G. 1956. Untersuchungen der Radioaktivität der Atmosphäre mit der Filtermethode. Archiv f. Meteorol. Geophys. & Bioklimatol., **9** (Ser. A): 204–23.

SIKSNA, R. 1949. Radioactivity induced from the atmospheric air, investigated with the photographic emulsion method. Arkiv f. Geofys., **1**: 128–47.

WILKENING, M. H. 1952*a*. A monitor for natural atmospheric radioactivity. Nucleonics, **10**:36–39.

———. 1952*b*. Natural radioactivity as a tracer in the sorting of aerosols according to mobility. Rev. Sci. Instr., **23**:13–16.

———. 1959. Daily and annual courses of natural atmospheric radioactivity. J. Geophys. Res., **64**:521–26.

YANO, N. 1961. Measurement of natural radioactive dust in the atmosphere by electric precipitator. Paper in Meteorol. & Geophys. (Tokyo), Vol. **12.**

By H. ISRAËL, *Aachen, Germany*

THE RADON-220 CONTENT OF THE ATMOSPHERE

Comparison of the expected and the measured values of the radon-220 content of the atmosphere discloses a considerable discrepancy:

Expected values: Considering that the natural radioactivity of the atmosphere is to be understood as an equilibrium state between the exhalation of radon-222 and radon-220 from the soil and their decay in the air, the concentration of both gases and their decay products in the atmosphere has to be related to the content of their long-living mother substances in the soil. The relation is given by the equation

$$c_0 = C\sqrt{D/K} , \qquad (1)$$

where c_0 is the concentration of the particular element in the atmosphere near the ground, C the concentration of the long-lived parent substance in the soil, D the diffusion coefficient of the emanation in the soil air, and K the "apparent diffusion coefficient" ("Scheindiffusionskoeffizient") in the atmosphere.* Since C has the following mean values for continental soils,

$$C_{U_{238}} = 1.0 \cdot 10^{-12} \text{ c/gm,}$$
$$C_{Th_{232}} = 1.1 \cdot 10^{-12} \text{ c/gm,} \qquad (2)$$

the relation of the mean radon-222 content and the mean radon-220 content in the continental atmosphere near the ground—given in c/cm³—is expected to be

$$(c_0)_{Rn_{222}} : (c_0)_{Rn_{220}} = 100 : 110 , \qquad (3)$$

that is, the radon and thoron concentrations should be of the same order of magnitude in the lowest layers of the atmosphere.

Measured values: In contradiction to this result, the measured values of the radon-220 content of the atmosphere given in the literature are in general much smaller than the values of the radon-222 content.

---

* H. Israël, "Die natürliche und künstliche Radioaktivität der Atmosphäre," in H. Israël and A. Krebs (eds.), *Nuclear Radiation in Geophysics* ("Kernstrahlung in der Geophysik") (Berlin: Springer Verlag, 1962).

Discussion: This discrepancy can be cleared up as follows:

In view of the short lifetime of radon-220, there is no possibility of enrichment. Therefore the radon-220 content usually is computed from the lead-212 (thorium B) content of the air probe assuming radioactive equilibrium between both elements, radon-220 and lead-212. Hence the results depend on the question whether or not this assumption is allowed.

As is well known, equilibrium between the elements of a radioactive series can be reached only when the top element is longer lived than all decay products or when the concentration of the top element of a part of a series is maintained by production. The first condition—fulfilled for radon-222 as the top element in view of its short-lived decay products—is not fulfilled for radon-220 as the top element. Also the second condition is not fulfilled for the following reason:

The radon-220 exhaled by the surface of the soil, as well as the elements produced by decay, is distributed in the atmosphere under the influence of the "austausch." The mean concentration of any decaying element for different altitudes is given by the equation

$$c_h = c_0 \cdot e^{-\sqrt{\lambda/K} \cdot h} , \tag{4}$$

where $c_0 =$ concentration near the ground, $c_h =$ concentration at the altitude h, $\lambda =$ decay constant, and $K =$ "apparent diffusion coefficient." Therefore the ratio of the radon-220 content to the lead-212 content varies with the altitude according to the equations

$$\left(\frac{c_{\mathrm{Tn}}}{c_{\mathrm{ThB}}} \right)_h = \left(\frac{c_{\mathrm{Tn}}}{c_{\mathrm{ThB}}} \right)_0 \cdot e^{-(h/\sqrt{K})\,(\sqrt{\lambda_1} - \sqrt{\lambda_2})} , \tag{5a}$$

$$\left(\frac{c_{\mathrm{Tn}}}{c_{\mathrm{ThB}}} \right)_0 = \sqrt{\lambda_1/\lambda_2} = 26.5 , \tag{5b}$$

where λ_1 and λ_2 are the decay constants of radon-220 and lead-212, respectively.

CONCLUSION

Consequently, radon-220 values derived from lead-212 measurements under the assumption of radioactive equilibrium are too small and have to be corrected according to equations (5a) and (5b). Doing so shows that the values published up to now come to the same order of magnitude as the values of the radon-222 content in the low atmosphere.

HENRY F. LUCAS, JR.

17. A Fast and Accurate Survey Technique for Both Radon-222 and Radium-226

ASSESSMENT OF THE RISK to man of a chronic small increase in the radiation dose is of great concern and interest not only because of the radiation resulting from fallout, but also because of our increasing awareness of the large number of areas and people who are being exposed to a higher than average level of natural background radiation. Since direct experimentation to determine the effect of radiation on man is not desirable or possible, one must utilize and study those people who reside in geographic areas having an unusually high radiation background. Several studies have been made to evaluate the risk from the variation in the external radiation dose (World Health Organization, 1959; Gianferrari et al., 1962). However, from a recent paper on the radioactivity of plants and animals from the high-radiation area of Brazil (Penna-Franca and Gomes de Freitas, 1963) and from our own study in the midwestern United States (Lucas, to be published), it appears that the radiation dose from the internally deposited radioelements is of the greatest importance.

The study in progress at the Argonne National Laboratory began with the discovery in 1948 of the high natural radioactivity in the drinking water of the adjoining municipal water supplies. Dr. A. F. Stehney later identified the radioactivity as resulting from radium-226 and radium-224 (Stehney, 1955). Other studies have defined the boundaries of the geographic region in which the high-radium waters are consumed (Lucas and Ilcewicz, 1958; Lucas, 1959a), the number of people involved (Lucas and Krause, 1960), and such metabolic factors as the relative contribution of both food and water to the whole body content of exposed individuals (Lucas, to be published; Lucas and Krause, 1960). These studies have provided the basis for the initiation

HENRY F. LUCAS, JR., is associate chemist, Radiological Physics Division, Argonne National Laboratory, Argonne, Illinois.

of an epidemiological survey to determine the effect of these internally deposited radioelements on man, which has recently been undertaken by the United States Public Health Service.

In the course of these studies, a fast, simple, and reliable method was developed for (*a*) a mail-order system of sample collection, (*b*) a γ scintillation method of radon-222 assay, and (*c*) an improved emanation method of radium-226 assay.

SAMPLE COLLECTION

In the original survey of the radium-226 content of well waters in Illinois (Lucas and Ilcewicz, 1958), the water samples were obtained by visiting each of the cities of interest, finding the person in charge of the well, inquiring about well use and construction, and, finally, obtaining the desired sample. This method of sample collection was not satisfactory for the larger study made, for the following reasons: (1) Samples were available only from 70 to 75 per cent of the desired sources at the time of the visit because of maintenance, repairs, or absence of anyone to start the pump. (2) The person in charge of the pumping station, except at the larger cities, was unable to supply or obtain any information about well depth, construction, or usage, and (3) on the average, 2–3 hours were required to obtain a sample from each of the municipalities of interest. While this includes travel time, approximately 1 hour was spent in locating someone to start the pump. From experience and discussion with the various water-department personnel, it became apparent that they were, in some cases, collecting weekly or monthly water samples for the various public health and state testing laboratories. All the requests and instructions were received by letter. Thus it appeared that a mail-order system of sample collection would be successful and would overcome many of the objections to the field method.

The system as developed is composed of the following material: (*a*) A 1-liter, small-mouth, screw-cap polyethylene bottle contained in a close-fitting cardboard mailing carton. (This carton, among five tested, was the only one remaining intact when containing a full bottle of water and thrown five times, as hard as possible, against a concrete wall. This test was just barely adequate, since about 1–2 per cent of the samples were received separated from their cartons or held together by wire or string.) (*b*) A letter (Fig. 1) explaining the reason for our interest in water from the particular well described and including instructions for collecting the sample. (*c*) A questionnaire to verify the existing well records and obtain information about the method and the fraction of the supply that is softened or otherwise treated. (*d*) A duplicate of the instructions and identification in the

Argonne National Laboratory

OPERATED BY THE UNIVERSITY OF CHICAGO
9700 SOUTH CASS AVENUE
ARGONNE, ILLINOIS

TELEGRAM WUX LB ARGONNE, ILL.

CLEARWATER 7-7711

TELETYPE TWX ARGONNE, ILL. 1710

April 17, 1963

Water Superintendent
Water Department
Someplace, Illinois

Dear Sir:

We are engaged in collecting water samples from water supplies throughout the Midwestern United States. These water samples will be analyzed for their naturally occurring radium content as part of a research program at this Laboratory to determine the natural radioactivity of man and his environment. The results of a survey of the "Natural Radium-226 Content of Illinois Water Supplies" have already been published in the November 1958 issue of the Journal of the American Water Works Association.

The development by this Laboratory of new ultrasensitive methods for detecting radioactivity permits the Laboratory to engage in this large scale effort to determine the distribution of the naturally occurring radioactive elements in man's environment. These elements - uranium, radium, carbon, potassium, etc., have always occurred in extremely small amounts in the earth on which we live, and thus in the foods and waters man consumes. It is a major purpose of this investigation to measure the rate at which man accumulates and stores these minute amounts of naturally occurring radioactive elements within his body.

Since we are unable to visit the many localities from which samples of water are needed, we are requesting your help in providing us with a sample of water from your well.

With this letter, I am sending a one (1) quart polyethylene bottle and a stamped, self-addressed return label.

a) We would greatly appreciate a water sample from your well identified in our records as:

  Well, South, Number 67, 2205' deep, drilled in 1897, and cased from surface to 1603:

b) Fill the bottle with raw (or chlorinated) water obtained from a tap at the pumping station. Pumps should be in operation one or two hours before sample is taken.

c) Fill out the questionnaire on the bottle as accurately as possible. If there is any question about it or about the sample collection, by all means please note on the questionnaire or on a separate sheet of paper.

d) Remove stamped, return-addressed shipping label and place over old label. Enclose water sample in shipping container. Seal and deposit in nearest Post Office.

Thank you for your assistance and consideration in this matter.

Sincerely yours,

HFL:lmc
cc: Reading file
encl.

Henry F. Lucas, Associate Chemist
Radiological Physics Division

FIG. 1.—Sample copy of form letter requesting a water sample

letter on a half-size sheet of paper fastened with tape to the bottle. (*e*) A return-addressed label with correct postage and three pieces of sticky tape for resealing the carton.

The package described above was sent via parcel post to the water superintendent of the water department of the city of interest. In the event that there was no municipal water department, the package was sent to the police chief of the city of interest. In nearly all cases the individuals reached have been most co-operative and eager to assist. Excluding incorrectly addressed samples, of the 600 evaluated, nearly 90 per cent were returned within three weeks; and a letter of inquiry has increased returns to about 94 per cent. While about 20 per cent of the water samples obtained are from sources other than those requested, this is no worse than was obtained by field collection, and in most cases these samples could be obtained at some known future date. Samples lost in the mails are less than 1 per cent and are not a serious problem.

This mail-order system of sample collection had several advantages over that of the field method. The information contained on the questionnaire appeared to be more precise, and, on checking with ten individuals selected at random, it was found that, rather than relying on memory, all but one had referred to written records when filling out the questionnaire. Samples were collected during the normal operation of the water system and were found to have the same radium-226 concentration as was found by field collection from the same well. Thus, this system appears to be unexpectedly reliable. Its ease and speed make it a very useful system, especially when used in conjunction with field methods of sample collection.

Radon-222 Assay

Perhaps the greatest problem in the assay of radon-222 in water is in obtaining a representative sample. Several special techniques and collection apparatus have been developed by the geologists. We, however, are interested in human exposure to radon-222 from the water. By having the exposed person collect the sample, perhaps he will superimpose the effect of his own individual habits on the geological factors affecting the radon-222 content. Thus, this method may provide a more realistic sample in terms of human exposure.

The radon-222 content of drinking water varies from nearly zero in water derived from lakes and rivers to about 0.4 μc/l from a private well in New Hampshire (Smith *et al.*, 1961). This well is not now in use, however. The average radon-222 in water from 228 wells in Maine and 26 wells in New Hampshire has been found to be 53,000 and 101,000 pc/l, respectively (Smith *et al.*, 1961).

These concentrations are very much higher than the maximum permissible concentration (MPC) of 1,000 pc/l suggested in 1955 (National Bureau of Standards Handbook 61). However, no recommendation was made in the subsequent handbook (No. 69, 1959). A provisional MPC of 35,000 pc/l has been suggested from measurements of the fraction of radon-222, which decays within the body (Mays *et al.*, to be published).

With radon-222 concentrations of this high level, a γ-counting system is the obvious choice for speed, ease of use, and freedom from chemical processing or manipulation. Abbatt *et al.* (1960) have shown that radon-222 is only slowly lost from a polyethylene bottle. The observed half-life for the γ activity of 3–3.3 days, when compared to 3.825 for radon-222, indicates a loss of radon-222 due to permeability of about 4 per cent per day. The 1-quart polyethylene bottles used in the mail-order system of sample collection have a thinner wall thickness than do the 5.6-liter bottles used by Abbatt *et al.* (1960), which permits a more rapid loss of radon-222. An apparent radon-222 half-life of 2.3–2.7 days was found.

It is essential for a reliable radon-222 γ-counting system that the counting efficiency be essentially the same for radon-222 in the liquid or the gas phase of the sample container. The counting configuration shown in Figure 2, position *A*, is very sensitive to the volume of the sample and to the temperatures, since the coefficients of solubility (ratio of the concentration in the water and the air) at 20° and 30° C. are, respectively, 0.250 and 0.195 (Jennings and Russ, 1948). This means that for equal gas and liquid volumes 80 per cent of the radon-222 is in the gas phase and 20 per cent is in the liquid phase. Thus, if the sample container is placed on the top of the crystal, as in position *A* in Figure 2, then small variations in volume will have a large effect on the radon-222 distribution and hence on the detection efficiency. If the sample is placed on its side, as in position *B* in Figure 2, the detection efficiency is relatively independent of distribution.

The counting system used in this work is 4×4 in. sodium iodide (thallium activated) crystal canned in 0.030-in.-thick stainless steel, and having a $\frac{1}{4}$-in.-thick quartz window. It is coupled to a 5-in. photomultiplier tube and placed in a 5-in.-thick iron shield. A background counting rate of 460 c.p.m. is obtained for all pulses above 0.060 mev. For a 1-quart sample, the counting efficiency for radon-222 in equilibrium with its daughters is 0.3 cpm/pc radon-222. Letting the minimum level of sensitivity be equal to 4 times the standard error of the count, then, for a 10-minute counting interval, the minimum detectable level of radon-222 will be 100 pc/l.

Measurements of the radon-222 concentration of water samples from several nearby wells were found to agree with similar measurements made by Scholl in 1917 and ranged from 60 to 500 pc/l (Scholl, 1917). Calibration was achieved by collecting duplicate samples, which were analyzed by the radon-222 emanation technique. Because of the low amounts of radon-222 available for this study, a critical evaluation of the technique was not made.

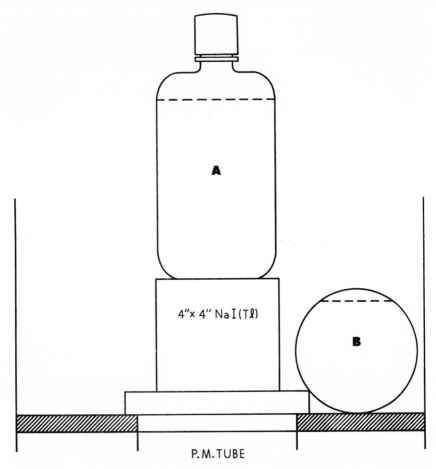

Fig. 2.—Sample configuration for radon γ counting system

Radium-226

EMANATION SYSTEM

Radon-222 from radium-226 is transferred by flushing from the emanation flask to charcoal cooled to —80° C. By heating and flushing of the charcoal, the radon-222 is then transferred to the very efficient and reliable α scintillation radon-222 counter. For a 20-minute counting interval, this procedure can readily detect 0.1 pc/l radium-226 from a 1-liter sample, or less from a larger sample. Samples as large as 20 liters can be conveniently used. Thirty or more (1-liter) samples may be easily processed by one man in an 8-hour day. The system used is shown in Figure 3 and involves the following operations:

1. Place the emanation flask (*A*) and charcoal (radon-222) trap on the system (Fig. 3).* Cool the water traps (*B*) and charcoal traps (*C*) with a solid carbon dioxide (dry ice) acetone mixture.†

2. Transfer the radon-222 from the emanation flask to the charcoal trap by flushing at 1 l/min for 12 minutes.

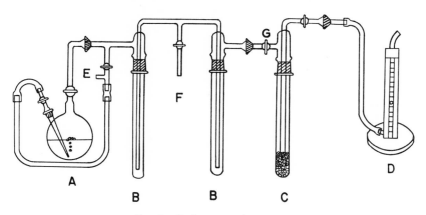

Fɪɢ. 3.—Radon emanating system

3. Seal emanation flask, close stopcock *E*, and evacuate radon trap with mechanical pump for 5 minutes. This operation may be automated with solenoid valves‡ and a sequence timer. Seal charcoal trap, remove from cold bath, and allow to warm.

4. Place charcoal trap (*B*) and α scintillation counter bottle (*H*) on transfer system shown in Plate I.

5. Place furnace (*A*) at 500° C. over charcoal, but do not open charcoal trap to line. Evacuate the counter bottle and the rest of the system. This operation requires about 1 minute.

6. Open charcoal trap to system and transfer radon by flushing with repeated addition of small volumes of helium from reservoir (*D*) to

---

* Two or three of these systems should be in use simultaneously.

† The following non-combustible, non-toxic solvent is very convenient: Freon, Formula TA, E. I. du Pont de Nemours & Co., Freon Products Division, Wilmington 98, Del.

‡ Solenoid valve, catalog No. V5D 5703R, 0.031-in. orifice, Skinner Electric Valve Division, New Britain, Conn.

PLATE I

System used to transfer radon from charcoal to radon counter bottle

the hot charcoal trap and pumping the helium and radon-222 with Sigma pump (*F*) to the counter bottle. Solenoid valves* and a timer† automate this operation.

Low radon-222 air is obtained by aging 220-cu.-ft. compressed dry air tanks for sixty days. Radon-222 concentrations ranging from 5×10^{-4} to 5×10^{-3} pc/l have been found in 12 tanks having a capacity of 220 cu. ft. at 2,000 p.s.i. (Stehney *et al.*, 1955). This radon-222 probably originates from radium-226 on the wall, since the concentration of radon-222 increased as the pressure and volume was reduced. However, the 12-liter flush needed for a 1-liter water sample will introduce only 0.06 pc. radon-222, and a correction can be conveniently made.

If radon-222-free air is desired, it can be obtained by absorbing the water on magnesium perchlorate (Anhydrone) or on Linde Molecular Sieve type 5A and the radon-222 on charcoal cooled to $-80°$ C. Copper tubing should be used between the tank and the system, since both water and carbon dioxide have a high permeability rate through rubber. No difficulty will be encountered, however, if the total length of exposed rubber in connectors and couplings does not exceed about 25 cm. Tubing must be replaced regularly or at first sign of cracking when bent sharply.

Removal of radon-222 at room temperature from the emanation flask shown in Figure 3 is independent of the shape, flow rate of the flush gas, bubble size or concentration of either hydrochloric, nitric, or perchloric acids. The fraction of the radon-222 removed is determined by the ratio of the flush gas and sample volumes. Successive samples of radon-222 from water samples ranging from 0.1 to 20 liters and for flushing rates from 0.5 to 2 l/min have shown that 50 per cent of the radon-222 is removed by a flush volume equal to twice the sample volume (Nelson, personal communication).

In another experiment, 2,000 pc. radium-226 was dissolved in 300-ml. 6 M nitric acid, placed in a 500-ml. emanation flask, flushed, and allowed to stand 30 days for radon-222 growth. This solution was then flushed at 1 l/min with successive 3-, 10-, 10-, and 10-liter volumes of air. The radon-222 content of each sample was 0.973, 0.025, 0.001, and 0.001 of the total. The radon-222 in the first 3-liter aliquot was slightly higher than the 96.9 per cent predicted, while the last two 10-liter portions correspond to the radon-222 produced in the 10 minutes required to collect the sample.

A simple, low-cost emanation flask can be made from a 1-liter narrow-neck bottle. Loss of radon-222 through the two-hole No. 4 rub-

* Solenoid valve, catalog No. V5D 5703R, 0.031-in. orifice, Skinner Electric Valve Division, New Britain, Conn.

† Timer, 3-gong, 1 r.p.m., catalog No. HR3-1, Herbach and Rademan, Inc., 1204 Arch Street, Philadelphia, Pa.

ber stopper is 2 ± 2 per cent. The stopper must be replaced often because even a slight cracking of the surface will result in an increased radon-222 loss. In addition, the per cent radon-222 loss will increase with a decrease in sample volume. New stoppers of polyvinyl* having a lower permeability coefficient may permit this type of flask to be used even for work of highest accuracy.

Contamination of the sample by room-air radon-222 trapped within the connections and tubing of the system is generally negligible. In the system described in Figure 3, the trapped volume is less than 25 cc. While room-air radon-222 concentrations of 5 pc/l have been found that were from five to ten times higher than for outdoor air, the average concentration for the midwestern United States of 0.2 pc/l will result in only 0.005 pc. radium-226 contamination. Lower radon-222 concentrations are found in air near the oceans.

Retention of radon-222 on the charcoal trap is greater than 99.5 per cent, provided that the flush gas is free of water and carbon dioxide (Stehney *et al.*, 1955). Quantitative removal of water from a high-velocity gas stream is extremely difficult because of the formation of ice crystals that are carried through the water freeze-out traps. If flush rates greater than 1 l/min are used, then additional water freeze-out traps must be added. Carbon dioxide is removed by the initial flushing of the acidic solutions.

The charcoal trap shown in Figure 3 was designed for flow rates to 10 l/min at which the pressure drop across the trap was 2 inches of mercury or less. Other types, such as a simple U-tube, have been used. Activated coconut charcoal, 6–8 mesh, from several sources has been found to work equally well. No loss of radon-222 occurs if the charcoal is evacuated while still at —80° C. and without warming to room temperature. If it is allowed to warm to room temperature, the absorbed gases vented, and then sealed, loss of radon-222 is less than 0.5 per cent (Stehney *et al.*, 1955).

The technique of absorbing radon-222 on charcoal has the advantage of permitting the assay of radon-222 from essentially any source. Measurements of radon-222 in the water (Lucas and Ilcewicz, 1958), radium-226 solutions (Lucas and Krause, 1960), atmospheric (Moses *et al.*, 1963) or stratospheric air samples (Machta and Lucas, 1962) or even soil air have all been made (Pearson, personal communication).

TRANSFER SYSTEM

The radon-222 on the previously evacuated (—80° C.) charcoal trap is placed on the system shown in Figure 4. The furnace (*A*) at 350–500° C. is placed over the charcoal-containing part of the trap,

---

* Bel-Art Products, Pequannock, N.J.

while the α scintillating radon-222 counter (*H*) and the rest of the line is being evacuated. After about 1 minute, the charcoal trap is opened to the line, helium from the reservoir (*D*) (a ⅛-in. NPT pipe tee) is admitted, and the Sigma pump (*F*) is started. The pressure-volume relationship of the reservoir is adjusted so that about 5 cycles are required to fill the α scintillation radon counter to 1 atmosphere. The gum-rubber tubing* is lubricated with castor oil and need not be replaced for 100 or more samples.

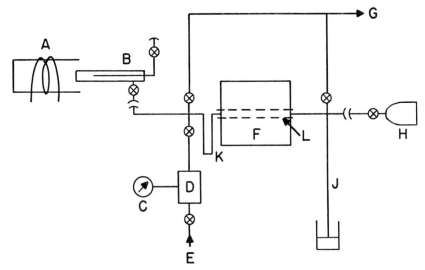

Fɪɢ. 4.—Radon transfer system. *A*, furnace at 350–500° C.; *B*, charcoal trap; *C*, vacuum pressure gauge; *D*, helium reservoir; *E*, helium supply regulated at 2–5 p.s.i.; *F*, Sigma motor pump, belt drive at 300 r.p.m.; *G*, mechanical vacuum pump; *H*, radon counter bottle; *J*, 1-mm.-bore Hg manometer; *K*, water freeze-out trap at −80° C.; *L*, gum-rubber tubing.

For one person to process 30 samples per day, transfer of radon-222 from the charcoal to the radon-222 counter chamber is automated by adding a 1-mm.-bore mercury manometer (*J*), a photoelectric switch, and a repeat cycle timer. The circuit diagram is shown in Figure 5. The three circuits to be activated are (1) fill reservoir with helium, (2) expand helium in reservoir to the charcoal trap, and (3) run Sigma pump to transfer helium in charcoal trap to counting chamber. A photoelectric relay on the manometer terminates transfer operation when the pressure in the counter bottle reaches atmospheric pressure less 2–3 cm.†

* Pure gum-rubber tubing (amber) ¼ in. I.D. × ⅛ in. wall, Amazon Hose and Rubber Co., 130 N. Jefferson St., Chicago 6, Ill.

† Available from W. H. Johnston Counter Laboratory, 3617 Woodland Avenue, Baltimore 15, Md.

The radon-222 content of the α scintillation radon-222 counter bottle* is determined by placing it on the end of a photomultiplier tube (Dumont 6292); covering it with a light-tight cover; turning on the high voltage; and counting. A 2–4-hour delay between filling the counter bottle and beginning the count may be used, but it is not necessary, since the counting rate of the radon-222 plus daughters

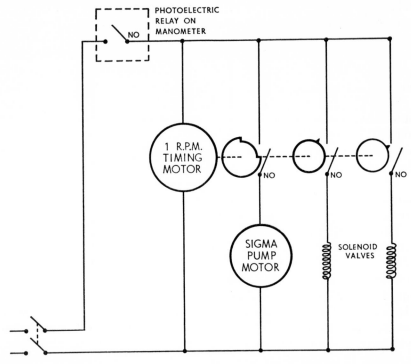

Fig. 5.—Control circuit for automatic transfer of radon from charcoal to α scintillation radon counter bottle.

follows the theoretical growth curve. The relative detection efficiencies for the radon-222, polonium-218, and polonium-214 α's are, respectively, 0.853, 0.895, and 0.912 counts per disintegration.

COUNTING SYSTEM

The counting efficiency of this system is unusually insensitive to either changes in amplifier gain, voltage on the photomultiplier tube, or trace impurities in the transfer gas. At present, eight counting systems are in use. Addition of 10 cc. air to 85 cc. helium changes the counting efficiency by less than 2 per cent. Typically, the counting

* Available from W. H. Johnston Counter Laboratory, 3617 Woodland Avenue, Baltimore 15, Md.

efficiency varies 0.05 per cent/per cent change in high voltage and 0.02 per cent/per cent change in amplifier gain. In these systems an amplifier gain of 1,000, voltage on the photomultiplier tube of 700 volts, and a discrimination level of 12 volts are used. A plateau obtained at fixed discrimination level and increasing voltage has a slope of about 5 per cent per 100 volts and extends for more than 500 volts. The background counting rate of the chamber (and electronics) must be determined over the same range. The lowest background and highest stability are obtained at an operating voltage just above the knee of the plateau. A simple transistorized unit has recently become available.* While not in use at this laboratory, an automatic sample changer is also available.†

Build-up of bottle background follows the theoretical growth equation and is negligible, provided that samples of the same relative activity are counted. Assuming that radon-222 is in the chamber for 4 hours, then a sample with 215,000 c.p.m. will be required to increase the background by 1 c.p.m. Further, this increase will grow in with a 138-day half-life. Contamination of the sample by a previous high sample is less than 0.1 per cent and can be reduced further by increased flushing of the line and this counting chamber.

This system has an over-all counting efficiency including the 5 per cent radon-222 loss in the bottle-filling system of 5.4 cpm/pc radon-222 and a background counting rate of 0.06 c.p.m. The high-efficiency counting system, combined with adsorption of radon-222 from gas, liquid, or solid samples on charcoal, produces a highly versatile system. Measurements can be routinely made of radon-222 concentrations as low as 1×10^{-14} curies or 1×10^{-16} c/l. Since contamination of the system or the counter is negligible, many of the charcoal traps and several of the a scintillation radon-222 counters have been in use since 1956. It is not necessary to rebuild any of the apparatus; thus, this system provides a sensitive, accurate, fast, and reliable technique for either radon-222 or radium-226 assay.

ACKNOWLEDGMENTS

The very considerable contributions by F. H. Ilcewicz, R. B. Burke, F. Markun, and T. P. Kinsella are gratefully acknowledged. Thanks are also due R. B. Holtzman for many helpful discussions and permission to use some of his data and also J. E. Rose and L. D. Marinelli for their interest and encouragement.

* Available from W. H. Johnston Counter Laboratory, 3617 Woodland Avenue, Baltimore 15, Md.

† Available from W. H. Johnston Counter Laboratory and Instrument and Development Products Co., Inc., 355 West 109th Place, Chicago, Ill.

REFERENCES

ABBATT, J. D., J. R. A. LAKEY, and D. J. MATHIAS. 1960. Natural radio-activity in West Devon water supplies. Lancet, **2**:1272–74.

GIANFERRARI, L., A. SERA, G. MORGANTI, V. GUALANDRI, and A. BONINO. 1962. Mortality from cancer in an area of high background radiation. Bull. World Health Organization, **26**:696–97.

HIGGINS, F. B., JR., W. N. GRUNE, B. M. SMITH, and J. G. TERRILL, JR. 1961. Methods for determining radon-222 and radium-226. J. Am. Water Works Assoc., **52**:63–74.

JENNINGS, W. A., and S. RUSS. 1948. Radon, Its Technique and Use. Pp. 13. Middlesex Hospital Press.

LUCAS, H. F., JR., 1959a. Further studies on the natural radioactivity of municipal water supplies. Argonne Nat. Lab., Radiol. Phys. Div. Semiannual Rpt. ANL-5967, pp. 54–57.

———. 1959b. Correlation of the natural radioactivity of the human body to that of its environment: Uptake and retention of Ra-226 from food and water. Argonne Nat. Lab., Radiol. Phys. Div. Semiannual Rpt. ANL-6297, pp. 55–66.

———. Epidemiological aspects of background exposure from radium. Oral presentation, July 25, 1962. (To be published.)

LUCAS, H. F., JR., and F. H. ILCEWICZ. 1958. Natural radium-226 content of Illinois water supplies. J. Am. Water Works Assoc., **50**:1523–32.

LUCAS, H. F., JR., and D. P. KRAUSE. 1960. Preliminary survey of radium-226 and radium-228 (MsThI) contents of drinking water. Radiology, **74**:114.

MACHTA, L., and H. F. LUCAS, JR. 1962. Radon in the upper atmosphere. Science, **135**:296–99.

MAYS, C. W., D. R. ATHERTON, B. J. STOVER, and S. R. BERNARD. Radon concentration guide for drinking water. (To be published.)

MOSES, H., H. F. LUCAS, JR., and G. A. ZERBE. 1963. The effect of meteorological variables upon radon concentration three feet above the ground. Air Pollution Control Assoc., **13**:12–19.

NATIONAL BUREAU OF STANDARDS. 1955. Regulation of radiation exposure by legislative means. National Bureau of Standards Handbook 61. Pp. 60.

———. 1959. Maximum permissible body burdens and maximum permissible concentrations of radionuclides in air and in water for occupational exposures. National Bureau of Standards Handbook 69. Pp. 95.

NELSON, N. Personal communication.

PEARSON, J. Personal communication.

PENNA-FRANCA, E., and O. GOMES DE FREITAS. 1963. Radioactivity of biological materials from Brazilian areas rich in thorium compounds. Nature, **197**:1062–63.

SCHOLL, C. 1917. Radioactivity in Illinois waters. Univ. Illinois Bull., Water Survey Series, No. 14, pp. 114–39.

SMITH, B. M., W. N. GRUNE, F. B. HIGGINS, JR., and J. G. TERRILL, JR. 1961. Natural radioactivity in ground water supplies in Maine and New Hampshire. J. Am. Water Works Assoc., **53**:75–88.

STEHNEY, A. F. 1955. Radium and thorium X in some potable waters. Acta Radiol., **43**:43–51.

STEHNEY, A. F., W. P. NORRIS, H. F. LUCAS, JR., and W. H. JOHNSTON. 1955. A method for measuring the rate of elimination of radon in breath. Am. J. Roentgenol., Radium Therapy & Nuc. Med., **73**:774–84.

WORLD HEALTH ORGANIZATION. 1959. Effect of radiation on human heredity: Investigations of areas of high natural radiation. World Health Organization, Tech. Rpt. Ser., No. 166, p. 1.

LUTHER B. LOCKHART, JR.

18. Radioactivity of the Radon-222 and Radon-220 Series in the Air at Ground Level

F OR A NUMBER OF YEARS the U.S. Naval Research Laboratory has obtained information on the concentrations of the short-lived daughter products of radon-222 and radon-220 (thoron) in the air near ground level by use of the technique described in detail elsewhere in this symposium (Lockhart and Patterson, 1963). Briefly, this procedure involved the estimation of the contribution of lead-214 plus bismuth-214 (RaB + C), lead-212 plus bismuth-212 (ThB + C), and gross fission products to the total β activity collected on air filters through measurement of the rate of decay of this activity during a 16-hour period.

Since this equipment was designed primarily for the detection of nuclear explosions through identification of fission products in the air, its original deployment in a network was based on this function. This equipment was also employed in field tests during the Greenhouse test series of 1951 in the Pacific; as a by-product, some information on natural activity levels in the mid-Pacific area has been obtained (Blifford *et al.*, 1956). Upon discontinuation of those programs, surplus equipment became available for use elsewhere; it has been transferred to various organizations, primarily in the Southern Hemisphere, where it has been operated on a co-operative basis for several years. The sites at which this equipment has been operated for extended periods and the co-operating agencies or institutions are listed in Table 1. Most of the information presented in this report comes from these stations.

LUTHER B. LOCKHART, JR., is head of the Physical Chemistry Branch, Chemistry Division, U.S. Naval Research Laboratory, Washington, D.C.

EXPERIMENTAL PROCEDURE

By measurement of the β activity on an air filter immediately after termination of the collection and again during hours 6 and 16 (or at two other times after disappearance of the lead-214 plus bismuth-214 activity but prior to decay of the lead-212 plus bismuth-212 activity), it is possible to determine the contribution of radon-222 daughters (lead-214 plus bismuth-214), radon-220 daughters

TABLE 1

SITES AT WHICH EXTENSIVE MEASUREMENTS OF ATMOSPHERIC RADIO-
ACTIVITY HAVE BEEN MADE THROUGH USE OF
NRL AIR MONITOR EQUIPMENT

SITE	LOCATION		PERIOD OF OBSERVATIONS	CO-OPERATING ORGANIZATION
	Lat.	Long.		
Wales, Alaska.	65°37′ N	168°03′ W	Jan., 1953–Oct., 1959	U.S. Navy
Kodiak, Alaska	57°45′ N	152°29′ W	Jan., 1950–Oct., 1960	U.S. Navy Electronics Lab.
Washington, D.C........	38°49′ N	77°01′ W	Jan., 1950–Dec., 1961	U.S. Naval Research Lab.
Yokosuka, Japan........	35°20′ N	139°35′ E	July, 1954–Dec., 1958	U.S. Navy
Lima, Peru...	11°40′ S	77°05′ W	May, 1959–Present	NASA Satellite Tracking Station, Ancon
Chacaltaya, Bolivia*....	17°10′ S	68°15′ W	Sept., 1958–Present	Laboratorio de Fisica Cosmica de Chacaltaya, Universidad Mayor de San Andres, La Paz
Rio de Janeiro, Brazil......	23°00′ S	43°25′ W	Aug., 1958–Present	Pontificia Universidade Catolica do Rio de Janeiro
			Jan., 1958–Apr., 1960	Instituto Nacional de Tecnologia
Little America V, Antarctica	78°30′ S	160°00′ W	Apr., 1956–Oct., 1958	U.S. Weather Bureau/ U.S. Navy
South Pole†...	90°00′ S	....	Feb., 1959–Present	U.S. Weather Bureau/ U.S. Navy

* Elevation 5,220 meters.   † Elevation 2,800 meters.

(lead-212 plus bismuth-212), and gross fission products to the initial radioactivity. These activity measurements can be related mathematically to the average activity concentrations of the materials in the air during the collection period (Lockhart and Patterson, 1963). Owing to the saturation effect, or equilibration of the rate of decay of lead-214 on the filter with its rate of collection from the air, the measured concentrations are characteristic of those prevalent during the last hour of the collection. Since the collections were generally terminated in the mid-afternoon, when turbulence in the lower atmosphere is at a maximum, the concentrations of radon-222 and its daughter products measured are representative of the minimum in the diurnal variation of this activity. The measured concentrations of lead-212, on the other hand, are more characteristic of the daily averages of this isotope in the air because of its longer effective lifetime on the filter.

TABLE 2

MONTHLY AVERAGES OF RADIOACTIVITY DATA COLLECTED
AT VARIOUS SITES IN DECEMBER, 1959*

| SITE | AVERAGE COUNTING RATES† | | | RATIO 6/16 | GROSS FISSION PRODUCTS IN HR.-16 COUNT (per cent) | CALCULATED ACTIVITY AT ZERO TIME | | | COUNTER EFFICIENCY (TOWARD PROTACTINIUM-234 [UX₂]) (per cent) | COUNTER BACKGROUND (c/m) |
	Initial	Hr. 6	Hr. 16			Lead-214+Bismuth-214 (RaB+C) (c/m)	Lead-212+Bismuth-212 (ThB+C) (c/m)	F.P.'s (c/m)		
Kodiak, Alaska.	88.7	3.1	2.2	1.43	53	84.7	2.8	1.2	19.3	30
Washington, D.C.	1,175	184	103	1.79	14	916	245	14	14.3	35
Lima, Peru	567	236	128.6	1.835	10	236	318	13	10.6	23
Chacaltaya, Bolivia	759	81.1	42.7	1.900	2.5	644	114	1.1	14.2	77
Rio de Janeiro, Brazil	1,160	473	251	1.886	4	488	662	10	14.8	40
South Pole	28.5	12	11.8	1.02	98	16.3	(0.6)	11.6	13.9	45

* Collections terminated at 1500–1600 local time; air flow averaged 20 or more c.f.m. (ambient) at all sites.
† Background subtracted.

The measured lead-214 air concentrations may be considered to be equivalent to those of radon-222 in the air, since, under the sampling conditions described, essentially complete secular equilibrium between radon-222, lead-214, and bismuth-214 should have been established. Lead-212, on the other hand, cannot be directly related to the air concentrations of the rare gas radon-220 because of the transient nature of this material (54-second half-life).

The monthly averages of the counting rates recorded during the decay measurements of samples collected at several sites during December, 1959, are shown in Table 2. The calculated contributions of radon-222 decay products (lead-214 plus bismuth-214), radon-220 decay products (lead-212 plus bismuth-212), and gross fission products to the initial counting rate of the filter are also shown, together with the counter background and the counter efficiency (including geometry factor) for the uranium oxide standard. The magnitude of the background count does impose a limit to the accuracy with which the determinations can be made, particularly in regard to the determination of low concentrations of fission products. The importance of the counting error is decreased significantly by reporting results as monthly averages; this method of reporting also smooths out the rather large day-to-day variations that occur in the concentrations of all three of these contributors to the radioactivity of the atmosphere.

Summary of Observed Air Concentrations of Natural Radioactivity

A summary of the average concentrations of lead-214 (or radon-222) and lead-212 in the air at ground level at several sites having distinctly different geographical and meteorological environments is given in Table 3. These results indicate that different areas of the

TABLE 3

SUMMARY OF MEASUREMENTS OF NATURAL RADIOACTIVITY
IN THE GROUND-LEVEL AIR

Site	Period of Observation	Radioactivity (pc/m³)		Activity Ratio Lead-214/ Lead-212
		Lead-214	Lead-212	
Wales, Alaska	1953–59	20	0.16	125
Kodiak Alaska	1950–60	9.9	0.04	250
Washington, D.C.	1950–61	122	1.34	91
Yokosuka, Japan	1954–58	56	0.48	117
Lima, Peru	1959–62	42	1.33	28
Chacaltaya, Bolivia	1958–62	40	0.53	72
Rio de Janeiro, Brazil	1958–62	51	2.54	20
Little America V, Antarctica	1956–58	2.5	<0.01	(>250)
South Pole	1959–62	0.47	<0.01	(> 50)

world, as would be expected, have quite different levels of background radiation owing to the natural radioisotopes in the air and, furthermore, that the concentrations of lead-214 and lead-212 do not necessarily parallel each other. From these long-term averages it is impossible to draw more than the obvious conclusion that sites located in the path of maritime air have generally lower natural atmospheric radioactivity than have those whose environmental air has had an overland route. This, of course, results from the fact that the sea is a negligible source of the rare gases radon-222 and radon-220, and, hence, in this environment, radioactive decay of these materials greatly exceeds their rate of replenishment in the atmosphere.

In Figure 1 information is presented on the monthly variations of lead-214 and lead-212 in the air at ground level at Yokosuka, Japan, over a 4-year period. It is evident that short-term measurements of atmospheric radioactivity cannot suffice to define the levels of airborne radioactivity, since seasonal effects are so great. Observed day-to-day variations are even greater. It might be pointed out that measurements made at a single time during the day, as these were, are also

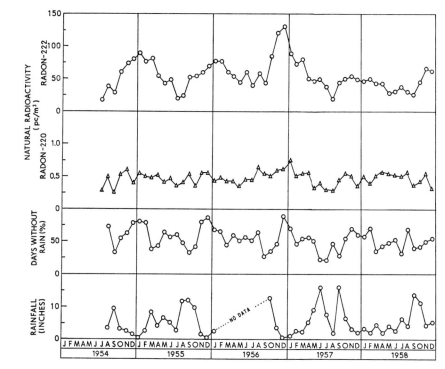

FIG. 1.—Natural radioactivity in the air at Yokosuka, Japan, 1954–58

subject to gross error in evaluating the average conditions prevalent at the site. Similar limitations may be expected to exist in data collected at any site. It is also apparent from Figure 1 that, while there is a generally inverse relationship between atmospheric radioactivity and rainfall, there is no simple correlation of these factors.

EFFECT OF GEOGRAPHICAL LOCATION ON
RADIOACTIVITY LEVELS

Before a presentation of the monthly and seasonal variations in natural activity levels that have been observed at all sites at which extensive measurements have been made, a brief description will be given of the effects of continental and oceanic environments on activity concentrations in general. For the purpose at hand the equilibrium concentration of lead-214 (or radon-222) over land is assumed to be 200 pc/m³ in the lower levels of the atmosphere, while that of lead-212 is assumed to be 2 pc/m³; it is recognized that these may not be realistic values, since an important variable, the effect of vertical diffusion, is being intentionally neglected. Under the set of conditions

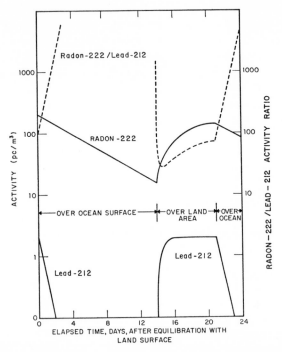

Fig. 2.—Effect of continental and oceanic environments on natural radioactivity levels in the air.

stated above, activity concentrations should follow the general patterns shown in Figure 2 during extended periods of residence over the different surfaces.

If the behavior of an air mass that leaves a land surface after an extended stay is followed as it progresses across an ocean (or an ice-covered land surface), its content of lead-212 will be seen to be rapidly depleted (with a half-life of 10.6 hours) and will be essentially zero in 2–3 days; lead-214, on the other hand, owing to its parent radon-222, with which it will be in secular equilibrium, will decay more slowly (3.8-day half-life) and should exist in measurable concentrations even after several weeks. Under such conditions the lead-214/lead-212 activity ratio will be exceedingly high.

Upon re-encountering a land mass, lead-212 activity would increase much more rapidly than that of lead-214, resulting in a low lead-214/lead-212 ratio. This latter pattern is what might be expected at a site near the seacoast while under the influence of a sea breeze. Later (within a day or two), after passage of the air mass farther inland, lead-212 should approach an equilibrium value, while lead-214 should continue to increase, as would the lead-214/lead-212 activity ratio.

This two-dimensional picture would be complicated by vertical motions of the air, which would dilute the activity of the surface air with air of lower lead-214 and lead-212 contents. The composition of the activity in these descending air masses would be governed by the time they had been out of contact with the earth's land surface in the same way that the composition of the air is affected by an oceanic environment. To obtain a good three-dimensional picture would require the collection of both radioactivity and meteorological data by an extensive network of stations. Such a project would be extremely worthwhile.

Seasonal Variations of Lead-214 (RaB) and Lead-212 (ThB) in the Air

In the next few figures the monthly averages of lead-214 and lead-212 in the air at essentially ground level are presented from data accumulated at several sites where the most extensive measurements have been made (Lockhart, 1959, 1960, 1962a, b). Undoubtedly, more information could have been obtained by a day-to-day analysis of the radioactivity data, particularly if they were interpreted in terms of the meteorological conditions prevailing at the time of collection. Such daily radioactivity data are available for more detailed analysis and interpretation in the future, if such are warranted; some have been presented elsewhere for the year 1957 (Lockhart et al., 1958).

In Figure 3, the monthly variations in natural activity at Washington, D.C., and Yokosuka, Japan, are presented. Note should be made of the fact that lead-212 concentrations have been multiplied by 100 so as to be reasonably presentable alongside the lead-214 values. The activity concentrations of both lead-214 and lead-212 are considerably higher at Washington than at Yokosuka, while the seasonal patterns of these two activities are distinctly different from each other at both sites. The lead-214 concentrations in the air at Washington are lowest in the spring months, when the air is usually most turbulent and intrusions of maritime air from the Gulf of Mexico are most common;

conversely, the concentrations are highest during periods of stagnation or inversion, which commonly occur in the late summer and fall. Lead-212 concentrations are highest by far in the summer; this apparently is the result of a temperature-induced increase in the rate of diffusion of radon-220 gas from the mineral sources in the soil or in the rate of fumigation of the soil which overrides the effects of the normal seasonal changes in the atmospheric mixing processes. Here,

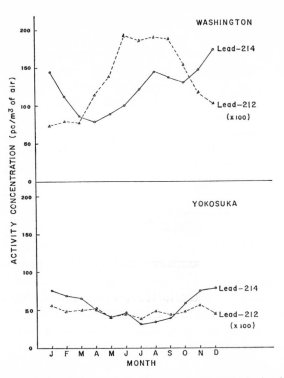

Fig. 3.—Seasonal variations in lead-214 (RaB) and lead-212 (ThB) in the air at Washington, D.C., and Yokosuka, Japan.

again, more detailed analysis on a short-term basis might prove highly enlightening as to the cause-and-effect relationships involved.

At Yokosuka, which is located on Tokyo Bay near its entrance to the Pacific Ocean, the lead-214 concentration is substantially lower than at Washington, in keeping with its insular location, but it undergoes a similar seasonal change. Lead-212 concentrations, on the basis of averages extending over a 4-year period, appear rather invariant with the season; however, significantly large day-to-day and month-to-month variations have been observed (Lockhart, 1959; Lockhart et al., 1958).

Similar data collected at several sites in South America are shown in Figure 4. Both Lima and Rio de Janeiro, though geographically in quite different situations, show increased concentrations of lead-214 in the air sampled during the winter months (June, July, August); at Chacaltaya, however, lead-214 appears rather uniform, with no observed seasonal effects. Lead-212 levels are of great interest, not so much because of the rather random changes that appear with time,

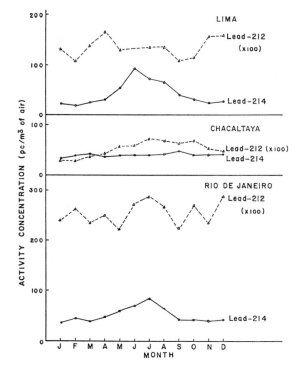

Fig. 4.—Seasonal variations in lead-214 (RaB) and lead-212 (ThB) in the air at several sites in South America.

but because of the magnitude of the concentrations of lead-212 relative to lead-214. Since both Lima and Rio de Janeiro are located on the coast, this effect may be the result of the presence of air at these sites that has had only a limited time of residence over the land, whereby the rate of lead-212 growth has exceeded that of lead-214 (see Fig. 2). This might also be true at Chacaltaya, with the exception that upper tropospheric air rather than maritime air is involved. Another cause to be considered is the possible difference in the uranium/ thorium ratio of minerals in the soils of South America (or at least those in the environs of these sites) when compared to the ratio in soils near Tokyo or Washington, D.C. High concentrations of thorium are known to exist in areas of Brazil not too distant from Rio de Janeiro; the radioactivity of these areas is presently being studied intensively (Roser and Cullen, 1961, 1963).

The concentration changes that have been observed in the lead-214 content of the air at two sites in Alaska (Wales, on the mainland, and Kodiak Island) and two in Antarctica (Little America V and South

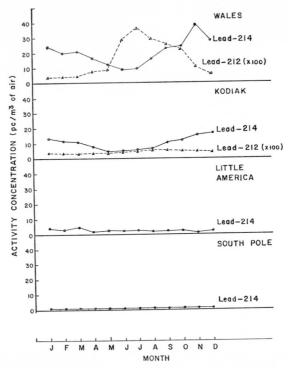

Fig. 5.—Seasonal variations in lead-214 (RaB) and lead-212 (ThB) in the air at the northernmost and southernmost sites.

Pole) are presented in Figure 5. Lead-212 values for the Alaska sites are also shown; the measured activity at the Antarctic sites was too low to be meaningful other than to confirm that no significant amounts of thorium minerals were exposed in this area. The lead-214 concentrations in Antarctica were also extremely low, as might be expected from the great distance of these sites from exposed land masses and the consequent decay of the 3.8-day radon-222 gas during its transport. As a matter of curiosity only, and ignoring some obvious objections, it can be calculated that the travel time for an air mass between Little America and the South Pole is about one week, based on observed differences in lead-214 (radon-222) at the two sites (ratio 0.2–0.3).

The Alaskan sites show the normal seasonal pattern of high lead-214 concentrations during the winter months in spite of the fact that this is a period when much of the surface is snow or ice covered. The lower concentrations of lead-214 (and also of lead-212) at Kodiak are consistent with its island location. The most interesting pattern observed is the tremendous increase in lead-212 concentrations at Wales, Alaska, during the summer and early fall months (June–October), when its concentration is 5 times as high as during the remainder of the year. Though this factor has not been investigated, it is assumed that removal of the snow and ice cover and thawing of the surface soil during this period are responsible. It should be noted that soil condition is extremely critical in regard to the quantity of radon-220 diffusing from the soil and entering the atmosphere, since the half-life of this gas is only 54 seconds; radon-222, with its 3.8-day half-life, is less sensitive to this parameter.

VARIATIONS IN THE LEAD-214/LEAD-212 ACTIVITY RATIO

Figures 6 and 7 show the ratios of the lead-214/lead-212 activity concentrations at the various sites during the course of an "average"

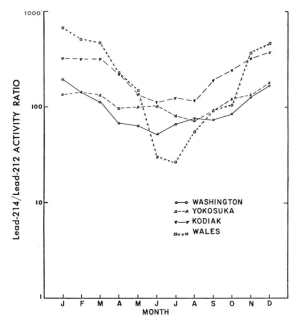

Fig. 6.—Seasonal changes in the lead-214/lead-212 (RaB/ThB) activity ratio at some Northern Hemisphere sites.

year. Apparently, each site has a signature that is intimately associated with the geography and meteorology of the local area, particularly as the latter relates to both large-scale and small-scale mixing and diffusion processes.

As shown in Figure 6, all the Northern Hemisphere sites show a summer minimum in this ratio, which is to be expected from the data presented above. Wales again is an extreme case because of the exceptional changes occurring in the lead-212 concentrations during the warmer months.

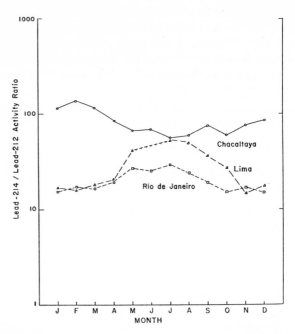

Fig. 7.—Seasonal changes in the lead-214/lead-212 (RaB/ThB) activity ratio at some Southern Hemisphere sites.

In Figure 7, it is evident that Lima and Rio de Janeiro show patterns that are similar to those presented in Figure 6, but that are six months out of phase, that is, a maximum in the ratio occurs in June–August, which is the Southern Hemisphere winter season. Chacaltaya, for reasons not yet understood, would fit better with the Northern Hemisphere than with the Southern Hemisphere sites.

CONCLUSIONS

The radioactivity of the atmosphere is the result of the interaction of a great number of factors, including the time of day, the season, the weather, and, for fission products in the air, the past history of nuclear testing. It is quite evident that the air concentrations of such materials as the naturally occurring radioisotopes that are being produced at a constant rate in the earth's crust vary so greatly, even when averaged over relatively long periods of time, that it becomes impossible to define the natural radioactivity background at a given site by any but an extended series of measurements.

The variability of the natural radioactivity of the air, on the other hand, and its obvious dependence on meteorological factors afford a means of studying the movement and mixing of air masses of different radioactivity contents and of deducing something of the past history of these masses. The simultaneous study of radon-222 with a half-life of 3.8 days and of lead-212 with a 10.6-hour half-life can give information on both long-term and short-term atmospheric processes. Unfortunately, at the present time the high background of fission products in the atmosphere makes more difficult the accurate evaluation of its lead-212 content.

Some work that is called for in the near future is a study of the vertical gradients of radon-222 and lead-212 in the lower atmosphere under various weather conditions, such as during frontal activity, in the presence of onshore and offshore breezes, and during subsidence of air from the upper troposphere. A comparison of the differences in the distribution of radon-222 with altitude over land and over sea far from land should provide useful information on the rates of vertical mixing or diffusion of this gas in the atmosphere.

REFERENCES

BLIFFORD, I. H., JR., H. FRIEDMAN, L. B. LOCKHART, JR., and R. A. BAUS. 1956. Geographical and time distribution of radioactivity in the air. J. Atmos. & Terr. Phys., **9**:1–17.

LOCKHART, L. B., JR. 1959. Atmospheric radioactivity levels at Yokosuka, Japan, 1954–1958. J. Geophys. Res., **64**:1445–49.

———. 1960. Atmospheric radioactivity in South America and Antarctica. *Ibid.*, **65**:3999–4005.

———. 1962a. Atmospheric Radioactivity at Washington, D.C., 1950–1961. Naval Research Lab. Rpt. NRL–5764.

———. 1962b. Natural radioactive isotopes in the atmosphere at Kodiak and Wales, Alaska. Tellus, **14**:350–55.

LOCKHART, L. B., JR., R. A. BAUS, R. L. PATTERSON, JR., and I. H. BLIFFORD, JR. 1958. Some Measurements of the Radioactivity of the Air during 1957. Washington, D.C. Naval Research Lab. Rpt. NRL-5208. Pp. 19.

LOCKHART, L. B., JR., and R. L. PATTERSON, JR. 1963. Techniques employed at the U.S. Naval Research Laboratory for evaluating airborne radioactivity. This symposium.

ROSER, F. X., and T. L. CULLEN. 1961. Radiation levels in selected regions of Brazil. Project Guaraparí. Institute of Physics, Pontifical Catholic University, Rio de Janeiro, Brazil.

———. 1963. External radiation levels in high background regions of Brazil. This symposium.

JOHN B. HURSH AND ARVIN LOVAAS

19. A Device To Measure Thorium-228 at Natural Levels

Tʜᴇ ᴍᴇᴀꜱᴜʀᴇᴍᴇɴᴛ of thorium-228 in concentrations found in natural materials can be made in a number of ways. Table 1 lists some of the general methods in use, each of which has special features that may apply to particular analytical problems. Method 1, although the most sensitive, requires chemical separations that are time-consuming and present opportunities for loss and for contamination. The method of γ spectrometry permits the use of a large sample. The identification is positive when the thallium-208 photopeak is measured, but the equipment required is rather expensive, and the sensitivity limits the types of natural materials that can yield reliable thorium-228 measurements. The very elegant method of α spectrometry developed by Hill (1961) has the manifest advantages of relatively simple preparation and the positive identification and measurement of all the long-lived α emitters in the sample. The sensitivity of the method may have been improved over the initial design, and this would extend its usefulness. The ionization chamber is of special design, and the electronic equipment is relatively expensive.

The fourth general method listed is specifically limited to measurement of thorium-228* and forms the subject of the present paper. The method as presented resolves itself into a measurement of the radon-220 (thoron) gas generated by the 3.64-day half-life radium-

JOHN B. HURSH and ARVIN LOVAAS are in the Department of Radiation Biology, University of Rochester School of Medicine and Dentistry, Rochester, New York.

Based on the work performed under contract with the U.S. Atomic Energy Commission.

* To the extent that the radium-224 content of the sample is significant, i.e., other than a reflection of thorium-228 content, this nuclide is measured as the immediate parent of radon-220.

224 daughter of thorium-228. The measurement of radon-220 incorporates the advantage of a relatively simple physical-separation step but presents the special problem of measuring a nuclide with a half-life of only 54.5 seconds. The difficulty has been solved by a variety of expedients. Evans (Aub *et al.*, 1952) described a procedure for collection of the 10.6-hour half-life lead-212 (thorium B) by electrostatic deposition on a disk centrally located in a 12-liter radon-220 gas-flow hold-up tank. After a short time for build-up, the counting rate of the α daughters of lead-212 was determined. Rundo *et al.* (1958) rapidly transferred the collected radon-220 into a previously evacuated

TABLE 1

DETECTION OF THORIUM-228

Method	Sensitivity (curies/gm ash)	Remarks
1. Chemical separation of thorium; allow Ra^{224} to grow in, separate Ra^{224}, and count in gas-flow or scintillation system	5×10^{-15}	Uses 10 gm. ash; potentially permits Th^{232} measurement as well
2. γ NaI (Tl) crystal spectrometry; sample must be sealed, stored until Ra^{224} is in equilibrium, and photopeak of Tl^{208} measured; requires analyzer	10^{-13}	Uses 100 gm. ash; permits simultaneous measurement of Ra^{228}
3. α spectrometry; special equipment such as designed by C. R. Hill (1961)	10^{-13}	Uses 1.5-gm. ash sample; yields simultaneous identification and measurement of all other α emitters in sample
4. Newly developed method; storing until Ra^{224} is in equilibrium; sweeping the thorium-228 granddaughter radon-220 from a solution containing the sample, adsorbing, and counting by scintillation techniques	5×10^{-14}	Uses 10 gm. ash dissolved in 200-ml. solution; Th^{228} only is measured

4-liter ionization chamber and counted radon-220 and polonium-216 (thorium A) as they decayed with approximately the half-life of radon-220. Zimmerman and Bouvier (1955) measured the thorium content of ores by dissolving the ore and continuously sweeping the solution with air, which then passed by concentrically-arranged, lucite-cylinder-supported zinc sulfide coatings. A photomultiplier tube placed at one end of this array detected the light pulses transmitted through the lucite. The system to be described derives from the Zimmerman-Bouvier device but achieves a sensitivity somewhat greater than the methods referred to above. The detector was designed to measure radon-220 in the respired breath. The description of the detector and its application in a breath assay procedure are described elsewhere (Hursh and Lovaas, 1962).

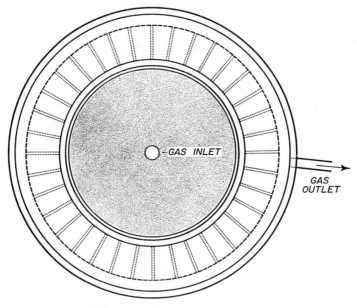

TOP VIEW
TO ILLUSTRATE MANIFOLD

GAS INLET

GAS
OUTLET

CHARCOAL PARTICLES

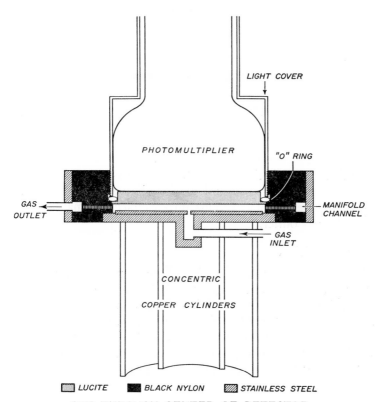

LIGHT COVER

PHOTOMULTIPLIER

"O" RING

GAS
OUTLET

MANIFOLD
CHANNEL

GAS
INLET

CONCENTRIC

COPPER CYLINDERS

LUCITE   BLACK NYLON   STAINLESS STEEL

CUT THROUGH CENTER OF DETECTOR

Fig. 1.—Schematic drawings of the detector as cut through the center and as viewed from the top with the photomultiplier, its light cover, and the lucite plate removed.

THE RADON-220 (THORON) MEASUREMENT SYSTEM

The detection system for radon-220 uses a scintillation technique, as does the Zimmerman-Bouvier system, but it retains the radon-220 in the sensitive counter volume rather than measuring it *en passant*. Figures 1 and 2 provide a schematic drawing of the detector. The radon-220 is adsorbed on the surfaces of small activated charcoal particles fixed by pressure adhesive to a steel plate, which forms the floor of a 3-inch-diameter by ⅛-inch-deep space bounded on top by a zinc sulfide–coated lucite plate. The face of a 3-inch-diameter photomultiplier tube is joined to the uncoated surface of the disk with

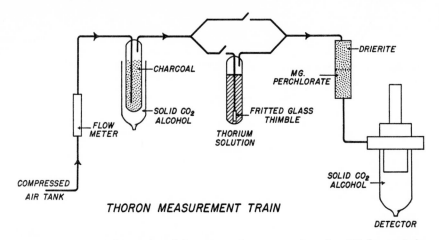

THORON MEASUREMENT TRAIN

Fig. 2.—Schematic illustration of the train used to convey the radon-220 from solution to the detector.

silicone fluid. The radon-220 gas carried by a flow of tank air enters the chamber through the central hole and passes over the charcoal. The unadsorbed radon-220 and carrier air pass out of the chamber through multiple peripheral vents into a collecting manifold and out through the gas outlet.

THE GAS TRAIN

Figure 2 is a schematic drawing of the train used to collect radon-220, to dry the gas, and to introduce it into the detector. Aged compressed air is used in order to minimize the interference from naturally occurring radon-222, which, like radon-220, will deposit on charcoal. A further radon-222 protective device is the insertion in the train of a cooled charcoal trap prior to the de-emanation vessel. The flow meter is of the Manostat type and has a stated accuracy of 2 per cent of full scale. For the experimental measurements reported, the gas flow was maintained at 1 liter per minute. The fritted-glass thimble serves to produce fine bubbles, which sweep the radon-220 from the solution. The drying trap employs indicator Drierite as a first stage, followed by solid magnesium perchlorate to reduce the water vapor to approximately 0.002 mg. per liter. This is necessary because the system counts α particles, and any slight amount of water condensate on the charcoal or zinc sulfide surfaces will reduce sensitivity markedly. From the drying trap, the gas passes into the detector. The copper cooling fins illustrated in Figure 1 are immersed in an alcohol-solid carbon dioxide slush so that the charcoal particles are cooled to about $-70°$ C., enabling an efficient adsorption of the radon-220 gas.

Counting Equipment

The accessory electronic equipment is of the usual kind, and measurements are made by recording counts shown on a scaler decade as a function of time.

Sensitivity

Since radon-220 has a 0.16-second half-life daughter, it would be expected that, for each picocurie of radon-220, there would be 4.4 disintegrations per minute available for possible detection. The decay rate of the adsorbed radon-220 would be equal to the thorium-228 decay rate (1) if in the experimental solution thorium-228 were in equilibrium with the radon-220 parent, radium-224, (2) if the radon-220 were completely removed from the solution, (3) if no radon-220 decay occurred en route to the detector, and (4) if all the radon-220 were adsorbed on the charcoal particles. The first two conditions can be readily satisfied, but an experimental compromise must be arranged in respect to conditions 3 and 4. High gas-flow rates minimizing radon-220 decay are associated with decreased partial pressure of radon-220 and a lesser fraction adsorbed. The compromise in the system reported here takes the form of setting the gas flow at 1 liter per minute and providing a dead space volume of 225 ml. between the 200-ml. solution and the detector. These conditions are associated with radon-220 decay to about 84 per cent of the initial activity and a factor of radon-222 adsorbed to radon-222 entering the detector equal to approximately 0.8.

In addition to these losses, a consideration of the counting arrangement makes it apparent that, at best, 2π geometry and 50 per cent counting efficiency might be expected. In the practical case, even though the charcoal particles are macroscopically small (300–325 mesh), the surface is highly irregular as seen by the α particle. Consequently, there is appreciable absorption of the α particles, and we obtain 35 per cent counting efficiency with the device illustrated. Application of these factors yields the expectation of about 1 count per minute per picocurie of thorium-228 in the solution. It is not surprising that this expectation is borne out by the calibration data presented below, in view of the circumstance that it is partly based on these data.

The background count is 0.15 counts per minute. Assuming that a measurement can be made of an amount of activity three times background and that 10 gm. of ash can be made soluble in 200 ml. of solution, a working sensitivity of approximately 5×10^{-14} c/gm ash may be calculated.

MEASUREMENT OF THORIUM-228 IN ROCK SAMPLES

As an illustration that this system may be used to measure natural radioactivity, we have performed analyses on 4 types of rock samples supplied and identified by Dr. William Basset, of the University of Rochester Geology Department. Dr. Robert Sutton, of the same department, has made available to us core samples of ocean-bottom material, which we have processed and measured.

DIGESTION

Ten gm. of the powdered rock was treated with hydrofluoric acid, then boiled with concentrated nitric and perchloric acid, and finally taken up in 200 ml. of dilute nitric acid. Samples were prepared in duplicate.

MEASUREMENT

As needed, individual sample solutions were transferred to the de-emanation vessels and measured in the counting system for three 10-minute periods. Since the accumulation of the 10-hour lead-212 daughter increases the background measurement, the samples were measured in the order of increasing activity. This order was established by preliminary brief measurement periods, after which a new charcoal-covered plate and zinc sulfide coating were prepared for the systematic counts.

CALIBRATION

In order to calibrate the system, solutions containing increasing, weighed amounts of 30–35-year-old thorium chloride salt were dissolved in a 200-ml. digest of 10 gm. dolomite (a rock that has a very low thorium-228 content). These standard solutions were then measured in the same way as were the experimental samples. The results plotted as counts per minute versus weight of thorium chloride yielded a straight line through the origin, permitting the calculation of an activity conversion factor of 1.07 counts/min/pc thorium-228.

RESULTS

Application of the calibration factor to the counting rate measured for the experimental samples yields the results listed in Table 2.

TABLE 2

THORIUM-228 CONTENT OF ROCKS AND SEDIMENT

Description	Th-228 Conc. (pc/gm)	Equivalent* Th-232 Conc. (p.p.m.)
Dolomite (Lockport)...................	0.03	0.27
Diabase (Palisades)...................	0.26	2.3
Shale, Dinsmore Creek, Rochester.......	0.61	5.1
Westerly granite, Westerly, R.I.........	5.4	48.6
Ocean bottom V7-58, 0–100 cm..........	1.16	10.4
Ocean bottom V7-58, 465–85 cm.........	0.70	6.3
Ocean bottom V7-57, 100–215 cm........	0.47	4.2

* Assuming radioactive equilibrium between Th-232 and Th-228. This is done for comparison, with no prejudice as to whether the assumption is or is not valid.

CONCLUDING REMARKS

A method has been presented applicable to investigation of the thorium-228 content of natural materials. Its advantages are relatively simple sample preparation, positive identification, and a sensitivity sufficient for measurements on rocks, some food materials, and perhaps the natural content of bone ash. We have not attempted to measure the last two categories and anticipate that bone ash will be either at or beyond the sensitivity of this approach.

In respect to the results on rocks and ocean bottom, it is apparent that interpretation would be facilitated by companion determinations of thorium-232. We hope to make such measurements. There is a temptation to speculate on the meaning of the decrease of thorium-228 in the ocean-bottom data from the 0–100-cm. core to the 465–85-cm. core. However, two considerations restrain us from pursuing this point: additional data are required, and speculations on this matter, as indeed on the interpretation of the other measurements on rocks, can be made with much greater authority by the professional geologists than by the authors of this paper.

REFERENCES

Aub, J. C., R. D. Evans, L. H. Hempelmann, and H. S. Martland. 1952. The late effects of internally-deposited radioactive materials in man. Medicine, 31:221–329.

Hill, C. R. 1961. A method of alpha particle spectroscopy for materials of very low specific activity. Nuc. Instr. & Methods, 12:299–306.

Hursh, J. B., and A. Lovaas. 1962. A device for measurement of thoron in the breath. Univ. Rochester Atomic Energy Project, Atomic Energy Commission Res. and Dev. Rpt. UR-619. Pp. 26.

Rundo, J., A. H. Ward, and P. G. Jensen. 1958. Measurement of thoron in the breath. Phys. Med. & Biol., 3 (No. 2): 101–10.

Zimmerman, J. B., and J. A. F. Bouvier. 1955. Measurement of thorium in ores by the thorium emanation method. Tech. Paper 14, Canada Dept. of Mines and Technical Surveys, Mines Branch, Ottawa. Pp. 24.

ELVIRA R. DI FERRANTE, ERIC GOURSKI,
AND RENÉ R. BOULENGER

20. Detector for Radon-222 Measurements at Very Low Level

T HE DETECTOR DESCRIBED HERE is useful for the determination of radium-226 by the emanation technique, especially in samples of very low content. Its construction presents no particular difficulties and is inexpensive. As with other radon-222 counters of analogous type (Damon and Hyde, 1952; Lucas, 1957; Malvicini, 1954; Van Dilla and Taysum, 1955), an α-scintillating inner surface is obtained by use of zinc sulfide. The low background of the detector is due to its form, which represents an optimization of the surface/volume ratio, the material used, and the counting system.

CONSTRUCTION

The detector consists of a cylindrical box with a spherical cavity of 48 cm.^{3} uniformly covered by zinc sulfide. The exact form and dimensions are shown in Figure 1. Lucite has been chosen as the material of construction because of the high transparency and the low radioactive contamination. Lucite can be easily shaped on a lathe, and for a mass production it can also be molded.

The cylindrical box is made in two parts, each with a hemispherical cavity. The cavity is polished and coated with the scintillator before the two halves are cemented together. The cement used is Araldite, type 123 B (Ciba), which adheres well to the lucite. The layer of zinc sulfide (RCA 33-Z-20-C) fixed on a very thin film of silicone grease (S.I.7 Dow Corning Corp.) is monogranular, with an average thickness of about 6 mg/cm^2.

ELVIRA R. DI FERRANTE is Euratom agent, ERIC GOURSKI is head of the Electronics Department, and RENÉ R. BOULENGER is director of the Health Physics Division, all at the Centre d'Étude de l'Energie Nucléaire, Mol-Donk, Belgium.

The outer surface of the box, except for the base (*B*), is left un-polished and covered with thick white paper (base material for the preparation of photographic paper). A small opening through the neck of the upper part, to which a micro-stopcock is attached by means of plastic tubing, permits the filling of the detector.

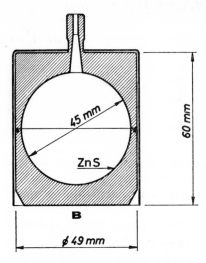

FIG. 1.—Detector in lucite for radon measurements

COUNTING SYSTEM AND DETECTOR CHARACTERISTICS

A 2-inch photomultiplier, RCA 6655A, and a cathode follower, which transmits the anodic signal (on 20 pf.) to a scaler having a minimum input level of —0.2 v., are the essential parts of the count-ing system. The elimination of an amplifier reduces the sensitivity to external electrical noise, while the insulator noise is avoided by using a photomultiplier that works at low voltage. During the counting the detector is placed on the head of the photomultiplier (Lucas, 1957).

The transparency of the zinc sulfide layer was studied by means of an α source (polonium-210) introduced on a narrow metallic support in the equatorial plane of the detector. Figure 2 shows the counting rate-voltage curves obtained with the source facing the bottom (*a*) and facing the top (*b*) of the detector. An attenuation factor of only about 2, due to a combined effect of transparency and distance, exists for the scintillation pulses of the upper part in comparison with those of the lower part. This fact justifies the elimination of a transparent window. Since the layer of zinc sulfide is monogranular, with the grains measuring 0.01–0.04 mm., it presents vacancies that affect the detector efficiency for the α particles hitting the scintillator almost perpendicularly. In spite of this, the value of the total efficiency of the method remains rather consistent.

The plateau of Figure 3 was obtained for the detector filled with radon-222 in an atmosphere of helium. The arrow indicates the operating voltage chosen, which gives the photomultiplier a low current amplification (lower than 10^5), insuring stability and absence of afterpulses. At the operating voltage the threshold corresponds to a signal of about 250 electrons emitted by the photocathode. This number of electrons corresponds to an energy absorption of 100–200 kev. in the

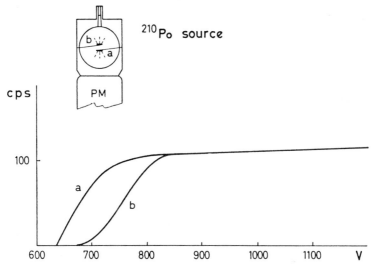

Fig. 2.—Counting rate-voltage curves obtained with a polonium-210 source placed in two different positions at the interior of the detector.

scintillator, while the energy absorbed by a grain of 0.01 mm. of zinc sulfide is 1–2 mev., according to the α-particle speed. Therefore, since grains of 0.01 mm. form layers of 4 mg/cm^2, the average thickness of 6 mg/cm^2 considered for the present work is a good compromise. In fact, a thicker layer would weaken the luminous signal and would make necessary the use of a higher voltage, resulting in an increased possibility of counting photomultiplier noise. This possibility is practically nonexistent in the region of the voltage used. Another effect of a thicker layer could be a higher background counting rate if a contamination is supposed to exist in the zinc sulfide rather than in the lucite.

At the operating conditions described, the background obtained for 18 detectors was 0.5 ± 0.1 c.p.h. Unfortunately the lucite presents a certain memory, especially after measurements of high counting rate. By means of repeated evacuations and refillings with helium, the contamination can be almost completely eliminated. During some of these flushing procedures it was observed that, after one evacuation followed by helium refilling, the counting rate was 0.3 per cent of that given by the sample; after the second flush, 0.05 per cent; after a third flush, 0.007 per cent. The countings were always performed not later than 3 or 4 hours after the end of the radon transfer to the detector; no appreciable loss of radon through the porous lucite (Van Dilla and Taysum, 1955) was observed.

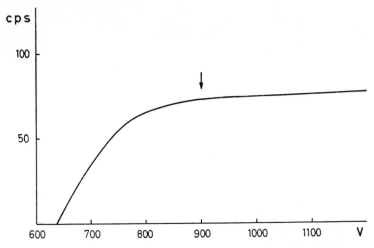

Fig. 3.—Counting rate-voltage curve obtained with radon in the detector

EXAMPLES OF MEASUREMENTS

Numerous environmental samples, such as water, milk, grass, animal bones and teeth, and also human teeth were analyzed for radium-226 content by the radon emanation technique. For all the analyses performed, the detector and the counting system described above were used. The radon removed with nitrogen from the solutions of the samples (previously ashed) was adsorbed on charcoal before being transferred with helium as carrier gas to the detector (Lucas, 1957, 1960).

The total efficiency of the method, determined by measuring the radon from a standard solution of radium, is 75 per cent of the α particles emitted (3 for each radon disintegration). The calibration of the system was repeated several times with different detectors, and the values found agreed to within ± 2 per cent.

As examples of the lowest levels measured, radium-226 content of human teeth from a normal environment is shown in Table 1. Corrections for the reagent-system blank, of the order of 0.002 pgm. radium-226, were made in the calculations.

TABLE 1

RADIUM-226 CONTENT OF HUMAN TEETH

Sample No.	Ash Weight (in gm.)	Ra-226 (pgm/gm ash)
107................	0.933	0.007
112................	2.011	0.015
137................	2.221	0.012
141................	1.669	0.005
142................	1.890	0.021
153-1..............	4.581	0.023
153-2..............	3.377	0.019
153-3..............	3.348	0.019
191................	1.047	0.021
199................	1.793	0.005
204................	2.205	0.005

Average values of two or more analyses; range of estimated error = ± 0.002–0.004 pgm/gm ash.

REFERENCES

DAMON, P. E., and H. I. HYDE. 1952. Scintillation tube for the measurement of radioactive gases. Rev. Sci. Instr., **23**:766.

LUCAS, H. F. 1957. Improved low-level alpha-scintillation counter for radon. Rev. Sci. Instr., **28**:680–83.

———. 1960. Correlation of the natural radioactivity of the human body to that of its environment. Argonne Nat. Lab. Radiol. Phys. Div. Semiannual Rpt. ANL-6297. Pp. 55.

MALVICINI, A. 1954. Camera a scintillazione per la misura dell'emanazione contenuta nell'aria. Nuovo Cimento, **12**:821–23.

VAN DILLA, M. A., and D. H. TAYSUM. 1955. Scintillation counter for assay of radon gas. Nucleonics, **13**:68–69.

MARVIN H. WILKENING

21. Radon-Daughter Ions in the Atmosphere

Nᴏᴛ ʟᴏɴɢ ᴀꜰᴛᴇʀ the discovery of radium by the Curies in 1898, it was observed that minute amounts of the short-lived decay products of the emanation gases were present in the atmosphere (Rutherford *et al.*, 1930). It was soon shown that the radioactive material in these "active deposits," as they were called, owed its origin to uranium and thorium minerals distributed in the earth's crust. Because of the inert nature of the emanation gases, they can diffuse into the atmosphere. Radon-222, radon-220 (thoron), and radon-219 (actinon) all are present in the atmosphere in measurable amounts, but, since the half-life of radon-222 (3.82 days) exceeds that of radon-220 (54.5 sec.) and radon-219 (3.92 sec.), the radon-222 has a much better chance of escaping into the air in spite of the fact that thorium is more abundant than uranium in the earth's crust. The activity due to the active deposit of radon-220 is only about 5 per cent of that due to the active deposit of radon. Additional details regarding the occurrences of these emanation isotopes in the atmosphere are given by Israël (1962). Radon-222 and its short-lived daughter products are the only elements considered in this report.

That portion of the uranium-radium series from radon-222 to the 22-year lead-210 isotope is shown in Figure 1. The short-lived daughter products include two α emitters, polonium-218 (RaA) and polonium-214 (RaC'), and two β emitters, lead-214 (RaB) and bismuth-214 (RaC). The branching that occurs at polonium-218 and at bismuth-214 is neglected. A very important feature with regard to the observation of radon-daughter products in the atmosphere is illustrated in the right-hand part of Figure 1. Radon-222 disintegrates with the emission of an α particle having an energy of 5.47 mev. At the instant of disintegration, the residual atom, polonium-218, has a recoil energy of 0.11 mev., which is sufficient to insure that most of the atoms will be

MARVIN H. WILKENING is head of the Department of Physics and Geophysics, New Mexico Institute of Mining and Technology, Socorro, New Mexico.

positively charged ions. These ions have a high probability of retaining at least a unit positive charge for some time after the disintegration event. Later, they may pick up an electron and become neutral atoms or they may attach to an aerosol particle in the atmosphere. Although the details of these processes with respect to their relative abundances are not well known, the existence of the positively charged radon-daughter ions is well established. It is these ions that form the "active deposit" reported by early investigators that was collected on negatively charged plates or wires.

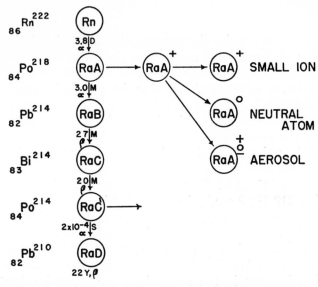

F IG. 1.—Portion of uranium-radium series showing the short-lived daughter products of radon-222. Isotopic symbols are given in the left-hand column. A positively charged polonium-218 (RaA) recoil ion is indicated on the right.

I ON C OLLECTION A PPARATUS

The so-called "induction method" of collecting an active deposit of radon-daughter products (Israël, 1951) has been modified in the present work so that the radon-daughter ions from a known volume of air can be collected and measured.

The apparatus shown in Figure 2 incorporates the features desired. Air is drawn through the tube at about 9 cubic meters per minute, so a normal collection will process 90 or more cubic meters of air. During use, the device is set out-of-doors with the axis of the tube perpendicular to the prevailing wind direction. The axis is arbitrarily set 0.95 meters above the ground. The No. 24 (0.56-mm. diam.) annealed copper wire is maintained at a negative potential of 2,000 volts, which is well below the level at which corona occurs. The high transverse electric field accelerates positive ions toward the wire. For the values of potential and air flow used, all ions having mobilities greater than about 0.3 cm.^{2} sec.$^{-1}$ volt^{-1} are collected. This is apparently sufficient to collect most of the radon-daughter positive ions, excluding those that might be attached to large molecular aggregates or to dust particles. It is of interest to note that when the collecting wire is made positive, the activity is reduced to only a few per cent of the negative-wire value.

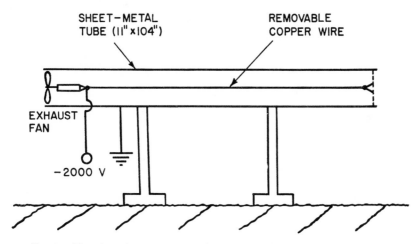

FIG. 2.—Negative-wire apparatus used to collect radon-daughter positive ions

At the end of a collection period, the wire, 2.25 meters in length, is wound onto a flat spiral for counting. Care must be exercised in this procedure lest some of the active deposit be lost by abrasion of the surface of the wire. The daughter-product activity is measured by means of a thin-window Geiger-Müller counter or an α scintillation counter. Routine procedure calls for a collection time of 10 minutes, a 3-minute period for preparation of the wire for counting, a 10-minute counting period under a Geiger-Müller counter, a background determination, and a second counting period of 5 minutes for the wire. The arbitrary activity level for the radon-daughter ions is taken as the mean counting rate over the first 10-minute counting period. The second count taken on each wire permits a comparison with the theoretical composite decay curve of radon daughters. The choice of a 10-minute collection time represented a lower practical limit on time resolution, and it also means that the contribution of the radon-220 (thoron) series to the total daughter-product activity is negligibly small, since the activity of the radon-220 daughters is governed by the

10.6-hour decay period of lead-212. The composite decay curve for the short-lived radon daughters in equilibrium with the parent, radon-222, has an effective half-life of about 36 minutes after the first 10 minutes.

THE THUNDERSTORM EFFECT

The radon-daughter ion-collection equipment was first used during the summer of 1958 at a station on top of 10,297-foot Mount Withington located some 50 miles southwest of Socorro, New Mexico. Collections were made in order to study the anomalously high values obtained by "induction method" collections made several years

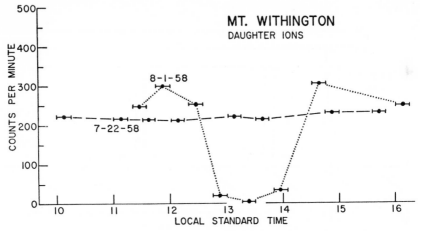

FIG. 3.—Radon-daughter ion collections showing the "thunderstorm" effect

earlier on the mountaintop and to investigate the possibility of studying downdrafts in thunderstorms using radon measurements.

A remarkable effect is illustrated by Figure 3. The curve of July 22 was typical of fair-weather days at the mountain station. On August 1 a thunderstorm passed near the station with the result shown. Although no precipitation occurred at the station, there was a heavy cumulonimbus build-up over the mountain. Strong electrical activity was recorded at the station, and some rain was observed a kilometer or more to the west. The activity due to the daughter-product ions dropped sharply to less than one-tenth of its prestorm value.

A case in which an active precipitating storm passed directly over the station is shown in Figure 4. Vonnegut and Moore (1963) were making extensive measurements at the mountain station at the same time and kindly provided the meteorological data shown. A comparison of the time of onset of the sharp decrease in the daughter-product ion activity with other characteristics of these and other storms studied during August, 1958, indicates quite conclusively that the sharp changes in daughter-product activity were produced by atmospheric electric-field changes associated with the thunderstorms. The effect of the storm, however, on the concentration of the parent radon was still unknown. Hence, before a return to the mountain for further

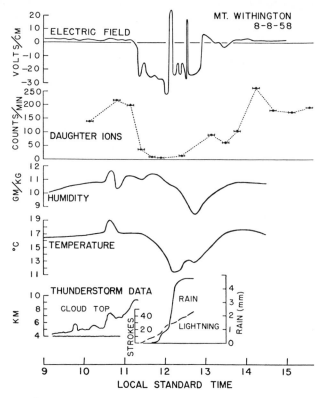

Fig. 4.—Identification of the decrease in daughter-ion content with the "negative" electric field of a thunderstorm.

studies of the thunderstorm effect, it was necessary to design and calibrate a device for making quantitative determinations of the radon content of the atmosphere.

Radon-222, Electric Field, and Conductivity Measurements

The concentration of radon-222 is determined by means of an apparatus similar to that described by Moses *et al.* (1960). Radon-222 gas is adsorbed from a sample air stream by 10 grams of activated charcoal contained in a copper-tubing trap held at a temperature of approximately −50° C. At the end of a collection period, argon gas is used to flush the radon from the charcoal trap into an α scintillation detection chamber. The trap is heated over a gas burner at about 800° C. during the exhausting process. Calcium chloride traps are used to prevent water from collecting in the charcoal trap and from entering the scintillation detector. Radon from a 1×10^{-11} curie standard radium solution was used in calibrating the apparatus. The scintillation detector consists of a 5-inch photomultiplier tube mounted above a 0.340-liter cylindrical detection chamber. The latter is coated with silver-activated zinc sulfide suspension.

A continuous record of the earth's electric field was provided by a recorder and amplifier from an electrostatic induction field mill. The instrument used is a modification of the alternating-current field meter described by Gunn (1954), which uses a phase-sensitive detector.

Equipment was also available for measuring the conductivity of the atmosphere. A Gerdien-type apparatus (Gunn, 1954) was operated to give continuous records of positive and negative conductivity. It is to be noted that the concentrations of small ions of high mobility to which this meter is most sensitive are the predominant carriers of electric charge at the mountain station. All rapid changes in the electric field records produced by lightning strokes in the vicinity of the station have been removed from the records in this study.

ELECTRICAL CHARACTERISTICS

Figure 5 shows a 24-hour record of radon concentration, daughter-product ion content, the electric field, and the total conductivity (sum of the magnitudes of the positive and negative values). The data were taken under fair-weather conditions. The radon-222 characteristically has a mean value of from 0.1 to 0.15 picocuries per liter of standard air. It is somewhat higher in the morning hours before the onset of vertical mixing by the "austausch" effect. The daughter-product ion activity shows an increase during the latter half of the day, a trend which is almost identical to that of the conductivity. This is especially significant, since it indicates that the radon-daughter positive ions do follow the same behavior as that of the total small ions of the atmosphere as measured by the conductivity meter. This fact has been corroborated by other data taken over many days of operation. Finally, the electric field is relatively constant at a level of about 150 volts per meter throughout the day with the exception of a small fluctuation near noontime. The weather was clear and cloudless with the exception of two hours in the late forenoon when fair-weather cumulus clouds were present over the mountain area. The irregularity shown in Figure 5 in the electric field record at 1200 was undoubtedly due to one of these small clouds.

Another example of the thunderstorm effect, together with electric field and radon concentration records is illustrated in Figure 6. Two storms were observed in rapid succession with a short period of electric field of the same sign as the fair-weather field between them. The

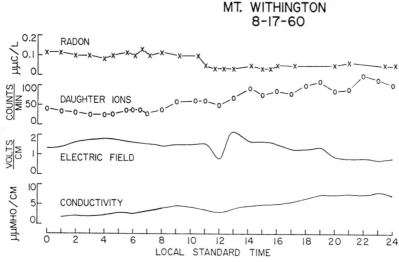

Fig. 5.—Radon-222, daughter ions, and electrical characteristics for a typical fair-weather day on the mountain.

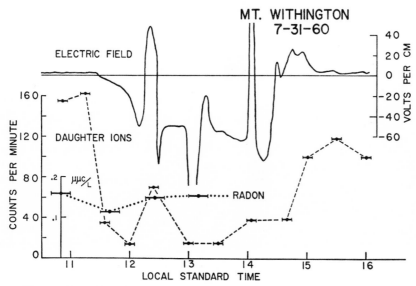

Fig. 6.—The effect of two storms passing over the station in rapid succession

PLATE I

An isolated thunderstorm system typical of those used in these studies

first field reversal occurred at 1130, and light precipitation fell at the station at 1210. This storm passed by at 1220 and was followed by 15 minutes of strong positive field. The 1225 collection unfortunately included only a part of this positive field interval, but, even so, a recovery in the negative-wire activity is noted. A second heavy storm engulfed the station from about 1330 to 1415. The late afternoon sun was shining by 1500. The radon-222 content is observed to decrease with the advent of the first storm and then remain relatively constant. Unfortunately, the radon-222 apparatus became inoperative during the progress of the second storm. These measurements were repeated for 16 storms; in all cases the change in daughter-ion activity was very dramatic compared with the relatively small changes in radon content.

A typical isolated storm of the type especially useful in studies of this type is shown in Plate I. This particular storm was photographed from Mount Withington at a distance of approximately 12 miles. The terrain elevation is 7,150 ft., the cloud base is at 12,700 ft., and the cloud top is at 35,000 ft.

Summary and Conclusions

The results of these investigations are summarized as follows:

1. The negative-wire apparatus measures only the radon-daughter small ions of positive charge and high mobility that result during the decay of radon-222 and its short-lived daughters. A thunderstorm passing over the station produces a reversal of the fair-weather electric field, changing in value from approximately 1.8 volts per cm. to as much as -340 volts per cm. This change is invariably followed by a sharp decrease in the radon-daughter ion content. When a storm passes by, the daughter-product ion content returns to nearly its normal value. The ion content responds to electric field reversals in a prompt manner, being of the order of a few minutes or less. The radon content remains relatively constant, with a maximum fluctuation of 40 per cent from fair weather to thunderstorm conditions. This change is always small compared with the decrease of 1,000 per cent or more in the radon-daughter ion content.

2. The anomalous decrease in the daughter-ion activity may occur in the total absence of precipitation at the station, provided that a storm center is sufficiently close to produce a reversal in the electric field.

3. The radon-daughter ion anomaly is not restricted to a mountaintop environment. The same effect has been observed on 6 occasions in the valley at Socorro (elevation, 4,620 ft.). It is more convenient to use the mountain station because of the greater frequency of isolated thunderstorms useful in these experiments.

These results show that the strong electric fields associated with thunderstorms are of the right polarity and of sufficient magnitude actually to sweep most of the positive radon-daughter ions upward, leaving a deficiency near ground level. When an electric field of the same sign as the fair-weather field but of greater magnitude occurs in a storm situation, positively charged daughter ions are brought down from higher levels in the atmosphere. No deficiency is observed in this case.

The net result of the thunderstorm effect is to provide an upward flux of radon-daughter ions from the ground to higher levels in the atmosphere. A similar flux could be expected from the top of the thunderstorm. This could account for some of the lead-210 (RaD) content of the stratosphere.

Since the daughter-ion content is found to correlate well with the total small-ion content of the atmosphere, as reflected by conductivity measurements, the measurement of radon-daughter ions becomes a useful tool in studying the atmospheric electrical environment.

ACKNOWLEDGMENTS

This work was carried out with the aid of a grant from the Atmospheric Sciences Section of the National Science Foundation. The help of Dr. Minoru Kawano and of the students who assisted with the data taking is gratefully acknowledged.

REFERENCES

CHALMERS, J. A. 1957. Atmospheric Electricity, pp. 128–30. New York: Pergamon Press.

GUNN, R. 1954. Electric field meters. Rev. Scient. Instr., **25**:432–37.

ISRAËL, H. 1951. Radioactivity of the atmosphere. *In* T. F. MALONE (ed.), Compendium of Meteorology, pp. 155–61. Boston: Am. Meteorological Soc.

———. 1962. Die natürliche und künstliche Radioaktivität der Atmosphären. *In* H. ISRAËL and A. KREBS (eds.), Nuclear Radiation in Geophysics, pp. 76–86. New York: Academic Press.

MOSES, H., A. F. STEHNEY, and H. F. LUCAS, JR. 1960. The effect of meteorological variables upon the vertical and temporal distributions of atmospheric radon. J. Geophys. Res., **65**:1223–38.

RUTHERFORD, E., J. CHADWICK, and C. D. ELLIS. 1930. Radiations from Radioactive Substances, pp. 153–58, 560–61. Cambridge: Cambridge University Press.

VONNEGUT, B., and C. B. MOORE. (Arthur D. Little, Inc., Cambridge, Mass.) Private communication, 1963.

S. GOLD, H. W. BARKHAU, B. SHLEIEN,
AND B. KAHN

22. Measurement of Naturally Occurring Radionuclides in Air

As part of the program of evaluating hazards to man from ionizing radiation, the U.S. Public Health Service is engaged in measuring concentrations of naturally occurring radionuclides. The range of radiation exposure commonly encountered can be computed from these concentrations for consideration with other sources of radiation. Significant differences in exposure values may then be used as a basis for epidemiological studies of their effect.

Reported here are measurements of naturally occurring radioactivity associated with air particulates in Cincinnati, Ohio. In recognition of the wide periodic variations in concentration, measurements were made daily for extended periods. At the same time, meteorological variables, such as wind speed and precipitation, were observed so that their influence on radionuclide levels could be evaluated. Samples were collected at a height above the ground at which air would be inhaled. The α particles from short-lived descendants of radon-222 and radon-220 were detected and the data used to compute radon-222 and radon-220 concentrations in air. In addition, β particles from the bismuth-210 daughter of 20-year lead-210 and γ radiations from 53-day beryllium-7 were measured.

Radon-222 and radon-220 gas, formed in uranium and thorium ore, respectively, reaches the atmosphere by emanating from the ground. Radon-220, because of its short half-life (0.9 minute), is detected only near its point of production. Radon-222, on the other hand, decays with

S. GOLD is chemist, Methods Development (Chemistry), Radiological Health Research Activities, H. W. BARKHAU and B. SHLEIEN are physical scientists, and B. KAHN is assistant chief, Radiological Health Research Activities, U.S. Department of Health, Education, and Welfare, Public Health Service, Bureau of State Services, Division of Radiological Health, Robert A. Taft Sanitary Engineering Center, Cincinnati, Ohio.

a 3.8-day half-life and can be found at a considerable distance from its source, following movement through the air. The major fraction of their radioactive daughters adheres to particles in the 0.005–0.04 μ range (Wilkening, 1952) and can be collected on efficient air filters. The measured radon-220 daughter decays with a 10.6-hour half-life and can therefore be found farther from its radon-220 precursor in time and location than can the short-lived radon-222 daughters, which decay with a 0.5-hour half-life.

Unlike radon-222 and radon-220 gas and their short-lived particulate daughters, lead-210 and beryllium-7 are not restricted by radioactive decay to the vicinity of their origin. Their movement is controlled by the factors affecting particle movement in air, such as diffusion, settling, winds, precipitation, and retention on the ground. Lead-210 is formed by the decay of the short-lived radon-222 daughters, and therefore a large fraction originates near the ground. Beryllium-7 is primarily produced in the stratosphere, where the cosmic radiation intensity is greatest.

RADON-222 AND RADON-220

PROCEDURE

Radon-222 and radon-220 daughters are collected by drawing air at the rate of 30 m³/day through a 4.7-cm.-diameter membrane filter.* The filter is shielded from precipitation by a 16-cm.-long glass tube. The apparatus and procedure have been described in detail (Setter and Coats, 1961). Figure 1 (after Anon., 1962) shows the collection cycle and the approach to equilibrium between daughters and parents. The filter is placed in the sampler at 8:00 A.M. and remains there for 24 hours except for a short α count at 3:00 P.M. After its removal, the filter is α-counted immediately to obtain the sum of the radon-222 and radon-220 daughter count rate. To measure the radon-220 daughter only, the filter is again counted after a 7-hour decay period. The count rate of the radon-222 daughter is calculated from the difference of the two measurements, with compensation for radioactive decay. Radon-222 and radon-220 count rates are then extrapolated to the end of the collection period and converted to radon-222 and radon-220 concentrations, based on assumption of radioactive equilibrium for the radon-222 daughters and uniform radon-220 levels throughout the 24 hours for the radon-220 daughters. The α count at 3:00 P.M. is converted to radon-222 activity without considering radon-220 daughter activity, because the latter constitutes less than 1 per cent of the total count rate at that time.

* No. AA, from the Millipore Filter Corp., Bedford, Mass. (NOTE.—Mention of commercial products is not to be construed as an indorsement by the Public Health Service.)

Experiments were performed to assess the effect of the exact location of the particulate collector and to compare the computed radon-222 concentrations with direct measurements of radon-222. Directional effects were evaluated by operating a second air-particulate collector beside the first one, pointing at 90° and 180° angles, and up and down, in turn. The second collector was also operated 4 feet below the original one for a period of time. To compare radon-222 and radon-222 daughter measurements, air was passed first through the

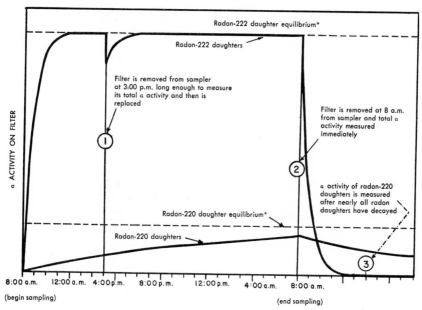

FIG. 1.—Idealized profile of α activity on filter during and after sampling

filter to collect the radon-222 daughters and then through tubes of activated charcoal at —78° C. to retain the radon-222. The tubes were sealed, stored for 3 hours so that the γ-emitting bismuth-214 daughter of radon-222 could grow to equilibrium, and then subjected to γ spectrometry (Shleien, 1963). With a calibrated spectrometer, the radon-222 activity in the charcoal tube was computed from the count rates of the 0.61- and 1.76-mev. γ rays of bismuth-214.

The variation of radon-222 daughter levels was observed continuously between April 17 and May 7, 1963, by collecting air particulates on a membrane filter immediately beneath an α radiation detector (Coats, 1960). Decay studies showed that the α activity was due to the polonium-214 daughter of radon-222. The filter was changed daily, and a continuous record of the count rate was maintained. Because of the rapid decay of the radon-222 daughters, the record reflects the levels at the time of counting. Recorded values were below actual levels immediately after a new filter was inserted, and above them during periods of rapid decrease.

Five soil samples, taken 2–12 inches below the surface of the ground near the collector, and 9 sections of a 20-foot soil core were analyzed for radium-226 content with the calibrated γ spectrometer.

Initially, wind speed and precipitation were measured by the U.S. Weather Bureau 1 mile southeast of the sampler. A second sampler has recently been operated at a U.S. Weather Bureau station 6 miles from the original sampler. At the new location, additional meteorological measurements, such as vertical temperature profiles and solar radiation intensities, are available.

RESULTS AND DISCUSSION

The effect of pointing the collector in various directions or lowering it by 4 feet was small. All radon-222 daughter concentrations on the two filters differed by less than 20 per cent, and over half the values agreed to within the counting error. Even the effect of horizontal movement was small; four-fifths of the radon-222 daughter measurements performed during the winter months at the two locations—one within the city and one near its outskirts, 6 miles away—agreed to within 20 per cent.

The radon-222 levels computed from its daughter activity are all lower than the directly measured values (Table 1). Either non-equilibrium conditions between radon-222 and its daughter in the air passing through the collector or incomplete particle retention on the filter could be the cause. The relatively lower activity obtained from radon-222 daughter measurements after high wind speeds or rain (runs 1 and 4)—conditions leading to non-equilibrium—suggests that the major factor is non-equilibrium. It is to be noted that equilibrium conditions between radon-222 and its daughters require constant radon-222 levels in air for more than 2 hours during collection, whereas no equilibrium can be reached for radon-220 and its daughters during the 24-hour collection period. The values in Table 1 agree with estimates that radon-222 values computed from its daughter may range between 50 and 100 per cent of actual radon-222 concentrations (Harley, 1953).

Daily morning radon-222 and radon-220 levels for 1962 are shown in Figure 2. The radon-222 levels pertain only to the 2 hours prior to collection at 8:00 A.M., while radon-220 values are influenced by air levels throughout the day. The fluctuation of radon-222 and radon-220 in Figure 2 is qualitatively similar except during January. The dissimilar pattern of radon-220 compared to radon-222 during January may be caused by the decreased diffusion rate through frozen soil, which affects radon-220 more seriously because of its short half-life.

A seasonal trend can be seen superimposed on the daily variations shown in Figure 2 and even more clearly by the average monthly

TABLE 1

COMPARATIVE RESULTS OF FILTER-PAPER AND CHARCOAL
ADSORPTION TECHNIQUES*

RUN No.	DURATION OF SAMPLING (MIN.)	TOTAL VOL. COLLECTED (M.³)	RADON CONCENTRATION, PC/M³, DETERMINED BY			RATIO OF FILTER-PAPER METHOD TO CHARCOAL ADSORPTION METHOD		CONDITIONS
			Filter Paper	Charcoal Adsorption		1.76-Mev. Peak	0.61-Mev. Peak	
				1.76-Mev. Peak	0.61-Mev. Peak			
1...	240	2	170 ± 20†	380 ± 30	400 ± 10	0.45 ± 0.06	0.42 ± 0.05	Haze first half-hour, thereafter clear; wind 4–8 knots
2...	120	1	850 ± 40	1,040 ± 50	1,050 ± 20	0.82 ± 0.05	0.81 ± 0.04	Ground fog and haze; wind calm to 3 knots
3...	120	1	590 ± 40	750 ± 40	670 ± 20	0.79 ± 0.07	0.88 ± 0.06	Ground fog and haze; wind calm to 3 knots
4...	240	2	90 ± 20	170 ± 20	170 ± 10	0.53 ± 0.12	0.53 ± 0.12	Some rain and fog; wind calm to 8 knots

* Reproduced in part from Table VI of Shleien, 1963.
† Counting error at 95 per cent confidence level.

values in Table 2 and for morning radon-222 in Figure 3. Minimum radon-222 concentrations occurred in March, and maximum values during August–October. The similarity for the 4 years is evident. Seasonal variations have been observed widely (Wilkening, 1952; Israël, 1951); for the Chicago area, a maximum has been reported in September–October and a minimum in April (Moses *et al.*, 1963). At Washington, D.C., a minimum was observed in the spring, and maxima in late summer and early winter (Lockhart, 1962). The second maximum was attributed to the long overland route of air masses during

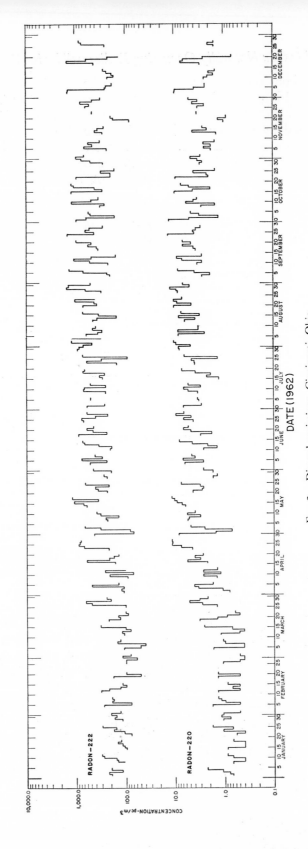

FIG. 2.—Diurnal variations, Cincinnati, Ohio

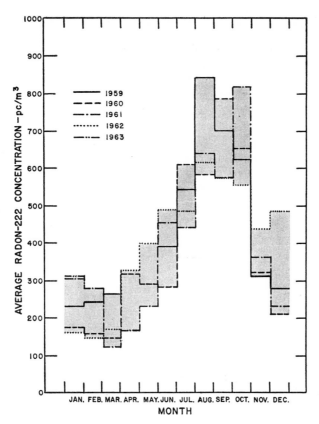

Fig. 3.—Average monthly morning radon-222 concentration

TABLE 2

AVERAGE MONTHLY MORNING AND AFTERNOON RADON-222 AND MORNING RADON-220 VALUES
(PC/M³)

Month	Morning Radon-222*					Afternoon Radon-222†					Morning Radon-220*				
	1959	1960	1961	1962	1963	1959	1960	1961	1962	1963	1959	1960	1961	1962	1963
January......	230	170	310	170	310	140	130	160	130	180		1.1	1.3	0.8	1.2
February.....	240	160	280	150	200	200	100	110	110	100		0.9	1.1	0.9	0.9
March........	270	150	120	170		90	70	80	80			0.7	0.9	1.5	
April........		320	170	320			70	70	80			4.6	1.8	4.4	
May.........		290	230	400			90	80	110			4	2.6	4.6	
June........	390	280	450	490		100	110	130	140		4.8	3.5	5	4.7	
July.........	540	610	440	480		80	140	130	150		5.3	5.7	3.5	4.5	
August......	840	580	640	610		160	160	160	180		8.8	5.7	5.4	7.1	
September....	700	790	570	570		150	230	130	170		7	7.8	5.1	5.1	
October.......	620	660	820	550		180	240	200	150		4.8	6	5.9	4.4	
November....	310	320	360	440		120	160	150	160		2.2	3.4	2.3	2.7	
December.....	280	210	240	490		140	130	170	230		1.9	2.4	1.6	3.4	
Average..		380	380	400			140	130	140			3.8	3.0	3.7	

* Collected at 0800.   † Collected at 1500.

winter, and the minimum to wind from the ocean, with its low radon-222 content. In part, the seasonal variation may be caused by changes in radon-222 diffusion through the ground. Freezing minimizes diffusion in winter. As the moisture content of the soil decreases with increasing temperatures, the emanation rate would be expected to increase.

To evaluate the effect of air stability, the daily morning radon-222 values were categorized in Figure 4 according to wind speed between

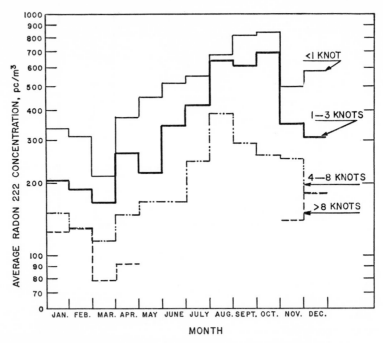

Fig. 4.—Effect of wind speed on morning radon-222 concentration

6:00 and 8:00 A.M. The graph shows a seasonal variation independent of wind speed and an increase of radon-222 levels with decreasing wind speed. The second effect is probably exaggerated by the above-mentioned non-equilibrium conditions between radon-222 and its daughters at higher wind speeds, resulting in the underestimation of radon-222 levels computed from the daughter activity. This induced error, however, is not so large as to affect the entire trend (Fig. 4). Thus, high radon-222 concentrations are associated with stable air. Early morning maxima in radon-222 concentrations have also been attributed to the greater stability of the air at that period of the day (Moses *et al.*, 1963). The relation between inversion and high radon-

222 levels has been noted repeatedly (Lockhart, 1962; Moses *et al.*, 1963); as with inversion, low wind speed at night and during early morning implies local ground-air stability.

Precipitation before and during sampling is also associated with low radon-222 levels (Israël, 1951; Gale and Peaple, 1958; Lockhart, 1962; Moses *et al.*, 1963). Rain would be expected to lower emanation from the earth by clogging pores and to wash radon-222 and its daughters from the air. It is generally accompanied by higher wind speeds so that lower radon-222 levels in a particular situation cannot be attributed with certainty to either wind speed or precipitation. The effects are separated in Table 3, comparing average daily morn-

TABLE 3

EFFECT OF WIND SPEED AND PRECIPITATION ON
AVERAGE MORNING RADON CONCENTRATIONS
(PC/M³)

6–8 A.M. WIND SPEED (KNOTS)	DAILY PRECIPITATION (INCHES)		
	<0.01	0.01–0.09	>0.09
<1............	630	430	310
1–3............	440	280	240
4–8............	200	190	150
>8............	150	100	110

ing radon-222 concentrations at the same wind speed in groups of less than 0.01, 0.01–0.09, and above 0.09 inches of rain during the 24-hour period prior to the end of collection. The 585 observations show that both increasing wind speed and rain independently decreased radon-222 levels, high wind appearing to be more effective than heavy precipitation.

Examples of continuous variation in radon-222 levels are given in Figure 5. Curves *A*, *B*, and *C* represent, respectively, the mean, highest, and lowest daily radon-222 concentrations during the period. Between 5:00 and 9:00 A.M., wind speeds were <1–4 knots for curve *A*, 7–9 knots for curve *C*, and less than 1 knot for curve *B*. Precipitation is indicated in Figure 5. Peak radon-222 levels occurred during the early morning hours, when air movement was least and inversions frequently occur. The mean of the 8:00 A.M. and 3:00 P.M. measurements approximated the average daily radon-222 concentration during this period. Assuming this to be true throughout the year, the average annual radon-222 concentration would be the average of these two measurements, or 260 pc/m³, according to Table 2.

Radium-226 in the ground—the probable source of the radon-222—has a concentration ranging between 0.3 and 0.7 pc/gm of moist soil. Values are the same within a foot of the surface and in the 20-foot core.

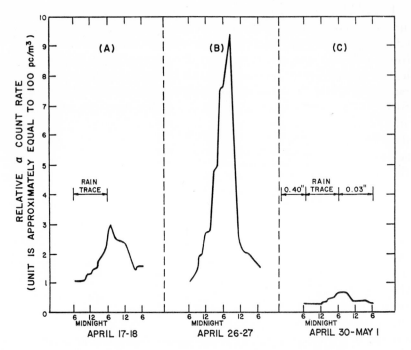

Fig. 5.—Continuous measurement of α activity in air particulates

Lead-210 and Beryllium-7

PROCEDURE

Lead-210 and beryllium-7, together with fission products from fallout, were measured in particulates collected on glass-fiber filters during April and May, 1961. The filters were inserted in high-volume air samplers, shielded from precipitation and the deposition of large particulates (Setter and Coats, 1961). Approximately 2,000 m.³ of air was filtered in 24-hour periods beginning at 8:00 A.M. Collection periods varied between 2 and 4 days.

At the time, measurement of the two naturally occurring radionuclides was relatively direct because fission product levels were low, with little interference by man-made fission products. Beryllium-7 was measured on the calibrated γ spectrometer by means of its 0.48-mev. γ ray, emitted in 12 per cent of its disintegrations. The 8 × 10-

inch filter was folded to one-quarter size and counted for 50 minutes with a 4×4-inch cylindrical sodium iodide detector. Corrections were made for interference of cesium-137, and small amounts of ruthenium-106–rhodium-106 from previous nuclear tests.

Lead-210 was separated chemically from the filter and its collected air particulates. Half the collected sample was used for analysis. After addition of lead carrier, the glass-fiber filter was fused with sodium hydroxide and leached with water. The basic solution containing the lead was acidified and saturated with hydrogen sulfide. The precipitated lead sulfide was metathesized to the sulfate by heating with sulfuric acid in the presence of bismuth holdback carrier, filtered, dried, and mounted for β counting. The ingrowth of the 5-day bismuth-210 daughter was followed by the increasing count rate of its relatively strong β particles. Activity of the lead-210 parent was computed from the bismuth-210 count rate, measured on a low-background β counter.

RESULTS AND DISCUSSION

Figure 6 shows that beryllium-7 concentrations for the 7-week period varied between 0.2 and less than 0.02 pc/m³. The average for the period, weighted for the length of each collection interval, was 0.096 pc/m³. The 2σ (sigma) error attributable to the γ counting and the correction for interfering γ radiation is ± 0.02 pc/m³. There is no simple relation of the variation in beryllium-7 concentration with measured local meteorological conditions. Changes in beryllium-7 values, however, parallel those for gross fission-product activity, as shown in Figure 6, and specifically those for the long-lived fission product cesium-137. Essentially all the fission products originated in nuclear tests carried out more than 2 years before April, 1961, and were therefore of stratospheric origin. The measured values of beryllium-7 would be expected to be above the annual average because the period of measurement occurred during the annual spring maximum for fallout of stratospheric origin. The average concentration is approximately twice as great as the monthly beryllium-7 values measured at Chicago in April and May of the previous year (Gustafson *et al.*, 1961).

For lead-210, also plotted in Figure 6, highest and lowest values are 0.015 and 0.0030 pc/m³, respectively, with a daily weighted average of 0.0082 pc/m³. The estimated analytical and counting 2σ error of these values is 10 per cent. If the lead-210 were formed entirely from local radon-222, an ingrowth period of 0.4 day would be needed to form 0.0082 pc/m³ from a constant radon-222 level of 200 pc/m³. However, because of its long half-life, lead-210 would be expected to

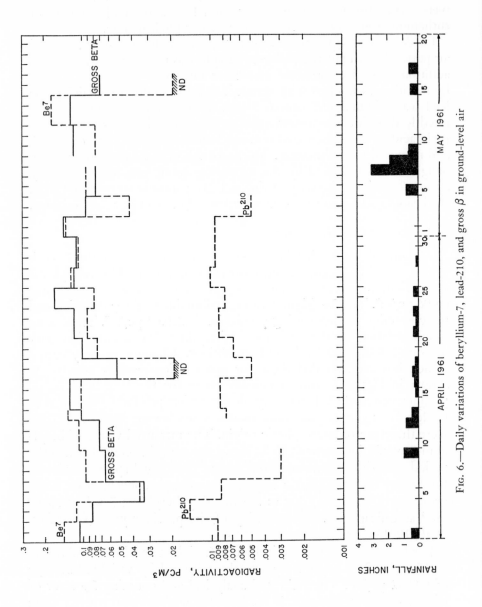

FIG. 6.—Daily variations of beryllium-7, lead-210, and gross β in ground-level air

come from remote sources as well as from the immediate vicinity. Because only early-morning radon-222 levels were measured, no correlation can be made between local daily radon-222 concentrations and these lead-210 levels. As indicated in Figure 6, lead-210 levels often varied in the same pattern as the other measured radionuclides, despite the different origins of beryllium-7, lead-210, and gross-fission products and their different distributions throughout the atmosphere. Lead-210 values lie in the range of previously reported data (List and Telegadas, 1961).

SUMMARY

The concentration and pattern of variation of radon-222 at Cincinnati are similar to those reported for a number of other locations. Seasonal variations show minimum early-morning radon-222 averages in March and maximum averages in September. Concentrations are affected by the degree of air stability and by precipitation. Radon-220 concentrations are approximately 1 per cent of radon-222 values and are parallel to them.

The measurement of radon-222 through its daughter is advantageous because of its simplicity. The relation between parent and daughter concentrations appears to be affected by meteorological conditions, and this relation must be determined to permit accurate indirect radon-222 measurement. Once this has been achieved, the factors affecting radon-222 concentrations in air will be studied further. The effect of radium-226 concentration, emanation from the ground, and meteorological factors at the point of radon-222 measurement will be considered.

These variables are to be observed to predict radon-222 and radon-220 levels for other population areas. Calculations would be based on a knowledge of normal radionuclide concentration patterns and on seasonal meteorological conditions and their effects. A gross annual radon-222 value of 260 pc/m³ was estimated for Cincinnati from the average 8:00 A.M. and 3:00 P.M. concentrations. It should be possible to obtain similar estimates from only a few radionuclide measurements performed at different seasons and under selected weather conditions.

The beryllium-7 and lead-210 concentrations measured during the relatively short period in the spring of 1961 were each fairly uniform. Of the 20 beryllium-7 values, 16 were within 55 per cent of the average of 0.096 pc/m³, and 12 of the 14 lead-210 samples were within 45 per cent of the 0.0082 pc/m³ average. The causes of daily variation have not been determined and should be examined in the future.

ACKNOWLEDGMENTS

We gratefully acknowledge the aid of the Nuclide Analysis (Physics) and Methods Development (Physics) groups in performing radioactivity measurements and the assistance of Mrs. A. Lett in performing lead separations. Meteorological data were supplied by the U.S. Weather Bureau, Cincinnati, Ohio.

REFERENCES

ANON. 1962. Natural and fission product radioactivity in surface air particulates at Cincinnati, Ohio. Radiological Health Data, 3:430–32.

COATS, G. I. 1960. A continuous automatic air sampler for the determination of radon and thoron. Health Physics, 4:192. (Abstr.)

GALE, H. J., and L. H. J. PEAPLE. 1958. A study of the radon content of ground level air at Harwell. Internat. J. Air Pollution, 1:103–9.

GUSTAFSON, P. F., M. A. KERRIGAN, and S. S. BRAR. 1961. Comparison of beryllium-7 and cesium-137 radioactivity in ground-level air. Nature, 191:454–56.

HARLEY, J. H. 1953. Sampling and measurement of airborne daughter products of radon. Nucleonics, 11:12–15.

ISRAËL, H. 1951. Radioactivity of the atmosphere. In T. F. MALONE (ed.), Compendium of Meteorology, pp. 155–61. Boston: Am. Meteorolog. Soc.

LIST, R. J., and K. TELEGADAS. 1961. The pattern of global atmospheric radioactivity—May 1960. USAEC Health and Safety Laboratory Quarterly Rpt., HASL-111, pp. 186–212.

LOCKHART, L. B., JR. 1962. Atmospheric Radioactivity at Washington, D.C., 1950–1961. Washington, D.C., Naval Res. Lab. Rpt. NRL-5764. P. 18.

MOSES, H., H. F. LUCAS, JR., and G. A. ZERBE. 1963. The effect of meteorological variables upon radon concentration three feet above the ground. Air Pollution Control Assoc. J., 13:12–19.

SETTER, L. R., and G. I. COATS. 1961. The determination of airborne radioactivity. Am. Ind. Hygiene Assoc. J., 22:64–69.

SHLEIEN, B. 1963. The simultaneous determination of atmospheric radon by filter paper and charcoal adsorptive techniques. Am. Ind. Hygiene Assoc. J., 24:180–87.

WILKENING, M. H. 1952. Natural radioactivity as a tracer in the sorting of aerosols according to mobility. Rev. Sci. Instr., 23:13–16.

R. L. PATTERSON, JR., AND L. B. LOCKHART, JR.

23. Geographical Distribution of Lead-210 (RaD) in the Ground-Level Air

LEAD-210 (RaD) is the longest-lived (22-year half-life) radio-active member of the radon-222 decay series, and its occurrence in the atmosphere both at ground level and in the stratosphere is well established (King *et al.*, 1956; Stebbins, 1961; Telegadas and List, 1961; Haxel and Schumann, 1955; Burton and Stewart, 1960; Blifford *et al.*, 1952).

It is, of course, recognized that this isotope is formed in the atmosphere by the normal process of decay of radon-222, which has diffused as a gas from the soil into the atmosphere. Moreover, all the radon-222 descendants are solids, initially electrically charged, and are rather rapidly adsorbed irreversibly onto air-borne dust particles. As solids, rather than gases, these products are subjected to removal and deposition processes different from those for the parent radon-222. Thus, whereas the main removal process for radon-222 is radioactive decay (this is also true for the short-lived radon-222 daughters, which rapidly approach secular equilibrium with the radon-222 parent), the principal removal process for lead-210 is rainout (Blifford *et al.*, 1952), with smaller quantities removed by dry deposition, or fallout.

The site of production of radon-222 is in the radium-226-bearing components of the earth's surface, primarily the land surface; the concentration of radon-222 is related to its source strength, its rate of diffusion into the air, and various meteorological activities that cause its dilution and dispersion in the atmosphere, in addition to radioactive decay. The site of lead-210 production, on the other hand, depends

R. L. PATTERSON, JR., and L. B. LOCKHART, JR., are with the U.S. Naval Research Laboratory, Washington, D.C., the former as head of the Radiochemistry Section, Physical Chemistry Branch, Chemistry Division, and the latter as head of the Physical Chemistry Branch, Chemistry Division.

solely on the location of the radon-222 at the time that it undergoes decay. Since information on the vertical distribution of radon-222 is lacking and since there is little information on its geographical distribution in the air at ground level, it is impossible to define in more than a general way a birthplace for the lead-210.

This report will consider some of the observed concentrations of lead-210 in the air as a function of time and space and attempt to correlate these results with the concentrations of radon-222 observed concurrently at the same sites.

EXPERIMENTAL DESIGN

During the past few years, the U.S. Naval Research Laboratory, with the support of the U.S. Atomic Energy Commission and various co-operating agencies and institutions in countries along the 80th meridian (west), has had under way a program of sampling radioactive isotopes in the air at ground level by air filtration. One of the isotopes of interest has been lead-210; the method employed in its collection and analysis has been described in detail elsewhere (Baus *et al.*, 1957; Lockhart and Patterson, 1963). The sampling sites are shown in Figure 1; they are located primarily along the 80th meridian (west), except for the Mauna Loa, Hawaii, station (155°36'W) which is at a high altitude (3,394 meters) and at a latitude equivalent to Chacaltaya (5,220 meters) in the Southern Hemisphere.

RESULTS AND DISCUSSION

The concentrations of lead-210 in the air at ground level determined through the radiochemical analysis of monthly or bimonthly composited samples from each site are listed in Table 1 in units of 10^{-15} curies per standard cubic meter of air. This information is shown in bar graph form in Figure 2 as a function of time and latitude.

The air concentrations of lead-210 at the subtropical and tropical sites (Miami to Antofagasta) show definite period changes, which occur at roughly comparable times of the year regardless of latitude or of season. The expected out-of-phase seasonal relationship between hemispheres is absent. This same pattern is observed at both high-altitude and sea-level sites. This pattern also does not seem related to periods of rainfall, at least at Panama, where the data have been examined in some detail. In previous years an inverse relationship of fission-product concentration in the air to the quantity of rainfall was noted (Lockhart and Patterson, 1960); however, during one of the two years of the current study, lead-210 concentrations were lowest during the dry season, while, during the other year, the concentra-

ATMOSPHERIC RADIOACTIVITY STATIONS
ALONG THE 80 TH MERIDIAN

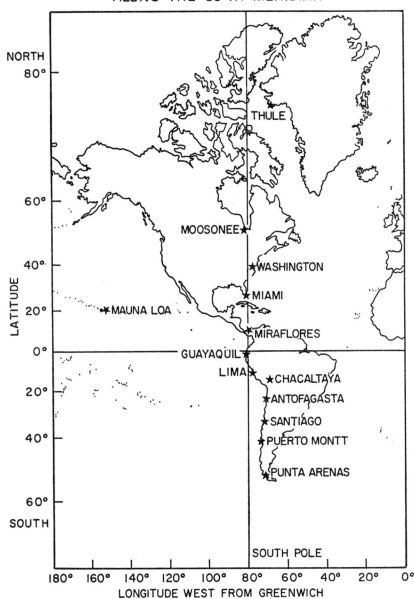

FIG. 1.—Location of air-sampling stations

TABLE 1

Measured Concentrations of Lead-210 in the Air at Ground Level

(Activity in units of 10^{-15} curies per standard cubic meter of air)

Station	1960		1961						1962					
	Sept.	Nov.	Jan.–Feb.	Mar.–Apr.	May–June	July–Aug.	Sept.–Oct.	Nov.–Dec.	Jan.–Feb.	Mar.–Apr.	May–June	July–Aug.	Sept.–Oct.	Nov.–Dec.
Thule	2.57	9.90	10.22	12.38	6.80	3.29	3.59	9.23	1.12	7.83	4.59	3.44	2.90	4.28
Moosonee	9.45	12.24	19.35*	21.87†	3.60	6.66	6.98	15.89	6.98	8.60	5.85	1.65	5.0	9.36
Washington	21.83	28.35	24.48	15.62	14.45	9.81	24.26	8.15	2.72	8.15	14.54	5.18	16.38	……
Miami	3.60	7.07	8.60	12.78	10.22	6.48	4.42	3.78	2.21	2.66	5.45	4.34	4.73	……
Mauna Loa	5.76	7.29	6.80	8.37	10.98	8.42	5.04	2.60	2.58	8.46	7.16	7.56	4.09	……
Panama	2.18	3.25	3.38	5.99	8.55	3.43	1.90	1.40	0.400	1.20	2.70	2.92	3.11	……
Guayaquil	3.69	6.08	10.13	12.11	8.10	6.98	6.93	10.67	9.14	10.62	……	13.00	10.89	……
Lima	1.73	3.29	2.48	0.797	3.71	3.17	2.85	1.83	2.91	4.06	4.50	2.82	1.76	……
Chacaltaya	11.70	10.71	8.37	9.05	11.61	17.14	13.19	9.41	7.47	7.92	11.79	13.23	13.86	10.7
Antofagasta	1.35	5.09	6.08	15.53	17.42	4.45	2.26	2.68	5.76	6.98	11.70	9.41	2.30	3.5
Santiago	4.41	5.99	5.76*	……	12.51‡	……	3.66	5.63	6.98	7.38	17.24	……	6.12	……
Puerto Montt	1.35	……	……	……	……	2.66	1.23	0.79	0.84	1.66	2.90	2.94	1.85	……
Punta Arenas	1.72	0.59	0.72	0.16	0.71	0.79§	0.66‖	0.57	1.30	0.82	1.62	2.32	1.12	……

* January only.   † March only.   ‡ May only.   § August–September.   ‖ October only.

tions were continuously increasing during the dry season, as might be expected.

At the other stations in the Northern Hemisphere, month-to-month changes are often large and rather erratic, with no discernible seasonal pattern. Fewer data are available from the comparable Southern Hemisphere sites, confusing any interpretation of possible systematic changes. The most outstanding factor is the decreasing concentration of activity in the more southerly regions. This is undoubtedly due to the lack of appreciable radon-222 sources in the area, since this portion of the earth has essentially no exposed land surfaces. Deposition processes operating on the lead-210 thus predominate over its formation.

The effect of latitude on the observed average concentrations of

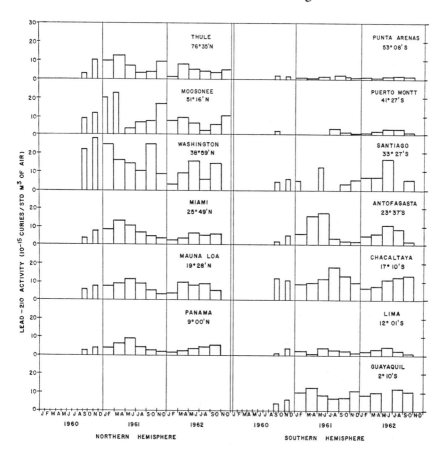

Fig. 2.—Average concentration of lead-210 (RaD) in the ground-level air as a function of time and latitude.

lead-210 in the ground-level air is shown in Figure 3 for two successive one-year periods. The vertical lines indicate the spread in values during the years covered (November, 1960–October, 1961, and November, 1961–October, 1962), while the profiles are drawn through the yearly averages of the values at each site. The profile presented is similar in its gross aspects, both in shape and in magnitude, to that of strontium-90 in the air, as measured at these same sites. Prior to resumption of nuclear testing in 1961, lead-210 concentrations exceeded those of strontium-90; at present, the reverse is true, at least for most sites. The place of major disagreement between the lead-210 and strontium-90 profiles is at Guayaquil, which rather consistently exhibits high lead-210 and low strontium-90 concentrations.

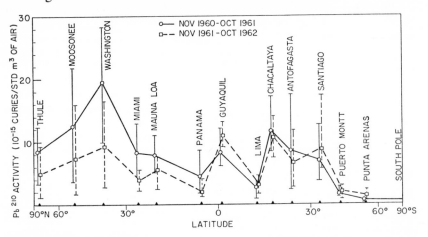

Fig. 3.—Latitudinal profiles of the average lead-210 (RaD) concentrations in the air during successive one-year periods.

In the Far South, the low concentrations observed at Puerto Montt and Punta Arenas are continued on to the South Pole, where the lowest concentration of all air-borne radioactive isotopes, both natural and fission product, are found.

In an attempt to interpret such information as is presented here, it is well to bear in mind that the meteorological conditions prevailing in the area of sampling may be of more significance at some sites than the actual geographical location or latitude.

RELATIONSHIP BETWEEN LEAD-210 AND RADON-222
CONCENTRATIONS

Studies have been made of the relative concentrations of lead-210 (RaD) and radon-222 in the air at ground level in an attempt to establish a residence time for the lead-210 in the atmosphere (Haxel and Schumann, 1955; Blifford *et al.*, 1952); however, these measurements are suitable to determine only an order-of-magnitude effect. An accurate determination would require the integration of concentrations of these isotopes to the top of the atmosphere, since both have large vertical distributions.

Some relationships of the concentrations of lead-210 on the filter to those of radon-222 in the air are shown in Table 2 for four sites at which parallel measurements of radon-222 and lead-210 have been made. First, it should be pointed out that the lead-210 resulting from the collection of the solid, short-lived radon-222 daughters, lead-214

TABLE 2

RELATIONSHIP OF LEAD-210 (RaD) ON THE FILTER TO RADON-222
AND LEAD-214 (RaB) IN THE AIR

STATION	DATE	RADON-222 CONTENT OF AIR* (d/m/m³)	LEAD-210 EQUIVALENT (calc.) From Lead-214 (d/m/m³)	LEAD-210 EQUIVALENT (calc.) From Radon-222 (d/m/m³)	LEAD-210 CONTENT OF AIR (d/m/m³)	LEAD-210 FROM LEAD-214 DECAY (per cent)†	LEAD-210 FROM RADON-222 DECAY (per cent)†
Washington...	Sept.–Dec., 1961	269	6.25×10^{-4}	12.7×10^{-2}	3.60×10^{-2}	1.7	350
Lima.........	Sept.–Dec., 1961	70	1.62×10^{-4}	3.3×10^{-2}	0.55×10^{-2}	2.9	600
Chacaltaya...	Sept.–Dec., 1961	87	2.01×10^{-4}	4.12×10^{-2}	1.47×10^{-2}	1.4	360
South Pole....	Jan.–Sept., 1960	1.13	0.026×10^{-4}	0.05×10^{-2}	0.124×10^{-2}	0.21	40

* At time of daily minimum.   † Relative to lead-210 (RaD) in the air.

and bismuth-214 (RaB and RaC), on the filters is negligible. The contribution from this source amounts to, at most, a few per cent in the examples presented; since the values shown are based on the measured minima in the concentrations of lead-214 and radon-222 in the air, more representative figures would be perhaps two to three times higher than shown, but still less than 10 per cent of the total lead-210 collected.

On the other hand, if an air mass containing the measured concentrations of lead-210 and radon-222 were isolated until the radon-222 decayed completely, lead-210 from this process would be three to six times as high as that actually found (perhaps ten to twenty times as high, if the sample were isolated during the morning, when radon-222 concentrations are highest). The exception is at the South Pole, where much of the radon-222 had already undergone decay by the time of its arrival there, so that completion of this decay process would add only about 40 per cent to the lead-210 content of the air, which was itself depressed due to washout during the transport process. Here, also, no diurnal variations in radon-222 concentration have been observed, though this would be difficult because of the low activity present; actually, however, no such variation would be expected, since radon-222 is not generated locally because of isolation of the surface from the soil by the thick ice cover.

The preponderance of freshly diffused radon-222 into the lower atmosphere will give a distorted picture of lead-210 residence times, in which the calculated residence time is much shorter than the true residence time. As an example, the ratio of the lead-210 from radon-222 decay to the total lead-210 in an air parcel after complete radon-222 decay can be employed to give an apparent, but erroneous, residence time of lead-210 in the atmosphere. These "residence times," calculated from data given in Table 2, are as follows for the periods indicated: Washington, 1.4 days; Lima, 0.8 days; Chacaltaya, 1.7 days; South Pole, 6.9 days. The times would be correspondingly lower if the true, average radon-222 activity rather than the radon-222 minimum activity had been employed.

LEAD-210 IN RAIN WATER

To complete the picture on the distribution of lead-210 (RaD) in the atmosphere, some data on its concentration in rain water at various sites (King *et al.*, 1956) are included in Table 3. The results are about what would be expected from the positions of the collecting sites relative to air masses from oceanic or continental environments; that is, island and coastal sites exhibit lower lead-210 concentrations in rainwater than do inland sites or sites that are in the path of con-

TABLE 3

CONCENTRATION OF LEAD-210 IN RAIN WATER
AT VARIOUS LOCATIONS

Location	Date Collected	Pb-210 Activity (pc/l)
Washington, D.C.	June–Nov., 1950	2.4
Glenview, Ill.	Jan.–July, 1950	2.5
Kodiak	Apr.–Dec., 1950	0.77
Panama	Mar., 1950–Jan., 1951	0.36
Hawaii	Mar.–Dec., 1950	1.1
Philippine Islands	May–Nov., 1950	1.1
Samoa	Jan.–Sept., 1950	0.23

tinental air masses. Samoa, far from large land masses and experiencing the washing effect of daily rains, exhibits the lowest concentrations of lead-210 in the rain water. In future work of this type it would be advisable to collect sufficient data so that the concentrations in rainfall could be integrated over the total rainfall for a given period and compared with parallel concentrations of lead-210 and radon-222 in the air.

Conclusions

The concentration of lead-210 in the air at ground level at any site is quite variable because of both the changing weather patterns in the area and the geographical location of the sampling site relative to radon-222 sources. As a consequence of the predominance of deposition processes over generation of lead-210 in the higher southern latitudes, concentrations of this isotope in the Antarctic region are extremely low.

It can be concluded from this study that short-term measurements of lead-210 in the air cannot suffice to define its average activity concentration in any area. Moreover, the lead-210 in a given air parcel at ground level bears little relation to the radon-222 in the same sample, since the radon-222 source of the lead-210 has been marked by the introduction of a preponderance of fresh radon-222 from the soil. Consequently, the lead-210/radon-222 ratio is generally unsatisfactory for the determination of the residence time of lead-210 in the atmosphere.

References

Baus, R. A., P. F. Gustafson, R. L. Patterson, Jr., and A. W. Saunders, Jr. 1957. Procedure for the Sequential Radiochemical Analysis of Strontium, Yttrium, Cesium, Cerium, and Bismuth in Air-Filter Collections. Washington, D.C., Naval Research Lab. NRL-Memo-758, Project NR-571–003. Pp. 24.

Blifford, I. H., Jr., L. B. Lockhart, Jr., and H. B. Rosenstock. 1952. On the natural radioactivity in the air. J. Geophys. Res., **57**:499–509.

Burton, W. M., and N. G. Stewart. 1960. Use of long-lived natural radioactivity as an atmospheric tracer. Nature, **186**:584–89.

Haxel, O., and G. Schumann. 1955. Selbstreinigung der Atmosphäre. Ztschr. f. Phys., **142**:127–32.

King, P., L. B. Lockhart, Jr., R. A. Baus, R. L. Patterson, Jr., H. Friedman, and I. H. Blifford, Jr. 1956. RaD, RaE, and Po in the atmosphere. Nucleonics, **14**:78–84.

Lockhart, L. B., Jr., and R. L. Patterson, Jr. 1960. Measurements of the air concentration of gross fission product radioactivity during the IGY, July 1957–December 1958. Tellus, **12**:298–307.

LOCKHART, L. B., JR., and R. L. PATTERSON, JR. 1963. Techniques employed at the U.S. Naval Research Laboratory for evaluating air-borne radioactivity. This symposium.

STEBBINS, A. K., III. 1961. Second Special Report on the High Altitude Sampling Program (HASP). Washington, D.C., Defense Atomic Support Agency Rpt. DASA-539B. Pp. 243.

TELEGADAS, K., and R. J. LIST. 1961. B-57 air sampling program (1960). U.S. Atomic Energy Comm. (NYOO) Fallout Program Quarterly Summary Rpt. HASL-105, pp. 150–61.

Discussion, Chapters 15–23

(Numbers in parentheses refer to chapter authored by discussant.)

Wilkening (21) asked Lucas (17) whether it was not desirable also to perform γ spectrometry in determining radon-222 and radium-226. Lucas (17) replied that both γ spectrometry and biased counting had been used and that the spectrometric method gave a 20–30 times better signal-to-noise ratio but that the spectrometric method was much more expensive.

Israël (Aachen, Germany) asked about the ratio of radon-222 to radon-220 in the atmosphere as measured by Kawano (16). Kawano (16) replied that this ratio varied from a minimum of 6 or 7 to a maximum of 15 or 16.

In response to a question by Gold (22), Lucas (17) stated that the charcoal trap in his apparatus was maintained at the temperature of solid carbon dioxide, which is in mixtures of acetone or carbon tetrachloride with chloroform. A 1:1 mixture of carbon tetrachloride and chloroform is preferred because solid carbon dioxide (dry ice) floats on this mixture, facilitating changing of the cold trap. As with many chlorinated hydrocarbons, one must protect one's self from the liver effects that sometimes occur.

Kamath (58) asked Hursh (19) whether he had encountered any corrosion problems due to the acidity of the thorium solutions used in the emanation technique. Hursh replied that he had not noticed any corrosion after repeated use of his chambers, even though approximately 1 N hydrochloric or nitric acid solutions were used.

In response to a question on the residual activity present on the charcoal, Hursh indicated that he believed that most of the background of 0.1 c.p.m. was due to the charcoal activity. Sieving through a 300–325 mesh makes the size sufficiently small to reduce adsorption considerably. A wide survey of available charcoals might be one way of reducing this activity.

C. R. HILL, R. V. OSBORNE, AND
W. V. MAYNEORD

24. *Studies of α Radioactivity in Relation to Man*

MANY OF THE STUDIES that have been reported hitherto of natural α radioactivity in man and his environment have been based on techniques specific for a single nuclide or element. Examples of these are the radon emanation method for estimation of radium-226 (Walton et al., 1959; Hursh et al., 1960; Lucas, 1957; Hallden et al., 1962) and the electrochemical deposition technique for polonium-210 (Black, 1961). This paper describes a complementary approach in which α activity is treated as a whole. Initially, a simple and sensitive technique for total α counting was developed, and this was later supplemented by techniques of α spectroscopy developed for use with the very low specific activities commonly found in biological materials.

TOTAL α COUNTING

For total α counting, Turner et al. (1958a, b) have modified the standard zinc sulfide scintillation counting technique in two important respects. In the first place, the sample material is arranged in a sealed capsule in intimate contact with its own zinc sulfide screen, so contamination background is almost completely eliminated. An advantage of this arrangement is that it makes possible identification of radium-226 by virtue of the growth of radon-222 in the sealed capsule. In a similar way the presence of polonium-210 (138-day half-life) can often be detected in cases in which radioactive equilibrium has been disturbed.

The other main feature of this technique is the facility for detection

DR. C. R. HILL, DR. R. V. OSBORNE, and PROFESSOR W. V. MAYNEORD are in the Physics Department of the Institute of Cancer Research, Royal Cancer Hospital, Surrey, England. Dr. Osborne is at present at the Biology and Health Physics Division, Atomic Energy of Canada Ltd., Chalk River, Ontario, Canada.

of fast pairs of α particles originating from both radon-220–polonium-216 (thoron–thorium A, 0.158 sec.) and radon-219–polonium-215 (actinon–actinium A, 0.0018 sec.). This provides a means for detecting the presence in a sample of the corresponding portions of the thorium and actinium series.

This technique is quick and reliable, and it is possible with it to make measurements of specific α activities down to 10^{-13} c/gm (which, in practice, covers almost all environmental materials) and simultaneously to obtain some information as to what groups of nuclides may be present. In this way a considerable quantity of information has been obtained on the total α activity of human tissues and the human environment (Turner *et al.*, 1958*a*, *b*; Mayneord *et al.*, 1960; Turner and Radley, 1960; Turner *et al.*, 1961).

LARGE-AREA α SPECTROSCOPY

In order to provide more precise identification of the nuclides responsible for the α activity observed in these materials, it was decided that an attempt should be made to carry out α-particle spectroscopy on them—if possible, without recourse to chemical separation techniques. With specific activities that may be as low as 10^{-13} c/gm, the achievement of a significant counting rate and the simultaneous avoidance of appreciable self-absorption requires the use of source areas on the order of 10,000 cm.², even in a device having high counting geometry. It seemed that only a pulse ionization chamber could accommodate sources of such a size. We therefore designed and built an instrument of this type in which the source material (having been finely ground and slurried) is sprayed onto a 90 × 180-cm. sheet of 0.125-mm.-thick aluminized Mylar or cellulose acetate, which can subsequently be rolled to form the outer electrode of a cylindrical ionization chamber.

The use of a sheet of organic material for the outer electrode (which is the main solid surface presented to the counting volume) minimizes α contamination background, while the other important source of background counts, radon emanating from the constructional materials of the chamber, can be reduced to a low level by continuous circulation of the counting gas over cooled charcoal. The instrument will accommodate 1.5 gm. of sample material, and it is possible with it to detect and identify nuclides present at concentrations down to 10^{-13} c/gm, while achieving resolutions down to 150 kev. A full description of this instrument has been given elsewhere (Hill, 1961).

An example of the type of information that this instrument can provide is given by the spectrum shown in Figure 1, which was obtained from the liver of a man who, more than forty years prior to his death, had been exposed for some four years to inhalation of uranium ore. The peak due to polonium-210 is found also in unexposed individuals, as discussed below, although in the present case the level is some 2–3 times above normal, probably as a result of transport to the liver of polonium produced by decay of the skeletal burden of radium-226

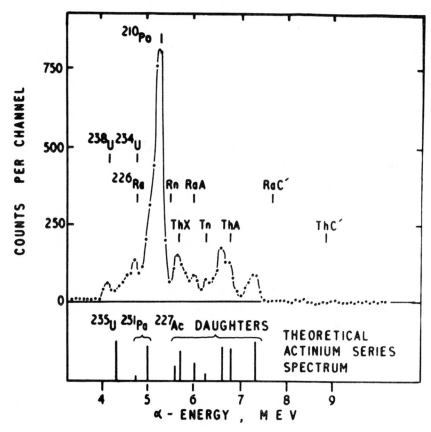

Fɪɢ. 1.—Alpha spectrum of human liver ash from a worker in a radium extraction plant

(5 × 10⁻⁹ gm., or some 300 times above normal). The remaining peaks in the spectrum can be identified as being due to the decay products of actinium-227. This finding was unexpected in view of the low natural abundance of the actinium series nuclides in uranium ore, but it has been confirmed, by the use of the fast-pairs technique referred to above, both in this case and in a similar but quite independent case that we have recently had the opportunity to examine. In the sample whose spectrum is shown in Figure 1 the activity due to

polonium-210 and that due to each of the five actinium-227 daughters present were 1.3 and 0.2 $\times$ 10^{-12} c/gm ash, respectively.

Several further examples of the application of this technique in the investigation of normal and other biological materials have been given in a published paper (Hill, 1962). Its application to certain geological problems is referred to in another contribution to this symposium (Cherry, 1963), and it has also been used for measurements of naturally occurring isotopic uranium-234 enrichment in water residues (Hill and Crookall, 1963).

One of the earliest and most striking findings was that of the high α activity frequently to be observed on foliage, which was shown to be due to α particles giving a peak in the energy region 5.1–5.3 mev. Furthermore, the amounts of this activity present were found to be rainfall dependent (Hill, 1960). From observations of half-life and by chemical analysis it was possible to show that most of this activity is due to polonium-210 (5.30 mev.), but it seemed important to discover how much, if any, plutonium-239 or plutonium-240 (5.15 and 5.16 mev., respectively) might be present. A particular reason for interest in this problem is that the occurrence of polonium-210 on foliage and other exposed surfaces may be attributed to deposition from the atmosphere of the long-lived decay products of atmospheric radon (Burton and Stewart, 1960), a "natural fallout" phenomenon. Both polonium and the artificially produced plutonium may therefore enter the biosphere in a similar manner, and this common factor, together with their very similar radiation characteristics, makes it necessary for great care to be exercised in attempts at their separate estimation in biological materials.

High-Resolution α Spectroscopy

With this and other possible requirements for higher resolution in mind, a modified version of our original α spectrometer has been built, again in the form of a cylinder, but rather smaller in size than the previous instrument and incorporating a Frisch grid, which effectively eliminates "positive ion" line broadening.

The design of this instrument is shown schematically in Figure 2. The gas chamber, which is constructed of stainless steel, incloses a light cylindrical open framework, to which the aluminized Mylar source backing sheet may be secured with Scotch tape. Concentric with this is a cylindrical grid of stainless-steel wires, each 9.14 $\times$ 10^{-3} cm. in diameter and spaced 0.246 cm. apart. The collector is a brass rod 0.2 cm. in diameter, the cylindrical grid and source diameters being 9.4 and 20 cm., respectively. Each electrode is 25 cm. long, and the effective source area thus provided is approximately 1,500 cm.². A standard source for energy calibration is mounted on one side of a rotatable strip, located in a slot cut lengthwise from the main source holder, and can thus be exposed to the counting volume at will by means of a control knob outside the gas chamber. Negative voltages are fed to the source and grid through a filter and potential dropping network on one end of the chamber, and output signals are taken to a head amplifier mounted on the other end. The head amplifier was

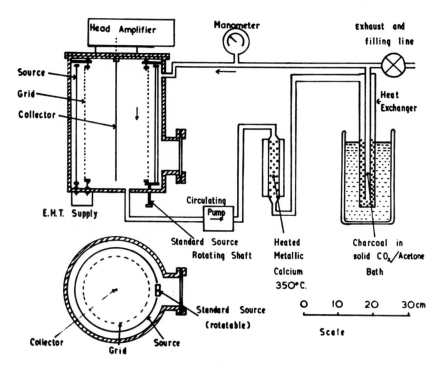

Fig. 2.—Cylindrical gridded pulse ionization chamber

specially constructed for the purpose and employs a modified version of a circuit published by Cottini *et al.* (1956).

In operation the chamber is filled with a 90:10 argon-methane mixture to a total pressure of 2 atmospheres, and the purity of the gas with respect to electronphilic contaminants and radon is maintained by continuous circulation over heated metallic calcium and cooled charcoal.

With this instrument it is possible to obtain spectra with line widths down to 40 kev., although the quality of the source is very critical if one is to attain this figure. It so happens, however, that, in the main use to which we have put the instrument so far—the simultaneous spectroscopic measurement of plutonium and polonium in atmospheric dust—it has been possible to obtain sources of high quality. In this case the sources have been formed by covering with aluminized Mylar one of the collecting plates of a large electrostatic precipitator. In this way the dust extracted over a period of a week from a total of some 4,000 kg. of air is deposited on the Mylar as a quite uniform and very thin film, and the whole sheet can then be transferred to the spectrometer and analyzed. Two spectra obtained in this manner are shown in Figure 3.

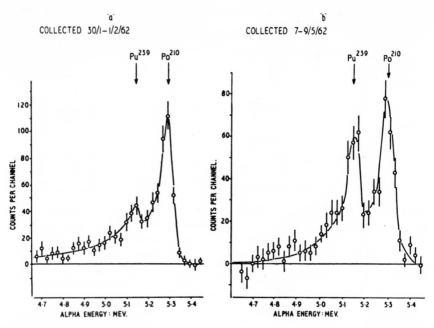

Fig. 3.—Alpha spectra from atmospheric dust samples collected at Belmont, Surrey, England.

The background counting rate of this instrument varies somewhat over the energy range of interest and is equivalent, for a good source giving a 40-kev. line width, to a source contamination of between 5 and 10×10^{-3} d.p.m. per nuclide. For a thicker source, which might typically contain 150 mg. of material, resulting in a degradation of line width to 120 kev., the equivalent background would be about three times the figure given above.

In order to be able to deal most effectively with samples that are essentially weightless, the chamber has been designed in such a way that the complete cylindrical electrode system can be easily removed and replaced with a gridded parallel-plate system fitted with a source changer capable of accommodating four 25 cm.² sources, which can be selected and counted in turn without opening the chamber. The line widths obtainable with this arrangement from a good source are again about 40 kev., but, owing to the smaller size of the counting volume, the background counting rate is appreciably lower than that of the cylindrical system, being equivalent to a source contamination of between 0.5 and 5×10^{-3} d.p.m. per nuclide (for a 40-kev. line width). The design of this electrode system is illustrated in Figure 4.

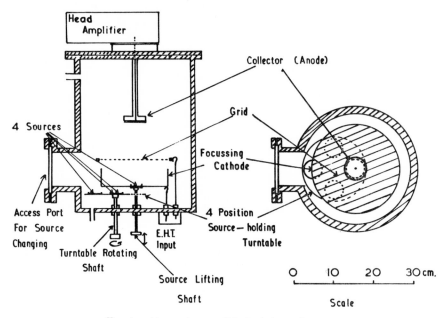

Fig. 4.—Alternative parallel plate electrode system

A fuller description of the complete instrument and its associated equipment will be given elsewhere.

It is found in practice that the spectroscopic and total α counting techniques are complementary: normally, samples are first total counted and then analyzed spectroscopically, and the two sets of counting rates and identifications so obtained are compared. Apart from the value of having two quite independent sets of measurements, this procedure provides a check on the introduction of contamination in the process of spectroscopic source preparation. In addition, the simplicity of the total counting techniques makes it possible to survey a much larger number of samples than could be examined in the spectrometers.

POLONIUM-210 IN THE BIOSPHERE

Two observations made quite early in the course of applying the techniques described above were, first, that radium-226 is almost always accompanied in biological materials by comparable activities of radium-228 and its daughters (Turner *et al.*, 1958*b, c*) and, second, that polonium-210 often occurs quite independently of its precursor, radium-226, and at considerably higher activities.

As a consequence of the second observation, a program was undertaken, and is still in progress, with the object of measuring the naturally occurring levels of polonium in human tissues, and particularly in

TABLE 1

α ACTIVITY AND RESULTING DOSE RATE DUE TO POLONIUM-210
IN NORMAL HUMAN TISSUES

TISSUE	No. SAMPLES	Po^{210} α ACTIVITY (pc/100 gm wet)		MEAN DOSE RATE* (mrem/yr)
		Range of Values	Mean	
Bone.............	6	0.5 –3.1	1.7	16
Liver.............	9	0.7 –1.8	1	10
Kidney..........	2	0.5 –0.9	0.7	7
Spleen..........	3	0.1 –0.6	0.3	3
Lung............	4	0.1 –0.4	0.3	3
Muscle..........	6	0.03–0.4	0.13	1.3
Testis...........	7	0.05–0.5	0.24	2.3
Ovary...........	1	....	0.7	7
Pancreas........	1	....	0.3	3

* Taking the energy of the polonium-210 α particle as 5.3 mev. and assuming a quality factor (RBE) of 10.

soft tissues, which appear to have been rather neglected hitherto in favor of bone (Osborne, 1963*a*). At the same time, it is hoped to obtain more information on the origin of this polonium (Holtzman, 1960) and, in particular, whether, as now seems likely, the majority of it in fact originates from the atmospheric deposition process referred to above. Because of the high volatility of polonium it is not possible to estimate it quantitatively by α spectrometry of ashed samples, and we have therefore found it necessary to return to the electrochemical deposition technique, followed by total α counting, for this purpose. Some preliminary results of such measurements are given in Table 1 (see also Osborne, 1963*a*). It should be noted that the presence of polonium-210 in a tissue may result either from the direct uptake of polonium or from the uptake of lead-210 or bismuth-210 and subsequent radioactive decay with retention of some of the resulting polonium-210 in the tissues. In the latter case it appears that the biological-ly effective dose from the lead-210 and bismuth-210 β and γ radiations is at least an order of magnitude less than that from the α radiation due to the associated polonium-210.

It is questionable what significance can be attached to mean dose rates resulting from very low concentrations of α emitters (where the distribution of absorbed energy, even over long periods of time, is ex-tremely non-uniform), but it is of interest that the values shown in the table range between 10 and 100 per cent of the internal dose due to potassium-40.

In parallel with the measurements of polonium in tissue, we are car-rying out periodic determinations of the concentrations of both polo-nium-210 and plutonium-239–plutonium-240 in the atmosphere (Os-borne, 1963b), using the electrostatic precipitation technique referred to above, and are also attempting to measure the plutonium-239–plu-tonium-240 content of human lungs (Osborne, 1963b) and other tis-sues. In view of the extremely low specific activities that are likely to occur in tissue and the serious possibility of confusion with other α emitters, particularly polonium-210, we have adopted the following procedure in this work. To a 10–30-gm. sample of tissue ash, a "spike" containing approximately 0.1 d.p.m. of plutonium-238 (α energy 5.50 mev., and known from atmospheric measurements not to be present to any appreciable extent in fallout) is added, and plutonium is then chemically extracted from the sample and assayed using the parallel-plate spectrometer. Any plutonium-239 and plutonium-240 present (5.15 and 5.16 mev.) is determined, the chemical yield is found from the counts in the plutonium-238 peak, and the effectiveness of the chemical extraction in discriminating against polonium-210 (5.30 mev.) can be checked.

Conclusion

It has been the purpose of this paper to describe in outline a set of techniques developed for the investigation of natural α radioactivity in man. Results of measurements have been given primarily to illus-trate the application of the techniques, and fuller accounts, together with more detailed descriptions of the individual instruments, are given in papers listed in the References.

By adopting an experimental approach that takes account of all α-emitting nuclides that may be present in a material, we have been able to build up what we believe is a fairly comprehensive picture of the main contributions to long-lived natural α radioactivity in man. One result of this approach has been to indicate the relative importance of certain nuclides that had previously attracted little attention. In particular, polonium-210, whose presence in the body is thought to originate largely from a process of atmospheric fallout of natural radioactivity, appears to be responsible for mean dose rates of several millirem per year in normal human bone and reproductive organs. This experimental approach has also provided a sensitive means for studying, in accidentally exposed humans, the metabolism of some of the potentially hazardous natural and artificial heavy radioelements.

ACKNOWLEDGMENTS

We are indebted to numerous of our colleagues, whose generous help and advice has made this work possible, and in particular to Drs. R. C. Turner and J. M. Radley, who carried out much of the earlier experimental work upon which ours is founded.

REFERENCES

BLACK, S. C. 1961. Low-level polonium and radiolead analysis. Health Physics, **7**:87–91.

BURTON, W. M., and N. G. STEWART. 1960. Use of long-lived natural radioactivity as an atmospheric tracer. Nature, **186**:584–89.

CHERRY, R. D. 1963. Alpha particle detection techniques applicable to the measurement of the natural radiation environment. This symposium.

COTTINI, C., *et al.* 1956. Minimum noise pre-amplifier for fast ionization chambers. Nuovo Cimento, **3** (Ser. 10):473–83.

HALLDEN, N. A., J. H. HARLEY, and I. M. FISENNE. 1961. Radium-226 in the diet in three U.S. cities. *In:* Argonne Nat. Lab. Rpt. ANL-6637, pp. 85–95.

HILL, C. R. 1960. Lead-210 and polonium-210 in grass. Nature, **187**: 211–12.

———. 1961. A method of alpha particle spectroscopy for materials of very low specific activity. Nuc. Instr. & Methods, **12**:299–306.

———. 1962. Identification of alpha emitters in normal biological materials. Health Physics, **8**:17–25.

HILL, C. R., and J. O. CROOKALL. 1963. Natural occurrence of U-234-enriched uranium. J. Geophys. Res. **68**:2358.

HOLTZMAN, R. B. 1960. Some determinations of the RaD and RaF concentrations in human bone. Argonne Nat. Lab. Rpt. ANL-6199, pp. 94–118.

HURSH, J. B., A. LOVAAS, and E. BILTZ. 1960. Radium in Bone and Soft Tissues of Man. Rochester, N.Y., Univ. Atomic Energy Project Rpt. UR-581. Pp. 11.

LUCAS, H. F. 1957. Improved low-level alpha scintillation counter for radon. Rev. Sci. Instr., **28**:680–83.

MAYNEORD, W. V., R. C. TURNER, and J. M. RADLEY. 1960. Alpha activity of certain botanical materials. Nature, **187**:208–11.

OSBORNE, R. V. 1963*a*. Lead-210 and polonium-210 in human tissues. Nature, **199**:295.

———. 1963*b*. Plutonium-239 and other nuclides in ground level air and human lungs during spring 1962. *Ibid.*, pp. 143–45.

TURNER, R. C., and J. M. RADLEY. 1960. Naturally occurring alpha activity of cigarette tobaccos. Lancet, 1197–98.

TURNER, R. C., J. M. RADLEY, and W. V. MAYNEORD. 1958*a*. The alpha-ray activity of human tissues. Brit. J. Radiol., **31**:397–406.

———. 1958*b*. Alpha ray activities of humans and their environment. Nature, **181**:518–21.

———. 1958*c*. The naturally occurring alpha ray activity of foods. Health Physics, **1**:268–75.

———. 1961. Naturally occurring alpha-activity of drinking waters. Nature, **189**:348–52.

WALTON, A., R. KOLOGRIVOV, and J. L. KULP. 1959. The concentration and distribution of radium in the normal human skeleton. Health Physics, **1**:409–16.

R. D. CHERRY

25. *Alpha Particle Detection Techniques Applicable to the Measurement of Samples from the Natural Radiation Environment*

THE USE OF α particle detection techniques in studies of the natural radiation environment has tended to be a relatively neglected field. In the last five years, however, as a result of the refinement of these techniques and their application by a small number of research groups, attention has been focused on the following simple facts.

a) Most naturally occurring radioactive elements *are* α emitters.

b) Although the short range of naturally occurring α particles results in their being of small consequence as external radiation sources, the presence of α emitters inside the human body is a feature of considerable importance.

c) The α-emitting nuclei are frequently γ emitters also. Thus identification and measurement of the concentration of a particular α emitter in a sample from the natural environment often enable information concerning the concentration of associated γ-emitting nuclei to be deduced.

d) Techniques for α particle detection possess certain definite advantages compared with comparable β and γ measurements. In particular, low background counting rates (and hence increased sensitivity) are a feature of α detectors, while α spectroscopic techniques are characterized by the lack of ambiguity in and the ease of the interpretation. Even the main disadvantage of α detectors—the inherent limitation of sample volume due to the short range of the α particles—falls away in cases in which only a small amount of sample is available.

R. D. CHERRY is lecturer in the Physics Department, University of Cape Town, Rondebosch, South Africa. He is currently on leave at the Geology Department, Rice University, Houston, Texas.

The purpose of this paper is to present a brief review of those α detection techniques which are of use in measuring samples from the natural environment. The subject will be discussed under two headings: (1) "total" α counting (i.e., no attempt at energy measurements) and (2) α spectroscopic techniques, where measurements are made of α particle energies in order to identify specific nuclides.

"Total" α Counting

THEORY

The basic theory of α particle emission from samples is well known for the case in which the α emitters are homogeneously distributed throughout the sample. This case has been discussed by a number of authors; probably the best reference is still the early paper of Finney and Evans (1935). Other useful treatments have been given by Evans and Goodman (1944), Nogami and Hurley (1948), Kulp *et al.* (1952), Turner *et al.* (1958a), and Keevil and Grasham (1943). The theory can be subdivided into two classes: "thick" and "thin" sources, "thick" sources being those whose thickness is greater than the maximum range of the α particles emitted by the sample. This range is typically about 25–30 microns in solid materials and is seldom greater than 90 microns; the volume of the sample that contributes to the α particle emission from the front face is thus very definitely limited.

The basic equation of thick-source α counting is

$$N_0 = \frac{NRA}{4},\tag{1}$$

where N_0 is the number of α particles emerging from an area A of the sample in unit time, N is the number of α disintegrations per unit volume of sample per unit time, and R is the range in the sample of the α particles concerned. N_0 is, of course, a measure of the counting rate; for 100 per cent efficient detection, N_0 *is* the counting rate. N and R are usually variables beyond the control of the experimenter (although N can sometimes be artificially enhanced by concentration techniques), and one must therefore use a sample and detector of large area if maximum count rates are to be obtained.

For thin sources, of thickness t, the basic equation is:

$$N_0 = \frac{NRA}{4}\left[\frac{2t}{R} - \left(\frac{t}{R}\right)^2\right].\tag{2}$$

The counting rate from a thin source is thus lower (since t is by definition less than R), and, moreover, it is necessary that t be known. In addition, uniform thin sources are considerably more difficult to prepare than are thick sources. Thin-source counting nonetheless possesses one important advantage; inspection of equation (2) shows that, *in the first order*, N_0 is independent of the range R. A less precise knowledge of R is thus required for the determination of N than is the case with thick-source counting, and the discussion of the range problem that is given below shows that this is by no means a trivial point. The extra effort involved in thin-source preparation is frequently worthwhile on this account, particularly if the quantitative precision of the results is of importance.

Both equations (1) and (2) must be modified to allow for (i) absorbers between the sample and the detector, (ii) the fact that α particles below a certain minimum residual range might not give rise to a response in the detector, and (iii) the fact that natural samples usually contain mixtures of α emitters, thus necessitating summations over different N and R. With modern experimental techniques (see below) modifications (i) and (ii) are usually negligible; (iii) is, on the other hand, often necessary and is a complicating feature. Since most of the naturally occurring α emitters are members of one of the three well-known natural radioactive series (see, e.g., Evans, 1955, chap. 16), questions of radioactive equilibrium arise, and some assumption concerning the degree of secular equilibrium attained by the sample is necessary. The favorite assumption is that secular equilibrium is complete, and all the references quoted above (Finney and Evans, 1935; Evans and Goodman, 1944; Nogami and Hurley, 1948; Kulp *et al.*, 1952; Turner *et al.*, 1958*a*) discuss this case in more or less detail.

EXPERIMENTAL TECHNIQUES

In the early work (Finney and Evans, 1935; Evans and Goodman, 1944; Keevil *et al.*, 1943) ionization chambers were used. Subsequently, the proportional counter became popular, and analyses of natural samples utilizing commercial proportional counting apparatus have been described by Adams *et al.* (1958) and others. A flow-type proportional counter designed specifically for measuring low-level α activity in biological materials has been described by McDaniel *et al.* (1956). This utilizes rectangular, screen-wall counters accommodating sources up to 141 cm.^{2} in area, with background counting rates of 3.8 counts/hr. More recently, Kiefer and Maushart (1961) have discussed proportional flow counters of area up to 700 cm.^{2} with backgrounds in the α pulse-height region of about 0.5 counts per hour per cm.^{2}. Mention should also be made of the high-quality proportional counters developed by the Glasgow group (Curran, 1955); although these have been applied primarily to problems of low-energy β spectroscopy, they could no doubt be utilized as α detectors to good effect.

Proportional counters are still in use in a number of laboratories and do possess certain advantages—these have been enumerated by Kiefer and Maushart (1961). The scintillation counter with an activated zinc sulfide phosphor, however, appears to be the best general-purpose low-level α detection instrument available today. Early examples of such apparatus were those of Reed (1950) and Kulp *et al.* (1952), but the most effective system is undoubtedly that devised by the group at the Royal Cancer Hospital (Turner *et al.*, 1958*a*; Mayneord *et al.*, 1958). This system is discussed in the references cited and elsewhere in the proceedings of this symposium (Hill *et al.*, 1963) and will not be described in detail here. It will suffice to say that it provides a simple, economical, sealed sample-phosphor system of great sensitivity and convenience. No emanation problems (such as are encountered with flow-type proportional counters) arise; in fact, measurement of the counting rate of the sealed system as a function of time allows sample disequilibria to be studied effectively. Sample areas are limited by the size of the photomultiplier tube; areas of just under 100 cm.^{2} (a $4\frac{1}{2}$-inch photocathode) are commonplace. With reasonable care (good quality zinc sulfide, selected photomultiplier tubes, and low cathode-to-first-dynode voltages [Pagano *et al.*, 1962]) background count rates of less than 0.01 counts per cm.^{2} per hour are obtainable. For a phosphor area of about 100 cm.^{2}, these rates correspond to activities of a few hundredths of a picocurie per gram in a typical thick-source solid sample or, in geochemical terms, to the activity of a typical rock sample containing 25 parts per thousand million of thorium and 8 parts per thousand million of uranium. The sensitivity of the method is thus apparent, although it must of course be borne in mind that extremely long counting times are necessary if the ultimate in sensitivity is to be attained. The associated electronics is, however, simple, and long-term stability is not a major problem.

THE PAIRS TECHNIQUE

The parameter obtained from a "total" α counting experiment is essentially the specific α activity, N, of the sample. It is often desirable to obtain further information as to the α-emitting nuclides present, and it is possible to do so to a limited extent without recourse to α spectroscopy by making use of the so-called "pairs" technique. This utilizes the fact that, in two of the three natural radioactive series, pairs of successive α particles are emitted within short intervals of time. These are due to the short-lived α emitters polonium-216 ($T\frac{1}{2} = 0.158$ seconds) in the thorium series and polonium-215 ($T\frac{1}{2} = 0.0018$ seconds) in the actinium series. Time analysis of the pulses from the α counter—easily accomplished by simple paralysis circuits—permits the separation of pulse pairs due to these particular α emitters; suitable equilibrium assumptions then enable the relative contributions of the three series to be assessed. This method appears to have been suggested first by Hurley and Shorey (1952), and later by Hirschberg (1954, 1955); it was the paper by Turner *et al.* (1958*a*) that first described its successful application in any detail. Subsequently, its application to the case of thorium and uranium determinations in geological samples has been discussed by Cherry (1963). The pairs technique, combined with the zinc sulfide scintillation counter previously discussed, has already resulted in the accumulation of a vast amount of data concerning the α activities of natural samples (Turner *et al.*, 1958*b*, *c*, 1961; Turner and Radley, 1960; Mayneord, 1960; Mayneord *et al.*, 1960; Marsden, 1960, 1963; Cherry, 1962), and its use is to be commended to workers in this field.

THE RANGE PROBLEM

From equations (1) and (2) it is obvious that a knowledge of the range, R, of the α particle in the sample is essential if a quantitative result is required. It is customary to use some empirical formula for calculating α particle ranges. The geological workers (Finney and Evans, 1935; Evans and Goodman, 1944; Nogami and Hurley, 1948; Kulp *et al.*, 1952) tend to favor the Bragg-Kleeman relation, which gives:

$$R\rho = 0.32(10^{-3})W^{1/2}R_a \text{ gm/cm}^2 , \qquad (3)$$

where ρ is the sample density, $W^{1/2}$ is the square root of the atomic weight calculated on the "atomic fraction" basis (Nogami and Hurley, 1948), and R_a is the range in centimeters in air of an α particle of the same energy. Turner *et al.* (1958*a*), on the other hand, replace $W^{1/2}$ by $Z^{2/3}$, where Z is the atomic number. For many common natural materials (both geological and biological) we have $W^{1/2} \sim Z^{2/3} \sim 4.6$, and the two expressions give very similar results.

The question of the accuracy of these expressions is of major importance. Nogami and Hurley (1948) claimed that the Bragg-Kleeman relation is accurate to within 5 per cent. Beharrell (1949) disputes this and claims that the Bragg-Kleeman law gives ranges that are about 10 per cent too low. Yagoda (1949) finds the reverse and says that the computed ranges are 10 per cent too high. A recent review article by Whaling (1958) presents the ranges of α particles of various energies in twelve gaseous and seven solid materials (all metallic). The ranges in gases agree fairly well with the Bragg-Kleeman law, but those in solids are definitely higher than those given by equation (3) in the low atomic weight range ($W^{1/2} = 5$ and less), which is applicable to most naturally occurring samples. Thus, if one draws smooth curves through Whaling's points, one finds that for $W^{1/2} = 4.6$ (which is generally equivalent to $Z^{\frac{2}{3}} = 4.6$) equation (3) gives ranges that are approximately 6 per cent, 13 per cent, and 25 per cent too low for α particles of energies 8, 6, and 4 mev., respectively. Whaling's values were based on experimental data available to him at the time (1958), and it is instructive to consider experimental data (unfortunately very meager) that have been published since then. Thus Garin and Faraggi (1958) determined the ranges of 4.5 mev. α's in uranium, gold, zirconium, and silicon. For these four elements they obtain ranges that were, respectively, 32 per cent, 28 per cent, 22 per cent, and 4 per cent higher than those given by the Bragg-Kleeman equation. Replacement of $W^{\frac{1}{2}}$ by $Z^{\frac{2}{3}}$ results in a considerable improvement; there is agreement with the Garin-Faraggi data to within 5 per cent for all four elements. Kamke and Kramer (1962) have determined α ranges in boron; in the region of 4.5 mev. their results are about 13 per cent higher than that given by the Bragg-Kleeman equation and 23 per cent higher than the same equation with $Z^{\frac{2}{3}}$ substituted for $W^{\frac{1}{2}}$. Finally, calibrations using the thorium series in equilibrium in thorium oxide (Cherry, 1963; Cherry, in press) indicate that equation (3) gives results that are 25 per cent low; $Z^{\frac{2}{3}}$ instead of $W^{\frac{1}{2}}$ gives results that are only 5 per cent low.

The whole subject of α particle ranges in solids is thus obviously in a state of some confusion and is in need of serious attention. A systematic study of α ranges as a function of particle energy and atomic weight would be most valuable and would help establish total α counting as a reliable *absolute* method of measurement. Meanwhile, it appears that the following very tentative conclusions can be drawn: (i) The Bragg-Kleeman equation (3) would appear to give ranges that are somewhere between 5 per cent and 30 per cent too low. (ii) Replacement of $W^{\frac{1}{2}}$ in equation (3) by $Z^{\frac{2}{3}}$ results in considerable improvement in the upper half of the periodic table but makes matters worse for the lightest elements. (Unpublished data by Osborne, 1962, appear to substantiate this latter conclusion.)

THE HOMOGENEITY PROBLEM

The derivation of equations (1) and (2) is based essentially on the assumption that the α activity is distributed homogeneously throughout the sample. For many natural samples this is not the case, and homogenization is necessary—a problem of which the geologists are particularly aware (Kulp *et al.*, 1952; Hurley, 1950). Grinding of the sample down to grain sizes very much less than the smallest α range encountered (in practice this requires grinding to micron size or less) and following this by thorough homogenization will of course take care of this problem, but it would seem desirable that the theoretical aspects be systematically investigated. Preliminary calculations (Cherry, in press) assuming two different types of inhomogeneity ("hotspot" and "surface coatings") indicate that (i) grinding down to at least one-tenth of the α particle range is necessary to insure less than 5 per cent error in formula (1) and (ii) the "pairs" count rate (see above) is distinctly more sensitive to inhomogeneities in the radioactivity distribution than is the "total" α count rate. Further investigations are proceeding.

CONCENTRATION OF THE ACTIVITY

As has been mentioned above, α counting techniques are so sensitive that they can frequently be used without any previous treatment of the sample concerned. Some environmental samples are, however, so low in radioactivity that some form of concentration is essential. Thus Turner *et al.* (1958c) ashed foodstuffs prior to counting and boiled drinking water (Turner *et al.*, 1961) to dryness before counting the residue. Chemical extraction is of course also feasible in most cases, although it does introduce an additional element of uncertainty and should be avoided where possible. In this connection it is perhaps worthwhile to emphasize the utility of modern ion-exchange techniques; two useful references that discuss the methods for extracting the trans-lead α emitters are those of Korkisch and Janauer (1962) and Choppin and Sikkeland (1960).

An intriguing and simple method of extracting the heavy elements from aqueous solution has been mentioned by Rosholt (1957) and Morken (1959). Zinc sulfide scintillation phosphor powder is simply added to vials containing the solution to be investigated, the vial is shaken vigorously, and the powder is allowed to settle. The zinc sulfide apparently adsorbs some (or all) of the heavy-element α emitters, and the scintillations emitted by the powder can be counted directly with a photomultiplier tube. Morken obtained over 95 per cent extraction for both polonium and plutonium under the right conditions of pH and considered that "an adsorption mechanism is indicated, in which case the extent of removal is probably independent of the element used." Turner *et al.* (1961) have used this technique effectively to determine radon in water, and further investigations would be of great interest.

GASEOUS SAMPLES

The treatment described above has tacitly assumed that the sample materials are solids or liquids. Gaseous samples will of course require special techniques of counter design, but the basic detection principles remain the same. Two selected references to the zinc sulfide type of scintillation detectors for gaseous samples will serve as examples. Lucas (1957) has described a radon counter, while Giffin *et al.* (1963) have applied the pairs technique to the case in which a gas is used in the detector chamber.

AUTORADIOGRAPHIC TECHNIQUES

Autoradiographic techniques have been discussed by Yagoda (1949), and their application to naturally occurring materials is reviewed elsewhere in this symposium (Ragland, 1963). They are listed here merely for the sake of completeness. As total α counters, they compare unfavorably with the techniques discussed above; as a tool for the identification and study of microsites of radioactivity, they are valuable and powerful.

α SPECTROSCOPIC TECHNIQUES

The amount of information obtainable from "total" α counting, even when the pairs technique is used, is strictly limited, and it is very often desirable to obtain information concerning the relative contributions of the specific nuclides responsible for α particle emission. In these cases recourse must be made to α spectroscopic techniques—these measure the energy of the α particles and thus provide a means of identification of nuclear species. The power of the technique is of course matched by increased complexity, particularly insofar as the associated electronics are concerned. A pulse-height analyzer (preferably of the multichannel variety) is an essential tool. The book by Chase (1961) is an excellent introduction to the electronic problems of nuclear spectroscopy. Source preparation, too, is a considerably more critical matter—sources not only must be "thin" but must be "spectroscopically thin," which is one or two orders of magnitude thinner (100 μgm/cm^2 is a frequently quoted upper limit for an acceptable α spectroscopic source). Source preparation, often a somewhat empirical art, has recently been reviewed by Jaffe (1962), and special methods for the preparation of the large-area sources frequently needed in the α spectroscopy of natural environmental samples have been given by Hill (1961), Korolev and Kocharov (1958), and Kocharov and Korolev (1961). A recently described (Bjørnholm and Lederer, 1962) cation-exchange foil technique is also worthy of special mention. We shall pass on to a brief survey of the detectors that can be used for low-level α spectroscopy.

ION CHAMBERS

The ion chamber is by far the most commonly used and most widely suitable tool for α spectroscopic studies. The parallel-plate, gridded-pulse ionization chamber (Wilkinson, 1950; Rossi and Staub, 1949; Herwig *et al.*, 1955; Cranshaw and Harvey, 1948; Bunemann *et al.*, 1949; Harvey *et al.*, 1957) is in turn the most popular type of chamber and is commercially available. A typical chamber of this sort possesses a sample area of about 10 cm.² and a detection efficiency of about 45 per cent of 4π and yields resolutions of about 60 kev. for a's of about 5-mev. energy. Ray and Hammond (1961) applied an instrument of this type to natural environmental samples without previous chemistry, and the work of Goldberg and Koide (1962) on thorium isotopes in ocean sediments provides a good example of high-quality results obtained after chemical extraction. The big disadvantage of this type of chamber lies in its limited source area, and the quest for greater sensitivity has led, in recent years, to the design of ion chambers capable of handling much larger sources. The properties of several of

TABLE 1

COMPARISON OF VARIOUS LARGE-AREA ION CHAMBERS*

Author	Configuration	Grid	R (kev.)	B (counts/ hr)	S (cm²)	S^2/B ($\times 10^{-6}$)
Hill (1961)...............	Cylindrical	No	110	4	15,000	56
Hill *et al.* (1963)...........	Cylindrical	Yes	40	0.3	1,500	7.5
Kocharov and Korolev (1961)..	Parallel plate	Yes	25	0.03	200	1.3
Nurmia (1962)..............	Parallel plate	Yes	100	0.03	300	3
Doke (1962) ⎫ Ogawa *et al.* (1961)⎰·········	Parallel plate	Yes (two)	70	0.01	150	2.3
Macfarlane and Kohman (1961)	Cylindrical	No	60†	0.3†	1,200	4.8
Lonati *et al.* (I) (1958).........	Parallel plate	Yes	300	1	110	0.01
Lonati *et al.* (II) (1958)........	Cylindrical	Yes	300	35	2,800	0.2
Lonati and Tonolini (1960)....	Parallel plate	Yes	300	0.18	110	0.07

* Numbers quoted are approximate in many cases.

† An estimate at an a energy of 2 mev. instead of 5 mev.

these are summarized in Table 1. The resolution R is the usual full width at half-maximum count rate, and the background B is taken over this same spectral line width; an a energy of about 5 mev. has been assumed. In general, the optimum figures quoted by the respective authors have been used; where no figures are quoted, the relevant parameters have been estimated from data given in the papers.

A study of Table 1 reveals the following essentials:

i) *Sensitivity.* A suitable "figure of merit" for comparing the sensitivity of the ion chambers listed would appear to be the quantity $(SE)^2/B$ where E is the (geometrical) counting efficiency expressed as a percentage of 4π. This figure takes into account the desirability of making the ratio (source count rate)²/background as high as possible in low-activity experiments (see Freedman and Anderson, 1952) and seems to be more suitable than the figure SE/B used by Doke (1962). Values of E are unfortunately not always quoted by authors, and thus the figure of merit cannot be calculated exactly. However, the variation in E is not great (from about 20 to 50 per cent for all the chambers quoted), and useful comparisons can be made on an approximate basis. It is easily seen that the ion chamber of Hill (1961), which has an efficiency close to 50 per cent, possesses by far the largest figure of merit, a fact due almost entirely to its very large surface area. This chamber and the results obtained from it are discussed in another paper at this symposium (Hill *et al.*, 1963) as well as in the literature (Hill, 1960, 1961, 1962; Mayneord and Hill, 1959; Hill and Jaworowski, 1961). It is, however, worth pointing out that its high sensitivity is obtained without recourse to electric background reduction techniques (e.g., anticoincidence shielding as used by Macfarlane and Kohman [1961] or utilization of the grid pulses as used by Kocharov and Korolev [1961] and Doke [1962]).

ii) *Resolution.*   A study of the energies of the α particles emitted by the naturally occurring α emitters (see, e.g., the article by Hanna [1959] or the isotope table of Strominger *et al.* [1958]) shows that in many cases two or more α "lines" will lie very close together. Thus the α particles from the common isotopes uranium-238 and thorium-232 are separated in energy by about 200 kev.; these α groups are *just* separated by a chamber of the Hill type, and sources containing both the species can accordingly be counted with fair accuracy without prior chemical treatment. Other cases are, however, not so favorable; a particularly important clash is that between thorium-230, uranium-234, and radium-226, which have their main α groups at 4.68, 4.77, and 4.78 mev., respectively. To separate these requires either prior chemical extraction or better resolution, and attention is drawn to the excellent resolving power of the instrument of Kocharov and Korolev (1961). Subject to the availability of sufficiently thin sources (a useful figure to remember is that a 100 μg/cm^2 source contributes about 80 kev. to the resolution), this chamber should be able to separate thorium-230 from uranium-234 and radium-226, although it will not be able to distinguish between the latter two isotopes. Optimum resolution with this chamber is, however, attainable only at the cost of a reduction in the efficiency E (down to about 10 per cent).

Whether one selects a chamber of maximum sensitivity or one with maximum resolving power depends, of course, on the problem at hand. The chambers of Hill *et al.* (1963), Nurmia (1962), Doke (1962), Ogawa *et al.* (1961), and Macfarlane and Kohman (1961) are all useful compromises between the two extremes. Those of the Italian group (Lonati, 1961; Lonati and Tonolini, 1960; Lonati *et al.*, 1958) appear to have been superseded, but they can claim the honor of having been the pioneers of the large-source ion-chamber technique as applied to naturally occurring α-active samples (Lonati, 1961; Lonati *et al.*, 1958; Facchini *et al.*, 1956).

MAGNETIC SPECTROMETERS

Magnetic spectrometers provide the ultimate in resolution (down to about 3 kev.) but possess such poor geometries (several orders of magnitude below the ion chamber) that their use in studying natural-environment samples is likely to be minimal. A good discussion of this type of instrument is given in the review of Baranov and Zalenkov (1959).

SCINTILLATION SPECTROMETERS

Scintillation counters are in many respects more convenient to operate than are ion chambers. As an α spectroscopic tool, however, they suffer from a major intrinsic defect—poor resolution. Thus typical resolutions obtained with scintillation α spectrometers have been 13 per cent (gaseous phosphor [Forte, 1956]) and 17 per cent (organic liquid phosphor [Seliger, 1960]). Inorganic crystals are rather better. Thus the early work of Broser and Kallman (1949) gave resolution better than 5 per cent for cadmium sulfide, while resolutions of 1.8 per cent (Martinez and Senftle, 1960) and 2.6 per cent (Fleury *et al.*, 1960) have been reported under optimum conditions for α's in cesium iodide. More typical of the sort of resolution likely to be obtained in practical applications (where one has larger area sources and no collimation) are the results of Gale (1962). Gale describes a simple cesium iodide α spectrometric system capable of measuring activities as low as picocuries and obtains a resolution of about 14 per cent using an 0.005-inch-thick crystal of 1.5-inch diameter and source diameters up to 1 cm. Belyaev *et al.* (1961) reported similar results for cesium iodide; resolutions from 11 to 22 per cent with 3–5½-cm.-diameter sources, but resolution as low as 4 per cent with 3-mm.-diameter sources. The potentialities of the cesium iodide (and possibly also the cadmium sulfide) spectrometer appear to be worthy of further investigation, even though they cannot hope to compete seriously, as to resolution, with the ion chamber. In this connection, the useful reference of Aniansson (1961) concerning the preparation of thin cesium iodide layers by vacuum deposition should be mentioned.

SEMICONDUCTOR DETECTORS

The recently developed semiconductor detectors possess many advantages for charged particle spectroscopy; excellent resolution (e.g., 17 kev. for a 1-cm.^{2} detector [Blankenship and Borkowski, 1961]), good linearity, good stability, and low background rates are the chief of these. They do, however, suffer from one major disadvantage as far as low-level detection is concerned, namely, their small area. Detectors with areas of several cm.² and resolutions approaching 50 kev. are now available commercially. If areas can be increased to above 10 cm.² without sacrificing resolution, such detectors will provide a serious alternative to the "conventional" (i.e., small area) Frisch grid-type ion chamber, and even now they are superior in resolution to the ion chamber if very small (millimetric dimension) sources are involved. Chetham-Strode *et al.* (1960) have discussed α particle spectroscopy with $\frac{1}{4}$ cm.² silicon-surface, barrier-type detectors; they emphasize the need for avoiding contamination of the detector by recoil nuclei if the excellent low-background characteristics of the detector are to be maintained. A general review of semiconductor detector devices has been given by Miller *et al.* (1962).

NUCLEAR EMULSIONS (YAGODA, 1949)

These suffer from the disadvantages of long exposure times, extremely tedious measurement, and poor resolution (about five times worse than a good ion chamber [Macfarlane and Kohman, 1961]). They are, however, very sensitive and have been employed in several experiments. Notable among these have been the searches for α radioactivity in the medium-heavy elements (e.g., Porschen and Riezler [1956] and Macfarlane and Kohman [1961] give a list of references to this sort of work) and the important experiments of Picciotto and Wilgain (1954) on thorium isotopes in deep-sea sediments. The technique used by the latter authors is so sensitive that it permits the measurement of as little as nanograms of thorium, provided it is sufficiently highly concentrated. Another emulsion technique that also possesses a degree of sensitivity of this order has been described recently by Geiger *et al.* (1961).

"THICK-SOURCE" SPECTROMETRY

In an effort to avoid the necessity for the preparation of spectroscopically thin sources, and/or to increase the sensitivity by increasing the volume of sample material, several authors have investigated the nature of the α spectrum from a "thick" source. Thus Graeffe and Nurmia (1961) have applied thick-source spectrometry to determinations of the energies of α's from the weakly α active medium-heavy elements, while Chudacek (1958) has suggested a semiempirical formula for the energy distribution of α's emerging from a thick source. A final useful reference is that of Abrosimov and Kocharov (1962), who discuss the effects of source thickness on the shape of the energy and directional distributions of the α particles.

Conclusion

In conclusion, two matters of general importance in low-level α counting techniques should be mentioned.

a) THE NATURAL α ACTIVITY OF MATERIALS

In the construction of α detectors it is important to insure that the surfaces exposed inside the detector possess a low natural α emission rate and a low emanating power (Hill *et al.*, 1963). Data for α emission rate from various materials have been published by several authors (McDaniel *et al.*, 1956; Bearden, 1933; Adams and Richardson, 1960). Features worth noting are the high levels of emission from solder (Bearden, 1933) (2,800 α's per 100 cm²/hr) and the extremely low level of emission from aluminized cellulose acetate (Hill, 1961) (about 0.3 α's per 100 cm²/hr). In general, materials should be checked on a total α counter before they are built into a low-level α detector.

b) COUNTING STATISTICS

Counting at low levels of radioactivity requires that particular attention be paid to problems of counting statistics. A simple matter, not often discussed, is the fact that for a low number of total counts the usual assumption that the Poisson distribution is equivalent to the normal distribution is invalid. Confidence limits based on the Poisson distribution itself must be used, and these have been given by Garwood (1936). The question of optimization of source-background ratios has been discussed by several authors (e.g., Jaffey, 1960; Freedman and Anderson, 1952) and leads to the generally accepted (source count rate)²/background figure of merit mentioned earlier.

A final problem that arises when counting members of the naturally occurring radioactive series is the question of the statistical distribution of counts from a sample containing both parent and daughter activities. The pioneer work concerning this complex question appears to have been that of Adams (1933); a more recent and more complete treatment is given by Huybrechts (1957), who applies his results to specific cases (the uranium and thorium families in equilibrium) for a specific time interval (thirty days).

ACKNOWLEDGMENTS

It is a particular pleasure to thank Dr. C. R. Hill, who introduced the author to large ion chamber experimental techniques and who criticized the manuscript constructively. The hospitality of the laboratories of Professor W. V. Mayneord (Physics Department, Royal Cancer Hospital, London) and Professor J. A. S. Adams (Geology Department,

Rice University, Houston) is also gratefully acknowledged, as is the award of a research fellowship by Rice University, during the tenure of which the present review was prepared.

REFERENCES

ABROSIMOV, N. K., and G. E. KOCHAROV. 1962. The effect of the thickness of the source on the shape of the energy distribution and angular distribution curves of alpha-particles. Bull. Acad. Sci. U.S.S.R., Physical Ser., **26**:237–44.

ADAMS, J. A. S., J. E. RICHARDSON, and C. C. TEMPLETON. 1958. Determinations of thorium and uranium in sedimentary rocks by two independent methods. Geochim. & Cosmochim. Acta, **13**:270–79.

ADAMS, J. A. S., and K. A. RICHARDSON. 1960. Radioactivity of aluminum metal. Econ. Geol., **55**:1060–63.

ADAMS, N. I., JR. 1933. An application of probabilities to the counting of alpha-particles. Phys. Rev., **44**:651–53.

ANIANSSON, G. 1961. Thin CsI(Tl) scintillating layers produced by vacuum deposition. Trans. Roy. Inst. Technol., Stockholm, No. 176. Pp. 20.

BARANOV, S. A., and A. G. ZALENKOV. 1959. Alpha-ray spectroscopy. Instr. & Exp. Techniques, No. 5, pp. 695–711.

BEARDEN, J. A. 1933. Radioactive contamination of ionization chamber materials. Rev. Sci. Instr., **4**:271–75.

BEHARRELL, J. 1949. Absorption of alpha rays in thick sources. Trans. Am. Geophys. Union, **30**:333–36.

BELYAEV, L. M., A. B. GIL'VARG, and V. P. PANOVA. 1961. CsI (Tl) scintillators for recording alpha-particles. Soviet Phys. Cryst., **6**:108–10.

BJØRNHOLM, S., and C. M. LEDERER. 1962. Thin cation-exchange foils. Nuc. Instr. & Methods, **15**:233–36.

BLANKENSHIP, J. L., and C. J. BORKOWSKI. 1961. Performance of silicon surface barrier detectors with charge sensitive amplifiers. IRE Trans. Nuc. Sci., **NS-8**:17–20.

BROSER, I., and H. KALLMAN. 1949. Measurements of alpha-particle energies with the crystal fluorescence counter. Nature, **163**:20–21.

BUNEMANN, O., T. E. CRANSHAW, and J. A. HARVEY. 1949. Design of grid ionization chambers. Canad. J. Res., **27A**:191–206.

CHASE, R. L. 1961. Nuclear Pulse Spectrometry, pp. 221. New York: McGraw-Hill Book Co.

CHERRY, R. D. 1962. Thorium and uranium contents of australites. Nature, **195**:1184–86.

———. 1963. The determination of thorium and uranium in geological samples by an alpha-counting technique. Geochim. & Cosmochim. Acta, **27**:183–89.

———. (In press.)

CHETHAM-STRODE, A., J. R. TARRANT, and R. J. SILVA. 1960. The application of silicon detectors to α-particle spectroscopy. IRE Trans. Nuc. Sci., **NS-8**:59–63.

CHOPPIN, G. R., and T. SIKKELAND. 1960. Scheme for the separation of the elements francium through uranium. *In* E. K. HYDE, The Radiochemistry of Thorium, pp. 60–61. Procedure 19, Nat. Acad. Sci., Nat. Res. Council, Nuc. Sci. Ser. 3004.

CHUDACEK, I. 1958. Energy spectrum of α particles emitted from sources of different thicknesses. Czech. J. Phys., **8**:396–403.

CRANSHAW, T. E., and J. A. HARVEY. 1948. Measurement of the energies of α-particles. Canad. J. Res. Ser. A., **26**:243–54.

CURRAN, S. C. Proportional counter spectrometry. 1955. *In* K. SIEGBAHN (ed.), Gamma-Ray Spectroscopy, pp. 165–83. New York: Interscience Pub.; Amsterdam: North-Holland Pub. Co.

DOKE, T. 1962. A new method for the reduction of alpha-ray background in a gridded ionization chamber. Canad. J. Phys., **40**:607–21.

EVANS, R. D. 1955. The Atomic Nucleus. New York: McGraw-Hill Book Co. Pp. 972.

EVANS, R. D., and C. GOODMAN. 1944. Alpha-helium method for determining geological ages. Phys. Rev., **65**:216–27.

FACCHINI, U., M. FORTE, A. MALVICINI, and T. ROSSINI. 1956. Analysis of U and Th minerals by alpha spectrum. Nucleonics, **14**:126, 128–31.

FINNEY, G. D., and R. D. EVANS. 1935. The radioactivity of solids determined by alpha-ray counting. Phys. Rev., **48**:503–11.

FLEURY, J., P. PERRÚN, M. BOGE, and J. LAUGIER. 1960. Spectromètre α a scintillation avec I Cs(Tl). J. Phys. Rad., **21**:480–83.

FORTE, M. 1956. Light pulses excited by α particles in argon: A gaseous scintillation detector. Nuovo Cimento, Ser. 10, **3**:1443–55.

FREEDMAN, A. J., and E. G. ANDERSON. 1952. Low-level counting techniques. Nucleonics, **10**:57–59.

GALE, H. J. 1962. The use of simple scintillation alpha spectrometry for contamination control in laboratories. Phys. Med. Biol., **6**:577–82.

GARIN, A., and H. FARAGGI. 1958. Parcours des alpha de 4.5 MeV dans l'uranium, l'or, le zirconium et le silicium. J. Phys. & Rad., **19**:76–78.

GARWOOD, F. 1936. Fiducial limits for the Poisson distribution. Biometrika, **28**:437–42.

GEIGER, E. L., A. N. TSCHAECHE, and E. L. WHITTAKER. 1961. Simplified autoradiography technique for α-emitters. Health Physics, **4**:302–4.

GIFFIN, C., A. KAUFMAN, and W. BROECKER. 1963. Delayed coincidence counter for the assay of actinon and thoron. J. Geophys. Res., **68**:1749–57.

GOLDBERG, E. D., and M. KOIDE. 1962. Geochronological studies of deep sea sediments by the ionium/thorium method. Geochim. & Cosmochim. Acta, **26**:417–50.

GRAEFFE, G., and M. NURMIA. 1961. The use of thick sources in alpha spectrometry. Ann. Acad. Sci. Fennicae, Ser. A, No. 77, **6**:1–14.

HANNA, G. C. 1959. Alpha-radioactivity. *In* E. SEGRE (ed.), Experimental Nuclear Physics, **3**:54–257. New York: John Wiley & Sons.

HARVEY, B. G., H. G. JACKSON, T. A. EASTWOOD, and G. C. HANNA. 1957. The energy of alpha particles from U-234, U-238, and Th-232. Canad. J. Phys., **35**:258–70.

HERWIG, L. O., G. H. MILLER, and N. G. UTTERBACK. 1955. Some characteristics of a gridded parallel-plate ionization chamber. Rev. Sci. Instr., **26**:929–36.

HILL, C. R. 1960. Lead-210 and polonium-210 in grass. Nature, **187**: 211–12.

———. 1961. A method of alpha particle spectroscopy for materials of very low specific activity. Nuc. Instr. & Methods, **12**:299–306.

———. 1962. Identification of alpha emitters in normal biological materials. Health Physics, **8**:17–25.

HILL, C. R., and Z. S. JAWOROWSKI. 1961. Lead-210 in some human and animal tissues. Nature, **190**:353–54.

HILL, C. R., R. V. OSBORNE, and W. V. MAYNEORD. 1963. Studies of α radioactivity in relation to man. This symposium.

HIRSCHBERG, D. 1954. Dosage de radioéléments par la distribution des intervalles entre désintégrations: Application au RdTh. Nuovo Cimento, Ser. 9, **12**:733–42.

———. 1955. Dosage de radioéléments par la distribution des intervalles entre désintégrations: Efficacité des estimations. *Ibid.*, Ser. 10, **1**:341–43.

HURLEY, P. M. 1950. Distribution of radioactivity in granites and possible relation to helium age measurement. Geol. Soc. America Bull., **61**:1–8.

HURLEY, P. M., and R. R. SHOREY. 1952. Discrimination of thoron alpha activity in presence of radon. Trans. Am. Geophys. Union, **33**:722–24.

HUYBRECHTS, M. 1957. Théorie statistique du comptage de particules alpha émises par des désintégrations successive d'éléments radioactifs. Nuovo Cimento, Ser. 10, **6**:811–31.

JAFFE, L. 1962. Preparation of thin films, sources, and targets. Ann. Rev. Nuc. Sci., **12**:153–88.

JAFFEY, A. H. 1960. Statistical tests for counting. Nucleonics, **18**:180–84.

KAMKE, D., and P. KRAMER. 1962. Energieverlust und Reichweite von α-Teilchen in Bor im Energiebereich von 0.2 bis 5.3 Mev. Ztschr. d. Phys., **168**:465–73.

KEEVIL, N. B., and W. E. GRASHAM. 1943. Theory of alpha-ray counting from solid sources. Canad. J. Res., Ser. A, **21**:21–36.

KEEVIL, N. B., A. R. KEEVIL, W. N. INGHAM, and G. P. CROMBIE. 1943. Causes of variations in radioactivity data. Am. J. Sci., **241**:345–65.

KIEFER, H., and R. MAUSHART. 1961. Large-area flow counters speed radiation measurements. Nucleonics, **19**:51–54.

KOCHAROV, G. E., and G. A. KOROLEV. 1961. Ionization α-spectrometer with high resolution. Bull. Acad. Sci. U.S.S.R., Phys. Ser., **25**:237–56.

KORKISCH, J., and G. E. JANAUER. 1962. Ion exchange in mixed solvents: Adsorption behaviour of uranium and thorium on strong-base anion-exchange resins from mineral acid-alcohol media: Separation methods for uranium and thorium. Talanta, **9**:957–85.

Korolev, G. A., and G. E. Kocharov. 1958. Thin α-emitting sources of large areas. Instr. & Exp. Techniques, No. 5, pp. 702–3.

Kulp, J. L., H. D. Holland, and H. L. Volchok. 1952. Scintillation alpha counting of rocks and minerals. Trans. Am. Geophys. Union, **33**:101–13.

Lonati, R. D. 1961. The presence of long life alpha emitters in the air. Energia Nucleare (Milan), **8**:217–20.

Lonati, R. D., J. U. Facchini, Z. Zori, F. G. Houtermans, and E. Tongiorgi. 1958. Study on α radioactivity in low concentration. Nuovo Cimento, Ser. 10, **7**:133–41.

Lonati, R. D., and F. Tonolini. 1960. Study of α-radioactivity in the soil and in biological material by spectrum determination. Energia Nucleare (Milan), **7**:107–10.

Lucas, H. F. 1957. Improved low-level alpha-scintillation counter for radon. Rev. Sci. Instr., **28**:680–83.

McDaniel, E. W., H. J. Schaefer, and J. K. Colehour. 1956. Dual proportional counter for low-level measurement of alpha activity of biological materials. Rev. Sci. Instr., **27**:864–68.

Macfarlane, R. D., and T. P. Kohman. 1961. Natural alpha radioactivity in medium-heavy elements. Phys. Rev., **121**:1758–69.

Marsden, E. 1960. Radioactivity of soils, plants and bones. Nature, **187**: 192–95.

———. 1963. Radioactivity of some rocks, soils, plants, and bones. This symposium.

Martinez, P., and F. E. Senftle. 1960. Effect of crystal thickness and geometry on the alpha-particle resolution of CsI(Tl). Rev. Sci. Instr., **31**:974–77.

Mayneord, W. V. 1960. Problems in the metabolism of radioactive materials in the human body. Clin. Radiol. **11**:2–13.

Mayneord, W. V., and C. R. Hill. 1959. Spectroscopic identification of alpha-emitting nuclides in biological material. Nature, **184**:667–69.

Mayneord, W. V., J. M. Radley, and R. C. Turner. 1958. The alpha-ray activity of humans and their environment. Proc. 2d United Nations Internat. Conf. on Peaceful Uses of Atomic Energy, **23**:150–55.

Mayneord, W. V., R. C. Turner, and J. M. Radley. 1960. Alpha activity of certain botanical materials. Nature, **187**:208–11.

Miller, G. L., W. M. Gibson, and P. F. Donovan. 1962. Semiconductor particle detectors. Ann. Rev. Nuc. Sci., **12**:189–220.

Morken, D. A. 1959. Integral-phosphor counting of α-emitters. Health Physics, **2**:77–78.

Nogami, H. H., and P. M. Hurley. 1948. The absorption factor in counting alpha rays from thick mineral sources. Trans. Geophys. Union, **29**:335–40.

Nurmia, A. 1962. Experimental techniques in low-level alpha spectrometry. Proc. Conf. Nuc. Electronics, I, Belgrade, pp. 297–302. Vienna: International Atomic Energy Agency.

OGAWA, I., TADAYOSHI DOKE, and M. TSUKUDA. 1961. The double-grid ionization chamber. Nuc. Instr. & Methods, **13**:169–76.

OSBORNE, R. V. Ph.D. thesis, London University, 1962.

PAGANO, R., C. J. S. DAMERELL, and R. D. CHERRY. 1962. Effect of photo-cathode-to-first dynode voltage on photomultiplier noise pulses. Rev. Sci. Instr., **33**:955–56.

PICCIOTTO, E., and S. WILGAIN. 1954. Thorium determination in deep-sea sediments. Nature, **173**:632–33.

PORSCHEN, W., and W. RIEZLER. 1956. Sehr langlebige natürliche α-Strahler. Zs. f. Naturforsch., **11a**:143–51.

RAGLAND, P. C. 1963. Autoradiographic investigations of naturally occurring materials. This symposium.

RAY, E. L., and S. E. HAMMOND. 1961. The use of an α pulse-height analyzer in environmental monitoring. Health Physics, **5**:50–56.

REED, C. W. 1950. An end-window alpha scintillation counter for low counting rates. Nucleonics, **7**:56–62.

ROSHOLT, J. N., JR. 1957. Quantitative radiochemical methods for determination of the sources of natural radioactivity. Anal. Chemistry, **29**:1398–1408.

ROSSI, B. B., and H. H. STAUB. 1949. Ionization Chambers and Counters: Experimental Techniques. New York: McGraw-Hill Book Co. Pp. 243.

SELIGER, H. H. 1960. Liquid scintillation counting of α-particles and energy resolution of the liquid scintillation α- and β-particles. Internat. J. Appl. Radiation & Isotopes, **8**:29–34.

STROMINGER, D., J. M. HOLLANDER, and G. T. SEABORG. 1958. Table of isotopes. Rev. Mod. Phys., **30**:585–904.

TURNER, R. C., and J. M. RADLEY. 1960. Naturally occurring alpha activity of cigarette tobaccos. Lancet, pp. 1197–98.

TURNER, R. C., J. M. RADLEY, and W. V. MAYNEORD. 1958a. The α-ray activity of human tissues. Brit. J. Radiol., **31**:397–406.

———. 1958b. Alpha ray activities of humans and their environment. Nature, **181**:518–21.

———. 1958c. The naturally occurring α-ray activity of foods. Health Physics, **1**:268–75.

———. 1961. Naturally occurring alpha activity of drinking waters. Nature, **189**:348–52.

WHALING, W. 1958. The energy loss of charged particles in matter. In S. FLÜGGE (ed.), Handbuch der Physik, **34**:193–217. Berlin: Springer Verlag.

WILKINSON, D. H. 1950. Ionization Chambers and Counters. Cambridge: Cambridge University Press. Pp. 266.

YAGODA, H. J. 1949. Radioactive Measurements with Nuclear Emulsions. New York: John Wiley & Sons. Pp. 356.

(Numbers in parentheses refer to chapter authored by discussant.)

In response to a question, Cherry (25) stated that his comments about the detection of α particles with nuclear emulsions were especially true in the measurement of different α particle energies. Even where the α emitters are buried in the emulsion, the resolution of different α energies would be much inferior to some of the chamber techniques. The main advantage of the emulsions is their great sensitivity to very minute amounts of radioactivity.

Spiers (55) inquired of Cherry (25) about the treatment of chance pairs in the pair method of α analysis. Cherry (25) replied that these chance pairs, in which two α particles come from separate nuclei, can be shown to have practically no effect because the rate of pairs of α particles from the same nucleus goes up at about the same rate as does the rate of chance pairs. Turner and co-workers have discussed the problem of chance pairs in the *British Journal of Radiology* and in *Geochimica et Cosmochimica Acta*.

Commenting on the remarks of Hill (24) regarding polonium-210 in marine organisms, Miyake (11) stated that his group had found marked increases of radium and uranium with depth in the ocean. Miyake (11) asked Hill (24) whether there were any data on the uranium and radium content of marine organisms. Hill (24) replied that he had none, but the unusually high α activity of some marine organisms is an interesting subject deserving of more attention.

Gold (22) asked Hill (24) to discuss the effect of sample thickness of the shape of α spectra, and Hill (24) replied that this was a difficult question to answer quantitatively. The question is discussed by Chudacek in the paper cited by Cherry (25). For practical purposes, the important effect is the development of a low-energy tail around each peak, and the shape of this tail is dependent on how smooth and homogeneous the source is made.

SERGE A. KORFF

26. Production of Neutrons by Cosmic Radiation

THE EARTH is continually being bombarded by cosmic radiation. This radiation is incident from all directions and arrives at all times. Some of it is sufficiently penetrating to reach sea level. The primary cosmic radiation is by definition a radiation originating outside the solar system—in the cosmos. The simple analysis is complicated by the fact that occasionally there are cases of solar injection of high-energy radiation, which is added to the cosmic radiation and which also occasionally penetrates to sea level.

Upon striking the atoms and molecules in the upper atmosphere, the primary radiation generates a vastly complicated secondary radiation, which contains all the entities known to physics and may, indeed, contain particles still undiscovered. Few primaries survive to reach as deep into the atmosphere as the tops of mountains. It is the μ mesons that are the principal agency penetrating to sea level.

Another part of the secondary radiation produced by the primaries is the neutron component. Since neutrons have a radioactive half-life of only about 1,000 seconds, even if they experienced large relativistic time dilations, they could not reach the earth from any point outside the solar system. They could reach the earth from the sun, and the problem of neutrons of solar origin will be discussed below. However, the majority of neutrons may be assumed to be of secondary origin, produced by the primaries within the atmosphere.

Properties of the Primary Cosmic Radiation

There are four principal properties of the primary radiation that are of concern for the purposes of the present discussion. These will be reviewed in turn.

SERGE A. KORFF is professor in the Department of Physics, New York University, University Heights, New York, New York.

Work supported in part by the National Science Foundation and in part by the Cambridge Air Force Laboratory, AFOSR, Hanscom Field, Bedford, Massachusetts.

427

The first important characteristic is the composition of the radiation. The primary radiation is, by number of entering particles, composed of about 85 per cent protons, 14 per cent α particles, and 1 per cent nuclei heavier than helium, from which all electrons have been removed. The exact fractions and the ratio of α's to protons are still being debated by some experts (Voyvodic, 1954), but, fortunately, for what follows this ratio is not crucial and the numbers given above may be taken as a good working guide, with the realization that some experts will make slightly different assignments. Measurements show that there are few electrons or γ rays present in the primary radiation. The most recent results indicated that perhaps 2 or 3 per cent of the radiation, at the bottom end of the spectrum, may be thought of as being γ ray or electron in type.

The second characteristic of this radiation is the energy distribution. The above-mentioned entities arrive with an energy per particle ranging from about 10^9 ev. up to perhaps 10^{19} ev. The upper end of the cosmic-ray spectrum (Greisen, 1956) is not very well known, and it will astonish few cosmic-ray physicists if in the next few years the limit is extended upward by an order or two of magnitude. The bottom energy limit is also confused, for at just about 10^9 ev. two effects happen. First, the primary radiation from great distances seems to diminish at lower energies, and, second, occasional bursts from the sun contain radiation from this value or, indeed, from a few times 10^9 ev. downward to a few kev. Further, the spectrum of the primary radiation is a power law of the type:

$$N(E)\,dE = AE^{-n}, \tag{1}$$

where $N(E)\,dE$ is the number of particles with energy E in the energy interval dE, and A and n are constants. The value of n is about 2, but, again, the exact number varies slightly according to different experts. Indeed, it is not known whether this number is constant over all the range of the spectrum, but it is a good approximate guide. The result of a spectrum such as this is that there are very few high-energy particles, and most of the energy is carried by the lower energy portion of the radiation.

The third important characteristic is the total energy, which can be obtained by multiplying the number of particles in each energy interval by the energy of that interval, that is, by integrating the spectrum and measuring the constants. This has been done by several observers, and the results may be summarized by saying that the total energy carried by the cosmic radiation is roughly equal to the total energy of starlight. In units of ergs per cm.^{2} per second, we find that about an average of 1, or slightly less, particle per cm.^{2} per sec. of an average energy of 2×10^9 ev. arrives at the top of the atmosphere. Thus the arriving flux has an energy of about 3×10^{-3} erg/cm²/sec or an energy density, dividing by c, of about 10^{-13} ergs per cc.

This total energy is incident from all directions. However, owing to the interaction between the earth's magnetic field and the charged incoming particles, there is a latitude effect. The net result of this effect is that the particles with the lowest energies reach the earth only in the vicinity of the geomagnetic poles, whereas those with energy in excess of about 10^{11} ev. can reach the earth anywhere. Because the low-energy particles are much more numerous, most of the cosmic-ray energy will be incident and most of the secondary production will take place in polar rather than equatorial regions. However, owing to the distribution of the resultant radioactive products by atmospheric circulation and winds, there is no observable latitude effect in the radiocarbon. This point has been carefully established by Libby (1952).

The fourth feature to be reviewed here is the constancy of the incoming radiation. Whereas actual radiation measurements of radiation intensities have existed only for the past thirty years or so, analysis of the radiocarbon content of archeological and paleontological specimens has shown that the cosmic-ray intensity has not altered by any great amount in the last 40,000 years, and probably for very much longer. The latter inference is drawn from studies of the longer-period products, such as beryllium-10, which are formed in very small amounts as spallation products in the atmosphere. We conclude, therefore, that to a first approximation we may regard the cosmic radiation as constant in amount. It is true that there appears to be a fluctuation (Dorman, 1957) of the order of a few per cent that shows an 11-year trend in phase with sunspot activity. On a long-term average, this disappears in integrating the total amount.

Short-time fluctuations (Dorman, 1957) occur that are quite spectacular. In the event of November, 1960, there were two increases at sea level of the order of 100 per cent and 85 per cent in the intensity. (See Fig. 4.) In the February 23, 1956, event the intensity at middle latitudes rose by a factor of 7 for a few hours, then returned to normal. However, integrating over a year, the contribution of these large events is still small.

NEUTRON EQUILIBRIUM IN THE ATMOSPHERE

In a consideration of the production of neutrons and their equilibrium in the atmosphere, it is realized at once that a long-term secular equilibrium must exist. In any such equilibrium the rate of production must equal the rate of disappearance, and in turn these two quantities are related to the equilibrium density and mean lifetime through

$$- dn/dt = dn/dt \qquad (2)$$

$$= \rho/t = q , \qquad (3)$$

where we designate the number of neutrons by n, the rate of production being the time derivative $(d/dt)n$, by t the mean lifetime, by ρ the density-equilibrium density of neutrons per cc., and define the rate of production as q for simplicity.

It will be recognized at once that the rate of production q is the sum of the number of neutrons produced in unit time by all processes and that likewise the rate of disappearance is the sum over all processes. If minor effects are neglected, one can state that the majority of the neutrons are produced in the atmosphere by the so-called "nucleonic cascade," in which the incoming particle produces secondary nucleons, each of which in turn has energy enough to produce still further nucleons. In nitrogen or oxygen the supply of protons and neutrons per nucleus is equal, so one would have, to a first approximation, a nucleonic cascade consisting of roughly equal numbers of protons and neutrons. The resulting neutrons are of all energies from an upper limit close to that of the incident primary particle, say a few or a few tens bev., down to about one or two mev., which is about the minimum energy with which a neutron can escape from a nucleus. Perhaps a majority of the neutrons will be produced as a result of the so-called "evaporation stars," which represent the terminal process in the nucleonic cascade, since at lower energies the process converges rapidly to zero. These neutrons will initially have energies of a few mev.

These neutrons are then slowed down by collisions with the nitrogen and oxygen nuclei in the atmosphere. The lowest level in the nitrogen nucleus being around 2.3 mev. and that for oxygen being 6.05 mev., according to the tables (National Research Council, 1962), any neutrons above 2.3 mev. will be rapidly slowed by inelastic collisions, while below this energy the neutrons will be slowed by elastic collisions. The mass of nitrogen being 14 times that of the neutron, the slowing-down process is not efficient, and many collisions are required for moderation of the neutron. The elastic scattering cross-sections for nitrogen are known (AEC, 1952) to be of the order of 10 barns, that of oxygen being smaller. The free path in S.T.P. air is of the order of $l = 1/2 \, N\sigma$, where the free path is designated by l, N is the number of atoms per cc. (the Loschmidt number, 2.7×10^{19} atoms or molecules per cc. at S.T.P.), and σ is the cross-section. The factor 2 occurs because the nitrogen (and oxygen) molecules each have 2 nuclei. The free path is then about 18 meters. A fairly slow neutron will take about 10^{-2} seconds to go such a distance. Thus the lifetime of a neutron in the atmosphere, during which time it is being moderated, is less than a second. No appreciable number of neutrons will be lost by decay.

During and after moderation, neutrons will be captured. Since the cross-sections are larger for nitrogen than for oxygen, and since there is four times as much nitrogen present in the atmosphere, only the nitrogen capture will be considered in this brief review. Examination of the cross-section curves (AEC, 1952) shows that the nitrogen nucleus has some large resonances in the general region of 0.4 mev. and a $1/v$ capture at low energies. One would therefore expect that some neutrons will be captured while being slowed down through the energy interval between 1 and 0.3 mev., and that those surviving this transition will be further moderated and then captured by the $1/v$ process when slow. The $1/v$ capture is essentially an (n,p) reaction, leading to the formation of radiocarbon, while the resonances lead both to the formation of radiocarbon and to tritium by an (n,T) process. Eventually, the carbon-14 β-decays back to nitrogen-14, so there is no new isotope built up on a cumulative basis. The product nucleus of the (n,T) reaction is likewise unidentifiable, since it merely constitutes a small addition to the already abundant carbon-12 in the atmosphere.

A fraction of those neutrons which are formed in the upper atmosphere will find themselves going upward, and, if they are within a mean free path of the outside, they may escape. As seen from outside, the earth will appear to be a source of neutrons, just as the target of an accelerating machine does. The neutrons that travel upward and escape are called "albedo neutrons," although the definition of the term "albedo" is not the same as that used by astronomers, referring, for example, to a planetary atmosphere. The escape velocity for neutrons, to permit them to escape from the earth's gravity, corresponds to a neutron energy of about 0.66 ev., so the low-energy neutrons will experience gravitational trapping, while those of larger energy will move outward and escape into space where they will eventually β-decay. It is not necessary to embark here upon a review of various theories and computations of the numbers of such albedo neutrons, which decay within the Van Allen radiation belts, for this subject has been extensively treated elsewhere. It is, however, pertinent to note that the probability P of any neutron decaying in a given cc. is

$$P(\mathrm{cc}) = \exp - t/T , \qquad (4)$$

where t is the length of time the neutrons spend in this cc., that is, reciprocally related to the velocity, and T is the neutron lifetime. For a neutron traveling at 10^{10} cm/sec, T being 1,000 seconds, P (cc.) is about 10^{-13}. This number, multiplied by the flux, is then the rate of production of charged particles, protons, and electrons.

The production and disappearance processes and their relation to the equilibrium can be summarized through Figure 1, in which are drawn the various principal processes that are operative. The diagram shows the equilibrium relationships. It also shows some of the properties of the equilibrium that have caused confusion and led to misunderstanding. These include the fact that there are two sources of radiocarbon, the slow $1/v$ capture and the fast-resonance capture. Also, the difficulty of fitting the tritium and helium-3 measurements is apparent, since some tritons are produced directly in the nucleonic cascade, in the evaporation stars, as also are some helium-3 nuclei.

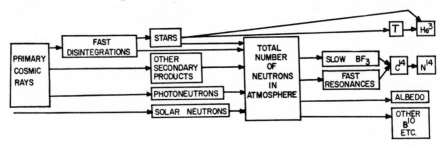

Fig. 1.—Neutron balance in the atmosphere. Diagram showing relationships between principal processes in the production and disappearance of neutrons.

Evaluation of the Various Measurements

Consider first the measurement of slow neutrons by boron trifluoride detectors. This was historically the first experimental system used for the quantitative study of cosmic-ray neutrons and has been the system from which the vast majority of the present data has been obtained. The boron trifluoride detectors operate using the (n,a) reaction in the boron-10 nucleus, measurement being made of the emerging a particle. The reaction is in the $1/v$ category. As was shown some twenty-five years ago by Bethe, Korff, and Placzek (1940), the counting rate of the boron trifluoride counters determines the density of neutrons, or the rate of production. For the density of neutrons the relation is

$$\rho = n/PVL\sigma_B v_B , \tag{5}$$

where n is the number of counts per sec. observed in a counter containing pure boron-10 trifluoride (i.e., corrected for whatever isotope ratio the gas used in the detector may have), P is the pressure in the counter in atmospheres, V the volume in cc., L the Loschmidt number, σ_B the cross-section of the boron-10 nucleus at velocity v_B. Inserting constants of nature, the relationship generally used is

$$\rho = 4.43 \times 10^{-5}(n/PV). \tag{6}$$

Similarly, to express the rate of production, q, and to relate this to the observed counting rate, n, one has

$$q = n(\sigma_A/\sigma_D)(780/PV), \tag{7}$$

where the cross-sections of the absorber, A, and the detector, D, are indicated by the sigmas with the appropriate subscripts. Inserting constants gives

$$q = 0.493(n/PV). \tag{8}$$

Next to be mentioned are the dimensions for the density and for the rate of production. In the foregoing the density has been defined as the number of neutrons per cc., and the rate of production as the number per cc. per second. If desired, instead, the rate of production can equally logically be expressed in units of per gram per second, and the density in units of neutrons per gram of atmosphere, a quantity that would not involve the variation of atmospheric density with altitude.

The actual boron trifluoride measurements give either density or rates of production in the atmosphere. If the measured values are plotted against depth in the atmosphere, one obtains a curve such as that shown in Figure 2. It will at once be noted that the curve has a maximum at roughly 0.1 atmosphere below the top, at which level the number of neutrons is a maximum. This point is often called the Pfotzer maximum, after the similar maximum in the intensity of the ionizing component of the cosmic-ray intensity. At various latitudes the curve exhibits a strong latitude dependence, and the family of curves all over the world is shown in Figure 2. The total number of neutrons produced in the entire atmosphere is obtained by integrating the area under the family in Figure 2. The family of curves in Figure 2 is a composite, in which Soberman (1956), of the New York University cosmic-ray group, put together all available data from this group and from other observers.

In deriving equations (5) and (7), it is assumed that both the detector and the absorber operate in the region of the $1/v$ law. The density of neutrons is therefore that of neutrons that have survived slowing-down processes and have entered the $1/v$ velocity interval. For these detectors, that means the region in which the energy is below

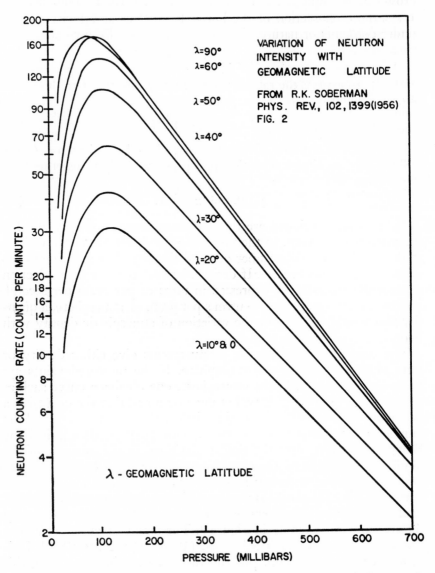

Fig. 2.—Neutron production throughout the atmosphere at various latitudes over the earth's surface. Observations of slow neutrons made with boron trifluoride counters flown in balloon experiments.

about 1 kev., since at about this energy the absorption and scattering cross-sections are equal, and at higher energies processes other than $1/v$ absorption dominate. Similarly, the rate of production must be understood to mean the rate of production of neutrons that survive slowing-down and get to the $1/v$ domain.

As indicated above, the assumption is made that the neutrons survive capture during slowing-down. The total number of neutrons that go into radiocarbon is therefore those which this experiment determined plus those which are captured at higher energies or by one of the resonances which also form radiocarbon.

The specific numbers that the observations of the New York University group (Soberman, 1956) give at the altitude of the Pfotzer maximum and in latitudes corresponding to the knee of the latitude-effect curve are the following. The neutron density is measured to be 6.5×10^{-7} neutrons produced at this elevation per cc/sec. This number implies that the average lifetime of a neutron in the atmosphere at this elevation is 0.7 sec., in agreement with the previous figure.

As has been pointed out, the figure for the rate of production cited is in units of per cc/sec. Since a level in the atmosphere at which the density is about 10^{-4} gm/cc is being dealt with, one would obtain, by converting units, a rate of production of about 9.3×10^{-3} neutrons per gram and per second.

The number of neutrons captured while being slowed down is given by

$$\exp\left[([3A + 2]/6) \int_{E_f}^{E_o} (\sigma_o/\sigma_t)(dE/E) \right], \tag{9}$$

where the cross-sections are designated by subscripts o for capture and t for total, the limits of integration are from an upper limit of around 0.4 mev., to a lower limit of 0.4 ev. (the cadmium cutoff), and the fraction preceding the integral is the fractional amount of energy lost per average collision, A being the atomic weight, in this case 14. This integral was derived by Flügge (1946) in his chapter in Heisenberg's book. Inserting numbers, the entire expression is found to be 1.77. Further estimating that about 20 per cent of the originally produced neutrons is captured in the resonances while at energies above 0.4 mev., one finally arrives at a factor 2.12, which is a multiplying factor for the number of neutrons measured with boron trifluoride counters, to convert them into the number of neutrons originally produced that will, after slowing down, give the number measured.

As mentioned before, by integrating the total number of neutrons produced at all altitudes, and at all latitudes, one can determine a world average of neutron production per sq. cm. column through the atmosphere and per second. This was done by Soberman (1956) and came out as about 1.1 neut/col sec. This number is that of the slow neutrons. Adding in the neutrons captured while still fast by multiplying by the factor 2.12 derived in the previous paragraph, one gets about 2.2 neutrons per sq. cm. column per sec. Now it is also possible to determine the amount of radiocarbon, which has been done by Libby (1952). He finds that the world totals correspond to rates of production of around 2.1 neutrons per sq. cm. column per sec. It is seen, therefore, referring to Figure 1, that the box representing the slow neutrons is measured with boron trifluoride counters, the radiocarbon box is measured by Libby, and the estimates for the box marked "fast neutrons," calculated from equation (9), agree reasonably well with the two observed quantities.

Next to be discussed is the tritium in nature. This substance has been identified by Grosse *et al.* (1951) and others and is today also used as a geophysical time measuring system. Here the simple determination of rate of production is complicated by the fact that not all the tritium is produced by cosmic-ray neutrons. Some tritons emerge directly from evaporation stars at the low-energy end of the nucleonic cascade. Further, it is not possible to infer the rate of production of tritium directly from the measurements of helium-3, for some helium-3 nuclei also emerge directly from the stars. Thus in the case of tritium, one knows roughly the amount in equilibrium at the present moment with a rate of production of the order of 0.2–0.5 per sq. cm. column per sec. This, in turn, is due to the sum of two processes, neither of which is well known, so the estimates are subject to some uncertainty. Very roughly, perhaps about 0.3–0.4 of the tritium comes from neutrons, and the remainder comes from stars.

As was pointed out some time ago by Korff (1956), when a process leads to the formation of an identifiable stable nucleus, the amount, N, of this substance produced over a time interval is given by

$$N = \int_{t_1}^{t_0} q\, dt,\qquad (10)$$

where q is the rate of production of this substance and the limits of integration cover the time interval under discussion. One limit is often the present time, and the other is that time in the past at which the process started, as far as the sample under study is concerned. In the simple case when q is constant, N is merely q times the time interval. This method has been applied to the helium-3 production. To give a numerical example, if one helium-3 atom is produced per cc. per second, then, if q stayed constant or were an average value over a period of

3×10^9 years, or 10^{17} seconds, there would be 10^{17} helium-3 atoms/cc in the sample. Since most substances that are solid at room temperature contain on the order of 3×10^{22} atoms/cc, helium-3 should be present in solids in the amount of a few parts per million. This is a maximum amount, for usually the value of q will be less, and seldom will the time interval be as long. Also, it assumes that no helium-3 escapes. Measurement of helium-3 in meteorites, which have been exposed to cosmic-ray bombardment in outer space, yields numbers smaller than those shown above.

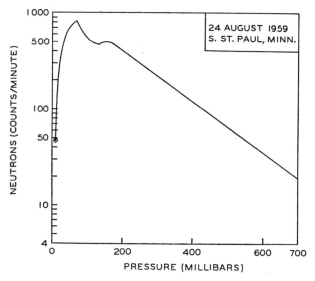

Fig. 3.—Neutron production at high balloon altitudes in the atmosphere. Neutron counting rate at 140,000 feet is determined with some accuracy, since flight floated for many hours.

Finally, there is the problem of the albedo neutrons, which are the neutrons escaping from the top of the earth's atmosphere. The high-altitude balloon flight measurement by Haymes and Korff (1960) shows clearly that the number of slow neutrons escaping upward is small. The balloon data number, extrapolated to the top of the atmosphere, is about 1 per cent of that at the Pfotzer maximum. (See Fig. 3.) If this number is translated to a flux by supposing that the velocity of the typical slow neutron corresponded to an energy of 0.2 ev., then the outward flux of slow neutrons is of the order of 0.03 neutrons per sq. cm. per sec. On the other hand, the flux of fast neutrons may be considerably greater. This problem is at present under study. A preliminary flight was made in November, 1962, by R. Mendell of the New York University cosmic-ray group. This flight determined the flux at altitudes up to 93,000 feet and was not high enough to permit a good evaluation of the number at the top of the atmosphere. Further flights are under preparation, and it is hoped that there will be a good number for this quantity before long.

Neutrons have also been measured in rocket flights above the earth's atmosphere. A model from which calculations can be made was proposed by Hess and his colleagues (1961), and a computed

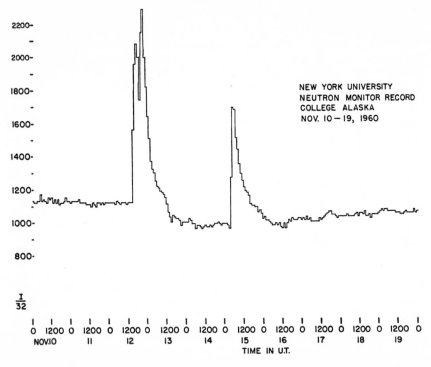

FIG. 4.—Neutrons observed at sea level, showing large fluctuations at times of solar disturbance, injection by flare of solar neutrons. Observations in November, 1960, by New York University neutron monitor at College, Alaska.

amount therefore exists. The rocket measurements do not agree with the computed amounts unless certain rather uncertain corrections are made. Therefore, at present there is no good measurement of the neutron flux above the atmosphere. The rocket flights all appear to show evidence for considerable amounts of local production of neutrons in the body of the rocket. Thus, for example, the rocket measurement by Reidy, Haymes, and Korff (1962) gave observed numbers that were higher than the computed values by a factor of 6 or so, and the flights of other observers (Bame *et al.*, 1963) are similarly high by varying amounts. The determination of really good values therefore still remains to be done.

It will be recognized at once from the foregoing that any space ship will, when in orbit, be permeated by a flux of neutrons produced by the cosmic rays that impinge upon it. Since the geometry and chemical composition vary markedly from ship to ship, it is impossible to compute precisely the levels to be expected. However, one can state that it is highly improbable that the number of neutrons could exceed that at the Pfotzer maximum, for at this atmospheric depth the incoming radiation is in equilibrium with its neutron secondaries. Therefore, the numbers cited in this paper for the Pfotzer maximum may be taken as a guide as to the upper limits.

REFERENCES

AEC NEUTRON CROSS SECTION ADVISORY GROUP. Neutron Cross Sections. 1952. AEC Neutron Cross Section Advisory Group, Rpt. AECU-2040.

ANDERSON, E. C., and W. F. LIBBY. 1951. World-wide distribution of natural radiocarbon. Phys. Rev., 81:64–69.

BAME, S. J., J. P. CONNER, F. B. BRUMLEY, R. L. HOSTETLER, and A. C. GREEN. 1963. Neutron flux and energy spectrum above the atmosphere. J. Geophys. Res., 68:1221-28.

BEGEMANN, F., and W. F. LIBBY. 1956. Continental Water Balance, Ground Water Inventory and Storage Times, Surface Ocean Mixing Rates and World-Wide Water Circulation Patterns from Cosmic Ray and Bomb Tritium. Univ. Chicago, Enrico Fermi Inst. for Nuclear Studies, Rpt. OSR-TN-56-561 (AD-110381). Pp. 38.

BETHE, H. A., S. A. KORFF, and G. PLACZEK. 1940. On the interpretation of neutron measurements in cosmic radiation. Phys. Rev., 57:573-87.

CRAIG, H. 1957. The natural distribution of radiocarbon and the exchange time of carbon dioxide between atmosphere and sea. Tellus, 9:1-17.

CRAIG, H., and D. LAL. 1961. The production rate of natural tritium. Tellus, 13:85–105.

DORMAN, L. I. 1957. Cosmic ray variations. Transl. from a publication of the State Publishing House for Technical and Theoretical Literature, Moscow. Pp. 736.

FLÜGGE, S. 1946. On the excitation of neutrons by cosmic rays and their distribution in the atmosphere. *In* W. HEISENBERG (ed.), Cosmic Radiation, pp. 144-58. New York: Dover Publishers.

FREDEN, S. C., and R. S. WHITE. 1962. Trapped proton and cosmic-ray albedo neutron fluxes. J. Geophys. Res., 67:25–29.

GREISEN, K. 1956. The extensive air showers. *In* J. G. WILSON (ed.), Progress in Cosmic Ray Physics, 3:1–141. Amsterdam: North-Holland Pub. Co.

GROSSE, A. V., W. M. JOHNSTON, R. L. WOLFGANG, and W. F. LIBBY. 1951. Tritium in nature. Science, 113:1–2.

HAYMES, R. C., and S. A. KORFF. 1960. Slow-neutron intensity at high balloon altitudes. Phys. Rev., **120**:1460–62.

HESS, W. N., E. H. CANFIELD, and R. E. LINGENFELTER. 1961. Cosmic-ray neutron demography. J. Geophys. Res., **66**:665–77.

HESS, W. N., H. W. PATTERSON, R. WALLACE, and E. L. CHUPP. 1959. Cosmic-ray neutron energy spectrum. Phys. Rev., **116**:445–57.

HESS, W. N., and A. G. STARNES. 1960. Measurement of the neutron flux in space. Phys. Rev. Letters, **5**:48–50.

KORFF, S. A. 1956. The effects of cosmic rays on the terrestrial isotope distribution. Ann. New York Acad. Sci., **67**:35–54.

LIBBY, W. F. 1952. Radiocarbon Dating. Chicago: University of Chicago Press. Pp. 124.

LORD, J. J. 1951. The altitude and latitude variation in the rate of occurrence of nuclear disintegrations produced in the stratosphere by cosmic rays. Phys. Rev., **81**:901–9.

MARTIN, J. P., L. WITTEN, and L. KATZ. 1963. Measurement of cosmic ray albedo neutron flux above the atmosphere. J. Geophys. Res., **68**: 2613–18.

NATIONAL RESEARCH COUNCIL. 1962. The Energy Levels of Light Nuclei. National Academy of Sciences, National Research Council, Nuclear Data Group. Pp. 339.

REIDY, W. P., R. C. HAYMES, and S. A. KORFF. 1962. A measurement of slow cosmic-ray neutrons up to 200 kilometers. J. Geophys. Res., **67**: 459–65.

SMITH, R. V., L. F. CHASE, W. L. IMHOFF, J. B. REAGAN, and M. WALT. 1961. Radiation measurements with balloons. Holloman AFB Rpt. ARL-TDR-62-2. Pp. 57.

SOBERMAN, R. K. 1956. High-altitude cosmic-ray neutron intensity variations. Phys. Rev., **102**:1399–1409.

VOYVODIC, L. 1954. Particle identification with photographic emulsions, and related problems. *In* J. G. WILSON (ed.), Progress in Cosmic Ray Physics, **2**:217–88. Amsterdam:North-Holland Pub. Co.

WHYTE, G. N. 1951. Cosmic-ray bursts and the nucleonic cascade. Phys. Rev., **82**:204–8.

JACOB KASTNER, B. G. OLTMAN, AND
L. D. MARINELLI

27. Progress Report on Flux and Spectrum Measurements of the Cosmic-Ray Neutron Background

PIONEERING MEASUREMENTS have been made by the Berkeley group (Patterson *et al.*, 1959) of the cosmic-ray-produced neutron flux as a function of altitude. This has aroused speculation that the neutron dose being received by man per year may be larger and a more important fraction of the total natural dose than had previously been supposed. Studies of the fast-neutron background (1–10 mev.) begun at Argonne centered on the "twin" scintillation fast-neutron spectrometer first described by Berlman and Marinelli (1956) and later improved by Grismore *et al.* (1961).

More recently, pulse-shape discrimination circuits have been reported (Owen, 1958; Forte, 1958; Broek and Anderson, 1960; Daehnick and Sherr, 1961) and have even become commercially available. These have proved to be far superior to the twin system for the unequivocal determination of the small but significant numbers of fast neutrons with which we are continuously being bombarded.

Our hope in developing these techniques is based on the use of large hydrogen-rich organic scintillators, most probably liquid, to provide the necessary high-efficiency and tissue-like characteristics. Furthermore, we wanted portability for eventual field studies.

The following is a presentation of the details of the apparatus and our progress to date.

JACOB KASTNER is associate physicist, B. G. OLTMAN is research technician, and L. D. MARINELLI is division director, in the Radiological Physics Division, Argonne National Laboratory, Argonne, Illinois.

Work done under the auspices of the U.S. Atomic Energy Commission.

441

PLATE I*a*

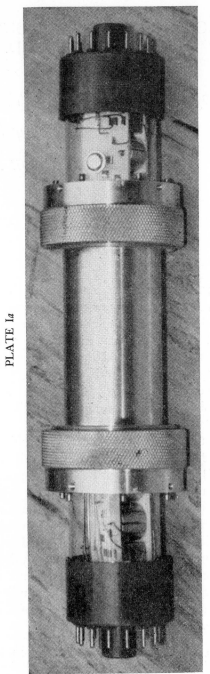

Twin scintillation detector for fast neutrons

PLATE I*b*

Twin scintillation detector for fast neutrons

TWIN SCINTILLATOR

The maximum response of the twin type of scintillation fast-neutron detector (see Pl. I) is obtained when one of the solutions has a relatively low proton-electron ratio and the other is rich in protons. The combination that we now employ is xylene and hexafluorobenzene (C_6F_6). The latter is a rare compound that only recently became available as a special product of 97 per cent purity from the Imperial Smelting Company in England. The pulse height of the raw material relative to xylene is about 25 per cent, which is quite good enough.

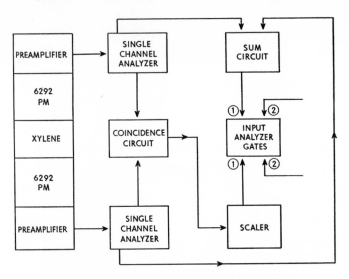

FIG. 1.—Block diagram of one-half of twin scintillator

Both solutions are nitrogen-quenched to improve the light yield. The cells are open-ended teflon cylinders with outer $\frac{1}{16}$-inch-thick aluminum walls. The two multiplier phototubes that close a cell are sealed by O-rings. The volumes are about 80 cm.[3].

Figure 1 shows the electronic system for pulse-height analysis of the two cells. Each of the solutions is viewed by a pair of multiplier phototubes operated in coincidence. The outputs from a pair are summed, amplified, and then fed to the input of a 400-channel pulse-height analyzer. The spectra from the two solutions are recorded in the opposing halves of the analyzer memory, thus permitting simultaneous recording of the data from both, neglecting coincidence between cells.

Cobalt-60 and cesium-137 were used to provide a γ-ray energy calibration of the luminescence from the two liquids. The proton-recoil calibration of the xylene was carried out using a Cockroft-Walton $\frac{1}{4}$-mev. accelerator to provide 2.5- and 15-mev. neutrons from D-D and D-T reactions (Fig. 2). The γ-ray-induced distribution in C_6F_6 must be corrected for its higher electron density before being subtracted from that of the xylene to get the net proton-recoil distribution.

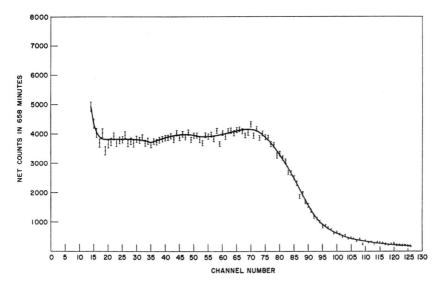

F<small>IG</small>. 2.—Proton recoil spectrum obtained from 14.65-mev. D-T neutrons with the "Twin" Scintillation Fast-Neutron Spectrometer.

P<small>ULSE</small>-S<small>HAPE</small> D<small>ISCRIMINATION</small> C<small>IRCUITS</small>

We have tested two different electronic circuits designed to discriminate between neutron- and γ-ray-produced scintillations. Both circuits utilize an RCA type 6810A 14-stage multiplier phototube. Stilbene has been employed with the circuit of Broek and Anderson (1960), wherein the discrimination pulses that gate the pulse-height analyzer to admit neutron events are taken from the last dynode of the multiplier phototube. The latter is operated with only a small accelerating potential between the dynode and the anode so that space-charge limitation of the anode current occurs with large pulses. Under the proper conditions of operation, proton recoil scintillations will produce positive pulses at the dynode that will gate the pulse-height analyzer. However, electron scintillations, in which the high-intensity, fast component of decay is of considerably greater importance, will produce there only negative pulses due to space charge repulsion and will thus be eliminated.

The second circuit is the Daehnick and Sherr (1961) type, commercially available.* The discrimination pulses that gate the pulse-height analysis of neutron events are formed by adding negative pulses from the multiplier phototube anode to the corresponding positive pulses from the thirteenth dynode. In operation, the relative amplitudes of the two that are to be combined are adjusted so that, after the addition and subsequent charge integration, the resultant pulses for electrons are as small as possible. These pulses, then, will not gate the pulse-height analyzer. However, since proton recoil

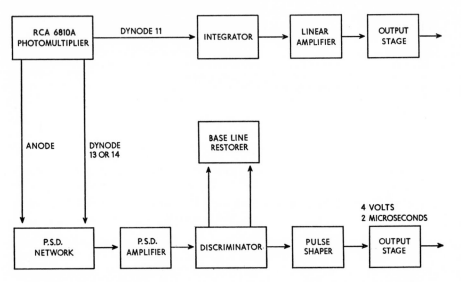

Fig. 3.—Block diagram showing circuit functions in pulse-shape discrimination probe

scintillations contain a relatively larger proportion of slow component than do electron scintillations, they will produce positive resultant pulses that will gate the pulse-height analyzer. The pulses to be analyzed are taken directly from a third element of the phototube.

Results and Observations

The twin scintillator system proved to be an excellent system for fast-neutron spectroscopy when the ratio of neutrons to γ's is not too small. It proved to be disappointing, however, as a tool for background studies simply because of the relatively large errors produced by internal radioactive contamination of the system. Numbers that made sense could be developed only after many subsidiary operations and manipulation of the data.

* Nuclear Enterprise Ltd., 1750 Pembing Highway, Winnipeg, Manitoba, Canada.

The pulse-shape discrimination technique, on the other hand, has proved most suitable to the particular problem of determining the fast-neutron component of the natural radiation environment.

Table 1 shows the data that we have obtained with 20 cm.^{3} of stilbene over a period of a month in our laboratory. These data are compared with data interpolated from a graph presented by the Berkeley group (Hess *et al.*, 1959). We also have begun to obtain data with 100 cm.^{3} of NE213 liquid scintillator and the commercial probe. The improvement in total count rate gives us good reason to be optimistic about obtaining greater precision in reasonable times.

We should point out the importance of calibrating these organic scintillators for proton energy versus pulse height (Brodsky, 1961)

TABLE 1

FAST-NEUTRON ENVIRONMENTAL BACKGROUND

Energy (mev.)	UCRL-8268 (n/cm² sec mev)	Argonne Data* (n/cm² sec mev)
0.75	0.0110	0.0101 ±0.0001†
1.00	0.0075	0.0094
1.25	0.0051	0.0054
1.50	0.0039	0.0039
1.75	0.0031	0.0021
2.00	0.0025	0.0018
2.25	0.0021	0.0017
2.75	0.0018	0.0016
3.00	0.0013	0.0014 ±0.0001

* Total time, 2.5 × 10⁶ sec.
† Standard error based on counting statistics only.

and also the need for careful adjustment of the pulse-shape circuit to give uniform discrimination efficiency over the energy range of interest (Grismore *et al.*, 1961). We have not as yet established the upper energy limits for pulse-shape discrimination.

At the moment, we have no explanation for the sharper rise at 1 mev. We plan to push the discrimination electronics to lower energies in any case, and this should serve to determine whether the rise is instrumental, a characteristic of our environment in the Argonne building, or a real description of spectral composition of cosmic-ray neutrons.

NOTE ADDED IN PROOF: Preliminary results from the experiments with a 1,600 cm.^{3} NE213 liquid scintillator give differential neutron flux values near sea level of approximately 40 per cent of those of Hess *et al.* (1959) between 1 and 8 mev. This is consistent with the total flux measurement of Kent (1963), giving about one-quarter of the UCRL total. Measurements are being continued in an effort to improve the precision of the data.

REFERENCES

BERLMAN, I. B., R. GRISMORE, and B. G. OLTMAN. 1960. Improved "twin" scintillation fast neutron detector. Rev. Sci. Instr., 31:1198–1200.

BERLMAN, I. B., and L. D. MARINELLI. 1956. "Twin" scintillation fast neutron detector. Rev. Sci. Instr., 27:858–59.

BRODSKY, ALLEN. 1961. The pulse distributions in stilbene produced by 1–15 MeV neutrons, corrected for end effect. Div. of Licensing and Regulations, Atomic Energy Commission, Rpt. TID-13075 (June). Pp. 68.

BROEK, H. W., and C. E. ANDERSON. 1960. The stilbene scintillation crystal as a spectrometer for continuous fast-neutron spectra. Rev. Sci. Instr., 31:1063–69.

DAEHNICK, W., and R. SHERR. 1961. Pulse shape discrimination in Stilbene scintillators. Rev. Sci. Instr., 32:666–70.

FORTE, M. 1958. Possibilities of discrimination between particles of different kinds by means of organic scintillator detectors. Proc. 2d United Nations Internat. Conf. on Peaceful Uses of Atomic Energy, 14: 300–304.

GRISMORE, R., O. J. STEINGRABER, B. G. OLTMAN, and L. D. MARINELLI. 1961. Improved apparatus for measurement of the cosmic-ray neutron background. Argonne Nat. Lab., Radiological Phys. Div. Semiannual Rpt., ANL-6297, pp. 40–44.

HESS, W. N., H. W. PATTERSON, R. WALLACE, and E. L. CHUPP. 1959. Cosmic-ray neutron energy spectrum. Phys. Rev., 116:445–57.

KENT, R. A. R. 1963. Cosmic-ray neutron measurements. Hanford Rpt. HW-SA-2870.

OWEN, R. B. 1958. The decay times of organic scintillators and their application to the discrimination between particles of differing specific ionization. IRE Trans. Nuc. Sci., NS-5:198–201.

PATTERSON, H. W., W. N. HESS, B. J. MOYER, and R. W. WALLACE. 1959. The flux and spectrum of cosmic-ray produced neutrons as a function of altitude. Health Physics, 2:69–72.

C. R. HILL AND D. S. WOODHEAD

28. *Tissue Dose Due to Neutrons of Cosmic-Ray Origin*

G REAT INTEREST is attached to the dose received by man from his natural radiation environment. One possible source of such dose is the neutrons that are produced near the top of the atmosphere as a result of collisions suffered by primary cosmic-ray particles. These neutrons possess very high energies, and they are known to penetrate to the earth's surface in measurable quantities.

Up to the present time, observations on these neutrons have been confined essentially to measurements of flux and flux spectra, with the majority of the detailed measurements being carried out at high altitudes.

In the present paper we discuss some of the difficulties that appear to exist in deriving estimates of tissue dose from flux measurements of this nature. We also describe an experimental program that has been undertaken with the ultimate object of making direct measurements of tissue dose due to cosmic-ray neutrons. Our primary intention is to stimulate discussion of a topic that is almost certainly the least explored aspect of the natural radiation environment; such conclusions as we have drawn are tentative only.

THE NEUTRON FLUX SPECTRUM AT SEA LEVEL

Patterson and Hess and their co-workers at Berkeley have recently measured the flux spectrum in the atmosphere of neutrons of cosmic-ray origin (Patterson *et al.*, 1959; Hess *et al.*, 1959). Their measurements show that an equilibrium condition exists at altitudes below about 12 km., in which the flux has a spectrum of constant shape and an intensity that is approximately halved by traversing 100 gm/cm^2 of air (equivalent to approximately 1 km. in altitude near sea level).

C. R. HILL, Ph.D., and D. S. WOODHEAD are in the Physics Department of the Institute of Cancer Research, Royal Cancer Hospital, Belmont, Surrey, England.

In their spectrum there is a peak at about 1 mev. due to evaporation neutrons resulting from collisions between very high energy neutrons and nuclei in the atmosphere. The authors have suggested that the biologically effective radiation dose arises largely from neutrons near this peak, and their calculation of its detailed energy dependence is reproduced in Figure 1 (in which we also show the theoretical response of the Hurst-type dosimeter, as discussed below). The magnitude of the total dose rate at sea level, in the absence of shielding,

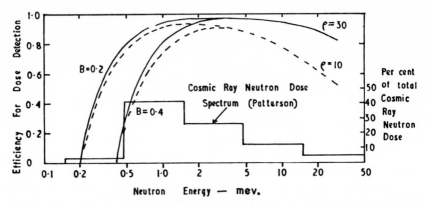

Fig. 1.—Comparison of Patterson's cosmic-ray neutron dose spectrum with theoretical response of the Hurst dosimeter. (B = bias setting in mev.; ρ = instrumental scale parameter.)

is estimated by Patterson and Hess (1959) to be 25 mrem. per year,[*] or about one-quarter of the total natural radiation background dose to man.

Neutron Dose in Finite Tissue Volumes

If we accept these data, which are probably the best of their kind available, it is necessary to recognize that a dose rate derived in this way from flux measurements in the atmosphere only properly applies in the case of a very small, isolated volume of tissue. As a first approach to the problem of estimating dose in a human body due to the neutron spectrum in the atmosphere described above, it is instructive to consider the simple case of an infinite slab of tissue (ignoring for the present any dose contribution arising from evaporation neutrons produced in the tissue). For this purpose it is necessary to take into account the directional properties of the flux in the atmosphere. The high energy component is clearly directional, but the evaporation component is more nearly isotropic, and, since this is thought to

[*] An RBE of 10 for heavy particles is assumed throughout the present paper.

contribute most of the dose, an assumption of isotropic flux appears to be appropriate as a first approximation.

Making these assumptions and using the results of Snyder's Monte Carlo calculations given in NBS Handbook 63 (National Bureau of Standards, 1957), we have calculated the dose rates at various depths in the infinite slab of tissue, with the results shown in Figure 2.

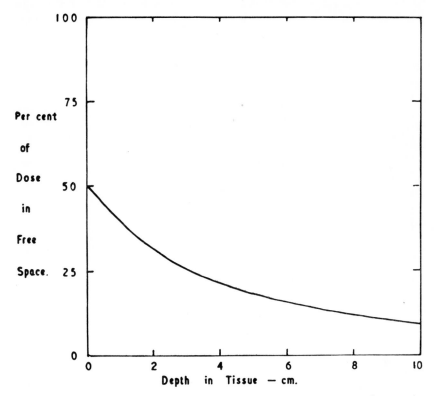

Fɪɢ. 2.—Dose in an infinite tissue slab due to isotropic cosmic-ray neutron flux entering the slab from the atmosphere. (Based on Snyder's Monte Carlo data and Patterson's neutron energy spectrum.)

The dose rate at the surface is half the free-space value, corresponding to the condition that the flux is in one direction only across the surface (the effect of backscatter is ignored). Within the slab, the dose rate falls off rapidly over distances of the order of a few centimeters, and it is thus clear that the infinite slab model is a useful first approximation to the case of a human being of mean diameter 20 cm.

The reduction in dose rate to about 10 per cent of the free-space value at a depth of 10 cm., in the absence of evaporation neutron production in the tissue, is to be compared with the reduction only to

94 per cent to be expected (according to Patterson and Hess) from interposition of 10 gm/cm² of air under equilibrium conditions for production of evaporation neutrons. It seems clear, therefore, that actual dose rates within a human being, or within any tissue volume having linear dimensions of more than a few centimeters, may be strongly dependent on the processes of generation and subsequent moderation of these secondary evaporation neutrons within the tissue. No attempt appears to have been made hitherto at the estimation of

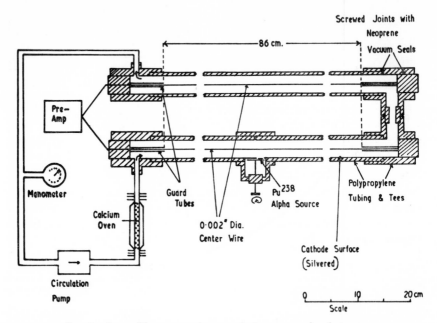

Fɪɢ. 3.—Large Hurst type, tissue-equivalent proportional counter

possible depth-dose variations arising in this way, and the dose contribution from the evaporation neutrons originating in the tissue has been ignored in the calculations made in the recent report of the United Nations Scientific Committee (1962). In fact, there do not yet appear to be adequate data available for this purpose on the cross-sections for production of evaporation neutrons under these conditions. The only available approach to the problem would therefore seem to be by way of the direct measurement of dose.

Experimental Approach to Dose Determination

As a possible approach to the direct measurement of cosmic-ray neutron dose and its variations, we have investigated the properties of the CH₂ wall–CH₂-filled proportional counter.

The application of tissue-equivalent proportional counters to fast-neutron dosimetry has been studied by Hurst (1954; Wagner and Hurst, 1958, 1959) and his co-workers at Oak Ridge and by Moyer (1952) at Berkeley. It has been shown in practice that proportional counters can be constructed and operated satisfactorily in which both the wall material and gas filling have the chemical composition $(CH_2)_n$, and it has also been shown theoretically that the first collision fast-neutron dose in $(CH_2)_n$ has essentially the same energy dependence as that in tissue (Wagner and Hurst, 1958). In principle, therefore, it is possible in this way to achieve a nearly perfect Bragg-Gray design of fast-neutron dosimeter.

In practice, in order to reduce the γ sensitivity of such an instrument to a satisfactory level, it is necessary to disregard all pulses of energy less than a certain value.* In this process a certain fraction of the neutron dose is inevitably lost, and it is important to know the magnitude of this loss. Transfer of energy to a hydrogenous medium from neutrons of energy up to 30 mev. is due principally to the recoil protons resulting from elastic collisions with hydrogen nuclei. We may conveniently consider in the following three groups the possible situations in which such recoil protons will contribute less than the bias energy B mev. to the filling gas of a counter:

1. Protons to which energy less than B is imparted by the collision.

2. Protons with initial energy greater than B, but which *either* start *or* finish their paths in the counter wall and therefore transfer less than B mev. to the gas.

3. Protons with initial energy greater than B, but which *both* start *and* finish their paths in the counter wall and therefore transfer less than B mev. to the gas.

Groups 1 and 2 have been considered by Ritchie (1959) for the case of a gap between two plane parallel slabs of hydrogenous material irradiated by a collimated beam of neutrons incident normally on the outer face of one slab. It appears that, except for very small counters, the fraction of dose lost is mainly determined by the ratio between B and neutron energy, and Ritchie's calculations may be taken as a reasonable approximation in the case of cylindrical counters in either collimated or isotropic flux.

The losses resulting from situation 3 do not appear to have been investigated hitherto: it is these, however, that influence the response of such counters at relatively high energies. In this case, in contrast

* It may sometimes be possible to distinguish between "gamma" and "neutron" pulses on the basis of their shape (Bennett, 1963), but improvement in response by this means can be expected only in the energy region below 1 mev., where protons have short ranges and high specific ionization.

to cases 1 and 2, the magnitude of the losses will be strongly dependent on the shape of the counter, and calculations appropriate to the case of an infinite cylinder are set out in an appendix to the present paper. A more rigorous derivation is understood to be in preparation by the National Bureau of Standards (Caswell *et al.*, 1963).

Comparison of the theoretical response of a Hurst counter, obtained in the manner described above, with Patterson's estimate of the energy dependence of cosmic-ray neutron dose can be made by reference to Figure 1. From this it will be seen that such an instrument may in principle be expected to record the total dose with an efficiency of between 75 and 80 per cent. In view of the considerable range of neutron energies encountered, it is unlikely that any other single device could be found that would be appreciably better in this respect.

CONSTRUCTION AND PERFORMANCE OF A LARGE HURST COUNTER

A large instrument of the Hurst type has recently been constructed in this laboratory. In order to achieve high sensitivity, it has a counting volume of 1.75 liters, made up of two parallel tubes, each effectively 86 cm. long and of 3.6 cm. internal diameter. The counter is filled with cyclopropane (C_3H_6) at atmospheric pressure and is constructed completely of commercially available, 6-mm.-wall polypropylene tubing and fittings, which are screwed together with neoprene "O" rings to form the vacuum system. An advantage of this method of construction is that it avoids the inclusion in the vacuum system of large solid-metal surfaces, which, in our experience, are liable to emanate a quantity of radon sufficient to produce an appreciable spurious response in high-sensitivity, heavy particle counters.

The anodes are 0.002-inch-diameter stainless-steel wires overlaid at each end by thin stainless-steel guard tubes that serve to define the counting volume. Cathodes were formed by chemically silvering the inner surface of the tubing to a thickness of not more than 50 micrograms per cm.^{2}, but this process is not to be recommended, since it has been found to give rise to serious gas contamination. In order to eliminate drift, it has been necessary in our present counter continuously to circulate the counting gas over metallic calcium heated to 350° C., which is an undesirable expedient in view of the possibility of releasing traces of radon into the counting gas. In operation, the counter is shielded by a $\frac{1}{16}$-inch aluminum cover, which has been found necessary to eliminate completely spurious pulses of electronic origin. A 100-channel pulse-height analyzer is used for recording the output spectrum.

Energy calibration of the counter and routine checking of its operation are effected by means of a 300 d.p.m. plutonium-238 α source that can be exposed through a small opening in the side of one of the counting tubes. Direct checks of the response of the counter to fast neutrons have been made using a calibrated 1-mc. actinium-227–α–beryllium sealed source, and these have shown agreement to within 30 per cent with dose rates calculated from Snyder's data (National Bureau of Standards, 1957). The neutron spectrum produced by such a source extends up to 14 mev., with a broad peak at about 5 mev., and thus largely overlaps that of the cosmic-ray neutrons.

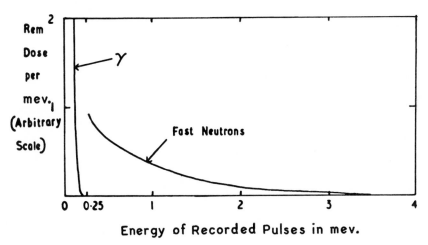

Fig. 4.—Response of counter to equal rem. doses of actinium-227–beryllium fast neutrons and cobalt-60 γ radiation.

Ability to discriminate between γ and neutron dose is illustrated in Figure 4. In this, the dose per mev. is plotted as a function of output pulse energy for equal rem. doses from the actinium-227–beryllium neutron source and a cobalt-60 γ source, respectively. A bias setting of 160 kev. provides a discrimination factor of 100:1 in favor of neutrons.

Since this instrument, when operated with an appropriate bias, is essentially responsive only to heavy particles, the main source of spurious response is likely to be internal α contamination. In order to determine the approximate shape of the output pulse spectrum that would arise from such contamination, a foil coated with a small quantity of thorium-228 was temporarily introduced into the system as a radon-220 (thoron) emanator. Radon-220 (thoron; 6.3 mev. α particles) was thus circulated with the counting gas and the α emitters polonium-216 (ThA; 6.8 mev.) and polonium-212 (ThC'; 6 and 8.8

mev.) deposited on the cathode, subsequently decaying with the 10.6-hour half-life of lead-212 (ThB) after the foil was removed. A spectrum obtained in this way and weighted to correspond to dose is shown in Figure 5. Included on the same diagram is a reproduction of the corresponding spectrum obtained from the actinium-227—beryllium neutron source and normalized to give equal areas under the two curves (and hence equal total dose).

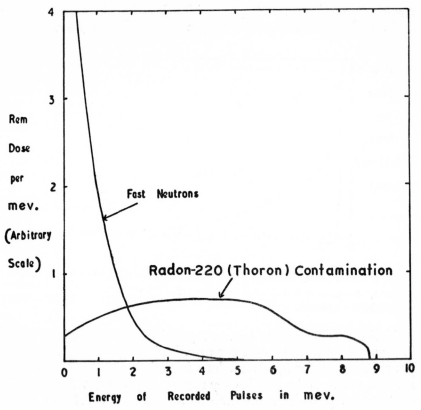

F_IG. 5.—Normalized energy spectra of actinium-227–beryllium fast neutrons and radon-220 (thoron) α contamination.

BACKGROUND MEASUREMENTS

In a preliminary attempt to measure cosmic-ray neutron dose with this instrument, output pulse spectra have been recorded under four different conditions:

1. In the basement of our laboratory near London, 105 meters above sea level, with an estimated 50 gm/cm² of concrete and masonry in the floors and roof above the counter.

2. In the same position but with the counter surrounded by approximately 10 cm. of paraffin wax.

3. In a horizontal gallery excavated from the chalk subsoil underneath the same room, providing about 250 gm/cm² of chalk shielding above the counter. According to Patterson's data, this might be expected to produce a reduction by a factor of 4 or 5 in the flux of high energy neutrons.

4. Under the same conditions as in 1, but with the cyclopropane (C_3H_6) counting gas replaced by argon.

In all four cases the spectrum recorded has been of the form shown in Figure 6. No significant decrease in counting rate was observed either when the counter was taken underground or when the counting gas was replaced by argon.

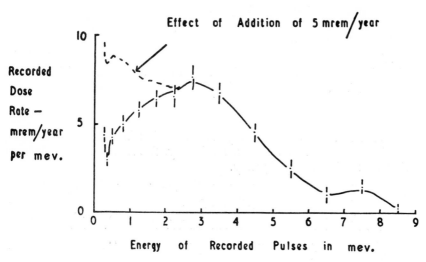

FIG. 6.—Background energy spectrum from large Hurst counter

These observations, together with the fact that the spectral shape is characteristic of α contamination rather than fast neutrons, appear to indicate that the recorded counts are due predominantly to α contamination. The actual counting rates are, however, sufficiently low that neutron dose rates of the magnitude predicted by Patterson would be expected to have a significant effect on spectral shape. In Figure 6 the broken line indicates the change of spectral shape due to superposition of a dose rate of 5 mrem. per year from actinium-227–beryllium neutrons. This compares with the figure of approximately 20 mrem. per year predicted from Patterson's data for the first of the experimental conditions given above.

Conclusion

This paper is intended primarily to discuss some of the apparent difficulties involved in estimating tissue dose to man due to cosmic-ray neutrons and to report the progress of a recently undertaken and still incomplete program of measurements. We have investigated the Hurst counter as a means of measuring both the absolute levels of cosmic-ray neutron dose to man and its possible variations. The instrument that we are at present using, although capable of further development, particularly with respect to the spurious response from internal α contamination, nevertheless is sufficiently sensitive to respond to the dose rates of the order of 25 mrem. per year that have been predicted. On the basis of the few measurements that we have been able to make so far, it appears that the actual dose rates may be appreciably less than that figure.

Acknowledgments

We are indebted to those of our colleagues who have helped and advised in various aspects of this work. In particular we wish to thank Professor W. V. Mayneord for his advice and encouragement.

References

BENNETT, E. F. 1963. Gamma ray discrimination in a proton recoil proportional counter. *In* Neutron Dosimetry, **2**:341–49. Vienna: International Atomic Energy Agency.

CASWELL, R. S., W. B. BEVERLY, and SPIEGEL, JR. 1963. Energy dependence of proportional-counter fast-neutron dosimeters. *In* Neutron Dosimetry, **2**:227–37. Vienna: International Atomic Energy Agency.

HESS, W. N., H. W. PATTERSON, and R. W. WALLACE. 1959. Cosmic-ray neutron energy spectrum. Phys. Rev., **116**:445–57.

HURST, G. S. 1954. An absolute tissue dosemeter for fast neutrons. Brit. J. Radiol., **27**:353–57.

MOYER, B. J. 1952. Survey methods for fast and high-energy neutrons. Nucleonics, **10**(5):14–19.

NATIONAL BUREAU OF STANDARDS. 1957. Protection against Neutron Radiation up to 30 Million Electron Volts. U.S. Dept. of Commerce, National Bureau of Standards Handbook 63.

PATTERSON, H. W., W. N. HESS, B. J. MOYER, and R. W. WALLACE. 1959. The flux and spectrum of cosmic-ray produced neutrons as a function of altitude. Health Physics, **2**:69–72.

RITCHIE, R. H. 1959. Calculations of energy loss under the bias in fast neutron dosimetry. Health Physics, **2**:73–76.

UNITED NATIONS, GENERAL ASSEMBLY OFFICIAL RECORDS. 1962. Report of the United Nations Scientific Committee on the Effects of Atomic Radiation, p. 208. New York: United Nations.

WAGNER, E. B., and G. S. HURST. 1958. Advances in the standard proportional counter method of fast neutron dosimetry. Rev. Sci. Instr.,
29:153–58.
———. 1959. Gamma response and energy losses in the absolute fast neutron dosimeter. Health Physics, **2**:57–61.

APPENDIX

Following is a calculation of energy loss under the bias corresponding
to protons that make more than one traversal of the boundary of the
gas volume in a long, tissue-equivalent proportional counter of the Hurst
type.

In this calculation we first determine the distribution of ranges in the
gas of protons that emerge isotropically from the wall of an infinite
cylinder, travel in a straight line and leave the gas volume at some other
point on the wall. For a particle entering the gas at a point A on the
wall of a cylinder of radius R, in a direction making an angle ϕ with the
plane containing A and the axis of the cylinder, elementary geometrical
considerations show that the probability that its range in the gas will be
less than d is given by

$$\delta N = \frac{\delta \phi}{\pi} \int_0^{\cos^{-1} G(\phi,\, r)} d\theta = \frac{\delta \phi}{\pi} \cdot \cos^{-1} G(\phi,\, r),$$

where

$$G(\phi, r) = \frac{2 \cos^2 \phi}{(r^2 - 4 \cos^2 \phi \sin^2 \phi)^{1/2}},$$

and

$$r = \frac{d}{R}.$$

Thus the total probability that the particle will have a range in the gas
that is greater than r times the cylinder radius is given by:

$$1 - N = 1 - \frac{2}{\pi} \int_0^{\pi/2} \cos^{-1} \left(\frac{2 \cos^2 \phi}{(r^2 - 4 \cos^2 \phi \sin^2 \phi)^{1/2}} \right) d\phi.$$

This expression may be evaluated numerically as a function of r and gives
the distribution shown in Figure 7.

We now take into account the variation of specific ionization along the
track of a proton of given initial energy and estimate, using Figure 7,
the probability that, in its passage through the gas volume, the proton will
have a combination of range and specific ionization sufficient to transfer
an energy of at least B mev. to the gas. In this way we may estimate the
fraction of protons of given energy that will be recorded.

Since we are interested in energy transfer (rather than proton flux), it
is necessary to take into account the fact that the mean energy of the
recorded pulses is considerably greater than the mean energy of the
pulses lost below the bias, and the fraction of dose recorded is correspondingly higher than the fraction of proton flux recorded. In order to esti-

mate the magnitude of this correction, it is again necessary to apply the
data of Figure 7 to separate portions of the proton track and to derive a
value, effective over the total length of the track, for the ratio between
the mean energy of recorded pulses and that of pulses falling below the
bias. If, for a given proton energy, this ratio has the value H, and F is the
fraction of protons recorded, the fraction of dose recorded is given by

$$\frac{HF}{HF+1-F}.$$

In all cases of practical interest H is found to have a value close to 4. We
can thus calculate, at any proton energy, the fractional loss of proton
dose detection efficiency attributable to group 3 above for an infinite

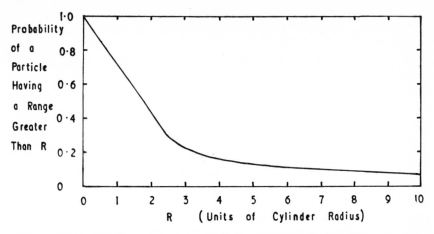

Fig. 7.—Range distribution in an infinite cylinder of an isotropic flux of heavy particles
that traverse the cylinder completely.

cylindrical counter of given design. For a counter of internal radius R,
containing a $(CH_2)_n$ gas filling at P atmosphere pressure and operating
at a bias of B mev., this is found to be a function only of

$$\rho = \frac{PRn}{B}.$$

We wish to calculate the efficiency of a particular counter for the
detection of dose due to fast neutrons of a given energy. This follows
from the proton dose efficiency versus energy relationships obtained as
given above and the known relationship between neutron energy and
corresponding recoil proton energy. For the present calculations, elastic
scattering has been assumed: this is believed to be an adequate approximation for energies up to 30 mev.

Fractional losses, f, in neutron dose detection efficiency, corresponding
to group 3 (protons that traverse the boundary of the gas volume twice),

have been calculated by the foregoing method for two values of ρ and are plotted (as functions of $[1 - f]$) in Figure 1. Also plotted in Figure 1 are values of $(1 - f)$ obtained from Ritchie's calculations (1959), corresponding to groups 1 and 2 above (protons whose tracks make more than one traversal of the boundary of the gas volume). In this case f is a function of both ρ and B, and values have been plotted for two values of each parameter. The gap width D in Ritchie's calculations for a plane parallel gap has been taken for the diameter $2R$ of the cylindrical system considered here. The values of ρ are chosen as being typical of counters of the type employed by Hurst (Hurst, 1954; Wagner and Hurst, 1958); $\rho = 10$ corresponds to a counter of 2-cm. radius filled with ethylene to a pressure of 0.5 atm. and operating with a bias equivalent to 0.2 mev., while $\rho = 30$ corresponds to the same instrument with a cyclopropane filling at 1 atm.

In deriving the data presented in Figure 1, it has been implicitly assumed that the thickness of the $(CH_2)_n$ wall of the counter is greater than the range of the most energetic proton. For smaller wall thicknesses, not only will the actual dose to the gas be reduced, but the efficiency of measuring the dose will also be less as a result of an enhanced proportion of protons traversing the gas volume at an early stage in their paths and having correspondingly low specific ionization. In practice, therefore, the foregoing calculations are only strictly valid where a counter is being used to estimate dose in tissue at a depth greater than that at which proton equilibrium has been produced (0.5 cm. at 15 mev. and 1.5 cm. at 30 mev.). The calculations also do not take any account of the dose contributions due to heavy recoil nuclei and inelastic collisions, both of which may become comparable with the contribution due to recoil protons at energies of the order of 30 mev., and both of which, for a given neutron energy, will be detected by instruments of this type with greater efficiency than will that from recoil protons.

HAROLD MAY AND LEONIDAS D. MARINELLI

29. Cosmic-Ray Contribution to the Background of Low-Level Scintillation Spectrometers

THE MEASUREMENT of low-level radioactivity by means of large sodium iodide scintillation counters has been the concern of our group at Argonne National Laboratory for some time, and much of this effort has been directed toward obtaining the maximum possible sensitivity. Although it is always more advantageous to increase the sample net counting rate, S, than to decrease the background rate, B—the figure of merit being S^2/B whenever $S << B$ (Jaffey, 1960)—sometimes this may not be possible. Such is the case when the sample size is limited (biological samples) or the sample activity cannot be increased with thick samples and large geometry because the flux at the crystal is limited by self-absorption. It is therefore often desirable to identify the various sources of background and their relative contributions in the energy region of roughly 100 kev.–3 mev. as a prerequisite to their reduction (Miller *et al.*, 1956). In some cases it is difficult to do this unambiguously; such is the case when one wishes to differentiate between the contribution of natural radioactivity (and possible fallout nuclides existing as contaminants) within shield materials and the low-energy γ rays originating therein from cosmic-ray interactions. Included in this category are numerous photons resulting from inelastic scatter and capture of those evaporation neutrons that are produced and moderated within the shield and surrounding structural members.

The principal components into which it is convenient to classify all cosmic-ray particles will be listed, and the major physical processes of interaction and decay giving rise to low-energy photons will be dis-

HAROLD MAY is associate physicist and L. D. MARINELLI is division director, Radiological Physics Division, Argonne National Laboratory, Argonne, Illinois.
Work performed under the auspices of the U.S. Atomic Energy Commission.

cussed. The way in which these depend upon the atomic mass number of the interacting elements is stressed in order to emphasize the relative merits of various shielding materials.

Cosmic Rays

The primary cosmic-ray particles are predominantly high-energy protons, plus a small fraction of α particles and heavier nuclei.

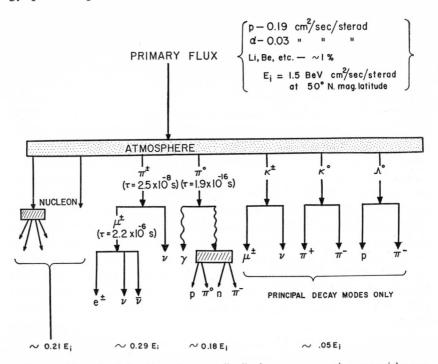

Fig. 1.—General relationship and energy distribution among cosmic-ray particles produced in the atmosphere.

Their energies range from roughly 1 gev. to at least 10^9 gev., the incident flux at energies less than 15 gev. being a strong function of geomagnetic latitude. Figure 1 depicts their principal modes of interaction with nuclei of the atmosphere and the secondary particles and radiation engendered therefrom.* The mean lifetimes and decay modes of unstable particles are shown, as well as an estimate of the fraction of incident energy carried by each mode.

The production of high-energy neutrons and protons by nuclear

* The figure is adapted from Puppi (1956); see also the excellent review by Peters (1958) for a somewhat more detailed diagram.

cascade, discussed by Dr. Korff in this symposium, accounts for roughly 20 per cent of the energy of the incident primary radiation. After repeated interactions within the atmosphere, these nucleons still possess sufficient energy at sea level to produce additional cascades in condensed materials. Production of charged pions, having a mass 273 times that of the electron and a mean life of 2.5×10^{-8} seconds, accounts for roughly 30 per cent of the energy; the muons resulting from their decay constitute the more penetrating or "hard" cosmic-ray component at sea level. Some 18 per cent of the incident energy goes into production of neutral π^0 particles which decay with a 1.9×10^{-16} second mean life and the emission of two high-energy quanta. These are the principal contributors to the electron-photon cascade shower, or "soft" component. Photonuclear reactions may transfer some energy to the nucleonic component. Production of kaons and hyperons is much less probable and, regardless of decay mode, leads only to production of additional pions, muons, electrons, neutrinos, and γ radiation.

Having thus very briefly reviewed the primary cosmic interactions within the atmosphere, we proceed to characterize the incident numbers and energy spectra of each component at sea level and their interactions in typical shielding materials.

The omnidirectional integral energy spectrum of the "soft" or electromagnetic (e.m.) component is shown in Figure 2, as obtained by Carmichael (1957). Above 100 mev. the ordinate is total number of electrons and photons, assumed approximately equal in number. In the lower-energy region the dotted line represents electron flux only; the total flux of electrons with energies greater than ~ 1 mev. amounts to 8.4×10^{-3} per unit sphere per sec.*

Attenuation of these high-energy electrons and photons occurs chiefly by radiation and pair production in the field of absorber nuclei. By making the simplifying assumptions that these are the *only* processes and by employing the asymptotic, high-energy cross-sections, the build-up and decay of a cascade shower in a thick medium may be calculated. A discussion of the results and limitations of shower theory may be found in the excellent volume by Rossi (1952). Since in the present context the photon energy range of particular interest is that in which Compton scattering and photoelectric absorption must also be considered, this theory cannot give meaningful estimates of the required shielding thickness. A computational method that includes all significant energy losses and realistic cross-sections,

---

* It should be noted that this value, and the spectrum, are calculated from the cascade theory approximations discussed below and hence are subject to the basic and considerable uncertainties that are inherent therein.

thus giving accurate results down to energies of ~5 mev., has been described by Olson and Spencer (1958).

The Monte Carlo approach is capable of representing all physical processes with great accuracy. An early study by Wilson (1952) employed a mechanical wheel of chance to simulate the statistical fluctuations of shower generation in lead. More recent work with high-speed electronic computers has been reported (Crawford and Messel,

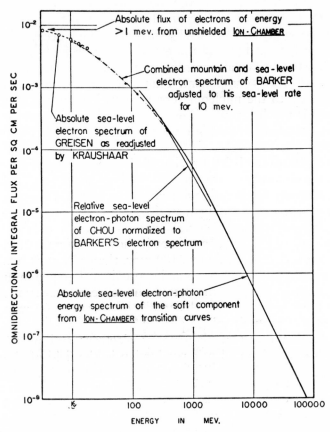

Fig. 2.—Omnidirectional integral energy spectrum of the electrons and photons of the soft component near sea level.

1962; Zerby and Moran, 1962) that appears to be in excellent agreement with the very limited experimental data. This work holds promise of direct application to several phases of the present discussion, but results in the media and energy range required are not available at present.

The "soft" cosmic-ray component received this classification, even

prior to realization of its true composition, on the basis of the observed dependence of the counting rate of a coincidence Geiger-Müller counter upon shield thickness. Such "transition curves" (see Rossi, 1952, pp. 318–29) indicate that some 10 cm. of lead (equivalent to 17 radiation lengths or to 30 cm. of iron) are required to suppress the shower electrons, whereas photons of a few mev. energy (in the broad minimum of the absorption coefficient characteristic of high Z materials) penetrate even deeper. In the absence of more clear-cut evidence bearing directly on low-energy photons in the "tail" of the shower, these figures appear applicable to the design of low-level shields. The attenuation of overlying structural materials may be appreciable.

The most penetrating cosmic rays consist of muons (no longer considered mesons in view of their weak nuclear interaction), which have a mass of 206.7 m_e and occur in both charge states. Those with positive charge are about 25 per cent more abundant, a direct consequence of the positively charged primary particles. The vertical intensity, I_v, under 167 gm/cm^2 of lead absorber is $0.82 \pm 0.01 \times 10^{-2}$ particles/cm^2/sec/sterad (Rossi, 1948), as measured at 40° N. latitude, and remains essentially constant north of this parallel. The muon energy spectrum, as derived from measurements by Pine, Davisson, and Greisen (1959) and others, is shown in Figure 3.*

The principal process by which incident cosmic-ray muons lose energy is that of ionization and collision, since radiation, pair production, and Cerenkov losses contribute to any extent only at incident energies above 40 gev. It is well to recall that, in the extreme relativistic case, total energy transfer to a knock-on electron within the absorber is permitted. Hence the ionization-collision losses cannot be dismissed entirely as low-energy events. The mean loss in iron, as averaged over the incident muon spectrum, is 1.70 mev cm^2/gm, of which some 0.134 mev cm^2/gm (or 8 per cent of the total) is lost to knock-on electrons having energies of 25 mev. or more—that is, above the critical energy, and hence having a good probability of producing radiation. The photon density which results, assuming at least one is produced by each knock-on, is equivalent to that of an activity of

---

* The direct experimental evidence on muons and protons is obtained with magnetic spectrographs detecting vertically incident particles only. Their intensity is known to vary with zenith angle θ as cos^n θ, where the value of n is energy dependent. For muons, a constant value of $n = 2$ is sufficiently accurate for most purposes, whereas the value for protons is in the range of 5–7. In order that these spectra be directly comparable with the neutron spectrum, which unavoidably was obtained with direction-insensitive detectors, the omnidirectional intensity J_2 of muons and protons is plotted, as calculated from the relationship $J_2(E) = [2\pi I_v(E)]/(n + 1)$.

2×10^{-15} c/gm. The fractional loss to electrons with energies greater than critical is substantially the same in lead.

When muons are brought to rest in condensed materials, they may either decay or be captured, these alternatives being represented by

$$\mu^{\pm} \rightarrow e^{\pm} + 2\nu . \tag{1}$$

$$\mu^{-} + Z^{A} \rightarrow (Z-1)^{A} \rightarrow n + (Z-1)^{A-1} + \nu . \tag{2}$$

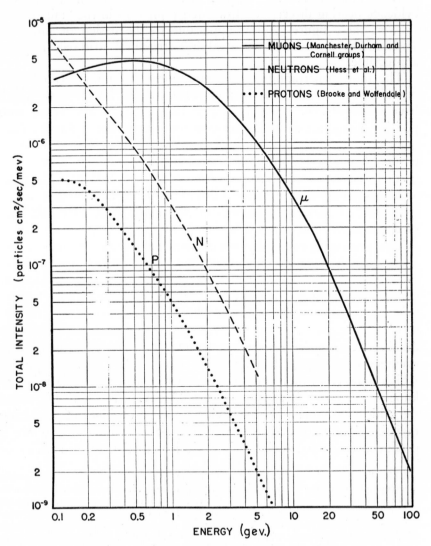

Fig. 3.—Omnidirectional differential energy spectrum of muons and nucleons near sea level and mid-latitudes.

Note that because of the Coulomb potential barrier only reaction (1) is possible for positive muons. The electron energy spectrum from muon decay in a liquid hydrogen target (Plano, 1960) is illustrated in Figure 4. The maximum and most probable energies are 52.8 and 46 mev., respectively, that is, most decay electrons are of sufficient energy to generate showers and, hence, contribute to the low-energy photon flux.

The probability of μ^- capture increases with the atomic number of the interacting medium, being equal to the decay probability at

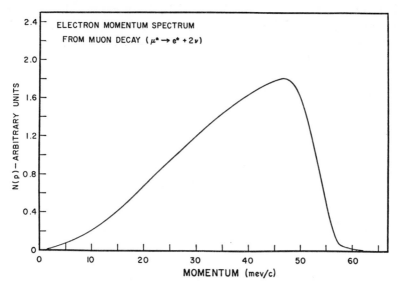

Fig. 4.—Differential energy spectrum of electrons resulting from muon decay in hydrogen.

about $Z = 11$. In this process, the muon is first trapped in a Bohr orbit. Mesic X-rays are emitted upon transition to inner orbits, the K_α (or $2p - 1s$) transition occurring roughly 80 per cent of the time. The orbital radii are much less than those of corresponding electron orbits due to the greater particle mass; in fact, the innermost orbit penetrates the nucleus for an appreciable fraction of time. The resulting nuclear capture releases about 100 mev., of which the largest portion is carried away by the neutrino. The remainder may be carried off by a single neutron but in heavy nuclei is normally shared by the entire nucleus. The "nuclear temperature" is increased, with subsequent "boil-off" of one or more neutrons. These "evaporation" neutrons have most probable energies of only a few mev. Several γ rays with energies in the range 1–10 mev. are emitted to complete the de-excitation of the residual nucleus. Table 1 summarizes for selected elements the capture probabilities, energies of K_α X-rays, and multiplicity $\bar{m}$, or mean number of evaporation neutrons per capture.

The contribution of the neutrons generated within shield materials to the low-energy photon background as a result of inelastic scatter and thermal capture therein will be discussed later. Consider instead a cubical shield having 1-m. outside dimensions and 20-cm.-thick iron walls—these dimensions being chosen as roughly typical of many installations in use. About 8,000 muons per minute intersect the top surface, and an equal number strike the sides, assuming negligible attenuation by overlying structural materials. In practice, this reduction factor may vary widely; we note only that 100 gm/cm² ($\approx$18 inches) of concrete reduces the incident flux by $\sim$10 per cent.

TABLE 1

MUON CAPTURE CONSTANTS FOR SELECTED ELEMENTS

Element	Z	Energy of K_α Mesic X-Ray (mev.)	μ Capture Probability	$\bar{m}$
Na......	11	....	0.46	1 $\pm$0.4
Al.......	13	0.344	0.61	....
Fe.......	26	1.26	0.91	1.15*
Ag.......	47	3.16	0.96	1.6 $\pm$0.18
Pb.......	82	6.00	0.98	1.64$\pm$0.16

* This value estimated from preliminary analysis of data by Kaplan *et al.* (1963).

It is clear that the actual path length of muons in the shield may range from zero to slightly over 100 cm., and, lacking any rigorous analysis of the actual path-length distribution, a mean path length of 40 cm. is assumed. From the range-energy relationship, and the incident spectrum, it can be calculated that some 16 per cent of the incident muons will lose all their energy in the shield. Hence, some 1,450 positrons are produced per minute from μ decay, each of which is capable of producing, by annihilation and cascade, several low-energy photons. An additional 1,150 mesic X-rays of 1.27-mev. energy arise from muon capture, as well as an indeterminate number of photons originating from the subsequent neutron capture, ranging in energy up to more than 9 mev. This photon production represents an additional "equivalent activity" of some 3×10^{-16} curies per gram.

High-energy nucleons, although numerically less abundant than muons, also contribute significantly to the low-energy γ-ray background. The proton spectrum of Figure 3 is a composite from many experimental studies (Mylroi and Wilson, 1951; Ogilvie, 1955;

Brooke *et al.*, 1962); that of the neutrons is due to Hess *et al.* (1959). The proton spectrum is known to much higher energies, and, by assuming that the neutron spectrum has the same slope above 10 mev., both spectra were graphically integrated in order to obtain the total incident particle flux. In the energy range from 100 mev. to 100 gev. a figure of 1.6×10^{-3}/cm^2/sec was found for neutrons; that of the protons in the same energy region, 2.17×10^{-4}/cm^2/sec. Hence the ratio of neutrons to protons is ~ 7.4, and of muons to nucleons ~ 9.5.

The processes by which high-energy nucleons interact in dense materials are similar in principle to the reactions in air that give rise to the nucleonic component but differ in detail because of the higher atomic number. These reactions are pion production, spallation, direct nuclear cascade, and emission of low-energy evaporation particles, predominantly accompanying de-excitation of the residual nucleus (Jackson, 1956). Many experimental studies of secondary spectra have been reported utilizing synchrocyclotron-accelerated protons ranging in energy from 30 to 350 mev.*

Monte Carlo calculations, which incorporate accepted theoretical concepts and experimentally determined cross-sections (Metropolis *et al.*, 1958; Bertini and Dresner, 1962), are invaluable in determining the systematic trends in multiplicity and energy distribution of emitted nucleons as a function of the atomic number of the absorber and the incident particle energy. Some results of such calculations for low, medium, and high atomic number materials are given in Table 2. The neutron multiplicity is found to vary with absorber material roughly as $A^{2/3}$.

Neutron emission from interactions of cosmic-ray nucleons has also been studied directly by several groups (Crouch and Sard, 1952; Geiger, 1956; Bercovitch *et al.*, 1960), and it is in some respects more significant in the present context, since it represents the effects of the total incident spectrum. On the other hand, one cannot readily separate the muon, neutron, and proton components and thus study their effects separately; moreover, the low incidence rates prevent accumulation of data with the desired statistical significance. Neutron multiplicities, inferred by analysis of data from multiple boron trifluoride counter arrays imbedded in paraffin moderator plus an absorber or "producer" material, are influenced by uncertainties in geometry, in detector efficiency (as determined from calibration sources differing in energy spectrum from that being measured), and corrections for gating period. The comparison given in Table 2 of Monte Carlo predictions

---

* A conveniently tabulated summary of many of these has been given by Mainschein (1962).

with experimental observations essentially illustrates an agreement between quantities that are not strictly comparable. Note that the calculations show a slightly higher multiplicity at all energies for incident neutrons as compared with incident protons, whereas the experimental work clearly demonstrates the opposite. As pointed out by Geiger (1956), this can readily be accounted for in terms of the much greater effective energy of the proton component, as contrasted to that of the neutrons (Fig. 3).

TABLE 2

EVAPORATION NEUTRON MULTIPLICITY IN NUCLEAR CASCADES

FROM MONTE CARLO CALCULATIONS*

INCIDENT PARTICLE AND ENERGY	TARGET ELEMENT		
	Al	Cu	Pb
Proton, 200 mev..............	0.60	2.69	8.10
Neutron, 200 mev.............	0.76	3.19	8.47
Proton, 300 mev..............	0.60	3	9.66
Neutron, 300 mev.............	0.75	3.22	9.50
Proton, 400 mev..............	0.61	3.11	10.60
Neutron, 400 mev.............	0.72	3.37	10.50

FROM COSMIC-RAY DATA

	Al	Fe	Pb
Nuclear component (Geiger, 1956).....................	1.9	2.9	6.5
Neutrons (Geiger, 1956).......	....	2.8	6.2
Protons (Geiger, 1956)........	....	3.7	11.0
Protons (300–800 mev.) (Berco-vitch *et al.*, 1960)...........	....	5–8 (Sn)	7–14

* Data from H. W. Bertini and Dresner (1962). Statistical uncertainties, as based on 1,000–2,500 case histories each, are not given.

The number of photons resulting from neutron inelastic scattering and from capture in the resonance and thermal energy regions may be calculated in principle from the transport and moderation properties and appropriate interaction cross-sections of the shielding material. The calculations are fairly straightforward for an infinite-slab geometry; however, neutron leakage at the shield surfaces is important in practical configurations, and the transport equations can be solved only approximately. Hence no quantitative estimate of these effects will be presented here. The thermal capture γ-ray spectra are well known for all elements of interest, whereas the inelastic scatter and resonance capture data are incomplete in some cases. Principal photons from thermal capture range in energy from 2.226 mev. in hydrogen to 7.40 and 7.64 mev. in lead and iron, respectively. Spectra of 4 shield materials are illustrated in Figure 5.

The average time required for a fast neutron to be slowed down, approach a Boltzmann distribution in the thermal energy region, and be captured is of definite interest. This time may be calculated for hydrogenous materials quite well and is known, somewhat less precisely, for lead. In the elements of medium atomic number, effects

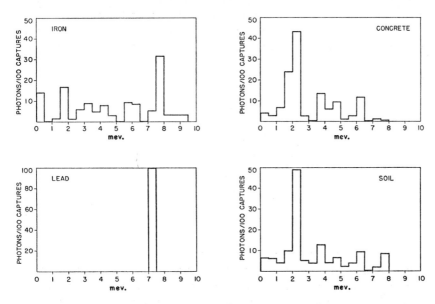

Fig. 5.—Energy spectra of γ rays resulting from the capture of thermal neutrons in selected shielding materials.

associated with the binding energy and thermal motion of the lattice are significant in the region of a few kev. energy and are not amenable to simple theory (Von Dardel and Sjostrand, 1958). Some values are given in Table 3 derived from theory and limited experiment. The time to reach an energy of 1 electron volt in iron was obtained by linear extrapolation of recent data (Isakov, 1962) to 4.9 ev., precisely in the region where crystal lattice effects predominate; hence, the figure may be in serious error. However, the orders of magnitude are undoubtedly correct, and they serve to indicate the long periods of gating required if anticoincidence techniques are to be effective.

Experimental investigations described below demonstrate the presence of evaporation neutrons in a typical low-level iron shield and the significant reduction of low-energy photons therein concurrent with their absence. The mean free path for high-energy nucleons is 135 gm/cm² in air, and somewhat less in concrete (Citron *et al.*, 1961); consequently their effect upon the background of a given shield will depend strongly upon the shield's location relative to overlying building materials. From the previous discussion it may be seen that complete attenuation of the nuclear-interacting cosmic-ray component may be most efficiently attained by an appropriate thickness of low atomic mass material. Of such *elements*, only carbon, sulphur, and aluminum are at all suitable from the standpoint of cost and chemical properties, and each has other drawbacks. Water is in many respects ideal, particularly if a boron compound is added in order to shift the thermal capture photon energy to 480 kev., where it can be

TABLE 3

TIME CONSTANTS FOR NEUTRON
TRANSPORT AND DIFFUSION

Material	Time To Reach 1 Ev. (μsec.)	Thermalization Time Constant (μsec.)	Diffusion Time Constant (μsec.)
H_2O	1.3	6	210
Fe	110	?	22
Pb	430	900	910

more readily absorbed. A further advantage is the ease with which radioactive contamination, whether arising from natural or fallout radioelements initially present or leached from the container walls, may be removed by ion exchange. The main disadvantage is its relatively low density, requiring large volumes and comparatively costly containers. Earth, rock, and concrete combine low cost, low effective mass number, and higher density, but their residual radioactivity is also high. The ultramafic minerals, olivine and dunite, are notable exceptions, since potassium and thorium concentrations approaching the current level of detection sensitivity have been reported (Wollenberg and Smith, 1962, 1963). Use of these might permit a sizable reduction in the amount of inner, low-activity shielding and a consequent cost reduction.

EXPERIMENT

The cosmic-ray-induced effects have been evaluated by a series of experiments conducted at Argonne National Laboratory. A low-activity 7-inch-diameter by 3½-inch-thick sodium iodide crystal, an iron shield having outside dimensions of $32 \times 32 \times 43$ inches and 8-inch-thick walls lined with ¼-inch lead sheet, and either 256- or 400-channel analyzers were employed. The relative contributions of each cosmic-ray component may be inferred from readings taken within the cave at 5 locations, affording a comparison between 4 different amounts of additional overhead shielding. The locations are as follows:

1. Two locations on the laboratory site, with only a thin metal roof (< 1 gm/cm²) overhead.
2. A basement laboratory, with some 80 gm/cm² of reinforced concrete (1 load-bearing floor, 1 roof).
3. An access tunnel to the ZGS proton synchrotron, with 1,460 gm/cm² (about 25 feet) of earth fill and concrete. (The synchrotron was under construction, so no activity from its proximity was possible.)
4. A Chicago water tunnel with 1.6×10^4 gm/cm² (about 250 feet) of rock, silt, and earth fill.

At times the crystal was surrounded inside the cave with plain or borated paraffin in order to demonstrate more clearly the presence of evaporation neutrons by reference to photons in the 2.23-mev. region. Further experimental details are to be found in our semiannual reports (May and Steingraber, 1958; May, 1962). The principal results at each of these locations are summarized in Table 4.

The average muon intensity at each location was determined from the counting rates observed between half-maximum points of the muon throughpeak, which extends from 32 to 57 mev. in our crystal. The relative rates were found to agree well with the coincidence telescope measurements of Wilson (1938).

The relative flux of evaporation neutrons versus location was determined by two independent measurements. In the first, the net increase in counting rate in the 2.23-mev. region was noted as the crystal was surrounded with some 17 cm. of paraffin within the cave. As tabulated above, a net increase of 4.2 ± 0.4 c.p.m. was observed in the unshielded location versus 2.8 ± 0.27 in the Building 203 basement. This represents 67 per cent of the increase at the first location. A direct measurement of neutron intensities at the two locations (but not within the iron shield) was made with a boron trifluoride counter and paraffin moderator—the usual "long counter" having approximately uniform sensitivity from 50 kev. to 5 mev. The possible presence of reactor- or accelerator-produced neutrons in the vicinity of the minimum shielding locations was investigated by also operating the detector at two remote locations known to be free of such influ-

TABLE 4

SUMMARY OF BACKGROUND COUNTING RATES OF SELECTED LOCATIONS

Location	ZGS, High Bay	203 Garage	203 Basement	ZGS Tunnel	Chicago Water Tunnel
Overhead shielding (gm/cm²)	~0	~0	80	1460	1.6×10⁴
Calculated relative intensity					
Muons	1	1	0.92	0.34	0.013
Nucleons	1	1	0.58	0.001	0

γ COUNTS/MIN (7 × 3½ IN. CRYSTAL, 20 CM. FE SHIELD)

Band	ZGS, High Bay	203 Garage	203 Basement	ZGS Tunnel	Chicago Water Tunnel
I Muon throughpeak	128	125	114	46	1.8±0.1
Muon throughpeak, Rel.	1	1	0.90	0.36	0.014
II 2.1–2.35 mev.					
a. No paraffin	9.0	9.8	7.1±0.25	3.53	2.06±0.1
b. With plain paraffin	12.7±0.6	14.3±0.2	9.9±0.1	3.24±0.1	....
c. With borated paraffin	7.7±0.02	9.2±0.2	6.6±0.1	3.1±0.1	....
III 60 kev.–1.575 mev.	556±15	569±14	435±10	340±6	380±10
	(Spring, 1962)	(Sept., 1962)	(1959–61)	(Spring, 1962)	(Fall, 1958)

FAST NEUTRONS, COUNTS/MIN (BF₃ LONG COUNTER, NET)

ZGS, High Bay	203 Garage	203 Basement	ZGS Tunnel	Chicago Water Tunnel
0.472±0.037	....	0.261±0.028 (in open)	0.003±0.017	....
0.400±0.014	(Garage, Hinsdale)	0.374±0.025 (in Fe room)		
0.383±0.042	(Bldg. 181)			

ence. The mean counting rate at the three locations listed is 0.42 ± 0.02 c.p.m. versus 0.261 ± 0.028 c.p.m. in the basement, the basement rate being 62 per cent of the mean.

These consistent experimentally determined ratios may be compared with the reduction of 42 per cent in high-energy nucleon intensity under 80 gm/cm², calculable from an interaction mean free path of 150 gm/cm². There is, therefore, a strong indication that under these experimental conditions the nucleonic component is by far the most important source of evaporation neutrons. This conclusion is strengthened by observations under 1,460 gm/cm² of earth, which is very nearly 10 mean free path lengths for the nucleonic component. Evaporation neutrons at this location were found to be ~5 per cent of their intensity in the basement laboratory, although, because of the meager statistical reliability of the results, this can be regarded as only an upper limit. Some evaporation neutrons are still present here from muon capture, but the extent of specifically muon-induced effects cannot be readily deduced from these data. Further studies at our laboratory await completion of improved facilities, now under construction, that will permit underground observations at any depth up to about 40 feet.

The integral counting rate at each location from 60 kev. to 1.575 mev. is also tabulated, as an indication of the practical import of these effects. Comparison of the 80 gm/cm² and 1,460 gm/cm² data reveals that a decrease of about 100 c.p.m., or some 23 per cent, accompanies the removal of the high-energy nucleons. The factor ranges from about 20 per cent at 300 kev. to 40 per cent at 3 mev. The apparent increase in the same region at the most heavily shielded site is most likely caused by changes in the residual radioactivity of the crystal during the interval of time that elapsed between the experiments.*

While elimination of the cosmic-ray interactions is thus shown to reduce the background appreciably, we should note that in order to effect a major order-of-magnitude improvement in sensitivity, this effort must be accompanied by a significant improvement in the local radioactivity of the crystal, photomultiplier, and shield. One advantage to be obtained by such measures will be an improvement in the stability of the background counting rate. When counting

---

* During this time, the crystal was recanned and a different photomultiplier tube was attached. The crystal is mounted in a double-walled, stainless-steel container designed to permit the introduction of mercury shielding into a close-fitting annulus. Strenuous measures were taken to insure removal of surface radioactivity at the time of fabrication, but it appears that repeated introduction and removal of the mercury, or perhaps simple decay of short-lived fallout nuclides, has reduced the inherent activity of the crystal package. The spectra distribution of the 40 c.p.m. difference yields no information as to the nature of these changes.

times of several hundred to a few thousand minutes are required, as is frequently necessary in low-level measurements, it is usually found that the background counting rate is not stable within the limits imposed by counting statistics alone. The principal causes of these variations are changes in ambient radon level (Moses *et al.*, 1963) and electronic drift or instability, as well as cosmic-ray intensity fluctuations. The quantitative identification and elimination of each is never a trivial problem, yet it is required if the full potential of the counter system is to be realized. Further discussion of these problems is outside the scope of this paper.

REFERENCES

BERCOVITCH, M., H. CARMICHAEL, G. C. HANNA, and E. P. HINCKS. 1960. Yield of neutrons per interaction in U, Pb, W, and Sn by protons of six energies between 250 and 900 mev selected from cosmic radiation. Phys. Rev., **119**:412–31.

BERNARDINI, G., G. CORTINI, and A. MANFREDINI. 1949. On the nuclear evaporation in cosmic rays and the absorption of the nucleonic component. I. Phys. Rev., **76**:1792–97.

BERTINI, H. W., and L. DRESNER. Monte Carlo Calculations on Intranuclear Cascades. Unpublished Rpt. ORNL-3383. Neutron Physics Div., Oak Ridge Nat. Lab., 1962.

BROOKE, G., P. J. HAYMAN, F. E. TAYLOR, and A. W. WOLFENDALE. 1962. The spectrum of cosmic ray muons and protons near sea level. J. Phys. Soc. Japan, **17** (Suppl. A-III): 311–72.

CARMICHAEL, H. 1957. Energy spectrum of the soft component near sea level. Phys. Rev., **107**:1401–9.

CITRON, A., L. HOFFMANN, and C. PASSOW. 1961. Investigation of the nuclear cascade in shielding materials. Nuc. Instr. & Methods, **14**: 97–100.

CRAWFORD, D. F., and H. MESSEL. 1962. Energy distribution in low-energy electron-photon showers in lead absorbers. Phys. Rev., **128**: 2352–60.

CROUCH, M. F., and R. D. SARD. 1952. The distribution of multiplicities of neutrons produced by cosmic-ray μ-mesons captured in lead. Phys. Rev., **85**:120–29.

GEIGER, K. W. 1956. Evaporation neutrons from cosmic ray nuclear interactions in various elements. Canad. J. Phys., **34**:288–303.

HESS, W. N., H. W. PATTERSON, R. WALLACE, and E. L. CHUPP. 1959. Cosmic-ray neutron energy spectrum. Phys. Rev., **116**:445–57.

ISAKOV, A. I. 1962. Nonstationary elastic slowing down of neutrons in graphite and iron. Soviet Physics, JETP, **14**:739–40.

JACKSON, J. D. 1956. A schematic model for (p, xn) cross sections in heavy elements. Canad. J. Phys., **34**:767–69.

JAFFEY, A. H. 1960. Statistical tests for counting. Nucleonics, **18**:180–84.

KAPLAN, S. N., D. HAGGE, J. BAIJAL, J. DIAZ, and R. V. PYLE. Lawrence Radiation Lab. Personal communication, 1963.

MAINSCHEIN, F. C. Unpublished Annual Rpt. ORNL-3360, p. 269. Neutron Phys. Div., Oak Ridge Nat. Lab., 1962.

MAY, H. A. Neutron production in massive shields and effect upon the low-energy gamma-ray background. Unpublished Semiannual Rpt. ANL-6646, p. 50. Radiological Phys. Div., Argonne Nat. Lab., 1962.

MAY, H. A., and O. J. STEINGRABER. Further studies on background counts in large NaI crystals. Unpublished Semiannual Rpt. ANL-5967, p. 145. Radiological Phys. Div., Argonne Nat. Lab., 1958.

METROPOLIS, N., R. BIVINS, M. STORM, J. M. MILLER, G. FRIEDLANDER, and A. TURKEVICH. 1958. Monte Carlo calculations on intranuclear cascades. II. High-energy studies and pion processes. Phys. Rev., **110**: 204–19.

MILLER, C. E., L. D. MARINELLI, R. E. ROWLAND, and J. E. ROSE. 1956. An analysis of the background radiation detected by NaI crystals. I.R.E. Trans. Nuc. Sci., NS, **3**:90–96.

MOSES, H., H. F. LUCAS, JR., and G. A. ZERBE. 1963. Effect of meteorological variables upon radon concentration three feet above ground. Air Pollution Control Assoc. J., **13**:12–19.

MYLROI, M. G., and J. G. WILSON. 1951. On the proton component of the vertical cosmic-ray beam at sea level. Proc. Phys. Soc. London A, **64**:404–17.

OGILVIE, K. W. 1955. Cosmic-ray experiments with a proton velocity selector. Canad. J. Phys., **33**:746–56.

OLSON, C. A., and L. V. SPENCER. 1958. Energy spectra of cascade electrons and photons. J. Res. Nat. Bur. Standards, **60**:85–96.

PETERS, B. 1958. Cosmic rays. *In* E. U. CONDON and H. ODISHAW (eds.), Handbook of Physics, Part 9, chap. 12, pp. 9-201–9-244. New York: McGraw-Hill Book Co.

PINE, J., R. J. DAVISSON, and K. GREISEN. 1959. Momentum spectrum and positive excess of μ-mesons. Nuovo Cimento, Ser. IX, **14**:1181–1204.

PLANO, R. J. 1960. Momentum and asymmetry spectrum of μ-meson decay. Phys. Rev., **119**:1400–1408.

PUPPI, G. 1956. Energy balance of cosmic rays. *In* J. G. WILSON (ed.), Progress in Cosmic Ray Physics, **3**:339–88. Amsterdam: North-Holland Pub. Co.

ROSSI, B. 1948. Interpretation of cosmic-ray phenomena. Rev. Mod. Phys., **20**:537–83.

———. 1952. High Energy Particles. Englewood Cliffs, N.J.: Prentice-Hall.

VON DARDEL, G., and N. G. SJOSTRAND. 1958. Diffusion measurements with pulsed neutron sources. *In* D. J. HUGHES *et al.* (eds.), Progress in Nuclear Energy, **2**, Ser. 1, 183. New York: Pergamon Press.

WILSON, R. R. 1952. Monte Carlo study of shower production. Phys. Rev., **86**:261–69.

WILSON, V. C. 1938. Cosmic-ray intensities at great depths. Phys. Rev., **53**:337–43.

WOLLENBERG, H., and A. R. SMITH. Earth Materials for Low-Background Radiation Shielding. Unpublished Rpt. UCRL-9970, Univ. of California Lawrence Radiation Lab., 1962.

——. 1963. Studies in terrestrial γ-radiation. This symposium.

ZERBY, C. D., and H. S. MORAN. 1963. Studies of the longitudinal development of high-energy electron-photon cascade showers in copper. J. Appl. Phys., **34**:2445–57.

(Numbers in parentheses refer to chapter authored by discussant.)

Adams (30, 34) asked whether there is any evidence as to how the cosmic-ray neutron component has changed in geologic time, in light of the possible effects of close supernova explosions (every 200 million years) or changes in the earth's magnetic field. Korff (26) replied that the net effect of the earth's magnetic field on total cosmic-ray intensity is not very large, affecting primarily the intensity distribution with latitude. The effect of the supernova phenomenon would depend on the time base; a large excursion in cosmic-ray intensity would not be significant unless there was a storage mechanism that provided a long-term enhancement. Such a mechanism does not appear to exist, and it is most likely that continuous generation of the existing cosmic radiation is taking place. Solar flares on a short timescale are observed from time to time. But one can infer from studies of meteorites and helium-3 that the general cosmic-ray intensity has not increased by any large amount. Eventually, beryllium-10 may provide an important dating tool.

May (29) and Korff (26) both remarked on the fact that the intensity of solar-flare events is quite different from place to place on the earth, and Korff (26) added the comment that the maximum energies of the particles vary from flare to flare.

Brar (31) asked Korff (26) whether the peak in the neutron intensity at 60–70 millibars in the atmosphere during solar-flare events was real and whether it varied with time. Korff replied that there were too few data to ascertain just what occurs near the top of the atmosphere during these events, except inferentially. It would be very useful to monitor the neutron or total ionization intensity at these altitudes, which will someday undoubtedly be done by means of a permanent satellite.

In response to a question by Spiers (47), Hill (28) stated the RBE that he used for his calculations was 10.

PART II

Environmental Radiation Measurements

Instrumentation and Techniques

Aerial Surveys

Ground Measurements and Surveys

An Address

J. A. S. ADAMS

30. Laboratory γ-Ray Spectrometer for Geochemical Studies

IN PRINCIPLE the natural α, β, and γ activities of the thorium-232, uranium-238, and uranium-235 series, as well as the γ and β activities of potassium-40, can be used in a variety of ways for the radiometric determination of these elements or any of their daughters. Analytical schemes based upon the ratio of γ to β activity, on the ratio of α activity to chemically determined uranium, and upon the time analysis of α emission (see Cherry, 1963) have all been described. Pulse-height analysis of α and γ activities has had wide application. The present work summarizes five years' experience with γ pulse-height or spectrometric determination of thorium, uranium, and potassium at the levels of concentration found in common terrestrial materials.

The γ spectrometric determination of thorium and uranium at ore levels of concentration was attempted early in the development of sensitive and somewhat stable γ spectrometers. Hurley (1956) described the application of the technique to common rocks, as did Adams et al. (1958). Some γ spectrometric analyses of common rocks have been reported by workers in Belgium (Brooke et al., 1959), the Soviet Union (Kartashov, 1961), France (Avan and Keller, 1961), the United Kingdom (Poole and Byrne, 1961; Bloxam, 1962), Japan (Sano and Nakai, 1961), and Canada (Gregory and Horwood, 1961). It should be noted that the same general methodology has been applied to ores (Mero et al., 1962), to whole-body counting (Marinelli et al., 1962), to meteorites (Van Dilla et al., 1960), and to fallout products (Gustafson et al., 1958). The application of γ spectrometry in the field has been described by Schneider and Schwerdtel (1960), Adams (1961), and Adams and Fryer (1963). Logging γ spectrometers for use down boreholes have been described

DR. ADAMS is professor of geology at Rice University, Houston, Texas.

by Brannon and Osoba (1956), Alekseev *et al.* (1959), and Johnstone (1963). The foregoing references are only representative, not exhaustive. The reader is referred to the recent summaries and review works of Crouthamel (1960) and Leddicotte (1962).

The determination of thorium, uranium, and potassium in terrestrial materials by γ spectrometry has the attraction of simplicity and relative freedom from self-absorption effects. Presently available detectors do not discriminate or resolve γ energies as well as the best α detectors discriminate α energies. For a relatively small percentage of samples available in limited quantity (for example, purified monomineralic separates) or uncommonly low concentrations (for example, ultrabasic rocks), present γ spectrometers do not have great enough sensitivity for routine analysis. The facts that uranium and thorium are determined indirectly by γ spectrometry and that thus one must assume secular radioactive equilibrium concentrations for measured daughters and parent thorium or uranium raise possible uncertainties, particularly where uranium-238 is determined by measuring a post–radon-222 daughter. Experimentally, however, clearly established cases of such radioactive disequilibria have been found but rarely in fresh rock samples. The high initial cost of γ spectrometers and associated facilities is a disadvantage when only a few determinations are made, but with full use of the equipment the cost per determination is far less than for wet chemical analysis, particularly for thorium (see Phair and Gottfried, 1963).

In the past five years nearly 10,000 γ spectrometric determinations of thorium, uranium, and potassium in powdered rock and soil samples, solid rock cores (right cylinders), irregularly shaped samples (tektites), and mineral separates have been made in this laboratory. Most of these data and their geochemical interpretation have been reported elsewhere (Pliler, 1956; Hamill, 1957; Adams *et al.*, 1958; Adams and Weaver, 1958; Murray and Adams, 1958; Whitfield, 1958; Adams *et al.*, 1959; Whitfield *et al.*, 1959; Adams and Richardson, 1960; Rogers and Ragland, 1961; Adams, 1962; Adams *et al.*, 1962; Heier, 1962*a, b;* Pliler and Adams, 1962*a, b;* Cherry and Adams, 1963; Heier and Rogers, 1963; Mahdavi, 1963; Richardson, 1963). The present paper is confined to the technique itself.

SHIELDING AND SURROUNDINGS

The laboratory for γ spectrometric analysis was specially designed as part of the Keith-Wiess Geological Laboratories at Rice University. The room is below ground level and isolated from other laboratories. The surrounding soil has nominal amounts of thorium (7–9 p.p.m.), uranium (2.5–3.0 p.p.m.), and potassium (0.8–1.2 per cent as the metal) (see Appendix 3, Rice Campus Station). The γ spectrometer room has a concrete baffle entrance, and all its sides, floor, and ceiling are 8-inch-thick concrete having the following apparent concentrations for cement and aggregate combined: thorium, 1.1 p.p.m.; uranium, 0.72 p.p.m.; and potassium, 1.6 per cent as the metal. These last results are in general agreement with those of Wollenberg and Smith (1962). It should be noted that this and most Gulf Coast cements are made from oyster shells in which the uranium series is deficient in radium-226, causing a lower activity than that found in most cements (*ibid.*). The low latitude (29.75° N.) and altitude (50 feet) decrease the cosmic-ray background slightly. The

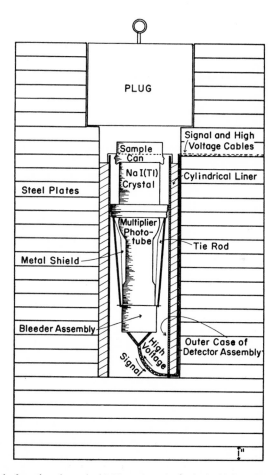

FIG. 1.—Vertical section through shield composed of 1-inch-thick and 18-inch-square iron plates; 3 × 3-inch sodium iodide (thallium activated) detector. The signal and high-voltage cables exit the shield along a curved path.

total overhead shielding from the overlying three-storied building amounts to about 2.5 feet of concrete in the vertical direction. The γ spectrometric laboratory has special equipment to maintain constant temperature and humidity.

A steel shield composed of 1-inch-thick and 18-inch-square steel plates proved satisfactory for γ spectrometry with vacuum tube electronics (see Fig. 1). However, with the greater stability of solid

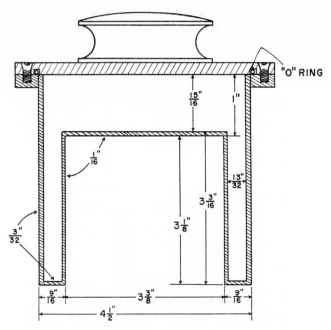

Fig. 2.—Vertical section of annular cylindrical cannisters constructed of aluminum and used to test shielding materials and maximum sample volumes.

state, multichannel pulse-height analyzers, the radioactive impurities in the steel shield proved to be the major limitation on the sensitivity of the method. Experiments with annular cannisters of aluminum (Fig. 2) with $\frac{13}{32}$-inch thickness of triply distilled mercury demonstrated that the greater part of the background was external to the detector components. Initial efforts to build a lower-level shield failed because of the inherent activity of the steel and, particularly, the welding material of the container for the mercury. A welding flux rod commonly contains rutile or some other form of titanium dioxide. If this titanium dioxide is obtained from placer sands, it can quite readily be contaminated with small, but significant, amounts of radioactive minerals, such as monazite. Steels alloyed with niobium, tantalum, or vanadium were found to have particularly unacceptable levels of inherent activities, including cobalt-60. Because the radio-activity of the shielding is a major limitation on sensitivity, the identification of low-level materials (Weller, 1963) will be most important for a more extensive geochemical application of γ spectrometry.

DETECTORS AND ELECTRONIC INSTRUMENTATION

Radiometric instrumentation evolves so rapidly that four distinctly different instrumental arrays have been in use in the past five years, and a fifth array is being put in operation (see summary in Table 1). The first array consisted of a 3×3-inch sodium iodide (thallium activated) scintillation crystal optically coupled with silicone grease to a Du Mont 6363 multiplier phototube. The multiplier

TABLE 1

SUMMARY OF EXPERIMENTAL ARRAYS

Array	Shield	Detector	Channels	Readout	Spectrum Reduction
1......	Iron plate	3″×3″ NaI(Tl)	1	Scaler	Desk calculator
2......	Iron plate	3″×3″ NaI(Tl)	256	Typewriter	Desk calculator
3......	Mercury in steel	3″×3″ NaI(Tl)	256	Typewriter	Desk calculator
4......	Mercury in steel	3″×3″ NaI(Tl)	128	Typewriter	Desk calculator
5*......	Mercury in steel	Twin 3″×3″ NaI(Tl)	128	Typewriter and punched-paper tape	Desk calculator or electronic computer

* With automatic programmer for sample changing, readout, and reset.

phototube was supplied with 900–1,100 volts by a commercially available, line-operated, vacuum tube high-voltage supply with internal stabilization (Tracerlab RLI-7). The signal from the multiplier phototube was analyzed with a commercially available single-channel vacuum tube pulse-height analyzer (Tracerlab RLA-4, RLA-5S, and RLS-2) after being amplified (Tracerlab RLA-1, RLA-3, and RLS-2). The number of pulses passing the pulse-height discriminators were recorded on a vacuum tube decade scaler (Baird-Atomic Scaler 131). Drift in the high-voltage supply and in the upper and lower discriminators was a major problem, even after adding a voltage stabilizer for the supply of critical circuits (Sorensen Regulator, Model 2505). The necessity of doing each analysis energy level in series intensified the problems of drift and introduced the possibility of missetting the discriminator levels. However, the down time and maintenance costs of the multichannel analyzers available five years ago were judged to negate any advantage that those multichannel instruments might have in principle. The first array had a resolution of 9.5 per cent with cesium-137. Interlaboratory calibration with similar γ spectrometers at Shell Development Company, Houston, Texas, and with wet chemical determinations at the U.S. Geological Survey yielded results that were general in agreement to within ±15 per cent at concentrations down to about 2 p.p.m. thorium and 1 p.p.m. uranium (Adams *et al.*, 1958; Murray and Adams, 1958).

PLATE I

Automatic sample changer for γ spectrometric determination of thorium, uranium, and potassium in solid-rock cores 1⅛ inches in diameter and 6 inches long. The core pieces are inclosed in plastic tubing. The sample tube at the left has just been picked up from the upper line of samples and will be deposited just in front of the cylindrical lead plug for insertion into the shield (*far left*) between two 3 × 3-inch NaI(Tl) detectors. After counting, the sample is placed with the lower line of samples.

The second instrumental array consisted of the same detector assembly and high voltage supply, but with the pulse-height analysis being done with a solid state, 256-channel analyzer with oscilloscope and typewriter readout (Nuclear Data Model ND-102). With the much greater stability of this array the major problem no longer was drift but was rather the inherent radioactivity of the detector components, the shield, and the surroundings. For the first time it was possible to identify positively uranium and thorium series photopeaks in the background spectrum and to test different shield materials (see above). With this array the standard diabase, W-1, could be analyzed for thorium, uranium, and potassium to well within ±10 per cent of the recommended values of 2.4 p.p.m. thorium, 0.52 p.p.m. uranium, and 0.65 per cent potassium oxide (Fleischer and Stevens, 1962).

The third array was designed to improve sensitivity by using triply distilled mercury as shielding (see above) and a 3 × 3-inch sodium iodide (thallium activated) scintillation crystal in an integral mount, giving 7.5 per cent resolution with cesium-137. However, there was relatively little improvement, for the reasons cited above, and both the original steel shield detector and the new array continued to be used. The fourth array was designed to make better use of the 256-channel analyzer by constructing a routing circuit that permitted the simultaneous pulse-height analysis of the signals from both γ detectors or from one γ detector and one α detector. This two-detector operation has proved quite satisfactory, and the 128-channel results are quite comparable to the 256-channel results in every respect.

The fifth array currently being put in operation consists of two opposing 3 × 3-inch sodium iodide (thallium activated) integral detectors with a combined resolution of about 8 per cent with cesium-137. These detectors are adjustable as to the distance between them and are mounted in a shield composed mainly of triply distilled mercury. The array has an automatic sample changer; a programmer for sample changing, data readout, and memory erasure; and a punched-paper-tape readout for electronic computer reduction of the γ spectrometric data. A variant of the fifth array is shown in Plate I.

SPECTRUM ANALYSIS

Potassium-40 emits only a 1.46-mev. γ photon, and that energy is the obvious choice for the determination of potassium in material also containing uranium and thorium, assuming a constant amount of potassium-40 in chemical potassium. By contrast, there is a complex of γ emitters in the thorium and uranium series (see Appendix 4). All these γ emissions may be used to calculate simultaneously for thorium, uranium, and potassium if an electronic computer is used for the data reduction (see Dean, 1963). This laboratory is converting to such data reduction in consideration of the greatly increased sample load, particularly of rock core. However, reduction of the spectral data, with or without electronic computers, must be considered in connection with (1) the size and density of crystal detector;

(2) the signal to background ratio, which in turn is dependent on sample size and sample concentrations of thorium, uranium, and potassium; and (3) the particular problem under study.

Aside from total reduction of the spectral data on an electronic computer, one may divide the simpler methods of reduction into two main types. The first type and the one used by Hurley (1956) and later workers uses the lower energy levels below the 1.46 mev. potassium-40 peak for the calculation of thorium and uranium. This method has the advantages of higher counting rates and permits the use of smaller scintillation crystals, including the well type. The low-energy method favors smaller samples to avoid self-absorption and

TABLE 2

EXPERIMENTAL CONSTANTS

	Potassium	Uranium	Thorium
Center of window in mev..............	1.46	1.76	2.62
Width of window in mev..............	0.23(0)	0.25(7)	0.28(0)
Background in c/m....................	7.4	2.4(5)	1.2(3)
Weight equivalent of total background in window.........................	1.7 gm.	0.57(5) mg.	1.1(2) mg.

TABLE 3

EQUATIONS FOR SIMPLE SPECTRAL ANALYSIS (350 GM. SAMPLE)

p.p.m. thorium* $\quad = 2.6$ (net counts at 2.62 mev.)
p.p.m. uranium* $\quad = 0.67$ (net counts at 1.76 mev. -0.75 net counts at 2.62 mev.)
6 per cent potassium* $= 0.068$ (net counts at 1.46 mev. -1.05 [net counts at 1.76 mev.
$\qquad -0.75$ net counts at 2.62 mev.] -0.84 net counts at 2.62 mev.)

* All elements calculated as uncombined metals, assuming secular radioactive equilibrium.

background buildup effects, and it requires the solution of three simultaneous equations (see Hurley, 1956). At these lower energies the resolution of photopeaks is somewhat less satisfactory and the corrections at each photopeak can be substantial. The higher-energy method involves the use of thorium and uranium series photopeaks above the 1.46 mev. potassium-40 peak and calls for larger crystals. The reduction of the data is simpler, but the counting rates are lower. For field instruments the higher energies are favored because of the larger volume of investigation (Adams and Fryer, 1963). The higher energy peaks have been used to date in this laboratory because (1) thorium is of prime interest and the 2.62-mev. peak from thallium-208 permits the determination of thorium-232 with minimum interference from uranium-238 and potassium-40; (2) the counting rates were acceptable until recently in view of the time required to do related chemical and petrographic studies; and (3) the calculations were somewhat quicker and simpler. Table 2 lists the background data for 350 gm. of purified sodium chloride in the 8-oz. seamless metal can used for most determinations in the past. Table 3 lists the formulation for the simple reduction of γ spectra obtained from these 8-oz. cans.

SAMPLE PREPARATION

In common rocks only potassium-40 can be determined directly by γ spectrometric methods; thorium and uranium must be determined indirectly, assuming that between the parent thorium or uranium and the γ-emitting daughter determined there exists the secular radioactive equilibrium ratio of parent to daughter. Thus, samples must be selected that are known or likely to be in secular radioactive equilibrium. In general, well-consolidated rocks have been found to be in equilibrium where independent chemical determinations have been made (see Pliler and Adams, 1962a; Richardson, 1963). Sealing samples and waiting for the buildup of radon-222 (Adams *et al.*, 1958) seldom indicates radon-222 loss from powdered samples or from samples composed of resistate minerals like those found in beach or placer sands. On the other hand, soils and other weathered, fine-grained materials are known to be out of secular radioactive equilibrium on the basis of the unsupported radium in ground and surface waters, on the basis of radon-222 and post–radon-222 daughters in the atmosphere and on the basis of the observation that field and laboratory determinations of uranium by γ spectrometry show the widest deviations from chemical determinations of uranium (see, e.g., a number of the contributions to this symposium).

For purposes of γ spectrometry it is unnecessary to grind samples at all except in cases in which it is necessary to homogenize them. Thus many solid samples—for example, fine-grained, homogeneous core and tektites (Cherry and Adams, 1963)—can be counted directly. Grinding to pass 60- or 80-mesh sieves is adequate to homogenize most coarse-grained samples. The major difficulty encountered in the preparation of samples is the problem of obtaining representative samples from many geologic environments. Thus, comparisons of field γ spectrometric measurements on some tens of kilos of sample with laboratory γ-spectrometric measurements on samples 2 orders of magnitude smaller yield consistent results only when large numbers of comparisons are made (see Adams and Fryer, 1963; Mahdavi, 1963). The natural stratified nature of beach sands can only be considered typical of the heterogeneity found in the field. It should be noted that still smaller samples are used for chemical and α pulse-height determinations. The size of sample depends upon how much sample is required to contribute at least 10 per cent or more of the amounts of equivalent thorium, uranium, or potassium in the background (see Table 2). In practical terms, 350 gm. of nearly all common terrestrial materials, with ultrabasic rocks and minerals being the major exceptions, are adequate. As little as 5–10 gm. of biotite or potassium feldspar are adequate for a potassium determination for the potassium-argon method of absolute geologic dating. Similarly, 5–10 gm. of zircon or monazite have proved adequate to determine thorium and/or uranium for uranium-lead and/or thorium-lead ages.

Standards

Cross-calibrations between field and laboratory γ spectrometric determinations of thorium, uranium, and potassium generally yield results that are in agreement to within 10 per cent of the amount present (Adams and Fryer, 1963; Mahdavi, 1963; Richardson, 1963). The agreement between γ spectrometric and chemical determinations is also to within about 10 per cent (Pliler and Adams, 1962a). Various laboratories can generally obtain agreement to within 10 per cent (see Appendix 3). In terms of signal-to-noise ratio and acceptable counting times it should be possible to obtain cross-calibrations that agree to within about 5 per cent at least. However, this can be accomplished only by a number of laboratories co-operating in the analysis by several different methods of carefully homogenized standards that cannot be biased by the packing-down or segregation of powdered material. To date, the best geochemical standard in all respects is probably the standard diabase W-1 (Fleischer and Stevens, 1962) discussed above.

Conclusions

The laboratory γ spectrometric determination of thorium, uranium, and potassium has proved reliable and expeditious for most terrestrial materials containing amounts equal to or slightly less than those listed as background equivalents in Table 2. Over 10,000 such spectrometric measurements have been made in this laboratory in the last five years and comparisons with other methods and other laboratories yield results that generally agree to well within 10 per cent of the amount present. Where more than a few hundred determinations are to be made, the γ spectrometer is considered to cost less per sample than alternative methods.

REFERENCES

ADAMS, J. A. S. 1961. Radiometric determination of thorium, uranium, and potassium in the field. (Abstract.) Geol. Soc. America Special Paper No. 68, p. 125.

———. 1962. Radioactivity in the lithosphere. *In* H. ISRAËL and A. KREBS (eds.), Kernstrahlung in der Geophysik, pp. 1–15. Berlin: Springer-Verlag.

ADAMS, J. A. S., and G. E. FRYER. 1963. Portable γ-ray spectrometer for field determination of thorium, uranium, and potassium. This symposium.

ADAMS, J. A. S., M.-C. KLINE, K. E. RICHARDSON, and J. J. W. ROGERS. 1962. The Conway granite of New Hampshire as a major low-grade thorium resource. Proc. Nat. Acad. Sci., **48**:1898–1905.

ADAMS, J. A. S., J. K. OSMOND, and J. J. W. ROGERS. 1959. The geochemistry of thorium and uranium. *In* Physics and Chemistry of the Earth, **3**:298–348. New York: Pergamon Press.

ADAMS, J. A. S., J. E. RICHARDSON, and C. C. TEMPLETON. 1958. Determinations of thorium and uranium in sedimentary rocks by two independent methods. Geochim. & Cosmochim. Acta, **13**:270–79.

ADAMS, J. A. S., and K. A. RICHARDSON. 1960. Thorium, uranium, and zirconium concentrations in bauxite. Econ. Geol., **55**:1653–75.

ADAMS, J. A. S., and C. E. WEAVER. 1958. Thorium to uranium ratios as indications of sedimentary processes: Example of concept of geochemical facies. Bull. Am. Assoc. Petrol. Geologists, **42**:387–430.

ALEKSEEV, F. A., S. A. DENISIK, V. V. MILLER, and V. P. ODINOKOV. 1959. Application of the method of γ-ray spectrometry to the investigation of boreholes. Inst. Petroleum, Acad. Sci. U.S.S.R., Moscow, Yadernaya Geofiz., Sbornik Statei, pp. 134–45.

AVAN, L., and P. KELLER. 1961. Determination of uranium and thorium in radioactive minerals by γ spectrometry. Compt. rend., **252**:1135–37.

BLOXAM, T. W. 1962. Quantitative determination of uranium and thorium in rocks. J. Sci. Instruments, **39**:387–89.

BRANNON, H. R., and J. S. OSOBA. 1956. Spectral γ-ray logging. J. Petrol. Technol., **8**:30–35.

BROOKE, C., E. PICCIOTTO, and G. POULAERT. 1959. Mesure directe de l'uranium et du thorium par spectrometrie gamma. Bull. Soc. Belge de Geol., **67**:315–28.

CHERRY, R. D. 1963. Alpha-particle detection technique applicable to the measurement of samples from the natural radiation environment. This symposium.

CHERRY, R. D., and J. A. S. ADAMS. 1963. Gamma-spectrometric determinations of thorium, uranium and potassium in tektites. Geochim. & Cosmochim. Acta, **27**:1089–96.

CROUTHAMEL, C. E. (ed.). 1960. Applied Gamma-Ray Spectrometry. New York: Pergamon Press. Pp. 443.

DEAN, P. N. 1963. Computer techniques in γ spectrometry. This symposium.

FLEISCHER, M., and R. E. STEVENS. 1962. Summary of new data on rock samples G-1 and W-1. Geochim. & Cosmochim. Acta, **26**:525–43.

GREGORY, A. F., and J. L. HORWOOD. 1961. A Laboratory Study of Gamma-Ray Spectra at the Surface of Rocks. Dept. Mines and Tech. Surveys, Ottawa, Canada, Mines Branch Research Rpt. R-85. Pp. 52.

GUSTAFSON, P. F., L. D. MARINELLI, and S. S. BRAR. 1958. Natural and fission-produced γ-ray emitting radioactivity in soil. Science, **127**: 1240–42.

HAMILL, G. S. The radioactivity, accessory minerals, and possibilities for the absolute dating of bentonites. M.A. thesis, Rice Institute, Houston, Texas, 1957.

HEIER, K. S. 1962*a*. Spectrometric uranium and thorium determinations on some high-grade metamorphic rocks on Langøy, northern Norway. Norsk Geol. Tidskr., **42**:143–56.

———. 1962*b*. A note on the uranium, thorium, and potassium contents in the nepheline syenite and carbonatite on Stjermøy, North Norway. *Ibid.*, **42**:287–92.

HEIER, K. S., and J. J. W. ROGERS. 1963. Radiometric determination of thorium, uranium, and potassium in basalts and in two magmatic differentiation series. Geochim. & Cosmochim. Acta, **27**:137–54.

HURLEY, P. M. 1956. Direct radiometric measurement by gamma-ray scintillation spectrometer. Parts I and II. Bull. Geol. Soc. America, **67**:395–412.

JOHNSTONE, C. W. 1963. Detection of natural γ radiation in petroleum exploration boreholes. This symposium.

KARTASHOV, N. P. 1961. The determination of traces of uranium, thorium, and potassium in rocks by means of γ-ray spectra. At. Energ. (U.S.S.R.), **10**:531–33.

LEDDICOTTE, G. W. 1962. Nucleonics. *In* Review of Fundamental Developments in Analysis. Anal. Chem., **34**:143R–171R.

MAHDAVI, A. 1963. Natural radioactivity of Gulf Coast and Atlantic Coast beach sands. This symposium.

MARINELLI, L. D., C. E. MILLER, H. A. MAY, and J. E. ROSE. 1962. Low level gamma-ray scintillation spectrometry: Experimental requirements and biomedical applications. *In* Advances in Biological and Medical Physics, **8**:81–160. New York: Academic Press.

MERO, J. L., G. M. GORDON, and L. E. SHAFFER. 1962. Ore analysis by gamma-ray spectroscopy. *In* G. B. CLARK (ed.), International Symposium on Mining Research, **1**:331–50. New York: Pergamon Press.

MURRAY, E. G., and J. A. S. ADAMS. 1958. Thorium, uranium, and potassium in some sandstones. Geochim. & Cosmochim. Acta, **13**:260–69.

PHAIR, G., and D. GOTTFRIED. 1963. The Colorado Front Range, Colorado, U.S.A., as a uranium and thorium province. This symposium.

PLILER, R. The distribution of thorium and uranium in sedimentary rocks and the oxygen content of the Precambrian atmosphere. M.A. thesis, Rice Institute, Houston, Texas, 1956.

PLILER, R., and J. A. S. ADAMS. 1962a. The distribution of thorium, uranium, and potassium in the Mancos Shale. Geochim. & Cosmochim. Acta, **26**:1115–35.

———. 1962b. The distribution of thorium, uranium, and potassium in a Pennsylvanian weathering profile. Geochim. & Cosmochim. Acta, **26**: 1137–46.

POOLE, J. H. J., and F. N. BYRNE. 1961. Measurement of the thorium content of natural materials by a γ-ray counting method. Nature, **191**: 62–63.

RICHARDSON, K. A. 1963. Thorium, uranium, and potassium in the Conway granite, New Hampshire. This symposium.

ROGERS, J. J. W., and P. C. RAGLAND. 1961. Variation of thorium and uranium in selected granitic rocks. Geochim. & Cosmochim. Acta, **25**: 99–109.

SANO, S., and J. NAKAI. 1961. A radiometric method of analysis of natural radioactive elements by gamma-ray spectrometry and application to granitic rock samples. J. Atomic Energy Soc. Japan, **3**:288–95.

SCHNEIDER, H., and E. SCHWERDTEL. 1960. Portable scintillometer with a differential discriminator. Atomkernenergie, **5**:278–81.

VAN DILLA, M. A., J. S. ARNOLD, and E. C. ANDERSON. 1960. Spectrometric measurement of natural and cosmic-ray-induced radioactivity in meteorites. Geochim. & Cosmochim. Acta, **20**:115–21.

WELLER, R. I. 1963. Low-level radioactive contamination. This symposium.

WHITFIELD, J. M. Uranium and thorium content of some granitic rocks. Ph.D. thesis, Rice Institute, Houston, Texas, 1958.

WHITFIELD, J. M., J. J. W. ROGERS, and J. A. S. ADAMS. 1959. Relation between the petrology and the thorium and uranium contents of some granitic rocks. Geochim. & Cosmochim. Acta, **17**:248–71.

WOLLENBERG, H. A., and A. R. SMITH. 1962. Portland Cement for a Low-Background Counting Facility. Univ. Calif. Lawrence Radiation Lab. Rpt. UCRL-10475. Pp. 29.

PHILIP F. GUSTAFSON AND SARMUKH S. BRAR

31. Measurement of γ-emitting Radionuclides in Soil and Calculation of the Dose Arising Therefrom

WITHIN THE LAST FEW YEARS, principally as a result of the publicity given to fallout from nuclear-weapons tests, there has been an increasing awareness of man's radiation environment. This environment consists of radiation due to radioactivity in the earth's surface, cosmic radiation, and radioactivity present within the human body itself. Within the last few decades the natural radiation environment of a significant portion of the world's population has been supplemented by the medical uses of radiation and, even more recently, by fallout and the industrial and technological uses of radiation sources. Concurrent with awareness of the radiation environment and increases therein has come concern with the possible injury to segments of a large population exposed to even minor increments in radiation dosage over that experienced from natural sources.

One approach to the assessment of the implications of relatively small increases in radiation levels for prolonged periods of time has been to study statistically large populations subject to different levels of environmental radiation. Since such studies are being undertaken at a time when the dose rate due to fallout may add substantially to that from natural sources, it is not possible to determine the true background dose to which a population has been exposed for eons by direct dose measurements alone.

This paper will describe a method using γ-ray spectrometry of soil cores to determine natural and fission-product radioactivity present. Furthermore, the γ-ray dose arising from these radionuclides can be calculated by a method that has been in use at Argonne for several

PHILIP F. GUSTAFSON is associate physicist and SARMUKH S. BRAR is associate electrical engineer, Division of Biological and Medical Research, Argonne National Laboratory, Argonne, Illinois.

Work was performed under the auspices of the U.S. Atomic Energy Commission.

years. Direct measurements of dose have also been made recently, and comparisons between that calculated and that measured directly will be presented and discussed.

Soil is sampled monthly by taking two cylindrical core samples, 6 inches in diameter and 8 inches deep, from each of two sites. The sites are approximately $1\frac{1}{2}$ miles apart, and samples within these areas are taken approximately 10 feet apart. Both sampling sites are flat, undisturbed areas covered with grass and weeds; there is no screening by buildings or trees within several hundred feet. Each of the four soil cores is analyzed separately, and the results are averaged to obtain the monthly activity levels.

Sample preparation consists of spreading fresh soil out on clean paper on a laboratory bench, removing stones, and air drying for several days. The dry soil and vegetation are pulverized in a ball mill. Intact pebbles are then removed, and approximately 2 kg. of mixed soil and vegetation are placed in a re-entrant cup for γ-ray analysis. The re-entrant cup allows a 1-inch-thick layer of soil to be placed around the circumference and over one end of a sodium iodide crystal 5 inches in diameter by 4 inches thick. The cup is made of $\frac{3}{32}$-inch stainless steel, with heli-arced seams. The stainless-steel cover plate is held in place with a strip of masking tape. The steel is sufficiently thick to stop 3.5-mev. β particles, the most energetic electrons encountered from radioelements normally found in soil.

The sodium iodide crystal is inclosed in electrolytic copper and is attached to a single 5-inch photocathode DuMont-6265 photomultiplier tube operated at 780 volts. An Argonne-type 256-channel analyzer manufactured by Radiation Counter Laboratories is used for pulse-height analysis. For most purposes a counting time of 100–150 minutes is sufficient. The energy region considered for purposes of spectral analysis extends from 0.1 to 2.8 mev.; thus it is possible to include cerium-141 and cerium-144 (0.144 and 0.134-mev. γ-rays, respectively) as well as the 2.62-mev. line from thallium-208. A typical γ-ray spectrum of a soil with the background spectrum removed is shown in Figure 1. The increased intensity at lower energy is due in part to the greater total absorption efficiency of the detector at lower energy and in part to the fact that appreciable Compton absorption and scattering occur within the soil itself.

Uranium, thorium, and potassium are determined by solving three simultaneous equations in the energy regions 1.35–1.55, 1.65–2.2, 2.2–2.8 mev. The coefficients required are determined empirically by counting sources of uranium and potassium of known activity. The uranium standard, for example, consists of 1 gm. of uranium ore thoroughly mixed with about 2 kg. of sodium phosphate (Na_2PO_4)

and placed in a re-entrant cup similar to those used for soils. The sodium phosphate is used as a soil blank or "mock soil" and has approximately the same density and effective proton number (Z) as that of local soils. In addition, it is essentially free from radioactive contamination in the amounts used. Similarly, the thorium standard is 1 gm. of thorium ore mixed with sodium phosphate, and the potassium source is 2 kg. of potassium chloride (chemical grade), both

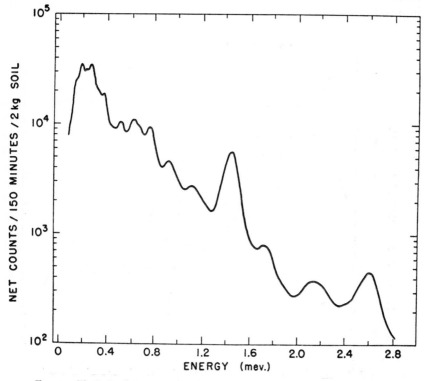

Fig. 1.—Typical γ-ray spectrum of soil with background spectrum removed

being contained in the re-entrant cups. These standards are used for frequent calibration of the entire system. A more detailed description, including the actual coefficients used and the numerical values obtained from a specific soil sample, is given in the Appendix to this paper. Solution of the appropriate equations now gives the concentration of natural activity in terms of grams of uranium, thorium, or potassium per gram of soil. Likewise, the contribution of the natural γ emitters to the total net γ spectrum arising from soil may also be determined, as shown in Figure 2, where the total spectrum is broken down into that due to these three components of natural activity.

An analogous procedure is now used to obtain the concentrations of various γ-emitting fission products also present in the soil. Known quantities of cerium-141, cerium-144–promethium-144, ruthenium-103, ruthenium-106, cesium-137, zirconium-95–niobium-95, and barium-140–lanthanum-140 are mixed with about 2 kg. of sodium phosphate. When placed in the re-entrant cups, these are suitable standards for determining the coefficients for use in the simultaneous equations and for obtaining the quantities of the various radionuclides present in a soil.

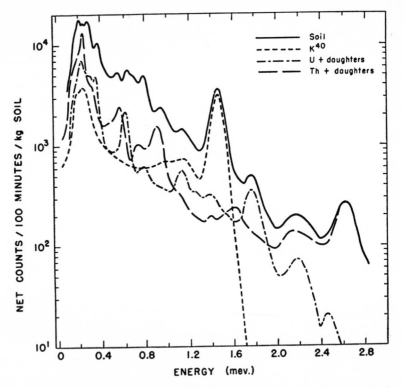

Fig. 2.—Gamma-ray spectrum of soil showing the contribution of uranium + daughters, thorium + daughters, and potassium-40.

If considerable time has elapsed (a year or longer) between the time of collection and the time of measurement of a soil specimen, the concentrations of cerium-141, ruthenium-103, and even zirconium-95–niobium-95 are greatly reduced and may be neglected (with caution). Similarly, if nuclear testing has not occurred within a year or longer, as was true from the beginning of 1960 to September, 1961, one may reasonably neglect these short-lived emitters. In such circumstances, the spectral contribution from uranium, thorium, and potassium-40, in the energy region 0.1–0.75 mev., is substracted. The resultant spectrum is the type shown in Figure 3, where the γ-ray lines of cerium-144, ruthenium-106, and cesium-137 are apparent. In the actual numerical solution of a complex spectrum, such stripping as illustrated in Figure 3 is not carried through, but rather the uranium, thorium, and potassium contributions in three regions—0.1–0.2, 0.42–0.58, and 0.58–0.75 mev.—are determined and removed. Concentrations of cesium-137, cerium-144, and ruthenium-106 are then determined by the solution of a 3×3 matrix (see Appendix, this article). An additional region, 0.2–0.42 mev., is also considered as a check region to provide an indication of the over-all validity of the solution.

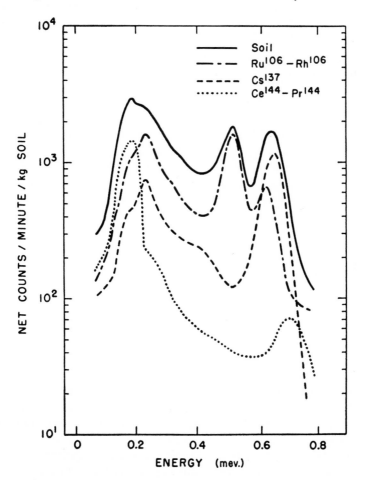

Fig. 3.—Gamma-ray spectrum of soil (uranium, thorium, and potassium-40 contribution removed) showing cerium-144, ruthenium-106, and cesium-137.

The sum of the contributions of uranium, thorium, potassium-40, cesium-137, cerium-144, and ruthenium-106 in the check region, as derived from the foregoing procedures, is compared with the net count in the region. If there is close agreement, the solution is considered reasonable. Lack of agreement may indicate errors in data manipulation, the presence of short-lived components that should be considered, or possibly the presence of other nuclides that have not been taken into account. Close agreement per se does not guarantee the accuracy of the determination, since there is no unique combination of intensities among the six sources considered that will lead to the observed net count. Such agreement does strongly suggest, however, that one has not grossly over- or underestimated any one component relative to another, nor have additional components of appreciable intensity been overlooked. Manganese-54 was found in soils collected in 1958 by means of such a check region and was verified by using an additional check region from 0.75–0.90 mev.

Because of the presence of antimony-125 in samples of surface air, one must look for it in soil. Analysis by γ spectrometry, considering now four regions and four unknowns after removal of uranium, thorium, and potassium, has failed to indicate conclusively the presence of this nuclide. However, the residual counts in the 0.2–0.42-mev. region may be used to place a limit on the amount of antimony-125 likely to be present. This approach indicated levels of antimony-125 $\leq$0.2 times those of cesium-137 in Argonne soils in August, 1961, with somewhat higher antimony/cesium ratios probable at this time.

A somewhat more lengthy process must be followed when cerium-141, ruthenium-103, zirconium-95–niobium-95, and possibly barium-140–lanthanum-140 are also present in soil. First, an initial spectrum is taken; the uranium, thorium, and potassium determined as before; and the contribution of these sources removed in the regions 0.1–0.2, 0.4–0.54, 0.54–0.68, and 0.68–0.80 mev. Two to three months later the same soil is again analyzed by the same procedure. The difference between initial and final values in the four regions yields a spectrum such as that shown in Figure 4. In this particular instance, cerium-141, ruthenium-103, and zirconium-95–niobium-95 are present, with no indication of barium-140–lanthanum-140. Because of the rather severe complication introduced by barium-140-lanthanum-140, soil cores are generally not counted until some 3–4 weeks after collection to allow for decay of a substantial amount of the radionuclide pair. The quantities of these short-lived fission products are now found by the solution of a 3 $\times$ 3 or 4 $\times$ 4 matrix, whichever is appropriate. For example, the amount of cerium-141 found by such a method is actually the quantity that has decayed in the interval between the initial and the final count; the quantity present initially is readily calculated, as is that present at any subsequent time. In like manner the initial amounts of zirconium-95–niobium-95 and ruthenium-103 are found. The contributions from cerium-141, zirconium-95–niobium-95, and ruthenium-103 in the initial and in the final spectrum are removed in the three energy regions 0.1–0.2, 0.42–0.58, and 0.58–0.75 mev., the thorium, uranium, and potassium-40 contributions already having been removed. One now is in a position to determine cerium-144, ruthenium-106, and cesium-137 as was done in the absence of the short-lived fission products. Agreement between initial and final values for the longer-lived components implies that the short-lived components have been determined correctly.

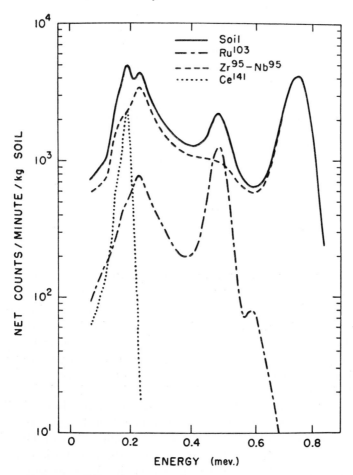

Fig. 4.—Gamma-ray difference spectrum of soil showing cerium-141, ruthenium-103, and zirconium-95–niobium-95.

One now has for a given soil sample the concentration of the natural and fission activities, and both may be expressed in activity per unit mass of soil. More frequently, uranium, thorium, and potassium are given in terms of grams of element per gram of soil, whereas fission products are expressed in curies per gram of soil. When one is ultimately interested in determining the dose arising from the γ emitters in soil, it is necessary to convert the concentration of fission products into activity *per unit area*. This may be done quite simply: the total mass and the diameter of the core are known, and one may

TABLE 1

FACTORS RELATING THE CONCENTRATION OF
URANIUM, THORIUM, AND POTASSIUM AND
THE AIR-DOSE RATE

$$D_U = 0.779 \times 10^6 \, S_U \ \mu r/hr,$$
$$D_{Th} = 0.375 \times 10^6 \, S_{Th} \ \mu r/hr,$$
$$D_K = 167.3 \qquad S_K \ \mu r/hr,$$

where concentration, S_i, is expressed in gm. of element per gm. of soil.

AVERAGE CONCENTRATIONS OF URANIUM, THORIUM,
AND POTASSIUM IN ARGONNE SOIL AND THEIR
DOSE-RATE CONTRIBUTIONS

Element	Concentration, S_i (gm/gm Soil)	Air-Dose Rate ($\mu r/hr$)
Uranium............	2.75×10^{-6}	2.14
Thorium............	5.33×10^{-6}	2
Potassium...........	1.40×10^{-2}	2.34
Total............	....	6.48

assume that the fission products have been uniformly mixed throughout this mass by the pulverizing process in the ball mill. Concentrations of fission-product deposition are usually expressed in mc/mi^2.

It is now possible to compute the γ-ray dose in air from the values found above. For the natural activities a method similar to that of Hultqvist (1956) and of O'Brien *et al.* (1958) is employed. In the latter method the energy absorbed per gram of air is presumed to be one-half that per gram of soil, since in air the solid angle is only 2π rather than 4π as in soil. The actual factors used are shown in Table 1. The air-dose rate (due to fission products) at 1 meter above the surface of a flat plane is derived from a modification of Dunning's (1957) calculations. This takes into account the distribution of activity in depth, which has been determined at the Argonne site, and introduces corrections for absorption and dose build-up in soil. In addition, air absorption has also been considered in integrating over the infinite plane. The factors used in obtaining the fission-product dose rate for a number of nuclides (Table 2) are in essence analogous to infinite-plane dose factors corrected for weathering.

Calculated air-dose rates for fission and natural radioactivity as derived from γ-ray spectrometry of soil cores are shown in Table 3. For the average topsoil at Argonne, the dose from uranium, thorium, and potassium-40 is 6.5 $\mu r/hr$; for clay about 20 feet below the surface, the dose rate was somewhat higher, 8.6 $\mu r/hr$. The fission-product dose rate has varied during the period in question from 5.3 to 7.1 $\mu r/hr$. In order to obtain the total open-field dose rate, it is necessary to include the contribution from cosmic rays; we have used the value of 4 $\mu r/hr$ determined by both Solon et al. (1960) and Kastner (private communication) at this latitude and elevation.

TABLE 2

FACTORS RELATING FISSION-PRODUCT CON-
CENTRATION (MC/MI²) AND AIR-DOSE
RATE (μR/HR) AT 1 METER

Radionuclide	Factor $\mu r/hr/mc/mi^2$
Cesium-137..................	3.4×10^{-3}
Zirconium-95................	4×10^{-3}
Niobium-95..................	4.2×10^{-3}
Ruthenium-103..............	2.75×10^{-3}
Ruthenium-106..............	1.3×10^{-3}
Cerium-141.................	0.35×10^{-3}
Cerium-144–promethium-144....	0.17×10^{-3}
Barium-140.................	1×10^{-3}
Lanthanum-140..............	16×10^{-3}

TABLE 3

CALCULATED AND OBSERVED AIR-DOSE RATE 1 METER ABOVE
GROUND, μR/HR, ARGONNE NATIONAL LABORATORY

DATE	CALCULATED				OBSERVED (TOTAL)
	Natural	Fission Products	Cosmic Rays*	Total	
9/27/62......	6.5	5.3	4	15.8	13.8
10/26/62.....	6.5	5.3	4	15.8	14
11/27/62.....	6.5	6.3	4	16.8	14.6
3/28/63......	6.5	7.1	4	17.6	17
3/29/63†.....	8.6	0	4	12.6	11

* See text.
† Soil collected and dose measured in excavation near Bldg. D203.

The calculated total dose rate has been compared with that measured directly by Kastner *et al.* (1963) with a muscle-equivalent ionization chamber, as described in this symposium. Dose measurements were made at the soil-collection sites, and the calculated doses were obtained from soil samples collected on the date of dose measurements. A series of measurements were also made in a recent excavation, over soil free from fission-product contamination; here the calculated and the measured natural and cosmic-ray dose rates can be compared without the additional uncertainties of fission-product dose estimates.

In all instances the calculated dose exceeds the measured dose by 10–20 per cent, as shown in Table 3. Considering the multitude of uncertainties, particularly in the calculated dose, the agreement is quite encouraging.

Further investigation is certainly indicated, particularly determination of the degree of radon-222 and radon-220 equilibrium in soil as a function of depth and moisture content. Since soils are not sealed in the re-entrant cups, radon-222 and radon-220 may escape, hence the concentrations of uranium-238 and thorium-232 inferred from the measurements may be too low. Present indications are that they are not greatly in error, but further evaluation of the absorption of soft γ radiation by the walls of the ionization chamber may prove otherwise. Much of the γ flux coming from the soil *in situ* is in the form of low-energy photons resulting from severe degradation of the relatively energetic primaries.

At present the method described herein appears to be wholly adequate as a means of determining the contribution of natural and fission-product γ-emitting radioactivity to the total dose rate measured *in situ*.

REFERENCES

DUNNING, G. M. 1957. Discussion of radiological safety criteria and procedures for public protection at the Nevada Test Site. *In* Hearings before the Special Subcommittee on Radiation of the Joint Committee on Atomic Energy, Congress of the United States, 85th Congress, First Session on "The Nature of Radioactive Fallout and Its Effects on Man." Washington, D.C.: U.S. Govt. Printing Office, Part 1, pp. 239–40.

HULTQVIST, B. 1956. Studies on naturally occurring ionizing radiations, with special reference to radiation doses in Swedish houses of various types. Kungl. Svenska Vetenskapsakademiens Handlingar, Vol. **6**, No. 3.

KASTNER, J. Personal communication.

KASTNER, J., F. R. SHONKA, and J. E. ROSE. 1963. Natural environmental radiation measurements utilizing muscle-equivalent ion chambers. This symposium.

O'BRIEN, K., W. M. LOWDER, and L. R. SOLON. 1958. Beta and gamma dose rates from terrestrially distributed sources. Radiation Res. **9**:216–21.

SOLON, L. R., W. M. LOWDER, A. SHAMBON, and H. BLATZ. 1960. Investigations of natural environmental radiation. Science, **131**:903–6.

APPENDIX

To show in more detail the actual procedures involved in determining the concentration of the various radionuclides in soil, a specific soil sample analysis will be described. The soil to be considered was collected in Argonne Park on September 27, 1962. The entire core had a dry weight of 4,900 gm. The first γ-ray counting was done 6 days after collection. The sample was counted again on March 18, 1963, after an interval of 166 days. The same mass of soil, 1,930 gm., was analyzed in both instances.

The concentrations of uranium, thorium, and potassium were determined in the three energy regions by means of the following equations:

1.35–1.55 mev:

$$\text{Net count}/100 \text{ min}/1.93 \text{ kg} = 29{,}183 = K + 0.37858 \text{ U} + 0.19364 \text{ Th.}$$

1.65–2.20 mev:

$$\text{Net count}/100 \text{ min}/1.93 \text{ kg} = 6{,}778 = 0.53248 \text{ U} + 0.31983 \text{ Th.}$$

2.20–2.80 mev:

$$\text{Net count}/100 \text{ min}/1.93 \text{ kg} = 4{,}404 = 0.08893 \text{ U} + 0.48652 \text{ Th.}$$

The solutions of these equations yielded

$$U = 8{,}192 \text{ counts}/100 \text{ min}/1.93 \text{ kg,}$$

$$Th = 7{,}555 \text{ counts}/100 \text{ min}/1.93 \text{ kg,}$$

$$K = 24{,}619 \text{ counts}/100 \text{ min}/1.93 \text{ kg,}$$

leading to concentration values of

$$S_U = \frac{8{,}192}{82{,}210} \times \frac{5}{100} \times \frac{1 \text{ gm. U}}{1.93 \times 10^3 \text{ gm. soil}} = 2.58 \times 10^{-6} \text{ gm U/gm soil ,}$$

$$S_{Th} = \frac{7{,}555}{37{,}440} \times \frac{5}{100} \times \frac{1 \text{ gm. Th}}{1.93 \times 10^3 \text{ gm. soil}} = 5.23 \times 10^{-6} \text{ gm Th/gm soil ,}$$

$$S_K = \frac{24{,}619}{39{,}670} \times \frac{5}{100} \times \frac{9.52 \times 10^2 \text{ gm. K}}{1.93 \times 10^3 \text{ gm. soil}} = 1.53 \times 10^{-2} \text{ gm K/gm soil .}$$

The standards were all counted for 5 minutes and the net counts/5 min in the three regions were used to normalize the soil readings. The figure of 9.52×10^2 gm. potassium is that amount of potassium in 4 lb. of chemical-grade potassium chloride.

The concentrations of uranium, thorium, and potassium were redetermined from the analysis made on March 18, 1963. The uranium, thorium and potassium contributions were subtracted in the energy regions 0.1–0.2, 0.40–0.54, and 0.68–0.80 mev. The March 18 results in these three regions were then subtracted from those obtained on October 3 to get the difference values shown below. The additional region for barium-140–lanthanum-140 is not included, since this radionuclide pair was not detected in this particular soil sample.

0.10–0.20 mev.:

$$\text{Counts}/100 \text{ min}/1.93 \text{ kg} = 16{,}495 = \text{Ce} + 0.6092 \text{ Ru} + 0.4349 \text{ Zr.}$$

0.40–0.54 mev.:

$$\text{Counts}/100 \text{ min}/1.93 \text{ kg} = 3{,}561 = 0.3908 \text{ Ru} + 0.0926 \text{ Zr.}$$

0.68–0.80 mev.:

$$\text{Counts}/100 \text{ min}/1.93 \text{ kg} = 11{,}747 = 0.4726 \text{ Zr.}$$

The solutions of these three equations were

$$\text{Zr}^{95} = 24{,}860 \text{ counts}/100 \text{ min}/1.93 \text{ kg.}$$
$$\text{Ru}^{103} = 3{,}224 \text{ counts}/100 \text{ min}/1.93 \text{ kg.}$$
$$\text{Ce}^{141} = 3{,}719 \text{ counts}/100 \text{ min}/1.93 \text{ kg.}$$

The values given above represent the following fractions of the October 3 counts:

$$\text{Zr}^{95} = 1 - e^{-1.7589} = 1 - 0.1722 = 0.8278 \,.$$
$$\text{Ru}^{103} = 1 - e^{-2.8726} = 1 - 0.0565 = 0.9435 \,.$$
$$\text{Ce}^{141} = 1 - e^{-3.4650} = 1 - 0.0313 = 0.0687 \,.$$

Therefore the values on October 3 were

$$\text{Zr}^{95} = 30{,}031 \text{ counts}/100 \text{ min}/1.93 \text{ kg} \,.$$
$$\text{Ru}^{103} = 3{,}417 \text{ counts}/100 \text{ min}/1.93 \text{ kg} \,.$$
$$\text{Ce}^{141} = 3{,}839 \text{ counts}/100 \text{ min}/1.93 \text{ kg} \,.$$

The difference value in the check region, 0.20–0.42 mev., was 9,240 counts/100 min, whereas the contributions from zirconium-95, ruthenium-103, and cerium-141 totaled 9,633, representing a surplus of 393 counts or a computed value 4.3 per cent higher than the observed difference.

The October 3 concentrations of zirconium-95, ruthenium-103, and cerium-141 were in $(10^{-10}$ c/kg):

$$Zr^{95} = 4.04.$$

$$Ru^{103} = 1.16.$$

$$Ce^{141} = 1.52.$$

Making use of the fact that the total core weight was 4.9 kg. and the diameter was 6 inches, the following activities in mc/mi² were computed for October 3 and corrected for decay to the time of collection.

	October 3	September 27
Zr^{95}............	281 mc/mi²	300 mc/mi²
Ru^{103}..........	81 mc/mi²	90 mc/mi²
Ce^{141}..........	106 mc/mi²	120 mc/mi²

The niobium-95 appeared to be initially present in 78 per cent of equilibrium in this sample. Hence on September 27 the corresponding niobium-95 activity was 510 mc/mi.²

The amount of cesium-137, ruthenium-106, and cerium-144 may now be determined using the three energy regions 0.10–0.20, 0.42–0.58, and 0.58–0.75 mev. This is illustrated for the October 3 measurements, where the contributions from uranium, thorium, potassium, zirconium-95 —niobium-95, ruthenium-103, and cerium-141 are also indicated.

Region	Soil Net	U	Th	K	Zr^{95}–Nb^{95}	Ru^{103}	Ce^{141}	Net'
0.1 –0.2 mev.	150,213	37,203	49,047	30,342	13,061	2,081	3,839	23,640
0.42–0.58 mev.	28,960	4,818	7,703	4,842	2,780	1,335	.....	7,482
0.58–0.75 mev.	54,581	10,724	9,801	7,817	15,381	.....	.....	10,858

The figures in the Net' column are presumed to be due to cerium-144, ruthenium-106, and cesium-137, the amounts of which may be determined from the following equations:

0.1–0.2 mev.:

$$23{,}640 = 0.89218\ Ce + 0.50881\ Ru + 0.41869\ Cs.$$

0.42–0.58 mev.:

$$7{,}482 = 0.02959\ Ce + 0.32575\ Ru + 0.08609\ Cs.$$

0.58–0.75 mev.:

$$10{,}858 = 0.07823\ Ce + 0.16544\ Ru + 0.49521\ Cs.$$

The solutions of these equations are

$$Cs^{137} = 14{,}327 \text{ counts/100 min/1.93 kg.}$$

$$Ru^{106} = 18{,}339 \text{ counts/100 min/1.93 kg.}$$

$$Ce^{144} = 9{,}314 \text{ counts/100 min/1.93 kg.}$$

The Net' value in the check region, 0.2–0.42 mev., was 9,635 counts/100 min, 8,965 counts of which were attributable to cesium-137, ruthenium-106, and cerium-144, leaving 670 counts or 7.5 per cent unaccounted for.

The concentration per kg. and the concentration per mi.² of cesium-137, ruthenium-106, and cerium-144 were found by the same method as that used for the short-lived emitters. The activities on September 27, 1962, were

$$Cs^{137} = 220 \ mc/mi^2.$$

$$Ru^{106} = 660 \ mc/mi^2.$$

$$Ce^{144} = 1,075 \ mc/mi^2.$$

The γ-ray dose rate due to the measured concentrations of natural and fission radioactivity may now be determined using the factors shown in Tables 1 and 2 of the text.

For uranium, thorium, and potassium-40 we have

$$D_U \ = 0.779 \times 2.58 = 2.01 \ \mu r/hr.$$

$$D_{Th} = 0.375 \times 5.23 = 1.96 \ \mu r/hr.$$

$$D_K \ = 167.3 \times 1.53 \times 10^{-2} = 2.56 \ \mu r/hr.$$

$$\text{Total} = 6.53 \ \mu r/hr.$$

And for the fission products, the rates on September 27, 1962, were calculated to be

$$D_{Cs}{}^{137} = 3.4 \ \times 10^{-3} \times 2.20 \ \times 10^2 = 0.75 \ \mu r/hr.$$

$$D_{Zr}{}^{95} \ = 4.0 \ \times 10^{-3} \times 3.00 \ \times 10^2 = 1.20 \ \mu r/hr.$$

$$D_{Nb}{}^{95} \ = 4.2 \ \times 10^{-3} \times 5.10 \ \times 10^2 = 2.14 \ \mu r/hr.$$

$$D_{Ru}{}^{103} = 2.75 \times 10^{-3} \times 0.90 \ \times 10^2 = 0.25 \ \mu r/hr.$$

$$D_{Ru}{}^{106} = 1.3 \ \times 10^{-3} \times 6.60 \ \times 10^2 = 0.86 \ \mu r/hr.$$

$$D_{Ce}{}^{141} = 0.35 \times 10^{-3} \times 1.20 \ \times 10^2 = 0.04 \ \mu r/hr.$$

$$D_{Ce}{}^{144} = 0.17 \times 10^{-3} \times 10.75 \times 10^2 = 0.18 \ \mu r/hr.$$

$$\text{Total} = 5.42 \ \mu r/hr.$$

The total obtained from this soil, including 4 $\mu r/hr$ for cosmic radiation, on this date was 15.9 μ/hr. The average dose rate measured with the muscle-equivalent chamber on the same date was 13.8 $\mu r/hr$. Hence the calculated dose rate is 1.15 times the observed value.

HAROLD A. WOLLENBERG AND ALAN R. SMITH

32. Studies in Terrestrial γ Radiation

O
UR STUDIES of terrestrial γ radiation were motivated by a specific problem: to determine the most suitable mineral materials to use in construction of the massive shield of a large low-background counting facility. These investigations prompted the development of techniques that have general application to environmental radiation studies and to certain geophysical problems. We have also collected a considerable body of data bearing on both the environmental γ-ray intensity and the concentrations of several γ emitters in a variety of rock formations.

We first discuss theoretical considerations, then describe our techniques for both field work and laboratory analysis. Finally, to illustrate applications of these techniques, we present some data organized to conform to the context of this meeting.

THEORETICAL CONSIDERATIONS: RADIOACTIVE ELEMENTS

Natural radioactivity in the earth's crust varies markedly with rock types. The principal radioactive elements that contribute to the radiation emanating from rocks are potassium-40, the uranium series, and the thorium series. These three elements all have very long half-lives, greater than 10^9 years, and all three emit γ rays as they undergo radioactive decay.

Potassium is present in almost all the earth's crustal material, and its isotope of mass 40 is radioactive. The radioisotope potassium-40 comprises 0.0119 per cent of natural potassium, and the ratio of potassium-40 to potassium-39, $1/8,500$, is considered to be fairly constant throughout the earth's crust. Potassium-40, with a 1.3×10^9-year half-life, has a branched decay scheme: about 89 per cent goes to calcium-40 with emission of a 1.3-mev. β particle; about 11 per cent goes

DR. WOLLENBERG is geologist and ALAN R. SMITH is physicist in the Health Physics Department, Lawrence Radiation Laboratory, University of California, Berkeley, California.

Work done under the auspices of the U.S. Atomic Energy Commission.

to argon-40 via electron capture and emission of a 1.46-mev. γ ray.

The principal potassium-bearing minerals are the feldspar ortho-clase and the muscovite and biotite micas. Hornblende and plagioclase may contain up to 1 per cent potassium. The clay mineral, illite, also has potassium as a principal constituent. Though not a principal constituent of the clay mineral, montmorillonite, potassium may be incorporated by cation exchange.

In igneous rocks, the concentration of potassium varies roughly with the abundance of silica, potassium being more prevalent in the acidic igneous rocks, such as granite, than in the ultramafics (peridotite, dunite, and serpentine). Rankama and Sahama (1950) show that because of the large ionic radius of potassium (1.33 Å) and its twelvefold coordination with respect to oxygen, the element is excluded from the early-formed crystallates of magmatic differentiation and becomes enriched in residual melts and solutions. Potash feldspars are therefore characteristic of the late crystallates, the igneous rocks most abundant in silica.

Ahrens (1954) gives the following concentrations of potassium for various igneous rock types:

In granite, the range is from 2 to 6 per cent.

In basalts, potassium concentrations vary with individual flows (potassium being sensitive to fractionation in a basaltic magma). A variation between 0.65 and 1.4 per cent was found by Ahrens *et al.* (1952) in Columbia River basalt. Daly (1933) reported 0.37 per cent in oceanic basalt and 0.65 per cent in plateau basalt.

Peridotite, pyroxenite, and dunite and their serpentinized forms have the lowest potassium concentrations of the igneous rocks—about 10 p.p.m.

The potassium content of sedimentary rocks depends largely upon the relative amounts of the feldspars, micas, and clay minerals that partially comprise the mineral-aggregate sediments. A sandstone derived from a close granitic source would contain an appreciable amount of feldspar and therefore exhibit a potassium content roughly that of its source granite. A pure quartz sandstone derived from a quartzitic source, or a sandstone at a distance from its granitic source great enough that the feldspars have been removed during the transport, would contain a relatively low potassium concentration.

Shales or argillaceous sediments with an abundance of mica and clay minerals contain appreciable potassium. Limestones are generally low in alkalies, though the presence of the authigenic feldspars and some argillaceous material filling cracks in the limestone (hydromica developing from detrital clays during diagenesis of the limestone) may increase the percentage of potassium over that of pure limestone.

Average potassium contents of the broad groups of sedimentary rocks (Ahrens, 1954) are: shales and argillaceous sediments, 3 per cent; sandstones, >1 per cent; and limestones, tenths of 1 per cent.

URANIUM AND THORIUM

Uranium-238 (half-life 4.49×10^9 years) and thorium-232 (half-life 1.39×10^{10} years) and their decay products are also major contributors of radioactivity in the earth's crust.

Uranium and Thorium in Igneous Rocks. As with potassium, the uranium and thorium contents in igneous rocks vary with the percentage of silica. Along with being potassium rich, the granites are usually uranium and thorium rich, while the ultramafics are quite lean in all three elements. Keevil (1944) gives the probable value for the thorium-uranium ratio in igneous rock as 3–3.5. Evans and Williams (1935) show an almost linear relationship of potassium oxide with uranium concentration in the volcanic rocks of the Lassen Peak region. J. A. S. Adams (1954) attributes the nearly linear relationship to systematic concentration of both uranium and potassium in the liquid phase of the magma as crystal fractionation proceeded. Thus, uranium and potassium oxide were excluded to a degree from the common minerals formed during crystallization of the Lassen magmas. Uranium and potassium oxide do not substitute readily for other common ions because of the different space requirements of potassium and uranium. Their ionic radii and coordination numbers with respect to oxygen do not permit their ready substitution by other common ions. This would suffice as a partial explanation of the higher uranium and potassium concentrations in the more acidic igneous rocks.

Larsen *et al.* (1956) show from their study of the uranium concentrations of three Mesozoic batholiths of the western United States that, where fractional crystallization is assumed to have been the major factor in magmatic differentiation, uranium is enriched in the youngest rocks, those being high in silica and potassium oxide and low in calcium oxide and magnesium oxide. A maximum enrichment of greater than 20 p.p.m. is found in differentiates very poor in calcium oxide. From chemical analyses of samples from the southern California batholith, Larsen *et al.* show that the uranium content of the common rock-forming minerals increases with total uranium in the bulk rock as the percentage of silica increases in the rock.

Davis and Hess (1949) have studied the concentration of radium in ultramafic rocks. Average values for igneous rocks range from 0.01 pgm/gm of radium in ultramafics to 1.0 pgm/gm in felsic rocks. The absence of uranium (parent element for the radium) in ultramafics is attributed to its strong concentration in residual liquids during magmatic differentiation, with uranium being apparently excluded from the crystal structures of the early-formed minerals, olivine, pyroxene, spinel, and plagioclase. Uranium is tetravalent at high temperatures with an ionic radius of 0.97 Å. This size may permit tetravalent uranium replacement of bivalent calcium or monovalent sodium, but an eightfold coordination is required with oxygen atoms. No such positions are available in the crystal structures of these early-formed minerals.

By comparing the concentration of radium in ultramafic rock samples with the per cent weight loss of the samples at 1,000° C. (representing essentially the percentage of water present in the original rock), Davis and Hess (1949) show that the parent uranium was probably contained in the interstitial liquid trapped between accumulated crystals in the cooling magma.

In examining the distribution of radium in individual minerals comprising a sample of dunite, Davis and Hess found that tremolite (1.2 per cent of the total sample composition) contributed 20 per cent of the total radium; serpentine, talc, and kammererite (4.8 per cent of the total sample composition) contributed 30 per cent; while olivine (92 per cent of the total composition) contributed 45 per cent.

Uranium and Thorium in Sediments. Koczy (1954) states that, even though uranium is highly insoluble, the uranyl ion is able to form complex compounds, which are generally soluble. Thorium is also highly insoluble but does not form soluble compounds. Therefore, uranium can be carried in an oxidizing environment, while thorium isotopes cannot. In river water, uranium-238 is transportable in both solution and suspension, while thorium-232 is chiefly contained in mineral resistates and is hardly transportable in the dissolved condition because of its tendency to hydrolyze. Traces of uranium are also present in resistates. Therefore, because of the general insolubility of resistates in sea water, thorium and some uranium are deposited as heavy elements, usually near the coast or in river-mouth sediments. When compounds containing the uranyl ion reach a reducing (sapropelic) environment, uranyl complexes become reduced, and uranium is precipitated or absorbed as a tetravalent ion. Such areas as the Black Sea and Norwegian fjords are examples of present-day sapropelic areas. A concentration of 50–100 p.p.m. uranium is reported in late Quaternary sediments from Norwegian fjords. The concentration of uranium in sea water reaches a maximum at a depth of about 1,000 m. because of the reducing environment present at this level.

Petterson (1954) attributes the deficiency of radium-226 in sea water (there should be about 5×10^{-16} gm/ml radium-226 in sea water, but only about one-sixth of that is present) to the precipitation of the uranium daughter product, thorium-230, from sea water along with ferric hydroxide. Therefore, a potential source of radium-226 is carried down to the surface of the sea floor. In the sea-floor sediments, the fall-off of the radium content with depth in the sediments is irregular. There are two sharp maxima between 0 and 20 cm., and very low radium values from 40 to 80 cm. Petterson proposes that radium migrates from its mother substance, thorium-230, so that radioactive equilibrium between the two elements is not maintained.

Bell (1954) states that the separation of uranium from aqueous solutions can take place in reducing environments in the presence of carbonaceous material and sulfides and in the absence of dissolved oxygen. Uranium is adsorbed by clays, carbon, aluminum, manganese, and silica and is accumulated with phosphatic marine sediments. The precipitation or adsorption of uranium is inhibited by the presence of the carbonate ion, which accounts for the generally low uranium concentration in limestones and dolomites.

The highest syngenetic concentrations of uranium occur in sapropelic marine shales. Large amounts of organic material and sulfides and a scarcity of carbonate are characteristic of these shales. All known uraniferous black shales are of pre-Mesozoic age. As opposed to the sapropelic sediments, the humic sediments (peats, lignites, and coals), which are characterized by low hydrogen and high carbon and oxygen contents, are generally low in uranium content. The presence of uranium in some Tertiary and Cretaceous lignites is attributed by some geologists to epigenetic deposition in the lignites, and by others to syngenetic deposition of uranium with the carbonaceous material.

The occasionally appreciable uranium contents of colloidal sediments, such as chert (formed as gelatinous or colloidal precipitates), may be due to the adsorption of the uranyl ion onto the silica-gel surface.

Uranium and Thorium in Accessory Minerals. Accessory minerals—primarily zircon and to some extent apatite, cassiterite, sphene, and rutile—carry appreciable but varied amounts of uranium and thorium. The phosphate monazite is the principal thorium ore mineral. Frondel (1956) gives the average concentration of thorite ($ThSiO_4$) as 10–12 per cent in monazite and up to 1 per cent in zircon.

Larsen *et al.* (1956) state that the igneous accessory minerals—zircon, sphene, and apatite—have uranium concentrations of over 300 p.p.m. The uranium concentration of zircon increases with the radioactivity of the host rock from gabbro to granite.

Rankama (1954) attributes pleochroic discoloration in such minerals as biotite, chlorite, amphiboles, andalusite, and quartz to the effect on crystal structures of α and β radiation from uranium and thorium. Pleochroic halos around minerals are most numerous in silicic igneous rocks and are practically nonexistent in the ultramafic igneous rocks. Radioactive minerals, such as zircon, rutile, cassiterite, and apatite, normally comprise the nuclei of the halos. The halos are usually made up of rings with characteristic radii. A ring in biotite due to uranium-238 α emission has a radius of 12.7 microns, while a ring due to thorium-232 in biotite has a 12.4-micron radius. Measurements of ring radii can thus be used to determine the range of α-particle penetration in crystals as well as the α particles' energy.

The effect of α emission on some minerals, notably zircon, results in production of nearly complete isotropy. The mineral appears in the amorphous state and is said to be metamictized. All zircons show some degree of radiation damage, and the primary cause of this metamictization is the presence of uranium and thorium in the mineral structure.

As well as being sources of radioactivity in igneous rocks, accessory minerals, such as zircon, sphene, rutile, etc., have an appreciable effect on the radioactivity of sedimentary rocks. These minerals have a high specific gravity (about 3.5), so they are concentrated with metallic minerals during deposition. When the radioactivity of zones of heavy-mineral deposition in the Ione formation of central California is correlated with their mineralogy (T. C. Slater, private communication), there is evidence that the zones of highest radioactive background correspond to the areas of heavy-mineral deposition.

FIELD PROCEDURES

Our field operations normally involve a threefold procedure: determination of the geologic setting of the rock or soil exposure under study, radiometric scanning with a portable scintillation counter, and sampling. A previous perusal of reports and geologic maps of the area studied aids us in the evaluation of the geologic setting of a site. Observations are made of nearby sources of contaminating material (e.g., higher radioactivity) that may influence the field counting rate of the subject soil or rock.

Sampling. Sampling procedures vary depending upon the nature of the material under study. On bedded or laminated outcrops pick-sample lines are cut perpendicular to the layered orientation. Homogeneous outcrops are chip-sampled with an attempt to obtain fresh material with as little fallout contamination as possible. Nearby outcrops of rocks with possibly different radiometric characteristics in the vicinity of a sampling site are also sampled to aid in the determination of their influences on the site's field counting rate. Rock samples are generally contained and transported in half-gallon cardboard ice-cream cartons.

Our work in soils has been mainly in conjunction with evaluations of the fallout component of radioactivity. Therefore, most of our soil samples are surface samples and incorporate varying amounts of vegetation. Soil sampling consists of taking the top 1–1½ inches of an area of roughly 2 × 2 feet. In some areas where we wish to determine the depth penetration of fallout we have also sampled the barren soil directly beneath the grass-root zone. Soil samples are generally contained in polyethylene "poultry bags" with the opening tied off to insure little, if any, moisture loss. Because of their light weight and small space consumption we sometimes use heavy-duty polyethylene bags for rock samples as well as soil.

Our general field-sampling philosophy is to get the most representative sample of a rock outcrop or soil site from a minimum amount of material; therefore, the sampling method varies somewhat from site to site.

PORTABLE INSTRUMENT

The radiation detection instrument used for the field work was designed and built at the Lawrence Radiation Laboratory; details of the circuitry have been published by Goldsworthy (1960). A brief description of the instrument is included here, followed by a general outline of the way it was used in this project.

The detector is a 3-inch-diameter by 3-inch-high thallium-activated sodium iodide (NaI[Tl]) scintillation crystal, viewed by a DuMont 6363 3-inch-diameter photomultiplier tube. The detector assembly is housed in a 5-inch-diameter by 12-inch-high thin-walled stainless-steel case and is connected to the indicator unit by 5 feet of coaxial cable. The indicator unit contains a Cockroft-Walton-type high-voltage supply for the phototube, a four-transistor linear pulse amplifier, an integral pulse-height selector circuit, and a multirange count-rate-sensitive indicator circuit. All electrical power for the instrument is supplied by a single, self-contained 10.75-volt mercury battery. A fresh battery can be expected to provide at least 300 hours of instrument operation time. The relatively large sodium iodide crystal provides a count rate high enough that we can always achieve a steady and reproducible reading from the count-rate meter; thus we do not have to resort to earphones, even when assaying the lowest intensities.

Four linear ranges are used in the count-rate circuitry to span the wide latitude of radiation intensity encountered. The ranges are calibrated to span intervals of 0–100, 0–500, 0–5,000, and 0–50,000 c.p.s., by the use of a single meter scale marked with 50 equal divisions. These calibrations are established by use of a variable-frequency pulse generator to determine correct adjustment of circuit values; both full-scale deflection and scale linearity are thereby checked.

After the electronic calibration is satisfactory, we calibrate the instrument in terms of radiation intensity. The detector unit is first connected to the 100-channel pulse-height analyzer, and we record the γ-ray spectrum produced by room background. Our original examination of this spectral shape, along with several other considerations, showed that the instrument threshold should lie above 100 kev.; we chose a value of approximately 120 kev. After the background count rate has been determined from pulse-height analyzer data, the detector is reconnected to the indicator unit, and the threshold control is adjusted until the meter indicates the correct count rate. Of course, these two operations are performed in quick succession, while measuring the same background intensity.

We then expose the detector to known intensities of γ rays from radium-226 sources certified by the U.S. National Bureau of Standards. Using count rates observed from these precisely known radiation intensities (corrected for background), we can establish calibrated ranges in terms of radium-226 (in equilibrium with its decay products).

This calibration procedure is performed infrequently, usually when there is evidence of erroneous instrument behavior. We employ a simpler method with a disk source of refined uranium for routine calibration checks. The correct count rate must be observed from this source when it is in a standard position (against the detector-case end plate nearest the crystal) if instrument readings are to be considered valid. The γ-ray spectrum from this check source has a steep slope at the energy corresponding to the instrument threshold; thus the test provides a sensitive means to verify correct instrument performance.

The radium-226 γ-ray calibrations of the four ranges are listed in Table 1. The most frequently used scale is the 0–0.0125 mr/hr range; each scale division corresponds to 0.00025 mr/hr. The most sensitive scale, used for lowest intensity measurements, provides scale divisions corresponding to increment of 0.00005 mr/hr, or 0.05 μr/hr.

Data taken with the 100-channel pulse-height analyzer show that, to a first approximation, normal mineral samples, unshielded natural background radiation, and radium-226 in equilibrium have spectra of the same gross shapes as viewed by our crystals. Thus the radium-226 calibration is very useful in practice, for it can be applied generally to convert field data into meaningful radiation-intensity values.

For field use, calibration checks are normally made twice during a measurement sequence: just after turn-on and just before turn-off of the instrument. One check may be sufficient for single-point measurements. For surveys lasting an hour or longer, the uranium source may be carried along by a second person so that frequent checks can be made (the main concern here is that damage to the instrument may result from vibration and shock inflicted during negotiation of rough terrain).

TABLE 1

RANGES FOR THE PORTABLE SCINTILLATION
COUNTER AND CALIBRATION WITH
RESPECT TO RADIUM-226

Scale Range (c.p.s.)	Scale Range (mr/hr)	Smallest Scale Division (mr/hr)
0–100	0–0.0025	0.00005
0–500	0–0.0125	0.00025
0–5000	0–0.125	0.0025
0–50000	0–1.25	0.025

FIELD USE OF THE INSTRUMENT

We now discuss field use of the instrument, particularly to clarify the significance of measurements taken in the field. The instrument is usually carried with the detector about 2 feet above ground level. Response time of the count-rate circuits is rapid enough that significant changes in radiation intensity are registered while the surveyor walks at a normal pace. Lowering the detector to the surface will permit a more precise measurement of the radioactive content in the volume immediately beneath.

Some survey work has been attempted with the detector carried inside a moving vehicle. We soon learned that the only quantity to be measured with confidence by this technique is the radioactive content of the roadbed; such information is generally of secondary importance. Information obtained while traversing narrow gravel or dirt roads may be the exception; these roads usually contain much more "local" material than do wide, paved highways. However, when unequivocally correct data are to be obtained, there is no substitute for literally getting into the field. Some improvement in quality of mobile data would be realized if the detector were positioned 10–15 feet above the surface; this approach has been used by British workers as a surveillance technique in areas adjacent to reactor sites (Cavell and Peabody, 1961).

Problems of interpretation arise when surveys are taken in the vicinity of, or across, areas of sharply contrasting radioactive content. In movement from an area of high activity into an area of low activity, readings will not decrease exactly in accordance with the changes in rock activity. For example, a survey taken from Franciscan sandstone across a sharp surface contact into serpentinized rocks will indicate a rather abrupt decrease in intensity just as the contact is passed but will also show an additional gradual decrease as one moves further into the ultramafic formation. An apparent activity decrease may be noted for several tens of feet into the serpentine. In reality, laboratory analysis may show that the entire activity change occurs across a distance measured in inches or fractions of an inch. The contribution of undesirable radiation in this situation can be reduced substantially if the detector is lowered to the surface. Additional protection from the high-intensity area can be provided by digging a shallow pit for the detector or by piling loose material around the detector to form a low barrier. Care must be exercised that such barriers are built of the same material as the formation to be assayed.

When the detector views a large area of uniform radioactive content, the measured intensity will remain nearly constant if the detector is lowered to the surface. Conversely, when the intensity changes, say, 10 per cent or more for this test, the measured intensity is probably not characteristic of the bulk formation at hand. Float from a nearby area may influence the reading; so may sloughing overburden or a variety of other geologic conditions. In short, one must always study the usual geologic and topographic features carefully to insure that radiation-intensity measurements are interpreted correctly.

We have encountered a surprisingly wide latitude of radiation intensity in our survey of various rock formations. For example, the highest intensity from surface material (not considered to be of radioactive-ore grade) was a reading of 1,400 c.p.s. observed over latite from the Table Mountain flows west of Jamestown, California. Readings across a sharp contact between Table Mountain latite and serpentine ranged from 1,400 c.p.s. over the latite to 160 c.p.s. over the serpentine. Here is a good example of the problems that arise in interpretation of field measurements; the serpentine is virtually free of radioactivity.

The lowest-intensity surface readings were 20–30 c.p.s., observed in serpentine at several localities. Some of these serpentine deposits are located within 2–3 miles of the Table Mountain formation. We note that lower readings are never observed at the surface, even over areas of freshly exposed serpentine—material that laboratory analysis shows to have essentially no measurable radioactive content. Thus we attribute a significant fraction of the 20–30 c.p.s. reading to air-borne activity and cosmic rays. The lowest intensity that we have observed was recorded in a magnesite mine, where serpentine is the host rock. A reading of 2 c.p.s. was noted at a position 1,200 feet in from the portal and under about 300 feet of cover. We hope to be able to obtain γ spectra at this location in the future. Laboratory experience shows that most of the residual count rate derives from radioactive content of the detector assembly itself. It is interesting to note that our detector reads 6–8 c.p.s. inside a "standard steel room" low-background facility at the Donner Laboratory, Lawrence Radiation Laboratory, Berkeley.

Some field scanning, primarily at limestone formations, was done in conjunction with a Precision Model 111 B scintillation counter. It was observed that readings were not as steady or reproducible on the Precision instrument as on the Lawrence Laboratory counter. However, the Precision instrument had sufficient sensitivity to detect contrasts between the carbonate rocks and surrounding higher-background materials. The Precision instrument uses a considerably smaller scintillation crystal than is used in the Lawrence Laboratory counter and therefore must operate with a much lower count rate; this factor alone can account for the different characteristics just mentioned. It can be assumed that a high-quality commercial portable scintillation counter that uses a crystal of size comparable to ours and is capable of registering radiation intensities as low as 0.0005 mr/hr can be successfully utilized for low-background γ-ray surveys.

FIELD SPECTROMETRY

We have just begun to study the application of γ-ray spectrometry to field work. Figure 1 shows a spectrum taken over soil at a location outside our laboratory building in March, 1963. Two important points are noted here. First, a very strong fallout peak is noted in the low-30's channel group. Second, the great intensity of low-energy air- and ground-scattered γ rays masks spectral structure below $\sim$0.6 mev. The upper spectrum on this figure was taken with the detector in its normal housing, as might be used in field work. The lower spectrum was taken at the same site, but with a ⅛-inch-thick lead absorber surrounding the detector assembly. The spectra are normalized to coincide beyond $\sim$1-mev. energy. Absorption data for ⅛-inch-thick lead show that 0.5-mev. γ rays suffer attenuation by a factor of 2, that lower energies are more strongly affected, and that higher energies are less strongly affected. Thus a thin lead absorber should produce clarification in the spectral region above $\sim$0.3 mev. without too great a sacrifice in count rate caused by attenuation. In particular, the strong 0.61-mev. γ ray from the uranium decay series should stand out clearly, when it is not overwhelmed by fallout peaks. In our example, the zirconium-95–niobium-95 peak does just such a masking job; however, this soil is not very rich in uranium. We intend to make two points by this example: that fallout is *now* a serious problem for surface spectral measurements and that a thin lead absorber, surrounding the detector, can add considerable clarity to field spectra. Absorber use may therefore aid significantly in identification of the

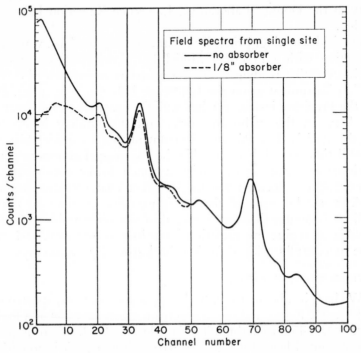

Fig. 1.—Field spectra from single site (*solid line*, no absorber; *dashed line*, $\frac{1}{8}$-inch Pb absorber)

isotopes present, particularly in the case of uranium, where other spectral areas are not so favorable for these purposes.

In the field, one expects a situation in many ways similar to that obtained in the laboratory by Gregory and Horwood (1961), namely, nearly all observed γ rays originate from the top 1-foot layer of ordinary earth material. In addition, there will be present a very strong, low-energy air-scattered component that obscures structure at the low end of the spectrum. Use of thin $\frac{1}{16}$–$\frac{1}{8}$-inch-thick lead absorbers may be very advantageous here, although such absorbers are not important for the simple count-rate scanning that we normally perform.

LABORATORY PROCEDURES

In the laboratory a γ-ray spectrometric analysis was conducted on each sample collected in the field. Usually one or more petrographic thin sections were prepared on representative samples from the field collection process. A description of the spectrometric equipment, evaluation and calculation procedures, and representative γ-ray spectra are included in the following sections.

PREPARATION OF SAMPLES

Samples are prepared for γ-ray evaluation by a standard procedure. Materials having pieces greater than 1–1½-inch diameter are reduced in size by crushing by hand to 1-inch diameter or smaller; materials already within this size range are used as received. Each sample is hand-packed in a thin-walled polyethylene container 4 inches in diameter and 3½ inches high. We use commercially obtainable Tupper refrigerator food containers. The weight of contained material ranges from 500 to 1,500 gm.; typical samples prepared from materials with bulk specific gravities of 2.5–3 and weight of 700–1,000 gm.

Great care is taken to insure that sample materials do not become contaminated with foreign radioactivity in any phase of the work. Samples are placed in closed containers immediately upon collection and are removed from these containers only when they are to be prepared for γ-ray analysis. Sample preparation, handling, and storage are carefully controlled in this respect. All samples thus far analyzed have been retained so that we can re-examine specimens as required.

EQUIPMENT

We use a γ-ray scintillation spectrometer in the laboratory to assay the radioactive content of sample materials. The detector unit consists of a 4-inch-diameter by 2-inch-thick sodium iodide scintillation crystal coupled optically to a DuMont 6363 phototube; the unit is constructed of special low-radioactivity materials. The detector and a sample to be counted occupy one compartment inside a lead shield. Interior dimensions of the detector compartment are $6 \times 8 \times 24$ inches. The shield itself is built of standard $2 \times 4 \times 8$-inch lead bricks and presents at least a 4-inch thickness of lead to all external radiation sources. A second identical shield compartment houses an identical detector, which is used exclusively to measure background radiation during each sample run. The background detector is shielded from the sample detector by a 4-inch thickness of lead. Both compartments are actually part of a single shield structure. The entire lead shield is covered with 0.030-inch cadmium sheet, which is in turn surrounded by an 8-inch thickness of low-activity concrete blocks; both these materials serve as neutron-absorbing elements.

A separate background detector is used because the background inside the shield is not constant. We need to measure sample activity that produces a net count rate that is only 1 per cent of the background count rate; therefore, our knowledge of the background must be exact. The only reasonable way that the background variation can be followed with sufficient precision is to use a second detector (identical to the sample detector) for background determination during each sample run. Problems related to background determination are more fully discussed below.

Output signals from each scintillation crystal are sent to identical UCRL Model-6 linear-pulse amplifiers. Amplifiers are operated in the double-delay-line clipping mode to minimize problems created when very large overload pulses must be handled (cosmic-ray mesons traversing crystals of this size may deposit 70–80-mev. energy).

Information from the sample detector drives a Penco PA-4 100-channel differential pulse-height analyzer. The gain of the system is adjusted so that each of the 100 equal-width channels is 20 kev. wide, and the total γ-ray energy interval spanned is 0.080–2.08 mev. The pulse-height analyzer accepts amplitudes in the 4–104-v. range from amplifier output pulses up to an amplitude of 150 v.

Sample data are also recorded on four integral scalers, as follows:

1. Total counts in channels 1–100.
2. Total counts above channel 100.
3. Total counts above channel 10 (0.28 mev.).
4. Total counts above channel 80 (1.68 mev.).

The background detector system is adjusted to have exactly the same gain as the sample system, and the resultant information is recorded on three integral scalers, as follows:

1. Total counts above channel 10 (0.28 mev.), on two units in parallel, to verify the data; this is *the* background correction number.
2. Total counts above channel 80 (1.68 mev.).

We count the fast-neutron flux on an eighth scaler, utilizing a moderated boron trifluoride counter as the detector.

At the completion of a run, stored pulse-height data are tabulated on a Victor Printer adding machine and are simultaneously plotted in a semilogarithmic mode to produce a curve of the counts per channel versus energy. Figures 3–9 are examples of curves derived from several runs.

The initial calibrations of energy versus channel number for both systems and the analyzer were performed with sources of known, different γ-ray energies. Routine calibration, performed at least once a week includes:

a) Center the peak from cesium-137 exactly in channel 29 for both crystals.

b) Adjust all integral scaler thresholds, using precision pulse generator and the pulse-height analyzer to determine accurately the correct pulse amplitude for each threshold.

Most sample spectra show pronounced peaks from the included radio-isotopes; the potassium-40 peak is particularly useful in this manner, and proper energy calibration of the sample crystal can usually be confirmed by inspection of plotted data. Amplifier output signals are always displayed on an oscilloscope to assist early detection of electronic troubles and thereby minimize collection of worthless data.

A Stabiline line-voltage regulator provides all a.c. power for this equipment. All d.c. power is derived from separately regulated supplies. The equipment is located in a fully air-conditioned laboratory and operates continuously except as required by servicing. Thus we have taken all reasonable steps to insure stable operation of equipment and acquisition of the greatest amount of valid data.

COUNTING PROCEDURE

Prepared samples are always counted with the container lid against the flat face of the detector crystal case; thus the counting geometry is constant. Sample counting time depends upon the amount of radioactivity present and the precision required for the data. Lower radioactive content requires longer count time for any specified accuracy; conversely, greater accuracy demands longer count time for any given radioactive content.

The most active samples, which produce count rates 2–10 times the background rate in our selected energy band, can be analyzed in 50–150-minute runs. Samples with intermediate activity can be analyzed in 200–500-minute runs. The lowest-activity samples are counted for 600–2,000 minutes; even the longest runs do not always provide the precision we desire but must be terminated because of practical time limitations and restrictions of present detection sensitivity.

SAMPLE THICKNESS AND SPECTROMETER RESPONSE

We cast a series of 4-inch-diameter disks from a high-activity, uranium-rich cement. These disks, of measured thicknesses, were then used to explore the response of our spectrometer crystal to various thickness samples. Figure 2 shows some results of these tests and indicates clearly that little is to be gained in our system by increasing sample thickness beyond the dimension now in use. The limitation here is mainly that due to a decrease in solid angle subtended by the more distant sample layers, rather than the effect of γ-ray absorption within the intervening sample material. The use of a larger-diameter crystal, or location of the entire sample at greater distance from the crystal, would show that layers at greater depth within a sample will

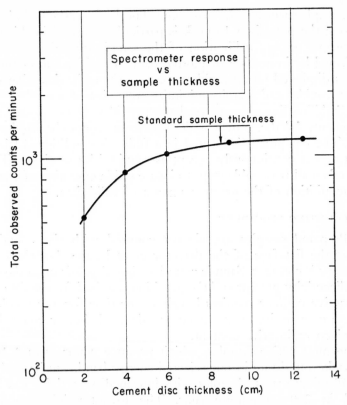

Fig. 2.—Spectrometer response *vs.* sample thickness

still contribute a significant count-rate increment. Gregory and Horwood (1961) found that their crystal-sample relationship required at least a 1-foot thickness of crushed material before the addition of more material gave no significant count-rate increase. In short, each different crystal-sample configuration will show a unique relationship between observed count rate and sample thickness. Thus, calibrations must be performed for each configuration if quantitative results are to be obtained. Because calibration procedures are relatively difficult, we have settled upon one configuration for all laboratory work; it is, at the same time, the one that provides the highest detection efficiency from a simple sample-container shape.

Our spectra show distinct sharp peaks for all prominent γ rays down to and including the 0.24-mev. member of the thorium series. This is a definite aid to identification of the radioactive components present. Gregory and Horwood (1961) also show that increasing sample thickness gradually obscures peaks at energies below 1 mev., until, at the 1-foot depth, these peaks may be little more than shoulders along the spectrum. Good spectral resolution and maximum count rate are seen to be incompatible, and each worker must make a compromise here.

BACKGROUND COUNT-RATE CONSIDERATIONS

The data from a sample run include both sample and background information. We infer the correct background to subtract from the sample count rate by use of data from the second detector and a set of background correction tables. These tables were derived from a number of long (about 1,000-minute) runs in which both detectors measured background. Background data for all energy intervals of interest were fitted by the least-squares method to linear equations, which were then used to generate tables that relate background detector data to sample detector background. By this technique, we can infer the correct background to within ±1 c.p.m., while the quantity we infer varies from 220 to 270 c.p.m.

The background variation is found to be related linearly to the slow-neutron flux at the detector. This variable neutron flux is produced by high-energy particle accelerators located about ½ mile from our detectors. The background count rate observed in the lead shield showed variation greater than a factor of 2; this was an unacceptable situation. We first covered the entire shield with a thin cadmium layer to reduce the slow-neutron effect. Even so, the background increased from 220 to 270 c.p.m. when the external slow-neutron flux increased from 0.005 neutron/cm²-sec (cosmic-ray intensity near sea level; Patterson *et al.*, 1959) to 0.030 neutron/cm²-sec (typical accelerator-produced intensity at our site). We then added an outer layer of low-activity concrete blocks to provide moderation and absorption of fast neutrons. The background variation has thereby been reduced so that the range is between 220 and 230 c.p.m. for various accelerator-operating conditions. The changes are now related to the external fast-neutron flux, although they do not follow so regularly as they did before addition of the concrete. The relationship between backgrounds in the two crystals does still follow as closely, though.

Other experiments with these crystals in the same close-fitting lead shield show that neutron-produced background effects are caused primarily by slow-neutron capture, either as a prompt capture radiation or as decay radiation from induced activation. By far the most important item for both these processes is the crystal itself. We have determined that the 25-minute half-life isotope iodine-128 is the most serious activity induced and that activation to equilibrium in a thermal neutron flux of 1 neutron/cm²-sec produces an observed decay count

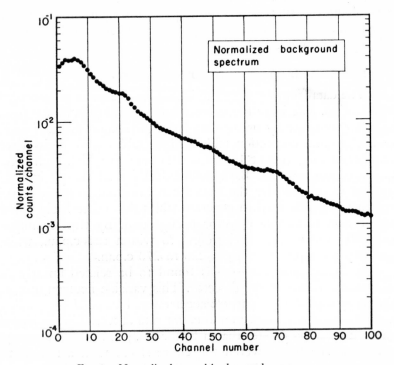

Fig. 3.—Normalized crystal background spectrum

rate of ~1,700 c.p.m. immediately following such an irradiation. Under equilibrium conditions during an irradiation in unit thermal flux, the total observed count rate is ~6,200 c.p.m. The implications of these facts to accurate low-level counting situations are obvious.

Figure 3 illustrates the sort of spectrum observed in a long background run with the sample crystal. The shape is rather undramatic; the only definite features are provided by backscatter effects in the low channels, a weak annihilation-radiation peak at channels 21–22, and a modest potassium-40 full-energy peak around channels 68–70. The backscatter effects derive principally from the small interior dimensions of our shield; the annihilation-radiation peak is associated with both cosmic rays and neutrons; the potassium-40 peak is produced by potassium in the glass phototube envelope. Other minor contaminants are probably masked by these predominant activities.

It is worth noting that, when a reasonably low background is obtained, successful measurement of small increments to this background depends more on constancy than on magnitude. Our concern for accurate background data is thus understandable.

DATA ANALYSIS

Data obtained from sample materials are analyzed with varying degrees of sophistication. We describe the analysis steps here and indicate the kinds of information derived from treatment of data by such methods. It will be assumed that data have been corrected for background if no mention of this correction appears in the text.

The simplest analysis involves use of only the total stored counts: a summation of channels 1–100. These data are reduced to values of specific activity for each sample, expressed as counts/min/gm. The single number so obtained is a measure of the total γ-ray activity of the sample detected by our crystal; however, we learn nothing about the identity of the radioactive isotopes from this simple analysis.

A second simple analysis technique involves inspection of the 100-channel differential spectrum, performed most effectively by use of the semilogarithmic data plot acquired for each sample run. We compare such a sample plot with plots of standard spectra obtained from materials that contain only uranium, thorium, or potassium, singly. The comparison shows which of these three is present in considerable abundance and can indicate relative amounts of each substance in favorable cases.

We obtain quantitative information for the three normal radioactive components (uranium series, thorium series, potassium-40) from a detailed analysis for each 100-channel spectrum. The process involved and an example of its application are set forth in the following paragraphs.

The calibration spectra of the three components (uranium, thorium, and potassium-40) were examined, and three energy groups were chosen that exhibited dissimilar shapes in each spectrum when compared with the other two. Each interval contains a non-zero fraction from its respective complete spectrum. These nine numbers are used as the constants of a set of three simultaneous equations in three unknowns (uranium, thorium, and potassium-40). Sample count rates from the same three intervals are inserted in the equations; solution of the equations then yields values for uranium, thorium, and potassium-

40 in the sample. The values so obtained are in terms of count contribution from each component present in the entire spectrum. The sum of the calculated components is compared to the observed sample activity to provide one check for validity of the process.

The three intervals selected are channels 41–50 (0.88–1.08 mev.), 51–60 (1.08–1.28 mev.), and 62–74 (1.30–1.56 mev.). The two lower intervals provide contrasting shapes in the uranium and thorium spectra, a condition required for obtaining unambiguous solutions from the equations. The highest interval includes the total absorption peak of potassium-40.

Several other considerations enter into selection of energy intervals for this analysis. The first factor is simply the necessity to acquire enough counts per interval to permit a meaningful solution of the equations with respect to statistical counting errors. We are thus encouraged to use low-energy intervals, thereby taking advantage of greater crystal efficiency at low energies plus Compton contribution in these channels from higher-energy γ rays. The second factor warns us to stay relatively high in energy to avoid backscatter problems and self-absorption within the sample of the low-energy γ rays. A third factor is important for those samples that contain "old" fallout. The highest γ-ray energy present in "old" fallout comes from cesium-137, at 0.661 mev., and produces a peak centered in channel 29 that extends into the low 30's. The effects of recent or fresh fallout are discussed in detail below. The lowest interval for the uranium, thorium, and potassium-40 determination must lie above this peak, or erroneous results may be obtained. Furthermore, choice of the three intervals above the cesium-137 peak permits us to calculate an approximate count contribution of "old" fallout to a spectrum. Thus we have standardized the three intervals for all samples, to be compatible with fallout-bearing samples, rather than to generate multiple sets of parameters for different classes of samples.

Early attempts to use two energy intervals above the potassium-40 peak position for uranium and thorium determination met with little success; we could not acquire enough counts in these intervals in reasonable run times to provide acceptable precision for calculated activity values. The problem of low count rate also applies to use of the 2.62-mev. total absorption peak for thorium determination, although a lower background and the presence of only the thorium component at this high energy do favor the use of this interval. A larger crystal would provide greater response at 2.62 mev. and therefore make this peak more useful for thorium assay. For the present, the 2.62-mev. peak region can serve as a check for thorium values computed from the three-interval method.

The set of equations from each sample run is solved by use of third-order determinants. Solutions are obtained by direct numerical computation or by use of a programmed digital computer. We first show derivation of the constants used in these equations and then work a sample calculation to illustrate the procedure.

$$C_1 = 1.0000 \ U_1 + 1.0000 \ Th_1 + 1.0000 \ K_1,$$
$$C_2 = 1.3016 \ U_1 + 0.9107 \ Th_1 + 0.4261 \ K_1,$$
$$C_3 = 1.3024 \ U_1 + 3.6796 \ Th_1 + 0.4389 \ K_1,$$

where C_1, C_2, and C_3 are the total counts (after background is subtracted) in channels 62–74, 51–60, and 41–50, respectively; and U_1, Th_1, and K_1 are the count contributions in the C_1 interval. The coefficients in these equations were determined from spectrometer runs of the standard uranium, thorium, and potassium samples. For convenience the coefficients are normalized to values of unity in the C_1 interval.

Third-order determinants were used to solve the foregoing equations in terms of U_1, Th_1, and K_1:

$$U_1 = (-0.4830) \ C_1 + (1.3398) \ C_2 + (-0.2004) \ C_3,$$
$$Th_1 = (-0.0067) \ C_1 + (-0.3570) \ C_2 + (0.3620) \ C_3,$$
$$K_1 = (1.4897) \ C_1 + (-0.9828) \ C_2 + (-0.1616) \ C_3,$$

where now U_1, Th_1, and K_1 are the contributions of uranium, thorium, and potassium to the C_1 interval, and C_1, C_2, and C_3 are the count contributions in each interval from a sample spectrum. These are the three equations used to compute the radioactive components of all samples. The U_1, Th_1, and K_1 values are then multiplied by 40.368, 69.593, and 5.2858, respectively, to calculate U_t, Th_t, and K_t, the contribution of each component to the total spectrum. These results can then be used with sample weight and carefully determined calibration factors to compute actual concentrations for uranium, thorium, and potassium-40 in each material. We illustrate, using as an example data for "Top Sand," run 806, 1,009 gm. weight, 140-min. run time.

Channel group	62–74	51–60	41–50
Counts	4,530	3,207	4,456
Counts/min	32.36	22.91	31.83
Background*	10.50	10.31	14.73
Net counts/min	21.86	12.60	17.10
	(C_1)	(C_2)	(C_3)

* Background data obtained from tables with background detector registering 259 c.p.m. this run.

U_1 = 2.898; multiply by 40.368 = U_t = 117 c.p.m.

Th_1 = 1.544; multiply by 69.593 = Th_t = 107 c.p.m.

K_1 = 17.418; multiply by 5.2858 = K_t = 92 c.p.m.

Sum = 21.860 (C_1 check)   316 c.p.m. calculated total

(329 c.p.m. observed total)

Uranium assay at 127,970 cpm/gm of U: concentration = 0.91 p.p.m.

Thorium assay at 51,526 cpm/gm of Th: concentration = 2.09 p.p.m.

Potassium assay at 15.845 cpm/gm of K: concentration = 0.58 per cent

Such calculations are easily performed in a few minutes with a desk calculator; it is our general procedure to compute each run in this fashion and to use the programmed digital computer for special situations.

Treatment of statistical errors as they appear in calculated components is rather cumbersome if exact expressions are used. However, if we make several simplifying assumptions, the errors can be computed rather easily to yield values that are approximately correct for most cases. These assumptions are as follows:

a) Errors in the nine parameters obtained from the uranium, thorium, and potassium-40 calibration spectra are very small.

b) Errors in the nine coefficients calculated from the foregoing parameters are small compared to errors of the C_1, C_2, C_3 quantities measured in sample runs; errors in these coefficients can therefore be neglected.

c) Errors in C_1, C_2, and C_3 are nearly the same in magnitude and can therefore be considered equal.

When these three assumptions are applied to our analysis method, the approximate errors become:

$$\sigma_U \cong 57 \, \sigma_{C_1},$$

$$\sigma_{Th} \cong 34 \, \sigma_{C_1},$$

$$\sigma_K \cong 9.2 \, \sigma_{C_1},$$

where the errors are standard deviations of the computed values for uranium, thorium, and potassium in the total spectrum, expressed in c.p.m., in terms of the error in C_1. (Exact treatment of errors in our IBM 7090 Computer program confirms that these expressions are good approximations.) Note that the absolute magnitudes of these errors are not dependent upon the amounts of uranium, thorium, and potassium; thus the relative errors (per cent errors) in the three components are inversely proportional to their abundances. Consequently, it is difficult to measure accurately a small amount of one component in the presence of a large amount of another component.

The errors discussed here are only those due to the statistical nature of the counting procedure and do not include such factors as inhomogeneity of activity distribution in samples or differences in γ-ray absorption within samples due to differences in atomic number and total weights. Such factors should be explored and their magnitudes determined so that the technique can be improved to yield more precise information.

We have enlisted the aid of an IBM 7090 digital computer in an effort to extract the maximum information from our 100-channel spectra. Although this program is just now in the development stage, some early results are suitable for discussion here.

The first computer analyses are designed to check the validity of our method, to calculate statistical errors of results, and to separate the fallout-component spectrum from any sample spectrum. Computer input data include 100-channel spectra from the uranium, thorium, and potassium calibration samples, a standard background, and the sample runs to be analyzed, along with other pertinent sample-run data. The computer first makes a background correction and then calculates uranium, thorium, and potassium quantities, using the three-equation method. Errors are computed for these quantities from exact expressions. The uranium, thorium, and potassium solution is next used to synthesize a 100-channel natural activity spectrum. This natural activity spectrum is subtracted from the background-corrected observed spectrum to produce a third spectrum. This residual spectrum, which we call "fallout," shows the differences between an observed sample spectrum and its calculated natural-activity spectrum and so includes statistical errors, any systematic errors, and the actual fallout spectrum. Computer readout includes solutions for uranium, thorium, and potassium concentrations with errors, the background-corrected observed spectrum, the synthesized natural-activity spectrum, and the "fallout" spectrum.

Two classes of samples have been examined in this fashion: those known to be free of fallout and those known to contain fallout. A study of the results from fallout-free samples has not yet revealed any systematic errors in our assay methods. In the strictest sense, this merely means that the sample materials produce spectra indistinguishable from linear combinations of the calibration spectra. No direct verification for our actual assays is obtained from this sort of analysis. The residual 100-channel spectra do not show regular shapes, such as would be the case if, for example, the calibration spectra were in error. One exception is noted, and that is found at analysis of the virtually non-radioactive ultramafic samples. Here we see a cluster of residual counts in the low-energy channels that show a regular shape, just that which should appear from backscatter effects.

Fallout-bearing samples produce residual spectra that we have come to recognize as true fallout spectra. We have concentrated mainly on soil samples, from which we seek to determine the fallout contribution. The residual spectra are smooth except where statistical errors are relatively large. We note one effect in a number of these spectra that may be a systematic error if it persists after a larger group of samples is analyzed. This effect appears as two artificially deep valleys in the fallout spectra and results from an overestimate of the uranium content.

We are expanding the scope of computer analysis to include a search for more favorable spectral intervals for uranium, thorium, and potassium determinations; lower-energy intervals would certainly improve the accuracy of our assays. A careful study will be made to determine the improvement in accuracy obtainable by imposing some limit conditions from various portions of the spectrum on the allowable solutions for uranium, thorium, and potassium. Other methods of solution will also be studied.

STANDARD SAMPLES

The uranium, thorium, and potassium contributions to the total γ-ray spectrum from a sample are calculated by use of coefficients developed from standard spectra. These standard spectra will be explained in detail here so that a clear understanding of our methods can be obtained. We have acquired separate assayed samples of uranium ore and thorium ore that are certified to be in secular equilibrium with their respective decay series. It is very important that equilibrium exists in the calibration samples; otherwise, distortion of spectra will result, especially if late members of the series are lacking in abundance. This lack of equilibrium would also lead to incorrect assays for experimental samples.

The assayed equilibrium ores of both uranium and thorium were obtained from the U.S. Atomic Energy Commission's New Brunswick Laboratory, New Brunswick, New Jersey. We have prepared samples for calibration purposes, using the standard polyethylene containers and two different bulk dilution materials. In each case, known amounts of the assayed ores were carefully mixed with known amounts of bulk material and then packed into the containers. The bulk materials used are (*a*) Jefferson Lake minus 50-mesh serpentine fines and (*b*) Clemco No. 24 silica sand.

No significant differences in either the spectral shapes or the count rates were noted when an ore was analyzed in these two matrix materials; however, since the particle sizes of ore and matrix are more nearly equal with serpentine fines than with sand, a better mixture is obtained by using serpentine as the matrix. Discussion of the results of calibration runs is limited to those obtained with the ore-serpentine mixture.

The New Brunswick Laboratory (NBL) ores are in the form of nominal 100-gm. samples of accurately known concentrations of each ore diluted in and thoroughly mixed with fine-ground dunite. The uranium ore used to make the calibration sample was the 0.50 per cent uranium concentration member of the NBL-42 series. This uranium ore is certified to be in equilibrium by measurement of the radium-to-uranium ratio and contains a negligible amount of thorium. A 50-gm. quantity of this ore was mixed with 850 gm. of serpentine fines and packed in a polyethylene container to constitute our uranium calibration standard. The thorium ore used was a 1.01 per cent thorium concentration member of the NBL-79 series; this ore contains 0.035 per cent uranium, a factor that we have taken into account when calculating the thorium calibration values. A 50-gm. quantity of this ore was mixed with 850 gm. of serpentine fines and packed in a polyethylene container to constitute our thorium calibration standard.

Eight days after these two mixtures were packed, spectrometer-calibration runs were taken to insure that any disturbance of equilibrium (loss of gaseous decay-series members) would be repaired and could therefore not affect our results. The potassium standard spectrum is obtained from a homogeneous mixture by using the pure chemical compound, potassium chloride. A 654-gm. quantity of this compound was packed in a polyethylene container to constitute our potassium calibration standard. The potassium calibration standard was also run at this time.

Each of the three calibration spectra is unique and shows characteristic peaks by which its identity can be verified in the mixtures encountered in sample analysis. Figures 4, 5, and 6 show calibration spectra from uranium, thorium, and potassium-40, respectively. The uranium spectrum shows a relatively flat distribution up to channel 14, then a sharp drop, a prominent peak at channels 26 and 27, a broad peak in the low 50's, another broad peak in the 80's, followed by a sharp drop in the 90's. Thorium shows a sharp peak at channel 8, rising from a plateau that drops sharply beyond channel 14, a modest peak at channels 25 and 26, a broad peak in the low 40's, followed by a dip in the 50's and a relatively flat distribution in the 80's and 90's. Potassium shows a typical monoenergetic γ-ray spec-

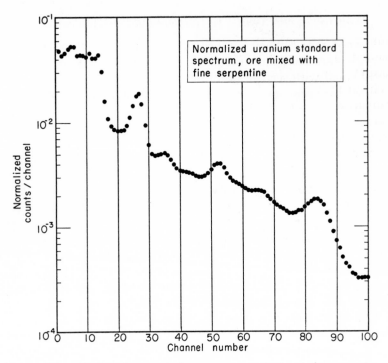

FIG. 4.—Normalized uranium standard spectrum; ore mixed with fine serpentine

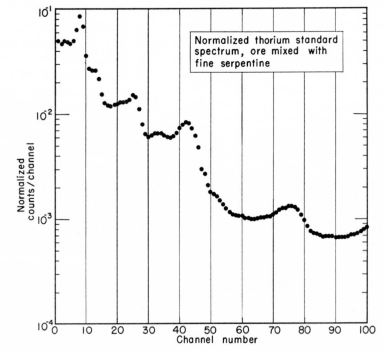

FIG. 5.—Normalized thorium standard spectrum; ore mixed with fine serpentine

trum, with a nearly flat Compton distribution extending to about channel 60, but modified by a rise at the low-energy end by various scattering interactions, and the single total-absorption peak centered at channels 69 and 70. Figures 7, 8, and 9 are spectra from samples of uranium-rich, thorium-rich, and potassium-rich rocks collected in the course of field examination; background has not been subtracted from these three spectra. A comparison between these sample spectra and the calibration spectra is instructive.

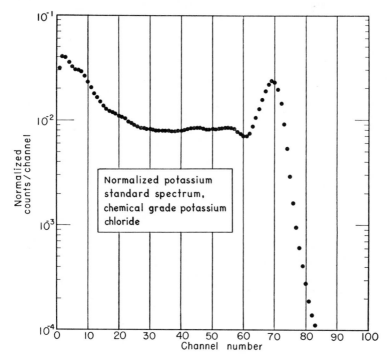

Fig. 6.—Normalized potassium standard spectrum, using chemical grade potassium chloride

Each of the three standards was counted until at least 200,000 counts were accumulated in each of the three intervals used for our calculations. We require such precision so that none of the parameters derived from these calibrations (subsequently used in solution of the equations) will contribute any appreciable error to the resultant analysis. In other words, only the statistical counting errors in the sample runs need to be taken into account when calculating precision of final results; such calculations are thereby simplified considerably. Table 2 lists the calibration parameters obtained from these runs. The last column at the right in Table 2 lists spectrometer response in terms

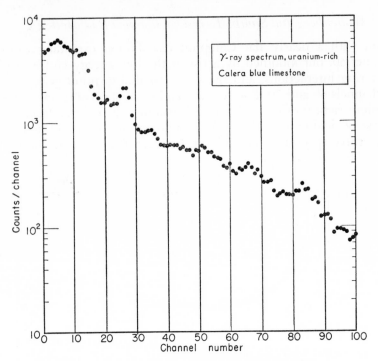

Fig. 7.—Gamma-ray spectrum, uranium-rich Calera blue limestone

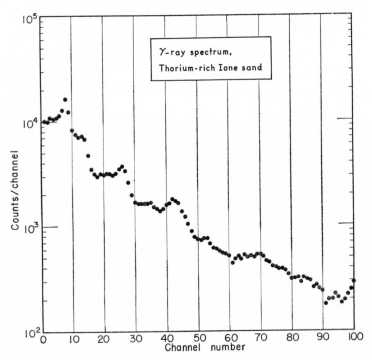

Fig. 8.—Gamma-ray spectrum, thorium-rich Ione sand

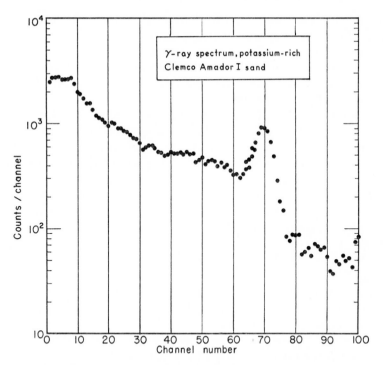

Fig. 9.—Gamma-ray spectrum, potassium-rich Clemco Amador I sand

TABLE 2

CALIBRATION PARAMETERS FOR URANIUM, THORIUM, AND POTASSIUM-40

	C_1 Channel Group 62–74	C_2 Channel Group 51–60	C_3 Channel Group 41–50	Total Spectrum Channels 1–100
U (250 min.)				
Observed counts.......	207,369	269,163	270,371	8,334,555
Normalized counts.....	1.0000	1.3016	1.3024	40.368
Uranium (cpm/gm)....	3,170	4,126	4,128	127,968
Th (490 min.)				
Observed counts.......	214,316	206,147	713,521	13,912,562
Normalized counts.....	1.0000	0.9107	3.6796	69.593
Thorium (cpm/gm)....	740.4	674.3	2,724	51,526
K (654 min.)				
Observed counts.......	704,554	303,684	315,379	3,844,474
Normalized counts.....	1.0000	0.4261	0.4389	5.2858
Potassium (cpm/gm)...	2.9976	1.2774	1.3158	15.845

of counts per minute in the total observed spectrum from 1 gm. of each of the three radioactive components.

No naturally occurring γ-ray emitters other than uranium, thorium, and potassium are likely to be encountered in sample analysis. However, we have detected fission-product activities (nuclear-weapons-test debris) in some samples, particularly in ultramafic rocks. A few samples collected before September, 1961, showed two long-lived γ emitters from "old" fallout: cesium-144 (a sharp peak in channels 3 and 4), and cesium-137 (a sharp peak at channel 29). Samples collected since September, 1961, particularly after the onset of winter rains, show several short-lived γ emitters from "fresh" fallout. Surface and near-surface materials are now contaminated to such an extent that we cannot always apply the methods described here without studying fission-product decay at the same time.

γ-RAY SCATTERING EFFECTS

Careful study of laboratory data taken on the lowest-activity materials, particularly the serpentines, points out one important problem associated with this sort of work: in general, the presence of any non-radioactive mass close to a detector will alter the count rate measured by the detector. When the detector is adequately shielded from external terrestrial radiation by a material that is not itself radioactive, as in our case, then the presence of the sample will invariably increase the observed count rate. The increase is expected to be small —only a few c.p.m.—and therefore cannot be verified when high-activity samples are counted.

As the simplest description of the situation, we say that the sample acts as a scatterer and that some of these scattered γ rays are directed into and detected by our crystal. Most of these secondary γ rays are produced from Compton interactions of primary γ rays in the sample. The only concentrated source of radioactivity known to be inside the shield is the phototube glass envelope, and the most important part of this envelope is the large, flat photocathode end plate, which rests against the crystal window. Some of the γ rays emerging from this glass pass through the crystal without interaction and then interact in the sample to send Compton-scattered γ rays back into the crystal; γ rays scattered through such large angles by the Compton process are generally termed "backscatter" γ rays and produce the characteristic spectral shape that we would observe in the low-energy channels. All other Compton-scattered photons also have energy less than that carried by the primary photons. Thus we expect the contribution from all Compton events to appear in the low-energy channels. We do not expect sharp structure in this spectrum because Compton-scattered photons can have a wide distribution of energies when derived from a single primary-photon energy.

The serpentine spectra show no increase above background for the three energy groups used in the quantitative analysis. Yet there is an increase noted above background when the entire 100-channel spectrum is used to assay sample activity. The excess counts are always at the low-energy end of the spectrum, in agreement with the proposed Compton-scattering effects.

We have attempted to determine the magnitude expected from such scattering interactions, using a small uranium-ore source and a standard serpentine sample in normal counting position. When the uranium source is placed as close as possible to the phototube end plate, we note a count increase of about 2 per cent when the sample is present, compared to the no-sample condition. If it is reasonable to treat the total background count rate in this fashion, then our serpentine activities are all 4–5 c.p.m. high; such a correction would reduce the detected activities of some serpentines below the point at which they have any statistical significance.

WATER-BEARING MATERIALS

Some of the lowest-activity materials that we have attempted to assay, the serpentines, illustrate another aspect of neutron interference. The serpentines are hydrated ultramafic rocks and contain about 10 per cent water by weight; this quantity of water is effective as a moderator for fast neutrons. In the presence of a fast-neutron flux, a serpentine sample then becomes a source of slow neutrons. It is important to point out that, although our outer-shield layer effectively absorbs the external fast-neutron flux, the inner lead shield acts as a fast-neutron-flux generator via cosmic-ray interactions. The cosmic-ray-produced fast-neutron flux is small, to be sure, but it is not negligible in the present context.

Thermal-neutron capture cross-sections for the major constituents of serpentine (magnesium, silicon, oxygen, and hydrogen) and the sodium-iodide crystal strongly favor capture in the crystal. Thus the crystal becomes slightly activated and reports information that appears to indicate radioactivity in the serpentine, while in reality these excess counts are due to the neutron-activation effect in the detector. Here again, the effect is measured in terms of a few c.p.m.; however, the apparent sample activity is of comparable magnitude.

LOWER LIMIT OF SENSITIVITY

Our present lower limit for detection of radioisotopes is determined by several interrelated factors. We are, in a sense, limited by the length of count time, but a fourfold increase in count time is required to reduce the statistical standard deviation (error) by a factor of 2. When count times of 2,000 minutes are already employed, it is evident that simply increasing the count time may not be the most satisfactory way to increase the detection sensitivity. It is also true that it requires a fourfold reduction in background to reduce the standard deviation by a factor of 2, if count time remains fixed. No such dramatic decrease in background can be expected beyond the point that we have reached. We see that both count time and background magnitude are related to detection sensitivity in a "square-root" fashion; it requires large changes in either quantity to effect significant improvement in the desired direction.

TABLE 3

LEAST DETECTABLE QUANTITIES OF URANIUM, THORIUM, AND
POTASSIUM-40 WITH PRESENT ANALYSIS SYSTEM

Radioisotope	Counts/min/gm	Grams To Produce 1 count/min	Concentration in 1,000-gm. Sample To Produce 1 count/min/gm
U in equilibrium....	1.28×10^5	7.8×10^{-6}	7.8×10^{-9}
Ra in equilibrium...	3.57×10^{11}	2.8×10^{-12}	2.8×10^{-15}
Th in equilibrium...	5.15×10^4	1.9×10^{-5}	1.9×10^{-8}
K.................	1.58×10^1	6.3×10^{-2}	6.3×10^{-5}

However, the standard deviation decreases in direct proportion to the increase in sample count rate, other parameters remaining constant. Thus the most fruitful approach to improved detection sensitivity is to increase the sample count rate, either by presenting a larger sample mass to the detector or by employing a more efficient (larger-volume) detector. We cannot employ either of these alternatives while working in the confines of the small lead shield; however, we will be able to take advantage of both soon, when a large shield structure has been completed.

Table 3 illustrates the performance of the present spectrometer-sample system in terms of the concentration of each component that produces 1 c.p.m. in the total observed spectrum. We assume that a 1,000-gm. sample is used and that a count time of 1,000 minutes is allotted so that the 1 c.p.m. difference is statistically significant. It should be noted that we lose the ability to identify the radioactive components as this lower limit of detection is approached.

We are now able to assay potassium down to the range of 50–70 p.p.m.; in the large shield structure, and with a larger detector, which views a greater sample mass, we should be able to reach the range of 1–2 p.p.m. potassium content. Improvement in detection sensitivity for uranium and thorium is expected to be of a magnitude similar to that quoted for potassium.

The problems discussed here are general for situations in which the sample count rate is but a small fraction of the background rate. We do not mean to imply that increasing count time and decreasing background rate are not worthwhile achievements; we wish only to point out the relative magnitudes of the effect produced when these parameters are varied. Obviously, the background cannot be too low; nor can a longer count time fail to improve precision of results. However, these factors must be viewed in proper perspective, and, in this context, sample count rate takes a pre-eminent position. Finally, so long as the background is high relative to the detected activity, this background must be known exactly if precise results are to be obtained. The background must remain extremely constant or, in our case, must be measured each time simultaneously when variations are encountered.

FALLOUT PROBLEMS

Our laboratory assay methods for uranium, thorium, and potassium-40 rest upon the assumption that no other γ-ray emitters contribute counts to the three selected spectral intervals. These intervals have been selected to lie above the spectral region to which cerium-144 and cesium-137 in "old" fallout contribute γ-ray events. In fact, the computer analysis permits us to measure the amount of these fallout isotopes present in contaminated samples. We can take account of "old" fallout rather easily.

The fission products generated from nuclear-weapons tests conducted since September, 1961—"fresh" fallout—are quite a different matter, however. This category of fallout can definitely interfere with the uranium, thorium, and potassium laboratory analysis; it can interfere very seriously with field survey work, and so we believe that the matter must be discussed here at considerable length.

At locations remote from the weapons-test sites, fission products fall to earth mainly in the form of dust-sized particles. If rainfall is frequent, occurring several times per month, most fallout is brought down by this precipitation; conversely, if dry spells of several months' duration occur, most fallout drifts slowly to the surface as dust, settling through the air. In most of California we have alternate periods for each mode of deposition every year: about 4–5 months of wet deposition and 7–8 months of dry deposition. Although the fallout particles are largely insoluble, their fate may depend strongly upon the mode of deposition. Dry deposition must be mainly and evenly on the surface and can deliver a considerable burden to leaves of grasses, shrubs, and trees. Rainfall deposition, however, may purge activity from growing vegetation, leaving it relatively clean, and may also scour fallout particles from other surface "dust-catching" facilities. Running water has great power to transport such small particles laterally; thus, when rainfall is great enough to produce surface run-off, fallout particles are transported by this agent. For example, field work generally reveals higher activity in ditches and other low places where runoff has collected; laboratory analysis confirms that the excess activity is indeed fallout. Furthermore, fallout particles may be carried into cracks and crevices in weathered or broken rock formations. We have not usually found fallout in samples taken from greater than 2 feet below grade level in undisturbed material; however, it is not possible to generalize on the matter; each situation must be evaluated separately. In summation, the extent of fallout interference at a particular field site may depend upon the length of time since the last rain, the intensity of the last rain, or perhaps whether trees and shrubs have leaves at the times in question.

These variables are of controlling importance for field work with an instrument of the type we use, a simple gross-activity indicator. Lowder *et al.* (1963) describe a technique to alleviate some of these difficulties. Our work with surface outcrops of ultramafic formations showed field count rates of 20–30 c.p.s. before September, 1961. Since that time, count rates as high as 250 c.p.s. have been observed at these same locations. Again, laboratory analysis shows the increase to be due to fallout. Field readings for higher-activity formations have shown similar or even greater absolute-count-rate variations, although the relative differences here are less. Figure 10 shows computer-derived spectra for a surface-soil sample collected at El Cerrito, California, in late October, 1962. The background-corrected observed spectrum appears as the top solid curve on this figure. The 100-channel synthesized natural activity spectrum is the next lower curve and is seen to merge with the observed spectrum beyond about channel 40. The "fallout" spectrum appears next as a set of points; it is a true fallout spectrum and extends only to channel 40. The computer shows 383 c.p.m. natural activity and 276 c.p.m. fallout for this sample. For purposes of comparison we show a fourth spectrum, appearing at the bottom of Figure 10; it is the spectrum from dried weeds in a ditch, collected at Felton, California, in April, 1962. The only detectable activities in the weed sample are fission products; the soil "fallout" spectrum is seen to be very similar. This figure also illustrates the sort of information that we derive from the computer analysis.

When in the field over unfamiliar formations, we cannot *now* assign a natural activity count rate; we might easily be in error by a factor of 2, or by nearly a factor of 10 if ultramafics are at hand. A qualitative test for the seriousness of the fallout problem can be performed by digging a pit for the detector to get it below grade level. When fallout is a problem, the pit position will show a lower count rate than will the normal surface position. We have not yet tried to make this test quantitative but have firmly established its qualitative validity.

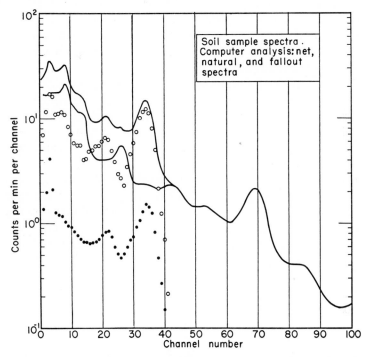

Fig. 10.—Soil-sample spectra analyzed for net, natural, and fallout contributions

At the present time, we are forced to rely upon laboratory analysis, even to interpret the simple data collected in the field; but the laboratory job is by no means clearly resolved for all samples. We now recognize two kinds of "fresh" fallout situation. The first case includes all samples that show lanthanum-140, a 1.60-mev. γ-ray emitter with an effective half-life of about 2 weeks; samples are likely to be in this class if they have been collected from the surface shortly after a rain and within a few weeks after large-scale atmospheric testing. The uranium, thorium, and potassium assays are not valid for these samples.

The second class includes the first-class samples after about a month's decay time and all other samples except those taken from deep underground or from other sufficiently protected sites. These samples may contain the medium-lived fission products cerium-144, rhodium-102, ruthenium-103, ruthenium-106–rhodium-106, and zirconium-95–niobium-95. Several of these isotopes have complex spectra, but they combine to show three prominent peaks: channels 3–4, 18–20, and 33–35 (zirconium-95–niobium-95). Such samples may also contain cesium-137, but its peak at channel 29 is usually masked by the zirconium-95–niobium-95 peak in the low 30's at this early age of fission products.

We can perform the uranium, thorium, and potassium assay for these samples. If the zirconium-95–niobium-95 peak is very intense, its upper edge may contribute slightly to the channels' 41–50 interval and thereby produce an artificially high thorium assay. However, this difficulty seems not to be serious in most cases and is mainly a problem for those surface samples collected explicitly for fallout measurement. It must be noted that "very intense" is a relative term as applied to the zirconium-95–niobium-95 peak, and ultramafic materials may *all* show trouble here because their natural activity is so low.

Results from our computer program have begun to show a consistent overestimate of uranium assay in some of our surface-soil samples. These samples contain a significant fallout component. It is possible that some of the relatively low-abundance high-energy γ rays from rhodium-102 cause this error. Additional computer analysis of such spectra must be accomplished before we can learn the truth of this matter.

The items discussed here may be only of short-term value; cessation of atmospheric nuclear-weapons testing would insure the virtual disappearance of these problems after a year or so. However, we can only speculate on the testing schedule, and these problems are certainly with us right now. It is generally true that most of the fallout does stay at or near the surface, but one cannot be certain about this matter. If most fallout at a site does lodge near the surface, then field scanning is subject to potentially great error, while subsurface collection for laboratory analysis is in a more favorable circumstance. Finally, anyone who has worked in the field will appreciate the difficulty of "mining" deep enough into a weathered surface outcrop to obtain a sample that is almost certainly free of fallout.

COMPARISON OF FIELD AND LABORATORY ANALYSES

A comparison of field-survey measurements with laboratory spectrometer analyses shows that the two kinds of data agree generally. There are some exceptions. Among the medium- to high-activity sample materials, these disagreements can usually be attributed to some constraint in the sampling environment: lack of uniformity of sample area, smallness of sample volume, or, recently, the presence of fresh fallout.

There is often very poor agreement between the two kinds of data for analyses of the materials with lowest activity—the ultramafic formations. There is good reason for this lack of agreement. For example, when we examine a surface deposit of serpentine, the portable instrument readings represent a relatively large amount of material, some of which may be foreign to the serpentine itself. However, a sample collected for laboratory study will contain only serpentine and *no* foreign materials, when we wish to assay serpentine activity. Therefore, field and laboratory results might disagree. Nearby outcrops of high-activity rocks can influence field measurements but would not affect laboratory data. Furthermore, in many cases the field instrument detects almost no radiation from the serpentine but records mainly air-borne activity and internal contamination. We cannot expect to distinguish a small increment to this base count rate with much precision, especially when the base count rate may vary with time or location. Laboratory conditions are much more favorable for determining these small increments accurately. With the field instrument we can always identify the ultramafic formations by their very low activity (except where there is appreciable "fresh" fallout); it is merely that we cannot distinguish the lowest from the low in the field.

Higher-activity rock types tend to show better field versus laboratory activity correlations. This is illustrated in Figure 15, where a fair correlation exists with cement-plant limestones and a good correlation exists with cement-plant "shales."

Results

During the last two years we have accumulated γ-ray spectrometric data from over 1,500 soil and rock samples. The impetus for the rock-sample study was furnished by the quest for low-radioactivity aggregate materials for our low-background counting room; the soil samples were collected during periodic fallout surveys in the San Francisco area. Soil samples were subsequently analyzed for the natural component of γ radioactivity. Rock-sample data are extensive enough that γ-radioactivity characteristics of several widely varying rock types can be compared; these include Oregon basaltic flows, central California ultramafics, California carbonate rocks, siliceous sands, and vein quartz.

An example of the practical application of our techniques is that of the analysis of portland cement raw materials and finished products. Our cement data are clearly of value to persons interested in low-background concrete ingredients; they should also be useful to persons in the cement industry concerned with raw-material ingredient concentrations in mill feed and with the ultimate distribution of raw-material chemical components in produced cements.

VOLCANIC ROCK

Fifteen samples were collected from the sequence of Tertiary and Recent volcanic flows, tuffs, and gravels, outcropping along State Highway 20, which traverses the Cascade Mountains west of Bend, Oregon. The sample with the lowest natural activity, 0.180 cpm/gm, consists of olivine basalt collected from the vicinity of Santiam Summit. The highest activity, 0.815 cpm/gm, occurs in a sample of feld-

TABLE 4

γ-RAY SPECTROMETRIC EVALUATION OF OREGON VOLCANIC ROCK

DESCRIPTION	CONCENTRATION			FIELD COUNTING RATE* (c.p.s.)	RATIO TH/U	NATURAL ACTIVITY COUNTING RATE (cpm/gm)
	U (p.p.m.)	Th (p.p.m.)	K (per cent)			
1. Andesite pebbles, Metolius Junction..................	1.16	1.43	0.56	195 av. over surface 160 surrounded	1.23	0.309
2. Lava Butte cinders..........	1.39	3.58	1.08	250	2.58	0.531
3. Pilot Butte cinders and soil..	0.625	2.25	0.66	175	3.60	0.300
4. Basalt, Santiam Junction....	0.85	2.0	0.59	175	2.36	0.304
5. Cinders, Santiam Junction stockpile..................	0.62	0.78	0.65	150	1.26	0.223
6. Basalt, vicinity, Laidlaw Butte	1.87	2.37	0.28	190	1.27	0.406
7. Cinders, Suttle Lake........	0.85	1.23	0.55	225	1.44	0.258
8. Solid basalt, Suttle Lake.....	0.70	0.70	0.41	185	1.0	0.190
9. Vesicular basalt, Santiam Summit...................	0.32	0.96	0.38	230	2.99	0.218
10. Solid basalt, Santiam Summit	0.57	0.79	0.42	230	1.39	0.180
11. Feldspathic basalt, Tombstone Pass......................	0.41	0.96	0.45	215	2.36	0.173
12. Feldspathic tuff, Rooster Rock	2.04	5.96	1.56	500	2.92	0.815
13. Fine-grained basalt, ≈1 mile east of confluence of Middle and South Santiam Rivers...	0.56	2.28	0.34	240	4.12	0.242
14. Columnar basalt, ≈2 miles west of Cascadia............	0.98	3.63	0.68	250	3.72	0.419
15. Volcanic pebbles............	1.84	3.07	0.68	300	1.67	0.498
Average values.............	0.99	2.13	0.62	....	2.26	0.338

* Includes fallout component.

spathic tuff collected on the western slope of the Cascades. Field counting rates average 234 c.p.s. with the extremes of 500 c.p.s. over the feldspathic tuff and 150 c.p.s. over stockpiled basalt cinders.

Ten samples (1–10 in Table 4) were taken from olivine basalt and andesite, which form the summit and east slope of the Cascades, an area described by Williams (1957). These lavas range in age from Pliocene to Recent, with the oldest, corresponding to samples 7–10,

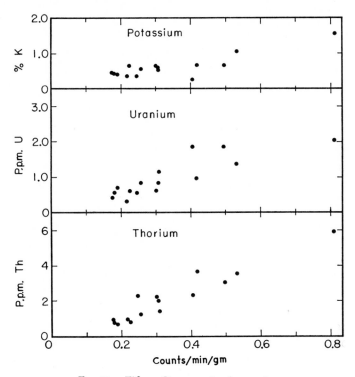

FIG. 11.—Fifteen Oregon volcanic samples

having undergone glaciation. Average isotope concentrations for the ten samples follow: uranium, 0.90 p.p.m.; thorium, 1.61 p.p.m.; potassium, 0.56 per cent.

The remaining five samples listed in Table 4 were collected in older (Oligocene to Pliocene) volcanic rocks, which crop out on the west slope of the mountains. Among these formations the Miocene Stayton lavas, corresponding to sample 13, are considered to correlate stratigraphically and lithologically with the Miocene Columbia River basalt; both are low in olivine. The five samples from the western Cascades average 1.17 p.p.m. uranium, 3.18 p.p.m. thorium, and 0.74 per cent potassium.

When the plots of potassium, uranium, and thorium concentration versus over-all counting rate are compared (Fig. 11), it is apparent that thorium is the principal determinant in the radioactivity of the volcanic-rock samples. Uranium and potassium concentrations also follow count-rate variation, but to a lesser extent than for thorium.

It is interesting to note that the feldspathic tuff, considered to be the most acidic member of this volcanic sequence, is highest in uranium, thorium, and potassium concentrations and is also highest in over-all count rate, while the basic basalts of Santiam Summit have the lowest radioisotope concentrations and the lowest over-all count rates.

Of further interest would be the possibility of a correlation between the alkalinity of the plagioclase feldspar and the radioactivities of these and other basalts.

ULTRAMAFIC ROCKS

Serpentinized ultramafic rocks were considered those most likely to fulfill the low-background requirements of aggregate for our counting facility. Possible sources of serpentine and magnesite in central and northern California were examined and sampled, with samples being subjected to γ-ray spectrometric analysis. Average values for 26 selected "uncontaminated" samples follow: over-all counting rate, 0.0197 cpm/gm; concentrations: potassium, 0.001 per cent; uranium, 0.11 p.p.m.; thorium, 0.07 p.p.m. Field portable scintillation-counter readings corresponding to 22 of the foregoing samples averaged 37 c.p.s. All readings were made prior to fallout in the autumn of 1961, so cesium-137 was the predominant fission product present.

Serpentine may contain uranium concentrations higher than those in unaltered ultramafic rocks (Davis and Hess, 1949), and it is therefore possible that pure peridotite or dunite may contain less radioactivity than we have measured in our serpentines. Economics dictated that we locate a source of aggregate fairly close to Berkeley; sources of pure peridotite or dunite were out of range on this basis. We have not yet obtained good data from such unaltered formations but plan to do so in the near future. Samples of forsterite brick, in the form of both raw material and finished products, have been assayed; results show these samples to contain considerably more radioactivity than do our serpentines. The source for this product is in an unaltered dunite from North Carolina; however, no special effort was made to keep radioactive contamination from the samples prior to our receipt of them. The high radioisotope content may be due to this factor. It is also possible that ultramafic rocks in other parts of the United States may not have the low radioactivity of those on the West Coast.

The marked contrast between the low activity of ultramafic rocks and the higher activity of adjacent rock types was evident wherever field readings were made. A good example is the contrast at the Phoenix Asbestos Mine near Napa, California: a reading of 300 c.p.s. was observed over sandstone, while 10 feet away, down a steep incline across the sandstone-serpentine contact, the reading over serpentine was 30 c.p.s.

The effect of an adjacent high-activity rock type on serpentine field-count rates is illustrated in the Sierran foothills, where Tertiary latite lava flows overcap Mesozoic serpentine. At one location, field readings were 1,400 c.p.s. over the latite and 160 c.p.s. over a nearby serpentine outcrop. Laboratory analysis indicates that the serpentine is virtually free of radioactivity, showing that the nearby latite produced the abnormally high reading over the serpentine.

There are areas near the coasts of California, Oregon, and Washington where similar low-activity ultramafic rocks constitute the bulk of surface outcrops. These areas should show an unusually low terrestrial component of natural radiation; in our experience this is always the case, wherever readings have been taken. The extremely low radioactive content of these formations, and the close proximity of some of them to quite high-activity formations, provide a fascinating opportunity for study. In the absence of fallout, only cosmic rays and air-borne activity contribute to the radiation field, save for contamination within a detector and people in the vicinity. Such locales should provide very favorable circumstances for the study of air-borne radioactivity and low-energy cosmic-ray phenomena. It is an interesting speculation that the lowest-intensity natural background outside the ice-covered polar regions might be found over an extensive outcrop of ultramafic rocks along the Pacific Coast, where the prevailing westerlies, having traversed thousands of miles of open ocean, should be devoid of radon. The radiation intensity at such a place should be lower even than that observed over the ocean.

LIMESTONE AND DOLOMITE SAMPLES

In the course of an investigation of cement-plant raw materials and possible low-background aggregates we have subjected samples of California carbonate rocks to γ-ray spectrometric analysis. In these samples (28 of which are considered fairly "pure" carbonate rocks, that is, with little or no interbedded shale material) uranium predominates over thorium and potassium as the principal contributor of radioactivity. This is illustrated by the frequency-distribution histograms in Figure 12, which show that in over 40 per cent of the samples uranium is responsible for 90–100 per cent of the total counting rate. This predominance of uranium can be explained in part by the fact that the activities ratio of uranium:thorium:potassium in cpm/gm of element is 128,000:51,500:16. However, the absolute abundance of uranium over thorium, and the scarcity of potassium in the carbonate rock samples, are illustrated by the maximum, minimum, and average values given in the accompanying tabulation and

	CONCENTRATION			RATIO Th/U	OVER-ALL COUNTING RATE DUE TO NATURAL RADIO-ACTIVITY (cpm/gm)
	U (p.p.m.)	Th (p.p.m.)	K (per cent)		
Maximum......	4.87	1.80	0.28	....	0.601
Minimum.......	0.03	....	....	....	0.023
Average........	1.25	0.35	0.05	0.28	0.186

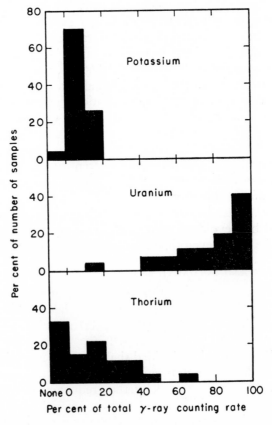

Fɪɢ. 12.—Frequency distribution of contributions of potassium, uranium, and thorium to the γ-ray counting rate in 28 carbonate-rock samples.

the plots of uranium and thorium concentrations versus over-all counting rate shown in Figure 13. Table 5 lists the γ-ray spectrometric evaluations of carbonate rocks.

In their discussion of thorium and uranium in carbonate rocks, Adams and Weaver (1958) state that about 80 per cent of the uranium in limestone is bound up in the calcite crystal lattice, while

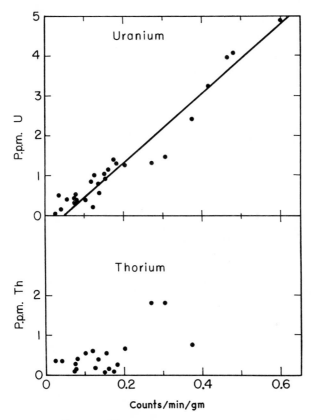

Fɪɢ. 13.—Limestone and dolomite samples

thorium and about 20 per cent of the uranium occur in the acid-insoluble detrital fraction. Therefore, the purity (magnesium or calcium carbonate content) of a limestone or dolomite should vary inversely with its thorium content as well as with the concentration of potassium, whose presence is due primarily to small amounts of clay and authigenic feldspars. A carbonate rock with a predominance of uranium over little or no thorium or potassium could be considered high purity as opposed to one with appreciable thorium and potassium, indicating the presence of clay and detrital minerals.

TABLE 5

γ-RAY SPECTROMETRIC EVALUATION OF CARBONATE-ROCK SAMPLES

RUN NO. AND LIMESTONE FORMATION	CONTRIBUTION TO TOTAL γ-RAY SPECTRUM (c.p.m.)			CONCENTRATION			OVER-ALL COUNTING RATE (cps/gm)	FIELD COUNTING RATE (c.p.s.)
	K	U	Th	K (per cent)	U (p.p.m.)	Th (p.p.m.)		
478 Calera, white......	4	3	14	0.03	0.03	0.36	0.023	50
721 Calera, blue.......	2	387	....	0.02	3.96	....	0.465	250
588 Gabilan...........	10	41	23	0.08	0.39	0.55	0.101	....
590 Sur series.........	4	122	3	0.03	1.03	0.07	0.149	....
609 Kernville (some biotite)..............	11	131	26	0.09	1.27	0.65	0.201	120
611 Kernville (biotitefree)..............	3	45	4	0.03	0.44	0.10	0.076	80
604 Bean Canyon high grade.............	12	145	....	0.09	1.30	1.80	0.268	....
634 Sparkuhle contact..	21	159	3	0.15	1.38	0.07	0.175	140
618 Sparkuhle.........	4	120	23	0.02	0.80	0.39	0.135	70
619 Sparkuhle, dolomitic.............	1	33	....	trace	0.26	......	0.034	70
620 Shay-Klondike.....	6	42	16	0.04	0.40	0.39	0.080	70
659 Black Mt. conglom.	32	270	33	0.23	2.40	0.76	0.376	120
642 Black Mt., low MgO	41	175	84	0.28	1.46	1.78	0.304	140
666 Furnace white.....	9	25	28	0.07	0.22	0.62	0.121	40
677 Furnace blue.....	1	50	12	trace	0.45	0.28	0.074	70
617 Jurupa, low MgO..	1	90	....	trace	0.83	....	0.117	50
676 Chino, footwall....	....	541	....	....	4.87	....	0.601	220
684 Chino, high lime...	7	377	....	0.04	3.23	....	0.416	190
678 Calaveras, high grade.............	2	55	....	0.02	0.39	....	0.054	60
687 Calaveras, dolomitic................	5	37	13	0.03	0.31	0.28	0.074	30
732 Calaveras, dark....	3	94	7	0.03	1.00	0.19	0.127	....
818 McCloud...........	....	126	6	......	1.14	0.14	0.161	....
808 Calera, Rockaway Beach.............	3	150	11	0.02	1.30	0.24	0.182	....
508 Gabilan dolomite, Natividad.........	9	25	21	0.05	0.16	0.33	0.041	80
800 Calaveras, dolomitic, Sonora.........	1	62	7	trace	0.51	0.14	0.077	....
864 Calaveras white dolomite.............	10	108	27	0.07	0.91	0.56	0.151	....
827 Calaveras gray dolomite.............	7	113	....	0.05	0.57	....	0.137	....
826 Calaveras white limestone.........	1	376	....	trace	4.05	....	0.479	110
Average values....	....	....	....	0.05	1.25	0.35	0.186	103 (for 19 samples)

HIGH-SILICA MATERIAL

In our investigation of possible sources of low-background aggregates several commercial sources of quartzose, high-silica sand were investigated in central California in the hope that suitable material for the minus $\frac{3}{16}$-inch fraction could be obtained. We also investigated sources of quartz for possible use in the coarser aggregate sizes. The results of the radiometric examinations of these materials are listed in Tables 6 and 7.

TABLE 6

γ-RAY SPECTROMETRIC EVALUATION OF SAND SAMPLES

RUN NO. AND DESCRIPTION	COUNTING RATE (cpm/gm)	CONCENTRATION			CONTRIBUTION TO TOTAL γ-RAY SPECTRUM (c.p.m.)		
		U (p.p.m.)	Th (p.p.m.)	K (per cent)	K	U	Th
899 Nortonville sand............	0.610	0.98	3.82	1.89	233	98	154
645 Corral Hollow sand, in old coal adit.......................	0.694	1.39	7.12	1.03	164	180	370
646 Corral Hollow sand, tailings dump......................	0.823	2.36	9.66	0.22	29	251	414
648 Corral Hollow sand, pile alongside hopper.................	0.548	1.07	5.24	1.32	184	120	237
649 Corral Hollow sand, hopper near adit portal..................	0.655	1.39	5.04	1.53	235	174	253
468 Ione sand, Buena Vista pit....	1.219	2.96	17.28	0.15	19	294	690
665 Clemco Amador I sand.......	0.415	0.21	1.35	2.01	275	24	61
673 Clemco II silica sand........	0.123	0.62	0.74	0.06	10	82	40
685 Monterey sand..............	0.785	0.03	3.96	3.21	520	4	209
672 Sierra Gem No. 12 sand......	0.097	0.31	0.56	0.15	23	39	28
580 Blackhawk (Ottawa, Ill.) silica sand.......................	0.047	0.19	0.37	....	....	26	20
664 Felton–Santa Cruz sand......	0.621	0.74	1.40	2.82	406	87	66
806 Contra Costa Ready-Mix Co. "Top Sand".................	0.306	0.91	2.06	0.58	92	117	107
532 Clemco No. 24 Silica Sand.....	0.088	0.19	0.96	0.06	10	27	55

With the exception of the Blackhawk silica sand (considered "Ottawa Standard Sand" by laboratories in the Bay area), the sand samples show generally high radioactivities when compared with the quartz samples. This can be attributed to the varying concentrations of grains of potassium feldspar and resistate minerals in the commercial sands, while vein quartz in its pure form has little or no radioactivity.

An interesting comparison is that between quartz and quartzite cobbles found in old placer tailings at Michigan Bluff, California. The quartzite is the product of the metamorphism of quartzose sandstone.

Radiometric evaluation of the quartzite (Table 7) indicates the presence of potassium, uranium, and thorium in amounts similar to those in a high-grade silica sand, while the quartz cobbles derived from stream-transported vein material are essentially free of radioactivity.

TABLE 7

γ-RAY SPECTROMETRIC EVALUATION OF QUARTZ SAMPLES

RUN NO. AND DESCRIPTION	COUNTING RATE (cpm/gm)	CONCENTRATION			CONTRIBUTION TO TOTAL γ-RAY SPECTRUM (c.p.m.)		
		U (p.p.m.)	Th (p.p.m.)	K (per cent)	K	U	Th
874 Michigan Bluff, Calif., placer quartz cobbles.................	0.011	......	......	0.02	3	......	......
797 Michigan Bluff, Calif., quartzite cobbles......................	0.151	0.47	1.46	0.14	17	46	57
875 Crushed "Clear Cat" vein quartz	0.020	0.11	......	......	......	11	....
885 Hornitos, Calif., vein quartz, with pyrite, and some mariposite.........................	0.012	......	......	0.05	6	......	......
815 Hornitos vein quartz..........	0.0045	......	......	......	......	......	......
907 Rogue River, Oregon quartz (crushed by Industrial Minerals and Chemical Co., Berkeley)...	0.018	0.08	0.12	......	......	9	5
873 Vein quartz, Coulterville, Calif..	0.001	......	......	......	......	......	......
872 Iron-stained, weathered, vein quartz, Coulterville, Calif......	0.006	......	0.05	trace	1	......	2

PORTLAND CEMENT STUDY

The usefulness of γ-ray spectrometry is demonstrated in a search that we conducted for a portland cement adequately low in radioactivity for use in the concrete of our low-background counting facility. The results of this study are fully reported by the authors in UCRL-10475. To determine the sources of radioactive contamination of cement, we visited all operating California plants and their adjacent quarries, sampled all raw materials and subjected them to γ-ray spectrometric analysis. As a result, we determined that uranium, the principal contributor of radioactivity to the limestone, is also the principal contributor to cement's activity, while thorium, the principal contributor in the siliceous-"shale" and aluminous-clay raw materials, is the secondary contributor in California cements.

With one exception, "shale" material at California plants has a higher radioactivity than has the corresponding limestone. The histogram in Figure 14 illustrates the higher activity of "shale" when compared to limestone. Laboratory counting rates for "shale" average 0.900 cpm/gm, while limestone averages 0.231. Field scintillation-counter readings average 314 c.p.s. in "shale" deposits, 103 c.p.s. in limestone. Plots of field versus laboratory counting rates for limestone and "shale" (Fig. 15) indicate good correlations where field count rates are representative of the material scanned.

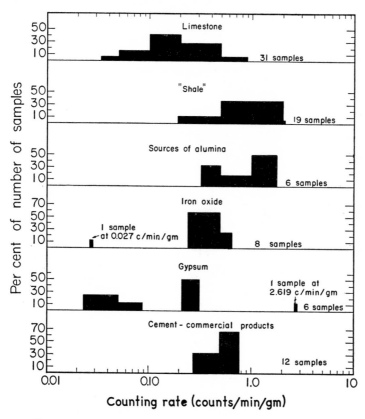

Fig. 14.—Frequency distribution of over-all γ-ray counting rates of cement raw-material ingredients.

On the basis of our California cement raw-materials study, we can state that, with some exceptions, limestone is the ingredient primarily responsible for determining the radioactivity of the finished product. Limestone usually makes up about 80 per cent of the kiln feed at cement plants, while "shale" material comprises only 10–15 per cent. However, at plants where "shale" has an activity 3–4 times that of the limestone, "shale" achieves parity or predominates in contributing to the cement's radioactivity. This is the case at some plants in southern

California where micaceous schists and metasediments are the sources of "shale." We emphasize, though, that low-activity cement can never be made from high-activity limestone. We expect that most workers who care about cement activity are attempting to obtain a low-activity product; thus the importance of limestone is evident.

California cement-plant "shale" materials are derived from widely

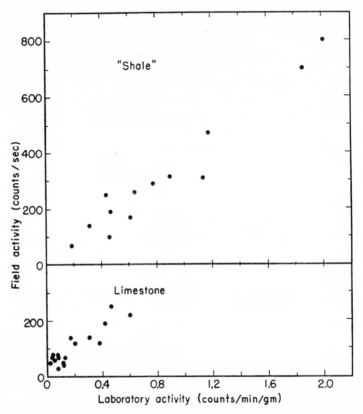

Fig. 15.—Field *vs.* counting rates for limestone and cement-plant "shale" samples

differing rock types, depending upon plant location. Besides the micaceous schists and metasediments mentioned above, southern California plants also draw on quartzites, lateritic clay, and interbedded shales, while plants in the northern part of the state utilize andesite, diatomaceous shale, phyllite, altered volcanics, and lateritic clay.

In other sections of the United States, it appears that limestones play a more dominant role in determining the radioactivities of their respective cements. Though we have not counted the raw materials from these plants, we suspect that a correlation exists between the counting rates of cements and of the most utilized limestones. This is illustrated in Table 8.

An example of the usefulness of γ spectrometry in determining the distribution of radioactivity in a cement plant's raw materials and finished product is illustrated by the study that we conducted on samples furnished by the Lafarge Cement Company of Vancouver, British Columbia. Along with the samples, the company sent us data indicating that there is an over-all weight loss (or ignition loss) of

TABLE 8

AVERAGE COUNTING RATES FOR CEMENTS DERIVED FROM
THE MOST-UTILIZED LIMESTONES

Limestone Formation	Geographic Location	No. Plants Reporting	Cement Av. Counting Rate (cpm/gm)
Oyster shells.........	Gulf and West Coast	5	0.416
Jacksonburg..........	Lehigh Valley, Pa.	6	0.478
Conosauga..........	Alabama and Georgia	3	0.541
Becraft, Coeymans, N. Scotland and Manilius..............	Appalachian region, Va., to Catskill Mts., N.Y.	5	0.524
Rogers City..........	Wisconsin, Michigan, and northern New York	4	0.680
Vanport.............	Western Pennsylvania and eastern Ohio	4	1.100
Austin Chalk.........	San Antonio and Dallas, Texas	2	0.692
Iola.................	Eastern Kansas	2	0.800
Metaline.............	Northeastern Washington	2	0.594
Annona Chalk........	Central Arkansas	2	0.635

36.4 per cent between kiln-feed slurry and clinker, owing to volatilization during calcining. Our spectrometric evaluation of calculated kiln feed and cement product indicated that during clinkerization, the potassium content is little changed; uranium, thorium, and over-all counting rate are increased by 38, 39, and 36 per cent, respectively. These increases are quite close to the over-all ignition loss of 36.4 per cent. This suggests that, at Lafarge, uranium and thorium are concentrated during calcining and are not among the volatile constituents of clinkerization. The potassium content remains unchanged during calcining, indicating that potassium is volatilized in an amount equal to the over-all ignition loss.

The following steps, illustrated in Table 9, were performed to achieve the results described above:

a) Calculate the percentages of raw materials in the finished product (clinker plus gypsum) by applying ignition-loss percentages to the amounts of raw materials in the slurry.

b) Calculate the contaminant concentrations and counting rates of the clinkerized ingredients by dividing measured uranium, thorium, potassium, and counting-rate values for each slurry component by 100 per cent minus respective ignition loss.

c) Calculate finished product's uranium, thorium, and potassium concentrations and over-all counting rate by weighting values from (*b*) above by percentage compositions of the raw materials in the clinker.

d) Comparison of hypothetical cement to measured values from sample of Type 1 cement shows excellent agreement, verifying the calculation procedure and the assumptions used in the process.

Conclusion

We have described both field and laboratory techniques for evaluation of the γ-ray component of environmental radiation; some applications for these techniques have been cited.

Laboratory analysis enables us to make quantitative assay for potassium, uranium, and thorium. Radiometric assay for potassium agrees well with chemical methods. Our assays for uranium and thorium are based only on the NBL calibration ores and have not been checked against chemical methods. Such cross-calibrations need to be performed. In addition, we need to try our assay methods on samples that have been assayed by other workers and would welcome any contribution of such samples for our tests.

Valid uranium and thorium assays require that the decay series have reached secular equilibrium. We have not made a thorough study of the errors introduced by lack of equilibrium. The relative insolubility of thorium and the half-lives of its decay series suggest that lack of equilibrium is unlikely here. However, uranium can be leached from its host material, there are long half-lived decay members, and the gaseous member can be lost to the air—three conditions which may lead to lack of equilibrium. Most of the uranium-series γ rays come from daughters of radium; therefore, some of our assays may really reflect only radium content. This situation would seem most likely to occur with soil samples; we are aware of these items but have not studied them thoroughly enough to know the real extent of the problems they pose.

FURTHER STUDIES

Until now, successful low-background shielding has been our principal goal. Recently, the Lawrence Radiation Laboratory gave us permission to expand our studies beyond the scope of low-background

TABLE 9

MEASURED AND CALCULATED RADIOISOTOPE CONCENTRATIONS AND COUNTING RATES FOR A TYPE I CEMENT

INGREDIENT	PER CENT SLURRY (BY WEIGHT)	MEASURED CONTAMINANT CONCENTRATIONS IN SLURRY INGREDIENTS			MEASURED OVER-ALL COUNTING RATE IN SLURRY INGREDIENTS (cpm/gm)	PER CENT CLINKER+GYPSUM (BY WEIGHT) CALCULATED FROM IGNITION LOSS	CALCULATED CONTAMINANT CONCENTRATION IN CLINKER INGREDIENTS			CALCULATED OVER-ALL COUNTING RATE IN CLINKER INGREDIENTS (cpm/gm)
		K (per cent)	U (p.p.m.)	Th (p.p.m.)			K (per cent)	U (p.p.m.)	Th (p.p.m.)	
Limestone, Hi Titer..........	54	0.04	1.50	0.33	0.216 (0.350 with fallout)	46	0.04	2.58	0.57	0.372
Limestone, Lo Titer..........	29	0.13	1.59	0.26	0.238 (0.463 with fallout)	27	0.13	2.56	0.42	0.384
Tailings..........	10	0.85	1.07	1.25	0.335 (0.402 with fallout)	15.5	0.85	1.10	1.29	0.345
Dunsmuir shale..........	4	1.05	1.82	4.24	0.619 (1.003 with fallout)	6	1.05	1.98	4.60	0.673
Cassidy shale (60 per cent combustible)..........	3	0.41	0.55	1.65	0.220 (0.248 with fallout)	1.5	0.41	1.83	5.50	0.734
Gypsum..........	....	....	....	....	....	4	0.05 (Measured)	0.35	0.13	0.059 (0.076 with fallout)
Weighted totals for slurry (calculated)		0.20	1.35	0.57	0.233	Weighted totals for clinker and gypsum (calculated)	0.20	2.09	0.92	0.367
					Measured Type I cement		0.18	2.17	0.93	0.365 (0.412 with fallout)

shielding to encompass the applications of γ spectrometry to the earth sciences and the study of the relationship between the γ radiation of natural raw materials and manufactured products.

Studies in Geoscience.   A project of particular interest to us is the application of γ-ray spectrometric determinations of natural radioisotopes in the study of heat generation in the earth's crust, heat flow, and rock metamorphism. Verhoogen (1956) and Birch (1954) present good synopses and bibliographies on recent and past earth-heat studies. At present there is active investigation (Lachenbruch, personal communication) dealing with the flow of heat from and through the earth's mantle and crustal layers whereby it may be determined whether the earth is gaining or losing heat. An attempt is also being made to establish the quantity of heat being transmitted from the earth's interior to the crust. Though heat transmission to the crust cannot be measured, it can be calculated if the crustal heat flow and heat generated by radioactivity are known.

Contemporary heat-generation data are also valuable in determining the heat generation, rapidity of heating, and maximum temperature attained in a sequence of rocks formerly buried at great depths. Metamorphosed rocks, particularly the blueschists of the Franciscan formation of the California coast ranges, are ready subjects for such a study. The Franciscan blueschists are considered by Bailey *et al.* (in press) to have been formed in an environment of abnormally high pressure relative to the temperature. Such low temperatures can result from rapid accumulation and burial or from abnormally low radioactivity in the rock. The parent material for the schists, Franciscan graywacke, generally contains little or no potassium feldspar (Bailey and Irwin, 1959); it is possible that they are also low in other potassium minerals as well as uranium and thorium.

With a sufficient concentration of radioactivity, the blueschists, if they remain in their environment of formation, should gradually be heated, converting them to the higher grade of regional metamorphism, the greenschist facies. The Franciscan blueschists have not converted; a knowledge of their radioisotope concentrations would indicate whether low radioactivity is responsible.

We are at present embarking on studies in conjunction with members of the U.S. Geological Survey in which field and laboratory γ-radiation data will be applied to such problems in crustal heating and regional metamorphism.

Studies of Structural Materials.   Besides continuing the investigation of cement raw materials, we have proposed that an investigation of steel should also be of interest. Iron ores that we have examined contain appreciable radioactivity, yet, in most cases, the finished-product steel is low in activity. Sampling and subsequent spectral analysis of the raw materials and products of the various stages of the steel-making process, for example, blast-furnace slag, open-hearth-furnace slag, and mill scale would document the purging of the initial radioactivity and also might furnish some insight into natural processes.

Our work will be carried on within the framework of the Health Physics Department at UCLRL, Berkeley, utilizing our recently constructed low-background concrete counting facility. We mean to pursue this work on a full-time basis and, in this respect, would like to establish and maintain contact with those who are also working in these areas. Such an interchange of information and ideas will certainly be mutually beneficial.

REFERENCES

ADAMS, J. A. S. 1954. Uranium and thorium contents of volcanic rocks. *In* H. FAUL (ed.), Nuclear Geology, pp. 89–98. New York: John Wiley & Sons.

ADAMS, J. A. S., and C. E. WEAVER. 1958. Thorium to uranium ratios as indications of sedimentary processes: Example of concept of geochemical facies. Bull. Am. Assoc. Petrol. Geologists, **42**:387–430.

AHRENS, L. H. 1954. The abundance of potassium. *In* H. FAUL (ed.), Nuclear Geology, pp. 128–32. New York: John Wiley & Sons.

AHRENS, L. H., W. H. PINSON, and M. M. KEARNS. 1952. Association of rubidium and potassium and their abundance in common igneous rocks and meteorites. Geochim. & Cosmochim. Acta, **2**:229–42.

BAILEY, E. H., and W. P. IRWIN. 1959. K-feldspar content of Jurassic and Cretaceous graywackes of northern Coast Ranges and Sacramento Valley, California. Bull. Am. Assoc. Petrol. Geologists, **43**:2797–2809.

BAILEY, E. H., W. P. IRWIN, and E. L. JONES. Franciscan and related rocks and their significance in the geology of the California coast ranges. Bull Calif. Div. Mines & Geology. (In press.)

BELL, K. G. 1954. Uranium and thorium in sedimentary rocks. *In* H. FAUL (ed.), Nuclear Geology, pp. 98–114. New York: John Wiley & Sons.

BIRCH, F. 1954. Heat from radioactivity. *In* H. FAUL (ed.), Nuclear Geology, pp. 148–74. New York: John Wiley & Sons.

CAVELL, I. W., and C. O. PEABODY. 1961. The Winfrith District gamma survey. United Kingdom Atomic Energy Authority, Research Group, Atomic Energy Establishment, Winfrith, Dorset, England, AEEW-R-62. Pp. 26.

DALY, R. A. 1933. Igneous Rocks and the Depth of the Earth; Containing Some Revised Chapters of "Igneous Rocks and Their Origin." New York: McGraw-Hill Book Co.

DAVIS, G. L., and H. H. HESS. 1949. Radium content of ultramafic igneous rocks. II. Geological and chemical implications. Am. J. Sci., **247**:856–82.

EVANS, R. D., and H. WILLIAMS. 1935. The radium content of lavas from Lassen Volcanic National Park, California. Am. J. Sci., Ser. 5, **29**:441–52.

FRONDEL, C. 1956. Mineralogy of thorium. U.S. Geol. Survey Prof. Paper 300, pp. 567–79.

GOLDSWORTHY, W. W. 1960. Transistorized portable counting-rate meter. Nucleonics, **18**:92–99.

GREGORY, A. F., and J. L. HORWOOD. 1961. A laboratory study of gamma-ray spectra at the surface of rocks. Canad. Dept. Mines & Tech. Surveys, Mines Branch Rpt. R-85. Pp. 56.

KEEVIL, N. B. 1944. Thorium-uranium ratios in rocks and minerals. Am. J. Sci., **242**:309–21.

KOCZY, F. F. 1954. Geochemical balance in the hydrosphere. *In* H. FAUL (ed.), Nuclear Geology, pp. 120–27. New York: John Wiley & Sons.

LACHENBRUCH, A. H. Personal communication, 1963.

LARSEN, E. S., G. PHAIR, D. GOTTFRIED, and W. L. SMITH. 1956. Uranium in magmatic differentiation. U.S. Geol. Survey Prof. Paper 300, pp. 65–74.

LOWDER, W. M., W. J. CONDON, and H. L. BECK. 1963. Field spectrometric investigations of environmental radiation in the U.S.A. This symposium.

PATTERSON, H. W., W. N. HESS, B. J. MOYER, and R. W. WALLACE. 1959. The flux and spectrum of cosmic-ray produced neutrons as a function of altitude. Health Physics, **2**:69–72.

PETTERSON, H. 1954. Radioactive elements in ocean waters and sediments. *In* H. FAUL (ed.), Nuclear Geology, pp. 115–20. New York: John Wiley & Sons.

RANKAMA, K. 1954. Isotope Geology. London: Pergamon Press.

RANKAMA, K., and TH. G. SAHAMA. 1950. Geochemistry. Chicago: University of Chicago Press.

VERHOOGEN, J. 1956. Temperatures within the earth. *In* L. H. AHRENS (ed.), Physics and Chemistry of the Earth, **1**:17–43. London: Pergamon Press.

WILLIAMS, H. 1957. Geologic Map of the Central Part of the High Cascade Range, Oregon, with Accompanying Text. Corvallis: Oregon Dept. of Geology and Mineral Industries.

RICHARD I. WELLER

33. Low-Level Radioactive Contamination

THE RADIOACTIVE CONTAMINATION of materials in the natural environment has presented a challenge to investigators of radiation since the discovery of radioactivity at the beginning of this century. The "contamination" referred to here is considered as any nuclear radiation emanating from a substance, regardless of whether it is natural in origin or a result of man's recent experiments with the atom. By "low level" is meant radiations that are usually considered inherent detector background activity. Although much radioactive contamination is considerably lower than the customary permissible radiation exposure levels established for health protection, sensitive detectors are affected by such materials.

At one time or another many scientists concerned with the measurement of weak activities have had to make measurements for suspected radioactive contamination in their detectors or shields. Yet, comparatively few systematic investigations have been reported.

A study of radioactive contamination of materials has been made by the Subcommittee on Radiochemistry of the Committee on Nuclear Science, National Research Council, National Academy of Sciences. DeVoe (1961) has written a comprehensive report on the results of an investigation of the problem performed by him and the committee. The report outlines the extent of radioactive contamination of materials and reagents, mechanisms of radioactive contamination, and effects of radioactive contamination on scientific research and suggests possible methods for obtaining uncontaminated substances. Included are data on radioisotopic and radiochemical contamination with man-made and naturally occurring radioisotopes, methods of assaying radioactivity, and a discussion of low-background detectors. In the appendixes to the report are reproduced extensive data on the reported measurements of a number of investigators on the radioactivity

RICHARD I. WELLER is professor of physics, Department of Physics, Franklin and Marshall College, Lancaster, Pennsylvania.

found in reagent chemicals and other materials. A considerable bibliography, as well as the inclusion of recommendations by the Subcommittee on Radiochemistry, makes the report an excellent reference base for future work in this field.

From time to time, the emission of a, β, and γ radiation has been measured and reported in other sources, usually when such data had been accumulated incidental to the problem of increasing the sensitivity of a detector by the reduction of background. To cite a few illustrations, Sharpe (1955) presented data on the rate of emission of a particles with energy greater than 100 kev. from various substances, and compared his results with those of Bearden (1933). Schumann (1962) presented a comparison of the a emission of miscellaneous substances as measured by Bearden (1933), Curtiss (1943), Sharpe (1955), and McDaniel *et al.* (1956). Background γ activity data for paints and varnishes, components used in multiplier tubes, sodium iodide crystals with various tube configurations, and a considerable number of other materials have been published by LeVine (1961). May and Marinelli (1962) have presented data on sodium iodide systems and the origin of background. A study of the internal contamination of steel with ruthenium-103 has been reported by Anderson and Rowe (1962).

PRESENT WORK

In the course of constructing sensitive radiation detectors, I found it necessary to measure the inherent radioactivity of materials to insure adequately low background. Some of the apparatus used in the educational program of Franklin and Marshall College seemed suitable, and this was augmented to a limited extent with additional equipment.

The low-level measurements laboratory of the college is located in the basement of a science building erected in 1900. The basement walls are constructed of stone and mortar and are at least two feet thick. The room was recently renovated for its present use and contains extremely well-regulated and filtered electrical supply lines as well as the required temperature controls.

The tabular results displayed below represent preliminary findings. This is the first phase of a survey to measure all (or most) of the elements and materials of interest to radiation scientists. The object of this survey is not only to point up low-activity materials, but to indicate the extent and level of radioactive contamination. The measuring instruments consisted of a comparatively large-area, thin-window gas flow counter, an internal 4π flow counter and a thallium-activated sodium iodide scintillation counter. All detectors could readily accommodate 5-inch-diameter samples. Only gross counting was undertaken, and the results are presented in counts per minute or per hour on a relative basis.

THIN-WINDOW FLOW COUNTER

The thin-window flow counter was a Sharp Laboratories low-beta counting system with two detectors. Each detector window was 5 inches in diameter and was completely surrounded by 1 inch of OFHC copper shielding. Around the copper were 4 inches of low-level lead. Directly above each detector was an 8-inch cosmic-ray guard detector connected in anticoincidence with the 5-inch detector to blank cosmic-ray counts from the readout of detector counts. A push-pull blanking system also eliminated electrical environmental noise from the background.

TABLE 1

THIN-WINDOW PROPORTIONAL FLOW COUNTER MEASUREMENTS

Sample Name	Description	Weight (gm.)	Area (cm.^{2})	Net α/100 cm^2/hr	Net β/100 cm^2/hr
Aluminum (54)....	Cast plate	154.2	98.06	5.76±1.86	348±8
Bismuth (26)....	99.999+% Pure	730	114.7	49.6±2.7	13.6±7.3
Cadmium (27)....	99.999+% Pure	620.8	112.8	—*	39.4±3.2
Copper (3)....	Powder	56.8	101	35.9±1.6	68.9±4.1
Gold (34)....	99.979% Pure	315.1	102.4	—	—
Graphite (44)....	Madagascar flakes, impure	13.2	101	74.6±1.9	1,130±8
Indium (28)....	99.999+% Pure	541.6	114.3	—	72.6±6.8
Iridium (35)....	99.9+% Pure	182.3	102.6	134±3	1,640±10
Nickel (5)....	Powder	68.5	101	26±2	114±6
Osmium (36)....	99.92% Pure	162.5	17.48	—	750±24
Palladium (37)....	99.865% Pure	195	102.3	—	70.8±4.7
Platinum (38)....	99.968+% Pure	347.9	102.3	—	68.4±4.1
Rhodium (39)....	99.940% Pure	199.4	101.6	20±1.6	415±6
Ruthenium (40)....	99.96% Pure	119.5	20.26	144±8	107±21
Silver (41)....	99.969% Pure	171.1	102.4	—	—
Tantalum (42)....	....	1.066	27.04	23.3±7.1	146±22
Tin (14)....	Pure, lead-free foil	1.203	62.95	182±5	1,260±14
Zinc (9)....	100% Pure	16.01	87.98	20.9±1.8	12.3±4.7
Lead [1] (15)....	Mexican origin	846.5	116.2	59.8±1.7	547±5
Lead [2] (16)....	Domestic origin	832.4	116.2	1,170±5	7,790±14
Lead [3] (17)....	Peruvian origin	845.1	117.5	728±3	5,430±8
Lead [4] (18)....	Domestic origin	842.2	116.8	17.8±2.2	238±8
Lead [4B] (19)....	....	812.5	113.2	45.4±2.7	296±7
Lead [5] (20)....	Domestic origin	841.8	116.7	58.1±2.8	216±7
Lead [6] (21)....	Australian origin	845.1	117.8	99.8±2.9	437±8
Lead [7] (22)....	Domestic origin	850	117.6	737±5	5,240±13
Lead [8] (23)....	Canadian origin	840.8	117.9	233±6	3,257±22
Lead [P] (24)....	Holland origin	816.1	112.9	—	59.5±7.4
Lead [S] (25)....	U. of C. shield	827.2	114.6	542±4	2,980±9
Lead (29)....	99.999+% Pure	842.3	114.2	298±4	2,910±13
Lead (4)....	Powder	99.6	101	263±10	4,630±41

* Dash indicates no detectable radioactivity above background.

The detectors had 800 μgm/cm^2 gold-coated Mylar windows. The maximum sample-to-window distance was $\frac{3}{64}$ inch, and the maximum sample thickness was $\frac{3}{8}$ inch. The gas flow detectors were operated in the proportional mode with a 90 per cent argon, 10 per cent methane counting gas mixture. Simultaneous α-β counting was employed using pulse-height discrimination. The manufacturer's claim that less than 10 per cent of the α's fall into the β channel was substantiated.

PLATE I

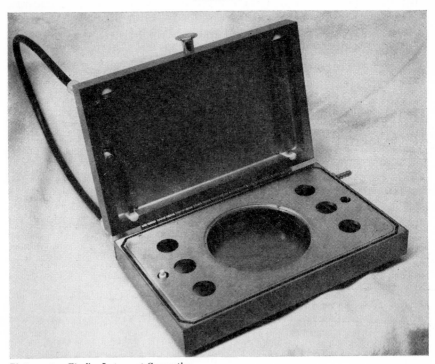

Photo courtesy Eberline Instrument Corporation

Internal 4π Flow Counter

The average α background was 0.3 c.p.m., with a standard deviation of ± 0.025 c.p.m. For β's, the background was 3.25 c.p.m., with a standard deviation of ± 0.12 c.p.m. In accordance with the customary statistical standards for nuclear counting, only counts that exceeded the background count by at least three times the standard deviation of the background count for the same counting time were considered detectable and reported. This criterion was followed with the measurements of all instruments.

Table 1 presents the measurements recently made with the thin-window proportional counter. The figures in the last two columns are the net counts (background subtracted) per 100 cm.² per hour with the standard deviation of the net counting rate for α and β radiation, respectively, based on counting statistics only. It is evident that activity was measurable from all samples except pure gold and silver. The lower portion of the table presents a study of a series of lead samples.

INTERNAL FLOW COUNTER

An internal 4π proportional flow counter was built for this work by the Eberline Instrument Corporation. The chamber, shown in

TABLE 2

INTERNAL 4π PROPORTIONAL FLOW COUNTER MEASUREMENTS

Sample Name	Description	Weight (gm.)	Area (cm.²)	Net α/100 cm²/hr	Net β/100 cm²/hr
Aluminum (54)	Cast plate	154.2	98.06	42.8±1.6	612±44
Copper (3)	Powder	56.8	101	55.2±3	1,010±146
Gold (34)	99.979% Pure	315.1	102.4	23.8±1.2	352±3
Iridium (35)	99.9+% Pure	182.3	102.6	138±9	2,164±64
Nickel (5)	Powder	68.5	101	61.8±3.4	1,010±142
Rhodium (39)	99.94% Pure	199.4	101.6	47.8±1.4	472±51
Ruthenium (40)	99.96% Pure	119.5	20.26	461±15	1,037±185
Silver (41)	99.969% Pure	171.1	102.4	13.2±2.01	—*
Tin (14)	Pure, lead-free foil	1.203	62.95	158±4	2,002±65
Lead [1] (15)	Mexican origin	846.5	116.2	146±3	775±26
Lead [2] (16)	Domestic origin	832.4	116.2	2,290±80	10,250±79
Lead [3] (17)	Peruvian origin	845.1	117.5	1,210±16	6,100±52
Lead [4] (18)	Domestic origin	842.2	116.8	86.6±2.7	539±40
Lead [4B] (19)	....	812.5	113.2	111±5	477±30
Lead [5] (20)	Domestic origin	841.8	116.7	121±6	694±58
Lead [6] (21)	Australian origin	845.1	117.8	128±4	789±27
Lead [7] (22)	Domestic origin	850	117.6	1,060±16	7,320±55
Lead [8] (23)	Canadian origin	840.8	117.9	459±9	3,842±63
Lead [P] (24)	Holland origin	816.1	112.9	30±1.4	531±26
Lead [S] (25)	U. of C. shield	827.2	114.6	1,126±33	3,770±65
Lead (29)	99.999+% Pure	842.3	114.2	373±10	4,440±98
Lead (4)	Powder	99.6	101	358±5	8,257±167

* Dash indicates no detectable radioactivity above background.

Plate I, consisted of hinged twin, facing probes with the sample mounted between them. The figure illustrates the 5-inch sample holder. It was found that sample thicknesses up to $\frac{1}{4}$ inch could be accommodated.

Long, flat plateaus were readily obtained with chemical-grade propane counting gas. A conventional preamplifier, power supply, and scaler were used with this detector. The chamber was surrounded by 2 inches of low-level lead on all sides and 4 inches on top. This was all the lead available. The mean α background for the reported data was 0.60 c.p.m., with a standard deviation of ±0.03 c.p.m. The mean β background was 360 c.p.m., with a standard deviation of ±1.0 c.p.m. The results in Table 2 for this instrument are in the same units as in Table 1. In all tables, a dash indicates that the radioactivity measured was not significantly above background. It should be noted that all samples (except powders) were cleaned prior to each measurement with acetone. The acetone used gave no observable measurement of α, β, or γ radioactivity. With this sensitive detector, radiation was measurable from all the samples tabulated.

SCINTILLATION COUNTER

The γ radiation counting system consisted of a 5-inch-diameter by 2-inch-thick thallium-activated sodium iodide crystal mounted in an aluminum container. The entrance window was of 0.025-inch-thick aluminum plus 20 mg/cm² of sprayed aluminum oxide (Harshaw). The scintillator was used with a single-channel pulse-height analyzer and a scaler. Gross counts were measured in the energy range 100 kev.–4 mev. For minimal background, the scintillator was mounted face down in a solid concrete-block shielding inclosure 32 inches thick. Four inches of low-level lead surrounded the scintillator-photomultiplier-preamplifier assembly within the block house. The average background for the measurements reported in Table 3 was 830 c.p.m., with a standard deviation of ±1.8 c.p.m. The results make evident the fact that, though this system is sensitive, the background is too high to provide low-level measurements comparable to the two preceding counters. Better electronics and shielding, as well as a larger crystal, are desirable.

Table 4 presents a composite tabulation of the lead samples measured with all the counters. All the figures are in counts per hour for the α and β radiations and in counts per minute for the γ radiation, as in the three preceding tables. The data in the last column represent measurements taken by Anderson *et al.* at Los Alamos Scientific Laboratory and are displayed here through the courtesy of that group for comparative purposes. Their specimens were $6 \times 6 \times 1$ inch in size. The samples were measured on a $7\frac{1}{2} \times 4$-inch sodium iodide crystal contained in a small lead shield, 4 inches thick. A $\frac{3}{8}$-inch lucite shield was placed between the crystal and sample to protect the crystal against impact and contamination. The crystal was canned in stainless steel. The counting rates listed were gross rates in the energy range 30 kev.–1 mev. Counting time was 1,000 minutes, so counting statistics were ±1 c.p.m. or about 0.1 per cent (one standard deviation). The spectrum of a reference sample, recorded on magnetic tape, had been subtracted. The group who made this study will elaborate upon it in a forthcoming report from the Los Alamos Laboratory.

TABLE 3

SCINTILLATION COUNTER MEASUREMENTS

Sample Name	Description	Area (cm.²)	Weight (gm.)	Net γ/kg min
Aluminum (54)....	Cast plate	98.06	154.2	156±10
Bismuth (26)....	99.999+% Pure	114.7	730.0	19.2±3.2
Cadmium (27)....	99.999+% Pure	112.8	620.8	—*
Copper (3)....	Powder	101	56.8	247±47
Gold (34)....	99.979% Pure	102.4	315.1	—
Graphite (44)....	Madagascar flakes, impure	101	13.2	758±114
Indium (28)....	99.999+% Pure	114.3	541.6	—
Iridium (35)....	99.9+% Pure	102.6	182.3	362±9
Nickel (5)....	Powder	101	68.5	263±41
Osmium (36)....	99.92% Pure	17.48	162.5	—
Palladium (37)....	99.865% Pure	102.3	195	35.9±13.3
Platinum (38)....	99.968+% Pure	102.3	347.9	—
Rhodium (39)....	99.94% Pure	101.6	199.4	1,043±10
Ruthenium (40)....	99.96% Pure	20.26	119.5	—
Silver (41)....	99.969% Pure	102.4	171.1	—
Zinc (9)....	100% Pure	125.4	441.7	29.4±4.8
Lead [1] (15)....	Mexican origin	116.2	846.5	0†
Lead [2] (16)....	Domestic origin	116.2	832.4	63.7±0.5
Lead [3] (17)....	Peruvian origin	117.5	845.1	54.5±0.3
Lead [4] (18)....	Domestic origin	116.8	842.2	7.12±0.24
Lead [4B] (19)....	....	113.2	812.5	9.85±0.25
Lead [5] (20)....	Domestic origin	116.7	841.8	1.19±0.05
Lead [6] (21)....	Australian origin	117.8	845.1	5.92±0.23
Lead [7] (22)....	Domestic origin	117.6	850	35.2±0.3
Lead [8] (23)....	Canadian origin	117.9	840.8	27.4±0.3
Lead [P] (24)....	Holland origin	112.9	816.1	6.13±0.26
Lead [S] (25)....	U. of C. shield	114.6	827.2	24.2±0.3
Lead (29)....	99.999+% Pure	114.2	842.3	26.1±0.3
Lead (4)....	Powder	101	99.6	—

* Dash indicates no detectable radioactivity above background.

† Minimum value considered as background. Other values for γ radiation listed for lead samples are with respect to this reference.

The results of Table 4 clearly indicate the higher sensitivity of the windowless counter relative to the thin-window counter. However, the well-known advantages of the thin-window counter should not be overlooked. The possible difficulties with charge effect, vapor effect, and counter contamination must be considered with internal windowless counters. Also, it should be noted that measurements could be made more rapidly with the thin-window counter because it required no delay for gas flushing. Furthermore, the thin-window counter utilized simultaneous α and β counting, resulting in a considerable saving in counting time.

The γ radiation tabulation shows that careful measurements do give a fair relative picture of comparative emanations. The superiority of the Los Alamos system is evident from the data presented.

TABLE 4

COMPOSITE TABULATION OF LEAD SAMPLES MEASURED ON ALL COUNTERS

LEAD SAMPLE	NET α/100 CM²/HR		NET β/100 GM²/HR		NET γ C.P.M.	
	Thin-Window Counter	4π Windowless Counter	Thin-Window Counter	4π Windowless Counter	Per Kg.	*
(4).........	263	358	4,630	8,257	—†	....
(15).........	59.8	146	547	775	0‡	7
(16).........	1,170	2,292	7,790	10,250	63.7	191
(17).........	728	1,205	5,430	6,102	54.4	35
(18).........	17.8	86.6	238	539	7.12	9
(19).........	45.4	107	296	461	9.85	....
(20).........	58.1	121	216	694	1.19	2
(21).........	99.8	128	437	789	5.92	26
(22).........	737	1,059	5,240	7,321	35.2	127
(23).........	233	459	3,257	3,842	27.4	56
(24).........	—	30	59.5	531	6.13	....
(25).........	542	1,126	2,980	3,770	24.2	0‡
(29).........	298	373	2,910	4,440	26.1	....

* From data by E. C. Anderson *et al.*, Los Alamos Scientific Laboratory.

† Dash indicates no detectable radioactivity above background.

‡ Minimum value considered as background. Other values for γ radiation listed for lead samples are with respect to this reference.

CONCLUSION

The object of this report was to call attention to measurements of radioactive contamination that have been made and that are currently in progress. The data presented are of a preliminary nature. Both the counters and the samples listed will be studied further before more definitive results and conclusions will be presented.

A new Subcommittee on Low Level Contamination of Materials (of which I am a member) has been formed by the Committee on Nuclear Science, National Research Council, National Academy of Sciences. It is anticipated that this group, under the chairmanship of E. C. Anderson, not only will give further consideration to the magnitude and specific nature of the problem but will also consider appropriate education, routine monitoring, and the possibility of stockpiling critical materials.

ACKNOWLEDGMENTS

This investigation was supported by U.S. Public Health Service Research Grant OH 00131-01 from the Division of Occupational Health. Many of the measurements were made by Douglas G. Smith. The U.S. Atomic Energy Commission provided much of the measuring apparatus. The 5-inch crystal scintillator was donated by the Budd Company. The following companies donated or loaned samples for measurement: American Smelting and Refining Company, Division Lead Company, Engelhard Industries, Mallinckrodt Chemical Works, Metals Disintegrating Corporation, and Williams Gold Refining Company.

REFERENCES

ANDERSON, E. C., and M. W. ROWE. 1962. Contamination of Steel by Ruthenium-103. Los Alamos Scientific Lab., U. of Calif. Rpt. LAMS-2780, pp. 163–73.

BEARDEN, J. A. 1933. Radioactive contamination of ionization chamber material. Rev. Sci. Instr., **4**:271–75.

CURTISS, L. F. 1943. Miniature Geiger-Müller counter. J. Res. Nat. Bur. Stand., **30**:157–58.

DEVOE, J. R. 1961. Radioactive contamination of materials used in scientific research. Washington, D.C., National Research Council, Committee on Nuclear Science. NAS-NRC-Pub-895. Pp. 149.

LEVINE, H. D. 1961. Advanced instrumentation for a gamma ray counting system to measure the radioactivity of humans. *In* H. v. KOCH (ed.), Instruments and Measurements, Vol. **2**. New York: Academic Press.

McDANIEL, E. W., H. J. SCHAEFER, and J. K. COLEHOUR. 1956. Dual proportional counter for low-level measurement of alpha activity of biological materials. Rev. Sci. Instr., **27**:864–68.

MAY, H. A., and L. D. MARINELLI. 1962. Sodium iodide systems: Optimum crystal dimensions and origin of background. Whole Body Counting Proceedings. Vienna. International Atomic Energy Agency.

SCHUMANN, G., II. 1962. Messmethoden. *In* H. ISRAËL and A. KREBS (eds.), Nuclear Radiation in Geophysics, pp. 295–342. New York: Academic Press.

SHARPE, J. 1955. Nuclear Radiation Detectors. New York: John Wiley & Sons.

J. A. S. ADAMS AND G. E. FRYER

34. Portable γ-Ray Spectrometer for Field Determination of Thorium, Uranium, and Potassium

In THE PAST FEW YEARS the development of transistorized circuits and miniature components has made it possible to construct portable radiometric instruments that are light in weight and low in power requirement, as well as stable and rugged in field use. The present work summarizes over two years' experience in the design, construction, and field use of portable γ-ray spectrometers to determine thorium, uranium, and potassium in the field.

The basic question is whether it is more efficient to take the sample to the instrument or the instrument to the sample. After more than 1,200 determinations made in the field (see Adams, 1961; Adams *et al.*, 1962; Mahdavi, 1963; Richardson, 1963), taking the instrument to the sample in the field has proved to have many advantages, particularly for thorium determinations. One major advantage in making the determination in the field is the knowledge of the concentration at one location before the next location is selected. This immediate knowledge makes it possible to develop quickly the most efficient sampling patterns and density. Furthermore, it is possible to advance and immediately test hypotheses regarding the relationship of thorium, uranium, or potassium to geologic structures, lithologies, or other factors.

Another advantage of the field instrument lies in the design choice of large scintillation crystals (3 × 3 inches) that permit the pulse-height analysis to be performed at the highest possible energies, namely, 2.62 mev. for the thorium-232 series, 1.76 mev. for the uranium-

J. A. S. ADAMS is professor of geology and G. E. FRYER is instrumentation specialist, Department of Geology, William Marsh Rice University, Houston, Texas.

Work performed under grants No. C-009 and No. K-054g from the Robert A. Welch Foundation.

238 series, and 1.46 mev. for potassium-40, assuming secular radioactive equilibrium in the two series and constant isotopic composition of chemical uranium and chemical potassium in nature. For a measurement centered on one of these γ-ray energies, the weight of rock investigated in an essentially 2π configuration ranges from about 50 kg. for potassium-40 to nearly 80 kg. for the thorium series. Although all parts of this system do not contribute equally to the measurement, the measurement may be considered quite representative relative to the grain sizes of rock and the usual size of samples taken for geochemical studies. The counting rates at the energies given above are such as to give acceptable statistics in 10 or 15 minutes in the 2.62-mev. window and in 5 minutes or less in the 1.46-mev. window. It should also be noted that the γ-ray energies cited are well above those of the more common and persistent γ-ray-emitting fallout products.

Absolute calibration of these instruments has proved difficult for the uranium-238 series and potassium-40. A major difficulty with the uranium calibration is the rarity of an essentially infinite and homogeneous source (approximately 80 kg.) of uranium-free thorium-232 that is in secular radioactive equilibrium with its daughters, which is needed to determine the contribution of the thorium series to the 1.76-mev. window used to determine uranium. Another practical difficulty with uranium-238 is that it is frequently out of secular equilibrium in soils and even in solid granite (Richardson, 1963). Potassium-40 can be determined only after corrections have been made for any thorium and uranium present; however, in most common geologic materials potassium-40 has such a relatively high counting rate that the corrections are often second order. The calibration difficulties with the present instruments can be reduced statistically, as has been done with sands (Mahdavi, 1963), where the detector was buried in the sample to achieve nearly 4π geometry. A similar reduction is in progress for the 2π case on solid rocks. Another difficulty with the present instruments has been the necessity to use a weighty shield for measurements on solid rock.

SHIELDING

Where the instrument can be buried in loose material in the field, as was the case with beach sands (Mahdavi, 1963), there is no need for shielding to achieve constant geometry. The instrument (Mark I, modified) used in the beach studies contained a 3 $\times$ 3-inch sodium iodide crystal and weighed 21.3 kg. By contrast, the lightest model with a shield weighed 36 kg., of which three-fourths was a lead shield. Without a shield or collimator, the detector is very sensitive to positive departures above a 2π plane. With the present shielded instrument (Mark II) a lead shield provides 90 per cent attenuation of 2.62-mev. γ rays coming from sources above the 2π plane. Operationally, it has been found that this gives very constant geometry, even within a few feet of a vertical wall. The configuration of the shield and instrument is illustrated in Plate I.

PLATE I

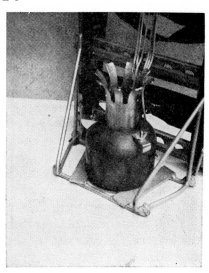

a   b

c

Views of the Mark II portable field γ spectrometer: *a*, in position to take field station; *b*, detail of the lead shield or collimator and packing-board mounting; *c*, the Mark II disassembled to show major subassemblies.

INSTRUMENT DESIGN CONSIDERATIONS

The instrument contains the same elements found in single-channel scintillation pulse-height analyzers used in the laboratory, namely, high-voltage power supply, detector, amplifier, differential discriminator, and scaler. The operation of these elements differs from conventional instruments in only two respects. First, the upper and lower discriminator levels are independently adjustable; second, the scaler readout is by means of a 4-digit mechanical register, which is preceded by a scale of 1, 10, 100, or 1,000 without interpolation indicators.

The need for battery operation, and hence low power consumption, as well as light weight and small size dictated a solid-state electronic design. Although more than one point on the γ spectrum was to be measured at each investigation site, it was thought that a single-channel instrument would be the most suitable for the immediate needs. While size, weight, and lower power consumption were factors in favor of a single channel, it was also thought that imperfect stability that might be expected from a somewhat hasty design of the semiconductor circuitry would complicate the operation of a multichannel instrument to the extent that little net advantage would be obtained over the single channel.

A battery was selected for the high-voltage power supply, as laboratory experience with a specially constructed battery pack had proved the stability of the method and the design of an electronic high-voltage supply was thereby avoided. The high-voltage batteries used have had service lives of over 1 year.

The amplifier and discriminators are analogous to those commonly found in vacuum-tube laboratory equipment. The upper discriminator level was made independent of the lower level to permit abnormal "window" widths. Since the instrument was not intended for measurements where the spectrum was to be scanned point by point, no serious inconvenience was introduced by this arrangement.

The choice of design for the scaler was necessitated by low power requirements. The absence of interpolation readout devices on the three decade counting units that precede the mechanical register was acceptable in view of the power saving.

Power for the entire electronic system, except the high voltage, is now obtained from a rechargeable nickel-cadmium battery pack. Disposable mercury batteries have also proved satisfactory, but their cost and unavailability at field locations led to the use of rechargeable batteries.

DETECTOR

The detector assembly of the most advanced design built and tested consists of a commercially available package containing a 2-inch-diameter photomultiplier tube and a 3-inch-high by 2-inch-diameter cesium iodide scintillation crystal. The detector package is mounted in a cylindrical housing attached to the bottom of the instrument case. In operation, the detector is inserted in the lead collimator or shield (see Pl. I). The use of a cesium iodide crystal was preferred to that of sodium iodide, since comparable sensitivity could be obtained in a smaller-diameter detector. The result was a significant weight-saving in the lead shield, which must be as light as possible for portability.

The photomultiplier is operated from a tapped high-voltage battery. The absence of a voltage divider for the photomultiplier dynodes virtually eliminates battery drain; 0.1-megohm resistors are connected in series with each 30-volt section of the battery to limit current. The battery pack is provided with three connectors to which the detector may be attached to obtain nominal total voltages of 700, 900, and 1100. The battery taps are arranged to provide conventional voltage ratios for the photomultiplier electrodes.

AMPLIFIER

The amplifier (Fig. 1) consists of a pair of 2-stage amplifier sections in cascade (*Q3–Q6*), preceded by a pair of emitter followers, which drive the coarse and fine gain controls (*Q1* and *Q2*), and followed by another pair of emitter followers (*Q7* and *Q8*), each of which drives one of the discriminators. The over-all gain of the amplifier is approximately 100, and the attenuator ratio is 16:1.

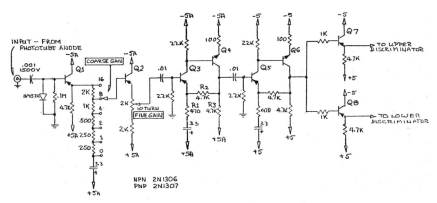

Fig. 1.—Circuit diagram of the amplifier

DISCRIMINATORS

Pulse-height analysis is accomplished by detecting non-coincident operation of a pair of discriminators biased at different levels. Pulses from the amplifier output that have sufficient amplitude to trigger the lower-level discriminator but not the upper-level discriminator are counted.

Each discriminator (Fig. 2) consists of a difference amplifier (*Q1*, *Q2*), a tunnel diode trigger circuit (*Q3*, *Q4*, *D1*), and a univibrator (*Q5*, *Q6*). In the quiescent condition *Q1* is cut off by virtue of the reverse bias established on the coupled emitters by *Q2*. The bias on

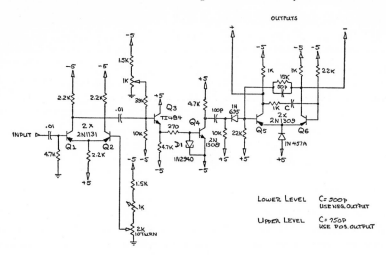

Fig. 2.—Circuit diagram of the discriminator

Q2 is adjustable by means of the discriminator level control. When a pulse occurs with an amplitude in excess of the bias level on the base of *Q2*, *Q1* turns on, and that portion of the pulse which exceeds the bias appears inverted and amplified on the collector of *Q1*. This pulse is coupled by means of an emitter follower to the tunnel diode. The bias on the base of the emitter follower can be adjusted slightly to alter the point at which the tunnel diode fires. This is in effect the zero adjustment of the discriminator.

In the absence of a pulse the current in the tunnel diode is small, and the voltage across the diode is less than 70 mv.—sufficiently low to hold *Q4* cut off. When a pulse occurs, if *Q4* were absent, the voltage across the tunnel diode would jump to 500 mv. when the current due to the pulse exceeds the peak-point current of the diode. However, this voltage is more than sufficient to turn on *Q4*, and with *Q4* connected the effect is to transfer a part of the diode current to the base of $Q4$. $Q4$ is then switched on, and a fast negative pulse appears on the collector. This pulse is used to trigger the univibrator.

Silicon transistors were used in the d.c. portions of the discriminator circuits, since feedback techniques were not applicable to the stabilization of this type of circuitry. The stability of the difference amplifier is enhanced by the balanced configuration where the thermal effects in $Q1$ and $Q2$ tend to cancel.

The univibrators of the lower- and upper-level discriminators generate pulses of 8 and 12 microseconds, respectively. The output is taken from the lower-level discriminator from the negative going collector of $Q6$. This pulse is differentiated, and the trailing edge is recovered as a short positive pulse. This operation is accomplished in one input stage of the anticoincidence circuit. The output from the upper-level discriminator is taken from the collector of $Q5$ and is fed to the other input stage of the anticoincidence circuit.

The time relation of the pulses of the lower- and upper-level univibrators is such that, if both discriminators are triggered, the trailing edge of the lower-level pulse occurs during the period of the upper-level pulse. The anticoincidence circuit treats these pulses so that an output is obtained if the lower-level discriminator is triggered but the upper-level one is not.

ANTICOINCIDENCE CIRCUIT

The anticoincidence circuit (Fig. 3) is a direct coupled transistor "AND" gate $Q3$ and $Q4$ with associated input and output stages. $Q1$ is an inverter and amplifier that handles the differentiated pulse from the lower-level discriminator. An emitter follower, $Q2$, couples to $Q4$,

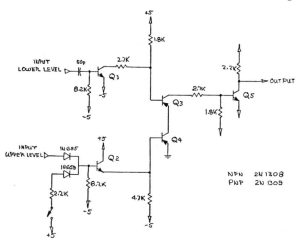

FIG. 3.—Circuit diagram of the anticoincidence gate

either the positive d.c. hold-off level supplied by the manually operated COUNT-ON-OFF switch or the positive univibrator pulse from the
upper discriminator. In the absence of a positive input at either of the
diodes on the base of *Q2*, *Q4* is held in saturation by the bias resistor
returned to the negative supply. In this condition, the occurrence of a
pulse at the base of *Q1* turns *Q3* on, and a negative output pulse is obtained at the collector of the output inverter *Q5*. If both lower- and
upper-level discriminators are triggered, a positive pulse appears at the
base of *Q2*, and *Q4* is cut off. Since the trailing edge of the lower-
level univibrator pulse occurs during the period of the upper-level
pulse while *Q4* is off, *Q3* cannot conduct and no output is obtained.
To interrupt the counting, the COUNT-ON-OFF switch is closed, continuously holding the gate cut off.

SCALER

The scaler (Fig. 4) consists of three decade-counting units and a
4-digit mechanical register. A scale factor selector switch permits the
use of scale factors of 1, 10, 100, or 1,000, depending on the number
of decade counting units in cascade prior to the register.

The decade counting units are totally encapsulated commercially
available devices.* The count of 10 is accomplished in these units by
progressively magnetizing a magnetic core in equal steps in such a
way that saturation of the core is achieved with the ninth step. The

* Sprague Electric Special Products Division, Type 73Z Decimal Counters.

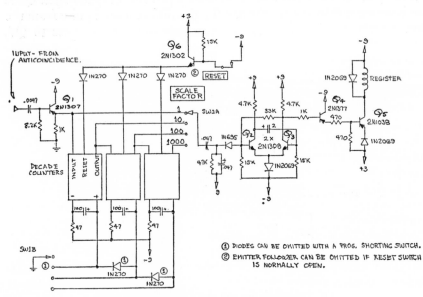

FIG. 4.—Circuit diagram of the scaler

impedance of the input winding of the core then drops, and the circuit configuration is such that the next pulse, that is, the tenth pulse, provides base drive for the reset transistor. This resets the core in preparation for the next cycle and produces an output pulse. Only 2 transistors are used in each decade unit: one to generate equal magnetizing pulses and the other to reset the core and generate output pulses—a considerable saving in components over conventional decade counters. An emitter follower (*Q1*) is provided to drive the low impedance input of the first decimal counting unit. The three decimal units are connected in series, but the output is taken from any point in the chain according to the position of the scale factor switch. The scale factor switch also switches power to the decimal units actually in use. This conserves power when low scale factors are used. Power is further conserved by operating at the lowest reliable voltage. Individual decoupling of the supply to the decimal units was found desirable for reliable operation. The output signal from the scale factor switch is coupled to the register driving univibrator (*Q2, Q3*), which has a period of 35 milliseconds. The register itself is driven by emitter follower *Q4* and driver *Q5*. The diodes associated with the power section of the scale factor switch avoid the necessity of a progressively shorting switch section for this purpose.

The reset of the decimal counting units is accomplished by means of the reset switch contained in the electromechanical register, which is actuated when the register itself is cleared manually. This permits clearing the entire scaler simultaneously. The transistor *Q6* is used in the reset circuit to permit the use of the normally closed reset switch that was incorporated in the register used. Diodes were placed in series with the reset inputs of the decimal units to isolate them except during reset action.

The decade counters are not equipped with interpolation indicators. Therefore, any count accumulated in the decades at the end of a counting period is lost. Thus it is necessary to accumulate a count on the register significant enough that the unknown count in the decades can be neglected. This is not a serious disadvantage, since the accumulation of three figures on the register limits the lost count to less than 1 per cent of the total.

The slight disadvantage resulting from a lack of interpolation indicators on the decades is more than offset by the saving in power consumption and size.

POWER SUPPLY

The power supply for the electronic package is an 18-volt nickel-cadmium battery pack tapped to provide unregulated voltages of +9, +3, −3, and −9 with respect to ground. In addition, regulated voltages of $+5$ and -5 are provided by a pair of emitter followers, the bases of which are held at fixed voltages by "zener" reference diodes.

The power ON-OFF switch has a third position that connects the three equal sections of the battery in parallel (with small current-limiting resistors) to the battery-charger connector. Parallel charging of the three battery sections was found desirable because the sections are not equally discharged during operation and because charging was to be accomplished from a 12-volt automobile battery if the need arose.

A meter is provided on the panel of the instrument to monitor battery current and voltage. Full discharge times are of the order of 10 hours. Longer operation can be obtained if low scale factors and low counting rates are used. For charging, a small external charging unit is connected to the instrument and to a source of power, either 115 volts a.c. or 12 volts d.c. In field operations the usual recharging time has been 1 or 2 hours. A simple stopwatch calibrated to 0.2 second and mounted on the instrument panel is used to determine total counting time.

INSTRUMENT DRIFT

Drift in the instrument arises from temperature changes and decline in the voltage of the nickel-cadmium batteries. The latter is quite systematic as to direction and rate. The "system gain" increases approximately 3 per cent with the normal decline in battery voltage from 21 to 18. The drift as observed is probably due to a combination of effects in the photomultiplier, the amplifier, and the discriminator. Measurements were made only of the over-all "system gain," which includes such factors as bias stability. Only large temperature changes produce noticeable drift and are not likely to occur during a normal counting interval. For example, a temperature change from 26° to 33° C. did not measurably alter the system gain. Although drift does occur to the extent indicated above, an experienced operator rarely has difficulty on this account. Furthermore, the standard procedure calls for pulse-height calibration before and after each station, and thus any drift that might occur can be noticed and corrected.

CALIBRATION

Gamma-ray lines at 0.663 mev., 1.46 mev., 1.76 mev., and 2.62 mev. were plotted using standard sources and discriminator windows of 1 per cent of the full discriminator span. The center points of these lines were plotted on a graph with energy as abscissa and discriminator setting as ordinate (see Fig. 5). The straight line drawn through the points confirmed the linearity of the system and indicated a zero error of 0.118 mev. in the discriminator.

In this procedure the system gain was arbitrarily adjusted to place the 0.663-mev. cesium-137 line at 17.5 on the discriminator dial. The cesium-137 peak was located by making a series of counts at three discriminator settings, 16–17, 17–18, and 18–19. The counting rates at the

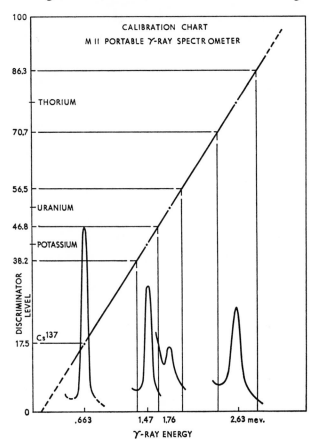

Fɪɢ. 5.—Calibration of the Mark II in the laboratory against known energies of γ rays in order to determine analysis windows.

upper and lower points were made equal by adjustment of the fine gain control, and the location of the peak was confirmed by obtaining a counting rate at the center setting 2–3 times that obtained at the upper and lower points. The exact counting rate at each of the settings was not considered important; only the position of the peak was considered. Highly variable background in this region of the spectrum and poor stability of the narrow window would in any event make absolute counting rates unreliable.

It was found that twofold difference in the counting rate between the upper and lower calibration points was indicative of a system gain change of less than 0.04 per cent. It was thus possible to control easily the system gain to the accuracy required by the relatively broad windows used in the measurements at field stations.

Having plotted the calibration curve showing the relationship between energy and discriminator settings, windows for the measurement of potassium-40, uranium-238 series, and thorium-232 series were selected. The calibration chart illustrated in Figure 5 shows windows centered on the three γ rays in question, with widths equal to plus and minus 10 per cent of the median energy. Other window widths are, of course, possible.

The equations used for determining thorium, uranium, and potassium from the counting rates are

$$\mathrm{Th} = C_t R_4 , \qquad (1)$$

where Th is thorium concentration in p.p.m., C_t is the thorium sensitivity constant, and R_4 is the counting rate at the thorium energy;

$$\mathrm{U} = C_u (R_3 - S_1 R_4), \qquad (2)$$

where U is uranium concentration in p.p.m., C_u is the uranium sensitivity constant, S_1 is the thorium-uranium stripping ratio, and R_3 is the counting rate at the uranium energy;

$$\mathrm{K} = C_k \left(R_2 - \frac{\mathrm{U}}{C_u}(S_2) - S_3 R_4 \right), \qquad (3)$$

where K is potassium concentration in per cent, C_k is the potassium sensitivity constant, S_2 and S_3 are uranium-potassium and thorium-potassium stripping ratios, R_2 is the counting rate at the potassium energy, and R_1 is reserved for the counting rate at another unused energy.

The "stripping ratios" are the factors that give the contribution to the counting rate at the energy in question, due to, or related to, higher energy radiation of other origin. In equation (3) the factor U/C_u is simply the counting rate due to uranium only at the "uranium" energy.

To determine the actual constants of each instrument and geometry for the solution of the concentration equations, it is necessary to use standards of known concentrations. Although it is, of course, desirable to use standards of density and volume similar to the materials to be measured, it is often not possible to find suitable standards of effectively infinite size. In such cases some success can be achieved by evaluating the "stripping ratios" from small samples. The assumption that the "stripping ratios" are relatively independent of the sample size and density was suggested by the great similarity observed between the spectrum from a small can of sample and that obtained from an effectively infinite source of the same material. This is illustrated in Figure 6, where the two curves were normalized for the peak counting rate of the 2.62-mev. γ ray. The detector-source geometries

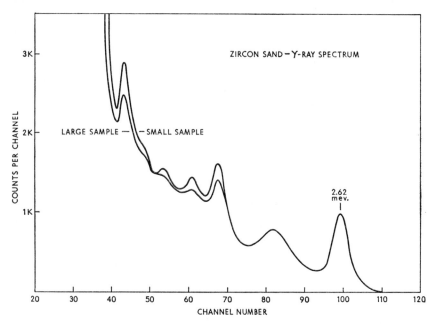

Fɪɢ. 6.—Comparison of spectra from infinite zircon sand source and small-can source or zircon sand.

are illustrated in Figure 7. According to this assumption, the "stripping ratios" were evaluated using small samples. The sensitivity constants were of necessity evaluated from effectively infinite standard sources. The sensitivity factor for thorium, C_t, is 2.14 for the small sample and 0.461 for the large sample, illustrating the strong dependence of this factor on sample size.

The data in Table 1 were used in the most recent calibration of the Mark II for the 2π or shielded case. In Table 1 the c.p.m. are taken with the Mark II; the thorium and uranium values are determined with the laboratory multichannel analyzer. The constants for the thorium, uranium, and potassium equations (1), (2), and (3) were determined from the data in Table 1, with the results given in Table 2.

The following procedures were used:

1. The thorium sensitivity constant, C_t, for a large sample, was evaluated by the use of the thorium window counting rate, R_4, for the large zircon sand sample for which the thorium and uranium contents had been previously determined by established laboratory radiometric procedures.

2. The thorium-uranium "stripping ratio," S_1, was evaluated using the two small samples and solving the two equations simultaneously.

3. The sensitivity constant for uranium, C_u, for a large sample was

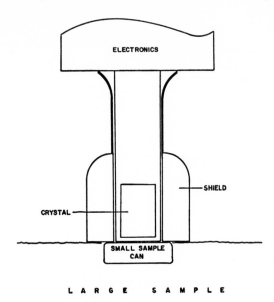

ELECTRONICS

SHIELD

CRYSTAL

SMALL SAMPLE CAN

L A R G E   S A M P L E

Fig. 7.—Counting configuration for spectra shown in Figure 6

then evaluated using the figure determined from field stations and the counting data for the large known zircon sand sample.

4. The potassium sensitivity constant, C_k, was determined from the counting rate obtained from a large potassium chloride pile.

5. The uranium-potassium and thorium-potassium "stripping ratios," S_2 and S_3, were determined by the simultaneous solution of the potassium equation set equal to zero for the counting rates obtained from the two small samples.

In the nearly 4π or buried detector case without shield, the heterogeneity of the beach sands was overcome by the comparison of nearly 100 field and laboratory measurements. These data for the Mark I (modified) yielded calibration constants that permitted the field determination of thorium, uranium, and potassium to within counting statistics (better than ±10 per cent) at all but the lowest concentrations, at which the cosmic-ray background became a factor (see Mahdavi, 1963). It is clear that similar accuracy at slightly higher concentrations can be obtained in the 2π case after a significant number of solid rocks have been measured in both the field and the laboratory so that the heterogeneities between the large field system and the small laboratory sample can be averaged out. To date, most of the

TABLE 1

CALIBRATION DATA

Probable Error

1. Zircon sand (essentially infinite source)

R4	1,470 c.p.m............	8 or 0.54%
R3	2,540 c.p.m............	12 " 0.47
R2	2,710 c.p.m............	12 " 0.45

Th = 679 p.p.m.
U = 228 p.p.m.

2. Zircon sand in laboratory counting can (4″ diam.×1.25″ thick; weight, 732 gm.)

R4	317 c.p.m............	4 or 1.2%
R3	572 c.p.m............	6 " 1.0
R2	584 c.p.m............	11 " 1.9

Th = 679 p.p.m.
U = 228 p.p.m.

3. Thorium standard. Monazite-dunite mixture in laboratory counting can (4″ diam.×1.25″ thick; weight, 406 gm.)

R4	100 c.p.m............	3 or 3 %
R3	85 c.p.m............	2 " 2.4
R2	82 c.p.m............	2 " 2.5

Th = 410 p.p.m.
U = 16 p.p.m.

4. Potassium standard (essentially infinite potassium chloride source)

R2	2,190 c.p.m............	19 or 0.9%

K = 52%

TABLE 2

EVALUATION OF CONSTANTS

1.	$C_t' = 0.461$		4.	$C_k' = 0.024$
2.	$S_1 = 0.72$		5.	$S_2 = 0.98$
3.	$C_u' = 0.154$		6.	$S_3 = 0.75$

measurements in the 2π situation have been for thorium alone, particularly in the Conway granite of New Hampshire (Adams *et al.*, 1962).

The thorium determinations on the Conway granite have been quite consistent and accurate, as shown in Table 3, where data are given for field, field core, laboratory core, and laboratory powder γ spectrometric measurements. These can also be compared with the γ spectrometric and wet chemical determinations on the same core at Oak Ridge National Laboratory (private communication). The over-all agreement to within a few per cent of the amount of thorium present is typical of the hundreds of such measurements. Such agreement with large numbers of measurements can be expected down to

TABLE 3

CROSS-CALIBRATION DATA

Depth	Rice U. No. Samples	Lab. γ-Ray on Core	Rice U. No. Samples	Field γ-Ray on Core (p.p.m. Th)	ORNL No. Samples	Thorium γ-Ray (p.p.m.)	Analyses Chemical (p.p.m.)
0–100....	64	49.50 p.p.m. Th 12.32 p.p.m. U 4.06 % K	29	47.6	1	57	58
100–200....	56	46.33 p.p.m. Th 12.97 p.p.m. U 3.94 % K	21	45.6	1	62	69
200–300....	63	46.56 p.p.m. Th 12.47 p.p.m. U 3.97 % K	23	44.9	1	45	45
300–400....	67	48.20 p.p.m. Th 10.72 p.p.m. U 3.96 % K	27	42.2	1	48	45
400–500....	70	47.66 p.p.m. Th 12.50 p.p.m. U 4.01 % K	27	48.2	....	....	....
500–600....	60	56.39 p.p.m. Th 14.71 p.p.m. U 3.98 % K	22	57	2	49	52

RICE UNIVERSITY LABORATORY CALIBRATION

No. Samples	γ-Ray Drill Core	γ-Ray Cans	Absorption Flame Photometry	
10..........	52.31 14.60 3.91	51.71 15.84 3.84	.... 3.92	p.p.m. Th p.p.m. U % K

thorium concentrations of a few p.p.m. before significant cosmic-ray background corrections must be made.

The present instrumentation is capable of determining the 1.76-mev. γ activity of bismuth-214 to an accuracy of a few per cent, but the uranium-238 concentration can be calculated from this determination only if secular radioactive equilibrium exists between uranium-238 and all its radioactive daughters. Where secular equilibrium has been examined in the upper few feet of the Conway core (Richardson, 1963), a deficiency of uranium-234 relative to uranium-238 was found. Even more significantly, Thurber (1963 and thesis) reports a surplus of uranium-234 relative to uranium-238 in sea water. These few observations, together with the movement of radon-222, lead-210, and bismuth-210 reported in several contributions to this symposium, all suggest that major disequilibria in the uranium-238 series may be the rule rather than the exception in the natural radiation environment of man. The resistate, sand-sized minerals in beaches represent one environment where the uranium-238 series is, as might be expected, in equilibrium.

The determination of potassium-40 is not fundamentally dependent upon complete radioactive equilibrium in the uranium-238 series. Thus, the counting rate in the 1.76-mev. window supplies a substantial, but not complete, correction for the uranium-238 series contribution to the 1.46-mev. window used for potassium. Furthermore, the potassium-40 counting rate is usually relatively so high that the potassium can be determined to within a few per cent in most terrestrial materials. It is indeed fortunate that most terrestrial materials do not depart widely from the average terrestrial ratio of 4 or so for both thorium-232/uranium-238 and thorium-232/potassium-40 (see Adams *et al.*, 1959; Adams and Heier, 1963). It should be noted that the foregoing estimates regarding accuracy apply for a number of measurements on different volumes of the material investigated. Any single potassium determination may be in error because of heterogeneities that make the composition of the thorium system different from that of the smaller uranium system and still smaller potassium system. Also, for the best accuracy, some care is required in the placing and operation of the portable γ spectrometer, as outlined in the procedures given below.

FIELD PROCEDURES

The field procedures used have evolved in the past few years as experience with the portable field γ spectrometers accumulated. The current procedures and their rationales are as follows:

1. Before departure for the field, the detector, powered by the field

high-voltage battery, is connected to a laboratory multichannel pulse-height analyzer. The cesium-137 emission is set in a certain channel by varying the gain in the laboratory instrument. The counting rate of the cesium-137 source attached permanently to the detector is then recorded together with the resolution of the scintillation crystal.

2. The nickel-cadmium batteries are usually charged at a low rate overnight, although the charging can be done in an hour or so. The low charge rate, combined with a few minutes of free running, puts the batteries on the flat part of the time-versus-voltage curve.

3. At the beginning of each day or traverse from base, the instrument is calibrated against a thorium-232 series source composed of finely divided pyrochlore cemented in plastic. The plastic cement prevents any mechanical segregation or packing-down of the mineral powder that might change its geometric relationship to the detector. The calibration is carried out by setting the lower discriminator at the bottom of the 2.62-mev. window determined in the laboratory (see above and Fig. 6). With the upper discriminator fully open, the fine gain is adjusted until a counting rate is obtained equal to that obtained in the laboratory under similar conditions. The upper-level discriminator is then lowered to the top of the 2.62-mev. window, and, if the count rate there is as expected, the instrument is considered calibrated and the thorium standard removed.

4. As a third check on the instrument, the discriminators are set on the cesium-137 window, keeping the gain constant. If the calibration has been properly performed, the expected count rate should be obtained. All the foregoing calibrations against the thorium-232 series and cesium-137 standards are done with 1-minute counts measured with a stopwatch. This interval yields statistics to better than 1 per cent and minimizes operator differences in simultaneously stopping both the count and the stopwatch. Both sources supply over 50 times the usual count from terrestrial sources in these windows.

5. The instrument is then moved to the field station. The Mark II (see Pl. I), complete with lead shield, is mounted on a packing board designed to place the major part of the 36-kg. load on the back of the hips. The instrument is not removed from the packing board to take a station. After a station is selected in the context of the geologic problem under investigation and with as close to 2π geometry as possible, the instrument cover is removed. The instrument requires no warm-up, and the cesium-137 calibration is then made. Counts are made successively in the 2.62-, 1.76-, and 1.46-mev. windows. In each case a count is continued until three significant figures have been obtained on the register. The 2.62-mev. count can be reduced to p.p.m. thorium while the 1.76-mev. count is being made. This makes it pos-

sible to continue the 1.76-mev. count long enough to obtain an optimum determination in terms of the amount of thorium present. The same is also true of the corrections for the potassium determination. The cesium-137 standard is checked at the end of the station, or more frequently if any drift is observed.

6. At the end of each traverse the instrument is checked once again against the thorium standard and then recharged.

7. At the end of each field trip the output of the detector is once again fed into a multichannel analyzer under exactly the same conditions as in the initial step. If the cesium-137 source attached permanently to the detector yields the same count rate as before and the resolution of the scintillation crystal is the same, the field data are considered acceptable.

GENERAL EVALUATION

The accuracy of the field γ spectrometric thorium, uranium, and potassium analyses has been discussed above. In general, the accuracy is quite comparable to other methods in use. It should be noted that the instrumentation and techniques described above were designed primarily for geochemical studies and not for environmental radiation measurements per se. Thus the results obtained in field comparisons with Lowder *et al.* (Fig. 2, 1963) and in Appendix 3 represent much smaller systems than those measured with most environmental radiation instrumentation mounted 3 feet off the ground. Despite these differences in system sizes, the agreement is generally quite satisfactory, and heterogeneities are believed to be responsible for most of the differences between various field and laboratory intercomparisons.

Hawkes and Webb (1962) considered field γ spectrometers delicate and demanding of very expert operation. Our experience has been quite the contrary; the present instruments have been operated continuously under very rugged field conditions. The only failures have been multiplier phototubes and the breakage on one resistor lead. In general terms, the present γ spectrometers are no more delicate or difficult to operate than magnetometers or gravimeters in common field use.

One man on foot by himself has maintained easily a sustained daily average of 20 thorium determinations with the Mark II in favorable terrain and weather. Although the 36-kg. weight of the Mark II is a drawback, it is estimated that to sledge out and bring back 20 equally representative samples would require much more time, effort, and even ton-miles. Furthermore, the grinding, splitting, and wet chemical determination of thorium in 20 such samples would require several man-weeks in the laboratory. In short, we estimate that, relative to other methods in use, the field γ spectrometers have been amortized long ago, particularly in consideration of the time and equipment of a skilled chemical analyst. Finally, we consider the major advantage of these instruments to be the great tactical advantage of obtaining many results while in the field. These immediate results make it possible to determine quickly the optimum sampling pattern and density, to formulate and test hypotheses immediately, and to accumulate data in sufficient quantities for detailed statistical analysis (Rogers and Adams, 1963).

REFERENCES

ADAMS, J. A. S. 1961. Radiometric determination of thorium, uranium, and potassium in the field. Geol. Soc. America Spec. Paper 68. Pp. 125.

ADAMS, J. A. S., M.-C. KLINE, K. A. RICHARDSON, and J. J. W. ROGERS. 1962. The Conway granite of New Hampshire as a major low-grade thorium resource. Proc. Nat. Acad. Sci., **48**:1898–1905.

HAWKES, H. E., and J. S. WEBB. 1962. Geochemistry in Mineral Exploration. New York: Harper & Row. Pp. 415.

HEIER, K. S., and J. A. S. ADAMS. 1963. The geochemistry of the alkalis. *In* L. H. AHRENS (ed.), Physics and Chemistry of the Earth, Vol. **5**. New York: Pergamon Press. (In press.)

LOWDER, W. M., W. J. CONDON, and H. L. BECK. 1963. Field spectrometric investigations of environmental radiation in the U.S.A. This symposium.

MAHDAVI, A. 1963. The thorium, uranium, and potassium contents of Atlantic and Gulf Coast beach sands. This symposium.

RICHARDSON, K. A. 1963. Thorium, uranium, and potassium in the Conway granite, New Hampshire, U.S.A. This symposium.

ROGERS, J. J. W., and J. A. S. ADAMS. 1963. Lognormality of thorium concentrations in the Conway granite. Geochim. et Cosmochim. Acta, **27**:775–83.

THURBER, D. L. 1963. Anomalous uranium-234/uranium-238 ratios in nature. *In* Nuclear Geophysics, Nuc. Sci. Ser. Rpt. No. 38, Pub. 1075 of the Nat. Acad. Sci.–Nat. Res. Council, Washington, D.C.

————. Natural variations of the ratio uranium-234/uranium-238 and an investigation of the potential of uranium-234 for Pleistocene chronology. Ph.D. thesis, Columbia University, New York, 1963.

WAYNE M. LOWDER, WILLIAM J. CONDON,
AND HAROLD L. BECK

35. Field Spectrometric Investigations of Environmental Radiation in the U.S.A.

Sɪɴᴄᴇ 1955, the Health and Safety Laboratory (HASL) has been conducting a study of the properties of the terrestrial γ and ionizing cosmic-ray components of the natural environmental radiation field, with particular reference to the exposure of the general population. Dose-rate measurements have been carried out in various parts of the United States with a wide variety of instrumentation (Solon et al., 1959, 1960; Shambon et al., 1963). These measurements have been supplemented by analytical calculations, based on the well-known properties of γ-ray propagation in matter, to provide interpretations of the measured quantities in terms of air, tissue, and gonadal dose (O'Brien et al., 1958). The present paper will describe the recent application of γ spectrometric techniques for determinations of the dose contributions of the individual components of the terrestrial γ radiation field and will summarize the results of a survey carried out in the western United States in late 1962.

Until 1961 we relied for dose determinations on readings obtained with 20-liter ionization chambers filled with air at atmospheric pressure (Shambon et al., 1963). The γ-ray calibration of these chambers was carried out using standard radium-226 sources, making suitable corrections for scattering in the laboratory. The ionization currents could be read with reasonable accuracy by means of vibrating-reed or Lindemann-Ryerson electrometers. However, difficulties were encountered when attempts were made to recheck earlier field readings. The lack of reproducibility observed at some locations was probably related to the fact that the chambers were not completely sealed. It is

WAYNE M. LOWDER, WILLIAM J. CONDON, and HAROLD L. BECK are physicists in the Radiation Physics Division, Health and Safety Laboratory, U.S. Atomic Energy Commission, New York, New York.

possible that sufficiently large changes in the amounts of α radioactivity in the chamber-filling took place to produce significant variations in the observed ionization current not related to the external γ field.

At this point HASL procured two high-pressure ionization chambers* filled with nitrogen and argon gas, respectively, to about 30 atmospheres (Shambon *et al.*, 1963). The pressure vessels are 8-liter cylinders with $\frac{1}{8}$-inch steel walls. The nitrogen chamber is connected by means of a 50-foot cable to an external vibrating-reed electrometer, while the argon chamber is a completely portable instrument, with its battery-operated electrometer mounted directly above the chamber itself. Separate γ and cosmic radiation calibration factors have been obtained, the latter from measurements at high altitudes in an airplane and near sea level in a small boat on several lakes. The

TABLE 1

SENSITIVITY OF IONIZATION CHAMBERS

Ion Chamber	Ion Current (amp/μr/hr)	
	Ra γ	Cosmic
20-Liter air, 1 atm..............	1.70×10^{-15}	1.70×10^{-15}
8-Liter nitrogen, 30 atm.........	1.03×10^{-14}	—
8-Liter argon, 30 atm...........	3.81×10^{-14}	4.5×10^{-14}

argon chamber has become our standard field instrument for total dose-rate measurements, combining the advantages of isolation from most environmental effects and a relatively large ionization current. A comparison of the radiation sensitivity of the three ionization chambers is given in Table 1.

Considerable information has become available in recent years concerning the significant contribution to the total environmental γ field from certain relatively short-lived fission products from nuclear-weapons testing. In particular, it has been shown that these isotopes, when deposited in the upper layers of the soil, can increase the γ background during and for many months after testing periods by 50 per cent or more at some locations (Peirson and Salmon, 1959; Spiers, 1959; Gustafson, 1960; Vennart, 1960; Collins *et al.*, 1961; Burch *et al.*, 1963; Gustafson and Brar, 1963). For those interested in natural radiation levels, this makes necessary the estimation of the additive effects of γ-emitting fallout on the total environmental radiation dose rates measured with ionization chambers.

* Reuter-Stokes Electronic Components, Inc., 18350 S. Miles Parkway, Warrensville Heights 28, Ohio.

DESCRIPTION OF SPECTROMETER

For this purpose, we have attempted to utilize a γ spectrometer to separate the radiation field into its various components. Insights into the applicability of this method have been obtained by reference to geological investigations carried out utilizing spectrometers for direct radiometric measurements in the field of the radioactive content of various rock types and geological formations (for example, see Berbezier *et al.*, 1958; Balyasnyi *et al.*, 1961). Such studies have indicated that considerable information could be derived from the pulse-height spectra obtained by placing a sufficiently large scintillation crystal at the same location relative to the sources as the ionization chambers, namely, at 1 meter above the ground.

The first field measurements were obtained by mounting an essentially unshielded 5-inch-diameter by 3-inch-high thallium-activated sodium iodide crystal on a small wooden tripod with the plane surface facing downward through a hole in the wooden platform. A 50-foot connection was run from the detector into a Corvan vehicle containing a Nuclear Data 256-channel pulse-height analyzer. A converter supplied 115-volt, 60-cycle a.c. from the 12-volt d.c. automobile battery. The field arrangement of the spectrometer and high-pressure argon ionization chamber is shown in Plate I. It was determined that the 20-minute readings would provide sufficient counts per channel over the energy range of interest (up to 3 mev.).

Two typical spectra, obtained during the survey trip to the western United States in late 1962, are shown in Figure 1. Three prominent total absorption peaks are visible, at 0.75 mev. from zirconium-95–niobium-95, at 1.46 mev. from potassium-40, and at 2.62 mev. from thallium-208 (thorium C″). Other small peaks are also apparent, of particular interest being the 0.5-mev. peak from several fission products (ruthenium-103, rhodium-106, barium-140–lanthanum-140) and the 1.76-mev. peak from bismuth-214 (radium C). The difference in the absolute counts per channel is a reflection of the differing terrestrial γ fields, as well as the altitude difference and the corresponding difference in cosmic-ray dose rates between Chicago and Colorado Springs. All the field spectra obtained in 1962 show approximately the same general features, but with significant differences in the relative sizes of the peaks.

DOSE CALIBRATION OF SPECTROMETER

Having determined that the field spectra provided sufficient detail for the identification of total absorption peaks, we then considered the problem of relating some property of these peaks to field

PLATE I

High-pressure ionization chamber and spectrometer as set up for a field measurement

dose rates. It was decided that, as a first step, the areas under the
various peaks would be determined by subtracting from the channel
data the straight-line fit between the continua on either side of each
peak when the data were plotted on semilogarithmic graph paper.
This method would minimize the effects of gain variation in the detector system during and between measurements and probably only
slightly underestimate the total peak counts. The assumption was then
made that the estimated total number of counts in each peak was

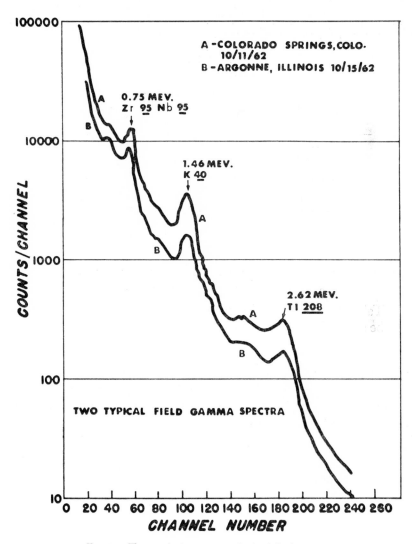

Fig. 1.—Two typical γ spectra obtained during 1962

proportional to the flux of primary photons of the particular energy incident on the detector.

The calibration of the detector was carried out by measurements in the laboratory of total absorption peak counts per unit incident primary flux as a function of energy, utilizing various point mono-energetic γ sources of known output. These source spectra were super-imposed on the normal laboratory background, thus providing a continuum closely approximating that obtained in typical field spectra. The response of the detector at each energy was first determined with the incident flux perpendicular to the plane face ($\theta = 0°$), and then again

TABLE 2

PRIMARY FLUX CALIBRATION OF DETECTOR

γ Energy (mev.)	Isotope	$R(0)$	α
0.51.......	Strontium-85	81,000	0.60
0.66.......	Cesium-137	69,000	0.59
0.75.......	Zirconium-95–niobium-95	62,000	0.53
0.89.......	Scandium-46	51,000	....
1.12.......	Scandium-46	42,000	....
1.28.......	Sodium-22	38,000	....
1.52.......	Potassium-42	34,000	....
1.69.......	Antimony-124	19,000	0.46
2.62.......	Thallium-208	17,000	0.39

at 15° intervals to $\theta = 90°$. The response function in units of peak counts *per unit flux* at $\theta = 0°$ [$R(0, E)$] is given as a function of energy in Table 2. It was found that the angular response of the detector could be approximated to within ±2 per cent by the equation

$$R(\theta, E) = R(0, E)\cos[\alpha(E)\theta], \qquad (1)$$

where the value for $\alpha(E)$ was separately determined for each source energy. The validity of the assumption of proportionality between our estimate of peak counts and the incident primary flux was verified by determinations at $\theta = 0°$ of the response of the detector with the various sources placed at several distances. With each source, the best fit of the plot of estimated peak counts versus incident primary flux was a straight line with a nearly zero intercept.

A. NATURAL EMITTERS

The primary flux per unit dose rate at 1 meter above the ground for any γ emitter uniformly distributed in the ground can be calcu-

lated from theory. In the case of the natural emitters, dose-rate calculations for unit sources in the half-space geometry have already
been carried out for uniform concentrations of potassium-40 and of
uranium-238 and thorium-232 and their daughters in equilibrium by
Hultqvist (1956) and O'Brien *et al.* (1958), using approximate rep-
resentations for the appropriate build-up factors to account for the
effects of the scattered γ photons. The primary flux incident on a
detector 1 meter above the air-ground interface for such sources can
be easily calculated for the 1.46-mev. γ ray from potassium-40 and
for any γ ray from the uranium and thorium series. We have chosen
the 1.76-mev. γ line from bismuth-214 and the 2.62-mev. γ line from
thallium-208 to represent these two radioactive series, on the basis of
their relatively clear presence in the field spectra.

Our method for inferring dose from the γ spectra depends on the
simple relationship

$$\frac{N}{I} = \frac{N}{\phi_p}\frac{\phi_p}{I}, \qquad (2)$$

where N/I is the total number of counts under a particular peak per
unit dose rate in the field measurement, N/ϕ_p is the total number of
counts in the field measurement per unit incident primary flux for the
particular γ ray, and ϕ_p/I is the primary flux per unit dose rate, the dose-
rate factor in each case taking into account all the γ rays from the isotope or series of isotopes under consideration. The primary flux in the
uniform half-space geometry can be obtained from the following equation:

$$\phi_p = S \int dV \frac{e^{-\mu_t r}}{4\pi r^2} = \frac{S}{2\mu_t}[e^{-t_h} - t_h E_1(t_h)], \qquad (3)$$

where S is the emission rate for the primary photons in $\gamma/cm^3/sec$,
μ_t is the total absorption coefficient in cm.$^{-1}$, t_h is the height of the
detector above the ground in units of γ mean free paths, and $E_1(t_h)$
is the well-known exponential integral function $[\equiv -E_i(-t_h)]$. If I is
given as the dose rate from uniform distributions of 1 per cent potassium metal, 1 p.p.m. uranium-238, or 1 p.p.m. thorium-232 in the
ground, ϕ_p/I can be calculated using equation (3) for each of these
emitters when the value for S is suitably chosen, depending on the
assumed decay schemes of the various isotopes.

The value for N/ϕ_p for any energy can be obtained by combining
equations (1) and (3) to account for the angular dependence of the
response of the crystal, suitably weighted according to the angular dis-

tribution of the incoming photon flux. This flux as a function of vertical angle is given by

$$d\phi_p = \frac{S}{2\mu_t} \sin\theta \, e^{-t_h \sec\theta} \, d\theta . \tag{4}$$

Then, since

$$N = \int_0^{\pi/2} R(\theta) \frac{d\phi_p}{d\theta} \, d\theta , \tag{5}$$

we get from (1), (3), and (4)

$$\frac{N}{\phi_p} = \frac{R(0) \displaystyle\int_0^{\pi/2} \cos a\theta \, \sin\theta \, e^{-t_h \sec\theta} \, d\theta}{\displaystyle\int_0^{\pi/2} \sin\theta \, e^{-t_h \sec\theta} \, d\theta} \tag{6}$$

$$\cong R(0) \left[\cos\left(\frac{a\pi}{2}\right) + \frac{a}{2!} \sin\left(\frac{a\pi}{2}\right) - \frac{a}{3!} \cos\left(\frac{a\pi}{2}\right) + \ldots \right] \tag{7}$$

for $t_h \cong 0$. Equation (7) gives the total absorption peak counts expected for unit primary flux at the detector in the field situation as a function of energy, assuming a uniformly distributed source in the

TABLE 3

DOSE CALIBRATION OF DETECTOR

	Potassium	Uranium	Thorium
Dose-rate contribution (μr/hr)....	1.71/%	0.76/p.p.m.	0.36/p.p.m.
γ Energy (mev.)...............	1.46	1.76	2.62
Isotope......................	K^{40}	Bi^{214}	Tl^{208}
ϕ_p/I	0.192	0.044	0.049
$R(0)$	34,500	18,500	17,000
N/ϕ_p	29,000	17,000	15,000
N/I (peak counts/μr/hr)	5,550	735	735

ground half-space. Equation (3) and the previously mentioned dose-rate calculations then provide the additional data needed for the dose calibration of the peaks under conditions of uniform half-space geometry. This procedure has been carried out for the chosen γ energies of 1.46 mev., 1.76 mev., and 2.62 mev., and the results are given in Table 3. The values indicated in the bottom row are the factors by which the peak counts in the field spectra are divided to obtain dose-rate estimates for the three main contributors to the natural radiation field.

The utility of the present method for relating the information content of our field spectra to mean concentrations of the various γ

emitters in the ground is indicated in Figure 2, where our determinations of the 2.62-mev. peak counts at certain locations are plotted against measured thorium concentrations in the ground. The thorium data were obtained at sites in California and Washington by Wollenberg and Smith (Lawrence Radiation Laboratory) from laboratory spectrometric measurements on soil samples (Wollenberg and Smith, 1962, 1963) and by Adams (Rice University) at locations in New Hampshire and Vermont by means of his field spectrometric technique for thorium determinations (Adams and Fryer, 1963)). The linear cor-

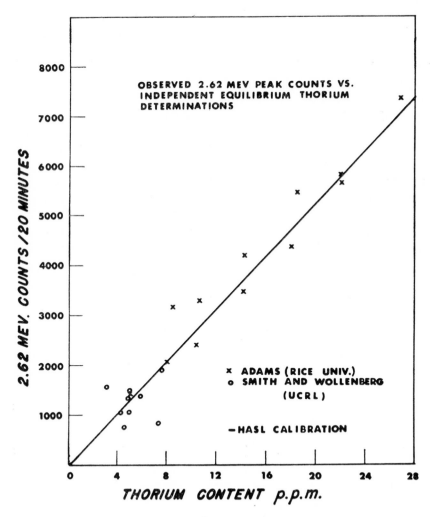

Fig. 2.—Experimental check of spectrometric thorium determinations

relation is quite evident from the plot, and the HASL calibration line, obtained independently as previously described, fits the data quite well. Insufficient experimental data on potassium and uranium contents exist at this time for similar validation of these calibration factors.

B. FALLOUT EMITTERS

The calculations of fluxes and dose rates at the detector for natural emitters were simplified by the assumption of uniform distribution in the ground half-space. When one considers the various γ-emitting isotopes contained in radioactive fallout, this assumption is no longer valid. In fact, two problems immediately arise, both related to the choice of a suitable model for depth distribution in the soil. The model chosen must be sufficiently realistic so that it represents an "average" situation, since the calibration factor, ϕ_p/I, is quite sensitive to the depth distribution. At the same time, the model must be amenable to reasonably straightforward mathematical techniques. Because of the inconsistencies in the available soil-depth data for the significant γ emitters, which are no doubt partly a reflection of large variations in actual depth distribution from place to place, it was recognized that the dose calibration factors for fallout would be considerably more uncertain than would be those for the natural emitters.

The assumption was made that the distribution of fallout γ emitters in soil could be reasonably approximated by an exponential function, that is, $S = S_0 e^{-az}$, where S_0 is the emission rate of primary photons at the surface, z is depth below the surface, and the constant "a" is to be chosen on the basis of available depth distribution data. The calculation of dose rates and primary fluxes from such a source was accomplished using the formalism of O'Brien *et al.* (1958), modified by the inclusion of the exponential source distribution. To simplify the integration, the build-up function was expanded binomially,* terminating the expansion at n terms, where n is determined as in the reference (O'Brien *et al.*, 1958). The results for $t_h \cong 0$ are then:

$$\phi_p = \frac{S_0}{2\mu_t} \frac{\ln(\gamma+1)}{\gamma} \qquad \text{photons/cm}^2/\text{sec};$$

$$I = \frac{S_0}{2\rho} \frac{\mu_e}{\mu_t} \frac{E_0}{\gamma} [\ln(\gamma+1) + A_0 - C_0] \qquad \text{mev/gm/sec};$$

* $b(t) = (1+at)^\beta = \sum_{i=0}^{n-1} B_i \frac{t^i}{i!}$.

where ρ is the density of air, $\gamma = a/\mu_t$, μ_e is the energy absorption coefficient in air for the primary radiation in cm.$^{-1}$, E_0 is the energy of the γ ray in mev.,

$$A_0 = \sum_{i=1}^{n-1} \frac{B_i}{i}, \qquad C_0 = \sum_{i=1}^{n-1} \frac{B_i}{i(\gamma+1)^i},$$

and the B_i are the coefficients in the binominal expansion of the build-up function. The constant "a" was chosen to be 0.33 cm.$^{-1}$ based on a rough fit of the available soil data (Gustafson, 1959; Friend, 1961; Telfair and Luetzelschwab, 1962; Walton, 1963). These results were then used to compute ϕ_p/I for zirconium-95–niobium-95 and several other significant fallout γ emitters. The final field calibration factors for these isotopes were obtained by combining ϕ_p/I with the angular and primary flux responses of the crystal in a manner analogous to that used for the natural emitters.

TABLE 4

FALLOUT CALIBRATION FACTORS

ISOTOPE	γ ENERGY (mev.)	PEAK COUNTS/μR/HR		
		Exponential*	Half-Space	Plane
Ba^{140}–La^{140}........	0.49	6,000	3,800	8,300
Ru^{103}–Rh^{103}.......	0.50	35,000	21,500	47,500
Ru^{106}–Rh^{106}.......	0.52	18,000	11,000	24,000
Zr^{95}–Nb^{95}.........	0.75 (av.)	20,500	13,000	26,000

* Assuming $S = S_0 e^{-0.33z}$.

These factors are given in Table 4 and compared with those obtained by assuming semi-infinite half-space or plane source distributions. The strong dependence of ϕ_p/I on the source distribution is evident, particularly as the constant "a" becomes very large, that is, for slight fallout penetrations into the ground. Consequently, the accuracy of the inferred fallout dose rates will depend somewhat on the time since deposition on the ground as well as the other factors that influence the penetration of the fallout particles into the ground.

In addition to the problems associated with non-uniform distribution of the fallout emitters in the ground, the dosimetric evaluation of the fallout contribution to the total γ field is complicated by two other factors. First, photopeaks characteristic of several significant fallout γ emitters are not observed in the field spectra. The low-energy cerium-141 and cerium-144 peaks are effectively buried in the continuum at the gain settings used for field measurements, and the cesium-137 peak (0.66 mev.) is hidden by the large zirconium-95–

niobium-95 peak (0.75 mev.). The dose-rate contributions from these isotopes would therefore not be adequately accounted for. Second, the dose calibration of the 0.5-mev. peak depends on the relative contributions of the isotopes listed in Table 4. In our present work, we have used the value given for ruthenium-106–rhodium-106 as a first approximation, both because it has a calibration factor intermediate between the others and because it has a long effective half-life, making it the dominant contributor to the 0.5-mev. peak after several months. It is possible that the ratio between the 0.5- and 0.75-mev. peaks may provide a rough criterion by which we can better evaluate the dose rate. For example, the 1.6-mev. lanthanum-140 peak has been noted at some locations where the 0.5-mev. peak is unusually high relative to the zirconium-95–niobium-95 peak, indicating that a lower value for the peak counts per unit dose rate may be appropriate.

The zirconium-95–niobium-95 and ruthenium-rhodium (103, 106) dose contributions can be estimated with a fair degree of accuracy by the peak methods. Total fallout γ dose rates can of course be inferred by subtracting the natural radiation levels estimated by means of the spectrometer data from the total dose-rate reading obtained with the ionization chamber. Both these methods are used routinely, and the agreement is generally quite good (± 0.5 $\mu r/hr$), indicating that the radioisotopes considered in the peak method are contributing most of the fallout γ dose rate at the time of measurement.

C. TOTAL γ DOSE RATE

A method of inferring total terrestrial γ dose rates from the spectrometer data has also been investigated. We have found that the total energy deposited in the sodium iodide crystal, excluding events below 0.15 mev. and above 3.4 mev., is a quantity closely proportional to the total γ dose. A calibration factor has been obtained using a standard radium-226 source in the laboratory. This experiment demonstrated the linearity of the response of the detector as a function of dose rate (from 4 to 40 $\mu r/hr$). After correction for the angular dependence of crystal response in the field situation, the dose conversion factor was applied to the data for a large number of field locations. The resulting inferred dose rates showed consistent agreement with the corresponding high-pressure ionization chamber readings.

The empirically determined linearity of crystal response with γ dose rate can be understood by the following considerations:

1. The well-known non-airlike response of sodium iodide at the lower energies is probably partially compensated for by the increased relative efficiency for higher-energy events in a detector of this size.

2. The photon energy spectra encountered in the field vary over a

limited range. Under the "worst" conditions in terms of a radium calibration, that is, high fallout levels with a greatly increased lower-energy component, no systematic deviations are observed in the total spectrometer energy readings in terms of dose rate as compared to the ionization chambers. (Such deviations are observed with count-rate meters.)

3. The low-energy cutoff was chosen to include the energy range where most of the air dose is contributed, but not to include the energy region where the sodium iodide response deviates most strongly from that of air. The dose contribution of the cerium isotopes in fallout is thereby neglected, but this generally is of the order of a few tenths of a μr/hr or less (Gustafson, 1959, 1960; Gustafson and Brar, 1963).

4. The high-energy cutoff at approximately 3 mev. includes essentially all events due to terrestrial γ-ray interactions in the crystal and renders negligible the energy contribution of cosmic radiation.

Further development of this method should provide an additional check on the dose inferences obtained by other methods, as well as a check on the reliability of the various instruments used for field surveys.

WESTERN UNITED STATES SURVEY

The dose rates for the individual components of the environmental γ radiation field have been calculated from the field spectra at many locations in the United States. The sum of these components usually agrees to within 5 per cent with corresponding high-pressure ionization chamber measurements. The level of agreement was maintained over a wide range of total natural dose-rate values and indicates the precision obtainable by the relatively simple methods described above. A summary of the data obtained in New England during July and August, 1962, is given in another symposium paper (Lowder *et al.*, 1963). In September and October of 1962, a survey trip was undertaken to the Pacific Coast. Spot readings were made at many locations in the western states during the course of the trip, with particular emphasis on three areas of interest. These included (1) the northern part of the Olympic Peninsula in the state of Washington, where the mean rainfall levels vary by almost a factor of 10 from east to west, resulting in a wide range of fallout deposition, (2) the San Francisco Bay area, where a number of fallout and natural background measurements had previously been made by the Health Physics Division at the University of California Radiation Laboratory (Stephens *et al.*, 1961), and (3) the Denver–Colorado Springs area, where earlier HASL surveys had indicated relatively high environmental radiation fields (Solon *et al.*, 1959,

1960). The readings obtained with the spectrometer on the Olympic Peninsula are given in Table 5. The natural radiation levels are quite low at all locations, and the fallout contributions increase from top to bottom in the table (east to west, geographically), as expected from the mean rainfall levels. The fallout levels calculated from (1) the analysis of the 0.5- and 0.75-mev. total absorption peaks and (2) the difference between the total γ dose rate as estimated from the total energy absorbed in the detector and the natural γ contribution as estimated from

TABLE 5

MEASUREMENTS ON THE OLYMPIC PENINSULA, WASHINGTON
OCTOBER 1–2, 1962

		γ DOSE RATES (μR/HR)						
TOWN	MEAN ANNUAL RAINFALL (IN.)*	K	U	Th	Zr-Nb	Natural	Total Fallout†	
							(1)	(2)
Sequim............	14	1.2	0.9	1.2	0.6	3.3	0.8	0.6
Sequim............	14	1.5	0.9	1.4	0.8	3.8	1.1	1.2
Port Angeles........	24	1.6	0.9	1.0	1.0	3.5	1.4	1.4
Port Angeles........	24	1.2	1.1	0.9	1.0	3.2	1.3	1.2
Joyce.............	54	1.7	1.2	0.8	1.8	3.7	2.4	1.8
Clallam Bay........	81	1.0	1.3	0.4	2.2	2.7	2.8	2.6
Forks.............	118	1.7	2.2	1.3	2.6	5.2	3.4	2.7
Forks.............	118	1.5	1.7	1.2	2.5	4.4	3.7	2.8
Forks.............	118	1.1	1.1	0.9	2.3	3.1	3.4	2.8

* These are 1960–62 values (L. Alexander, private communication).

† (1) From photopeak calibrations; (2) from total energy measurement, with natural component subtracted.

peak analyses were in good agreement, except at Forks. Here, there had been recent heavy rainfall, and substantial recent deposition of fallout was indicated by the presence of observable barium-140–lanthanum-140 peaks in the spectra at 1.6 mev. The assumed calibration factors (N/I) for the 0.5-mev. and 0.75-mev. peaks would then tend to be too low since recent deposition implies a more nearly plane source.

The California locations showed more typical natural γ radiation levels and relatively little fallout (approximately 1 μr/hr), due probably to the relatively little rainfall during the previous months.

Table 6 lists the readings obtained with the spectrometer in eastern Colorado. The high natural γ readings are evident, due primarily to high potassium ($\sim$3 per cent) and thorium (12–20 p.p.m.) contents of the soils. The differing results of the two fallout estimates at the

various locations point up the problem associated with taking differences (method 2).

The detailed data from the entire survey trip will be published in a
future Health and Safety Laboratory report. It should be noted that
the areas for which the data are given in Tables 5 and 6 are both
somewhat anomalous in terms of natural radioactivity. An examination of the data from more than 200 locations in the United States
provides more realistic estimates of typical potassium content of surficial soil as well as typical equilibrium quantities of uranium and
thorium: approximately 1.5 per cent potassium (2.6 μr/hr), 1.8 p.p.m.

TABLE 6

MEASUREMENTS IN EASTERN COLORADO, OCTOBER 10–11, 1962

TOWN	γ DOSE RATES (μR/HR)						
						Total	
	K	U	Th	Zr-Nb	Natural	Fallout	
						(1)	(2)
Fort Collins...........	2.7	1.7	4.2	1.7	8.6	2.1	1.7
Denver................	5.2	1.8	7.1	1.5	14.1	1.7	1.1
Denver................	3.4	1.2	6.2	1.5	10.8	1.8	1.7
Denver................	5.2	2.1	6.0	1.3	13.3	1.5	1.2
Denver................	5.2	2.3	6.9	1.1	14.4	1.3	0.1
Denver................	3.5	1.8	6.9	1.3	12.2	1.7	0.8
Colorado Springs......	5.2	1.3	3.7	2.8	10.2	3.3	2.2
Colorado Springs......	5.9	1.4	5.0	2.1	12.3	2.3	1.7
La Junta..............	3.5	1.4	4.0	1.8	8.9	2.0	1.9

uranium (1.4 μr/hr), and 9 p.p.m. thorium (3.0 μr/hr).* The potassium and thorium figures agree very closely with the available estimates
of mean soil contents in various parts of the world (Vinogradov, 1959;
Belousova and Shtukkenberg, 1961). The uranium figure, which
corresponds to a mean radium-226 soil content of 0.63×10^{-6} p.p.m.,
is somewhat less than the mean values found by previous investigators
(Vinogradov, 1959), but well within the range of $0.1–1.0 \times 10^{-6}$
p.p.m. indicated by unpublished data from a few points in the United
States (Harley *et al.*, 1956; Hardy, 1962). Very few of the locations
on the 1962 surveys exhibit radium concentrations in excess of 1.1×10^{-6} p.p.m., even in areas with unusually high thorium and potassium
contents, such as Denver, Colorado (see Table 6). The absence of

* It should be emphasized that our spectrometer calibration is in terms of dose *rate*,
and that our *soil* refers to earth material *in situ*, i.e., including the water content. Thus,
the true soil activity would tend to be slightly higher than our figures would indicate.

appreciable total absorption peaks from the uranium series can be observed in both spectra given in Figure 1, where peaks at 1.12 and 1.76 mev. would be quite noticeable for uranium dose rates comparable to those from potassium and thorium.

Radiation from the Uranium Series

In general, it has been observed that potassium and the thorium series contribute about equally to the natural γ radiation field, while the contribution of the uranium series appears to be seldom more than 25 per cent of the total dose rate and is usually much less than this in typical or high radiation areas. The relatively small contribution of the radium-226 daughters to the total γ radiation field can be tentatively explained by the additive effect of two processes. First, uranium and, to some extent, radium have a tendency to migrate out of the upper layers of the soil during the soil-forming and weathering processes, whereas potassium and thorium are more resistant to leaching (Vinogradov, 1959; Hansen *et al.*, 1960). Although far from invariably the case, such a displacement would effectively remove part of the parent elements of the γ emitters in the uranium series from locations where their decay would influence the γ radiation field at the ground surface. Second, the production of the gaseous daughter, radon-222, and its subsequent emanation into the soil air and migration into the atmosphere or deeper into the ground before decay provides another mode of removal of the γ-emitting daughters of radium-226 from the upper layers of the soil. A similar process takes place in the thorium series with the production of radon-220, but its short half-life (52 seconds) reduces the effect of its movement within the soil to negligible proportions. Calculations carried out using one-dimensional diffusion theory to account for the migration of free radon-222 indicate that under normal conditions almost the entire dose from the uranium series in the ground is contributed by the daughters of radon-222 trapped within the soil particles. Therefore, our estimate of radium-226 soil content depends on the emanation coefficient of the soil and, for a typical coefficient of 50 per cent (Baranov, 1957, Vinogradov, 1959), would be approximately one-half the actual content. Since one would not expect the emanation coefficient to vary significantly with depth in the upper few inches of soil under most conditions (Delwiche, 1958), the uranium series calibration factor of the spectrometer, calculated for a uniform source, would still be appropriate for normal situations.

It can be concluded from the foregoing discussion that the dose contribution of the uranium series depends strongly on the radon-222 emanation coefficient of the soil as well as on the radium-226 content of the upper layers, both of which vary considerably from place to place. In addition, a substantial reduction in the diffusion rate of soil gas, which might occur, for example, under conditions of wet or frozen ground, could result in a significant increase in the normally small dose contribution from the daughters of free radon in the soil. This phenomenon, along with variations in the atmospheric radon near the ground (O'Brien, 1958), the natural "fallout" of radon-222 daughters from the atmosphere (Thompson and Wiberg, 1962), and the shielding effect of water on or in the ground (Burch *et al.*, 1963) will produce changes in the natural radiation field with time that have as yet been little investigated (see Foote, 1963).

FUTURE PROGRAMS

It is clear that it would be quite useful to undertake spectrometric analyses of the environmental radiation field at locations where precise information as to the content and distribution of the natural and fallout radionuclides in the ground is available. The relatively few independent thorium determinations that we were able to obtain at spectrometer sites during 1962 provide an insight into the potential information content of the field spectra (see Fig. 2). HASL is now considering setting up chemical and γ spectrometric procedures in the laboratory for accurately determining the potassium-40, radium-226, thorium-232, and fallout contents of soils as a function of depth at a number of locations. By calibrating the portable spectrometer against these known soils, one could obtain more accurate dosimetric calibrations for the field spectra as well as considerable information on the nature and distribution of radioelements in soils.

In summary, HASL has developed a spectrometric technique for measuring the mean concentrations and dose contributions to population exposure of the various γ emitters in the ground. This method combines some of the advantages of soil-sampling and aerial surveys in that it provides considerable information for a relatively large sample volume. Sufficient detail can be obtained from γ spectra obtained at 1 meter above the ground for important dosimetric inferences to be made, which can be related to a source volume large enough to reduce greatly the potential difficulties associated with small-scale irregularities and anomalies in radioisotope distribution. This consideration is particularly important when measurements representative of a much larger area are required. The refinement of our present method will depend on the anticipated development of detectors with greater peak resolution, improved methods of spectrum analysis, and adequate analytical techniques for determining the radioactive contents of standard soils.

ACKNOWLEDGMENTS

The authors wish to thank James E. McLaughlin, director, Radiation Physics Division, for his continuing support of these investigations as well as direct assistance in the conduct of the survey trip to the Pacific Coast. Robert Sanna and Stephen Samson, of RPD, also participated in the survey trip. The soil data obtained and supplied to the authors by John A. S. Adams, of Rice University, Alan R. Smith and Harold Wollenberg, of the Lawrence Radiation Laboratory, University of California, and Lyle T. Alexander, of the U.S. Department of Agriculture, for a number of locations have been of great value in interpreting the results of our measurements.

REFERENCES

ADAMS, J. A. S., and G. E. FRYER. 1963. Portable γ-ray spectrometer for field determinations of thorium, uranium, and potassium. This symposium.

BALYASNYI, N. D., L. I. BOLTNEVA, A. V. DMITRIEV, V. A. IONOV, and I. M. NAZAROV. 1961. Aerial determination of radium, thorium, and potassium content of rocks. Soviet J. Atomic Energy, **10** (No. 6): 621–24. English transl. of Atomnaya Energiya, **10** (No. 6): 626–29.

BARANOV, V. L., as quoted in A. G. GRAMMAKOV. 1957. Field emanation method. *In* Radiometric Methods in the Prospecting and Exploration of Uranium Ores. Moscow: State Scientific-Technical Publishers of Literature of Geology and Mineral Resources Conservation. In English transl., USAEC Rpt. AEC-tr-3738, Books 1 and 2 (1959).

BELOUSOVA, I. M., and Y. M. SHTUKKENBERG. 1961. Natural Radioactivity. Moscow: State Publishing House of Medical Literature.

BERBEZIER, J., B. BLANGY, J. GUITTON, and C. LALLEMANT. 1958. Methods of car-borne and air-borne prospecting. The technique of radiation prospecting by energy discrimination. Proc. 2d United Nations Internat. Conf. on Peaceful Uses of Atomic Energy, **2**:799–814.

BURCH, P. R. J., J. C. DUGGLEBY, and F. W. SPIERS. 1963. Studies of environmental radiation at a particular site with a static γ-ray monitor. This symposium.

COLLINS, W. R., JR., G. A. WELFORD, and R. S. MORSE. 1961. Fallout from 1957 and 1958 nuclear test series. Science, **134**:980–84.

DELWICHE, C. C. 1958. Weathering of great world soil groups as related to general atmospheric radioactivity. Proc. 2d United Nations Internat. Conf. on Peaceful Uses of Atomic Energy, **18**:551–56.

FOOTE, R. S. 1963. Time variation of terrestrial γ radiation. This symposium.

FRIEND, J. P. (ed.). 1961. The high altitude sampling program. Vol. 5. Supplementary HASP Studies. Defense Atomic Support Agency Rpt. DASA-1300, Part III, chap. 3.

GUSTAFSON, P. F. 1959. Measurement of soil radioactivity and calculation of the dose therefrom. *In* Argonne Nat. Lab. Rpt. ANL-5967. Pp. 156–63.

———. 1960. Assessment of the radiation dose due to fall-out. Radiology, **75**:282–88.

GUSTAFSON, P. F., and S. S. BRAR. 1963. Measurement of γ-emitting radio-nuclides in soil and the calculation of the dose arising therefrom. This symposium.

HANSEN, R. O., R. D. VIDAL, and P. R. STOUT. 1960. Radioisotopes in soils: Physical-chemical composition. *In* R. S. CALDECOTT and L. A. SNYDER (eds.), Radioisotopes in the Biosphere, chap. 2, Minneapolis: Univ. Minnesota Center for Continuation Study of the General Extension Div.

HARDY, E. P. JR. (HASL). Private communication, 1962.

HARLEY, J. H., *et al.* (HASL), quoted in W. M. LOWDER and L. R. SOLON. 1956. Background Radiation: A Literature Search. Health and Safety Lab. Rpt. NYO-4712. Pp. 43.

HULTQVIST, B. 1956. Studies of Naturally Occurring Ionizing Radiations, with Special Reference to Radiation Doses in Swedish Houses of Various Types. Kungl. Svenska Vetenskapsakad. Handl., **6**, Ser. 4, No. 3. Pp. 125.

LOWDER, W. M., A. SEGALL, and W. J. CONDON. 1963. Environmental radiation survey in northern New England. This symposium.

O'BRIEN, K. 1958. Some variable contributors to natural background. USAEC, Health and Safety Lab. Rpt. HASL-27. Pp. 8.

O'BRIEN, K., W. M. LOWDER, and L. R. SOLON. 1958. Beta- and gamma-dose rates from terrestrially distributed sources. Radiation Res., **9**: 216–21.

PEIRSON, D. H., and L SALMON. 1959. Gamma-radiation from deposited fallout. Nature, **184**:1678–79.

SHAMBON, A., W. M. LOWDER, and W. J. CONDON. 1963. Ionization Chambers for Environmental Radiation Measurements. USAEC, Health and Safety Lab. Rpt. HASL-108. Pp. 39.

SOLON, L. R., W. M. LOWDER, A. SHAMBON, and H. BLATZ. 1959. Further Investigations of Natural Environmental Radiation. USAEC, Health and Safety Lab. Rpt. HASL-73. Pp. 34.

———. 1960. Investigations of natural environmental radiation. Science, **131**:903–6.

SPIERS, F. W. 1959. Measurement of background γ-radiation in Leeds during 1955–1959. Nature, **184**:1680–82.

STEPHENS, L. D., H. W. PATTERSON, and A. R. SMITH. 1961. Fallout and natural background in the San Francisco Bay area. Health Phys., **4**: 267–74.

TELFAIR, D., and J. LUETZELSCHWAB. 1962. Penetration of fallout fission products into an Indiana soil. Science, **138**:829–30.

THOMPSON, T., and P. A. WIBERG. 1962. Some observations of variations of the natural background radiation. UN Doc. A/AC.82/G/L. 753, submitted to UN Scientific Committee on the Effects of Atomic Radiation.

VENNART, J. 1960. Increases in the local gamma-ray background due to nuclear bomb fall-out. Nature, **185**:722–24.

VINOGRADOV, A. P. 1959. The Geochemistry of Rare and Dispersed Chemical Elements in Soils, chap. 16. 2d ed., rev. and enl. Transl. from Russian. New York: Consultants Bureau, Inc. Pp. 212.

WALTON, A. 1963. The distribution in soils of radioactivity from weapons tests. J. Geophys. Res., **68**:1485–96.

WOLLENBERG, H. A., and A. R. SMITH. 1962. Earth Materials for Low Background Radiation Shielding. Univ. California, Lawrence Radiation Lab. Rpt. UCRL-9970. Pp. 126.

———. 1963. Studies in terrestrial γ radioactivity. This symposium.

PHILLIP N. DEAN

36. Computer Techniques in γ Spectrometry

T HE ADVENT of the transistorized multichannel analyzer has greatly expanded the field of γ spectrometry. It has speeded up the process
of taking data and has thereby complicated the data analysis. Whereas
in the past only a few energy bands of a pulse-height spectrum were
measured, it is now possible to measure 400 or more energy bands
simultaneously on a single spectrum. With the multidimensional
analyzers, one can measure 256 individual spectra of 256 channels
each. This means that data are being accumulated much faster than
analyses can be accomplished by hand. The application of computer
techniques to the analysis of complex pulse-height spectra promises to
relieve the situation.

Gamma-ray spectra can be either continuous or discrete. The resolving of continuous γ-ray spectra is very difficult and not usually
encountered. The discussion here will be confined to γ-ray spectra
that have a primary line structure related to individual γ rays and in
which the continuum results only from scattering. This problem and
some of the methods developed to handle it are discussed in some
detail in the literature (Hubbell and Scofield, 1958; Barrus, 1960;
Mollenauer, 1961; Heath, 1960, 1962). A detailed discussion of the
methods used in the computations is not intended here but merely a
presentation of the techniques used for the different types of analysis
and an indication of the accuracy of the methods. The type of computer used in making the calculations described in this paper was the
IBM 7090. Smaller computers may be used, with the major requirement being a Fortran coding capability and a 32,000-word memory.
This is the standard memory for the larger computers. When such a
computer is not available, time can be purchased on the larger ma-
chines at commercial computing centers.

PHILLIP N. DEAN is a staff member at the Los Alamos Scientific Laboratory,
University of California, Los Alamos, New Mexico.
Work performed under the auspices of the U.S. Atomic Energy Commission.

TECHNIQUES

Computer programs utilized in γ spectrometry can be divided into two basic categories: quantitative and qualitative. By "quantitative" is meant determining the absolute amounts of each component of a multiple-component spectrum. By "qualitative" is meant determining the number and energy of all photopeaks in a complex spectrum. Since the approximate areas of the photopeaks can be determined in the latter method, a quantitative estimate of the relative magnitudes of the components is possible. The quantitative method is inherently more accurate in computing relative magnitudes, but the components of the composite spectrum must be known. Library or standard spectra of each component must also be available. Analysis is then reduced to adding the proper amount of each standard spectrum that will best represent the composite spectrum. This type of computer analysis is especially useful in the fields of radiochemistry and routine low-level counting of such items as milk samples, steel samples, etc., where the contaminant is known. The qualitative method is used when the components of the spectrum are unknown. This method furnishes the researcher with the energy of each photopeak present, from which he can deduce what mixture of isotopes is contained in the unknown. Then, after obtaining standard spectra of each of the components, the quantitative program can be used to determine the amount of each isotope in the mixture.

TECHNIQUE 1: QUALITATIVE METHOD

The proper use of the qualitative method of analysis requires an extensive knowledge of the response of the detector system to monoenergetic γ rays. The usual approach used to obtain this information is to measure experimentally the response of the detector to a series of monoenergetic γ rays available from radioactive nuclides. Interpolation is then used to predict the response to any other γ ray. There are many complications inherent in this type of procedure, which are too involved to discuss here. The basic procedure in the analysis of the complex spectrum is as follows. The spectrum is first searched by the computer for all the photopeaks present, and the energy of each photopeak is determined. The height and width of each peak are also computed. The width of the peak, along with its energy, gives an indication as to whether or not two or more Gaussians should be fitted to the peak. Spectra corresponding to monoenergetic γ rays are then obtained from the library spectra or computed by interpolation as mentioned above. Having a full spectrum available for each photopeak, the approximate relative amounts of each constituent are then determined by the weighted least-squares method.

An alternative method is as follows (Summers and Babb, unpublished). The unknown spectrum is first searched for the highest energy photopeak. The peak is then fitted with a Gaussian curve, or sum of Gaussian curves if the peak is too wide compared with resolution to be single, by the method of weighted least squares. Then, for each peak determined to be present, the total spectrum for that energy γ ray is computed. This spectrum is then subtracted from the unknown spectrum. The operation is repeated as many times as required until the residual spectrum is insignificant. The result of this type of analysis consists of the energy and area of each photopeak present in the unknown spectrum. This qualitative type of code is not sensitive to gain changes in the electronics unless such changes occur during the counting of a sample. An absolute energy scale is required to identify properly the energy of each photopeak. The photopeaks are usually also identified by their peak channel so that the relative energies of the various peaks are known if the absolute calibration is incorrect.

Technique 2: Quantitative Method

The quantitative method is the most frequently used type of program. It has been discussed extensively in the literature (Hubbell and Scofield, 1958; Barrus, 1960; Mollenauer, 1961; Heath, 1960, 1962). A statistical analysis of this type of computation has also been published (Pasternack, 1962). The computational technique is as follows. All components of the mixture must be known. A very accurate spectrum, statistically, of each of the radionuclides is required. This set of spectra is referred to as the "library." A set of equations is generated for the unknown spectrum, one for each channel. The total number of counts in each channel is a linear combination of the counts in that channel from each component. Therefore, one obtains a set of n equations, where n is the number of channels, of the form,

$$Y_i = \Sigma_j f_j E_{ij},$$

where Y_i is the count rate in channel i, f_j is the fraction of radionuclide j present in the sample, and E_{ij} is the count rate in channel i due to a known amount of the radionuclide j. The E_{ij} are obtained by counting the radionuclide j alone and comprise the standard spectrum. This set of equations is then solved for the f_j by the weighted least-squares method. This method is discussed in some detail in the literature (Moore and Zeigler, 1959).

An absolute energy scale is not necessary for this type of code. What is required is that the energy be exactly the same when the mixture is counted as it was when the standards were counted. An example of what effect gain shifts can have on the analysis is shown in Figure 1. The figure shows the residual spectrum from an analysis of a cobalt-60/cesium-137 mixture. There was a gain shift of approximately 1 per cent between counting the standards and the mixture. The amplitude of the residual peak is 10 per cent of the amplitude of the main peak. These gain shifts are usually due to instability of the

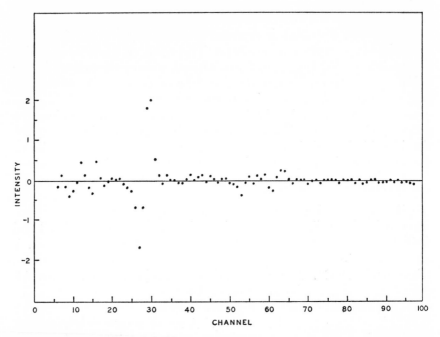

Fig. 1.—The residual spectrum from an analysis of a mixture of cobalt-60 and cesium-137 when there was a 1 per cent change in gain between counting the standards and the mixture.

electronics. This problem is being studied intensively, and R. Dudley of the International Atomic Energy Agency has developed a very promising system that can stabilize the gain to ±0.01 channel (Dudley, 1962). Since these codes depend on differences in structure of the library spectra, enough channels must be used in a spectrum to identify clearly all the fine structure.

DATA

Figure 2 shows the results from an analysis of a complex spectrum by the qualitative method. The mixture was found to contain 8 photopeaks. Each of these photopeaks plus its Compton distribution is shown on the graph. The energies and areas for these peaks are listed in Table 1. Computing time on this problem was under 30 seconds.

Several types of spectra encountered in routine low-level counting have been investigated. Figure 3 shows the experimental and computed spectra, along with the library spectra of a mixture of cobalt-

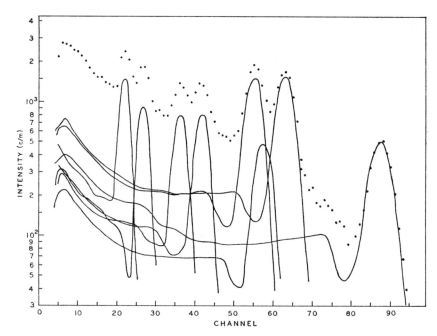

Fig. 2.—Composite spectrum of a mixture of four radionuclides (sodium-22, cesium-134, cobalt-60, and yttrium-88). The computed Compton distribution associated with each photo-peak detected is also plotted.

TABLE 1

RESULTS OF A QUALITATIVE ANALYSIS OF THE
SPECTRUM SHOWN IN FIGURE 2

Photopeak	Area (c/m)	Height (c/m/ch*)	Width	Position
1	3,354	503	6.26 ch. = 0.128 mev.	87.43 ch. = 1.817 mev.
2	8,416	1,540	5.13 ch. = 0.105 mev.	63.13 ch. = 1.335 mev.
3	2,634	496	4.99 ch. = 0.102 mev.	58.48 ch. = 1.240 mev.
4	7,793	1,524	4.80 ch. = 0.098 mev.	54.95 ch. = 1.171 mev.
5	3,601	824	4.10 ch. = 0.084 mev.	41.63 ch. = 0.907 mev.
6	3,282	804	3.83 ch. = 0.078 mev.	36.15 ch. = 0.798 mev.
7	3,271	943	3.26 ch. = 0.066 mev.	26.61 ch. = 0.608 mev.
8	4,676	1,459	3.01 ch. = 0.061 mev.	21.97 ch. = 0.513 mev.

* The abbreviation ch. = channel.

60, strontium-85, sodium-22, cesium-137, and manganese-54. The difference between the two spectra is so small that it is encompassed by the size of the plotted point. Table 2 shows the results of the analysis.

The greatest error is seen to be in resolving the strontium-85 and sodium-22. Both isotopes have a strong 0.511-mev. line. Computing time was 25 seconds. The components of the mixture were all standards of relatively high count rates, giving good counting statistics.

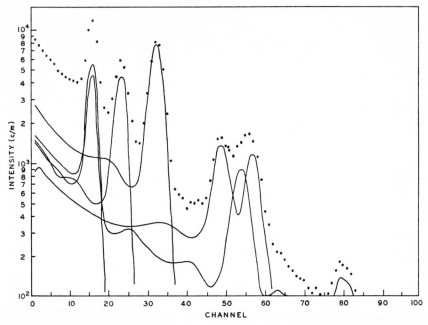

Fig. 3.—Composite spectrum of a mixture of cobalt-60, strontium-85, sodium-22, cesium-137, and manganese-54 (*dots*). The library spectra are also shown (*solid lines*).

TABLE 2

RESULTS OF A QUANTITATIVE ANALYSIS OF THE
FIVE-COMPONENT SPECTRUM
SHOWN IN FIGURE 3

Isotope	Computed Fraction	Actual Fraction
Cobalt-60............	0.994±0.003	1.000
Sodium-22............	1.022±0.006	1.000
Strontium-85.........	0.981±0.006	1.000
Cesium-137..........	0.999±0.003	1.000
Manganese-54........	0.999±0.001	1.000

A slightly different type of problem, in which counting statistics are not so good, is illustrated in Figure 4. This is the spectrum of a routine milk sample that was analyzed for iodine-131, cesium-137, and potassium-40. Both the experimental and computed spectra are plotted. The library spectrum for the potassium-40 was obtained by dissolving 100 grams of potassium chloride in 1 gallon of water. It should be emphasized that the standards, which furnish the library spectra, and the sample should be counted in as nearly identical

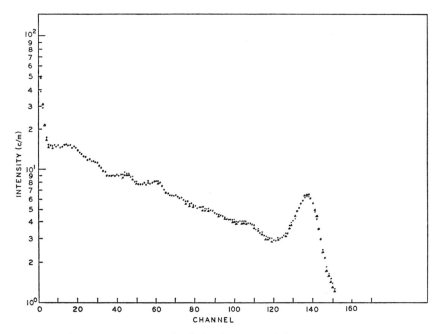

Fig. 4.—Composite spectrum of a fresh-milk sample. The computed spectrum is represented by a triangle point when it does not coincide with the milk spectrum.

TABLE 3

Results of Quantitative Analysis of a Milk Sample for
Iodine-131, Cesium-137, and
Potassium-40 (Fig. 4)

Isotope	Machine Computed	Hand Calculated
Iodine-131...................	21.8 ±18.6 pc.	Insignificant
Cesium-137...................	0.173± 0.021 nc.	0.164 nc.
Potassium (chemical)..........	6.86 ± 0.34 gm.	7.07 gm.
Background...................	0.95 ± 0.01	....

geometries as possible. The results of the analysis are shown in Table 3, along with hand-calculated values. The background here is treated as a component of the sample, as indeed it is. It contributes 95 per cent of the gross count rate. The computing time was 18 seconds.

Another type of mixture of radionuclides that is of interest in low-level counting is illustrated in Figure 5. This is a plot of the experimental and computed spectra of a mixture of radium-226 and thorium-228. The results are shown in Table 4. The computing time was

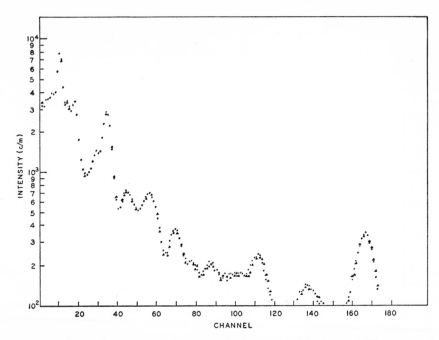

FIG. 5.—Composite spectrum of a mixture of radium-226 and thorium-228. The computed spectrum is represented by a triangle point when it does not coincide with the experimental spectrum.

TABLE 4

RESULTS OF QUANTITATIVE ANALYSIS OF A MIXTURE OF RADIUM-226 AND THORIUM-228 (FIG. 5)

Isotope	Machine Computed	Actual
Radium-226..............	1.01±0.03	1.00
Thorium-228..............	0.96±0.02	1.00
Background..............	1.00±0.01	1.00

15 seconds. When the two sources were counted together, the thorium-228 source was placed above the radium-226. A rough calculation gives 4 per cent as the absorption caused by the radium-226 source.

Summary

Computer codes used in γ spectrometry are of two general types: qualitative and quantitative. Codes written for qualitative analysis usually merely identify the photopeaks and give their width and area. Identification of the radionuclides is then left to the investigator. These types of programs are very accurate in determining the number of photopeaks present in a complex spectrum and their energies and are much more accurate than are hand techniques. With reasonably good statistics, photopeak energies can be determined to ±0.2 kev.

Codes written for quantitative analysis are used when the radionuclides comprising the mixture that gives the complex pulse-height spectrum are known. Very accurate standard spectra of these components are required, as well as very close control of the energy calibration. The accuracy of the technique depends mostly on the care of the investigator in taking the data and can be within counting statistics. With good counting statistics, five or more different radionuclides can be resolved from a complex spectrum. The codes have been proved to be more accurate than computations by hand and are much faster.

With the two types of computer codes described, most complex γ spectra encountered in the field of γ spectrometry can be efficiently and accurately analyzed.

References

Barrus, W. R. 1960. Unscrambling scintillation spectrometer data. IRE Trans. Nuc. Sci., NS, **7**:102–11.

Dudley, R. 1962. Paper presented at 2d Symposium on Radioactivity in Man. Northwestern University, September, 1962. (In press.)

Heath, R. L. 1960. Data processing techniques for routine application of gamma-ray scintillation spectrometry. Paper 16 of Proc. Total Absorption Gamma-Ray Spectrometry Symposium, Gatlinburg, Tennessee, May 10–11, 1960, USAEC Rpt. TID-7594, pp. 147–58.

———. 1962. Recent developments in scintillation spectrometry. IRE Trans. Nuc. Sci., NS, **9**:294–304.

Hubbell, J. H., and N. E. Scofield. 1958. Unscrambling of gamma-ray scintillation spectrometer pulse-height distribution. IRE Trans. Nuc. Sci., NS, **5**:156–58.

MOLLENAUER, JAMES F. 1961. A computer analysis for complex sodium iodide gamma spectra. Univ. Calif., Berkeley, Lawrence Radiation Lab. Rpt. UCRL-9748. Pp. 85.

MOORE, R. H., and R. K. ZEIGLER. 1959. The solution of the general least squares problem with special reference to high-speed computers. Los Alamos Scientific Lab., New Mexico, Rpt. LA-2367. Pp. 94.

PASTERNACK, B. S. 1962. Linear estimation and the analysis of gamma ray pulse-height spectra. Technometrics, 4:565–71.

SUMMERS, D. L., and D. D. BABB. Unpublished paper by the Dikewood Corp. Work done for Advanced Research Projects Agency and Air Force Special Weapons Center under Contract AF29(601)-4569.

M. H. SHAMOS, A. LIBOFF, AND J. SIDEROWITZ

37. *Thin-Wall Ion Chambers for Low-Level Radiation Environment*

O<small>NE OF THE PROBLEMS</small> associated with low-level ionization measurements, particularly of the natural radiation environment, has been the residual ionization due to radioactive contamination in the chamber walls. In some cases this residual ionization, caused chiefly by a contamination, can seriously limit the accuracy of such measurements. Various methods have been used in the past to minimize this effect or in some way to compensate for it. The latter methods generally involve measuring the residual ionization and correcting the data accordingly. Hess and Vancour (1949) used two such techniques to compensate for the residual ionization in their measurement of the absolute cosmic-ray intensity. They determined the wall effect by using four separate cylindrical chambers, identical in construction except for size,* and extrapolating the observed ionization currents to a chamber of zero diameter, which yields the residual current. Their second method has more frequently been used by other investigators. This involves successively reducing the chamber pressure, provided that the a particles terminate within the gas volume, and extrapolating to zero pressure to obtain the residual a current. Both methods yielded essentially similar results, but the second is obviously simpler to use with chambers that can withstand reduced pressures. While these techniques permit one to compensate the chamber cur-

M. H. SHAMOS is chairman and A. LIBOFF and J. SIDEROWITZ are research assistants, Department of Physics, New York University, Washington Square, New York.

Research supported by the Division of Biology and Medicine, U.S. Atomic Energy Commission.

* The ratio of length to diameter was kept constant ($l/d = 2$) as the diameter was changed.

rents* for the *mean* residual ionization, it is clear that fluctuations in the residual ionization cannot be taken into account in this way.

A more commonly used method of minimizing the residual ionization is the high-pressure ion chamber. Both Gray (1931) and Sievert (1932) pointed out some time ago that increasing the pressure in a chamber increases the ionization from external radiation more favorably than it does that due to wall contamination. In fact, in the ideal case, if the chamber is saturated, the ionization due to external radiation should increase linearly with pressure, while that due to α radiation from the walls should not increase beyond the point at which all the α particles spend their range within the chamber volume.

A number of investigators (Millikan, 1932; Millikan and Neher, 1936; Clay and Jongen, 1936; Compton and Turner, 1937; Clay and Clay, 1938; Burch, 1952, 1954; Hultqvist, 1956; Neher, 1957) have used pressurized chambers to measure the absolute cosmic-ray intensity, beginning with the early work of Millikan (1932) and extending up to the present time, with somewhat less than total agreement among them. The lack of agreement, according to Burch (1954), stems largely from differences in interpretation of high-pressure ionization chamber data.

Others have used such chambers for environmental or similar low-level radiation studies with varying results. Burch and Spiers (1953), using commercial nitrogen under pressures of 13.5–27 atmospheres, claimed that such nitrogen had a high recombination coefficient, such that nearly 90 per cent recombination of the α-induced ionization was effected.† Sievert and Hultqvist (1957) have used large (250 liter) pressurized chambers (20 atm. nitrogen or carbon dioxide) for measuring body radioactivity, apparently with good repeatability, although it is not clear to what extent the chamber wall influenced the absolute accuracy. Spiers (1956) measured the background γ radiation in two different locations with Geiger-Müller tubes inclosed in 1-mm. brass and found that some dose rates were $\sim$20 per cent higher than those previously obtained with a high-pressure chamber having a wall of 6.3-mm. steel. Both Clay (Clay and Jongen, 1936) and Burch (1954) have commented on the fact that, entirely apart from the residual ionization, the wall effect in high-pressure chambers—that is, attenuation of the incident radiation and production of secondaries—is the most likely source of discrepancy among different

---

* In practice one generally makes the correction numerically, although in principle the chamber current could be compensated automatically by means of an appropriate balancing network.

† Since commercial-grade nitrogen frequently contains water vapor, it is likely that the latter was responsible for the recombination.

investigators. If one seeks to eliminate this effect by decreasing the wall thickness, it is no longer possible to pressurize the chamber.

A third method of reducing the residual ionization is by the use of gridded chambers. If a wire-mesh screen is mounted a distance away from the chamber wall greater than the range of the α particles and is made one of the electrodes (the other being the central conductor), it can be seen that a marked reduction in residual ionization can be achieved. In fact, the reduction to be expected is roughly in the ratio of the area of exposed wire to total area of screen, since the residual radiation now comes essentially from the wire mesh. Such chambers have been used by Dumond and Hoyt (1930) and by Doke (1962), among others. The latter used a double-gridded chamber in pulse operation to measure the spectra of weak α activities.

Pressurized chambers have the advantage, of course, of greater sensitivity for a given volume. There are instances, however, where thin-wall chambers may be required. Such is the case, for example, if one wishes to measure the total environmental radiation, including the very low energy portion of the spectrum, or if one wishes to avoid the cumbersome wall corrections associated with thick-wall chambers. Under such conditions the residual ionization must be minimized by one of the other methods described above or, as in the present work, by the method of induced columnar recombination described below.

Residual Ionization

Assuming that the chamber walls and gas filling contain negligible amounts of contemporary carbon or other β emitters, which is relatively simple to achieve in practice,* the chief source of residual ionization is α emission from radioactive high atomic number impurities in the chamber walls. While certain materials are known to be better than others in this respect, all are contaminated to some extent. For brass, for example, Hess and Vancour (1949) report the value $n_a = 0.07$ $\alpha/cm^2/hr$, and point out that the lowest α contamination reported prior to that date was $n_a = 0.03$ for steel (Bearden, 1933). Wilkinson (1950) indicates a slightly lower value, $n_a = 0.01$, attainable with clean steel. On the other hand, Sharpe and Holton, whose unpublished data are tabulated by Sharpe (1955) along with those of Bearden, find a somewhat higher value for steel, $n_a = 0.05$,

---

* The carbon-14 content of contemporary carbon can contribute several per cent to the background ionization in a chamber constructed of this material. However, such plastics as lucite, polystyrene, or polyethylene, being natural gas or petroleum derivatives, contain essentially non-contemporary carbon; and one can prepare "dag" conductive coatings of non-contemporary carbon. Similarly, carbon-containing gases can be prepared free of carbon-14.

possibly owing to the fact that their detector had a lower energy cutoff than that employed by Bearden (100 kev. versus 250 kev.). The same tabulation indicates that, while an acrylic resin similar to lucite or plexiglass (Perspex) exhibits little or no α contamination, Aquadag, which is widely used as a conductive coating in plastic chambers, shows the value $n_a = 0.07$.

Taking $n_a = 0.05$ as representative of typical materials employed in ion-chamber construction, the effect of the α contamination may be estimated in the following way.

Assuming a spherical chamber of radius R, the ratio of area to volume is $A/V = 3/R$, and the relative contribution of α-induced ionization is

$$\frac{I_a}{I_0} = \frac{k A n_a}{n_0 V} = \frac{3 k n_a}{n_0 R},$$

where k is the mean ionization per α particle and n_0 the ionization density (ion pairs $\times$ cm.$^{-3}$ $\times$ sec.$^{-1}$) produced in the gas by external radiation. For background radiation at sea level, $n_0 \simeq 2$ in air at 1 atmosphere. Taking $k = 1 \times 10^5$ ion pairs per α, the ratio of ionization currents is $I_a/I_0 \simeq 2.1/R$, or, expressed in terms of chamber volume,

$$\frac{I_a}{I_0} \simeq \frac{3.4}{V^{1/3}},$$

where V is in cm.^{3}. Since A/V is a minimum for a sphere, this puts a lower limit on the effect of residual ionization.

It is evident that for a small chamber the effect of wall contamination is quite pronounced. In a 1-liter chamber, for example, the α contribution is roughly $\frac{1}{3}$ the ionization due to external background radiation. In a larger chamber the relative effect is naturally smaller; for one of the chambers (73.6 liters) used in the present investigation the estimated α contribution is found, from the estimate given above, to be about 8 per cent of the background radiation. The actual α contribution, as determined by an extrapolation technique, was somewhat higher, $\sim$11 per cent of the background ionization. Hence one or both of the values assumed for n_a and k may be slightly low. If n_a is taken to be 0.07, the value cited for Aquadag (Sharpe, 1955), the estimated α contribution becomes $\sim$12 per cent, in better agreement with the observed value.

Thus, even in large chambers the effect of α contamination can be significant when one is measuring background ionization levels. One can, by the extrapolation techniques described by Hess and Vancour

(1949), determine the *mean* residual ionization and correct the data accordingly. But, as noted earlier, this procedure does not take into account the influence of the α-induced ionization on the statistical fluctuations in ionization current. This effect can be rather large, as may be seen from the following analysis, due primarily to Evans and Neher (1934).

If we have n_a α particles per unit time, each producing k ion pairs, and n_β β particles, each producing j ion pairs on the average, the mean ionization current is

$$i_m = e(n_a k + n_\beta j),$$

where e is the electronic charge.

This is the mean expected value of the current after equilibrium has been established in the R-C network associated with the ion chamber; that is, after time $t_0 \gg RC$ following initial collection of charge or an abrupt change in radiation level. Ionization currents, particularly of the small magnitudes considered here, are generally measured with high-impedance, potential-operated instruments that record the potential difference across a parallel combination of R and C. Thus,

$$i_m = \frac{V_m}{R} = \frac{Q_m}{RC},$$

where Q is the charge on the associated capacitance. Further time lags may be introduced by electrical or mechanical damping of the meter (or chart recorder), but we assume here that this effect can be made small compared with RC.

A suitable measure of the effect of statistical fluctuations is the standard deviation of a single instantaneous reading of i_m, given by

$$\sigma(i_m) = \frac{e(n_a k^2 + n_\beta j^2)^{1/2}}{(2RC)^{1/2}}. \tag{1}$$

For the 73.6-liter chamber mentioned earlier, exposed to the sea-level radiation environment, the constants in equation (1) have approximately the following values:

$n_a \simeq 0.12$ sec.$^{-1}$,

$k \simeq 10^5$ ion pairs/α,

$n_\beta \simeq 75$ sec.$^{-1}$,

and

$j \simeq 2 \times 10^3$ ion pairs/β (assuming an average specific ionization of 70 pairs/cm).

Hence

$$i_m \simeq e(12 \times 10^3 + 148 \times 10^3)$$

$$= 2.56 \times 10^{-14} \text{ amp.},$$

and

$$\sigma(i_m) \simeq \frac{e}{(2RC)^{1/2}}(12 \times 10^8 + 2.96 \times 10^8)^{1/2}$$

$$= 0.62 \times 10^{-14}(2RC)^{-1/2} \text{ amp.}$$

The first term in each case is due to the α contamination, while the second represents the ionization produced by the external radiation. It should be noted that, while the residual α-induced ionization is only a small fraction of the total current (~ 8 per cent), it is nevertheless largely responsible for the statistical fluctuations in current. For an RC time constant of 1 sec., which corresponds roughly to that frequently employed with an electrometer in this current range,* the fractional standard deviation becomes

$$\frac{\sigma(i_m)}{i_m} = \frac{(n_\alpha k^2 + n_\beta j^2)^{1/2}}{(2RC)^{1/2}(n_\alpha k + n_\beta j)} \simeq 0.17.$$

This severely limits the precision that can be obtained with such chambers, particularly if one is interested in short-term variations in the radiation environment. It is generally possible, of course, to smooth out the fluctuations by increasing RC or otherwise damping the recording system, but this may not always be feasible in view of the nature of the measurement. If there were no α contamination, the fluctuations would be due to the β's alone, which in the foregoing example would result in a fractional standard deviation of ~ 0.08. Thus, while contributing only some 8 per cent to the total current, the residual α radiation more than doubles the statistical fluctuations.

Thin-Wall Ion Chambers

Except for the use of commercial nitrogen by Burch and Spiers (1953), which was possibly contaminated with water vapor, the tendency has been to use the rare gases, or other gases having negligible attachment coefficient, and to operate the chambers at high voltages so as to assure complete collection of the ionization. However, to discriminate against the residual ionization, it turns out that one should take quite the opposite approach, that is, electron attaching gases should be used with low collecting fields. This follows from

---

* The present measurements were made with vibrating-reed electrometers (Applied Physics Corp. Models 30 and 31). However, any electrometer having a sensitivity in the millivolt region would yield similar results.

the fact that at the very low ionization densities that result from the normal environmental radiation the only significant type of recombination is the so-called intratrack or columnar recombination. Treating the problem according to the columnar theory developed by Jaffe (1913) and extended by Zanstra (1935) yields the fraction of ions F that escape recombination:

$$F = \frac{i}{i_s} = \frac{1}{1 + g \cdot f(x)},$$

where $g = aN_0/8\pi D$, D = ionic diffusion coefficient, a = recombination coefficient, N_0 = linear ion density along track, i = measured ionization current, i_s = theoretical saturation current, $x = [(bk \sin \theta X)/2D]^2$, k = ionic mobility, X = electric field strength,

TABLE 1

IONIZATION RECOMBINATION IN AIR

X (volts/cm)	$f(x)$	F (β) $(N_0 \simeq 125/\text{cm})$	F (a) $(N_0 \simeq 4 \times 10^4/\text{cm})$
300........	1.16	1	0.92
100........	2.68	1	0.83
30........	4.88	1	0.73
10........	7.03	1	0.65
1........	11.63	0.998	0.53
0.1......	16.24	0.995	0.44

θ = angle between track and field, b = empirically derived constant, and $f(x) = \epsilon^x (i\pi/2) H_0^1(ix)$, where $H_0^1(ix)$ is the Hankel function of the first kind.

Applying this to an air-filled chamber results in Table 1.

Thus, as long as volume recombination is negligible, it is evident that the chamber is effectively saturated for β's (and hence for γ's) over the entire range of collecting fields from 0.1 to 300 volts/cm, while for a particles appreciable recombination occurs at the lower values of field strength. For gases that are more electronegative than air, and hence capture electrons more readily, it should be expected that the effect will be enhanced accordingly. Similarly, if the track density is increased, as in a heavy gas such as Freon-12, or if the diffusion coefficient is decreased, the recombination probability is further increased.

We have found Freon-12 (CCl_2F_2) to be an ideal gas for this purpose. It is extremely electronegative and has a high density (5.54×10^{-3} gm/cm³). The first chamber to be used in this fashion was the 73.6-liter chamber described earlier. This is a cylindrical plexiglass chamber having a wall thickness of 1,130 mg/cm² and permanently filled with Freon-12 to a pressure of about 1 atm. (Fig. 1). The chamber has been in continuous operation for several years as a background radiation monitor, with excellent results. The improvement over an air filling in the same chamber is shown in Figure 2, where the amplifier sensitivity has been adjusted to give the same deflection in both cases. It is clear that a marked reduction in fluctuations for a given field strength is achieved with the Freon-12. Moreover, while an increase in average current is apparent in air when going from 22.5 volts (field strength ≃ 1 volt/cm) to 90 volts, no such change is found with Freon-12. The increased fluctuations observed between

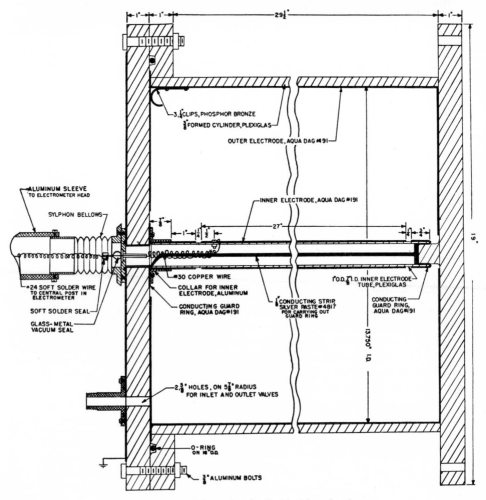

Fig. 1.—Schematic of cylindrical ion chamber

22.5 and 90 volts are due to the greater agitation energy of the electrons at the higher fields. The effect is relatively small in Freon-12, indicating low electron temperature, probably because of the very low excitation and dissociation potentials of this molecule. It follows, therefore, that the optimum collecting field for such an ionization chamber is the lowest value for which there is complete collection of the background ionization.

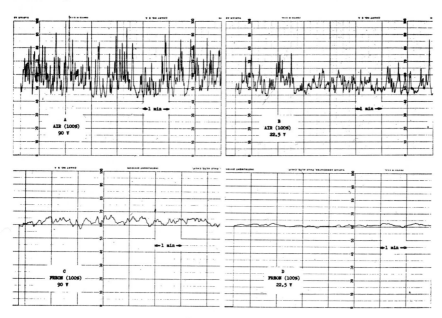

Fɪɢ. 2.—Current fluctuations in air and in Freon-12

In order to compare the ionization in Freon-12 with that in air for a given radiation field, two series of measurements were made nearly a year apart, using both background radiation and weak γ sources. Both sets of data gave the same result for equal pressures at ∼300° K., namely,

$$\frac{\text{Ionization in Freon-12}}{\text{Ionization in air}} = 3.90 \pm 0.005 \,.$$

ULTRA-THIN WALL CHAMBER

An ultra-thin wall chamber based on the same principle has been used in connection with a measurement of the very soft sea-level radiation (Shamos and Siderowitz, 1962). This is a flat cylindrical chamber 76 cm. in diameter by 8.4 cm. deep, with a volume of 38.1 liters and a window area of 4.54×10^3 cm.^{2} (Fig. 3). The cylindrical wall

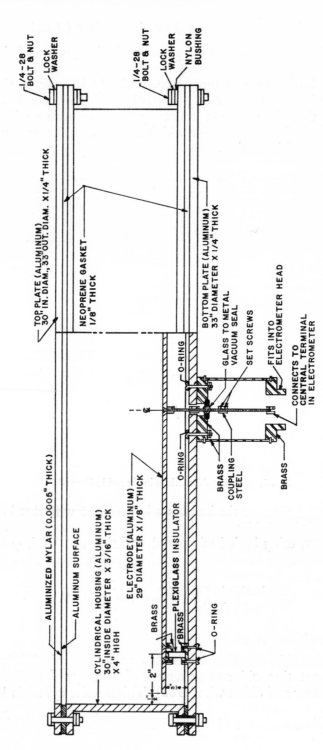

1/4-28 BOLT & NUT

LOCK WASHER

1/4-28 BOLT & NUT

LOCK WASHER

NYLON BUSHING

TOP PLATE (ALUMINUM) 30" IN. DIAM., 33" OUT. DIAM. X1/4" THICK

NEOPRENE GASKET 1/8" THICK

BOTTOM PLATE (ALUMINUM) 33" DIAMETER X 1/4" THICK

GLASS TO METAL VACUUM SEAL

SET SCREWS

FITS INTO ELECTROMETER HEAD

CONNECTS TO CENTRAL TERMINAL IN ELECTROMETER

O-RING

O-RING

BRASS

COUPLING

STEEL

BRASS

ALUMINIZED MYLAR (0.0005" THICK)

ALUMINUM SURFACE

CYLINDRICAL HOUSING (ALUMINUM) 30" INSIDE DIAMETER X 3/16" THICK X 4" HIGH

ELECTRODE (ALUMINUM) 29" DIAMETER X 1/8" THICK

BRASS

PLEXIGLASS INSULATOR

BRASS

O-RING

O-RING

2"

1/2"

FIG. 3.—Cross-section of thin-walled ionization chamber

636

PLATE I

Top view of thin-wall chamber showing Mylar window

PLATE II

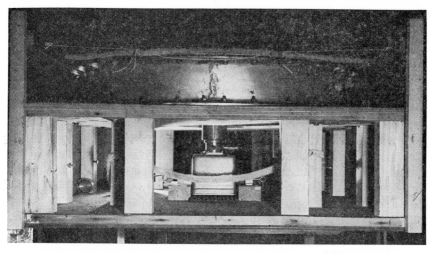

Side view of thin-wall chamber showing electrometer head

and collecting electrode are of aluminum, and the chamber window is of 0.0005-inch (1.77 mg/cm²) aluminized Mylar. Plate I is a view of the chamber showing the Mylar window, while Plate II shows the chamber with the electrometer head in place. The chamber was filled with a known mass of Freon-12 to a pressure of about 1 atm. and operated with a collecting voltage of 22.5 volts (field strength = 2.7 volts/cm), at which field the chamber is completely saturated for β particles but collects less than 2 per cent of the α-induced ionization. Because of the dead space below the collecting electrode in this chamber, which may contribute some extracameral ionization, it was not possible to determine the active volume accurately. Hence this chamber was used only for difference measurements, not for absolute ionization density.

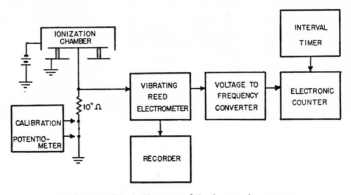

FIG. 4.—Block diagram of the integrating system

When used over extended periods of time, the chamber output is generally integrated electronically, as shown in the block diagram of Figure 4. The output signal from the vibrating-reed electrometer is fed to a voltage-to-frequency converter (Dymec Mod. 2210), which converts the d.c. signal from the electrometer to a proportional audio frequency. The audio frequency then feeds an electronic counter (Hewlett Packard Mod. 521 AR), which provides a direct-reading digital display of the d.c. voltage level averaged over the gate time of the counter. The latter is controlled by an electric clock that can be preset to any desired time interval. The electrometer signal is also fed to a chart recorder to provide a continuous visual record. The over-all system time constant is approximately 5 seconds.

TEMPERATURE EFFECT

Such a thin-wall chamber gives quite reliable results, provided that certain precautions are observed. Because of the extreme flexibility of the Mylar window, it responds very readily to small changes in temperature of the gas. As the temperature increases, the gas causes the Mylar to balloon out. This, in turn, changes the capacitance of the system, which results in a current being drawn through the 10^{11} Ω resistor. It is easy to show that the current thus produced is proportional to the rate of change of temperature, and for the present chamber it was estimated to be ∼1.6 per cent/° C/hr. Small changes in battery voltage due to temperature changes also result in spurious currents through the high-megohm resistor. This effect is again rate dependent, amounting to ∼0.007 per cent/° C/hr, considerably less than the capacitance effect. Consequently, the chamber temperature was kept constant by inclosing it in a wooden box whose top consisted of a rectangular frame covered with a 0.0005-inch sheet of Mylar on each face of the frame, a 1.5-inch air space between the two serving as a thermal insulator. Heating tape was wrapped around the top of the chamber near the Mylar window, and the temperature was maintained by a thermistor bridge at a value some 10° C. above ambient.

Changes in atmospheric pressure also affect the position of the Mylar window and hence the capacitance of the system. This effect, which is also rate dependent, was estimated to be approximately 3.9 per cent/cm mercury/hr, a negligible value for ordinary pressure variations.

Operating Characteristics and Errors

The current sensitivity of the thin-wall chamber is about 1.30 × 10^{-14} amp/μr/hr. As a result of the temperature control, the error from this source could be kept to within 0.15 per cent. The average deviation for repeated readings was of the order of 0.5 per cent, which is believed to stem primarily from inherent statistical variations in the radiation. The greatest source of error lies in any uncertainties in the mass of Freon-12 filling and in the value of the high-megohm resistance. In the present case the latter was known to within ±1 per cent, while the Freon-12 mass was uncertain by about ±3 per cent. Hence, for greater absolute precision it would be necessary to improve the Freon-12 mass determination. In the larger chamber (73.6 liters) all these quantities have been determined to better than ±1 per cent, which means that the absolute background ionization density can be measured with this sort of precision. A series of such measurements, taken at sea to eliminate the ground shine and radon-222 and radon-220 concentrations, gave the value 2.00 ± 0.03 ion pairs/cm³/

sec (3.45 ± 0.05 μr/hr) for the absolute cosmic-ray ionization density.*

Summary

Ion chambers for measuring the environmental radiation have excellent characteristics if filled with electronegative gases, such as Freon-12, and operated at low collecting fields so as to induce columnar recombination of the ions in the α tracks arising from radioactive contamination of the chamber walls. Freon-12 has the added advantage of high density and low electron temperature, giving the sensitivity of a pressurized chamber of roughly 4 atm. and reducing the noise fluctuations in the gas. Several such chambers have been constructed, some with ultra-thin windows, for operation at atmospheric pressure, and used with good success over a period of years.

* *Note added in proof:* This value should be corrected for the ionization due to γ radiation from the potassium-40 content of sea water. We have estimated this to be about 1.5 per cent, which reduces the cosmic-ray ionization to 1.97 ± 0.03 ion pairs/cm^3/sec.

References

BEARDEN, J. A. 1933. Radioactive contamination of ionization chamber material. Rev. Sci. Instr., 4:271–75.

BURCH, P. R. J. Ph.D. thesis, University of Leeds, 1952.

———. 1954. Cosmic radiation: Ionization intensity and specific ionization in air at sea level. Proc. Phys. Soc., 67A:421–30.

BURCH, P. R. J., and SPIERS, F. W. 1953. Measurement of the γ radiation from the human body. Nature, 172:519–21.

CLAY, J., and P. H. CLAY. 1938. The absolute intensity of cosmic radiation on sea-level. Physica, 5:898–900.

CLAY, J., and H. F. JONGEN. 1936. Absolute intensity of cosmic radiation at sea-level. K. Akad. Amsterdam Proc., 39:1171–73.

COMPTON, A. H., and R. N. TURNER. 1937. Cosmic rays on the Pacific Ocean. Phys. Rev., 52:799–814.

DOKE, T. 1962. A new method for the reduction of alpha-ray background in a gridded ionization chamber. Canad. J. Phys., 40:607–21.

DUMOND, J. W. M., and A. HOYT. 1930. Design and technique of operation of a double crystal spectrometer. Phys. Rev., 36:1702–20.

EVANS, R. D., and H. V. NEHER. 1934. The nature of statistical fluctuations with applications to cosmic rays. Phys. Rev., 45:144–51.

GRAY, L. H. 1931. Scattering of hard γ-rays. II. Proc. Roy. Soc. London A, 130:524–41.

HESS, V. F., and R. P. VANCOUR. 1949. New methods of determining the absolute intensity of cosmic rays in the atmosphere and the residual ionization in ionization chambers. Phys. Rev., 76:1205–8.

HULTQVIST, B. 1956. Studies on naturally occurring ionizing radiations, with special reference to radiation doses in Swedish houses of various types. Kungl. Svenska Vetenskapsakademiens Handl., Vol. **6**, No. 3. Pp. 125.

JAFFE, G. 1913. Columnar ionization theory. Ann. der Physik, **42**:303–44.

MILLIKAN, R. A. 1932. Cosmic-ray ionization and electroscope constants as a function of pressure. Phys. Rev., **39**:397–402.

MILLIKAN, R. A., and H. V. NEHER. 1936. A precision world survey of sea-level cosmic-ray intensities. Phys. Rev., **50**:15–24.

NEHER, H. V. 1957. Gamma rays from local radioactive sources. Science, **125**:1088–89.

SHAMOS, M. H., and J. SIDEROWITZ. 1962. Ionization due to the very soft sea level radiation. Interim Technical Report, AEC AT(30-1)-1704, New York University.

SHARPE, J. 1955. Nuclear Radiation Detectors. London: Methuen.

SIEVERT, R. M. 1932. Eine Methode zur Messung von Röntgen-Radium-und Ultrastrahlung nebst einige Untersuchungen über die Anwendbarkeit derselben in der Physik und der Medizin; mit einem Anhang enthaltend einige Formen und Tabellen für die Berechnung der Intensitätsverteilung bei γ Strahlungsquellen. Acta Rad. Suppl., **14**:1–179.

SIEVERT, R. M., and B. HULTQVIST. 1957. High pressure ionization chambers. Brit. J. Rad. Suppl., **7** (Pt. 1), 1–12.

SPIERS, F. W. 1956. Radioactivity in man and his environment: Presidential address. Brit. J. Radiol., **29**:409–17.

WILKINSON, D. H. 1950. Ionization Chambers and Counters. Cambridge: Cambridge University Press.

ZANSTRA, H. 1935. Ein kurzes Verfahren zur Bestimmung des Sättigungsstromes nach der Jaffe'schen Theorie der Kolonnenionisation. Physica, **2**:817–24.

A. WENSEL

38. *Measurement of Environmental Radiation with Plastic Scintillators*

T HIS PAPER is concerned with car-borne equipment for the measurement of low γ dose rates of natural or artificial sources. The artificial radioactive substances released during the normal operation of a reactor are of special interest. They cause an external γ irradiation of the population living in the vicinity of the reactor station. For most reactor stations the distance between the release point of the gaseous activity and the nearest populated area is so large that certain emergency limits are not exceeded in case of an accident. The doses to the population due to releases of radioactive gases (e.g., argon-41) during normal operation are therefore very low, in some instances of the order of the natural environmental radiation.

Measurements of the natural γ-radiation levels around a reactor station are necessary as a base for the determination of the dose rate produced by artificial isotopes. Readings during reactor operation correspond to the sum of both components. For these measurements car-borne equipment is desirable, permitting the survey of large areas in a short time.

The requirements for an instrument to be used for measurements of environmental γ radiation are the same as for determination of the γ radiation produced by artificial isotopes in a certain distance from the release point. In both cases the fundamental quantity for the assessment of radiation hazards to man is the dose rate. The energy distribution of the γ radiation is in both cases complex. A large part of the natural γ radiation originates in the soil and is scattered several times before reaching the surface. The gaseous isotopes released by a reactor diffuse in a large cloud, where multiple scattering occurs.

The car-borne equipment should meet the following specifications:

DR. A. WENSEL is head, Health Physics Department, Institut für Kernphysik, Der Johann Wolfgang Goethe Universität, Frankfurt am Main, Germany.

a) The measuring instrument should have a response (observed value/unit dose rate) which is energy independent over a wide range. For this range a lower limit of the order of 0.05 mev. is tolerated. The photon flux per roentgen is very high in this region.

b) The dose rates to be measured are very low (in the order of 10 $\mu r/hr$). The instrument should have a high sensitivity, allowing a short time constant so that readings can be taken in a moving vehicle. The time constant of the instrument determines the maximum possible speed of the car.

c) The instrument should be insensitive to mechanical vibrations and to changes of ambient temperature.

For the measurement of low dose rates, ionization chambers, Geiger-Müller counters, or scintillation counters can be used. Ionization chambers seem to be an ideal choice because their response curve is energy independent for a very large energy range. But they are suitable only for stationary measurements, since the time constant is very long.

Geiger-Müller counters have also a nearly flat response curve over a limited range (Hine and Brownell, 1956), but the count rate is so low that long time constants are required for accurate measurements.

Therefore, a scintillation counter with a large-volume plastic scintillator (17.5 cm. diam. $\times$ 15.5 cm.) was selected for the van used for measurements in the vicinity of the reactor station at Frankfurt am Main, Germany (FRF) (Type SP4, S.A. MESCO, Vauves [Seine], France). This instrument used as a normal pulse counter has a very high sensitivity, permitting very short time constants when the pulse rate is measured with a rate meter driving a recorder. The scintillation counter also meets the specification. The disadvantage of a counter of this type is that the response curve does not have any constant portions that would be suitable for dosimetric applications (Hine and Brownell, 1956). The response curve of a large plastic scintillator operated as a normal pulse counter is given in Figure 1.

There are several ways to overcome this disadvantage of the plastic scintillation counter. The simplest would be a measurement of the mean anode current of the photomultiplier. This arrangement works quite well for higher dose rates (Boulenger and Gourski, 1961; Wensel and Hasl, 1962), where the signal current is large compared to the dark current but presents difficulties for lower dose rates. The dark current will represent an appreciable fraction of the total current when dose rates of the order of the natural γ background are to be measured. The dark current depends strongly upon the ambient temperature, so that this method is not applicable for mobile equipment.

A second method using electronic means is proposed in this paper.

The equipment described has a flat response curve for a certain energy range and meets all other requirements for an instrument for the measurement of low dose rates mentioned above. The response for this instrument has been measured with several radioactive isotopes. Using experimental pulse-height spectra, a calculation has been performed to determine the influence of the discriminator bias on the performance of the system. To study the variation of the response versus energy resolution of the counter used, pulse-height distributions were calculated by Monte Carlo methods.

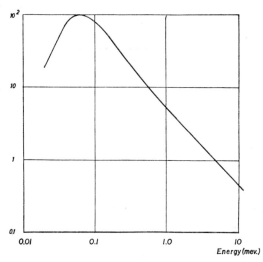

Fig. 1.—Response curve of a large plastic scintillator (17.5 cm. diam. × 15.5 cm.). (Wensel and Hasl, 1962.)

PULSE INTEGRATOR

A method to obtain a flat response curve for a plastic scintillation counter that is not influenced by the dark current of the photomultiplier is the use of a "pulse integrator." This device was developed for neutron dosimetry with proton-recoil proportional counters (Glass and Hurst, 1952; Hurst and Wagner, 1961). Each pulse of the photomultiplier that is higher than the bias of a discriminator is given a weight according to the pulse height. Figure 2 illustrates the principle of the pulse integrator with 4 discriminators.

A, B, C, and *D* are pulse-height discriminators biased so that output pulses appear for inputs larger than *a, 2a, 4a,* and *8a* volts, respectively; *a* can be chosen according to the type of discriminator used, that is, $a = 5$ corresponds to 5-, 10-, 20-, 40-volts bias. The 4 discriminators feed into 4 count stages of a binary scaling system. It can be shown that, on the average, 5-v. pulses give a net count of 1, 10-v. pulses a count of 2, 20-v. pulses of 4, and 40-v. pulses of 8. For neutron dosimetry a system with 7 discriminators instead of 4 has been used (Glass and Hurst, 1952).

Since the pulse-height distribution produced in a scintillation counter with a plastic scintillator is very broad, it was thought that a 4-discriminator pulse integrator should give sufficiently accurate results. For the measurements described in a later section of this paper, a

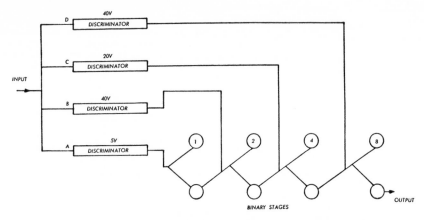

FIG. 2.—Pulse integrator with 4 discriminators. (Glass and Hurst, 1952)

4-discriminator arrangement with vacuum tubes was constructed. The method of coupling the discriminators into the stages of the binary counter given in Glass and Hurst (1952) was used.

The total count rate measured with a pulse integrator with appropriate bias setting is proportional to the current that would be measured wih no dark current present. Therefore, the calculations in Hine and Brownell (1956) for the anode current produced with organic scintillators are directly applicable, and it can be expected that the relative response is flat over a certain energy range. Because the dark-current pulses have to be eliminated by the lowest discriminator, it is not possible to design a counter based on this principle to measure γ rays with energies lower than about 100 kev. But normally the contribution of the γ rays of natural or artificial sources with energies below this limit should be small and proportional to the total dose, so this instrument should be well suited for survey measurements.

Measurements with a 4-Discriminator System

The response curve for a large plastic scintillator with a thickness of 15.5 cm. was determined with a 4-discriminator pulse integrator for two different bias settings. To cover the energy range of interest, 5 radioactive isotopes were used. The activity of these isotopes was known either by absolute calibration (cesium-137) or, for

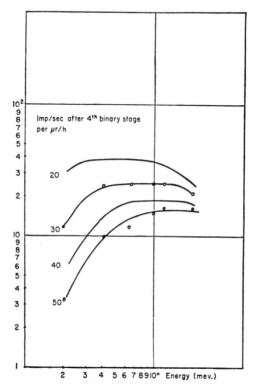

Fig. 3.—Response curves of an SP4 plastic scintillation counter. Circles give values from measurements with a 4-discriminator system. Curves are computed from experimental spectra.

the isotopes activated in the reactor (mercury-203, gold-198, cobalt-60, and sodium-24), by calculation. The dose rates were calculated from the activities and the distances from the sources to the geometrical center rather than to the unknown efficiency center of the scintillator. Since the distances were large, the error introduced by this method may be neglected. In addition, the dose rates were measured with an ionization chamber. The accuracy of the dose-rate determinations is about 20 per cent.

The results of these measurements are given in Figure 3 (measured values indicated by circles). The ordinate gives the count rate per $\mu r/hr$ after the fourth binary stage.

The response curve for the lower discriminator setting has a flat section between 0.3 and 2 mev. For the higher bias setting, this flat region is shifted to higher energies.

In addition to the isotopes mentioned above, the count rate was determined for radium as an emitter of a complex spectrum. The mean energy is assumed to be 1 mev.

CALCULATION OF RESPONSE CURVES FROM EXPERIMENTAL SPECTRA

To determine the influence of the bias setting on the shape of the response curve for a 4-discriminator pulse integrator, experimental spectra of the radioactive sources mentioned in the previous section of this paper were taken with a 400-channel pulse-height analyzer. The response curves for the 4-discriminator instrument, as well as for a continuous pulse integrator, were calculated with the IBM 704 computer at the DRZ, Darmstadt, Germany, using a code (EUSI) written in FORTRAN.

To obtain the count rate for a 4-discriminator pulse integrator, background was subtracted and the channel content multiplied by 1, 2, 4, or 8, depending on the ratio of the channel number to the bias setting.

Then the contributions of all channels above the bias setting were added. The count rate that would be measured by a continuous pulse integrator was computed in the same manner, multiplying the channel content by the channel number.

The response curves for the 4-discriminator device are shown in Figure 3. The shape of the curves depends upon the bias setting of the lower discriminator. For very low bias settings the response is constant for low energies and falls off for high energies because the pulse-height range that can be handled by this method is only about 1:8. For higher discriminator settings a marked drop in low-energy response can be observed. Between these extremes a bias setting can be found where dose rates of γ emitters in the energy range between 0.3 and 2 mev. can be measured with an accuracy better than 20 per cent.

In practice, the lower discriminator must be biased according to the height of the dark-current pulses of the photomultiplier. The bias should be set so that the contribution of the dark-current pulses is small, taking into account the fact that dark-current pulses contribute only with a small weight to the total count rate of the background.

Experimentally, it was found that, for the counter used, bias settings around 0.05 mev. give satisfactory results. For very high ambient temperatures the bias should be set somewhat higher.

The count rate calculated for a scintillator with a continuous pulse integrator was found to be proportional to the dose rate within the experimental errors introduced by the determination of the dose. The principle of a continuous pulse integrator is given in Glass and Hurst (1952). The pulses pass a linear gate that is operated by a discriminator without changes of the pulse height. After being stretched,

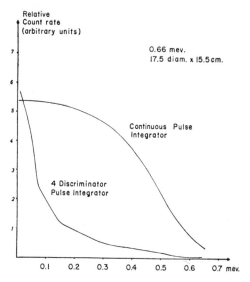

Fig. 4.—Discriminator curves for continuous and 4-discriminator pulse integrator

these pulses operate a rate-meter circuit. This method has several advantages. The energy range covered by an instrument with an integrator of this type should be larger than for a 4-discriminator circuit.

The discriminator curve for a continuous integrator is nearly flat at the low-energy end, whereas the curve for a 4-discriminator system is sharply increasing, as shown in Figure 4. A transistorized version of the continuous pulse integrator is under construction. Some transistorized circuits that can be easily modified to operate as a continuous pulse integrator are described in the literature (Emmer, 1962).

Monte Carlo Calculations of Pulse-Height Spectra in Plastic Scintillators

To study the influence of the properties of the scintillation counter (e.g., energy resolution) on the performance of a pulse inte-

MESIS

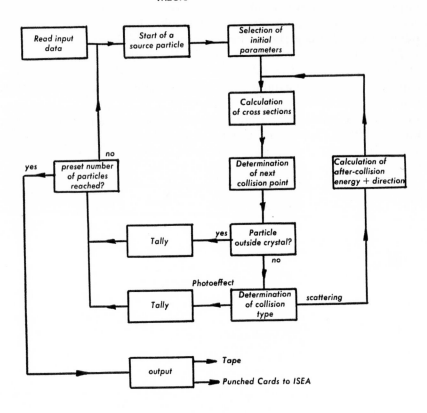

ISEA

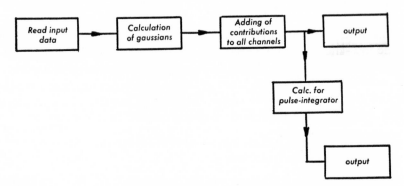

Fig. 5.—Flow chart of the Monte Carlo program MESIS

grating system, Monte Carlo calculations were made to yield the energy distribution of the electrons produced by γ rays. For this purpose a code (MESIS) was written in FORTRAN for calculations on the IBM 704 computer.

Only a general outline of this code is given here; a detailed description will be published elsewhere. A simplified flow chart for this program is shown in Figure 5. The scintillator is assumed to be cylindrical. Source particles start from the center point of one face of the cylinder in the direction of the axis. The interaction cross-sections of the γ rays are calculated by approximate formulas (Fessler and Wohl, 1961) in order to reduce the calculation time.

Energy distributions were calculated for two crystal sizes, 5 cm. diam. × 5 cm. and 17.5 cm. diam. × 15.5 cm., with 10,000 source

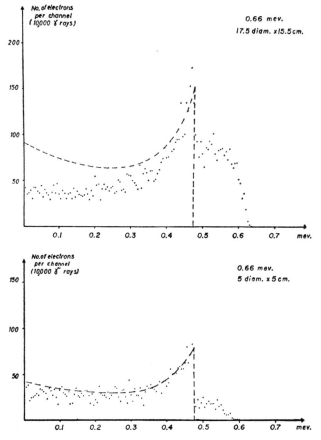

Fig. 6.—Electron distributions for two scintillator sizes calculated by the Monte Carlo method (MESIS). *Dashed curves*, single interaction distribution.

particles for each case. The following source energies were used: 0.1, 0.27, 0.66, 1.3, and 2.5 mev. As an example, the electron energy distribution for 0.66-mev. γ rays for the two crystal sizes is given in Figure 6. The corresponding distribution for single scattering of the same number of interacting particles is given by the dashed curves. Even for the smaller crystal, there is a contribution of electrons scattered several times.

A second code (ISEA) was written in FORTRAN for Gaussian broadening of the electron distribution, taking into account the spectral resolution of the scintillation counter. Each channel is replaced in

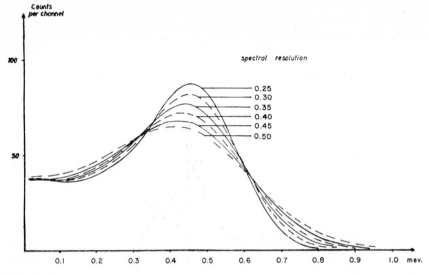

FIG. 7.—Theoretical pulse-height distributions (0.66 mev., 17.5 cm. diam. × 15.5 cm.)

turn by a Gaussian curve whose standard deviation is assumed to be inversely proportional to the square root of the energy. The contributions of the Gaussian curves are added in all channels. This method yields a smooth curve even if the input spectrum is distributed statistically. The intrinsic efficiency of the scintillator is assumed to be constant.

Figure 7 shows a set of pulse-height distributions for the larger crystal, with the spectral resolution as a parameter obtained with this code from the electron distribution given in Figure 6.

The values of the spectral resolutions given in Figure 7 correspond to the energy of 1 mev. The code ISEA was modified in the same way as described in the previous section to calculate the response curves for the continuous and the 4-discriminator pulse integrator. The dose corresponding to each spectrum can be easily determined from the number of impinging γ rays (10,000 in this case). The count rate for a continuous pulse integrator was found to be proportional to the dose rate for the 4-discriminator system. The results of the calculations are given in Figure 8. The shapes of the curves correspond to those obtained from the experimental spectra. The curves with corresponding bias settings are less widely spaced. This can be explained by the fact that the experimental spectra show a sharper increase at the low-energy end than do the theoretical spectra.

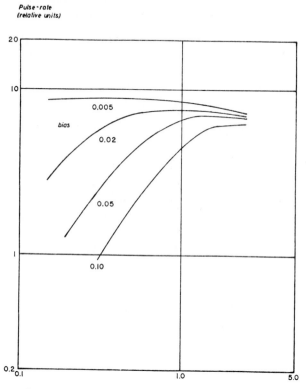

Fig. 8.—Calculated response curves for a 4-discriminator pulse integrator

The spectral resolution of the scintillation assembly has practically no influence upon the performance of a pulse integrator. The count rate varies less than 1 per cent if the energy resolution at 1 mev. changes from 0.25 to 0.5.

Conclusion

It is demonstrated that a large plastic scintillation counter in connection with a 4-discriminator pulse integrator is adequate for a carborne survey of γ radiation from natural and artificial sources.

The counter can be made energy independent for the energy region that is of primary interest to both cases. The sensitivity for a 17.5 cm. diam. × 15.5 cm. plastic scintillator (about 25 c.p.s. per μr/hr after the fourth binary stage) is high enough that a rate meter with a short time constant can be used and operation in a moving vehicle is possible.

REFERENCES

BOULENGER, R., and E. GOURSKI. 1961. Portable dosimeter with a scintillator. *In* Selected Topics in Radiation Dosimetry, pp. 219–25. Vienna: International Atomic Energy Agency.

EMMER, T. L. 1962. Nuclear instrumentation for scintillation and semiconductor spectroscopy. IRE Trans. Nuc. Sci., **9**:305–13.

FESSLER, T. E., and M. L. WOHL. 1961. Monte Carlo Studies of Gamma-Ray and Neutron Transport in Infinite Homogeneous Media. Cleveland: National Aeronautics and Space Administration Rpt. NASA-TN-D850. Pp. 50.

GLASS, F. M., and G. S. HURST. 1952. A method of pulse integration using the binary scaling unit. Rev. Sci. Instr., **23**:67–72.

HINE, G. J., and G. L. BROWNELL (eds.). 1956. Radiation Dosimetry. New York: Academic Press. Pp. 932.

HURST, G. S., and E. B. WAGNER. 1961. Special counting techniques in mixed-radiation dosimetry. *In* Selected Topics in Radiation Dosimetry, pp. 409–17. Vienna: International Atomic Energy Agency.

WENSEL, A., and G. HASL. 1962. Zur Messung der Umgebungsstrahlung mit plastischen Szintillatoren. Kerntechnik, **4**:118–19.

JACOB KASTNER, FRANCIS R. SHONKA,
AND JOHN E. ROSE

39. A Muscle-Equivalent Environmental Radiation Meter of Extreme Sensitivity

T HE RECENT DEVELOPMENT by one of us (Shonka, 1962) of a vibrating quartz fiber electrometer has made possible the construction of an uncomplicated, unpressurized, portable environmental radiation meter of extreme sensitivity. (See Plate I.)

The electrometer is routinely operable at a sensitivity of better than 500 divisions per volt. The sensitivity and balance adjustments may be made in the presence of a d.c. signal on the fiber. The electrometer has an inherent capacitance of 1–2 picofarads and a rapid response with no detectable anisotropy. Thus the instrument is an ideal detector for null measurements. A three-pound transistorized power supply capable of operating the system continuously for 150 hours is in use for field measurements.

A 16.5-liter pseudosphere with 6-mm.-thick walls was welded from six molded sections of conducting muscle-equivalent plastic (Fig. 1) (Shonka et al., 1958). The entire assembly contains no metal and has a polyethylene guard-ring insulator and a molded polystyrene insulator supporting a thin central collecting rod terminated by a thin-walled, hollow, tissue-equivalent, 4-cm.-diameter sphere. The ionization chamber was filled to 760 mm. mercury at 15° C. with a muscle-equivalent gas, recently formulated by Shonka, consisting by volume of 41.11 per cent neon, 39.59 per cent ethylene, 16.17 per cent ethane, and 3.13 per cent nitrogen. In this way the Bragg-Gray cavity principle applies without the usual stopping-power corrections; furthermore, the cavity-size restrictions were essentially removed. A similar

JACOB KASTNER is associate physicist and JOHN E. ROSE is division director at the Radiological Physics Division, Argonne National Laboratory, Argonne, Illinois; FRANCIS R. SHONKA is head, Physical Sciences Division, Physical Sciences Laboratory, St. Procopius College, Lisle, Illinois.

Work performed under the auspices of the U.S. Atomic Energy Commission.

ion chamber was molded as a sphere of 3-liter volume and 2-mm. walls. Additional caps were molded for this chamber of 2.5 and 6 mm. thickness. Studies of attenuation as a function of photon energy are under way.

Calibration of the chambers was carried out using a 1.11 mg. N.R.C. of Canada certified radium needle in 0.5-mm. platinum, with the usual checks for saturation, scattering, and inverse-square law behaviors. An extremely precise and accurate determination of the ratio

PLATE I

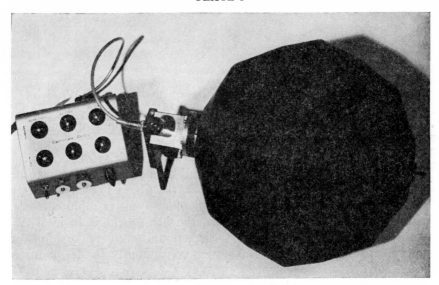

Environmental radiation meter

between muscle gas and air of "W," the energy required to create an ion pair, was graciously made for us by Dr. W. P. Jesse of St. Procopius College, Lisle, Illinois. The volume and capacitance of the chamber were measured to better than 0.5 per cent. The ionization rate calculated for muscle gas was compared with the ionization actually obtained. The agreement was good enough to assure that the ratio of electron stopping powers for our wall and gas is essentially unity.

The portability of even the 16-liter system and its extraordinary sensitivity of 0.32 mv/sec for one μr/hr enabled us to make measurements of environmental background in a skiff on Lake Michigan, on the top of a 40-meter ranger-type tower, and at various land sites and within buildings in the Chicago area.

The value obtained for cosmic-ray background is compared with that of other published determinations in Table 1. Corrections have been made for the height of Lake Michigan (about 580 feet) and for radon content in the air. The agreement is reasonable, especially when one considers the greater response of our chamber to low-energy radiation.

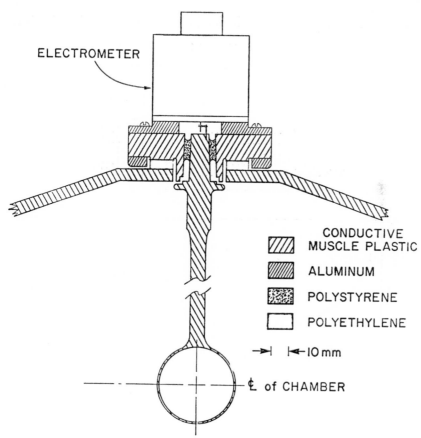

ELECTROMETER

|///| CONDUCTIVE MUSCLE PLASTIC

|///| ALUMINUM

|:::| POLYSTYRENE

|　| POLYETHYLENE

→| |←10 mm

₵ of CHAMBER

Fig. 1.—Muscle-equivalent ion chamber

Daily measurements in a first-floor laboratory over a period of months yielded a mean of 6.8 μr/hr with a coefficient of variation of better than 0.5 per cent.

The total radiation level at points outside the building corresponded to about 14–15 μr/hr, which is a shade higher than the values of 12–13 μr/hr reported for this area by the Health and Safety Laboratory of the Atomic Energy Commission (Solon *et al.*, 1959). The material of our building walls evidently attenuates the outdoor radiation considerably.

Measurements were carried out over a period of months in an 8-inch-wall iron room. The mean of very reproducible readings was 1 mv/sec, corresponding to about 3 μr/hr. Thus the cosmic-ray transmission by the iron room is about 67 per cent, which agrees very well with measurements made by Dr. F. W. Spiers (1960) in Leeds, England.

Figure 2 displays the data obtained on the tower. Each point is the

TABLE 1

SEA-LEVEL COSMIC-RAY IONIZATION

Observer	μr/hr	Ion Pairs (cc/sec)
Burch (1954)................	3.3	1.92
Neher (1952)................	4.7	2.74
Hess (1951).................	3.4	1.96
Shamos *et al.* (1963).........	3.4	1.97
This experiment............	4.2	2.44

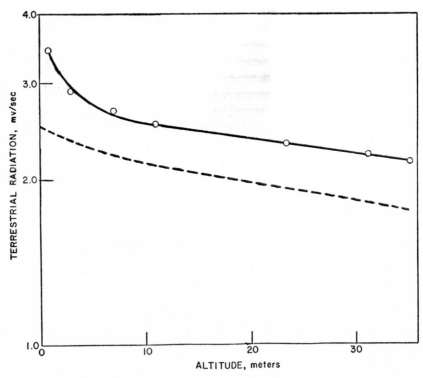

FIG. 2.—Air attenuation of terrestrial radiation. *Solid line,* experimental data; *broken line* theoretical. (After Hultqvist [1952] and O'Brien [1958].)

mean of about 15 measurements made at intervals of several weeks. The coefficient of variation of the data at each altitude is less than 4 per cent. The ordinate has been reduced to net terrestrial radiation by subtraction of the cosmic-ray contributions. The attenuation by air of terrestrial radiation has been calculated both by Hultqvist (1952) and by O'Brien *et al.* (1958) on the basis of assumed concentrations of uranium, thorium, and potassium. After filtration of the soft components by the first 10 meters of air there seems to be substantial agreement between the observed and calculated attenuation factors.

Earth samples were taken by Dr. P. F. Gustafson at sites that were concurrently surveyed by means of our meter. He analyzed the samples by γ-ray spectroscopy and derived a radiation output for the terrestrial component due to uranium, thorium, and potassium and fission products, if any (Gustafson *et al.*, 1958). The agreement is excellent (see Gustafson and Brar, 1963), and thus the two systems suitably complement each other.

The smaller 3-liter system has proved to be quite reasonable in its sensitivity for background measurements, and with it and the caps we plan to continue our Lake Michigan measurements in an attempt to determine quantitatively the soft component of cosmic-ray ionization.

To conclude, the Shonka electrometer, with its extreme sensitivity, has made possible a system for the reproducible measurements of absorbed dose rates of fractions of a microroentgen per hour. The apparatus is quite rugged, exhibits no geotropism, and is easily portable and battery operated. The electrometer and muscle-equivalent ion chamber seem to provide the simplest system (requiring the fewest corrections) for studying the dose to man from external, natural, and man-made environmental radiation.

ACKNOWLEDGMENTS

We must point out that much of the initial impetus for this work was provided by the late Dr. G. Failla. We are grateful to Mr. William Prepejchal for helping both in the accumulation of data in odd places and in making the auxiliary apparatus more compact. We are grateful also to Mr. Richard York for his valuable contributions in the design and construction of the electrometer and chamber.

REFERENCES

BURCH, P. R. J. 1954. Cosmic radiation: Ionization intensity and specific ionization in air at sea level. Proc. Phys. Soc. (London), Ser. A, **67**: 421–30.

GARRETT, C. 1958. Modification of the basis for roentgen calibrations between 0.5 and 3 mev. Canad. J. Phys. **36**:149–50.

GUSTAFSON, P. F., and S. S. BRAR. 1963. Measurement of gamma-emitting radionuclides in soil and the calculation of the dose arising therefrom. This symposium.

GUSTAFSON, P. F., L. D. MARINELLI, and S. S. BRAR. 1958. Natural and fission-produced gamma-ray emitting radioactivity in soil. Science, **127**:1240–42.

HESS, V. F., and G. A. O'DONNELL. 1951. The rate of ion formation at ground level and at one meter above ground. J. Geophys. Res., **56**: 557–62.

HULTQVIST, B. 1952. Calculation of the ionization due to radioactive substances in the ground. Tellus, **4**:54–62.

NEHER, H. V. 1952. Recent data on geomagnetic effects. *In* J. G. WILSON (ed.), Progress in Cosmic Ray Physics, **1**:243–314. Amsterdam: North-Holland.

O'BRIEN, K., W. M. LOWDER, and L. R. SOLON. 1958. Beta and gamma dose rates from terrestrially distributed sources. Radiation Res., **9**:216–21.

SHAMOS, M. H., A. LIBOFF, and J. SIDEROWITZ. 1963. Thin-wall ion chambers for low-level radiation environments. This symposium.

SHONKA, F. R. 1962. Vibrating quartz fiber electrometer. Radiology, **78**:112.

SHONKA, F. R., J. E. ROSE, and G. FAILLA. 1958. Conducting plastic equivalent to tissue, air and polystyrene. Proc. 2d United Nations Internat. Conf. on Peaceful Uses of Atomic Energy, **21**:184–87.

SOLON, R. L., W. M. LOWDER, A. SHAMBON, and H. BLATZ. 1959. Further Investigations of Natural Environmental Radiation. U.S. Atomic Energy Comm., Health and Safety Lab. Rpt. HASL-73. Pp. 34.

SPIERS, F. W. 1960. The measurement of natural environmental gamma radiation. Strahlentherapie, **111**:65–74. (In German.)

WERNER N. GRUNE, JOSEPH H. MEHAFFEY,
CHARLES H. KAPLAN, AND TOM L. ERB

40. *Feasibility Studies for the Establishment of Parameters for Background Radiation Measurements*

THE EVALUATION OF HUMAN EXPOSURE to naturally occurring radioactivity has become of public health interest and importance only during the last few years, although the existence of natural activity has been known for over 60 years. With the increased concern over problems of radioactive fallout and the monitoring of areas in the vicinity of nuclear reactors, extensive radiologic surveys have become necessary to determine the existing types and quantities of artificial and natural radioactives. The most important environmental factors affecting the selection of sites for atomic energy plants are the geology, hydrology, and meteorology of the area (Gorman, 1957). In some instances, the levels of natural activity discovered have been sufficiently high to cause concern. Because of the new problem, relatively few methods of analysis were suited to the requirements of field sampling programs.

DETERMINATION OF NATURAL RADIOACTIVITY IN WATER

The purpose of these studies was to develop a simple, reliable and reproducible field method for the detection and the analysis of naturally occurring radioactivity. The procedures were to be applied to the analysis of water supplies, predominantly ground waters, during a relatively extensive field sampling program. Therefore, the re-

WERNER N. GRUNE is professor of sanitary engineering, and JOSEPH H. MEHAFFEY, CHARLES H. KAPLAN, and TOM L. ERB are graduate research assistants, at the Sanitary Engineering Research Laboratories, School of Civil Engineering, Georgia Institute of Technology, Atlanta, Georgia. Joseph H. Mehaffey is currently employed as senior electronic systems design engineer, Lockheed Georgia Company, Division of Lockheed Aircraft Company, Marietta, Georgia.

quirements differed from those of many existing analytic methods. The equipment had to be compact and portable. To achieve significant results during a time-limited field survey covering a 2,500-square-mile area in Maine and New Hampshire, it was desirable that the analytic procedures be as rapid as possible without loss of accuracy.

INTRODUCTION OF THE PROBLEM

Following the original discovery, late in 1957, of radioactivity in a pressure tank fed from a deep-well water supply, a comprehensive radioanalysis over an extensive area was indicated (Grune *et al.*, 1960*a*). It was necessary to identify the chemical form and physical state of the radioactive material, whether present in the soluble, colloidal, or gaseous state, and the predominant naturally occurring radioactive series.

METHODS FOR RADON-222 ANALYSIS

Two approaches to analytic procedures may be employed to determine radon-222. Either the radon-222 itself may be isolated from the sample or the immediate daughters may be extracted and the radon-222 concentration calculated. Radon-222 isolation is to be preferred, but it is very difficult to collect, transport, or store a sample without radon-222 loss. The daughter products are easily collected and may be chemically precipitated, but it is necessary to free the sample from radon-222 at the time of collection. Analysis must be made soon thereafter, since the daughter activity decreases at a rate based on a half-life of approximately 35 minutes. Radon-222 collection produces a more sensitive test, since for a given amount of radon-222 the activity increases with time, while the daughter activity decreases if isolated.

Initially, studies were made of the precipitation of the polonium, lead, and bismuth daughters of radon-222 as sulfides, using a lead carrier and hydrogen sulfide or ammonium sulfide. A de-emanation method was next investigated and is described below.

The analytic procedure studied in great detail, and later employed in the field studies, involved the de-emanation of radon-222 from solution and the collection of the gases in a 125-ml. Erlenmeyer flask coated with a powdered, silver-activated zinc sulfide screen. This procedure was followed by α scintillation counting. Water samples were collected in specially designed glass bubblers, as shown in Figure 1 and Plate I. De-emanation was achieved by means of the evacuated zinc sulfide coated Erlenmeyer flask connected to the bubbler. The vacuum was used to produce an upward flow of air through the bubbler, resulting in radon-222 removal from the water and collection in

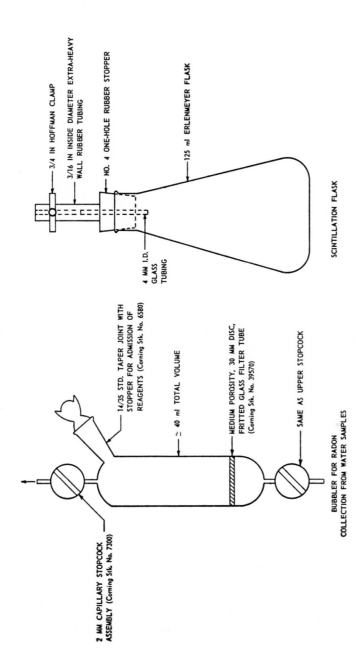

3/4 IN HOFFMAN CLAMP

3/16 IN INSIDE DIAMETER EXTRA-HEAVY WALL RUBBER TUBING

NO. 4 ONE-HOLE RUBBER STOPPER

125 ml ERLENMEYER FLASK

4 MM I.D. GLASS TUBING

SCINTILLATION FLASK

14/35 STD. TAPER JOINT WITH STOPPER FOR ADMISSION OF REAGENTS (Corning Stk. No. 6580)

~ 40 ml TOTAL VOLUME

MEDIUM POROSITY, 30 MM DISC, FRITTED GLASS FILTER TUBE (Corning Stk. No. 39570)

SAME AS UPPER STOPCOCK

2 MM CAPILLARY STOPCOCK ASSEMBLY (Corning Stk. No. 7300)

BUBBLER FOR RADON COLLECTION FROM WATER SAMPLES

Fig. 1.—Sketch of bubbler and scintillation flask

the flask. The analysis was completed by α scintillation counting of the flask, with the use of a specially designed detection unit.

To provide the scintillation screen, the insides of the flasks were coated with a layer of zinc sulfide crystals in accordance with the procedure outlined by Harris, LeVine, and Watnick (1957). The inside end of the stopper was coated, but the inside bottom of the flask was not. The clear flask bottom provided a window for a clear light path to the photomultiplier tube.

PLATE I

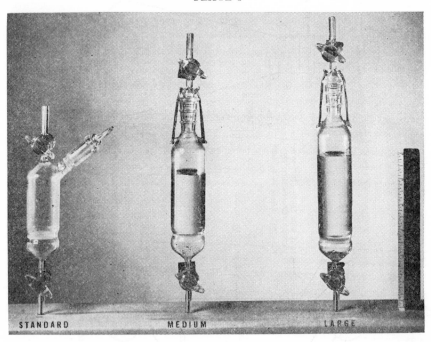

Original and redesigned glass bubblers

The collection and analysis of a sample were divided into four major steps:

1. *Bubbler and Scintillation Flask Preparation.*   After having been cleaned, the bubbler was prepared for sampling by insertion of 1 ml. of concentrated nitric acid through the tapered joint. When the sample (25–30 ml.) was added, the resulting acid concentration of approximately 0.5N hydrochloric acid helped to keep dissolved materials in solution and reduced contamination of the bubbler. The final preparatory step before sample collection was to weigh the bubbler for subsequent determination of sample volume.

The scintillation chamber was a 125-ml. Erlenmeyer flask coated with zinc sulfide. The sides of the flask were swabbed with a solution of approximately one part silicone grease to ten parts chloroform by volume. The chloroform evaporated rapidly and left a layer of grease to help produce a uniform coating of zinc sulfide powder. The coated surface extended from the base of the neck to the edge of the flat bottom.

To conserve flasks and to maintain fresh zinc sulfide coatings, each flask was used first for a background count before use with a sample and then recoated.

2. Sample Collection. Samples were collected directly from faucets to represent as nearly as possible the radon-222 concentration exposure of the consumer. Before a sample was collected, the faucet was turned on for a period of at least one minute to insure that the sample came from the pressure tank or the well. Since radon-222 escapes rapidly from water, any turbulence in the flow from the faucet results in a partial loss of radon-222. The bubblers were filled directly under the stream, quickly removed, and stoppered. The maximum time required for the entire operation did not exceed 3–4 seconds. This time was critical, since turbulence in the bubbler during filling would release radon-222 rapidly.

3. Radon De-emanation. Radon from the sample was extracted from the bubbler to the scintillation flask. The upper capillary tube of the bubbler (see Fig. 1) was connected by rubber tubing to the scintillation flask, previously evacuated with a high-vacuum pump. The clamp on the rubber tube and the upper stopcock of the bubbler were opened, and the lower stopcock was used to regulate the vacuum-produced flow of air.

The air flowing through the bubbler was dispersed by the porous glass plate into fine bubbles that scrubbed the radon-222 from the sample. After bubbling had ceased, the clamp on the rubber tubing was tightened to seal off the radon-222 in the flask, and the bubbler was stored for transportation to the laboratory. To maintain a constant counting efficiency, the pressure in the flask was equalized to atmospheric pressure.

Air was used to remove the radon-222 from the water and entered the scintillation flask. Therefore, it was necessary to account for the radon-222 concentration of the air. The problem was readily solved by taking an air sample at each radon-222 sampling point, using only a scintillation flask. The air sample was counted in the laboratory and its activity subtracted as a background count for absolute calculations.

4. Laboratory Analysis for Radon-222. The analysis of the sample in the laboratory proved to be a rapid and simple procedure. The bubblers were reweighed to determine the precise sample volume. Both sample and background flasks were counted to complete the analysis.

It was found to be essential to take special care during two steps of the laboratory analysis: (*a*) the flask should be centered on the photomultiplier tube to insure constant geometry, and (*b*) the sample should be separated at least 4 hours before counting is started. This time interval, as may be seen in Figure 2, insured that the radon-222 and daughters are in equilibrium and are changing only slowly with

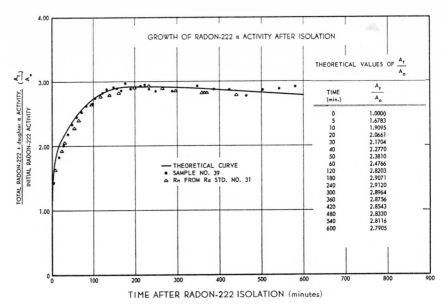

FIG. 2.—Build-up of radon-222 and daughters

the radon-222 half-life. Therefore, during the normal counting periods employed, no correction for decay was necessary because of the slow rate of change in the total activity. A preset counting time of 20 minutes was used for all radon-222 analyses, with background counts of 5 minutes' duration.

SCINTILLATION DETECTION UNIT

To obtain data in the field, a special scintillation-type detection unit was designed and assembled, as shown diagrammatically in Figure 3. The detection unit consisted of four major parts: (*a*) a photomultiplier tube, (*b*) a preamplifier, (*c*) a pulse amplitude discriminator, and (*d*) a low-voltage power supply.

A Model 6363 Dumont 10-stage, 3-inch photomultiplier tube was used. The high voltage necessary for operation of this tube was furnished by a Model 186 Nuclear-Chicago scaler unit.

To increase the output signal voltage of the photomultiplier to a more usable level and to lower the output impedance of the photomultiplier tube to a level suitable for driving a length of coaxial cable, a preamplifier was installed in juxtaposition to the photomultiplier tube.

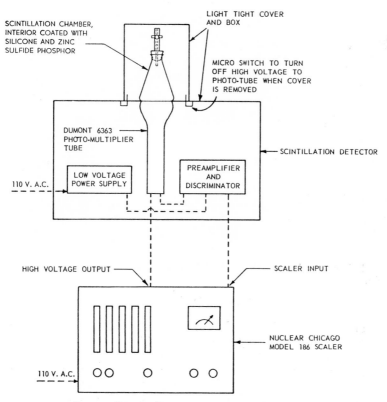

Fig. 3.—Sketch of scintillation detector and scaler

The amplifier discriminator performed two tasks. The first was to amplify further the signal from the preamplifier before the signal was introduced to the discriminator section. The other task was to pass signals above a preset level and to reject those below that level. This base-level discriminator allowed the random noise pulses, which are usually much below the average signal level, to be removed from the input signal.

The low-voltage power supply was required because the scaler power supply was inadequate to power all the components of the detection system. This unit was voltage-regulated to obtain reproducible readings. As an additional safeguard, a Sola transformer-regulator was used in the a.c. line to the power supply. This combination produced a most stable regulation under all conditions encountered in the field. A photograph of the analysis system in operation at the field laboratory in Raymond, Maine, is reproduced as Plate II.

PLATE II

Analysis system in operation at field laboratory (scintillation detector and scaler on bench).

REPRODUCIBILITY OF METHOD

The de-emanation method was found to have a reproducibility of ±10 per cent at the 95 per cent confidence level. This reproducibility was adequate when compared to the ±14 per cent at 95 per cent confidence for the variation in activity of a well with time. Thus, the confidence interval of the variation in activity with time exceeds that of the reproducibility of the method. The activity of a well in South Paris, Maine, was found to vary with time during a continuous pumping test, reaching a maximum about 15 minutes after the start of pumping and declining gradually thereafter.

PROCEDURE FOR RADIUM-226 ANALYSIS

Radium-226 was also analyzed for, using the de-emanation procedure, with a few simple changes in procedure. After de-emanation for radon-222 analysis, the water sample was thoroughly purged of any remaining radon-222. The bubbler was then sealed for 8–12 days to allow for build-up of the radon-222 daughter from the radium-226 in the sample. Following storage, a final radon-222 analysis was performed on the sample, and the radium-226 content was calculated from the known state of partial equilibrium between the two radioelements. Because of the small sample size available, limited by the size of the glass bubblers, it was found that reliable results were obtained only from samples with a minimum radium-226 content of 50 pc/l.

SUMMARY AND RESULTS

Approximately 350 water samples were collected in Maine and New Hampshire and analyzed for radon-222. Approximately 85 samples were also analyzed for radium-226. The following conclusions were reached:

1. Over 99.2 per cent of all water samples from drilled wells in Maine analyzed for radon-222 + daughters had activities exceeding 2,000 pc/l.

TABLE 1

SUMMARY OF RADON-222 CONCENTRATIONS IN WELLS

LOCATION	WELL TYPE	No. SAMPLES	ACTIVITY* (pc/l RN-222+DAUGHTERS)		
			Minimum	Average	Maximum
Maine..........	Drilled	128	1,120	87,600	884,000
Maine..........	Dug	76	0	7,870	31,400
Maine..........	Springs	18	0	18,800	113,000
New Hampshire..	Drilled	17	2,560	142,000	1,130,000
New Hampshire..	Dug	3	6,180	30,100	68,000

* The maximum permissible concentration (MPC) of radon-222 plus daughter products is 2,000 pc/l (National Bureau of Standards, 1955).

2. Approximately 84 per cent of all dug-well samples in Maine contained radon-222 + daughters with activities exceeding 2,000 pc/l.

3. It was established that radon-222 and radium-226 were not in equilibrium with each other, as is true for most natural waters.

4. Studies of the variation of activity with duration of pumping revealed that the radon-222 activity varies with the rate of pumping and the length of the resting period between pumpings.

5. No definite boundary of a radioactive zone could be established in either Maine or New Hampshire, and it appeared that some radon-222 is probably present in all wells. This widespread occurrence of radon-222 has been substantiated by similar discoveries in Tennessee, Colorado, and New York, and in other New England States.

6. Extremely high activities were found scattered over the entire 2,500-sq.-mi. area sampled. These findings indicate that hot spots may

be the result of intrusions, but no positive evidence to support this possibility could be obtained. The generalized results of the 1960 field survey are shown in Table 1.

A more complete treatment of the methods for determining radon-222 and radium-226, design and development of analysis system and operation, as well as the detailed results from the field studies, have been previously presented (Higgins *et al.*, 1961; Smith *et al.*, 1961; Grune, 1961, 1962).

DEVELOPMENT OF IMPROVED DETECTION UNIT

The equipment used during the initial field studies, shown in Figure 3 and Plate II, proved too bulky for convenient field use, and

PLATE III

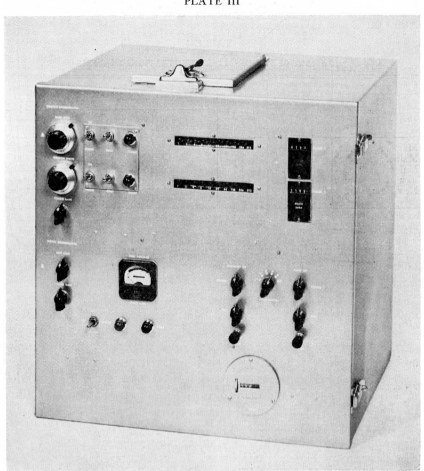

Transistorized detector and scaler unit

the need for a.c. power precluded its mobility in the field. To eliminate these difficulties, a new portable battery-operated unit, employing transistorized circuitry, was designed and developed at the Georgia Institute of Technology.

The unit, shown in Plate III, was transistorized insofar as possible to reduce weight, bulk, and power requirements. Another feature of this unit was the incorporation of a planchet counting chamber. This

PLATE IV

Two binary scalers

feature provided the capability for counting two samples simultaneously: a gaseous sample contained in an Erlenmeyer flask and a 2-inch planchet sample placed in a separate chamber.

The principal detector in the unit is a zinc sulfide α scintillator, coated on the inside of an Erlenmeyer flask. The clear bottom of the flask is placed in contact with the 5-inch photomultiplier tube within the light-tight chamber in the unit. The flask is filled with the sample to be analyzed by evacuating the flasks and taking samples at various locations.

Provision was made to substitute a sodium iodide crystal for the zinc sulfide detector in the flask chamber. The input to this scaler may be put through a window discriminator. This modification allows the unit to be used in conjunction with a built-in manual window discriminator to obtain pulse-height information, thus increasing the versatility of the unit.

The planchet chamber may be used with a β or γ scintillation crystal mounted in a removable tray.

The flask chamber consists of a light-tight, sheet-metal inclosure within the main cabinet. The 5-inch photomultiplier tube protrudes

PLATE V

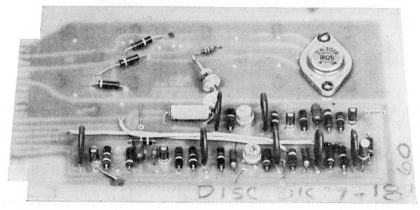

Upper, window discriminator; *lower*, single discriminator and low-voltage regulator

into the bottom of this chamber, in which up to 1-liter sample vessels may be accommodated.

Details of the electronic circuit design have been presented elsewhere (Grune *et al.*, 1960*b*). Printed circuit techniques were used throughout, and the salient circuit boards are shown in Plates IV and V. They represent the initial efforts of our laboratory electronics staff in preparing printed circuit boards.

Studies of Parameters for Background Radiation Measurements

During the fall of 1960 this project underwent a change in direction to obtain more information on environmental radiation. Specifically, the purpose of these investigations has been to determine the feasibility of obtaining field measurements of background radiation to estimate man's total radiation exposure. Initially, it was necessary to establish parameters and techniques for radiation measurements in the environment.

Interaction of cosmic rays with the atmosphere and terrestrial radiation was studied, and components of environmental radiation were summarized. Emphasis was placed on the development of a 3-channel γ-energy analyzer. Radiation flux and air-sample activity measurements were conducted utilizing the instrumentation developed under this study.

In addition, theoretical and experimental studies were conducted to determine a method for the absolute calibration of a sodium iodide scintillator with γ energies ranging from 0.1 to 1 mev. The completion of the over-all study will provide an inexpensive and reliable γ, β, and α monitoring system capable of unattended operation at remote locations. In case of atomic attack or nuclear excursion, a network of these systems would allow precise mapping of the fallout pattern. In addition to this objective, knowledge regarding the effects of chronic low-level exposure to ionizing radiation would be increased.

AUTOMATIC AIR-SAMPLING UNIT

During subsequent work a need existed to determine radioactivity present in the air in particulate form. It was desired to examine automatically each air sample immediately after deposition on a filter paper to aid in detecting any short-lived elements present. To implement this requirement, an automatic filter-tape air-sampling unit* was modified for radiation measurement. The tape sampler draws air through a filter tape for a preset time, then automatically moves the tape and takes another sample. An end-window Geiger-Müller tube was installed over the tape in such a manner as to count the sample just taken. A schematic diagram of the automatic air-sampling unit to count particulate matter is shown in Figure 4. The output of the Geiger-Müller tube was amplified and sent via coaxial cable from the roof, where the sampler was located, to a scaler inside the building.

* Model M, Hi-Flow Tape Sampler, manufactured by the Research Appliance Company, Allison Park, Pa.

The time required for a preset number of counts was recorded by a printing timer, and the scaler automatically reset. As soon as the collection of the next sample was completed, it automatically moved into position for counting.

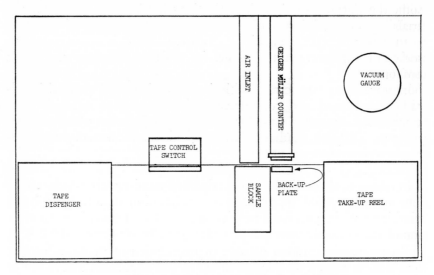

Fig. 4.—Schematic drawing of automatic air-sampling unit

DESIGN AND OPERATION OF THE γ-DETECTION SYSTEM

More recently the work in the area of instrument development work has been carried out in conjunction with feasibility studies to design suitable monitoring stations for the continuous mapping of background radiation patterns. Gamma radiation is a product of most nuclear reactions, and its large mean distance between collisions in the atmosphere increases the feasibility of detecting low levels of radiation. One of the problems encountered when attempting to measure γ radiation from unknown sources derives from the many types of conversions that γ radiation may undergo during the detection process. The three γ interactions that may occur in a detector are (*a*) Compton, (*b*) photoelectric, and (*c*) pair production. Each of these processes has an energy-dependent probability of occurrence and a different conversion efficiency. In addition, it is extremely difficult to determine which of the three processes produced a particular scintillation. Therefore, any scintillation detector employing a single photomultiplier-crystal combination can produce little quantitative energy *versus* activity data for a completely unknown source.

THREE-CHANNEL PULSE-AMPLITUDE ANALYZER

A compact transistorized 3-channel analyzer was designed and constructed. Plate VI shows two of the transistorized boards for the 3-channel analyzer. This unit employs a 2-inch diameter $\times$ 2-inch sodium iodide crystal and is the unit demonstrated during the field trials. A photograph of the unit is shown in Plate VII.

In the Georgia Institute of Technology laboratory setup, the sodium iodide scintillation detector was located on the roof of the building. The detector, shown in Plate VIII, was housed in a wooden box, which was copper-clad to shield out interference from a nearby radio tower and temperature-controlled to prevent changes in the gain of the photomultiplier tube. The signal was preamplified in the detector box and transmitted to the 3-channel pulse analyzer, inside the building, via coaxial cable.

The analyzer section divided the pulse heights into three ranges with gated discrimination. A readout consisted of the counts occurring in each of the three ranges and a total. The boundaries between these ranges are variable and were set by the use of a cesium-137 γ standard.

The γ analyzer has two independent output systems. The first system is used when only battery power is available and consists of four decimal scalers. Each scaler uses two "Nixie" readout tubes, driven by magnetic beam-switching tubes. A mechanical register makes possible a total count of 999,999 in each channel.

The second output mode, which can be used whether or not the decimal output is used, is analog rather than digital, and the data are recorded on a strip-chart potentiometer recorder. The outputs of each of the 3 channels and the totaled output are sampled periodically by a count-rate meter and fed into the recorder. After each sampling cycle, the count-rate meter records its zero reading and then samples a 60 c.p.s. input for calibration of the chart scale. Each cycle was timed to take exactly 30 minutes to facilitate data analysis.

The background β flux is to be counted with two plastic scintillators. One of these detectors is exposed to β and γ radiation, and the pulses are counted by a reversible scaler. The other detector, which is identical in geometry and sensitivity adjustment to the first, is shielded from β radiation and thus counts only γ photons and other highly penetrating radiations. The pulses from this detector are fed into the "subtract" input of the reversible scaler, leaving the difference, or net β count, showing on the readout. A block diagram of the β and γ systems is shown in Figure 5.

The reversibility of the scaler is obtained by driving each "Nixie"

PLATE VI*a*

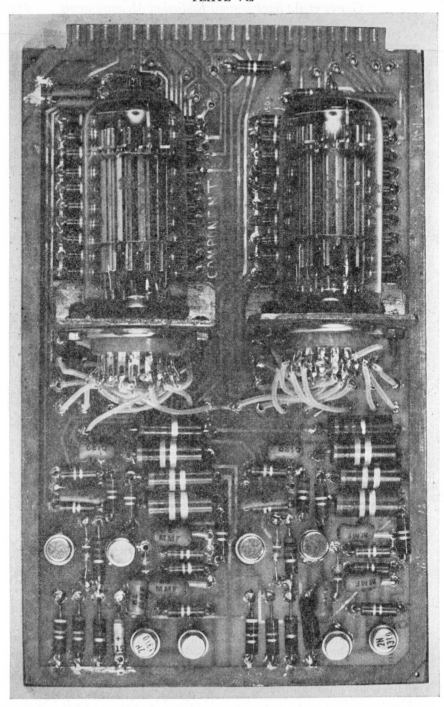

Two-decade, decimal scaling unit

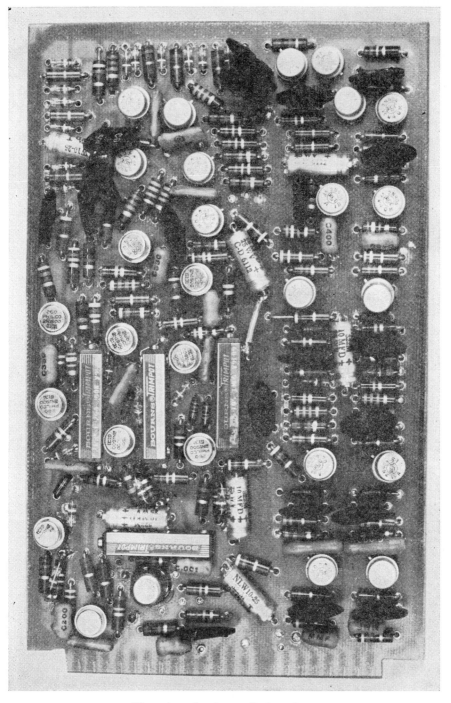

Three-channel, pulse-amplitude analyzer

PLATE VII

Complete γ-analyzer system

PLATE VIII

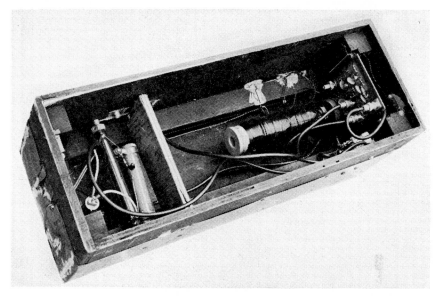

Gamma-detection unit

γ DETECTION SYSTEM

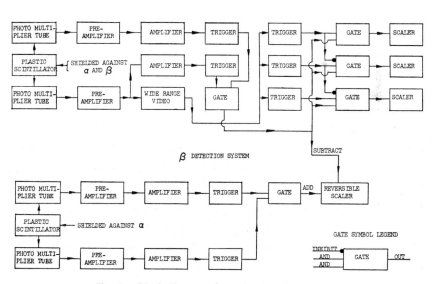

Fig. 5.—Block diagram of β and γ detection systems

readout tube with two beam-switching tubes, one wired to count forward, the other backward (i.e., 9–8–7–6, etc.). Thus, when the forward tube is pulsed, counts are added, and when the reverse tube is pulsed, counts are subtracted.

THREE-CHANNEL γ ANALYSIS

Typical data from the 3-channel γ analyzer, with a fourth channel giving total dose rate, are presented in graphic form in Figure 6, which shows a rather definite relationship between the 3 energy ranges from natural background radiation.

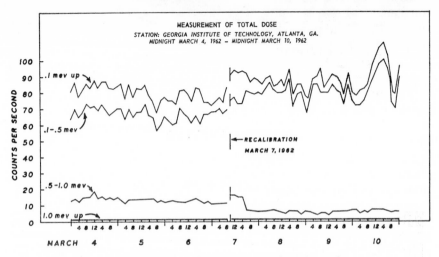

Fig. 6.—Measurement of total dose by 3-channel γ analyzer

COMPTON SHIFT ELIMINATOR STUDY AND ASSEMBLY

The scintillation detector is useful for isotope identification because a primary standard sample can be used for calibration and one may study its photoelectric peak with an analyzer. With the energy spectrum established, an unknown sample can be evaluated by comparison.

However, there is no primary "standard" with which to compare a background spectrum. The spectrum obtained on single or multichannel analyzers has a photopeak that corresponds directly to the energy of the photons emitted from the samples. The remainder of the "spectrum" comes from Compton shifted photons and/or pair production. The last two conversions cannot be directly correlated to γ photon energies.

The results of a thorough study of this complex problem of eliminating Compton shifted photons and the description of an experimental Compton shift eliminator assembly have been reported elsewhere (Grune *et al.*, 1962). It was determined from experimental observations with low-level equipment, developed under the research contract, that very few γ photons from the background have energies exceeding 3 mev. As the photoelectric effect provides almost a 1-to-1 γ energy to light energy conversion, it becomes evident that eliminating Compton γ interactions aids considerably in improving energy resolution of a detector for completely unknown samples. Therefore, plans for reducing the effect of Compton interaction with a "Compton shield" were initiated. Latest information supports the contention that such a shield is feasible, and in fact a similar assembly to the one proposed during our study was designed and constructed at Oak Ridge National Laboratory for Compton elimination in γ spectroscopy studies (Davis *et al.*, 1956).

The determination of the energy distribution of background γ flux by the use of a single detector does not appear to be practical. There is no technique with which to determine whether a pulse of a given height resulted from a photon producing a photoelectric effect, in which case the pulse height would be proportional to the γ-ray energy, or from a photon that had some undetermined higher energy undergoing Compton scattering or pair production. Since the last two events predominate over the photoelectric effect over most of the γ energy band, the pulse-height information from a sodium iodide crystal is almost useless for determining the actual energy distributions of cosmic radiation.

To overcome these difficulties, a system was proposed that will disregard most of the pulses resulting from γ reactions other than the photoelectric effect. The reaction of main concern is the Compton effect. In the Compton effect, part of the γ-ray energy is absorbed by an electron in the sodium iodide crystal, and the remainder may escape the crystal in the form of a scattered photon. It may be assumed that the maximum scattering angle is not much greater than 90 degrees and that the probability for scattering at greater than 90 degrees is low. Under these assumptions, photons that do scatter at angles greater than 90 degrees would have lost so much energy in the process that they would probably undergo a photoelectric reaction before escaping the crystal. Thus, the desired result would be obtained anyway.

In the proposed scheme, a shielding scintillator would surround the sodium iodide and almost all the scattered photons that escape the sodium iodide would enter the shield surrounding the sides and bottom of the crystal. This shield would be of sufficient thickness to insure a very good probability of some sort of light-producing inter-action with the scattered γ rays. By observing both the shield and the sodium iodide with separate photomultiplier tubes and rejecting all time coincident pulses from the two tubes, one effectively rejects most of the Compton pulses from the sodium iodide detector. This interaction can also take place in reverse, that is, the scattered photon going from the shield into the sodium iodide. However, these pulses are still coincident and therefore will be rejected. This system would eliminate many of the pair-production pulses, although laboratory studies have shown that there are not many background photons of sufficient energy for pair production. The pulses remaining after co-incidence elimination would consist principally of photoelectric pulses and could be pulse-height analyzed for γ-ray energy.

The over-all counting rate would be reduced, but most of the counts lost would have produced Compton shifts in the sodium iodide scintillator and therefore would have been useless from an energy-measurement standpoint.

A theoretical study was made utilizing the Klein-Nishina formula (Klein and Nishina, 1929) for the differential cross-section per elec-tron.

Some studies, based on the theory described above, were conducted with experimental apparatus constructed at this laboratory. A photo-graph of the assembly is shown in Plate IX. The initial experiments made with this apparatus did not produce the expected results.

The curves in Figures 7 and 8 show the percentage of the inter-acting photons that undergoes the Compton effect and scatters at angles of θ greater than $\pi/2$ and $3\pi/4$, respectively, for various ener-gies. It may be observed from Figure 8, that for photons with an energy of 0.1 mev., only 14.6 per cent of those undergoing the Compton effect will be scattered more than $\theta = 3\pi/4$. At this energy about 7.8 per cent of the interacting γ radiation undergoes the Comp-ton effect. Therefore, a Compton eliminator system with a shield around 10.7 steradians—equivalent to $\theta = 3\pi/4$—will eliminate all but 14 per cent of the 7.8 per cent initially interacting via the Compton effect. Therefore, over 98 per cent of the observed interactions would be photoelectric and usable for energy analysis.

As another example, consider a photon energy of 1 mev. For $\theta = 3\pi/4$, a Compton eliminator system would eliminate all but approximately 7.4 per cent of photons undergoing the Compton effect, thus bringing the effective Compton cross-section for single interactions into the range of the photoelectric cross-section. Since in a thick crystal the scattered photons are likely to undergo several

PLATE IX

Experimental Compton eliminator assembly and single-channel γ analyzer

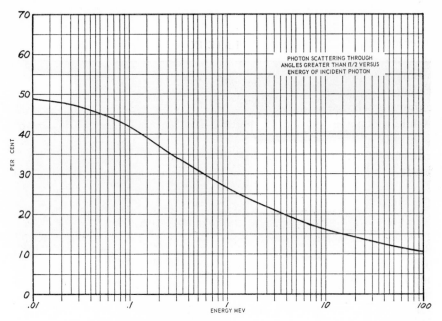

Fig. 7.—Photon scattering through angles greater than $\pi/2$ vs. energy of incident photon

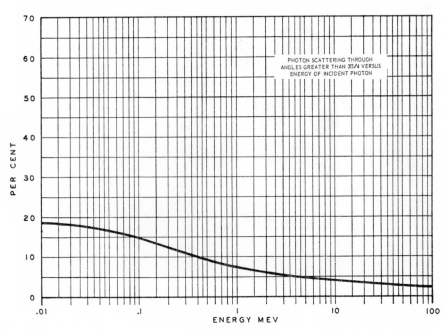

Fig. 8.—Photon scattering through angles greater than $3\pi/4$ vs. energy of incident photon

interactions, each of which decreases their energy and thus increases their absorption cross-section, the over-all probability for total energy absorption of an interacting γ photon is much better than the preceding thin-target analysis would reveal. Therefore, the Compton eliminator shield, in conjunction with a thick sodium iodide crystal, would yield quite useful energy information in the range of interest.

Future investigations should be conducted using larger and more efficient scintillation crystals than were available for the preliminary study.

Acknowledgments

The authors wish to express their appreciation for the assistance and generous hospitality extended to the field party by Dr. and Mrs. C. D. Brown at Raymond, Maine, and Mr. and Mrs. S. W. Hodgson at Nottingham, New Hampshire. Special recognition is due Messrs. F. B. Higgins and B. M. Smith, graduate students in sanitary engineering, through whose tireless efforts the field survey was conducted during the summer of 1959.

For the generous support of the field party, special mention is due Mr. Roy C. Peek, radiochemist, Sanitary Engineering Research Laboratories, Georgia Institute of Technology, who devoted many hours to the planning and logistics of the sampling program both prior and during the survey, and Messrs. J. G. Lee, E. H. Friedman, J. O. Etheridge, M. H. West, and T. D. Hughey, who assisted in the development and construction of the radiation detection equipment.

Without the excellent co-operation, sincere interest, and assistance provided by a number of U.S. Public Health Service officials and staff members of both the Maine and New Hampshire health departments, these investigations could not have been carried out. Acknowledgment is given of the splendid co-operation on the part of a number of staff members of the Division of Radiological Health, in particular to Mr. J. G., Terrill, Jr., Mr. D. J. Nelson, Mr. E. S. Weiss, Mr. E. C. Anderson, and Mr. R. D. Grundy. Valuable guidance and technical service were also rendered by several staff members of the Health and Safety Laboratory, U.S. Atomic Energy Commission, New York, New York.

The authors also acknowledge the active support of this work by the Engineering Experiment Station and the School of Civil Engineering, Georgia Institute of Technology, especially during the initial stages of the research program.

References

Davis, R. C., P. R. Bell, G. G. Kelley, and N. H. Lazar. 1956. Response of "total absorption" spectrometers to gamma rays. IRE Trans. Nuclear Sci., NS, **3**:82–86.

Gorman, A. E. 1957. Selection of sites for atomic energy plants. Am. Soc. Civil Engineers J. **83**:1175–10.

GRUNE, W. N. 1961. Natural radioactivity in ground water. Parts 1 and 2. Water & Sewage Works, **108**:409–11, 449–52.

———. 1962. Natural radioactivity in ground water. Part 3. *Ibid.,* **109**: 25–29.

GRUNE, W. N., F. B. HIGGINS, and B. M. SMITH. 1960*a*. Natural Radioactivity in Ground Water Supplies in Maine and New Hampshire. Complete Scient. Rpt. to Div. of Radiological Health, USPHS, Washington, D.C., Contract No. SAph-73551, February 1.

———. 1960*b*. Natural Radioactivity in Ground Water Supplies in Maine and New Hampshire. Final Rpt. Project A-473 to Div. of Radiological Health, USPHS, Washington 25, D.C., Contract No. SAph-73551, October 26.

GRUNE, W. N., J. H. MEHAFFEY, and T. L. ERB. 1962. Development of Instrumentation for Background Radiation Measurements. Final Rpt. Project A-530 to Div. of Radiological Health, USPHS, Washington 25, D.C., Contract No. SAph-76183, August 30.

HARRIS, W. B., H. D. LeVINE, and S. I. WATNICK. 1957. Portable radon detector for continuous air monitoring. A.M.A. Archives of Ind. Health, **16**:493–98.

HIGGINS, F. B., JR., W. N. GRUNE, B. M. SMITH, and J. G. TERRILL, JR. 1961. Methods for determining radon-222 and radium-226. J. Am. Water Works Assoc., **53**:63–74.

KLEIN, O., and Y. NISHINA. 1929. Über die Streuung von Strahlung durch freie Elektronen nach der neuen relativistischen Quantendynamik von Dirac. Ztsch. f. Phys. **52**:853–68.

NATIONAL BUREAU OF STANDARDS. 1955. Regulation of Radiation Exposure by Legislative Means. Nat. Bur. Standards, Nat. Committee on Radiation Protection, Handbook 61. Pp. 60.

SMITH, B. M., W. N. GRUNE, F. B. HIGGINS, JR., and J. G. TERRILL, JR. 1961. Natural radioactivity in ground water supplies in Maine and New Hampshire. J. Am. Water Works Assoc., **53**:75–88.

JOHN E. HAND

41. Instrumentation for Aerial Surveys of Terrestrial γ Radiation

INVESTIGATIONS of the relative terrestrial γ-ray environment over large areas of land can be successfully conducted with air-borne measuring systems. Aerial measurements of surface radioactivity were undertaken in the United States as early as 1948. At that time flights were made in order to investigate the problems associated with locating radioactive ore deposits from an air-borne vehicle. The results obtained on these missions indicated that surficial materials containing 0.01 per cent uranium could be detected from an aircraft flying at 500 feet above the ground. Subsequent efforts were directed toward improving the detection, measuring, and recording characteristics of the air-borne apparatus and toward establishing systematic aerial survey procedures. It is now commonplace to perform aerial measurements of the natural terrestrial γ-ray activity over soils and rocks containing less than 10 parts per million of potassium-40, thorium, and uranium combined.

The Aerial Radiological Measuring Surveys (ARMS) program conducted by Edgerton, Germeshausen and Grier, Inc. (EG&G) is sponsored by Civil Effects Test Operations, Division of Biology and Medicine, U.S. Atomic Energy Commission. The Division of Biology and Medicine has the responsibility of obtaining information concerning the radiation levels existing on the ground near those AEC installations whose activities present a potential radiological hazard to neighboring districts. Since it is possible that radioactive aerosols escaping from such installations can be transported large distances before deposition occurs, the area of concern surrounding each site is on the order of 10,000 square miles.

JOHN E. HAND is senior scientist at Edgerton, Germeshausen and Grier, Inc., Goleta, California.

In order to accomplish a systematic coverage of such a large area, established procedures are followed during presurvey planning, survey operations, and postsurvey data handling. The AEC site is generally taken as the center of the area to be surveyed. The boundaries of the area form a square 100 miles on a side; the orientation of the boundaries with respect to the compass points is dictated by the prevailing topographic features and the proximity of population centers.

Proposed survey flight lines are drawn on maps of the area prior to the departure of the field crew from the laboratory. The flight lines are drawn to scale approximately 1 mile apart and run parallel to two of the area boundaries. During aerial traverses along these lines, reference to the flight maps enables the navigator to remain cognizant of the position of the aircraft with respect to topographic and cultural features of the terrain.

In earlier instrumentation systems, the geographic position of the aircraft was determined using the method of visual navigation described above, in conjunction with aerial photographs of the ground that were taken as the aircraft progressed along a survey line. The method gave quite accurate results. Reduction and handling of the data were performed later at the home laboratory, where it was soon discovered that the effort required to correlate radiation-intensity values with map locations of where the radiation measurements were made was extremely laborious. As a consequence, a graphic presentation of the areal patterns of terrestrial γ-ray intensities was not available to the interested agencies for quite long periods after the completion of survey flight activities.

The procedure followed by our survey group to determine the aircraft position during flight utilizes visual navigation in conjunction with distance values supplied by an air-borne radar system. During flight, radiation-intensity values and ground-position values are recorded simultaneously.

INSTRUMENTATION

The instrumentation was constructed with solid-state circuit components wherever it was possible in order that the weight, volume, and power requirements of the system could be held to minimum values. In addition, pulsed-current circuit configurations are used in many of the modules to stabilize the system response against electrical transients and temperature variations.

Figure 1 shows a block diagram of the instrumentation in which the system is classified into three functional subsystems: (1) radiation detection and measurement, (2) aircraft position, and (3) data recording. System operation can be understood by an examination of each of the subsystem functions.

ARMS-II SUBSYSTEMS

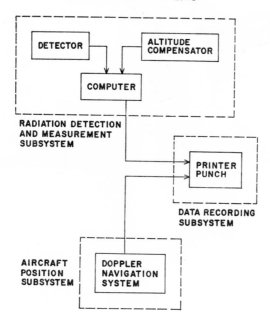

Fig. 1.—ARMS-II subsystems

RADIATION DETECTION AND MEASUREMENT SUBSYSTEM

γ-Ray Detectors. Dual γ-ray transducers provide the system with two detection-sensitivity ranges. The high-sensitivity-range detector is a sodium iodide (thallium activated) crystal 9 inches in diameter and 3 inches thick. It is optically coupled to a 12-inch diameter, flat-faced photomultiplier tube (EMI 9545B). Plate I shows the PM tube–crystal assembly in a laboratory rack.

The second detector is a Harshaw integral assembly and consists of a $\frac{3}{4} \times \frac{3}{4}$-inch right cylindrical crystal coupled to a Dumont 6262 photomultiplier tube. Both the high- and low-range detectors are shielded against extraneous magnetic influences. The two detectors are approximately a factor of 200 different in total detection sensitivities, so the range of detection is from 50 c.p.s. to 10^7 c.p.s. when given in terms of the equivalent large crystal count rate.

Each detector has a power supply and preamplifier, so overloading of either detector has no effect on the response of the other. Plate II

PLATE I

Scintillation crystal assembly

PLATE II

Radiation survey instrument panel

shows the assembled detector package as used in the aircraft, along with other components of the subsystem, as initial tests on system behavior were being performed in the laboratory.

Amplifier. The amplifier unit consists of 4-stage voltage amplification with an emitter follower output. Pulse-shaping and base-line discrimination circuitry are also housed in the amplifier box.

Arithmetic Computer. The arithmetic computer is a small special-purpose computer that performs both measuring and corrective functions on the gross radiation rates. Plate II shows the computer as it is operated from the system control panel. The functions performed on the data by the computer are the following:

1. It measures the incoming count rate and converts it to a decimal format display.

2. It performs altitude compensation on the data according to a signal from a radar altimeter.

3. It generates 1-second-sampling-period signals.

4. It provides cosmic and extraneous background corrections to the gross count rate.

5. It categorizes the net radiation intensity into channels of a pre-determined count-rate width.*

6. It converts the radiation channel number data to binary format and supplies these signals as the input for a decimal printer.

7. It generates command-to-print signals to the printer. A brief description of the computer operation is as follows: Since the unit is basically a digital computer, the incoming pulses from the amplifier are accumulated for a preset period of time. At the end of this period, a switching pulse is generated which transfers the contents of the input counting circuits to two encoder circuits connected in parallel. The input counting circuits are reset to zero just prior to the end of the switching period. At the close of the switching period a sampling function is again generated and incoming pulses are accumulated once more. The computer accumulates pulses as the previous-period pulse

---

* A typical example is as follows:

Channel No.	Count Range (hundreds of counts/sec)	Channel No.	Count Range (hundreds of counts/sec)
1	0–1	11	12–14
2	1–2	12	14–16
3	2–3	13	16–18
4	3–4	14	18–25
5	4–5	15	25–40
6	5–6	16	40–80
7	6–7	17	80–150
8	7–8	18	150–300
9	8–10	19	300–500
10	10–12	20	500–1000

accumulations are being converted to the required output functions. Thus, there is a one-sample-period delay before an accumulated count is displayed and before it is available for printout. The standard sampling time has been selected as 1 second at a survey altitude of 500 feet above terrain.

AIRCRAFT POSITION SUBSYSTEM

It is necessary that the values recorded for the radiation intensity be associated with a corresponding ground location. To accomplish this correspondence, an electronic navigation system is employed.

The principle of operation of the system is based on the Doppler effect as exhibited by electromagnetic radiation. High-frequency radar signals are directed at the ground in a dual-beam geometry. The signals strike the ground fore and aft of the aircraft and are reflected back to a receiver. The main frequency of the reflected radiation is shifted from the transmitted frequency by an amount that is proportional to the velocity of the aircraft. The navigation system measures the frequency displacements and converts the values into transverse and radial components with respect to a compass reference course set into the unit by the navigator.

The components of the Doppler navigation system consist of a transmitter-receiver unit, a frequency tracking unit, a computer, and the control units. An analog signal is taken from the computer position indicator circuitry and converted to an equivalent binary format through an analog-to-digital converter unit. Binary coded information is then routed to a decimal printer, which, upon command, records the along- and across-track components of the aircraft position on paper tape.

DATA-RECORDING SUBSYSTEM

The information generated by the radiation measurement and the aircraft-positioning subsystems is fed simultaneously to the data-recording subsystem. The nature of the circuitry is such that the prevailing values of the radiation level and aircraft position are always impressed on the recording subsystem input channels. Consequently, it is only necessary to generate a command-to-print signal in order to record the survey data. The data are recorded on two media: (1) decimal printout on paper tape and (2) binary coded information on punched tape. The decimal tape provides a rapid visual evaluation of the data during postflight data editing; the binary tape serves as the input data source to an automatic data-processing system. The two units are operated serially, the decimal tape being printed first. Figure

2 shows the data format of each device. The punched-tape entry corresponds to the decimal information of two prints.

The printer-punch combination was manufactured by Clary Corporation. The printer can be operated without the perforator if desired, but during normal survey usage both types of data entry are employed. Data are recorded every three seconds along a traverse line, which provides survey information approximately every 0.1 nautical mile.

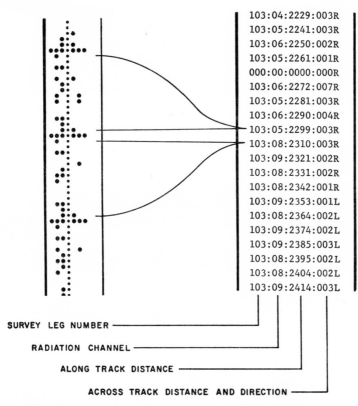

```
103:04:2229:003R
103:05:2241:003R
103:06:2250:002R
103:05:2261:001R
000:00:0000:000R
103:06:2272:007R
103:05:2281:003R
103:06:2290:004R
103:05:2299:003R
103:08:2310:003R
103:09:2321:002R
103:08:2331:002R
103:08:2342:001R
103:09:2353:001L
103:08:2364:002L
103:09:2374:002L
103:09:2385:003L
103:08:2395:002L
103:08:2404:002L
103:09:2414:003L
```

SURVEY LEG NUMBER ─────────

RADIATION CHANNEL ─────────

ALONG TRACK DISTANCE ─────────

ACROSS TRACK DISTANCE AND DIRECTION ─────────

SENSITIVITY PRINTOUT COLOR: HIGH (BLACK) LOW (RED)

Fig. 2.—Information printout

SYSTEM INSTALLATION

Plate III is a photograph of the aircraft in which the survey instrumentation has been installed. The plane is a Beechcraft Model E-50 Twin Bonanza. Gross weight is 7,000 pounds. With the normal survey gear and crew members aboard, sufficient fuel is carried for a 5-hour survey mission with a 45-minute reserve. The aircraft speed

PLATE III

Survey plane

PLATE IV

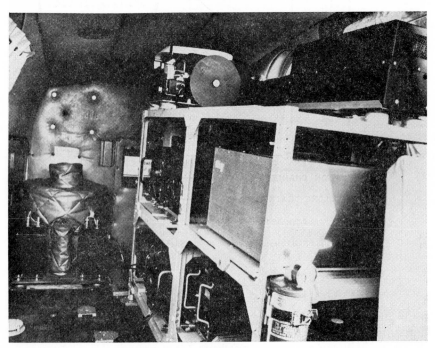

Detector and instruments in airplane

during survey activities is about 150 m.p.h. at an altitude of 500 feet above terrain.

At the time that the system was designed and assembled, this aircraft was selected as the survey vehicle for two principal reasons: (1) In the light, civilian class of aircraft the Twin Bonanza possessed the highest inherent structural strength. (2) The arrangement of cabin and baggage-compartment space in the basic design of the aircraft required the least amount of structural modifications for installation of the survey equipment.

PLATE V

Pilot's instrument panel

The aircraft has been equipped with heavy-duty 100 ampere, 24-volt d.c. generators to provide sufficient power for all the normal aircraft requirements in addition to the survey apparatus requirements. As a protective emergency measure, an auxiliary rocket engine has been installed in each engine nacelle.

Plate IV shows the equipment installation in the aircraft aft of the pilot's seat. Plate V shows a view of the pilot's instrument panel. The Doppler navigation system indicators and controls are seen to the immediate right of the engine-controls quadrant.

CALIBRATIONS PERFORMED ON THE SYSTEM

Each subsystem in the installation consists of components with quite complex circuitry. To insure that proper operation of the system is obtained during survey activities, calibrations are performed periodically during flight. These consist mainly of tests with the radiation subsystem and map checks of the Doppler apparatus. Prior to becoming an operational system, laboratory and field calibrations were performed on each component and on each subsystem as a whole. A description of the more important procedures is given.

CALIBRATION AND STANDARDIZATION OF THE RADIATION-MEASUREMENT SUBSYSTEM

The transistorized components of the radiation-measuring subsystems were standardized against laboratory counting equipment of proved merit. The base-line discriminator circuit components were adjusted to provide a 1:1 correspondence of γ-ray energy cutoff with the discriminator dial reading. Thus, with the base-line discriminator

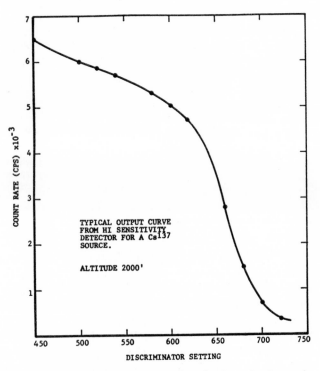

TYPICAL OUTPUT CURVE
FROM HI SENSITIVITY
DETECTOR FOR A Cs^{137}
SOURCE.

ALTITUDE 2000'

Fig. 3.—Count rate *vs.* discriminator setting. The inflection point due to the cesium-137 photopeak occurs at a discriminator setting of about 665.

control set at 662, the pulses arising from crystal γ-ray interactions of energy less than 662 kev. are rejected prior to entry into the arithmetic computer. The energy–dial-indication equivalence reduces the probability of confusion during the daily inflight calibration procedures by the navigator.

The crystal-detector assembly was examined for directional detection characteristics in the laboratory. During these tests various orientations of the assembly with respect to the magnetic field of the earth were used. If preferred directions of detection exist in the assembly, either from inadequate magnetic shielding or from inhomogeneities in the crystal or photomultiplier tube photocathode, they exhibit an effect of less than 1 per cent of the average counting rate.

The measuring accuracy of the arithmetic computer counting and encoding circuits was determined using a laboratory standard pulse generator. The generator also served as the input signal source during calibrations of the background subtraction circuits.

The periodic inflight radiation calibration procedure consists of exposing a small cesium-137 γ-ray source to the crystal and adjusting the amplifier gain to obtain a predetermined count rate with the baseline discriminator set at the cesium-137 γ-ray photopeak energy. Figure 3 shows a typical integral discriminator curve taken in flight. A count rate of approximately 2,500 c.p.s. occurs at the value of the source γ-ray energy. Hence, by duplicating this count rate with the cesium-137 source at a discriminator setting of 662, consistent survey data are obtained from day to day.

ALTIMETER AND SAMPLE GATE CALIBRATION

The radar altimeter was calibrated in flight at 500 feet over Yucca Flat at the Nevada Test Site.

The aircraft was photographed with an accurately positioned camera pointing vertically as the plane passed directly overhead. With knowledge of the focal length of the camera and the aircraft wing span, the altitude could be calculated. As a result of these runs, the altimeter was adjusted to give an uncertainty of ± 2 feet in the indicated altitude.

At the time the altimeter was calibrated, an experiment was performed to determine the extent of the error introduced by the survey pilot as he starts the Doppler navigator while flying over a visual ground check point. Two-way radio communication was set up between the pilot and the ground camera operator. Upon receipt of a signal from the pilot at the instant that he thought he was directly overhead, the camera operator photographed the aircraft. All aircraft position points but one fell within a 75-foot-radius circle of the camera station. Since visual aids off to the side were not used by the pilot in this experiment, as is done in survey work, it is thought that in practice a positioning error of less than ±50 feet is caused by pilot judgment.

The performance of the radiation detection and measurement subsystem altitude compensation circuitry was evaluated from data collected over a broad array of radioactive sources at the Nevada Test Site. Four-hundred small sources were placed on 100-foot centers to form a 2,000-foot square on a broad, fairly level area of the site.

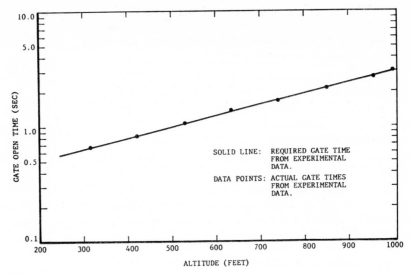

SOLID LINE: REQUIRED GATE TIME
               FROM EXPERIMENTAL
               DATA.

DATA POINTS: ACTUAL GATE TIMES
               FROM EXPERIMENTAL
               DATA.

FIG. 4.—Gate-open time *vs.* altitude

Cobalt-60 and cesium-137 γ-ray sources were used at different times. The altitude compensation circuitry is designed to provide counting-rate data collected at altitudes between 200 and 1,000 feet above terrain that is the equivalent to 500-foot altitude, 1-second-sampling-period count rate. This function is accomplished through a specially wound potentiometer on the altimeter output servo shaft. The signal appearing on the wiper arm of the potentiometer is routed into the arithmetic computer and is made to control the radiation sampling period. Hence, at altitudes between 500 and 1,000 feet above terrain, the sampling time is increased to provide an equivalent 500-foot level 1-second count. Similarly, at altitudes between 200 and 500 feet, the sampling period is shortened to provide the 500-foot equivalent count rate. Figure 4 shows the results of the measurements over the cesium-

137 array. Data were collected at 100-foot intervals from 200 to 1,000 feet in the compensated and uncompensated computer modes. Plates VI and VII show two views of the aircraft during recent recalibration procedures over the source arrays. Figure 5 shows a typical altitude versus radiation-intensity curve over an uncontaminated area. The effective height that terrestrial activity penetrates is readily seen.

Data from the array measurements were used to obtain an estimate

PLATE VI

Survey airplane over Nevada Test Site; calibration array

PLATE VII

Survey airplane over Nevada Test Site

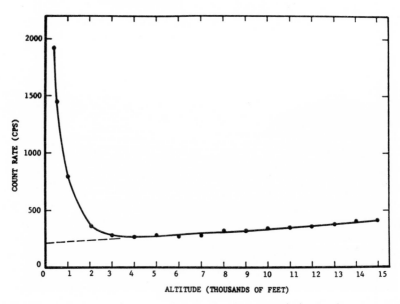

FIG. 5.—Cosmic count rate *vs.* altitude over land with no calibrating source in aircraft

of the lower limit of sensitivity of the large crystal detector. The assumption was made that at 500 feet above terrain the source array appeared to be nearly an infinite field. Also, considering the sources as being homogeneously distributed, the detector viewed a cobalt-60 field with an energy intensity of 4.8 $\mu c/m^2$; that of the cesium-137 appeared to be 16.15 $\mu c/m^2$. Average count rates at the center of the array were 4,400 c.p.s. for cobalt-60 and 5,600 c.p.s. for cesium-137. At 500 feet, then, the lower limit of ground concentration appears to be about 0.1 $\mu c/m^2$ for positive detection above terrestrial background.

DOPPLER RADAR CALIBRATION

The performance of the Doppler navigation system was evaluated by a series of flights over closed triangular courses. The lines were laid out in such a manner that a course was flown every 30° of the compass quadrant. The length of the flight lines ranged between 35 and 45 miles. The average values of the closing errors were ±0.33 per cent along track and ±0.35 per cent across track of the distances flown. Since the Doppler data are to be correlated with known map positions, it is necessary to apply corrective values to the indicated readings for instrumental error accrued during the traverse. From the along- and across-track values recorded over the traverse end point, the magnitude of the closing error is obtained. Corrections are then applied to the intermediate data points along the line in proportion to the along- and across-track distance values recorded at each point.

System Uncertainties

The uncertainties existing in the data collected by the ARMS installation are divided into two categories: (1) aircraft positioning errors and (2) errors in the recorded radiation values.

AIRCRAFT POSITIONING UNCERTAINTIES

Observational results indicate that a pilot positioning uncertainty exceeding ±50 feet at each end of a traverse leg is unlikely.

The maximum uncertainty of associating a radiation level with a geographic position is 1 radiation sampling period. At an altitude of 500 feet above terrain this corresponds to 1 second or about 200 feet of ground distance at an aircraft speed of 150 miles per hour. (If the survey altitude is higher, the error is increased according to the sampling gate time.)

The average error values for the along- and across-track distances are taken as those listed in the previous section, that is, 0.33 and 0.35 per cent, for the along- and across-track errors, respectively, of the distance flown. A tabulation of the uncertainties in the indicated position data for a 50-nautical-mile leg gives:

1. Pilot position error, ±100 ft.
2. Sampling period error, ±100 ft.
3. Instrumental error
   a. Along track, ±502 ft.
   b. Across track, ±532 ft.

The total uncertainty is then ±744 feet or ±0.12 nautical mile. If the uncertainty of point locations on USGS 4-mile-to-the-inch quadrangle maps is of the same order of magnitude as the error given above, then an uncertainty of about ±0.16 nautical mile will be encountered when associating a data point on the map with the actual ground location.

RADIATION-LEVEL UNCERTAINTIES

The uncertainties present in the recorded radiation values are considered to be due mainly to four causes: (1) meteorological effects; (2) radiation sampling time variations; (3) instrumentation response; and (4) radiation statistics.

The effect of meteorological changes during survey activities is to vary the density of the air column between the aircraft and the terrain. Our major area of concern is the uncertainty introduced into the data from relative meteorological changes, since it is not the purpose of the survey to obtain absolute ground concentrations of the γ-ray emitters. The normally occurring variations in temperature, pressure, and humidity during survey activities are fairly small, but if one considers a temperature variation of ±10° F., a pressure change of ±8-mm. mercury, and a 50 per cent change in relative humidity as the limits to be encountered over the survey area, then an uncertainty of about ±3 per cent is to be expected in the recorded radiation data.

In flight, some gustiness and turbulence are always encountered, so during any particular radiation sampling period, the altimeter varies the gating time. If it is assumed that a maximum rate af altitude change occurring during any sampling period is ±1,000 feet per minute, or about 17 feet per second, then (coupled with the uncertainty of the photographic calibration of the altimeter) an uncertainty of ±3.4 per cent is introduced into the radiation data.

Periodic calibrations of instrument response are performed during the survey flights using the cesium-137 source aboard the aircraft. If the calibrations are performed once an hour, the drift rate in instrumentation response amounts to an error of 5.6 per cent in the recorded count rate. In practice, calibrations are performed approximately every 30 minutes. For a maximum uncertainty determination the value of ±2.8 per cent is used.

The statistical error inherent in the detected radiation is due chiefly to three sources: (1) earth radiation, (2) cosmic plus air-borne radiation, and (3) cesium-137 source leakage through the shield.

The average background rate determined from many readings at 3,000 feet above terrain is about 1,200 counts per second, which consists of cosmic rays, calibrating source, and extraneous radiations sources. Although the background subtraction circuit removes the average value of this activity, the total statistical fluctuation is present with the net earth radiation. Considering an average net count rate of 500 counts per second, the standard deviation would be ±8.3 per cent.

The uncertainties in the recorded radiation levels are summarized as follows:

1. Meteorological effects, ±3 per cent
2. Altimeter sampling time, ±3.4 per cent
3. Instrument variations, ±2.8 per cent
4. Radiation statistics, ±8.3 per cent

The total uncertainty associated with the radiation values is, then,

$$\sigma_{tot} = \pm(3.0^2 + 3.4^2 + 2.8^2 + 8.3^2)^{1/2}\%$$

and

$$\sigma_{tot} = \pm 9.8\%.$$

The magnitude of the uncertainties listed applies to data taken during a single survey day at one site. To match data between days of widely different meteorological conditions, the proper corrections must be applied.

SUMMARY

The Aerial Radiological Measuring Surveys instrumentation constructed by EG&G has been installed in a light Twin Bonanza aircraft. Successful operation of the system over the last two years has provided net counting-rate data over large areas of the United States. The system allows altitude compensation of the data to be made and undesirable contributions to the gross counting rate to be subtracted in flight so that the recorded data are background-corrected and normalized to the value corresponding to 500 feet above terrain, 1-second count rate. The equivalent detection range of the system is from 50 c.p.s. to 10^7 c.p.s. at 500 feet above terrain. Flights over source arrays of known strength indicate that the lower limit of detection is about 0.1 $\mu c/m^2$ of nuclide concentration on the ground.

Uncertainties in the aircraft position data are on the order of $\pm\frac{1}{2}$ per cent of the distance flown, and uncertainties in the radiation data are nominally taken as ±10 per cent.

REFERENCES

DAVIS, F. J., and P. W. REINHARDT. 1957. Instrumentation in aircraft for radiation measurement. Nuc. Sci. & Engineering, **2**:713–27.
—— (eds.). 1962. Extended and Point-Source Radiometric Program. U.S. Atomic Energy Commission, Civil Effects Tests Oper., Rpt. CEX-60.3. Pp. 64.
HAND, J. E., R. B. GUILLOU, and H. M. BORELLA. 1962. Aerial Radiological Monitoring System. II. Performance, Calibrations and Operational Check-out of the EG&G ARMS-II Revised System. U.S. Atomic Energy Commission, Rpt. CEX-59.4, Part II. Pp. 66.
MERIAN, R. F., J. G. LACKEY, and J. E. HAND. 1960. Aerial Radiological Monitoring System. I. Theoretical Analysis, Design, and Operation of a Revised System. CEX-59.4, Edgerton, Germeshausen and Grier, Inc., and Div. of Biology and Medicine, AEC. Pp. 54.

ROBERT B. GUILLOU

42. The Aerial Radiological Measuring Surveys (ARMS) Program

THE CIVIL EFFECTS TEST ORGANIZATION (CETO), Division of Biology and Medicine (DBM), U.S. Atomic Energy Commission, is sponsoring a program of Aerial Radiological Measuring Surveys (ARMS) to determine the γ radiation background in various parts of the United States (Fig. 1). In addition to satisfying the radiological requirements of the AEC, the data have a potential use in public health, geologic, and other environmental investigations. The purpose of this report is to describe the program in sufficient detail to develop an understanding of what is measured and how, so that, as extensive ARMS data become available, workers in the natural radiation environment field can use them in their studies.

HISTORY OF ARMS

The nationwide ARMS program was started in July, 1958, by CETO in co-operation with the U.S. Geological Survey (USGS). The purpose of the program is to obtain γ background data that can be used as a basis for appraising changes in environmental levels of radiation brought about by nuclear testing programs, operation of reactors and other nuclear facilities, and radiation accidents. The requirement for data of this type was indicated by the experience of the British in studying the Windscale accident in 1957. An aerial survey using air-borne equipment and techniques developed to prospect for uranium deposits was able to outline rapidly and efficiently the areas of major iodine-131 contamination (Williams *et al.*, 1957). The concentration of iodine-131 on grass that produces the maximum permissible concentration of iodine-131 in milk, however, produces a γ count rate at 500 feet above the ground that is smaller than

ROBERT B. GUILLOU is senior scientist at Edgerton, Germeshausen and Grier, Inc., Santa Barbara, California.

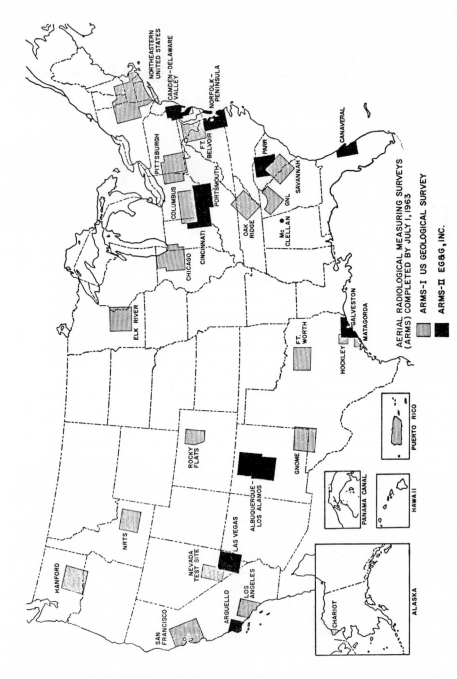

Fig. 1.—Civil Effects Test Operations ARMS Program

the changes in count rate that are due to variations in natural γ background in many areas. A concentration on the ground of 1 microcurie of iodine-131 per square meter will produce a count rate of about 360 counts per second on ARMS equipment at 500 feet (Davis and Reinhardt, 1962), while natural γ background commonly ranges from 100 to more than 1,000 counts per second. In order to delineate areas of maximum permissible concentration of iodine-131 contamination, therefore, it is necessary to have a preaccident knowledge of areal radioactivity levels.

The Geological Survey used two similarly equipped DC-3-type aircraft for ARMS work in 1959 and 1960. One aircraft was used on a part-time basis during 1961, 1962, and 1963. In 1959, because of the expanding requirement for ARMS data and the desire of the Geological Survey to limit its efforts to areas of geological interest, Edgerton, Germeshausen and Grier, Inc. (EG&G) was asked to develop ARMS instrumentation for CETO. This system, ARMS-II (the USGS system being called ARMS-I), became operational in May, 1961. To date, ARMS surveys totaling more than 225,000 square miles have been completed in 31 areas. See the Appendix to this article for a list of ARMS reports. In addition to these aerial surveys the ARMS-I and -II systems have been used on special missions, such as mapping fallout from nuclear tests at the Nevada Test Site.

DESCRIPTION OF ARMS-I INSTRUMENTATION

The ARMS-I γ radiation detection equipment was designed in 1951 for uranium exploration by the Health Physics Division of the Oak Ridge National Laboratory. It has been described in detail by Davis and Reinhardt (1957), who give the sensitivity of the equipment as "the count rate for a dose rate of one microroentgen per hour due to radium γ rays is 225 counts per second" (p. 717). A gyro-stabilized, continuous-strip-film camera records the flight path of the aircraft, and the distance of the aircraft from the ground is measured by a continuously recording radar altimeter. An electromechanical edge-mark system provides a common reference by making fiducial marks on the radiation and altimeter data tapes and on the camera film. The flight observer actuates this system when the aircraft passes over a recognizable feature on the ground.

The USGS scintillation detection element consists of 6 thallium-activated sodium iodide crystals 4 inches in diameter and 2 inches thick and 6 photomultiplier tubes connected in parallel. The signal from the detecting element is fed through amplification stages to a pulse-height discriminator, which is usually set to accept only pulses originating from γ radiation with energies greater than 50 kev. The signal is then fed to a circuit that records total count on a graphic milliammeter. The signal from the other rate meter is recorded by a circuit that includes a variable resistance that is controlled by the radar altimeter servomechanism, thereby compensating the data for deviations from the nominal 500-foot surveying altitude. A correction for cosmic background is also made in the compensated rate meter. Continuous profiles, or analog data, are recorded on chart paper by the USGS.

Description of ARMS-II Instrumentation

The EG&G ARMS-II instrumentation is installed in a Beechcraft Model 50 Twin Bonanza. The apparatus consists of 3 subsystems: (1) the radiation detection and measurement subsystem, (2) the aircraft space positioning subsystem, and (3) the information printout subsystem. The functions of these subsystems and their components are described in detail in Part 2 of the USAEC Report CEX-59.4 (Hand *et al.*, 1962).

The main detection element utilizes a 9-inch-diameter, 3-inch-thick thallium-activated sodium iodide crystal and a 12-inch photomultiplier tube. A $\frac{3}{4} \times \frac{3}{4}$-inch crystal is used in a low-sensitivity detection element. The radiation amplifier unit contains a voltage amplifier, a pulse-shaper, and an energy base-line discriminator. The discriminator is set to reject pulses due to γ rays with energies below 50 kev. for routine surveys and 662 kev. during calibration procedures using a cesium-137 source. The arithmetic computer performs a cosmic background correction, the compensation of the data for deviations from the nominal surveying altitude, the classification of the count rate into channels, and gives print command signals to the information printout subsystem. Compensation of the data for deviations from the nominal surveying altitude is accomplished through control of the sampling period by introducing a signal from the radar altimeter. The normal sampling period is 1 second. The sampling period is less than 1 second when the aircraft is below 500 feet and greater than 1 second when the aircraft is more than 500 feet above the ground. The count rate is normalized to 500 feet above the ground in the range 300–900 feet above the ground. The arithmetic computer classifies the count rate in digital channels of predetermined width. In the range of most natural materials, between 0 and 2,000 counts per second, the channel width is narrow. Above 2,000 counts per second, a progressively wider channel-width is used.

The position of the aircraft is determined by a modified General Precision Laboratory Doppler navigation system. The J-4 compass system establishes a reference line against which the actual path of the aircraft is compared, the heading information being held by either a driven gyro or a magnetically slave gyro. The Radan 500 Doppler radar unit determines the ground speed and drift angle of the aircraft relative to the J-4 reference line. Signals from the J-4 compass and the Radan 500 go to the TNC 50 (Track Navigation Computer), where the along-track and the across-track distances relative to an initial ground point are computed. The along-track and the across-track distance signals then go to the analog-to-digital converter. Upon receipt of a print command from the radiation computer these outputs go to the printer and are recorded.

The information printout subsystem consists of two data recorders, a decimal printer, and a binary tape punch. The data recorded are survey leg number, channel number, along-track distance, across-track distance and direction, and detector sensitivity.

Conduct of Surveys

The standard ARMS area is a square 100 miles on a side (10,000 square miles) centered on a nuclear facility. Mountainous terrain, location of a site near the ocean, overlapping of 100-mile squares, and special requirements at a particular site cause deviations from the standard-sized area. The size of ARMS areas ranges from 1,500 square miles for ARMS-I Matagorda to 28,500 square miles for ARMS-I Northeastern United States, which included seven facilities. The general technique in surveying an ARMS area is to fly equally spaced parallel lines across the area at an altitude of 500 feet above the ground. A spacing of 1 mile between flight lines is normal, but in some areas the spacing is 2 miles. Detail surveying on $\frac{1}{2}$–$\frac{1}{4}$-mile-spaced lines is done in the immediate vicinity of a nuclear facility if it is requested by the health physicists at the site. The flight lines are oriented generally across the trend of the geologic units in the area because this gives the best definition of the aeroradioactivity units, which usually parallel the geologic units.

Complete coverage of an area can be obtained over flat or moderately rolling terrain, where the pilot can keep the aircraft between 300 and 900 feet above the ground. In mountainous terrain, where it is not possible to fly parallel flight lines at the nominal surveying altitude, as much of the area as possible is surveyed on ridge and valley traverses.

ARMS Data

COMPILATION OF DATA

A brief description of the ARMS compilation and interpretation procedures is necessary for an understanding of ARMS data. Although the final product is similar, the USGS and EG&G use different compilation procedures because they have different equipment and objectives. It should be noted that both groups obtain more detailed data than can be presented on small-scale (1:250,000) maps.

COMPILATION OF ARMS-I DATA

The objective of the Geological Survey is to organize the ARMS-I compilation procedures so that compiled data that satisfy AEC (gross features) and Geological Survey (detailed features) requirements are obtained without duplication of effort. This is easily accomplished by making the initial compilation on maps having a

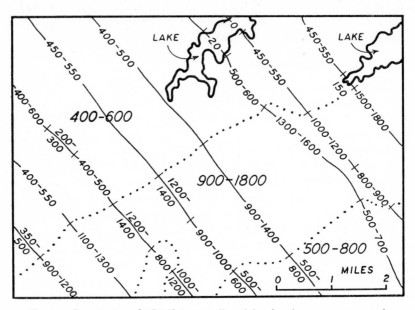

Fig. 2.—Compilation of ARMS-I aeroradioactivity data in counts per second

scale of 1 mile = 1 inch, which keeps the data at the scale necessary for geological studies. A brief outline of the USGS compilation procedures includes the following: (1) The traces of the flight lines are plotted on compilation base maps, scale usually about 1 mile = 1 inch, from the film record of the flight path. (2) Aeroradioactivity profiles are nested, and similar features or levels appearing on adjacent profiles are chosen. (3) The boundaries and count rates of similar features are plotted on the flight lines on the compilation base, and initial aeroradioactivity units are delineated on the map (Fig. 2). (4) The compilation base maps are reduced photographically to 1:250,000 scale (4 miles = 1 inch). (5) Final aeroradioactivity units are drawn on the publication base map (1:250,000 scale) from the photo reductions.

COMPILATION OF ARMS-II DATA

The objective of EG&G in compiling ARMS data is to get the data on a 1:250,000-scale map as rapidly, as accurately, and as economically as possible. Consistent with this philosophy, EG&G has developed automatic data-processing (ADP) techniques for handling ARMS-II data.

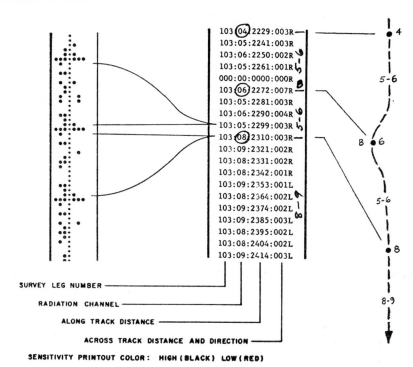

FIG. 3.—Manual compilation of ARMS-II data

During the first one and one-half years of operation, however, ARMS-II data were compiled manually. The following steps were used (Fig. 3): (1) The decimal data tape was edited by selecting the data points that divide the flight line into segments having similar aeroradioactivity and the points that indicate known locations or major changes in flight path. (2) The selected data points were plotted on tracing paper at map scale (1:250,000) using the recorded uncorrected Doppler distances. (3) Flight lines were corrected for instrumental error, and the true positions of the data points were plotted on the compilation map. The correction consisted of graphically proportioning the error between map locations and the segment end points. (4) The final step was the delineation of aeroradioactivity units or the selection of boundaries for those areas having similar count rates on adjacent flight lines.

The automatic data processing of ARMS-II data utilizes an IBM 704 computer to convert Doppler position data to geographic coordinates and then to X-Y plotter coordinates. Each data point, about 9 per mile, is plotted in its proper geographic position on a 1-mile = 1-inch map by a 30 × 30-inch EAI X-Y plotter. Aeroradioactivity units are delineated at this compilation scale, then photographically reduced to publication scale (1:250,000) and drafted on the final map. ADP has been used to process data from the Portsmouth, Cincinnati, Canaveral, and Parr surveys. In addition to the 1:250,000-scale maps that will be published in the CEX reports, the ARMS-II data for each of these areas are available on 1-mile = 1-inch plots and in a tabular listing showing radiation channel and geographic coordinates for each data point.

AERORADIOACTIVITY UNITS

Delineation of aeroradioactivity units is necessary to put the ARMS data in a form that can be readily understood because the natural background radioactivity is commonly complex, and the complexity is easily recorded by the ARMS instrumentation. The size of the units represents a compromise between the narrowest possible range in count rate and the largest possible area within one unit. Dissimilar count rates on adjacent lines, as well as fluctuations along the flight lines, contribute to the width of the range of count rate for a particular aeroradioactivity unit. The upper limit of one unit may be the lower limit of an adjacent unit, or the range of one unit may overlap the range of an adjacent unit. Since the data are prepared for presentation on a map of scale 1:250,000, or 4 miles = 1 inch, most of the units should be more than 2 miles wide (along the flight line) and encompass more than 4 survey lines. Aeroradioactivity units as narrow as $\frac{1}{2}$ mile are shown on the map if they differ substantially from adjacent units.

An example of complex aeroradioactivity data and units is shown in Figure 4 to illustrate the philosophy and problems connected with delineating aeroradioactivity units. The numerical values listed represent γ count rates in hundreds of counts per second. It can be seen that, in many places, the position of the unit boundary is unique and obvious, such as at *A*, *B*, and *C*. At other places, such as *D*, the placement of the boundary is arbitrary, and the dashed line could have been used. The selected boundary was chosen to indicate a unit with slightly lower radioactivity to the right of the line. A single unit (3–7) instead of three units (4–7, 3–6, 4–7) could have been used in this region. At *E*, the unit boundary was drawn through a fairly uniform segment (8–10), but sometimes this is necessary in order to simplify the shape of units while holding the range of units as narrow as possible. The 8–25 unit to the right of *E* is an example of a complex area that requires a wide range in count rate to avoid a multitude of small units.

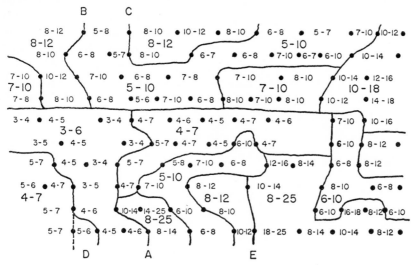

Fig. 4.—Aeroradioactivity data and units; count rate in hundreds of counts per second

THEORETICAL CONSIDERATIONS

SOURCES OF γ RAYS

The γ-ray activity at 500 feet above the ground that is measured by ARMS equipment has two principal sources: terrestrial sources, which are the radionuclides in the surficial materials of the earth, and non-terrestrial sources, which include cosmic radiation, the radionuclides in the air, and sources within the ARMS aircraft and instrumentation (fission-product contamination, calibration sources, etc.). Since it is not possible to measure directly the relative contribution from each source at a particular time while surveying, the calibration procedures for the ARMS instrumentation include measuring and subtracting from the gross γ count rate the count rate that is produced by non-terrestrial sources. This is accomplished by calibrating over water, or at 3,000 feet or more above the ground, where the activity from terrestrial sources is negligible. A fundamental assumption in this procedure is that the atmospheric sources, natural and artificial, have a generally uniform horizontal and vertical distribution. Except for a few days immediately after a nuclear test or during periods of severe inversion conditions, this assumption is probably valid. The correctness of the calibration can be checked where flight lines cross wide rivers or lakes.

TERRESTRIAL SOURCES OF γ RADIATION

A terrestrial component of γ radiation found at 500 feet above the ground comes from the radionuclides in the surficial 12 inches of earth materials. Radionuclides in soil and, to a lesser extent, in rock are the major sources of γ rays. The present distribution of the surficial material and the concentration of the natural radionuclides in it are determined by the original content and form of the radioactive material in the parent rock and by changes brought about by geologic and pedalogic processes.

TABLE 1

POTASSIUM-40, THORIUM, AND URANIUM IN IGNEOUS AND SEDIMENTARY ROCKS
(In p.p.m.)

	IGNEOUS ROCKS		SEDIMENTARY ROCKS		
	Basaltic	Granitic	Shales	Sandstones	Carbonates
Potassium-40*					
Average	0.8	3.0	2.7	1.1	0.3
Range	0.2– 2.0	2.0– 6.0	1.6– 4.2	0.7–3.8	0.0–2.0
Thorium					
Average	4.0	12.0	12.0	1.7	1.7
Range	0.5–10.0	1.0–25.0	8.0–18.0	0.7–2.0	0.1–7.0
Uranium					
Average	1.0	3.0	3.7	0.5	2.2
Range	0.2– 4.0	1.0– 7.0	1.5– 5.5	0.2–0.6	0.1–9.0

* Chemical potassium contains 0.0119 per cent potassium-40.

The principal natural γ-producing radionuclides found in soils and rock are potassium-40 and the members of the uranium and thorium series. The content of these radionuclides in various rocks is shown in Table 1. There is a general tendency for the content of natural radionuclides in igneous rocks to increase with increasing silica content. Among the sedimentary rocks, shales are generally more radioactive than sandstones and carbonate rocks. The radioactivity of metamorphic rocks, unless radionuclides were added or removed during metamorphism, reflects the potassium, uranium, and thorium content of the original sedimentary or igneous rock.

The concentrations of the natural radionuclides in soils are probably similar to the concentrations in sedimentary rocks because both are produced by the breakdown of pre-existing rocks. The fine clayey and silty soils are generally more radioactive than is coarser, sandy soil because much of the soil radioactivity results from radioelements that are fixed or adsorbed on clay. The interaction of various soil-forming processes, however, sometimes produces a concentration of radioactive accessory minerals and therefore an increase in total radio-activity in some sandy soils.

Artificial radionuclides are generally concentrated in the surficial inch or two of material and probably have had little effect on the distribution of aeroradioactivity units in ARMS areas. The contribution of fallout to the total radioactivity of soil must be small because the minimum aeroradioactivity levels in many parts of the United States are less than 300 counts per second. In some parts of the western United States the increased radioactivity of dry lake beds may be due in part to the concentration of fallout particles by running water.

CONVERSION OF COUNT RATE TO DOSE RATE

ARMS data are reported, as recorded, in counts per second normalized to 500 feet above the ground. For several reasons it is not possible to derive an accurate general conversion from count rate at 500 feet above the ground to dose rate at 3 feet above the ground. The

TABLE 2

FACTORS IN THE CONVERSION OF COUNT
RATE TO DOSE RATE

	DOSE RATE AT 3 FT.	COUNT RATE AT 500 FT.	
		ARMS-I	ARMS-II
Cobalt-60	1 $\mu r/hr$	18	22
Cesium-137	1 $\mu r/hr$	25	25

source of most natural γ radiation is a mixture of potassium-40 and the γ-producing members of the uranium and thorium series, but information is generally not available concerning the concentrations of these radioelements. The γ rays produced by these elements have different energies and are not attenuated equally in a 500-foot air column. Another important consideration is that the γ-ray intensity measurement at 500 feet above the ground represents the radiation coming from an area on the ground that is 800–1,000 feet in radius. A dose rate derived from this count rate is at best an integrated dose rate for this area and, in general, will not coincide with individual ground measurements. The common practice of not measuring, or removing, cosmic background when making ground measurements contributes to the conversion problem.

The EG&G ARMS-II instrumentation was designed to give data that were compatible with the data of the existing USGS ARMS-I equipment, and both units were flown over the Extended Source Calibration Area (ESCA) at the Nevada Test Site for cross-calibration purposes. The results of the flight measurements have been reported in CEX-60.3 (Davis and Reinhardt, 1962). The calibration range consisted of 400 equal-valued sources spaced on 100-foot centers to form a square that was 2,000 feet on a side. Cobalt-60 and cesium-137 sources were used, and the dose rate at 3 feet above the ground from both arrays of sources was about 0.2 mr/hr. A single conversion factor is recommended for use with ARMS data: 25 c.p.s. at 500 ft. = 1 μr/hr at 3 ft. This conversion factor should be used with caution in view of the uncertainties mentioned previously.

Uses of ARMS Data

The documenting of present levels of environmental γ radiation through the acquisition of ARMS data is an important contribution to the studies in at least three fields. The primary purpose of the ARMS program is to obtain data for the radiological requirements of the AEC, because knowledge of existing levels of environmental radiation is essential to an appraisal of any changes in level brought about by nuclear testing programs, operation of nuclear facilities, and radiation accidents. ARMS data have a geologic potential, as indicated by investigations of the correlation of aeroradioactivity data and areal geology by the USGS during the last eight years. Because they document one aspect of man's natural radiation environment, the data should play an important part in studies of the long-term biological effects of low-level radiation. It is to be expected that, as the data are more widely disseminated and become available for more areas in the United States, the value and uses of the data will increase.

GEOLOGIC USE OF ARMS DATA

The USGS is using ARMS-I data to develop a better understanding of the areal geology in many areas in the United States. Excellent correlations between aeroradioactivity data and areal geology may be found in CEX publications dealing with ARMS-I surveys and in other papers at this symposium.

Although a detailed correlation of aeroradioactivity data and geology is not included in EG&G's scope of work for CETO, we are responsible for a brief appraisal of the data. Interesting correlations have been seen in several areas. In the Las Vegas area distinctly dif-

ferent aeroradioactivity levels occur over several rock units (Fig. 5).
A small area of Precambrian metamorphic rock on the west side of
Frenchman Mountain produces a distinct aeroradioactivity unit of
400–800 c.p.s. Two small areas of volcanic rock separated by carbonate rock near Sloan cause the following sequence of units: 700–1,400,
300–500, and 800–1,400 c.p.s. Moderate count rates south of Henderson, 1,000–1,800, are associated with Cretaceous and Tertiary intru-

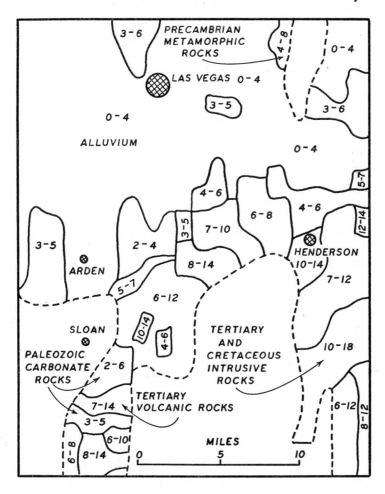

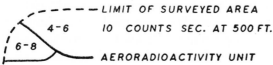

- - - - - LIMIT OF SURVEYED AREA

4-6   10 COUNTS SEC. AT 500 FT.

6-8   AERORADIOACTIVITY UNIT

FIG. 5.—Correlation of aeroradioactivity data and geology, Las Vegas area, Nevada

sive rocks. The western part of the Las Vegas valley, which is covered with alluvium derived from clastic and carbonate rocks, has a low (0–400 c.p.s.) aeroradioactivity.

It must be remembered when using ARMS data for geologic purposes that the radionuclides constitute a very small proportion of the surficial material and that only the distribution of the radionuclides is measured. Boundaries of rock types or geologic formations will not be reflected by changes in aeroradioactivity unless they differ from adjacent rock units in content of radionuclides. This means that a granite and a shale, a diabase and a limestone, or a Precambrian quartzite and an unconsolidated recent alluvium could produce single aeroradioactivity units, while the limestone and the silty limestone facies of a single formation could produce two distinct aeroradioactivity units. Since the concentrations of the radionuclides and their host minerals are generally controlled by the rock-forming and sedimentation processes, the trends of aeroradioactivity units usually parallel geologic structural trends. The delineation of aeroradioactivity units produces a framework upon which data from geologic field observations may be placed. In this way, ARMS data can furnish an excellent guide for field mapping and can determine lithologic continuity between outcrops. The technique is of most value in areas of low to moderate topographic relief, residual soil, and poor outcrop, such as the Piedmont. These are the areas where it is most difficult to determine the distribution and continuity of lithologic units by ordinary field methods.

ENVIRONMENTAL USE OF ARMS DATA

It is generally accepted that any radiation is genetically undesirable, since any amount of radiation can induce harmful mutations. Although voluminous data exist on the detrimental effects to man and other living organisms of moderate- to high-level ionizing radiation, there is a paucity of information on the long-term effects of low-level radiation. This lack of information is not surprising when one considers that the two terms, "long-term" and "low-level," are applied to a recently recognized problem.

Estimates of the effect of background radiation on the incidence of congenital malformation have ranged from negligible to dominant. Most estimates have been based on experimental work with fruit flies and mice, but some recent studies have used public health records. The major problem is determining the levels of environmental radiation. Wesley (1960) assumed that background radiation was proportional to the world-wide distribution of cosmic-ray energy flux and concluded that most fatal congenital malformations are caused by background radiation. His assumption concerning the cause of background radiation is not substantiated by studies of cosmic rays or the natural radioactivity of the environment.

Gross geologic information was used by Gentry *et al.* (1959) and Kratchman and Grahn (1959) to estimate low-level environmental radiation in their studies of congenital malformation in New York State and the entire United States, respectively. Both studies indicated a relationship between rates of congenital malformation and levels of environmental radiation. The Gentry study (1959) reported an average malformation rate of 15.1 per 1,000 live births in townships where the presence of extensive quantities of materials with relatively high concentration of the radioactive elements was "probable" and a rate of 12.8 per 1,000 live births in "unlikely" townships. The data from the preliminary study of Kratchman and Grahn (1959) suggest that mortality incidence from congenital malformation may be higher in those geologic provinces of the United States, particularly the Colorado Plateau, that contain major uranium-ore deposits, uraniferous waters, or helium concentrations. They inferred that these provinces have a higher than average level of environmental radiation.

ARMS data can make a valuable contribution to studies requiring a knowledge of environmental radiation because they show the detailed distribution of areas with different levels of γ radiation. It must be remembered that only the γ component of the background radiation is recorded by the ARMS equipment at 500 feet above the ground. A complete study of environmental radiation would include external and internal emitters and all components of the radiation field.

Future ARMS Studies

During the past year the emphasis in the ARMS program has been shifting from routine areal surveys to research-type activities. The routine surveys will be continued, but at a reduced rate. A significant accomplishment this year has been the development of ARMS-III, an aerial capability for obtaining γ spectral data. The major elements of this system are a Technical Measurements Corporation (TMC) Model 401, 400-channel pulse-height analyzer and a Harshaw matched-window detector assembly consisting of a 6-inch-diameter, 4-inch-thick sodium iodide crystal, and three 3-inch photomultiplier tubes. Excellent aerial spectra have been obtained with this new system. Future research with ARMS-III is directed toward developing the potential of a system that incorporates the mobility and speed of an aircraft and the extreme utility of spectral data. The long-range objectives of these investigations are the determination of ground concentrations of radionuclides and ground dose rates from aerial spectra.

A thorough study of the conversion of count-rate data at 500 feet above the ground to dose rate at 3 feet above the ground is another objective of the ARMS program for next year. A more effective conversion factor, or factors, will greatly enhance the utility of ARMS count-rate data. The most direct approach to the conversion-factor problem is to compare aerial count-rate data and ground dose-rate data. Experiments to obtain these data are planned for areas with a wide range in background dose rate and in concentrations of uranium, thorium, and potassium. In addition to this direct comparison, spectral data obtained with ARMS-III in the air and on the ground at these locations will be studied.

The ARMS program, which was started by CETO in 1958 to satisfy the radiological requirements of the AEC, has resulted in measuring the background γ radioactivity in more than 225,000 square miles. The many excellent correlations between aeroradioactivity data and areal geology indicate that small variations in the concentrations of uranium, thorium, and potassium-40 can be detected by ARMS instrumentation. Although ARMS data are measured at 500 feet above the ground, they constitute a wealth of data concerning terrestrial γ radiation. The purpose of this paper has been to describe this relatively untapped source of data so that students of the natural radiation environment will become aware of, and help exploit, the potential value of ARMS data.

REFERENCES

DAVIS, F. J., and P. W. REINHARDT. 1957. Instrumentation in aircraft for radiation measurement. Nuc. Sci. & Engineering, **2**:713–27.
—— (eds.). 1962. Extended-and-Point-Source Radiometric Program. U.S. Atomic Energy Commission Rpt. CEX-60.3. Pp. 64.
GENTRY, J. T., E. PARKHURST, and G. V. BULIN, JR. 1959. An epidemiological study of congenital malformations in New York State. Am. J. Pub. Health, **49**:497–513.
HAND, J. E., R. B. GUILLOU, and H. M. BORELLA. 1962. Aerial Radiological Monitoring System. II. Performance, Calibration, and Operational Checkout of the EG&G ARMS-II Revised System. U.S. Atomic Energy Commission Rpt. CEX-59.4, Part II. Pp. 66.
KRATCHMAN J., and D. GRAHN. 1959. Relationships between the Geologic Environment and Mortality from Congenital Malformation. U.S. Atomic Energy Commission, Division of Raw Materials and Division of Biology and Medicine, TID-8204. Pp. 23.

WESLEY, J. P. 1960. Background radiation as the cause of fatal congenital malformation. Internat. J. Radiation Biol., **2**:97–112.

WILLIAMS, D., R. S. CAMBRAY, and S. C. MASKELL. 1957. An airborne radiometric survey of the Windscale area, October 19–22, 1957. United Kingdom Atomic Energy Authority, Research Group. England: Atomic Research Establishment. Pp. 26.

APPENDIX

CIVIL EFFECTS TEST OPERATIONS
ARMS PROGRAM REPORTS

ARMS-I	CEX No.*
Chariot	No report
Chicago	59.4.13
Columbus	59.4.3
Elk River	61.7.1
Fort Belvoir	59.4.17
Fort Worth	59.4.9
Georgia Nuclear Lab	58.4.8
Gnome	59.4.24
Hanford	*59.4.11*
Hockley	No report
Los Angeles	*59.4.16*
Matagorda	No report
Nevada Test Site	No report
Northeastern United States—South	58.4.6
Northeastern United States—North	59.4.14
Nuclear Reactor Testing Station	59.4.10
Oak Ridge	*59.4.15*
Pittsburgh	59.4.12
Puerto Rico	61.7.2
Rocky Flats	....
San Francisco	58.4.5
Savannah	*58.4.2*

ARMS-II	
Albuquerque–Los Alamos	61.6.2
Arguello	62.6.3
Camden-Delaware Valley	61.6.3
Canaveral	63.6.1
Cincinnati	62.6.5
Galveston	*62.6.1*
Las Vegas	61.6.1
McClellan	No report
Norfolk-Peninsula	61.6.4
Parr	63.6.2
Portsmouth	62.6.6

CEX-59.4, Parts I, II, ARMS-II System

CEX-60.3, "Extended-and-Point-Source Radiometric Program" describes cross-calibration of ARMS-I, ARMS-II, and three other monitoring systems.

* Italics indicate report published, remainder in preparation.

JAMES A. PITKIN, SHERMAN K. NEUSCHEL,
AND ROBERT G. BATES

43. *Aeroradioactivity Surveys and*
Geologic Mapping

T HE U.S. GEOLOGIC SURVEY, on behalf of the Division of Biology
and Medicine, U.S. Atomic Energy Commission, has made background γ-radioactivity surveys in the vicinity of major nuclear installations in the United States during the period July, 1958–December,
1961. Figure 1 shows major areas surveyed as part of the Aerial Radiological Measurement Survey I (ARMS I) program.

The ARMS I surveys are part of a nationwide program (Guillou,
1963) to obtain data on the existing γ radioactivity for areas in and
adjacent to nuclear facilities. These data provide information that can
be used to detect any future variations in radioactivity that might
result from nuclear testing, reactor or other Atomic Energy Commission operations, or accidents involving radioactivity. The U.S. Geological Survey is using the ARMS I data in its study of the correlation of aeroradioactivity with areal geology.

AERORADIOACTIVITY SURVEYING

Surveys are made with continuously recording scintillation detection equipment (Davis and Reinhardt, 1957) installed in a twin-
engine aircraft. The aircraft follows a pattern of parallel, equally
spaced flight lines (usually spaced 1.6 kilometers apart), which, topography permitting, are oriented normal to the general geologic
trend of the area being surveyed. An approximate altitude of 150
meters above the ground is maintained, and the flight path is recorded
by a gyrostabilized continuous-stripfilm camera. A continuously recording radar altimeter measures distance of the aircraft from the
ground.

JAMES A. PITKIN, SHERMAN K. NEUSCHEL, and ROBERT G. BATES
are geologists, U.S. Geological Survey, Washington, D.C.
Publication authorized by the director, U.S. Geological Survey.

723

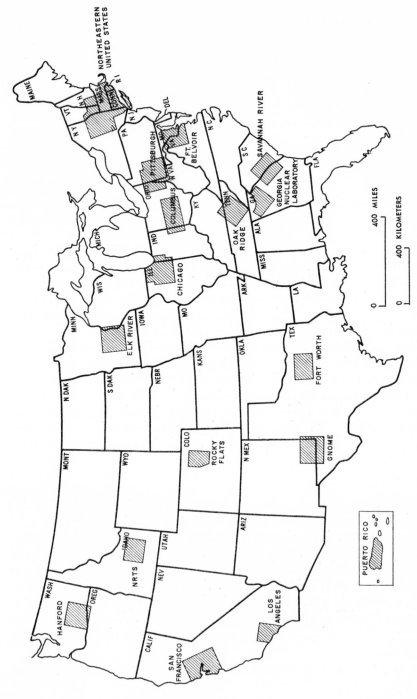

Fɪɢ. 1.—Aeroradioactivity surveys made by the U.S. Geological Survey, 1958–61, on behalf of the U.S. Atomic Energy Commission

Aeroradioactivity Detection Equipment

Figure 2 is a diagram of the U.S. Geological Survey aeroradioactivity equipment, which is calibrated in counts per second (c.p.s.). The sensitivity of the equipment can be described in several ways, one being "the count rate for a dose rate of one microroentgen per hour due to radium γ rays is 225 c.p.s." (Davis and Reinhardt, 1957, p. 717). Also, for simulated plane sources: "The count rates at 500 ft. (150 meters) equivalent to a ground reading of 1 $\mu r/hr$ for cesium-137 and cobalt-60 plane sources are 25 and 18 counts/sec, respectively" (Davis and Reinhardt, 1962, p. 239).

The detecting element consists of six sodium iodide (thallium-activated) crystals, each 10 centimeters in diameter and 5 centimeters thick, and six photomultiplier tubes connected in parallel. The signal from the detecting element is fed through amplification stages to a pulse-height discriminator set to accept only pulses originating from γ radioactivity with incident energies greater than 50 kev. The signal is then fed to two rate meters. One rate meter feeds a circuit that records total radioactivity on a graphic milliammeter. The signal from the other rate meter is recorded by a circuit that includes a variable resistance controlled by the radar altimeter servomechanism, which compensates the data for deviations from the 150-meter surveying altitude. Reasonable compensation is produced within a 30–270-meter range of altitude above the ground.

The effective area of response of the scintillation equipment at an altitude of 150 meters above the ground is approximately 300 meters in diameter, and the radioactivity recorded is an average of the radioactivity received from within the area (Sakakura, 1957). Periodic calibrations made with a cesium-137 source assure uniformity of equipment response.

The γ-ray flux at 150 meters above the ground has three principal sources: cosmic radiation; radionuclides in the air, which are mostly radon-222 daughter products; and radionuclides in the surficial layer of the ground (Gregory, 1960; Moxham, 1960, 1963). The cosmic component is determined twice daily by calibrations at 600 meters above the ground and is removed from the altitude-compensated rate-meter circuit. The component due to radionuclides in the air is difficult to evaluate. It is affected by meteorological conditions, and a tenfold change in radon-222 concentration is not unusual under conditions of extreme temperature inversion. The only valid measurement of the air component is obtained by flight over large bodies of water. Daily flying of a test line, part of a regular survey line, provides data that give an estimate of air-component variation. If inversion condi-

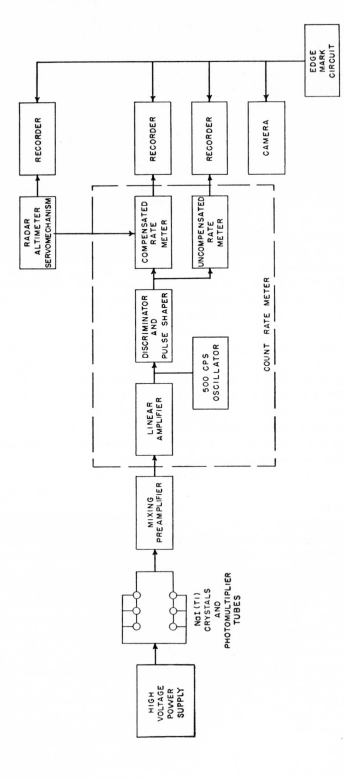

Fig. 2.—Diagram of U.S. Geological Survey aeroradioactivity equipment

tions are avoided, the air component may be considered to be fairly uniform on a given day in a particular area and will not affect interpretation of the aeroradioactivity data.

The ground component comes from the upper few centimeters of the ground. It consists of γ rays from natural radionuclides, principally members of the uranium and thorium radioactivity decay series and potassium-40, and from fallout of nuclear fission products. The majority of the ARMS I data was obtained during the period of the nuclear moratorium, from November, 1958, to September, 1961. Consequently, fallout has rarely complicated interpretation of the data.

CORRELATION OF AERORADIOACTIVITY WITH AREAL GEOLOGY

The usefulness of aeroradioactivity data as an aid to geologic mapping varies widely from area to area. Of all the ARMS I areas, the best results for geologic interpretation have been obtained in those including some part of the Piedmont physiographic province of the eastern United States. These areas include Fort Belvoir, Georgia Nuclear Laboratory, and the Savannah River Plant (Fig. 1). The province is an approximately linear belt of varied sedimentary, igneous, and metamorphic rock that extends from Georgia northeastward into eastern Pennsylvania and New Jersey. Bedrock is deeply weathered and covered by thick accumulations of residual soil, outcrops are sparse, and topographic variations are minor.

R. G. Schmidt (1961*a*), in his interpretation of the Savannah River area, reports: "Field examination and comparison with geologic maps indicate that the natural radiation level is closely related to the type of soil or rock at the surface of the ground. This relationship is locally so good that geologic contacts may be reasonably mapped from changes in radiation level." He also notes: "Several rock units in the Piedmont, irrespective of weathering, may be mapped by their characteristic radioactivity" (Schmidt, 1962*a*, p. 5). Interesting features in this area include erosional windows of Piedmont metamorphic and igneous rock inclosed by sedimentary rocks of the Coastal Plain. The windows are well defined by changes in level of radioactivity. Another feature is distinctive radioactivity zoning in Carolina Slate Belt rock, which field reconnaissance failed to explain, since the surface material was homogeneous in gross characteristics (Schmidt, 1962*a*, p. 32). Recent work in North Carolina shows that the Slate Belt there consists of a number of mappable units within a sedimentary basin. A careful field survey would reveal similar units in the Savannah River ARMS I area (J. S. Watkins, Jr., oral communication, 1963).

J. A. MacKallor (1963*a*, p. 34), discussing the Georgia Nuclear

Laboratory area, states: "The pattern of radioactivity reflects much of the geology, and many of the local changes in the regional trend of the rocks are marked by corresponding changes in the radioactivity." Several structural features are accurately defined by radioactivity. These include the Cartersville fault and the Brevard shear zone, the former being especially well marked by changes in radioactivity level, which range from 100 to 800 c.p.s. Field reconnaissance revealed that the radioactivity data would be a welcome guide to the geologist in this area of deep residual soils and infrequent outcrops (MacKallor, 1963*b*).

The Fort Belvoir ARMS I area (Fig. 1) has proved to be an excellent locale in which to correlate geology with aeroradioactivity. Figure 3 shows an area in the Piedmont of Maryland, approximately 25 kilometers northwest of Washington, D.C. The principal bedrock is shale, crystalline schist, and phyllite, which is intruded by numerous mafic (magnesium-iron) bodies. The diabase, gabbro, and serpentine intrusives are well delineated by low radioactivity levels of 200–400 c.p.s. These levels contrast sharply with the higher radioactivity of 400–500 c.p.s. of the shale and the 500–600 c.p.s. of the schist and phyllite. The schist and phyllite consist of varying assemblages of quartz, chlorite, muscovite, albite, and oligoclase. Also notable on Figure 3 is the well-defined low over the alluvium and water of the Potomac River.

Griscom and Peterson (1961) used aeroradioactivity data in mapping the Rockville quadrangle, Maryland, part of which is included in the eastern one-third of Figure 3. By using radioactivity lows and corroborative aeromagnetic data, they located several mafic intrusives that had not been previously recognized.

Figure 4 shows an area of the Piedmont approximately 20 kilometers west of Baltimore, Maryland. The principal rocks are crystalline schist, phyllite, and quartzite schist. These rocks exhibit a very uniform and moderate radioactivity of 400–500 c.p.s. in the northwestern half of Figure 4 and 500–600 c.p.s. in the southeastern half. Cutting these rocks are three northeast-trending lenticular bodies of gneiss and a small granite intrusive shown at the eastern edge of Figure 4. The gneiss is typically a banded granitoid gneiss composed of quartz, the potash feldspar microcline, the soda-lime feldspar oligoclase, and biotite. Radioactivity over the gneiss varies from 500 to 800 c.p.s. but is generally more than 600 c.p.s. The abrupt change in radioactivity is present nearly everywhere at the contact of the gneiss with the surrounding schist. In Figure 4 virtually all radioactivity higher than 600 c.p.s. is limited to gneiss and granite. Radioactivity over granite and granitic gneiss is usually moderate to high because of the presence of abundant potassium (and radioactive potassium-40) in one of the principal mineral constituents, the potash feldspar.

The serpentine dike of Figure 4 is part of a northeast-striking dike that extends a continuous distance of 29 kilometers in northeastern Maryland. Over most of its extent, the dike is only 0.2–0.4 kilometer in width, yet every flight profile over the dike made during the Fort

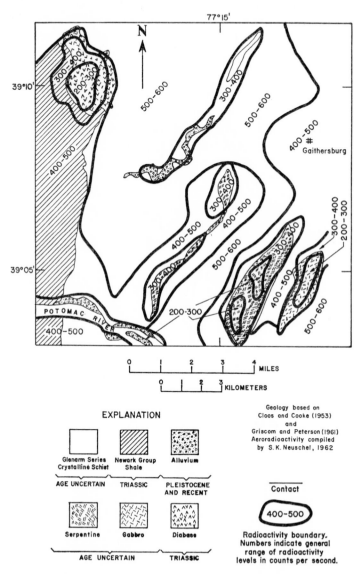

FIG. 3.—Geology and aeroradioactivity of an area in Montgomery County, Maryland

Belvoir ARMS I survey shows a distinct radioactivity low. The radioactivity of the dike is 200–300 c.p.s.; that of the surrounding schist is
400–500 c.p.s.

Meaningful radioactivity data may also be obtained in areas of sedimentary rocks, especially where the strata have an appreciable dip
from the horizontal. Such an area is shown in Figure 5, an area around
Athens, Tennessee (Bates, 1962*a*, pp. 37–39), about 80 kilometers
southwest of Knoxville in the Valley and Ridge physiographic province of the Appalachian Highlands. The bedrock consists of folded
and faulted sedimentary rocks of Paleozoic age. Delineation by radio-

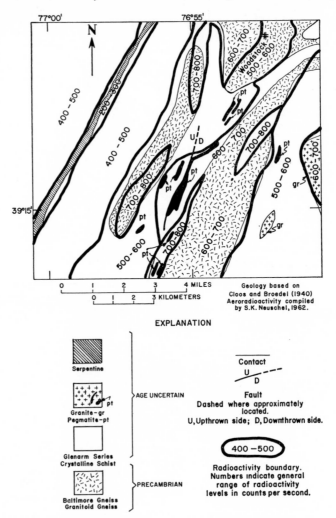

Fɪɢ. 4.—Geology and aeroradioactivity of an area in Howard County, Maryland

activity of these tilted strata is quite good; some units may be traced by radioactivity for 80–160 kilometers along strike (Bates, 1962*b*).

The Valley and Ridge province contains numerous subparallel faults that strike northeast. Many of these faults have thrust the radioactive argillaceous Rome formation or the Conasauga shale upon the less-radioactive Knox dolomite (Fig. 5). Thus the northwestern edge of the high radioactivity unit associated with the Rome or the Conasauga generally coincides with a fault trace (Bates, 1962*c*, p. 225).

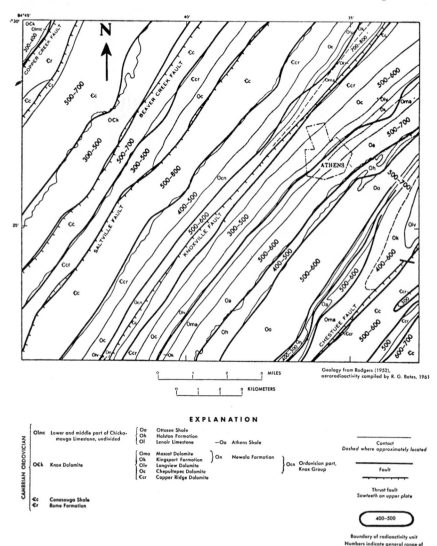

Fig. 5.—Geology and aeroradioactivity of an area near Athens, Tennessee

Some faults can be traced by radioactivity for 50 kilometers or more along their strike.

The importance of outcrop width of a formation to good definition by radioactivity data is well illustrated in Figure 5. The Holston formation is exposed on both limbs of a doubly plunging syncline. When the outcrop width of the formation is sufficiently wide, the low radioactivity units associated with the formation on both limbs of the syncline clearly outline the structure. The formation narrows to the northeast on the southwestern limb of the structure until it is not discernible by radioactivity because of its limited extent relative to the

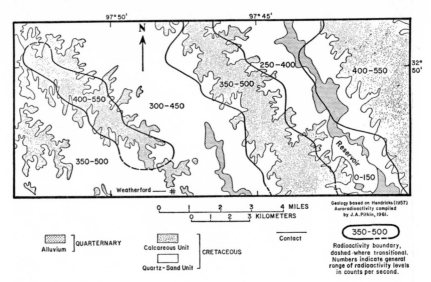

Fɪɢ. 6.—Geology and aeroradioactivity of an area in Parker County, Texas

area of response of the scintillation crystals. The Athens shale also narrows near this point and loses definition. The radioactivity units associated with these two formations merge into a broader unit encompassing six formations, and all definition of the southwestern limb of the syncline is lost.

Interpretation in sedimentary terrains becomes more difficult when the strata have an essentially horizontal attitude. Such a situation is shown in Figure 6, an area of sedimentary rocks of early Cretaceous age in Parker County, Texas, approximately 50 kilometers west of Fort Worth. The geologic units are almost horizontal in attitude, thin, and generally homogeneous in composition. Radioactivity correlation is difficult in such an area; however, a gross relationship is apparent between the calcareous unit and the quartz sand unit. The slightly higher radioactivity of the calcareous unit is due to its clay content (in marl or calcareous shale), whereas the less radioactive quartz sand unit contains little clay. The higher radioactivity is probably due to potassium-40 in the clay. These two units each contain several geologic formations; combination by the criterion of gross lithologic character was necessary for radioactivity interpretation.

East of the area of Figure 6, around Dallas, Texas, the geologic setting is much the same, except that the respective units are much broader in breadth of outcrop. Hence, radioactivity delineates the contacts of the Woodbine sand, Eagle Ford shale, Austin chalk, and Taylor marl—all of late Cretaceous age. The Eagle Ford shale is particularly well defined and exhibits consistent moderate levels of radioactivity.

Figure 7 shows an area in the Coastal Plain of Maryland, approximately 16 kilometers east of Washington, D.C. Bedrock consists of unconsolidated sediments, generally sands, silts, marls, and clays that dip gently eastward. A sharp radioactivity boundary marks the contact of the Patapsco and Monmouth formations, both of which are relatively unconsolidated sands and silts. The Monmouth contains an abundance of the mineral glauconite, a hydrous potassium iron silicate; the Patapsco is essentially non-glauconitic. The Monmouth (400–600 c.p.s.) is higher in radioactivity than the Patapsco (250–300 c.p.s.), probably because of potassium-40 in the glauconite. The higher radioactivity of the Monmouth and the abrupt change of radioactivity level at the Monmouth-Patapsco contact are distinct throughout Figure 7, except where masked by alluvial material of the Chesapeake Group. The Aquia formation, a greensand which overlies the Monmouth to the east, is also glauconitic. The Aquia is less glauconitic than the Monmouth, as shown by their respective radioactivity—350–500 c.p.s. for the Aquia, 400–600 c.p.s. for the Monmouth.

Aeroradioactivity data obtained over alluviated areas are of little value to geologic mapping because most alluvial material does not overlie parent rock. K. G. Books (1962*a, b*) could make only broad generalizations in his interpretation of the data obtained over the Los Angeles Coastal Plain, San Fernando Valley, and Oxnard Plain of southern California. The same situation is true in the Sacramento-Stockton part of the Great Valley of central California (K. G. Books, oral communication, 1963). In areas covered by eolian deposits, correlative geologic value is similarly lacking. Along the Columbia River area of southern Washington, the eolian cover is thick enough to mask the radioactivity of the underlying volcanic rock and prevent effective correlation (Schmidt, 1961*b*, 1962*b*).

Useful radioactivity data have been obtained over volcanic flows in the Snake River Plain of southern Idaho where eolian cover is sparse or lacking. Distinctive patterns of radioactivity measured over recent basalt flows require field investigation for proper explanation. Initial interpretation indicates that blocky (aa) lava is slightly higher in level of radioactivity than ropy (pahoehoe) lava.

In many glaciated areas, surface material has been transported some distance from its source area and consequently does not reflect the composition of the underlying rock. The heterogeneous composition of glacial deposits commonly gives a fairly uniform level of radioactivity, and boundaries of many glacial units are covered by post-

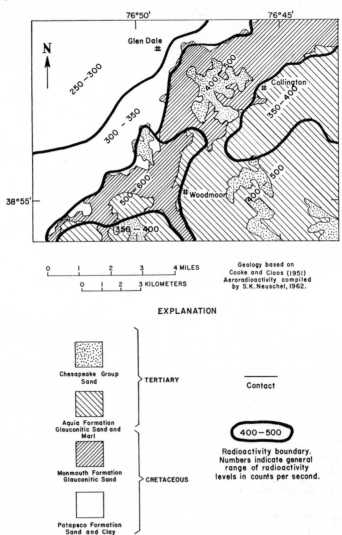

Fig. 7.—Geology and aeroradioactivity of an area in Prince Georges County, Maryland

glacial eolian deposits. Any direct correlation with bedrock geology is rare, and correlation of the uniform radioactivity levels with the glacial-eolian material at the surface is usually lacking (Flint, written communication, 1963).

An exception to the preceding conditions is found in southern New England, where most of the glacial material has not traveled far from its source area and reflects the composition of the underlying rock. Thus, many reasonable correlations of radioactivity with bedrock geology can be made. Rocks of the Oliverian Plutonic Series of New Hampshire, consisting chiefly of plagioclase gneiss, and the surrounding hornblende gneiss of the Bronson Hill anticline, may be traced across New Hampshire, Massachusetts, and Connecticut by their characteristic radioactivity levels despite the glacial cover. Distinctive radioactivity lows are also associated with several Pleistocene lake beds (Popenoe, 1963*a, b*).

REFERENCES

BATES, R. G. 1962*a*. Aeroradioactivity Survey and Areal Geology of the Oak Ridge National Laboratory Area, Tennessee and Kentucky (ARMS-I). CEX-59.4.15, Civil Effects Study, Civil Effects Test Operations, U.S. Atomic Energy Commission. Pp. 42.
———. 1962*b*. Natural gamma aeroradioactivity of the Oak Ridge National Laboratory area, Tennessee and Kentucky. U.S. Geol. Survey Geophys. Inv. Map GP-308.
———. 1962*c*. Airborne radioctivity surveys—a geologic exploration tool. Southeastern Geol., **3**:221–30.
BOOKS, K. G. 1962*a*. Aeroradioactivity Survey and Related Surface Geology of Parts of the Los Angeles Region, California (ARMS-I). CEX-59.4.16, Civil Effects Study, Civil Effects Test Operations, U.S. Atomic Energy Commission. Pp. 25.
———. 1962*b*. Natural gamma aeroradioactivity of parts of the Los Angeles region, California. U.S. Geol. Survey Geophys. Inv. Map GP-309.
———. Personal communication, 1963.
CLOOS, E., and C. H. BROEDEL. 1940. Geologic map of Howard and adjacent parts of Montgomery and Baltimore Counties (Maryland). Md. Geol. Survey. Scale 1:62,500.
CLOOS, E., and C. W. COOKE. 1953. Geologic map of Montgomery County (Maryland) and the District of Columbia. Md. Dept. Geology, Mines, and Water Res. Scale 1:62,500.
COOKE, C. W., and E. CLOOS. 1951. Geologic map of Prince Georges County (Maryland) and the District of Columbia. Md. Dept. Geology, Mines, and Water Res. Scale 1:62,500.
DAVIS, F. J., and P. W. REINHARDT. 1957. Instrumentation in aircraft for radiation measurements. Nuc. Sci. & Eng., **2**:713–27.
———. 1962. Radiation measurements over simulated plane sources. Health Physics, **8**:233–43.

FLINT, G. M., JR. Written communication, 1963.

GREGORY, A. F. 1960. Geological interpretation of aeroradiometric data. Canada Geol. Survey Bull. 66. Pp. 29.

GRISCOM, A., and D. L. PETERSON. 1961. Aeromagnetic, aeroradioactivity and gravity investigations of Piedmont rocks in the Rockville quadrangle, Maryland. U.S. Geol. Survey Prof. Paper 424-D, pp. 267–71.

GUILLOU, R. B. 1963. The aerial radiological measuring surveys (ARMS) program. This symposium.

HENDRICKS, L. 1957. Geology of Parker County, Texas. Univ. Texas Pub. 5724. Pp. 67.

MacKALLOR, J. A. 1963*a*. Aeroradioactivity Survey and Areal Geology of the Georgia Nuclear Laboratory Area, Northern Georgia (ARMS-I). CEX-58.4.8, Civil Effects Study, Civil Effects Test Operations, U.S. Atomic Energy Commission. Pp. 36.

———. 1963*b*. Natural gamma aeroradioactivity of the Georgia Nuclear Laboratory area, Georgia. U.S. Geol. Survey Geophys. Inv. Map GP-351.

MOXHAM, R. M. 1960. Airborne radioactivity surveys in geologic exploration. Geophysics, **25**:408–43.

———. 1963. Natural radioactivity in Washington County, Maryland. *Ibid.*, **28**:262–72.

POPENOE, P. 1963*a*. Aeroradioactivity of parts of east-central New York and west-central New England. U.S. Geol. Survey Geophys. Inv. Map GP-358.

———. 1963*b*. Aeroradioactivity Survey and Areal Geology of Parts of East-Central New York and West-Central New England (ARMS-I). CEX-59.4.14, Civil Effects Study, Civil Effects Test Operations, U.S. Atomic Energy Commission.

RODGERS, J. 1952. Geologic map of East Tennessee with explanatory text. Tenn. Dept. Conserv., Div. Geol. Bull. 58.

SAKAKURA, A. Y. 1957. Scattered gamma rays from thick uranium sources. U.S. Geol. Survey Bull. 1052-A., pp. 1–50.

SCHMIDT, R. G. 1961*a*. Natural gamma aeroradioactivity of the Savannah River Plant area, South Carolina and Georgia. U.S. Geol. Survey Geophys. Inv. Map GP-306.

———. 1961*b*. Aeroradioactivity of the Hanford Plant area, Washington and Oregon. U.S. Geol. Survey Geophys. Inv. Map GP-307.

———. 1962*a*. Aeroradioactivity Survey and Areal Geology of the Savannah River Plant Area, South Carolina and Georgia (ARMS-I). CEX-58.4.2, Civil Effects Study, Civil Effects Test Operations, U.S. Atomic Energy Commission. Pp. 41.

———. 1962*b*. Aeroradioactivity Survey and Areal Geology of the Hanford Plant Area, Washington and Oregon (ARMS-I). CEX-59.4.11, Civil Effects Study, Civil Effects Test Operations, U.S. Atomic Energy Commission. Pp. 25.

WATKINS, J. S., JR. Personal communication, 1963.

44. *Some Aerial Observations on the Terrestrial Component of Environmental γ Radiation*

G EOGRAPHIC VARIATIONS in the cosmic component of environmental γ radiation are well documented. Much less is known of the terrestrial component, though in many places it is the most important radiation source. Aeroradiometric surveys, though limited in resolution and accuracy in comparison to surface methods, can uniquely depict regional variations in radiation from which some estimates may be made of terrestrial radioelement content and of surface radiation intensities.

A very general relationship exists between radioelement content and lithologic character of common rock types, as shown in Table 1. Columns 1–3 give the average uranium, thorium, and potassium content. It is convenient, for evaluation of total γ emission from rocks, to reduce these chemical components to a total γ-ray equivalent quantity. Multiplication of the chemical quantities of thorium and potassium by their respective equilibrium quantity constants yields their γ-ray equivalents (columns 4 and 5, Table 1), which are summed in column 6. The constants used here represent a compromise among the most consistent published values (Moxham, 1963).

Table 1 illustrates, in a very general way, the substantial variation in radioelement content among common rock types. It would be exceedingly unwise, however, to infer from a geologic map the relative variations in the terrestrial component in a specific area, except in the most qualitative fashion. For nearly any given rock type, substantial masses exist whose total radioelement content departs widely from the values given in Table 1. Terrestrial emission, moreover, is a surficial

R. M. MOXHAM is chief, Branch of Theoretical Geophysics, U.S. Geological Survey, Washington, D.C.

Publication authorized by the director, U.S. Geological Survey.

TABLE 1

URANIUM, THORIUM, AND POTASSIUM CONTENT AND
γ-RAY EQUIVALENTS OF COMMON ROCK TYPES*

ROCK TYPE	CHEMICAL COMPOSITION			γ-RAY EQUIVALENTS		
	U (p.p.m.) (1)	Th (p.p.m.) (2)	K (per cent) (3)	$e\mathrm{U_{Th}}$ (p.p.m.) (4)	$e\mathrm{U_K}$ (p.p.m.) (5)	Total $e\mathrm{U}$ (p.p.m.) (6)
Granite...........	5	18	3.8	8	9.5	22.5
Shale.............	3.7	12	1.7	5	4.2	12.9
Limestone.........	1.3	1.1	0.2	0.5	0.7	2.5
Sandstone........	0.45	1.7	0.6	0.8	1.6	2.8
Basalt...........	0.5	2	0.5	0.9	1.2	2.6

* $e\mathrm{U_{Th}}$ = 0.45 chemically determined Th; $e\mathrm{U_K}$ = 2.5 chemically determined K. Limestone composition from Rankama and Sahama (1950); U and Th content of igneous rocks from David Gottfried, U.S. Geological Survey (oral communication); U and Th content of shale and sandstone from Adams *et al.* (1959); K content (except limestone) from Green (1959).

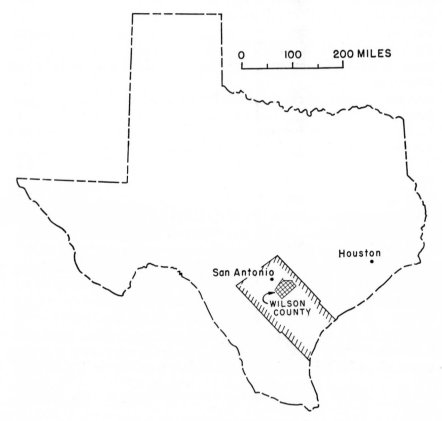

FIG. 1.—Index map of Texas showing location of aerial survey

phenomenon. Most geologic maps give no hint of the chemical and physical changes that bedrock undergoes in forming a residual soil mantle, though the radioelement concentration may be substantially altered in the process. Lateral soil transport may superpose foreign radioelements or absorbers on a bedrock unit, masking the intrinsic radioactive character of the underlying strata. Many soils maps, on the other hand (and from the geologist's prejudiced point of view), seem related to the bedrock inadequately to provide unequivocal data on the terrestrial sources. But, together, the geologic and the soils maps permit a more coherent analysis of the magnitude and distribution of radioelements at the earth's surface.

Some of the problems of radioelement distribution are illustrated in an aeroradiometric survey of the southeast Texas Coastal Plain made in 1956 (Fig. 1). The region is underlain by moderately to poorly consolidated clastic sediments of Tertiary and Quaternary age that dip gently southeast, toward the Gulf of Mexico. Flight lines were oriented southeast, normal to the regional structure. A 6-crystal detector was carried in a DC-3 aircraft at a nominal terrain clearance of 500 feet. No altitude compensation was used owing to the relatively flat topography. Surface measurements were made in 1957 with carborne and with hand-portable scintillation counters. Preliminary results of the aeroradiometry, and a description of the equipment used, have been given by Moxham and Eargle (1961) and Davis and Reinhardt (1957).

EFFECTS OF SOIL ON RADIATION

One of the most persistent radiometric patterns in the Coastal Plain is a low that coincides with the Carrizo sand that crops out along a strip 1–3 miles wide and about 70 miles long. The Carrizo sand is bounded on the northwest and southeast by more clayey strata of the Wilcox group and the Mount Selman formation, respectively. In most places, radiation is distinctively lower over the Carrizo sand than over the adjacent rocks, as expected. But in northwest Wilson County (Fig. 2), the low-radiation zone extends far beyond the Carrizo boundary on the southeast, overlapping the adjacent clayey members of the Mount Selman formation. The Carrizo weathers to very sandy soil designated *Norfolk sand* on a map by Lyman and Schroeder (1908). The *Norfolk sand* has apparently moved southeastward, down the topographic slope, covering the underlying clayey bedrock (Fig. 3). The radiation low closely follows the limits of the sandy soil, and emission from the underlying bedrock is effectively absorbed.

The opposite effect, that is, radiation enhancement through soil formation, has been described in Washington County, Maryland (Moxham, 1963), where radioelements, chiefly potassium, have been concentrated in residual soil through carbonate depletion of high-potassium limestones.

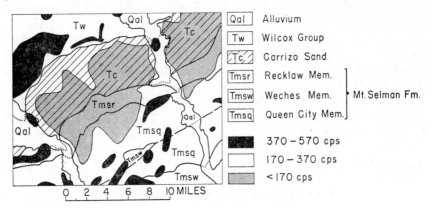

Fig. 2.—Geologic and aeroradiometric map, northern Wilson County, Texas

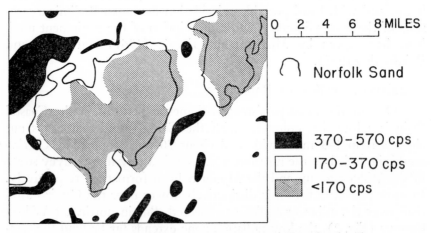

Fig. 3.—Soil and aeroradiometric map, northern Wilson County, Texas

RELATION OF RADIOELEMENT CONTENT AND RADIOMETRY

Soil samples collected in the Coastal Plain were analyzed chemically for uranium, thorium, and potassium (Table 2). The γ-ray equivalents were determined as described above. If it is assumed that detector altitude is constant, that the emitting source fully covers the field of view, is fairly homogeneous, and is in radioactive equilibrium, and that the spectral distribution is constant, then the radiation intensity at the 500-foot level will be directly proportional to the sur-face-radiation intensity and thus to the total eU content of the emit-ter. In Figure 4 the total calculated eU content of the samples has been plotted against the net (total minus cosmic) aeroradiation intensity at these places. The resulting curve, by least squares, is

$$S = 2.2 \times 10^{-2} I_a ,$$

where

$$S = \text{total calculated } eU(\text{p.p.m.}),$$

$$I_a = \text{net aeroradiation (c.p.s.)} .$$

TABLE 2

CHEMICAL ANALYSES AND RADIOMETRIC DATA, TEXAS COASTAL PLAIN

GEOLOGIC UNIT	SAMPLE No.	CHEMICAL ANALYSIS			eU (CALC.)			AERO-RADIO-ACTIV-ITY NET (C.P.S.)	NET SUR-FACE RADIO-ACTIV-ITY (μR/HR)
		U (p.p.m.)	Th (p.p.m.)	K (per cent)	Th* (p.p.m.)	K† (p.p.m.)	Total (p.p.m.)		
Wilcox gr. (Tw)....	62	5	n.d.‡	1.30	......	......	......	450	4.5
Wilcox gr. (Tw)....	63	2	6	0.80	3	2	7	300	3
Wilcox gr. (Tw)....	64	4	4	1.46	2	3.6	10	340	4.5
Jackson gr. (Tj)....	66	15	5	1.31	2	3.3	20	770	12.5
Goliad sand (Tg)..	68	10	n.d.	0.75	......	......	......	270	2.5
Goliad sand (Tg)..	69	4	3	0.60	1	1.5	6	270	1.5
Wilcox gr. (Tw)....	70	8	26	1.36	12	3.4	23	880	19.5
Carrizo sand (Tc)..	71	2	2	0.44	1	1.1	4	130	1
Cook Mountain fm. (Tcm)..........	72	6	10	0.89	5	2.2	13	570	8
Jackson gr. (Tj)....	73	4	8	1.74	4	4.3	12	500	8.5
Wilcox gr. (Tw)....	75	4	5	1.42	2	3.5	10	470	5.5
Mount Selman fm. (Tms)...........	76	7	16	0.60	7	1.5	16	830	10.5
Cook Mountain fm. (Tcm)..........	78	12	n.d.	0.84	......	......	......	410	4.3
Jackson gr. (Tj)....	79–80	5.5	10	1.80	5	4.5	15	750	10.5

* $eU_{Th} = 0.45$ chem. Th.   † $eU_K = 2.5$ chem. K.   ‡ Not determined.

The RMS error is 2 p.p.m. eU. A similar curve for data obtained in the Washington County, Maryland, survey (Moxham, 1963) has a coefficient 2.3×10^{-2}. The curve labeled "theory" is from Saka-kura's (1957, p. 10) extended source equation.* The difference be-

* Sakakura (1957, p. 10) gives an equation (7) which permits radioelement con-tent to be determined from aeroradiation. This equation contains a constant (3.19×10^7) based upon calibration over a point source, using the Geological Survey 3-crystal detector. Six crystals were used on the survey of Texas (Moxham and Eargle, 1961), and, lacking calibration at that time, we arbitrarily doubled the value of C to obtain the quantitative data given in that report. Later calibration of the 6-crystal detector gave a constant of 6.18×10^7, upon which the theoretical curve in Figure 4 is based.

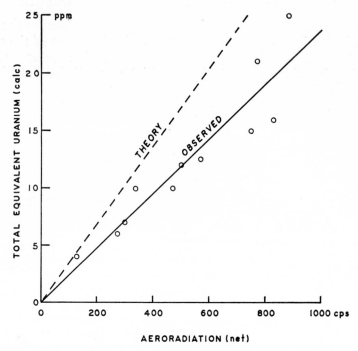

Fig. 4.—Total calculated *e*U content of extended sources *vs.* aeroradiation intensity

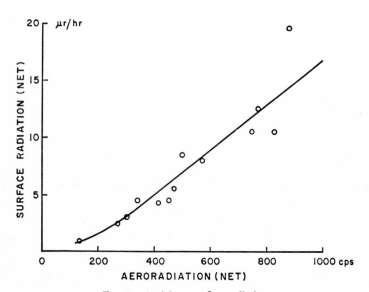

Fig. 5.—Aerial *vs.* surface radiation

tween the theory and the observed is within the accuracy limits given by Sakakura (1957, p. 10).

Surface radiation intensity was measured with a hand-portable scintillation counter at most of these locations and in Figure 5 is plotted against net aeroradiation. Surface measurements are given in microroentgens per hour ($\mu r/hr$), as is customary in most geologic studies, though the scintillation-counter measurement is effectively energy independent. Degradation of the primary spectrum by scattering in the source and in air is, however, a mitigating effect. A well-calibrated scintillation counter probably gives a fairly accurate measure of ionization for low-level, severely scattered spectra, but this is admittedly conjecture and should be studied further. The author's scintillation counter showed a γ intensity at sea level over water of about 2.5 $\mu r/hr$, which is probably 15–25 per cent lower than generally accepted sea-level cosmic-ray intensity, so absolute values referred to below may be subject to the same error. The quantity 2.5 $\mu r/hr$ was subtracted from all field observations to obtain the net surface radiation.

REGIONAL VARIATIONS IN RADIOELEMENT CONTENT AND TERRESTRIAL EMISSION

The aeroradiometric profile in Figure 6 was selected to show the relative radiation levels among the Coastal Plain formations. The line extends from the Midway group southeast to Aransas Bay on the Gulf Coast, passing about 5 miles southwest of Goliad. Intraformational variations are shown on an aeroradiometric contour map of the Coastal Plain survey area (Moxham and Eargle, 1961) compiled from 90 similar profiles. In the following discussion Figure 5 is used to estimate net surface radiation intensities from the net aeroradiometry; total calculated eU values, from equation (1), are given (in parentheses) for the associated radiation intensities.

The pre-Miocene part of the section shows relatively high amplitude variations ranging from about 1 $\mu r/hr$ (3 p.p.m.) over the Carrizo sand to about 9 $\mu r/hr$ (14 p.p.m.) over argillaceous, glauconitic, and ferruginous strata of the Wilcox, Mount Selman, and Jackson. The Jackson group locally contains highly abnormal concentrations of uranium that are quite apparent on the surface profile (Fig. 6), but much less so on the aerial record. The localized occurrences of uranium in such amounts are sufficiently rare that they need not be considered in this discussion of the regional radioactivity of the Coastal Plain. Nevertheless, the tuffaceous clays of the Jackson group are generally the most radioactive parts of the central Coastal Plain section.

In the strike direction, normal to the profile of Figure 6, the Mount

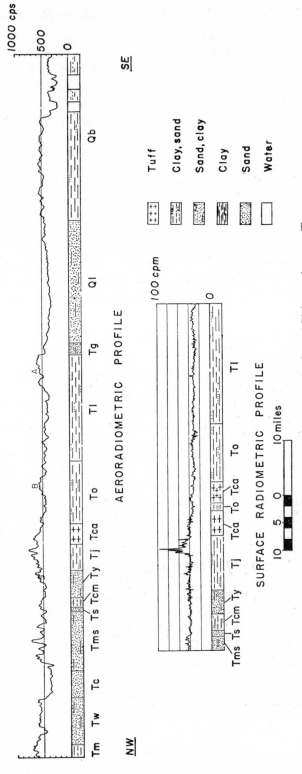

Fig. 6.—Radiometric profiles across the Coastal Plain, southeastern Texas

Selman and Wilcox strata locally show variations (Moxham and Eargle, 1961), in only a few miles, from 2 μr/hr (5 p.p.m.) over relatively pure quartz sands to off-scale anomalies over ferruginous sandstones estimated to peak at 15–17 μr/hr (22–24 p.p.m.).

The post-Miocene Oakville-Lissie strata give rise to a low, uniform 1–3 μr/hr (2–7 p.p.m.) pattern, extending for more than 50 miles and broken only by local highs (points *A* and *B*, Fig. 6) where the flight line crosses or skirts the San Antonio River. These features are probably caused by unmapped alluvial terraces that exist along most of the San Antonio River valley. From the top of the Lissie formation to the Coast, radiation increases slightly over the Beaumont clay, then drops sharply to the cosmic background level over Aransas Bay.

We have detailed information on intraformation variations in the post-Miocene only to about the base of the Lissie. In this area the same radiation uniformity over wide areas exists in the strike direction, so hundreds of square miles show variations of only a few μr/hr.

The foregoing results indicate that for most of the Coastal Plain the terrestrial component ranges from about 30 to 70 per cent of the total, with local contributions up to 90 per cent, the remainder being the cosmic component. Despite similarities in lithology, the older rocks of the Coastal Plain have a more heterogeneous distribution of radio-elements and a generally higher level of radioactivity than have the younger Coastal Plain sediments. This age-radioactivity relationship was also observed by Schmidt (1962) in a survey of the Atlantic Coastal Plain in South Carolina. The heterogeneity of the older sediments may reflect a more locally diverse depositional environment, but the slight regional difference in radiation level between the upper and lower Coastal Plain probably relates to depositional environment and rock weathering rather than to the age of the rocks per se.

REFERENCES

ADAMS, J. A. S., J. K. OSMOND, and J. J. W. ROGERS. 1959. The geochemistry of thorium and uranium. *In* L. H. AHRENS (ed.), Physics and Chemistry of the Earth, **3**:298–348. New York: Pergamon Press.

DAVIS, F. J., and P. W. REINHARDT. 1957. Instrumentation in aircraft for radiation measurements. Nuc. Sci. & Eng., **2**:713–27.

GREEN, J. 1959. Geochemical table of the elements for 1959. Geol. Soc. America Bull., **70**:1127–83.

LYMAN, W. S., and F. C. SCHROEDER. 1908. Soil Survey of Wilson County, Texas. U.S. Department of Agriculture.

MOXHAM, R. M. 1963. Radioactivity in Washington County, Maryland. Geophysics, **28**:262–72.

Moxham, R. M., and D. H. Eargle. 1961. Airborne radioactivity and geologic map of the Coastal Plain, southeast Texas. Geological Survey, Geophys. Investigations Map GP-198. Washington, D.C.

Rankama, K., and Th. G. Sahama. 1950. Geochemistry. Chicago: University of Chicago Press. Pp. 912.

Sakakura, A. Y. 1957. Scattered gamma rays from thick uranium sources. Geol. Survey Bull. 1052-A, pp. 1–50.

Schmidt, R. G. 1962. Aeroradioactivity Survey and Areal Geology of the Savannah River Plant Area, South Carolina and Georgia (ARMS-I). U.S. Atomic Energy Commission, Civil Effects Test Oper. Rpt. CEX-58.4.2. Pp. 41.

45. *Atmospheric Attenuation of γ Radiation*

$\mathbf{T}$HE MAJOR SOURCES contributing to the γ-ray spectrum of the earth at low altitudes are potassium-40 and some of the radioactive daughter products of uranium-238 and thorium-232. In addition to these naturally occurring elements, fallout from atmospheric tests of nuclear weapons has produced detectable amounts of zirconium-95–niobium-95 and barium-140–lanthanum-140. The daughter products of thorium-232 and uranium-238 exist in the soil and to some extent in the atmosphere. The radioactive fallout atoms are localized near the surface of the soil, probably distributed exponentially in depth, with a relaxation distance on the order of 1–2 inches.

The γ-ray spectrum of the earth is the composite sum of the photons from these source distributions and the scattered photons resulting from interactions between these primary photons and the constituents of the soil and the atmosphere. The result of these interactions is to change the discrete energy spectrum of the γ rays that are characteristic of the energy levels of radionuclides to a continuous energy spectrum that is characteristic of the soil and the atmosphere as well as the radionuclides themselves.

This paper is primarily concerned with the effects of the atmosphere on the γ-ray spectrum of the earth. Two effects are considered, first, the change in the spectrum as a function of distance above the surface and, second, the change at a fixed elevation due to variations in the mass density of the atmosphere. No particular attention will be given to the time variation of the γ-ray spectrum other than that due to variations in the meteorological conditions.

THEORY

The parts of the γ-ray spectrum that can be interpreted most easily are the various photopeak counting rates, that is to say, the

A. E. PURVIS is assistant professor, Department of Physics, University of Dallas, Dallas, Texas, and R. S. FOOTE is chief, Geonuclear Section, Texas Instruments Incorporated, Dallas, Texas.

parts of the spectrum that are produced by primary γ rays that arrive at the detector without interacting with the soil or the atmosphere and whose energy is totally absorbed by the detector. The analysis of the spectrum is complicated by the fact that the usual scintillation crystal does not absorb the total energy of every photon incident on its surface and the photomultiplier tubes do not produce exactly the same electrical pulse height when the total energy of the photon is absorbed.

The dependence of the photopeak counting rate of the γ ray having energy E_i on the distance above the surface of the earth, the mass density of the atmosphere, and the primary γ-ray energy can be determined from the equation,

$$d^4 N_i = [\lambda_i n_i d^3 V \, dt] \left[\frac{A_D}{4\pi r^2} \right] [\, e^{-\mu_e \rho_e r_e}] [\, e^{-\mu_a \rho_a r_a}],$$

where $d^4 N_i$ = number of γ rays having energy E_i that originate in the volume element $d^3 V$ in a time interval dt and are totally absorbed by the detector, λ_i = decay constant for the radionuclide emitting the γ ray whose energy is E_i, n_i = number of radioactive atoms per unit volume that emit the γ ray having energy E_i (this number usually depends on some parent radionuclide in the decay series), A_D = effective area of the detector for total absorption of a photon having energy E_i, μ_e = mass absorption coefficient of the earth at energy E_i, μ_a = mass absorption coefficient of the atmosphere at energy E_i, ρ_e = mass density of the earth, ρ_a = mass density of the atmosphere, r = total distance from between volume element $d^3 V$ and the detector, r_e = distance the γ ray must travel in the earth, r_a = distance the γ ray must travel in the atmosphere, $r = r_e + r_a$.

The bracketed terms in this equation represent (1) the number of γ rays leaving $d^3 V$ in dt, (2) the fraction that starts toward the effective area of the detector, (3) the fraction that escapes from the earth without scattering, and (4) the fraction that does not scatter in the atmosphere before impinging on the detector.

Considering the number of γ rays counted per unit time and integrating over the volume of the source, we have

$$\frac{d N_i}{d t} = \frac{\lambda_i A_D}{4\pi} \underset{\text{volume of the earth}}{\int \int \int} n_i \, e^{-(\mu_e \rho_e r_e)} e^{-(\mu_a \rho_a r_a)} \sin \theta \, d\theta \, d\phi \, dr \,.$$

Assuming that the earth is an infinite slab and that the source is distributed uniformly in the earth, the ϕ and r integrations can be carried out directly, leaving

$$\frac{d N_i}{d t} = \frac{\lambda_i n_i A_D}{2 \mu_e \rho_e} \left\{ e^{-(\mu_a \rho_a z)} - (\mu_a \rho_a z) \int_1^{\infty} \frac{e^{-(\mu_a \rho_a z) x}}{x} \, dx \right\},$$

where $x = (\cos\theta)^{-1}$ and z is the distance from the detector to the surface of the earth. This last integral is in a standard form and can be represented by an infinite series. Using the series expansion for the integral gives

$$\frac{dN_i}{dt} = \frac{\lambda_i n_i A_D}{2\mu_e\rho_e}\left\{ e^{-(\mu_a\rho_a z)} + [\mu_a\rho_a z]\left[\ln(\mu_a\rho_a z) + \gamma - \frac{(\mu_a\rho_a z)}{1\cdot 1!} + \ldots\right]\right\}, (1)$$

where γ is Euler's constant $0.5772157\ldots$ This expression represents a theoretical description of the photopeak counting rate with an infinite slab source distribution, which closely approximates the physical situation for the naturally occurring radioisotopes.

The photopeak counting rates due to the atmospheric concentration of radioisotopes can be determined from the expression:

$$d^4 N_i = [\lambda_i n_i d^3 V\, dt]\left[\frac{A_D}{4\pi r^2}\right][e^{-\mu_e\rho_e r}].$$

In this expression all symbols have the same meaning as in equation (1). Assuming that n_i is constant in the atmosphere and that the volume of the detector is negligible, the r and ϕ integrations can be carried out to give

$$\frac{dN_i}{dt} = \left\{\frac{\lambda_i n_i A_D}{2\mu_a\rho_a}\right\}\left\{2 - e^{-(\mu_a\rho_a z)} + (\mu_a\rho_a z)\int_1^\infty \frac{e^{-(\mu_a\rho_a z)x}}{x}\,dx\right\}. \quad (2)$$

A similar expression for the counting rate due to a plane source can be obtained by integrating the expression

$$d^3 N_i = [\lambda_i G_i d^2 a\, dt]\left[\frac{A_D}{4\pi r^2}\right][e^{-\mu_a\rho_a r}]$$

over the surface of the earth. In this expression G_i = the number of radioactive atoms per unit area on the surface of the earth, $d^2 a$ = differential element of area on the surface, and all other symbols retain their same meaning.

The integral form of this expression, assuming that the surface of the earth is an infinite plane, is

$$\frac{dN_i}{dt} = \frac{\lambda_i G_i A_D}{2}\int_1^\infty \frac{e^{-(\mu_a\rho_a z)x}}{x}\,dx.$$

Using the infinite series expression for this integral gives

$$\frac{dN_i}{dt} = -\frac{\lambda_i G_i A_D}{2}\left\{\gamma + \ln(\mu_a\rho_a z) - \frac{(\mu_a\rho_a z)}{1\cdot 1!} + \frac{(\mu_a\rho_a z)^2}{2\cdot 2!}\ldots\right\}. \quad (3)$$

This expression describes the photopeak counting rate due to a plane source distribution on the surface of the earth. It is assumed that this closely approximates the radioactive fallout distribution on the earth's surface.

Equations (1) and (3) involve the factors $(\lambda_i n_i A_D)/(2\mu_e\rho_e)$ and $(\lambda_i G_i A_D)/2$, which determine the absolute magnitude of the photo-peak counting rates. However, it is the dependence of the photopeak counting rates on elevation and mass density that are of interest, not their absolute magnitudes. To avoid evaluating the absolute magnitudes, one can consider the normalized photopeak counting rate, that is, the ratio of the counting rate at an elevation z to the counting rate at a standard elevation, z_0. This has the added advantage that the standard elevation can be chosen sufficiently high that it will reduce the effects of the earth's surface irregularities and irregularities in the source distributions within the earth.

TABLE 1

MASS ATTENUATION COEFFICIENTS FOR AIR

Gamma-Ray Energy (mev.)	Attenuation Coefficient (cm²/gm)
2.614	0.0395
1.76	0.0470
1.46	0.0515
0.75	0.0730

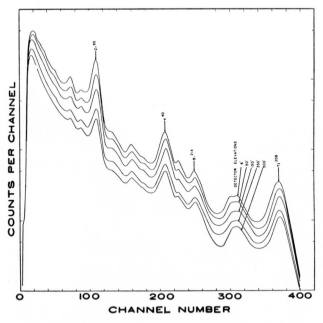

FIG. 1.—Surface γ spectra as a function of detector elevation

The normalized theoretical photopeak counting rates as a function of elevation are shown in Figures 2–5, for the 2.614-mev. γ ray from thallium-208, the 1.76-mev. γ ray from bismuth-214, the 1.46-mev. γ ray from potassium-40, and the photopeak due to the 0.757-mev. and 0.724-mev. γ rays from zirconium-95, and the 0.768-mev. γ ray from niobium-95. The first three are calculated from equation (1), the last is calculated from equation (2). The density of the atmosphere was taken as 33 grams per cubic foot; the absorption coefficients are shown in Table 1.

EXPERIMENTAL RESULTS

The γ-ray spectrum of the earth was measured at several elevations up to 500 feet with two sodium iodide scintillation crystals, one 5 inches in diameter by 5 inches thick, the other $11\frac{1}{2}$ inches in diameter by 4 inches thick. Figure 1 shows the spectra obtained with the $11\frac{1}{2} \times$ 4-inch crystal at 6, 150, 250, 350, and 500 feet.

These spectra were analyzed by two computational techniques to

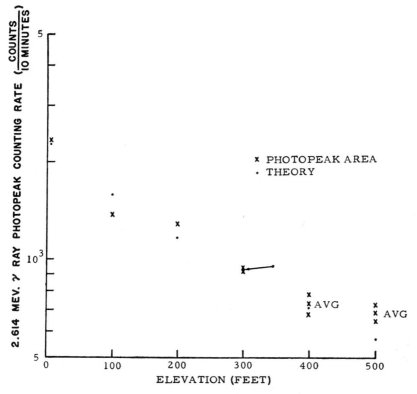

FIG. 2.—Thallium-208 counting rate, 5 × 5-inch detector, as a function of elevation

determine the photopeak counting rates. Calibration spectra for the response of the $11\frac{1}{2} \times 4$-inch detector to thorium, uranium, potassium, and zirconium-niobium sources have been obtained (Foote, 1963). These calibration spectra were used to form a response matrix which when inverted can be multiplied by the appropriate channel group summations from the spectra to yield numbers proportional to the photopeak counting rates. The spectra from both detectors were

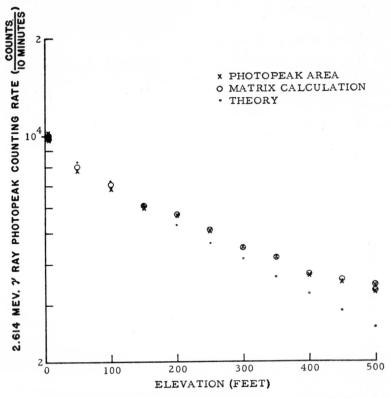

Fig. 3.—Thallium-208 counting rate, $11\frac{1}{2} \times 4$-inch detector, as a function of elevation

analyzed by assuming that the shape of the photopeaks is a Gaussian curve and that the rest of the counts in the spectrum in the energy interval of the photopeak can be represented by a straight line. This technique (Burrus, 1960) provides a good estimate of the actual number of totally absorbed primary photons. The major difficulty with the matrix calculation of the photopeak counting rates is that the matrix itself does not consider the change of the spectrum with elevation. The major difficulty with the Gaussian curve superimposed on a straight line assumption is not primarily the assumption but the sensitivity of the calculated value of the statistical variations of the number of counts in the channels at the ends of the energy interval.

Figures 2–6 show the theoretical and experimental photopeak counting rates. The theoretical values were obtained by assuming

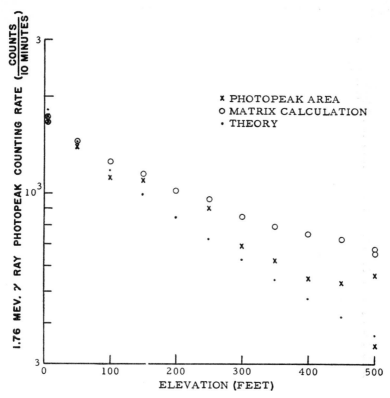

FIG. 4.—Bismuth-214 counting rate, $11\frac{1}{2} \times 4$-inch detector, as a function of elevation

that the experimental values, E_i, could be represented by some constant, a, times the theoretical values, T_i. Alpha was determined such that

$$\frac{\partial}{\partial a}\left[\sum_i \{E_i - aT_i\}^2 \right] = 0 .$$

DISCUSSIONS AND CONCLUSIONS

There are three disparities between the assumptions used to obtain the theoretical counting rates and the experiment itself. The first is that the probability that a photon will be totally absorbed by the detector does depend upon where the photon started from, even if it does arrive at the detector without scattering. This is most important for the 11½ × 4-inch detector at low elevations. At high elevations the primary flux tends to be a collimated, broad beam into the bottom of the detector. The second is that the thallium-208 and bismuth-214 concentrations in the atmosphere have not been included in the theoretical values. This is most important for bismuth-214 because of the

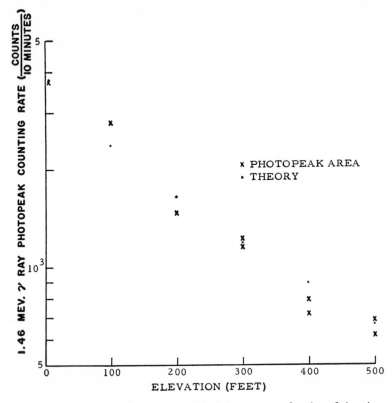

Fig. 5.—Potassium-40 counting rate, 5 × 5-inch detector as a function of elevation

relatively large concentration of radon in the atmosphere. The third is that the attenuation coefficient has not been reduced to compensate for the photons that have not been attenuated sufficiently to fall outside the photopeak due to the primary photons. This results in the theory's being approximately 5 per cent low at 500 feet.

This theoretical description of the flux of primary γ rays from the earth appears to be inadequate, since for only two of the counting rates, thallium-208 and potassium-40 within the 5 × 5-inch detector, is the shape of the theoretical curve consistent with the experiment

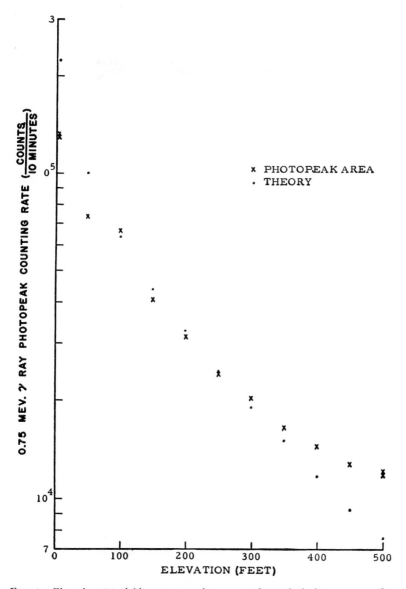

Fig. 6.—Zirconium-95–niobium-95 counting rate, $11\frac{1}{2} \times$ 4-inch detector, as a function of elevation.

data. These data demonstrate that accurate measurement of the γ-ray spectrum of the earth must include the atmospheric concentration of bismuth-214 and that spectral data to 500-foot elevation can be evaluated to provide surface amounts of specific radionuclides.

REFERENCES

BURRUS, W. L. 1960. Unscrambling scintillation spectrometer data. IRE Trans. Nuc. Sci., NS-7:2–3.

EVANS, R. D. 1955. The Atomic Nucleus. New York: McGraw-Hill Book Co. Pp. 972.

FOOTE, R. S. 1963. Time variation of terrestrial γ radiation. This symposium.

46. Time Variation of Terrestrial γ Radiation

E ACH γ-RAY–PRODUCING RADIOISOTOPE has its individual characteristic electromagnetic energy spectrum, which can be measured by the
use of detectors capable of totally absorbing the energy of the incident photon. If the counting rate as a function of the photon energy
is known for the shape and size of the detector used within the identical field measuring geometry, the amounts of each variable input can
be determined (Burrus, 1960). The detection system for the measurements to be discussed uses an $11\frac{1}{2}$-inch-diameter by 4-inch-thick
sodium iodide crystal whose pulses are analyzed by a 200-channel
pulse-height analyzer. Data have been taken every 47 minutes since
December 22, 1962, on punched paper tape and separated by computer operation to determine the amounts of each contributing radioisotope.

METHOD

Exact calibration spectra for all statistically recognizable inputs
in the composite γ-ray field spectra are necessary. If the exact calibration spectra for each input radioactive source are known, a matrix can
be formed by the inversion of the original matrix composed of calibration information. Since 9 inputs are involved, a 9×9-square
matrix is necessary, where each calibration spectra is separated into
identical sets of 9 channel groupings. These inputs are: thallium-208
(thorium-232 decay series), cerium-144–praseodymium-144, bismuth-
214 (uranium-238 decay series), barium-140–lanthanum-140, potas-
sium-40, zirconium-95–niobium-95, cesium-137, 0.51 mev. (source
complex), and iodine-131. Of these 9 inputs, the initial six are
statistically valid and are included. Calibrations for all inputs, except
for barium-140–lanthanum-140, were performed in large solid-angle
geometry, using low-activity silica sand in a tank contaminated with

ROBERT S. FOOTE is chief, Geonuclear Section, Texas Instruments, Inc., Dallas,
Texas.

each respective input source. Measurement of the barium-140–lanthanum-140 calibration was accomplished by distributing 100 μc. of source over an area ~2,500 square feet with the detector at an elevation of 4 feet to obtain the proper sum of peak intensity. Precise energy calibration is required prior to matrix operation on the field data. It is very difficult to maintain precise energy calibration of the

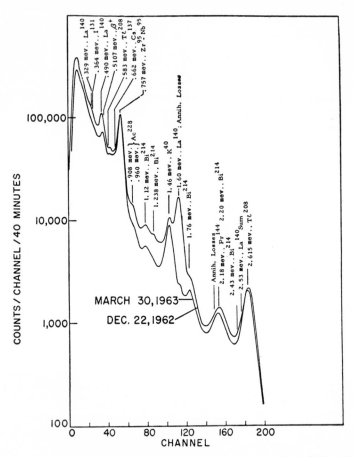

Fig. 1.—Gamma-ray field spectrum using 11½-inch-diameter × 4-inch-thick sodium iodide crystal at an elevation ~4 feet.

spectral analysis equipment in a remote location. Furthermore, in the average multichannel analyzer the calibration of zero energy is also variable. Therefore, prior to any operation on the data gathered, each spectrum must be corrected to proper gain and zero energy position. Channel correction is accomplished by using two known energy lines within the energy spectrum to provide sufficient information whereby recalibration of each channel and its associated number of counts can be made. With proper calibration, field spectra can be separated with accuracy using matrix separation techniques.

RESULTS

The environmental field γ-ray spectra of December 22, 1962, and March 30, 1963, are shown in Figure 1. The major difference observable is the large decay of lanthanum-140 at 1.60 mev. The locations of the major energy lines of all inputs are indicated. An example of one of the calibration spectra, the uranium series (bismuth-214), is seen in Figure 2. If the proper calibration information is given and the gains and resolutions of the detection system are constant or correctable, the matrix output can be used to give the individual amounts of

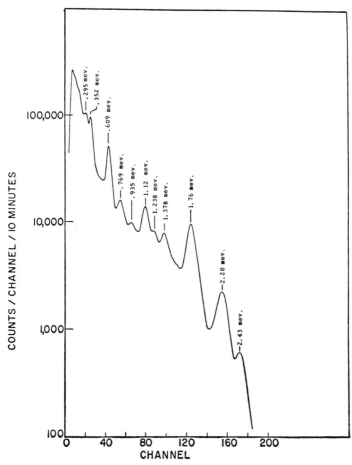

FIG. 2.—Uranium decay series—γ-ray field spectrum

each material. If the normalized spectra of all matrix inputs are then added, a composite spectrum will be created that should equal the field spectrum of the original input. If instrumental gain, resolution, linearity, as well as the matrix input calibration spectra, are not constant and correct, the difference between this created spectrum and the field spectrum will not be zero. Also, if other γ-ray inputs are present and not accounted for, the difference spectrum will not be zero. An example of the difference spectrum for the initial data point, December 22, 1962 (Fig. 1), is shown in Figure 3. The lanthanum-140

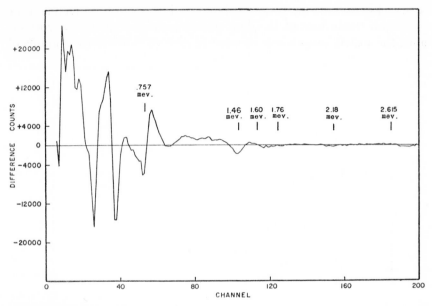

Fig. 3.—Difference spectrum; field spectrum less synthesized spectrum formed by matrix outputs and calibration spectra.

contribution to this spectrum is high and can be equated to ~40 nc/ft². The time variation of bismuth-214, along with temperature, barometric pressure, average surface wind velocity, and relative humidity, is shown in Figure 4. The effective surface bismuth-214 is seen to be highly variable, with an apparent relationship with all weather variables. In general, the minimum bismuth-214 is associated with temperature and surface wind velocity maxima. Maximum surface bismuth-214 is associated with minimum surface temperature and winds. Maximum-to-minimum bismuth-214 effective surface values oscillate up to 30 per cent within a daily cycle without the addition of rain. Radon-222 daughter product contribution to the surface amounts during rains (Foote and Humphrey, 1962) is seen, as well as the diurnal variations, which are highly variable. The variation of lanthanum-140 is observed in Figure 5, which shows the storm washout of lanthanum-140. The lanthanum-140 decays with the half-life of 12.8 days, except for rainfall additions. The amount of rain does not appear to be correlated with the amount of fallout. Figure 6 indicates the modulation of thallium-208, cerium-144–praseodymium-144, potassium-40, and zirconium-95–niobium-95 for the same period of time. Diurnal variations, if they exist, are very small.

Identical inputs were evaluated for the period of time from February 20, 1963, to March 6, 1963. Bismuth-214 and weather variations are indicated in Figure 7. The large diurnal variations in bismuth-214 are again seen. Variations of 5 other inputs are listed in Figure 8.

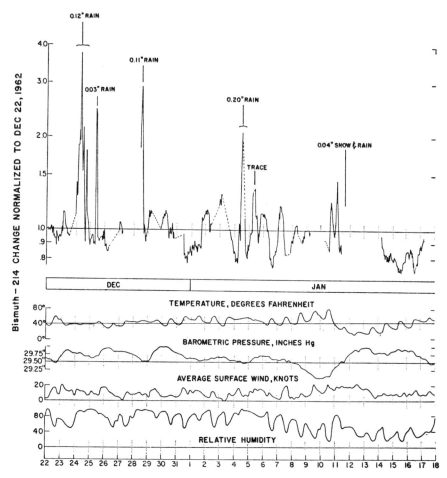

FIG. 4.—Modulation of effective surface bismuth-214, December 22, 1962–January 17, 1963

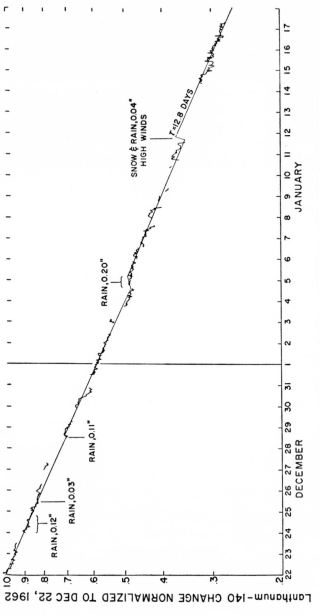

Fig. 5.—Decay of effective surface lanthanum-140, December 22, 1962–January 17, 1963

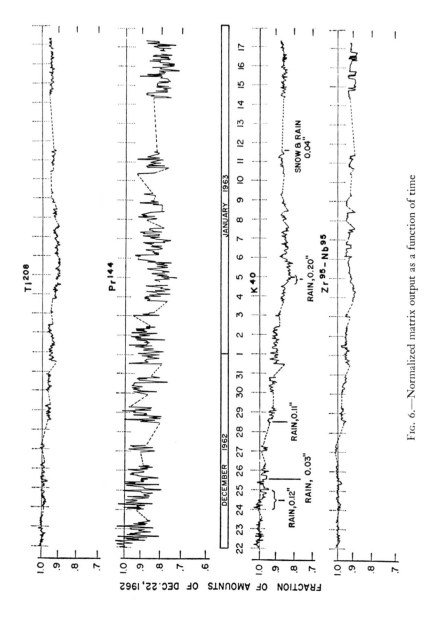

FIG. 6.—Normalized matrix output as a function of time

763

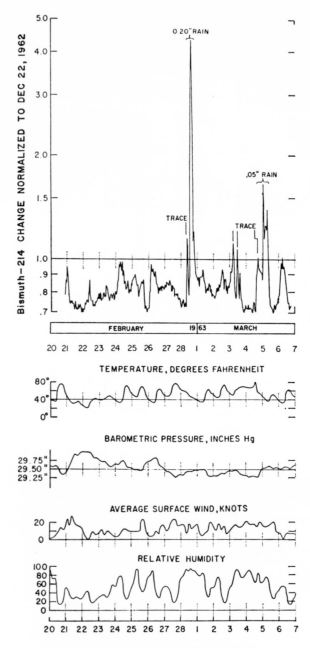

Fig. 7.—Effective surface bismuth-214

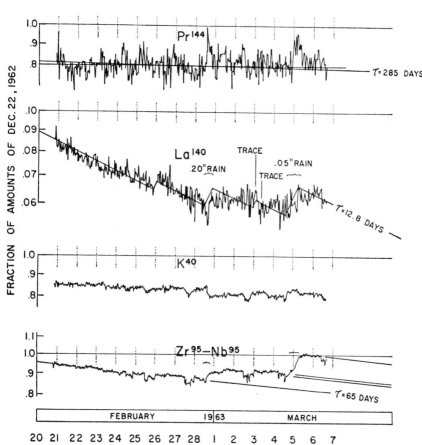

FIG. 8.—Normalized matrix material output

Radon-220 daughter products can be seen with the rain of February 28, 1963. The amounts of cerium-144–praseodymium-144 are small, providing a statistically more highly varying output. The downward modulation of potassium-40 with rainfall is observable, while lanthanum-140 and zirconium-95–niobium-95 are seen to build up with each rainfall. One per cent changes in zirconium-95–niobium-95 appear detectable. Each rain has brought down sufficient zirconium-95–niobium-95 to cause the average intensity to remain nearly constant since December 22, 1962.

The results of handling large numbers of spectra have given confidence in the ability to understand the radioactive material involved in the γ-ray field spectrum. It has given confidence in the ability to operate a remote spectral monitoring station, where the data obtained can be automatically recorded, corrected for energy calibration, and separated into normalized amounts of each radioactive input by a relatively simple computer operation. Although the spectral inputs used for each radioactive source are not completely identical with those obtained in field geometries, where the radioactive source is varying from an infinite plane to an infinite slab, they are sufficiently accurate to allow large variations in individual inputs not to affect the matrix mathematical outputs of the remaining inputs, such as the large variations created by rainfall addition of lead-214 and bismuth-214. Study of the variations of bismuth-214 with weather conditions is continuing, and it is believed that an empirical relationship using measurable weather factors can be developed to predict the modulation of the effective surface bismuth-214.

REFERENCES

BURRUS, W. L. 1960. Unscrambling scintillation spectrometer data. IRE Trans. Nuc. Sci., NS-7:2–3.

FOOTE, R. S., and N. B. HUMPHREY. 1962. Natural atmospheric radioactive fallout. Am. Physical Soc. Bull. **7** (Ser. 2): 321. (Abstract.)

P. R. J. BURCH, J. C. DUGGLEBY,
B. OLDROYD, AND F. W. SPIERS

47. Studies of Environmental Radiation at a Particular Site with a Static γ-Ray Monitor

MOST SURVEYS of natural environmental radiation have been concerned with variations in its intensity as a function of geography. At high altitudes we know that the cosmic-ray intensity increases, and we are equally familiar with the importance of geology in determining natural γ radiation levels. The latter are low over chalk, for example, but high over granite, and higher still over monazite sand. Although the time variations of cosmic-ray intensity at a fixed site have been intensively studied, rather less attention has been paid to the time variation of the natural γ-ray intensity in a given environment. The apparatus described in this article was designed to investigate this latter phenomenon.

SPECIFICATIONS FOR APPARATUS

It was decided to select a site remote from intense artificial sources of radioactivity and free from other abnormal disturbing influences. Accordingly, the apparatus was installed in a hut in an uncultivated field, at least 150 feet from the nearest tree or other surface encumbrance.

It was required that the apparatus should function continuously over long periods of time—years rather than months—and hence long-

P. R. J. BURCH is deputy director, J. C. DUGGLEBY is research assistant, B. OLDROYD is junior technical officer, and F. W. SPIERS is professor and head, Department of Medical Physics, Medical Research Council, Environmental Radiation Research Unit, Department of Medical Physics, University of Leeds, General Infirmary, Leeds, England. Mr. Duggleby's present address is Australian Atomic Energy Commission, Research Establishment, Private Mail Bag, Sutherland, N.S.W.

term stability was of paramount importance. Because of the continuous nature of the observations, it was necessary to design some form of automatic or semiautomatic recording capable of registering both short- and long-term variations. Any investigation of background radiation is complicated by variations in the cosmic-ray intensity, and hence it was necessary to determine the latter separately in order that variations in other components of the background radiation could be properly assessed. A 3-channel apparatus was therefore planned in which one channel would measure the γ radiation from the upper hemisphere (roof and sky), the second would record the cosmic-ray intensity, and the third would measure the γ radiation from the lower hemisphere (ground). To determine this last component, the apparatus had to be installed at an appreciable height above ground level; the floor level of the hut housing the apparatus is in fact 10 feet above ground level. The hut is supported on steel pillars, and access is by steel steps.

With respect to accuracy of recording, it was considered that a statistical error of less than ± 1 per cent (S.D.) for a 1-hour observation should be aimed at.

DESIGN OF APPARATUS

A simplified section through the apparatus is shown in Figure 1. It consists essentially of four large disk-shaped steel high-pressure ionization chambers. The upper chamber (*1*) is shielded from the sky by its own wall thickness and the thin roof of the hut only. The second chamber (*2a*) is separated from the top one by 2 inches of lead. The third chamber (*2b*) is contained within the same "shell" as *2a*, the active volumes being separated by a thin steel plate. The fourth chamber (*3*) is separated from *2a* and *2b* by another 2-inch layer of lead. The whole pile of chambers is mounted within a 4-inch-thick cylindrical steel shield to prevent γ radiation from the horizontal direction from affecting chambers *2a* and *2b*.

Chambers *1* and *2a* are connected differentially. Most of the hard component of cosmic radiation traversing chamber *2a* will also traverse chamber *1*, and, when the gas pressures are suitably adjusted, the differential current will be insensitive to changes in the intensity of the hard component of cosmic radiation. This may be regarded as an anticoincidence device. Changes in differential current will arise from variations in the soft component of cosmic radiation (which, in general, will be inversely correlated with barometric pressure) or from variations in the γ radiation component from the roof of the hut and from the sky. In practice, the deposition and wash-off of fission products on the roof of the hut have been the most important factors affecting the differential response of chamber *1*.

Chamber *2b* records the hard component of cosmic radiation. Most of the particles traversing *2b* will also penetrate chamber *3*. On subtracting the response of chamber *2b* from that of chamber *3*, then, provided that pressures or nominal response is suitably adjusted, an accurate estimate of the ground γ radiation affecting chamber *3* may be obtained. Again, anticoincidence properties are exploited.

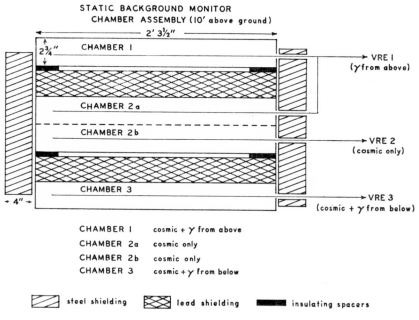

FIG. 1.—Simplified section through apparatus

The "active volume" of each chamber is $2\frac{3}{4}$ inches deep and $27\frac{1}{2}$ inches in diameter. Chambers are constructed of $\frac{3}{8}$-inch mild steel to withstand the working gas pressure of about 20 atmospheres. The lid of each chamber is removable and is secured by 32 nuts and threaded studs, the latter being screwed and welded to the chamber base. A gas seal is provided by lead washers between the lid and the studs and a copper washer between the securing nut and lid. A $\frac{1}{8}$-inch-diameter leather band, soaked in a latex adhesive, provides a gas seal around the flange.

To test for gas leaks, the chambers are filled to their working pressure and immersed in a large tank of water. Nuts are tightened until gas bubbles are eliminated. The chambers are tested initially with commercial nitrogen, but filled finally with about 20 atmospheres of commercial argon.

The outside, or shell, of the chambers is connected to a positive or negative 90-volt dry battery supply. Chambers (except *2a* and *2b*) are insulated from one another, and all are insulated from the shield by mechanical supports. Ionization current is determined through the rate of voltage drift (dV/dt) of the central electrode system; a vibrating-reed electrometer* (VRE) is used for this purpose (one for each "channel"). The central electrode system consists of a rigid outer ring held in position by three insulated supports (with guard-ring arrangement), one of which is connected through the chamber

PLATE I

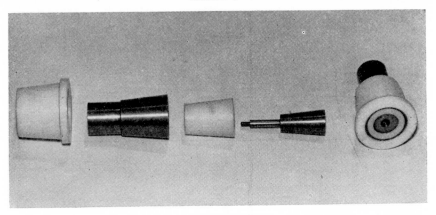

Photograph of dismantled electrode support system (*left*) and complete assembly (*extreme right*).

wall to the input of the electrometer. A network of thin stainless-steel wires is stretched between the studs and anchored to the rigid outer ring.

The electrode support assembly fitting into the cylindrical walls of the chambers is illustrated in Plate I; PTFE ("teflon") insulates the central electrode from the guard ring and also insulates the latter from the chamber wall. Coning of the assembly prevents the extrusion of insulators under pressure and encourages a secure gas seal; a 10° taper has been found to be satisfactory for this purpose.

The differential current from chambers *1* and *2a* (Fig. 1) is measured by VRE 1, VRE 2 measures the ionization current produced by cosmic radiation in chamber *2b*, and VRE 3 measures the current from chamber *3*.

* Ekco 1079.C.

The output from each of the three VRE's operated under voltage-drift conditions is fed into its appropriate recording channel. Three modes of recording are used. The first is a pen recorder that samples the output from each VRE in turn at 54-second intervals and prints a dot of the color appropriate to the channel with the same periodicity. By this method, one can time an abrupt change of radiation intensity

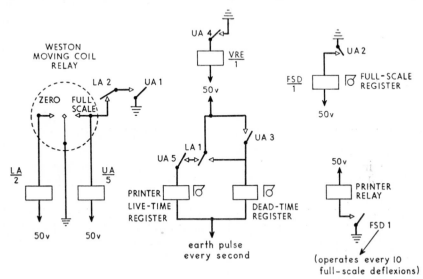

Fig. 2.—Simplified schematic diagram of control circuit. During transit between zero and full-scale output from the electrometer (*VRE*) the live-time register records live time in seconds. On attaining full-scale output, relay *UA/5* operates, locks itself via contact *UA1*, disconnects the live-time register at *UA5*, starts the recording of dead time through *UA3*, and registers one full-scale deflection at *UA2*. Meanwhile, *UA4* causes the input to the *VRE* to be short-circuited. On the collapse of the input to the *VRE*, the Weston moving coil relay returns to zero and relay *LA/2* operates. *LA/2* releases *UA/5*, which now restores the input to the *VRE* at *UA4*. *LA1* maintains the recording of dead time until "zero" output is exceeded. At that instant, *LA/2* is released and the recording of live time proceeds via *LA1* and *VA5*. At every ten full-scale deflections, *FSD1* closes and the "printer relay" operates to record the live time (in seconds) for that number of deflections. The accumulated live time is automatically reset to zero at the end of the printing cycle. On/off control switches and spark-quench circuits are omitted from the diagram.

by noting from the chart when a change occurs in the voltage increment between successive dots.

Additionally, the output from each VRE is also fed into a voltmeter relay* the contacts of which operate subsidiary relays and registers (see Fig. 2). In this control circuit the following operations are performed:

i) When the output from the VRE is less than the preset zero or

* Weston Moving Coil Meter Relay type S. 54.

greater than the preset full-scale deflection, an electromechanical register* records the "dead time" in seconds.

ii) When the output from the VRE reaches the full-scale deflection, a contact on the relay voltmeter causes the input to the VRE to be short-circuited. Another register also operates to record a full-scale deflection.

iii) The short-circuiting of the VRE input cancels its output, and, when the relay voltmeter returns, its "zero contact" operates, causing the input of the VRE to be opened again.

iv) When the output of the VRE rises to the preset "zero level," the "zero contact" of the voltmeter relay opens and the recording of "dead time" (which started with the closure of the full-scale deflection contact) is terminated.

v) Noting the total time for an observation and subtracting the "dead time" give the effective or "live time" t. Noting the total number of full-scale deflections that have been recorded gives the total voltage increment V during the observation. The average rate of voltage drift, is of course, $\dot{V}/t$, and this is proportional to the ionization current.

This control and recording system has been extremely reliable and satisfactory in service. It is used in conjunction with a third method of recording. Relay contacts are used to control an electromechanical printer that accumulates live time (in seconds) and prints the accumulated total for every 10 full-scale deflections. (This time is typically of the order of 1 hour.) In this way, a simple indication is given of changes in background intensity over, say, a 24-hour period. It is merely necessary to glance at the column of printed figures. Unfortunately, various mechanical troubles have been experienced with these printers, and we are planning to replace this type of recording by an "instantaneous" dV/dt display on a pen recorder.

RESULTS

The record of the ionization current in chamber *2b*, which responds to the hard component of cosmic radiation, is shown in Figure 3 for the period April–July, 1961. It is not possible, without information on the variation of atmospheric conditions with height, to interpret the data in detail, but some features can be identified. The record shown in Figure 3 has been corrected for the barometric effect, using a coefficient $\beta = -3.1$ per cent per cm. mercury (based on the mean regression line of ionization in chamber *2b* versus barometric pressure). The residual variations are due in part to atmospheric condi-

---

* Sodeco 4-digit manual reset register.

tions and in part to extra-terrestrial events. Two marked decreases in the record, on April 13 and July 12, were identified with large Forbush decrements accompanying solar flares. These were detected simultaneously on cosmic-ray neutron counters in the Physics Department of the University of Leeds. A fall of about 2 per cent in the average level of the hard component is evident in Figure 3, and this

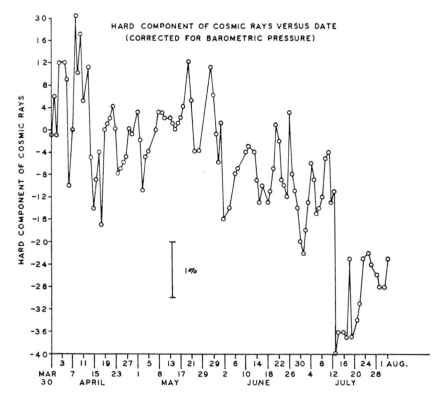

Fig. 3.—Record of ionization current in chamber *2b* from April to July, 1961. (Proportional to intensity of hard component of cosmic radiation.)

accords quantitatively with the known variation of cosmic-ray intensity for winter and summer conditions.

Measurements with a portable high-pressure ionization chamber (Spiers *et al.*, 1963) have shown that the fall in dose rate observed when the ground was covered with snow did not recover immediately after the snow melted. The initial dose rate was re-established only after a period in which presumably the surface water had drained away. This phenomenon is illustrated several times in the record of chamber *3* during the period April–July, 1961. The data in Figure 4

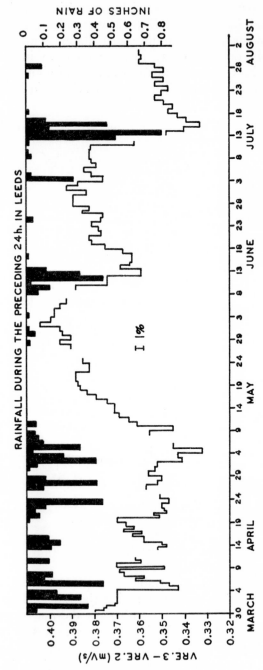

CORRELATION BETWEEN RAINFALL AND BACKGROUND CHANGES

FIG. 4.—Record of (i) relative intensity of ground γ radiation and (ii) rainfall, during a period of low fallout deposition, April–July, 1961

give (i) the response from ground γ radiation, that is, the current in chamber *3* minus that in chamber *2b*, and (ii) the rainfall in the 24 hours preceding each measurement. During this period of observation fallout deposition was very low, and the meteorological conditions of a typical English summer demonstrated convincingly the decreased dose rate following rain and the approximately exponential recovery as the ground subsequently dried out. Comparison of the periods June 7–13 (rainfall 1.16 inches) and July 10–17 (rainfall 2.08 inches) suggests that, for the type of field below the apparatus, the first inch of

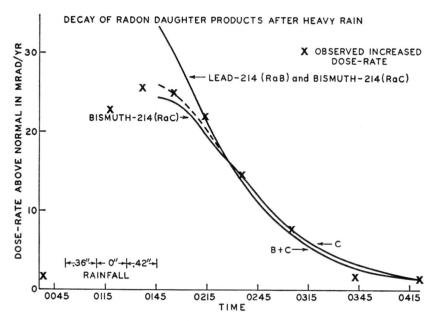

Fig. 5.—Decay of γ-ray dose rate from ground-deposited radon daughter products, following heavy summer rainstorm.

rain reduced the dose rate by nearly 8 per cent and the second inch of rain caused a further reduction of nearly 5 per cent. The initial recovery rate for these particular soil conditions was about 1 per cent per day.

Although there has been some discussion of the precise effects of rainfall on the escape of radon from the ground, the general finding seems to be that radon escapes more freely from dry soil than from wet soil. The decreased γ radiation over waterlogged ground (also noted by Marsden and Watson-Munro, 1944) is readily accounted for by the considerable extra self-absorption introduced by the presence of the water. Of course, if the rain is heavily contaminated with radioactive materials, the absorption effect can be completely out-weighed by γ radiation from the deposited radioactivity.

It is well known that following a dry spell the air can become dust-laden, and an equilibrium deposit of radon decay products becomes established on the dust particles. Sudden heavy rain, as in a thunder-storm, then brings down the decay products, lead-214 (RaB) and bismuth-214 (RaC), which produce a temporary rise in the ground γ radiation. An example of this phenomenon is shown in Figure 5, which records the ground γ radiation for a short period during a thunderstorm. The rise in γ-ray dose rate after two short, heavy downfalls (0.36 inches and 0.42 inches over a period of 55 minutes)

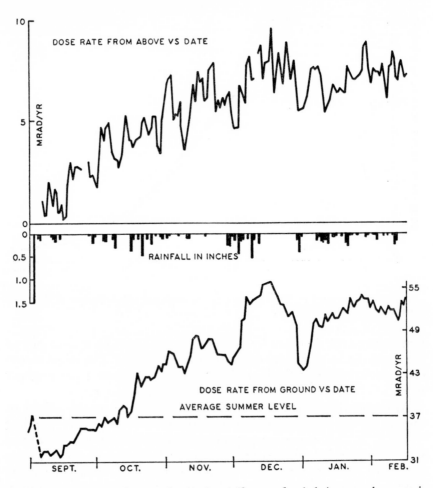

Fig. 6.—Gamma-ray dose rate in chamber *1* (from roof and sky), γ-ray dose rate in chamber *3* (from ground), and rainfall. From September, 1961, to February, 1962.

is succeeded by a decrease that closely follows the expected decay of the lead-214 (RaB) and bismuth-214 (RaC) activities.

The records of the responses of chamber *1* and chamber *3* (corrected for the hard cosmic-ray component) are shown in Figure 6 for the period September–December, 1961. The sharp increases in the radiation received by the top chamber indicate the passing radioactive clouds from the Russian test series of that period. Identification with

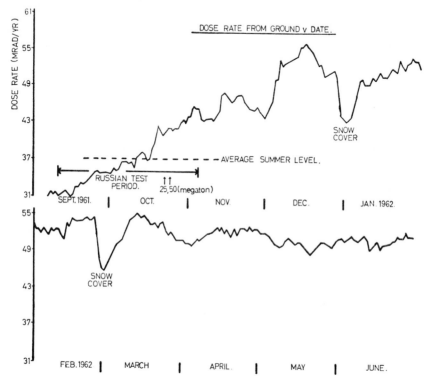

Fig. 7*a*.—Record of γ-ray dose rate in chamber *3* (from ground), from September, 1961, to June, 1962.

particular bombs is difficult, and the response to the fallout is determined chiefly by the activity deposited on the roof of the hut.

The record of the ground γ-ray dose rate began on September 1 at a level lower than the summer mean because of heavy rain at the end of August; the dose rate rose, however, almost continuously over this period with a number of steplike increases that depended on the activity at different heights in the atmosphere and the incidence of rain. The whole record up to the end of April, 1963, is shown in Figure 7. Various decrements due to snow are observed, together with a fall in the summer of 1962 before the additional tests during last autumn. Two striking increases characterize the last months of 1962, one of which might be linked with the 25-megaton bomb exploded on December 24, 1962. The effect of the snow cover lasting throughout January and February is seen, followed by recovery toward the end of April.

Fig. 7*b*.—Record of γ-ray dose rate in chamber 3 (from ground), from July, 1962, to April, 1963.

SUMMARY

The construction and performance of a large static background radiation monitor are described. The apparatus consists of four large disk-shaped high-pressure ionization chambers mounted on top of one another and placed in an open-ended cylindrical steel shield. The chambers are positioned in a hut supported on a framework so that the bottom chamber is 10 feet above grassland, which surrounds the hut to at least a distance of 150 feet.

The top chamber is shielded below by 2 inches of lead but, apart from the light roof of the hut, is unshielded from radiation from the sky. The bottom chamber is unshielded from γ radiation, but it is shielded on the top and sides by lead or steel and responds only to hard cosmic rays. The top chamber is connected differentially to the one below it so that together they respond to soft cosmic rays and to γ radiation from the sky. The third chamber measures the hard cosmic-ray ionization, and the bottom chamber measures the ground γ radiation plus the hard cosmic-ray component. The ionization currents are continuously recorded.

Results are given to show the dependence of the ground γ-ray dose rate on the water content of the soil, and records are shown of the variation in γ radiation from the ground and the sky following the resumption of nuclear-weapons testing in 1961.

ACKNOWLEDGMENTS

The authors are indebted to Mr. R. L. Corry and Mr. K. Dobson for the construction of parts of the apparatus and assistance in its assembly.

REFERENCES

MARSDEN, E., and C. WATSON-MUNRO. 1944. Radioactivity of New Zealand soils and rocks. New Zealand J. Sci. Technol., B, **26**:99–114.

SPIERS, F. W., M. J. McHUGH, and D. B. APPLEBY. 1963. Environmental γ-ray dose to populations: Surveys made with a portable meter. This symposium.

W. HERBST

48. Investigations of Environmental Radiation and Its Variability

FOR THOUSANDS OF YEARS all existing life has been exposed to a certain quantity of ionizing radiation from natural sources. A knowledge of this background radiation, its nature, its quantity, and, last but not least, its variability is, for obvious reasons, of considerable radiobiological interest. This knowledge is essential for statistical research on possible pathological effects on human beings caused by chronic exposure to natural background radiation, and it also provides a basis for a biological evaluation of additional radiation burdens from artificial radiation sources. A systematic measurement of this background-radiation dose and of the natural radioactivity seems indispensable in order to determine rapidly the extent of radioactive contaminations in the environment for any reasons whatever. Finally, profounder knowledge of this background radiation allows one to draw conclusions on the content of natural radioactive nuclides in geological strata.

Our surveys (with G. Hübner) of this background were initially confined to measuring the doses of the exterior penetrating radiation and to relating these doses to the distribution of the population. We paid special attention to the variability of this background radiation as it depended on the different natural conditions and those brought about by civilization factors.

The geological variability of Switzerland presented a good opportunity to study intensively the connections between the terrestrial radiation dose and geological structure. This work was carried out in collaboration with the Swiss Federal Public Health Department (Eidgenössisches Gesundheitsamt) in Bern (section chief, G. Wagner; collaborator, A. Mastrocola) and with the Institute of Mineral-

W. HERBST is professor at the Radiologisches Institut der Universität Freiburg, Freiburg, Germany.

ogy of the University of Bern (director, T. Hügi; collaborator, E. Halm).

These measurements were supplemented by corresponding measurements in the Federal Republic of Germany.

Of the man-made factors responsible for the variability of the background radiation, the influences of mining, road-making, housing conditions, agriculture, and forestry, as well as fallout from nuclear weapons tests, were studied.

It is considered that a reasonably exact picture of the radiation environment and of the radiation burden of man can be obtained only by taking into account the variability and diversity of the radiation background.

In the following we give a short survey of our findings.

METHOD

Our measurements of the dose rate of exterior penetrating radiation were carried out with the aid of an ionization chamber. To guarantee the best possible conditions for a comparison of our measurements with the measurements of this background radiation carried out in 1958 by Solon *et al.* in Switzerland and in the Federal Republic of Germany, we partly followed these authors in the construction of the chamber (Solon *et al.*, 1958).* This chamber proved useful also in our other measurements.

The housing of our chamber consists of a 25-liter polyethylene bottle. The electrodes were well isolated with amber and coated with a special graphite lacquer. The interior of the chamber was coated in the same way. The chamber is air-equivalent and is surrounded by 1.04 gm/cm^2 of absorbing material (polyethylene and aluminum, 3.2 mm. thick).

The voltage across the chamber is 30 v., about 10–15 v. higher than the saturation voltage.

Interferences of electrolytic and thermal contact voltages were eliminated by averaging the readings obtained with chamber voltages of opposite sign.

The basic calibration of the chamber was carried out with radium. To eliminate the influences of varying atmospheric pressures and temperatures, a further special calibration of the chamber with radium was combined with each measurement.

We estimate the accuracy of the measurements to ±10 per cent

---

* EDITOR'S NOTE: For a more recent discussion of the 1958 instrumentation see A. Shambon, W. M. Lowder, and W. J. Condon, "Ionization Chambers for Environmental Radiation Measurements," Health and Safety Laboratory (USAEC) Report HASL-108, February, 1963.

PLATE I

Automobile with instruments for radiation-level measurements

because of calibration with several γ emitters. A detailed description of the chamber has been published (Herbst and Hübner, 1961).

The measuring instruments were installed in a car. However, for each measurement the chamber was set up outside the car (Pl. I).

MEASUREMENTS IN CONNECTION WITH
THE WORK OF SOLON *et al.*

To compare our own method and values with those of Solon *et al.* (1960) and their results of corresponding measurements in the area of Lac Leman (Switzerland) as well as in Offenburg and Wesel (Federal Republic of Germany), special measurements were made in these places also. It must be mentioned that particulars about exact sites of the measurements are not stated by Solon *et al.* Therefore, we were unable to choose exactly the same places for parallel measurements.

With this reservation, these comparisons of the dose rates of the terrestrial component of environmental radiation yielded the following results:

1. Area of Lac Leman (Switzerland):
   a. Solon *et al.* (autumn, 1958): mean value of 4 measurements: 8.8 μr/hr
   b. Present author (summer, 1961): mean value of 4 measurements: 7.8 μr/hr
      The interpretation of these results must also take into consideration the fact that at the time of the measurements of Solon in autumn, 1958, a dose rate from fallout might have been effective giving at least 10 per cent higher values than at the time of our own measurements in summer, 1961.
2. Federal Republic of Germany:
   Offenburg:
   a. Solon *et al.* (autumn, 1958): 11.6 μr/hr
   b. Present author (June, 1960): 12.1 μr/hr
   Wesel:
   a. Solon *et al.* (autumn, 1958): 10.4 μr/hr
   b. Present author (May, 1962): 11.9 μr/hr
      It must be considered that the average increase of the dose rate from fallout at the time of our own measurements in May, 1962, was 3.1 μr/hr (Herbst and Hübner, 1962). At the time of the measurements of Solon *et al.* in autumn, 1958, this increase might possibly have ranged only between 1 and 2 μr/hr.

Considering the specific influences of fallout on the dose rate at the times of the measurements, our own values appear to correspond closely with those of Solon *et al.*, the accuracy ranging within ±10 per cent.

PERCENTAGE OF COSMIC RADIATION IN DOSE RATE
OF TOTAL BACKGROUND RADIATION

In order to determine the dose rate of the exterior penetrating radiation from terrestrial sources only, the percentage of cosmic radiation dose within the total values of the measurements, which were made up to heights of 2,600 meters above sea level, had to be taken into account.

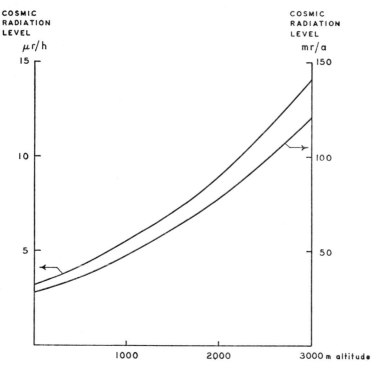

FIG. 1.—Cosmic radiation intensity as a function of altitude

We determined the contribution of the cosmic radiation to a first approximation by measuring the dose rate on a larger inland lake for the given height in areas between 46° and 50° N. latitude.

For the intermediate ranges we interpolated the dose rate dependence with height from corresponding results by Solon *et al.* (1960), Hultqvist (1956), Sievert (1958), and others.

After interpolation we obtained for cosmic radiation alone a dose rate at sea level of 3.2 μr/hr, corresponding to 28 mr/yr, and altogether a dependence of the cosmic dose rate on height as shown in Figure 1.

Our procedure therefore reads:

Dose rate (terrestrial) = dose rate (total) measured with the ionization
chamber minus dose rate of cosmic radiation obtained from Figure 1.

VARIABILITY OF TERRESTRIAL RADIATION DOSE RATE WITH RESPECT TO GEOLOGICAL STRUCTURES

The values of terrestrial dose rates obtained mainly in Switzerland with respect to defined geological structures and geological exposures are shown in Figure 2, *a–l*.

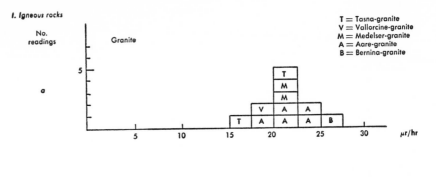

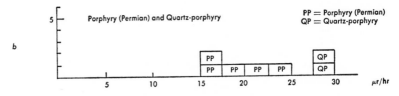

Fig. 2.—Part *A*—Terrestrial dose rates above geological formations

Naturally, for a specific geological formation the values show a
scatter depending on the natural radionuclide content of the particular rocks. In view of the low number of values in several series of
measurements, only a rough mean value can be obtained in these cases
(Halm *et al.*, 1962).

In correspondence with the results of other authors, we observed
the highest dose rates on acid igneous rocks, especially granite (17–26
μr/hr), quartz-porphyry (28–30 μr/hr), and porphyritic rocks of
different nature (14–24 μr/hr), and, in addition, above some mica-
schists (10–25 μr/hr) and metamorphic gneisses (10–32 μr/hr) with
relatively high values in the presence of feldspar (32 μr/hr). The
higher the chlorite content, generally, the lower the dose rate. Lower
values were found for amphibolite (12–20 μr/hr), greenschist (8–10
μr/hr), and serpentine (6 μr/hr).

In exposures of limestones we found only low dose rates, between 2 and 15 μr/hr. The dose rates above limestone-mica-schist (11–18 μr/hr), moraines of rock fragments (3–11 μr/hr), sands (11–12 μr/hr), and *Nagelfluh* (conglomerate) (8–12 μr/hr) were inversely proportional to the limestone content.

Medium-sized values were observed for dolomite (10–13 μr/hr), marl and marl schist (8–17 μr/hr), and clay-schist (9–20 μr/hr). The dose rates in open lodes with variegated sandstone originating from

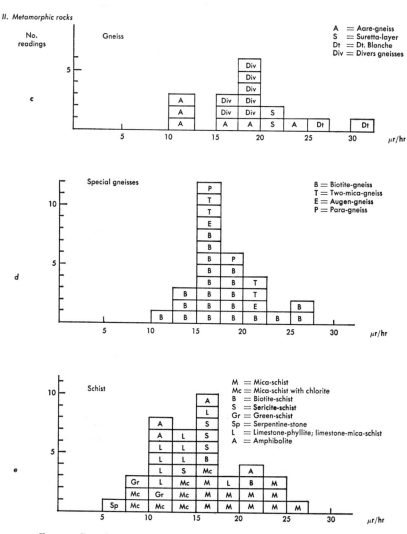

Fig. 2.—Part *B*—Terrestrial dose rates above geological formations

continental sedimentations (15 μr/hr) were higher than Molasse-sandstone originating from marine sedimentation.

As a general rule, sericite-quartzite of the Permo-trias formation (12–20 μr/hr) had values higher than those of quartzite in the Trias formation (7–10 μr/hr). Quartzites are partly mineralized with uranium. Exposures and barren rocks with relatively higher uranium content had chiefly higher dose rates (52–200 μr/hr).

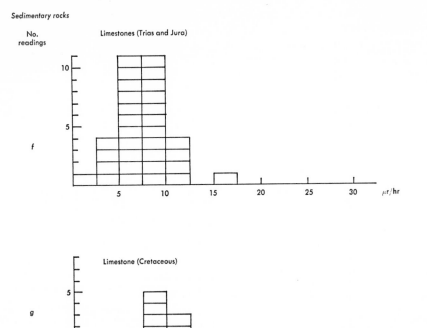

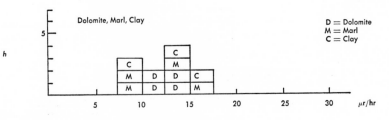

Fig. 2.—Part *C*—Terrestrial dose rates above geological formations

In a stock of potash fertilizer (60 per cent potassium) we investigated the influence of the potassium-40 upon the dose rate at different distances from the stock (Fig. 3).

Natural Radiation Burden within a Population

On the basis of the results of our measurements, the Swiss Federal Public Health Department (Eidgen. Gesundheitsamt), Bern (G. Wagner, W. Rottenberg, A. Mastrocola), evaluated the natural radiation burden within the population of Switzerland (Halm *et al.*, 1962). This analysis represents a population of 5,429,061 individuals, the entire population of Switzerland. Here, the mean dose rate from

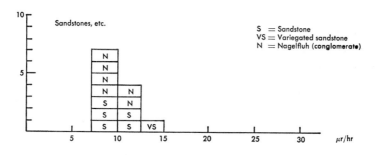

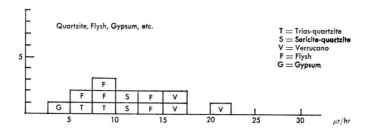

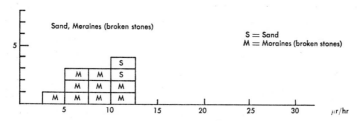

Fig. 2.—Part *D*—Terrestrial dose rates above geological formations

the earth's crust in air is 76.1 mr/yr. The corresponding mean dose rate from cosmic radiation is 38.4 mr/yr. Therefore, the mean natural background dose rate in air to which individuals are exposed is about 114 mr/yr.

By making corrections to account for habits of living, the mean dose rate from cosmic radiation may be reduced to 31 mr/yr and the mean dose rate from terrestrial environment may be increased to 91 mr/yr. Altogether, the mean total dose rate of external penetrating

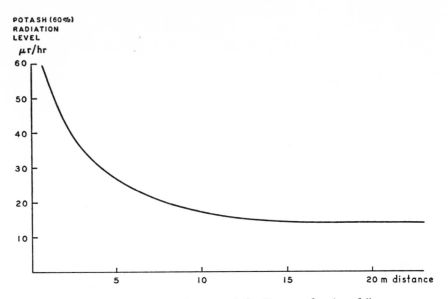

Fig. 3.—Radiation dose rate above potash fertilizer as a function of distance

radiation to individuals in Switzerland is 122 mr/yr. This value is only 7 per cent lower than 130 mr/yr, the best mean value of all countries reported in the United Nations report (1962). The situation is similar in the southwestern part of Germany.

Figures 4 and 5 indicate the distribution of dose rates of terrestrial and cosmic radiation and its percentages exceeding a certain value in Switzerland. It may be seen from Figure 4 that 4 per cent of the population will be exposed to more than 150 mr/hr, 10 per cent to more than 100 mr/yr, and 28 per cent to more than the mean value (79.1 mr/yr) of the dose rate from terrestrial sources. Figure 5 indicates that levels of the mean dose rate from cosmic sources (39.2 mr/yr) will be exceeded by 35 per cent, the level of 50 mr/yr by 5.7 per cent, and the level of 60 mr/yr by 1 per cent of the population.

AREAS OF HIGHER NATURAL RADIATION

It is considered that dose rates from natural sources exceeding the general mean value of background radiation (about 130 mr/yr) more than 150 mr/yr, that is, the mean additional maximum dose rate recommended by the International Commission on Radiological Protection (ICRP) for populations, are of interest from a biological and technical standpoint.

Smaller areas with more than $130 + 150 = 280$ mr/yr due to larger than normal amounts of naturally radioactive material are found sometimes in the gneiss, granite, and porphyry formations of our mountains. We found such higher natural radiation areas in

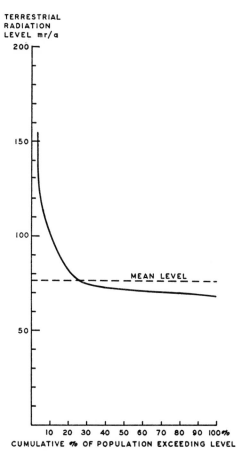

FIG. 4.—Cumulative per cent of the population of Switzerland exceeding value of terrestrial radiation level (with Eidgen. Gesundheitsamt, Bern).

Switzerland (up to 300 mr/yr) and in the Black Forest (Germany), with dose rates up to 500 mr/yr and in one uninhabited location up to 1,800 mr/yr. Such a Black Forest situation is shown in Figure 6.

VARIABILITY OF RADIATION LEVEL BY REASONS OF CIVILIZATION

I. MODIFICATIONS OF ENVIRONMENTAL RADIATION BY MINING

By mining, material with a specific content of natural radio-active nuclides is brought up from the interior of the earth's crust.

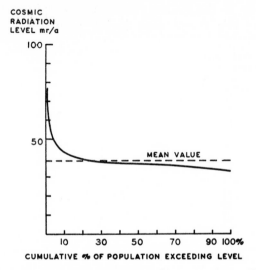

FIG. 5.—Cumulative per cent of the population of Switzerland exceeding value of cosmic radiation level (with Eidgen. Gesundheitsamt, Bern).

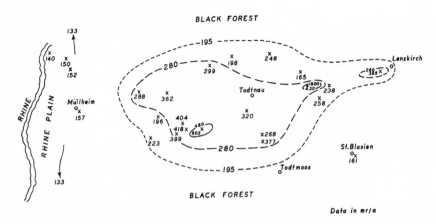

FIG. 6.—Example of a "hot" region in southwestern Germany

The blast-furnace process more or less concentrates this radioactive material. As a result, the slag stones often have increased radiation levels. An example of such a situation is shown in Figure 7 from Duisburg (with W. Altvater), comparing the normal values of radiation dose in the town with the dose rates in regions where slag stones are used for road-making, etc.

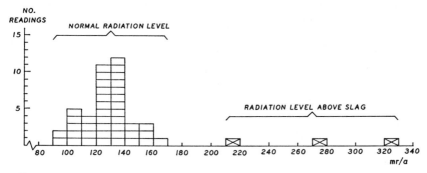

Fig. 7.—Normal terrestrial dose rate compared with the dose rate above slag in a town (Duisburg).

2. INFLUENCE OF ROAD-MAKING AND HOUSING CONDITIONS
ON BACKGROUND LEVEL OF RADIATION

Generally, road-making modifies the dose rate of environmental radiation. In the lower regions of southwestern Germany, the road metal often has a content of natural radioactivity higher than that of the environment. Therefore, on the average, the dose rates here are 35 per cent (−10 to +97 per cent) higher above the roads than in the vicinity. But the special layers of the Autobahn decrease the dose rate on the average by 17 per cent in these regions (Herbst and Hübner, 1962).

The opposite situation is observed in many mountain regions of the Black Forest, where the regional dose rate is relatively high, and the road material decreases the dose rate locally, on the average, by 26 per cent (17–41 per cent).

Owing to the different radioactive content of the building material, the ventilation, etc., large variations of the relative dose rate inside buildings are found in Germany and in Switzerland. There is a need for further measurements to have better information about the levels of radiation to which man is exposed indoors. At the present time, our data indicate that for the investigated situations in these regions, on the average, the terrestrial dose rate in air indoors may be 20 per cent higher than in air outdoors.

3. MICROCLIMATE OF BACKGROUND RADIATION IN AGRICULTURE AND FORESTRY

Decomposition and cultivation of the soil, the presence of potash fertilizers, physiological processes in the plants, etc., involve the possibility of local variations of the background-radiation dose rate above the soil surface. For these reasons, substantial variations of terrestrial radiation levels (between —25 per cent and +25 per cent) are occasionally found within small agricultural areas.

In forestry there are similar findings. Furthermore, tree tops have a filtration effect vis-à-vis the air-transported radioactive material. And on the weather side of the trees we found two to five times more fallout radioactivity than is found on the opposite side. With regard to occasionally higher dose rates from fallout (May, 1962: 3.1 µr/hr, max. 5.7 µr/hr), such a filtration process may influence also the local distribution of radiation levels.

NEUTRON COMPONENT OF ENVIRONMENTAL RADIATION

The more recent discussions of background radiation take into account also the dose rate due to the neutron component of cosmic rays. The cosmic radiation tissue dose rate is difficult to determine. For middle latitudes, the estimations vary between 25 and 88 mrem/yr at sea level and between 100 and 300 mrem/yr at an altitude of 3–4 km. (United Nations, 1962).

We are interested in the additional flux and dose rate due to the neutron component from spontaneous fission in the earth's crust. Again in collaboration with the Swiss Federal Public Health Department (Eidgen. Gesundheitsamt) in Bern (G. Wagner, W. Rottenberg, A. Mastrocola) and the Institute of Mineralogy of the University of Bern (T. Hügi, E. Halm), we started measurements to determine this component. We used (with H. Dresel) the simple procedure of placing nuclear track emulsion film badges at various locations and of counting the number of proton-recoil tracks resulting from the neutrons after long exposures.

With the exception of special geological structures, the calculated additional neutron flux from terrestrial sources remains low in open air (lower than 10^6 neutrons/yr/cm²). However, above rocks and in buildings with larger than normal amounts of natural radioactivity and especially in tunnels, in the midst of rocks with a thickness of 300–1,100 m. and with a high background of the ionizing radiation component (gneiss, granite, porphyry), we obtained a neutron flux up to 40×10^6 neutrons/yr/cm².

Taking into account a mean neutron energy from fission of 1–2 mev., we estimate in these exceptional locations, according to Snyder (1961) and Handbook 75 of the National Bureau of Standards (1961), a terrestrial neutron dose rate up to 100 mrad/yr and a tissue dose rate of several hundreds mrem/yr. These studies will be continued.

SUMMARY

Measurements of the dose rate of external penetrating environmental radiation were made with the aid of an ionization chamber (25 l., air-equivalent, calibrated with radium) in several regions of Germany and, in collaboration with the Swiss Federal Public Health Department, Bern, in Switzerland. The measurements are satisfyingly in accord with 1958 readings by Solon *et al.* in several locations in the two countries. The principal aim of the measurements was to determine the variability of the terrestrial component of the environmental radiation. The population burden of the natural radiation level, with terrestrial and cosmic component separated, in larger residential areas and in the whole of Switzerland was determined. The variability of the terrestrial dose rate was measured with respect to more than 25 defined geological formations. An example of a geological "hot" spot in the Black Forest is given. Attention was paid to variations of the dose rates by reasons of civilization and industry:

1. Settlements and housing conditions influence the radiation level.

2. Mining brings up to the earth surface subterraneous material containing specific natural radioactivities.

3. Road-making involves significant variations of the local background radiation.

4. Agriculture and forestry influence the "microclimate" of the radiation.

Preliminary estimations of the neutron dose from terrestrial sources are given.

REFERENCES

HALM, E., W. HERBST, and A. MASTROCOLA. 1962. Messung des natürlichen Strahlenpegels in der Schweiz. Beilage B, No. 6/1962 to Bull. des Eidgen. Gesundheitsamtes, Bern. Pp. 35.

HERBST, W., and G. HÜBNER. 1961. Untersuchungen über die durchdringende äusseren Umgebungsstrahlung. Atomkernenergie, **6**:75–81.

———. 1962. Zur Variabilität der terrestrischen Komponente der durchdringenden äusseren Umgebungsstrahlung im Freien. *Ibid.*, **7**:481–86.

HULTQVIST, B. 1956. Studies on naturally occurring ionizing radiations, with special reference to radiation doses in Swedish houses of various types. Kgl. Svenska Vetenskapsakademiens Handl. 6, Ser. 4, No. 3. Pp. 125.

NATIONAL BUREAU OF STANDARDS. 1961. Measurement of absorbed dose of neutrons, and of mixtures of neutrons and gamma rays. National Bureau of Standards, Handbook 75. Pp. 86.

SIEVERT, R. M. 1957. *In* transl. of pp. 11–21 of Effect of Radiation on Human Heredity. World Health Organization, Sept., 1957. Pp. 168. Informe del Grupo de Estudio sobre los Efectos Geneticos de las Radioacienes en la Especie Humana. *In* Boletin de la Oficina Sanitaria Panamericana (Washington), 45 (No. 3, Sept. 1958): 187–95.

SNYDER, W. S. 1961. Estimation of dose distribution within the body from exposures to a criticality accident. *In* Selected Topics in Radiation Dosimetry, pp. 647–56. Vienna: International Atomic Energy Agency.

SOLON, L. R., W. M. LOWDER, A. SHAMBON, and H. BLATZ. 1960. Investigations of natural environmental radiation. Science, 131:903–6.

SOLON, L. R., W. M. LOWDER, A. V. ZILA, H. D. LEVINE, H. BLATZ, and M. EISENBUD. 1958. External environmental radiation measurements in the United States. Proc. United Nations Internat. Conf. on Peaceful Uses of Atomic Energy, 23:159–64.

UNITED NATIONS. 1962. Report of the United Nations Scientific Committee on the effects of atomic radiation. New York: United Nations, General Assembly Official Records, 17th Sess., Suppl. 16. Pp. 442.

J. E. HOY AND L. F. LANDON

49. Background Radiation Measurements in the Environs of the Savannah River Plant, 1952–63

T HE SAVANNAH RIVER PLANT has conducted an 11-year survey of the natural γ radiation of its environs so that any changes in the natural level, resulting from plant operation, could be evaluated promptly and accurately (Reinig, 1963). The results of this survey and descriptions of the instruments that were used are presented in this paper.

The plant, which was built and is being operated for the Atomic Energy Commission by E. I. du Pont de Nemours and Company, occupies 320 square miles on the South Carolina side of the Savannah River approximately 30 miles southeast of Augusta, Georgia. Figure 1 shows the location of the plant and some of its major production units (five reactor and two chemical separation areas). Environmental radiation data were obtained at (1) "on-plant" stations—within 1 mile of production areas; (2) "plant perimeter" stations—near the plant perimeter several miles from the areas; and (3) "25-mile" stations—more distant monitoring points about 25 miles from the plant.

SURVEY OF ENVIRONMENTAL RADIATION

The Savannah River Plant (SRP) is located in the Atlantic Coastal Plain physiographic province, which extends east from Augusta, Georgia, to the Atlantic Ocean. The Atlantic Coastal Plain is composed of unconsolidated sediments ranging in thickness from

J. E. HOY and L. F. LANDON are with E. I. du Pont de Nemours and Company, Savannah River Plant, Aiken, South Carolina.

The information contained in this article was developed during the course of work under Contract AT(07-2)-1 with the U.S. Atomic Energy Commission.

zero along the inner boundary (the fall line) to more than 1,200 feet on the down-dip side along the coastline (Siple, 1960). Except for several exposed placer deposits of low-concentration uranium and thorium, the soils are mostly sandy clays of fairly uniform composition. The plant elevation ranges between 200 and 300 feet above mean sea level.

Background radiation levels were measured as part of an over-all program to establish the abundance of natural radioactivity in the environment, including that in air, water, vegetation, and living or-

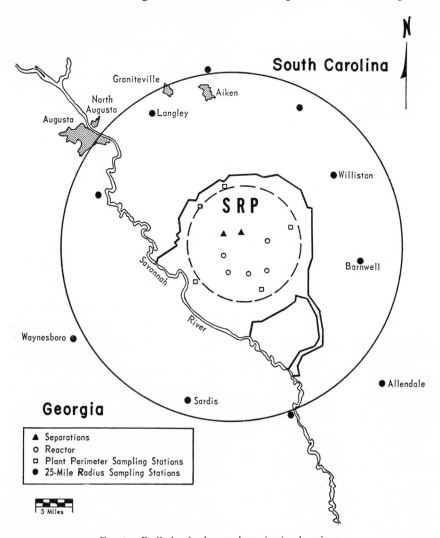

Fig. 1.—Radiation background monitoring locations

ganisms. From June, 1951, to early 1952, techniques of measuring background radiation levels were investigated, instruments were tested, and calibration work was completed. At this time ionization chambers with direct-current amplifiers and recorders were recognized as the most accurate instruments to measure radiation background. However, the high cost of these instruments and the previous

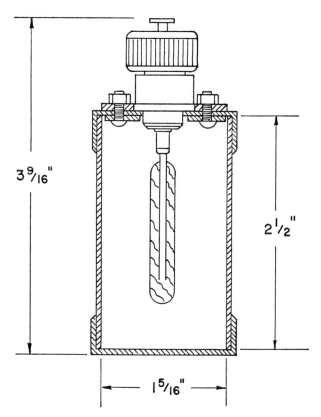

3⁹⁄₁₆"

2¹⁄₂"

1⁵⁄₁₆"

FIG. 2.—S-22 type of ionization chamber

favorable experience at Hanford with small rechargeable ionization chambers (S-22) led to their adoption (Singlevich *et al.*, 1951).

Starting in February, 1952, almost two full years before plant start-up, radiation at fixed locations on the plant site was measured with S-22 chambers (Fig. 2). This chamber has a volume of 120 cm.³, a surface density of 85 mg/cm² (which absorbs β particles of less than 300 kev.), and a full-scale response of 13 mrad. Chambers were calibrated against a radium standard with a rejection level set at ±6

PLATE I

Louvered station

PLATE II

Screened stand

per cent of the given exposure when charged and read on a Keleket minometer.

Environmental radiation was measured by exposing two S-22 chambers for weekly intervals in louvered monitoring stations (Pl. I) or screened stands (Pl. II) about 4 feet above ground. Chamber-leakage-induced errors were reduced by accepting the lower of the two readings. S-22 chambers were used from the start of the monitoring program through September, 1958. They were finally abandoned when the program was expanded because maintenance was a problem (in 1952 there were 24 monitoring stations; by 1958 there were 64).

INVESTIGATION OF OTHER SURVEY INSTRUMENTS

In late 1958 the feasibility of replacing S-22 chambers with commercially available Landsverk (L-65) pocket dosimeters (Pl. III) was investigated. The L-65 dosimeter normally ranges from 0 to 200 mr., but was changed to 0–75 mr. when a special minometer with a lower charging voltage was provided. Large quantities of these dosimeters were readily available, since they were used for routine personnel monitoring. L-65 chambers were specifically designed to

PLATE III

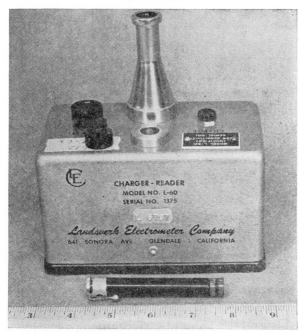

Landsverk charger reader

measure γ exposures and have a flat energy response (± 10 per cent) from 100 kev. to 1.2 mev., a volume of 5 cc., and a surface density of 238 mg/cm² (153 mg/cm² more than the S-22 chambers). Beta particles of <0.7 mev. are completely attenuated by the chamber wall. To determine the difference of the two chambers' responses to natural radiations, each type was exposed for 1-month periods at 54 routine monitoring stations. Data from this test indicated an S-22/L-65 ratio

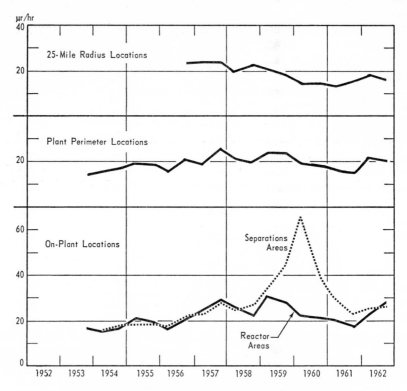

Fig. 3.—Environmental γ radiation levels (6-month averages)

of 1.97 ± 0.74 (90 per cent confidence level), implying that β and/or low energy γ radiations contributed about half the dose rate that was measured with the thinner-shelled S-22 chamber. Because primary interest was in the more penetrating component, earlier data were adjusted and the results expressed in consistent units ($\mu r/hr$). Data from the latter half of 1958 through 1962 were obtained with L-65 chambers.

Figure 3 summarizes the results of continuous measurements from 1952 to 1962 and represents the combined results of both cosmic and terrestrial components. These data are grouped according to distance from operating units and represent over 10,000 readings. (Although not of prime interest in this discussion, the bottom graph shows radiation exposure rates at less than 1 mile from radiochemical separation processes and reactor facilities.) Radiation exposure rates (environmental radiation) increased from an average of 9 μr/hr in 1952 to a maximum of 25 μr/hr in 1957. During 1961 at the end of the 3-year moratorium on testing of nuclear weapons, the radiation intensity approached the original 1952 level (average 13–15 μr/hr). This decrease and the rise in 1962 after nuclear-weapons testing was resumed provided strong evidence that fallout caused the intensity variations that were observed. Radiation levels at the plant perimeter locations and 25-mile-radius locations were generally unaffected by plant operation. The 1958 results agree well with the measurements of Solon *et al.* (1960) at Charleston, S.C. (17.2–18.2 μr/hr), and Columbia, S.C. (18.9–19.1 μr/hr), for the same period.

A chief disadvantage of rechargeable ionization chambers is that, when they are charged or read, an electronic transient is introduced. An average error of ±2.5 mr. (90 per cent confidence level) will occur in L-65 chambers during these operations; this is equivalent to ±3.5 μr/hr for a 30-day exposure. The significance of observed variations is difficult to interpret unless large numbers of readings are averaged; errors in these averages become smaller as the number of measurements increases.

Insulator leakage causes up-scale readings and produces errors that are additive. Even though the chambers are leak-tested before they are used, they may possibly break down under local climatic conditions. When pairs of chambers are exposed and only the lower reading is used, the leakage error is minimized but not completely eliminated.

SELF-READING DOSIMETERS

Bendix Model 862 self-reading dosimeters (0–200 mr.) were evaluated for possible use in the environmental monitoring program. Their rugged construction and hermetic seal reduce insulator leakage, and the direct-reading aspect reduces errors of reading and charging. Data from a Bendix dosimeter and L-65 chambers at three environmental monitoring stations for a 4-month period indicated that the Bendix dosimeter readings averaged 4.8 μr/hr lower than the L-65 results. This may be attributed to (1) less leakage in the Bendix dosimeter, (2) differences in chamber thickness, or (3) less error in reading the Bendix dosimeter because it was charged only at the beginning of the test.

SCINTILLATION AND GEIGER-MÜLLER SURVEY INSTRUMENTS

Scintillation instruments are widely used as radiation detectors because their sensitivity and efficiency for γ photons are high. Such an instrument (Technical Associates Model FS-11) was tested to determine whether it could measure environmental γ radiation levels after first being calibrated against radium. Figure 4 shows the results relative to a standard ion chamber (Victoreen Model 208). The response was extremely high at photon energies below 600 kev. and decreased to less than 1 relative to the ion chamber at higher energies.

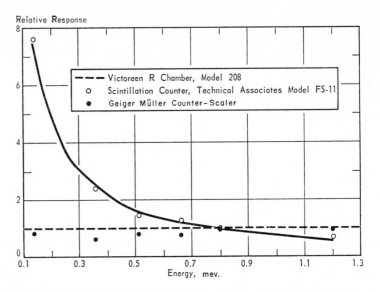

Fig. 4.—Energy dependence of background detectors relative to radium

Unless the energy of the radiation that is being measured is known, accurate dose rates could not be determined with this instrument. Figure 4 also shows the energy response of a shielded Geiger-Müller tube-scaler combination (Victoreen 1B85 shielded by 660 mg/cm² of stainless steel). Except for a 40 per cent low response at 360 kev., the curve was relatively flat (± 25 per cent) over the energy range shown. Radiation measurements with a shielded Geiger-Müller detector-scaler combination should be reasonably accurate and economical.

Of interest is the fact that the ratio of the response of a scintillation counter to that of a Geiger-Müller counter (or ion chamber) is sensitive to photon energy. This feature can be used to register significant changes in the background γ spectrum.

DOSIMETRY FILM

Personnel monitoring film has long been used to measure radiation exposures. The minimum exposure that can be detected by most types of monitoring film is between 10 and 20 mr. (equivalent to 14–28 μr/hr for a 30-day exposure). This limit is too high to measure background rates unless it is exposed for extended periods. Attempts to use personnel monitoring film (Du Pont Type 555) in the environmental program have shown that local climatic conditions restrict the exposure to a maximum of 30 days.

In the summer, films fog at high temperatures and often show positive results that are not verified by ion-chamber measurements. Additional errors may be introduced by latent-image fading if the

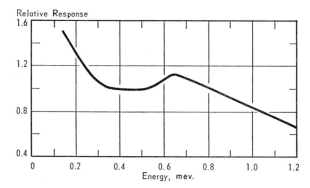

Fig. 5.—Energy response of film scintillator combinations at constant dose rates

films are not kept dry or by directional sensitivity of the film and its holder. The latter results from calibrating the film dosimeter in a unidirectional γ flux but exposing it to background radiation that is a volume-distributed source. Interpretation of background levels may be as much as 50 per cent low if directional response of the film and its holder is not considered.

Directional effects can be minimized and sensitivity limits lowered by combining dosimetry films with scintillators. Eastman X-ray Film Type KK between a pair of 1½-inch × ¾-inch sodium iodide scintillators will register radiation rates as low as 8 μr/hr in 14-day exposure, but it has the disadvantage of being dependent on rate of exposure (O'Brien *et al.*, 1958).

Identical total exposures with different dose rates do not produce equal film densities; for example, a 20 μr/hr dose rate produced a film density twice that on films without scintillators; at a 70 μr/hr rate the density of the film with a scintillator was 11 times that of regular film. This technique is useful, however, when the dose rate is constant; but when the rate changes or is unknown, exposures cannot be determined accurately. At constant dose rates the energy response of the film scintillator is flat (± 10 per cent) between 300 kev. and 900 kev. (Fig. 5) and has characteristics similar to the scintillation counter response at high and low energies.

CONCLUSIONS

The rechargeable ionization chambers that were used for environmental measurements were adequate to detect long-term trends in radiation levels. The absolute values of the results may have been too high because insulator leakage introduces some error. Charging and reading errors prevent the detection of minor short-term trends that may exist between successive readings.

Other techniques to measure background radiation were investigated and evaluated; none, except perhaps a shielded Geiger-Müller tube-scaler combination or Bendix self-reading ion chambers, was economically suited to replace the rechargeable ionization chamber.

The continuous survey with the existing instruments provides reliable data for surveillance of the environs and an accurate evaluation of any change resulting from plant operation.

REFERENCES

O'BRIEN, K., L. R. SOLON, and W. M. LOWDER. 1958. Dose-rate dependent dosimeter for low-intensity gamma-ray fields. Rev. Sci. Instr., **29**: 1097–1100.

REINIG, W. C. 1963. The 1951 preoperational environmental survey for the Savannah River Plant: In retrospect. Health Physics, **9**:83–85.

REINIG, W. C., R. E. GOSLINE, E. L. ALBENESIUS, and R. S. WILLIAMS. 1953. Natural Radioactive Contents of the Savannah River Plant. U.S. Atomic Energy Commission Rpt. DP-27.

SINGLEVICH, W., J. W. HEALY, J. H. PAAS, and F. E. CAREY. 1951. Natural Radioactive Materials at the Arco Reactor Test Site. U.S. Atomic Energy Commission Rpt. HW-21221 41.

SIPLE, G. E. 1960. Piezometric levels in the Cretaceous sand aquifer of the Savannah River Basin. Georgia Mineral Newsletter, Vol. **13** (No. 4, Winter).

SOLON, L. R., W. M. LOWDER, A. SHAMBON, and H. BLATZ. 1960. Investigations of natural environmental radiation. Science, **131**:903–6.

50. *Radioactivity of Some Rocks, Soils, Plants, and Bones*

THE PRESENT COMMUNICATION reports further observations, mainly of α radiation, on the substances referred to in earlier papers on the same subject (Marsden, 1959, 1960) and also from wheat samples reported by Marsden and Greer (1961). Owing to the wide range of materials dealt with, the work has been an extended reconnaissance directed toward the discovery of exceptional levels of activity rather than a detailed systematic investigation of particular classes of material. This was largely conditioned by the apparatus and facilities available, although, thanks to the kind help from colleagues in several institutions, the range of observations has been extended.

Although in many directions firm conclusions require more extensive experimentation and data, it is considered desirable to place on record such observations as have been made to date. The main interest centers in the Pacific island of Niue, but observations of material obtained elsewhere will first be described, since they help in our understanding of the situation in Niue.

Moreover, evidence is presented of the possible effects on humans of natural foods ingested containing significant quantities of α-radiating nuclides.

PHOSPHATE ROCKS AND THEIR USE IN FERTILIZER

In an earlier paper (Marsden, 1959) data were given for the α activities of various sources of phosphate rock as used in the manufacture of superphosphate for agricultural purposes, showing Florida and Pacific island phosphates as about 200 pc/gm total α activity and Peruvian phosphate as 4.5 pc/gm for old and 0.8 pc/gm for fresh guano. Samples have since been obtained from Nauru Island at

SIR ERNEST MARSDEN is guest research worker at Dominion Physical Laboratory, New Zealand.

regularly spaced intervals of 6 feet across the flat source area under extraction and at successive 3-foot depths into the spaces between the underlying coral pinnacles. The outstanding result was that, although the phosphorus content was approximately the same, that is, 16 per cent, at all points, both laterally and in depth, the α activity from uranium and daughter products in equilibrium varied sixfold laterally from point to point on the surface. In all cases, the value decreased with depth. A typical case was a diminution from 175 pc/gm at one point on the surface progressively to 60 pc/gm at 15 feet in depth. It may be that the raw material is a residual leached product and has in the past been enriched with uranium, etc., by sea spray (Wilson, 1959). An interesting feature is the absence of thorium, whereas the North African phosphates, obviously of marine origin at some stage of their history, as witness the presence of sharks' teeth, show only 55 pc/gm, with half their activity due to thorium. It may be that the Pacific island phosphates do not arise from sea-bird droppings as in Peru and as has sometimes been supposed. These considerations may have relevance to the accumulation of activity of Niue soils to be described later.

It is of some interest to investigate the question as to what extent, if any, the application of superphosphates to soils increases their radioactivity and that of the crops grown thereon. Superphosphates, because of the manufacturing process involved, have only about half the activity of the parent rock. In New Zealand the superphosphate activity is about 100 pc/gm. It is applied to permanent grasslands at an average rate of, say, 3 cwt. per acre, with a rainfall of about 45 inches per year fairly evenly distributed throughout the year. Considering the large rate of growth of grass and the relatively small amount of radon daughter products in natural fallout in the Southern Hemisphere, with its large proportion of ocean to land, it appears that the α activity of the grass does not arise appreciably from polonium, compared with the total α activity taken up from the soil. It so happened that an area of the Grasslands Research Station at Palmerston North was laid out in replicated plots some sixteen years ago. One section has received no fertilizer in the meantime, and another was given superphosphate at the very high rate of 16 cwt. per acre each year. The soil is fairly active naturally, approximately 12.5 pc/gm, with about half from thorium daughters. Samples of soil from depths of 0–$1\frac{1}{2}$ inches, $1\frac{1}{2}$–3 inches, and 3–6 inches were taken after the sixteen years referred to. There was found to be little variation of activity with depth, and the average fertilized soil was somewhat less than 5 per cent greater in activity than the non-fertilized soil. The γ-ray activity showed a similar ratio. In terms of the normal practice of using approximately 3 cwt. per acre, this difference after so many years is practically insignificant. Plots to which large dressings of gypsum and phosphate had been applied showed slightly enhanced activity. There was, however, a larger proportional activity in the ash of rye grass grown on the fertilized area. This may be related to possible decrease in pH caused by the fertilizer. It is hoped to repeat and extend the measurements to elucidate this question. It may perhaps be mentioned that the annual yield in dry matter on the superphosphate area was 11,700 lb. per annum as against 9,000 lb. on the control, and the phosphorus content was 0.46 per cent and 0.40 per cent, respectively.

This question is one facet of the general question of whether our natural environment and foodstuffs have become increasingly active over the last generation, which has shown tremendous increase in the use of superphosphates as well as several other sources of radiation, for example, tetra-ethyl lead. For instance, samples of old wheat collected in England, South Australia, and New Zealand thirty years ago and tested recently showed less activity on the average than present wheat in these countries, even, apparently, when the decay of radium-228 "pickup" is taken into account. This may be related to the varieties of wheat used at present, bred for rust resistance and higher yields on the more acid soils to which wheat-growing has been extended, or possibly in small part to the practice that has grown up of drilling in superphosphate with the seed wheat. An interesting case was found from the examination of a activity of bones of sheep raised on the University Farm at Cambridge, England. Forty years ago this was fertilized with basic slag, which is relatively very inactive and also tends to increase the exchangeable calcium in the soil. Sheep bones collected then and preserved in a clean condition showed only about one-quarter the activity of modern sheep bones from the area that has now been fertilized for a few years with superphosphate. There are still smallish but significant amounts of thorium in the old bones.

Bones of lambs raised in the area last year were secured immediately after killing, it being expected that, in accordance with results obtained in New Zealand, these would show appreciably lower activity than sheep bones. This was not the case, however, and the epiphyseal ends of the femurs showed activity about twice that of the bone shaft. It was subsequently found that the lambs had been "trough fed" for two months on a proprietary mixture including dried beans from North Africa, which proved to have pronounced specific a activity in the ash. One of the lambs had been put back on grass only, that is, without supplementary food, for two months, and the activity of the bones proved appreciably lower, with no difference between epiphysis and shaft. Unfortunately, a set of sample bones large enough for better statistics could not be obtained, but the tests seem to show that α-emitting nuclides in the diet, chiefly radium, move very quickly in and out of bones. On my return to New Zealand last year, metacarpel bones of 8–10-year-old cows of the same breed were secured from some 20 farms on different soil types with varying fertilizer practices and rainfall, but in all cases the cows had eaten growing grass only, that is, no foodstuffs from outside the farm. The bones gave activities ranging systematically from 2 to 6 pc/gm bone ash. The measurements are being repeated at intervals to provide estimates of the nuclides concerned, and the results await analysis, but in general the activity is lower on limestone soils.

These various measurements do, however, indicate that care is necessary in interpreting results of natural α activity of food products and bones and the migration of radioactive nuclides inside the body.

Wheat

Marsden and Greer (1961) gave results illustrating the dependence of the α activity of wheat on variety and soil. For different varieties on the same soil the variation was 2.5:1, and for the same variety of wheat on different soils the variation was 8:1, being highest on acid soils. Attention was also drawn to the high activity of the ash of a bulk shipment of wheat from Western Australia, which gave 15 pc/gm, as compared with 2.5 pc/gm for other parts of Australia and also New Zealand, only about 1 pc/gm for wheat from the United Kingdom and certain parts of Canada, and 8 pc/gm for two samples of wheat used for breakfast foods in the United States. Samples from wider areas in Canada have since shown values somewhat higher than the original ones, while wheat from two soil types in Southern Rhodesia gave a value of 1 and 8 pc/gm, respectively, and Irish wheat about the same. It is not suggested that these amounts constitute anything in the nature of a hazard.

However, by the courtesy of appropriate authorities, samples of wheat were secured from the various experimental trial areas in Western Australia, together with soil samples at two depths to 15 inches. In one case, representative of a fairly large area measured in tens of square miles, the variety "Gabo" showed the highest pickup and gave a result of 150 pc/gm ash and similar enhanced values in the bran and flour produced therefrom. Only one other of the trial areas gave values approaching this figure, and four were normal. Other varieties of wheat grown on the same soil gave about 75 per cent of this high value. Incidentally, it is interesting to note that nearly all Australian wheats give an ash content on the order of 1.25

per cent, as against 1.7–1.8 per cent for English wheats. The samples of soil referring to the above-mentioned wheat gave activities on the order of 150 pc/gm, interestingly enough with a higher value at 8–15-inch depth than at 0.6 inches. Two lots of sheep bones collected from abattoirs in the area gave values of 80 pc/gm ash, a hen bone gave 80 pc/gm ash, and the ash of trees also showed high activity. Samples of granite, the parent rock of the soil, showed very high α activity with a background γ activity of 0.1 mr/hr. The area in question is sparsely populated and hardly suitable for population genetic studies.

In considering the possible significance of such high α activity in wheat, regard must be given to the Ra/Ca ratios in the rest of the diet, since radium and calcium have somewhat similar chemical properties, though some of the salts show different solubilities. The mean daily intake in mg. of calcium in the human diet in the United Kingdom is as follows (ARCRL, 1960):

	Mg/Day
Milk and cream	496
Cheese	95
Vegetables and fruit	87
Flour and cereals	47
Other, eggs, meat, fish, etc.	56
Drinking water	60
Creta preparata	243
Total	1,084

Creta preparata is calcium carbonate compulsorily added to flour by law, but that used is fortunately low in activity. The contribution from milk and cream corresponds to about one pint of milk per day. It will be noted that the contribution of flour and cereals is 47/1,084 = 4.3 per cent. (For Indian agriculturists the figure is over 25 per cent.) Consider those people who consume only one-fifth pint of milk per day, for example, do not eat breakfast foods but possibly do eat whole-meal bread, and live in those countries where creta preparata is not added to flour. Then the proportional contribution of calcium from flour and cereals would be about 2.5 times as much, and, if the activity of the wheat concerned were 150 pc/gm, this would involve a few-hundred-fold increased intake of α activity from this source and an intake into the food of many times that of the normal Western diet. It may be that, in considering radiation hazards, we pay too much heed to γ radiation and to average α-radioactive intake rather than giving attention to the 5–10 per cent of the population subject to conditions very different from the average. More such cases will be given later in this paper, but it may be mentioned here in passing that the content of radon in potable water was found to vary from 0.3 pc/l for some New Zealand surface waters to over 1,000 pc/l for several of the artesian waters so far tested. Of course, the radioactive effect of this radon is far less than that of radium, etc., but the radon unexpectedly caused considerable trouble in apparatus used in some earlier measurements, since lead-214 and bismuth-214 were deposited on copper surfaces washed with tap water containing radon.

Before the subject of plant products is left, mention may be made of observations of the activities of wood ashes and tobaccos, since both show wide variations that may have significance in the latter case. Normal wood gives an ash content on the order of 0.3 per cent, with activity of, say, 2 pc/gm, but certain varieties of exotic timbers grown on one common soil type with high altitude and rainfall in Southern Rhodesia gave ash contents up to 1.9 per cent and α activities of ash up to 90 pc/gm. As to tobaccos, raw leaf was collected from many parts of the world, and ash contents varied from 7 per cent to 24 per cent, while α activity of ash varied from 3 pc/gm to over 100 pc/gm. Samples of seed showed very little activity. Mr. M. A. Collins, of this laboratory, measured the unstable and the stable active radicals in the smoke condensates from some of these samples of tobacco, and there appears to be a probability of some measure of relationship. The high activity tobaccos grew on soils derived from geologically old granites. These observations are dealt with in a separate paper (Marsden and Collins, 1963).

RADIOACTIVITY OF PLANKTON AND SHELLFISH

As a result of some observations of A. T. Wilson (1959) that nitrogen and phosphorus were air-borne to and deposited on land, particularly by the high westerly winds in New Zealand, some of the sea foam in a heavy wind was collected and dried. It was found to carry significant α activity, with little or no contribution from thorium products, and, in view of his observation of nitrogen and phosphorus content, it is presumed that this might arise from fragmental plankton. Consequently, tests were made on mussels, which use plankton as food, and these also showed similar and high activity. At the same time, speculations regarding the radioactivity of Niue indicated the possibility of a similar origin, at any rate in part. A plankton net was therefore sent to Niue, and the authorities there kindly arranged to have the net trolled from a canoe for the collection of plankton. This material was found to be highly radioactive, and the activity change with time indicated that it was not radium or thorium. Plankton was collected early in 1961 during a sea voyage from New Zealand to the United Kingdom, by allowing the deck hose to pass water from about a 17-foot depth through a small plankton net, and this radio-active property of plankton was shown to be general all across the Pacific and Atlantic Oceans. I have no knowledge of plankton types but, from the variations in times of day of collection, deduced that it was a common property of phyto- and zoöplankton. The yield varied considerably with locality. The specific activity was on occasion extraordinarily high, at several places 80 pc/gm dry matter at 110° C. There is an admixture of thorium products in plankton when collected in off-shore waters. However, there was a very definite variation from place to place in the relative amounts of lead-210 and polonium-210 in the plankton. This was shown by measurements showing increases in the α activity in the months following collection. In many samples the α activity showed a few-fold increase in the 8 months following collection; in other cases there was less than 10 per cent increase, in others an actual decrease in the 4 months following collection. Similar amounts of activity were also found in samples collected near the Azores through the kind courtesy of Dr. Currie of the "Discovery."

On arrival in the United Kingdom, Dr. C. R. Hill of the Institute of Cancer Research kindly examined one of the specimens in an α-particle spectrometer and found that the α activity was almost wholly due to polonium. The observations on the plankton collected were continued on the return sea voyage from the United Kingdom to New Zealand via South Africa and the Indian Ocean. Very large yields of plankton were obtained off the western coast of Africa (Gulf of Guinea), but there were small yields in the southern Indian Ocean. At Cape Town, I discussed the matter with R. D. Cherry, of the University of Cape Town, and he subsequently reported that he had made measurements on plankton collected there recently and also on some samples collected in 1900 and found that the latter showed only $\frac{1}{8}$ the activity of the former, in agreement with calculations based on the 20-year half-life of lead-210, the parent of polonium-210. It would appear, therefore, that plankton can scavenge large amounts of lead-210 and polonium-210 from sea water, but whether arising from products of uranium and radium present in the water or from the natural fallout from radon-222 in the atmosphere is not known. The latter might be indicated from earlier measurements made on ash from grass samples of areas with higher altitude and rainfall yet with slow growth of grass, these samples having been made available to me by Dr. Scott Russell and reported by Marsden (1959, 1960). In some of the plankton samples there were indications of the possible presence of very small amounts of strontium-90, but more extended tests would be necessary to establish and evaluate their presence.

Recently, observations were made on the bones of plankton-feeding birds, shearwaters, made available to me by the New Zealand Dominion Museum. These bones were relatively of very low activity and partly thorium, indicating that the plankton radionuclides were not particularly bone-seeking. Cockles, mussels, and pipis collected in New Zealand waters have also been examined, and they showed considerable polonium activity. The flesh of the paua was also examined and showed practically no radioactivity. I found later that the paua feeds on small seaweed and not on plankton. The activity of the foregoing three types of shellfish was on the order of 6 pc/gm dry matter, 22 per cent wet weight. C. R. Hill has obtained polonium values of 0.5–1.3 pc/gm for cockles, and also 0.1 pc/gm wet weight. Measurements are being continued to enable an estimate to be made of the ratio of lead-210 to polonium-210 and the contribution of thorium.

These observations of the α activity of plankton and shellfish raise two interesting speculations. In the first place, it is possible that the action of the intense α radiation on the plankton may be responsible for oil production in the sediments from a shallow ocean, as reported by K. E. Emery (1960). After 100 years this would leave no trace of activity. Some of the possible source beds in New Zealand may well be similar in nature. Another source of such transient radiation but with longer half-life is the large amount of thorium-230 that has been noted in deep sediments.

The second speculation relates to genetic characteristics of Polynesian people, for example, the New Zealand Maoris. Polynesians have always eaten considerable amounts of shellfish. It was dried and stored during their voyages. They live mostly near the sea, and in the case of the Maoris, for example, I am informed that it is a fairly low estimate to assume that they consumed, on the average, an equivalent of about 6 medium-sized mussels per day. This would correspond to some 150 pc. of polonium α activity per day, probably in more assimilable form than that of the assumed value of 5–10 pc. per day intake of all forms of α activity in the European diet of, say, 25 years ago. It is possible that the several-fold-higher incidence of such congenital impairments as clubfoot, albinism, etc., may be somehow related to mutagenic effects of this enhanced activity. I am indebted to Dr. Kennedy Elliott (personal communication, 1963) for unpublished figures indicating that the excess of clubfootedness in Maoris compared with Europeans is sixfold, and similarly for Hawaiians. Joseph Banks, who called with Captain Cook, mentions in his diary the frequency of albinism in Tahiti. Although the suggestion is speculative only, it is relevant to the Niuean situation described later. It might be interesting to ascertain the proportion of such defects in sections of Polynesians who lived inland, for example, the Urewera tribe in New Zealand, provided there were enough persons available from a statistical point of view.

NIUE

In the course of a survey of the α activities of the soils of New Zealand and the South Pacific islands, a sample from Niue was provided by the New Zealand Soil Bureau, which had already noticed some years ago that it possessed noticeable β-ray activity. When this sample was measured for α radiation, however, it gave very large values, and a study of the rise of activity after sealing showed that the activity arose mainly from radium. A γ-ray energy spectrum confirmed this with no evidence of thorium products. Recently, the Research Branch of the Division of Radiological Health, U.S. Public

TABLE 1

SOIL AND BONE ANALYSES FOR NIUE SAMPLES

	TOTAL URANIUM			RADIUM-226 (d/m/gm)	THORIUM-230 (d/m/gm)
	µg/gm	d/m/gm	(d/m/µgm U)		
Soil, 1.............	36	55	1.5	743	17
Soil, 2 (total α, 1,800 d/m/gm)...	36.5	56	1.53	415	59
Bone ash (hen).....	<0.04	<0.06		18	2

Health Service, has kindly supplied the analyses (in Table 1) of a fairly radioactive soil sample submitted to them, and also of hen bone (moderately active, compared with some more active ones available later). A different sample, examined by C. R. Hill and kindly reported to me, gave little uranium activity, compared with thorium-230 and radium, but considerable polonium. The latter may have arisen from air-borne fragmental plankton.

Niue (19° S. 170° W.) is an isolated island 13 miles wide and 11 miles long, surrounded by great ocean depths with no suggestion of past geological relationship or land connection to other lands. It is presumably a large volcanic cone completely covered to an unknown thickness with limestone in the form of emerged reefs, forming an old raised basin at about 200 feet, and other intermediate terraces. Niue soils consist of residue from weathered limestone considerably enriched by volcanic-ash-shower material of unknown but probably far distant origin.* An average soil has the percentage composition given in Table 2 (from Schofield, 1959). Mineralogically, the soil contains gibbsite, crandallite, goethite, and magnetite, etc. Crandallite is particularly noteworthy as a scavenger of certain nuclides.

The soil is mostly in pockets between limestone reaching to or near the surface and in the best portions (*A* in Fig. 1) up to 3 feet deep. It is very fine grained and porous, and nowhere does the 77 inches per annum of rainfall lodge. There are no ponds or streams, but the soil is quite moist at 1 foot. Thus the soil may possibly filter out the uranium, radium, etc., and fragmental plankton in the air-borne salt spray that accompanies the rain on occasions. The soil is less evident on the terraces nearer the coast; its activity when dry and in equilibrium to bismuth-214 is up to some 3,000 pc/gm (no thorium-232), found in one special somewhat undulating, but slightly depressed, area of approximately one-half square mile near *A* (Fig. 1). It is freely radon-emanating, samples of air near the surface giving about 5 pc/l, and a

TABLE 2

PARTIAL ANALYSIS OF NIUE SOIL

Per Cent		Per Cent	
0.8.	SiO_2	1.3.	CaO
36.	Al_2O_3	0.17.	MgO
27..equivalent..	Fe_2O_3	0.05.	K_2O
1.5.	TiO_2	0.17.	Cr_2O_3
2.45.	P_2O_5	0.2.	MnO

sample of air drawn from the bottom of a hole freshly dug to 18 inches depth giving 1,500 pc/l. The average γ activity over the whole island, except for the fringing coral terrace, is on the order of 0.05 mr/hr and at area *A* (Fig. 1) up to 0.30 mr/hr, varying with time and meteorological conditions, and some 15 per cent less over cultivated soil than over a grass track across the soil between the planted area. It is as though the escaping radon and γ-radiating bismuth-214 were absorbed in the grass mat. For comparison, the background activity at Samoa is on the order of 0.005 mr/hr; Tonga, 0.0025; Fiji, 0.003; Aitutaki, 0.01–0.02; most of the Cook Islands, 0.0025, the highest being 0.02 over Black Rock, Rarotonga. No other South Pacific islands show α or γ activities nearly comparable with that of the agricultural land of Niue.

It is difficult without a thorough examination of the rainwater and

---

* By courtesy of Dr. Bruce C. Heezan of the Lamont Geological Observatory I have recently secured samples of sea bed cores taken at 2,600 fm. off Niue at 19° S. 170°40' W. These show the presence of considerable volcanic ash. The nature and activity are being investigated.

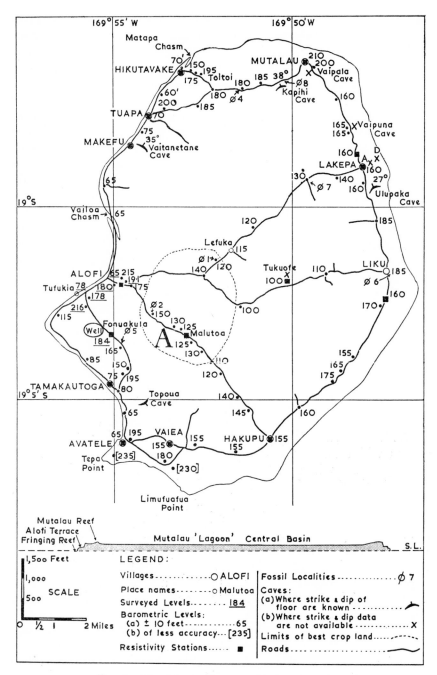

FIG. 1.—Niue Island and topographical cross-section

spray to estimate whether it is responsible for the disequilibrium as mentioned above, although the large proportion of polonium shown in the α spectrometer examination by Dr. Hill may be considered as evidence of such a source through the fragmental plankton. There seems no sign of radioactive nuclides rising from below the ground, although Schofield (private communication) adduces some possible evidence of hydrothermal activity from a supposed caldera below. The thickness of the coral to the underlying volcanic rock is not known, although a well hole was sunk 190 feet continuously through coral limestone to a water table just above sea level and at the position indicated on Figure 1. The underlying rock is presumably basalt, though there is a possibility of its being a form of the much more radioactive phonolite. Phonolite rocks, said to be of late Tertiary age, were collected from several places in the South Pacific, for example, the so-called "Black" rock at Rarotonga, Chatham Islands, an area near Dunedin, N.Z., and one from the Antarctic. The rocks are fine grained and massive. They show an activity of 65 pc/gm, over half from thorium daughters, but possibly an enhanced amount of actinium is indicated by a surplus of double scintillations from the radon-219 and polonium-215 decays. A sample of soil from Aitutaki, presumably phonolitic and collected recently from a good agricultural area, gave 80 pc/gm of α activity, with 80 per cent due to thorium products.

Two lots of sample profiles of Niue soil were taken down to 2 feet. There was not much change in the total activity of about 2,500 pc/gm, but in the lower levels there appeared to be somewhat more actinium. Moreover, samples of water were taken from the well mentioned above. This showed 90 p.p.m. of solid residue carrying 12 pc/gm of radium with an excess of unsupported radium-223, as was indicated by the decrease with time to almost zero of the ratio of double scintillations in the six weeks following collection.

At a later stage, an oxydized yellow palagonite-cemented tuff was discovered from the collection of a late medical officer and naturalist, Dr. Alan Berry, and presumed from his notes to be from a shelf some 10 feet above sea level in the northwestern part of the island. This tuff contained much calcium carbonate and appeared to be possibly slightly permeable, though the exposed part had a hard, thick, dark-brown crust. Its activity was 90 pc/gm total α activity and was interesting in relation to the soil in that it also contained no thorium, but the uranium and radium were approximately in equilibrium, as deduced from the amount of growth of radon, that is, very different from the soil, whose activity is so very largely derived from radium only. The most interesting feature of the tuff, however, was the apparent large excess of actinium, and this causes excess uranium-235 or protoactinium to be suspected, although in Niue soil the d/m/gm relative to the uranium content was near to normal, that is, 1.5 d/m/gm uranium (see Table 2). Larger samples of the tuff are to be collected so that this question can be examined. Gamma-ray spectra of the various materials were secured, including the tuff, using a sodium iodide scintillator and a multi-channel analyzer. Although these γ spectra verified the observations cited above and deduced from the examination of α particles in showing the uranium-235 line at 0.185 mev. prominently in all cases, there is a neighboring line arising from radium, and the general background activity in the laboratory was very large in relation to the effect arising from the tuff, so estimations could not be made with sufficient certainty. It is hoped to concentrate the uranium fractions and test these separately in the near future.

It is appreciated that this method of estimating actinium by radon-219 doubles (Marsden, 1960) is possibly affected by the state of subdivision and surface area of the particles of the specimen. Yet the importance of the question would seem to warrant further effort, particularly as the effect is shown also, though to a very minor extent, by the phonolites. In the latter case, a certain number of doubles would arise from any longer-period polonium-216 that may be present because over half the activity of the phonolite arises from thorium products. Nevertheless, comparison with a sample of thorium sulfate separated more than twelve years ago indicated that the effect was probably too large to be attributed wholly to thorium. Sample observations expressed in doubles per 1,000 counts per hour are plutonium, 1.1; uranium mineral, 1.6; uranium oxide, 1.2; Niue hen bone, 1.2; taro ash, 1.2–3 (according to variety); thorium sulfate, 3.8; phonolite, 5.0; Niue tuff, 9; and Niue soil, 1.28. The theoretical probability number is about 1.5, not allowing for relaxation time and doubles due to polonium-216. The figures given are for intercomparison only and include probability doubles. Similar measurements were made on material from a sea-bed core kindly supplied by Bruce Heezan of Lamont and collected from 2,693 fathoms at 19°03′ S., 170°37′ W., just off Niue and 5 cm. from the top. This material gave an activity of 30 pc/gm and an actinium figure for doubles, determined as above, of 25 doubles per 1,000 counts per hour. This indicates a high value of protoactinium.

The foodstuffs grown on the active Niue soil will now be considered. Observations have already been reported on most of these (Marsden, 1960) and indicate that the more interesting foodstuff, with the largest pickup of activity, is taro, of which the Niueans consume per head an average of 14 ounces per day (wet weight). A large collection was made of some half-dozen varieties of taro from all over the island, together with the soils on which they were grown. It is evident from the native names for different varieties that they have been collected in past voyages from islands far and wide, and it was possible, for instance, to identify two of the same varieties on a visit to Samoa. But the α activities there were of low order compared with those of Niue. Several varieties grown on the more active soils (near A in Fig. 1) gave astonishingly high results both on dry weight and as ash (0.9 per cent); for instance, two varieties gave 600 pc/gm ash, more than 90 per cent due to radium (in equilibrium with bismuth-214). There may also be a further contribution from polonium lost in the ashing process. This corresponds to a daily total α activity intake of over 2,000 pc/day, which is 100 times the average European intake on a diet of medium-low-activity carbohydrate content. In addition, there is a contribution, roughly 10–15 per cent, from the consumption of yam, cassava, etc., which are also very active, and there is a certain amount of activity from inspired radon according to circumstances. The water supply is usually rain water of low specific activity. A general account of the observations with different varieties of taro will be given at a later date, when the results and growth curves with age are fully analyzed, though in passing it may be stated that samples of the so-called giant "famine" taro showed little activity; but this variety is only consumed in periods of food crises. We may make a very rough estimate, however, that the people in a few villages drawing appreciable foodstuffs from the special active area have an intake of, say, one-fourth of the figure stated, and the islanders as a whole, say, one-twentieth, but even quantities as thus calculated are large compared with normal European daily dietary intakes. As to bones, it has not yet been possible to obtain human bones for examination. Teeth, not specially taken from those consuming the more active foodstuff, gave activities roughly 8 times those in the United Kingdom, for example. Bones of so-called "bush hens," nondescript descendants of hens introduced sixty years ago and ranging freely on Niue soil, gave very high activity, nearly all from radium, with, say, 10 per cent thorium-230. In the case of the only one shot on the special area, the activity was 300 pc/gm ash (in equilibrium with bismuth-214). The same content for a human skeleton would give a value some 2.25 times the accepted maximum permissible body burden (0.1 μgm. radium). Hens fed on wheat mash from New Zealand for egg production on an area with one-fifth the soil activity described above gave several times less activity in the bones. Eggs from bush hens gave values, in terms of ash, only of the order of 5 per cent of the bones of the mother hen. This observation may be compared with that reported earlier (Marsden, 1959), that the bones of newly born lambs and calves contain only about one-eighth the radium/calcium ratio of the bones of the mothers, indicating the discrimination of radium compared with calcium by the placenta. It was noted that fungus growing on neighboring radium-impregnated coral sands gave, after being washed and dried, a particularly high activity.

Possible Radiation Effects

When we look for possible radiation effects on the population, apart from observation problems, we are faced with great statistical difficulties due to the small total population of 4,600 with, say, 10 per cent sharing in the high-activity effects of the more special area. One important feature, however, arising from the system of land inheritance, is that the population is not migratory. Moreover, it is possible to obtain information on pedigrees and relationships from local leaders and from the records. It had been hoped to search for cytological studies of chromosome anomalies in the blood cells, since these might be expected to be much more in evidence than visible genetic impairments, but those with the necessary observational experience and techniques are rare, and none was available. However, the numbers of experienced technicians are increasing rapidly and such examination should soon be possible. One hopes that the work will not be long delayed, since certain exposure and dietary conditions are now changing.

When I first visited the island, forty years ago, there were dirt roads overhung with foliage, so inhaled soil dust and radon values must have been high. Now the roads are all covered with inactive coral, and fine chocolate-colored dust is no longer raised—a dominant feature formerly. Moreover, large areas of soil are being heavily disked and mixed with coral lime so that crop areas may be available nearer to the villages. There is probably less consanguinity in the last thirty years because the better roads enable the people to mix more. Also, there is the mixing influence of a recently established school system. More important, however, is the consumption of large quantities of milk powder, which even thirty years ago was distributed freely to expectant mothers and has for the past eleven years been given to all school children each day. The calcium in such reconstituted milk from New Zealand is of very low α activity and dilutes the higher-activity calcium accompanying radium intake from the local staple diet. Imported flour is being used in greater quantities. In addition, instead of houses with a proportion of soil in the floor, most of the islanders, because of reconstruction programs following two hurricanes, now have houses having inactive coral concrete floors and sides and allowing ample ventilation in more open conditions, which will keep down the inspired radon and local γ radiation. The removal of the villagers from Fatiau to Vaiea in 1952 with its beneficial results appears to be a case in point (Marsden, 1960). Nevertheless, chromosome anomalies will persist throughout life, and the elder people may still show such, apart from those of genetic origin.

The reported deaths from cancer in Niue are a little higher, but not significantly, than European figures, but proper pathological diagnostic service was not available. However, two cases of limb amputations due to bone cancer have been reported by a recent medical officer. In a previous paper (Marsden, 1960), particulars were given of the former high incidence of stillbirths and infertility. Last year's census figures show that 35 per cent of the 378 women over age fifty had never had a child. Among 239 of the age group thirty-five to forty-four the figure was 13 per cent. Stillbirths are not taken into account. The average number of children in the families of these groups were 3.9 and 5.9, respectively, and the same trends are shown in all near age groups. Of children under nine, there were 953 males and 817 females on the island.

The Niueans in general are pure Polynesians and may be compared with the Tongans, with whom they are most akin in normal blood grouping, and with the Samoans. The Niueans are, on the average, some 2 inches less in stature than either. This could be expected, perhaps, from past radiation exposure, considering the results of experiments with animals and findings following Rongelap, though of course it could be due more to different standards of infant feeding, protein probably being low in Niuean diet. Cataract was recently shown to be much in evidence in Niue, and, of 205 old and infirm people examined who had been left in the villages during the day as being unfit to go to the family plantation to work, 120 showed evidence of cataract condition. The ratio was somewhat higher in the villages near the high-activity family cropping areas. Following the survey and the spread of information that the medical officer was interested, a number of more seriously affected semiblind patients came forward and are included. As to genetic impairments, absolute numbers are statistically small and have been difficult to obtain reliably. There were eight albinos (2 related), or 1 in 600, in Niue, as against 1 in 6,500 in Tonga and similar ratios in Samoa and Fiji. Cases of extra digits and talipes showed correspondingly high ratios. In the case of extra digits, 3 cases were officially known, or 1 in 1,600. This compares with a figure nearer to 1 in 20,000 for the original Berlin report. Mr. A. C. S. Wright, the soil surveyor, reported verbally to the writer that he had seen 6 cases in a gang of 20 Niuean workers. These may, of course, have been in related persons. In the case of clubfootedness, although several past medical officers have remarked on this condition, a larger relative number of cases were reported to me by a Niuean teacher, and, because of the introduction of compulsory primary education, teachers may now be in a better position to give such estimates. On the whole, I consider that such impairments as those described above may well be 6 times as prevalent in Niue as in comparable islands, such as Tonga, Samoa, etc., whose chief medical officers kindly supplied me with figures.

It may possibly be relevant to mention that at the island of Aitutaki, with a population of 2,200 and where a visiting medical officer has reported a high incidence of talipes, the background γ radiation was measured as 0.02 mr/hr. A soil sample recently obtained from the collection of the New Zealand Soil Bureau and taken from a highly fertile area gave an α activity of 80 pc/gm, with some 70 per cent contribution from thorium products. The soil was well supplied with calcium and magnesium, with 92 per cent base saturation. It may well be that the soil is phonolitic in origin.

In considering the estimate of sixfold incidence of these visible genetic impairments in Nuie compared with most other South Pacific islands, we must realize that the latter may also be affected by the typical Polynesian diet rich in shellfish that have high α activity. Furthermore, the possible influence of consanguinity must be kept in mind, and estimates of its occurrence and significance must be made. Isolated communities, with frequent intermarriage, tend to concentrate such abnormalities.

On the whole, however, particularly in view of the relative figures for Tonga and Samoa, etc., and in spite of the statistical problems associated with small numbers, one gets the strong impression that radiation plays an important role in the lives of the people of Niue; but a better medical assessment is certainly called for before any definite conclusions can be reached. As to mental qualities, the Niueans, like the Tongans and Samoans, show a high degree of natural intelligence. The Niueans, in particular, are a lovable, happy people of high character. They contain a good proportion of highly intelligent members who occupy positions as pastors, teachers, nurses, officials, etc. It seems desirable that research be undertaken on the "spread" of intellect, using a suitable battery of intelligence tests. The suicide rate is somewhat high, however, and, in regard to the committed feeble-minded, it is noticeable that a high proportion appears to come from those who work on and consume the products of the more radioactive areas. On the whole, it is considered that there appears to be a definite radiation effect, probably arising more from ingested α-radiating foodstuffs than from external γ radiation.

ACKNOWLEDGMENTS

I wish to thank all those many kind friends and colleagues who have contributed to the work recorded, in particular Mr. H. Nemaia, assistant medical officer, who accompanied and guided me in my field observations; Mr. W. H. Ward, director of the Dominion Physical Laboratory, New Zealand, who gave me all facilities as guest worker; and the staff, particularly Mr. R. A. Morris, who took such pains in tropic-proofing and transistorizing my electronic apparatus that it gave me no instrumental troubles throughout. With their help I hope to continue the work.

REFERENCES

AGRICULTURAL RESEARCH COUNCIL RESEARCH LABORATORY, UNITED KINGDOM. 1960. Pub. No. 5. Pp. 20.

ELLIOTT, J. KENNEDY. Personal communication, 1963.

EMERY, K. O. 1960. The Sea off Southern California: A Modern Habitat of Petroleum. New York: John Wiley & Sons. Pp. 366.

FRONDEL, C. 1958. Geochemical scavenging of strontium. Science, **128**: 1623–24.

HILL, C. R. 1962. Identification of alpha emitters in normal biological materials. Health Physics, **8**:17–25.

MARSDEN, E. 1959. Radioactivity of soils, plant ashes and animal bones. Nature, **183**:924–25.

————. 1960. Radioactivity of soils, plants and bones. *Ibid.*, **187**:192–95.

MARSDEN, E., and M. A. COLLINS. 1963. Alpha-particle activity and free radicals from tobacco. Nature, **198**:962–63.

MARSDEN, E., and E. N. GREER. 1961. Alpha activity of wheat and flour. Nature, **189**:326–27.

SCHOFIELD, J. C. 1959. Geology and Hydrology of Niue Island. New Zealand Geol. Survey Bull. 62. Pp. 28.

WILSON, A. T. 1959. Surface of the ocean as a source of air-borne nitrogenous material and other plant nutrients. Nature, **184**:99–101.

F. X. ROSER AND T. L. CULLEN

51. External Radiation Levels in High-Background Regions of Brazil

Two types of high-background regions have been studied in Brazil. The first is the region of the monazite sand along the Atlantic Coast. The second is the zone of volcanic alkaline intrusives in the inland states of Minas Gerais and Goias. The geology of these regions is described in this article; their location is indicated in Figure 1. Reports of previous field trips have been published (Roser and Cullen, 1958, 1962a, b), and the present report will summarize the results.

INSTRUMENTS

To measure the external γ-radiation field, a combination of ionization chamber and portable scintillometers was used. The chamber, designed by V. F. Hess (Hess and O'Donnell, 1951), was a square one, $30 \times 30 \times 7.7$ cm., with one of the broad faces of thin aluminum foil to permit measurement of β radiation as well. The current was read with a vibrating-reed electrometer. The chamber, calibrated with a standard milligram source of radium, was used to establish base points of reference with readings at 1-meter height. Portable scintillometers were calibrated frequently at these points and then moved out quickly to measure wider areas.

Since anyone in his daily rounds moves through differing γ-ray fields, a device is needed to integrate the 24-hour dose he receives. Recently O'Brien, Solon, and Lowder (1958) described the development of a sodium iodide crystal with a Du Pont 508 X-ray film in a light-tight box as a sensitive dosimeter. The crystal intensifies the light

F. X. ROSER is director and T. L. CULLEN is assistant professor of physics at the Institute of Physics, Catholic University, Rio de Janeiro, Brazil.

Study supported by the National Research Council of Brazil, the Brazilian Nuclear Energy Commission, and the U.S. Atomic Energy Commission.

with an amplification factor of 1,000, and thus the dosimeter is sensitive to doses of the order of 1 milliroentgen.

Air filters were used to collect atmospheric radioactivity, and the activity of the filters was measured with the ionization chamber.

REGION OF MONAZITE SAND

The decomposition of the archeogneisses in the mountain range that parallels the Atlantic Coast has produced a natural separation and concentration of insoluble minerals such as monazite. Ground to fine

FIG. 1.—Regions of high natural radioactivity surveyed

particles and carried downstream by the many streams that empty into the ocean, the monazite underwent a second concentration process by specific gravity in the churning interplay of river and ocean currents.

This process has resulted in black deposits of strongly radioactive material, abrupt gradients, a mottled aspect on the present beaches, and occasional anomalies on older beaches inland.

Three towns built over monazite sand deposits were surveyed in detail: Guaraparí, Meaipe, and Cumuruxatiba. The first of these, Guaraparí, together with its sister city across the river, Muquiçaba, has a stable population of 5,600 and sees an annual influx of 10,000 vacationers.

The results of previous aerial studies made by two companies, Lasa and Prospec, were used in guiding the ground studies. An aerial photograph of Guaraparí is shown in Plate I, with isoradiometric lines superimposed on the picture. A map of Meaipe is given in Figure 2, with the ground scintillometric readings indicated.

On the several field trips some 59 ionization chamber measurements were made in these three towns, and the data are given in summary

TABLE 1

SUMMARY OF IONIZATION CHAMBER READINGS IN
MONAZITE SAND REGION: TOWNS OF GUARAPARÍ,
MEAIPE, AND CUMURUXATIBA

Range of Radiation Levels (mr/hr)	No. Measurements
1.00–0.50	3
0.50–0.30	2
0.30–0.15	7
0.15–0.10	10
0.10–0.07	10
0.07–0.05	12
0.05–0.03	5
0.03–0.02	6
0.02–0.01	4

form in Table 1. The points were distributed geographically through the towns as well as from the lowest radiation level to the highest. The chamber, however, was not brought inside the monazite separation plant or onto the beaches.

Because of the manner in which these readings were taken, they must be understood as representative but not as average readings of the towns. The average of the 47 readings taken in Guaraparí (0.08 mr/hr) is too low, while those taken in Meaipe (3 readings, 0.65 mr/hr) and Cumuruxatiba (2 readings, 0.33 mr/hr) are too high.

The streets of the three towns were surveyed in detail with portable scintillometers. Measurements were made by walking slowly down the street, continuously reading the scintillometer, held at a height of 1 meter, and writing down a reading for every 16 meters. Small anomalies were explored with care, and on streets that had a

PLATE I

Guarapari: Aerial map showing isoradiometric lines. (Courtesy of Lasa Co.)

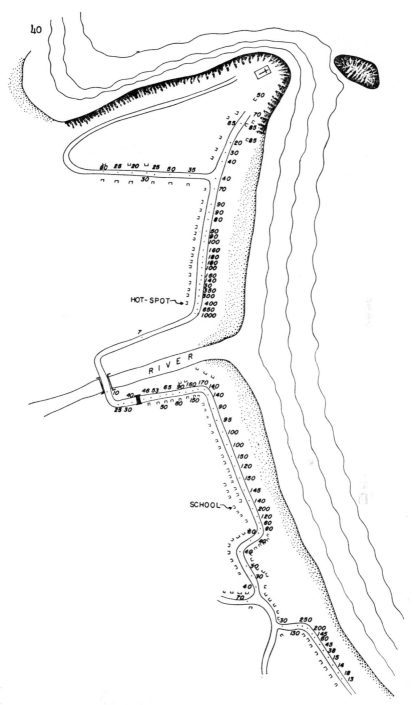

Fig. 2.—Meaipe. Scintillometer readings given in microroentgens per hour

higher population three traverses were made, one down the center and one on each side of the street.

A summary of the data is presented in Table 2. The first two principal streets of Guaraparí, rather densely populated, contain a number of hot spots, ranging from 1 to 10 square meters in area. They are distributed randomly through the streets, in gardens, and under houses. The maximum reading on these streets was due to a small anomaly next to a house, with a reading of 0.4 mr/hr.

The intensity level within buildings and homes is important because of the length of time spent by people indoors. Two types of

TABLE 2

SUMMARY OF PORTABLE SCINTILLOMETER MEASUREMENTS
IN STREETS OF GUARAPARÍ, MEAIPE,
AND CUMURUXATIBA

	Radiation Level (mr/hr)	No. Measurements
Guaraparí		
Principal streets		
I.	0.093	93
II.	0.130	120
III.	0.076	50
IV.	0.075	92
V.	0.060	88
19 Other streets.	0.057	336
Meaipe		
All streets.	0.133	94
Cumuruxatiba		
All streets.	0.053	68

homes are found in Guaraparí. The older type is a mud house with no floor. The modern type is constructed of hollow tiles and covered with surface concrete.

There was no indication of contaminated material in the mud houses, nor did there seem to be any shielding effect due to the walls. The readings within the homes simply reflected the level due to the earth below. It is evident, however, that radioactive sand is sometimes used in the construction of the tile homes, although the amount varies widely. Readings within these homes are frequently of the order of 0.05 but can be as high as 0.23 mr/hr.

The monazite sand separation plant, Mibra, presents occupational levels. There are some 500 men in the city who have worked for some years at the plant.

The highest levels of radiation intensity are encountered on the beaches. The monazite sand is concentrated at selected spots by the churning surf. These spots of *areia preta*, or black sand, are readily recognized because of the black ilmenite with which monazite is associated. These are preferred spots for sun and sand bathing and show levels up to 2 mr/hr.

A pilot program was run to measure population exposure data with the sodium iodide crystal dosimeters. A series of calibration experiments was run, tailored to fit the doses received in Guaraparí (Roser and Cullen, 1962*b*). Some dosimeters were left 24 hours at spots measured with the ionization chamber, and excellent correlation was obtained. Then some were distributed to people chosen at random. The results are given in Table 5. This work will be extended to a study of the levels received by sections of the population grouped according to the number of years they have lived within the city.

TABLE 3

RADIATION LEVELS WITHIN BUILDINGS, GUARAPARÍ

	Radiation Level (mr/hr)	No. Buildings
Hotels and boarding houses.........	0.075	7
Homes, tile and concrete...........	0.103	52
Homes, mud.....................	0.072	9
Mibra, monazite separation plant		
Separation room, bags...........	4	
Storeroom, bags................	4	
Rejected sand bin..............	3.05	
Office........................	0.30	
Playground, rejected sand........	0.25	

TABLE 4

RADIATION LEVEL ON BEACHES OF GUARAPARÍ

	Radiation Level (mr/hr)	No. Measurements
South Beach		
Inland edge............	0.185	63
Water's edge...........	0.090	63
Center................	0.070	65
Areia preta—up to.......	2.000	
North Beach		
Inland edge............	0.135	13
Water's edge...........	0.037	13
Center................	0.120	10
Areia preta—up to.......	1.850	

* *Areia preta*, or black sand, occurs in patches of 1–4 square meters.

Two air filters were exposed at a height of 6 meters at either end of the main street. They indicated a radon-220 concentration of 0.02 pc/l and a radon-222 concentration of 0.1 pc/l. These figures can be compared with the averages of three years of measurements at Catholic University, Rio de Janeiro: radon-222 concentration, 0.007 pc/l; radon-220 concentration, 0.005 pc/l.

TABLE 5

SUMMARY OF CRYSTAL FILM DOSIMETER
DATA; DOSE RATE RECEIVED BY PEOPLE

Range of Dose Rate (mr/hr)	No. People
More than 1	3
1 –0.30	3
0.30–0.20	7
0.20–0.10	6
0.10–0.06	7
0.06–0.02	6
0.02–0.01	2

REGION OF VOLCANIC INTRUSIVES

An outstanding example of a non-explosive volcanic intrusive is found in the region of Poços de Caldas. Here a protruding alkaline plug, 35 km. in diameter, has been raised to a height of 400–500 meters above the surrounding granitic substratum. Weathering and erosion lowered the inner portion of the plug, leaving intact the periphery of hard rock.

In the inner zone an intricate net of fracture lines permitted subsequent mineralization in many places. At Morro do Ferro, for instance, close to the center of the inner zone, there appeared two well-developed dikes of magnetites with admixtures of titanium and residual manganese. These are accompanied by secondary fractures yielding rare-earth oxides and a high percentage of thorium oxides (0.5–1.8 per cent) and traces of uranium.

Morro do Ferro is a steeply rising hill, 300 meters high. The radioactive region covers 0.35 km.², and borings indicate a depth of 100 meters. It is perhaps the largest known thorium deposit. More intense radiation levels are encountered in a smaller area of 0.15 km.², where the readings are up to 2–3 mr/hr.

The four measurements with the ionization chamber average 1.58 mr/hr, as indicated in Table 6. Two traverses of the hill were made with the portable scintillometers. The results of these traverses are given in Figure 3. Four filters were exposed on the hill. Two of them, run over an open trench 80 meters above the mine shaft, showed a radon-220 concentration of 10 pc/l, while filters exposed within the mine shaft indicated a concentration of 3,000 pc/l.

Cascata is a small town inside the Poços de Caldas intrusive, located near the uraniferous section of the region. The town was surveyed in detail with portable scintillometers. The average of the readings was 0.08 mr/hr, indicating some contamination. An air filter was exposed in the town and indicated a radon-222 concentration of 14 pc/l and a radon-220 concentration of 0.03 pc/l.

TABLE 6

SUMMARY OF IONIZATION CHAMBER READINGS
IN ZONE OF VOLCANIC INTRUSIVES

Locality	Radiation Level (mr/hr)	No. Measurements
Morro do Ferro.............	1.580	4
Araxá......................	0.324	4
Tapira.....................	0.200	3
Poços de Caldas............	0.027	2

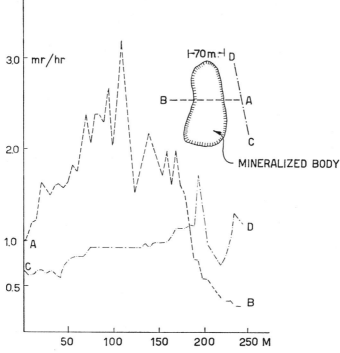

FIG. 3.—Radiometric traverse of Morro do Ferro. (Courtesy of Resk Frayha)

The city of Poços de Caldas itself is situated on the edge of the intrusive that is called by its name. Although the city is known as a spa resort with hydrothermal radioactive springs, the general level within the city is low, with an average of about 0.02 mr/hr. Two ionization chamber measurements are shown in Table 6.

Pedro Balão is a tourist-attraction spot just north of the city of Poços de Caldas and is situated on the crater rim. Here the alkaline rock is permeated with zirconiferous minerals in which uranium occurs. Over a wide region levels ranged from 0.02 to 0.03 mr/hr, and within a smaller area (0.5 km.²) from 0.04 to 0.05 mr/hr.

Taquarí is another hill within the Poços de Caldas intrusive similar in topography to Morro do Ferro but situated in the uraniferous section. In this region veins of zirconium minerals (10 per cent zirconium) with associated uranium (0.5 per cent) pervade the entire hill. Within a circle of several hundred meters the levels are between 0.1 and 0.2 mr/hr and within limited patches are between 0.3 and 0.4 mr/hr.

Araxá is a city in Minas Gerais located near another intrusive similar to that of Poços de Caldas. Here an interesting and important drainage pattern produces radioactive springs and thermal baths at the tourist center of hotels outside the city. Over an extensive pasture region, of 10 km.², the background is six to eight times normal. In smaller patches the level is as high as 0.4 mr/hr.

Tapira, in the state of Minas Gerais, is a small town of 350 within a circular plug and located near its drainage system. The intrusive is considerably smaller than Poços de Caldas, but there is rock evidence that explosive activity took place there. Possibly the violence took off much of the more radioactive rock, for the levels are lower. The interesting aspect is that the remaining radioactivity is found primarily over fertile farm land. Over a greater area, of 12 km.², the levels are three times normal, but in smaller areas as high as 0.3 mr/hr. The average for the three chamber measurements is given in Table 6.

Onça, in the state of Minas Gerais, shows an interesting example of gneissic rocks cut across with dikes of minerals rich in thorium oxides and silicates. The dikes are from 6 to 8 m. wide and up to 10 km. long. Over an area of 100 km.² the average levels are four times normal. Over the smaller area of 1 km.² the readings are between 0.04 and 0.2 mr/hr. Directly over the dikes, however, the radiation levels were between 2 and 2.5 mr/hr.

Extended Field Work in Monazite Zone

A 6,000-km. field trip was made by jeep through the northern part of the state of Rio de Janeiro, the state of Espírito Santo, and the southern part of the state of Baía. Old beaches are found inland, and, although they are usually covered with relatively inactive material, the radiation levels can be several times normal.

The purpose of these measurements was twofold: to search for populations that received radiation four or five times the normal background and to look for control populations that received normal levels. Each town was surveyed first with the portable scintillometer in all the streets of the town. An average was rapidly estimated, and an ionization chamber reading was made at a point that gave this average reading. The exposure levels for the 40 municipalities surveyed are given in Table 7.

TABLE 7

SUMMARY OF POPULATION EXPOSURE LEVELS IN
STATE OF ESPÍRITO SANTO

Range of Radiation Levels (r/yr)	Population
0.09–0.13	15,000
0.13–0.17	4,000
0.17–0.22	29,000
0.22–0.35	6,000
ca. 0.50	300
ca. 0.95	6,000
ca. 1.15	350

DISCUSSION

The significance of the radiation levels discussed in this report can be seen by comparing them with representative levels measured by Solon *et al.* (1960) in different cities of the United States. In New York City the external radiation levels are between 0.008 and 0.015 mr/hr, of which approximately 0.004 mr/hr is due to cosmic radiation. This would result in a yearly dose of 0.09 r.

During the calibration experiments the ionization chamber was placed in an iron house with walls 4 inches thick. Thus the ionization due to chamber contamination and hard cosmic radiation was measured. Since this reading was subtracted from all subsequent readings, only the levels due to external γ radiation have been given.

See Eisenbud *et al.* (this symposium) for a discussion of the internal sources of radiation.

REFERENCES

HESS, V. F., and G. A. O'DONNELL. 1951. On the rate of ion formation at ground level and at one meter above ground. J. Geophys. Res., **56:** 557–62.

O'BRIEN, K., L. R. SOLON, and W. M. LOWDER. 1958. Dose-rate dependent dosimeter for low-intensity gamma-ray fields. Rev. Sci. Instr., **29**:1097–100.

ROSER, F. X., and T. L. CULLEN. 1958. On the intensity levels of natural radioactivity in certain selected areas of Brazil. *In* Radioactive Products in the Soil and Atmosphere. Rio de Janeiro: Instituto Brasileiro de Bibliografia e Documentação.

———. 1962*a*. Environmental Radioactivity in High Background Areas of Brazil. Rio de Janeiro: Catholic University.

———. 1962*b*. Sodium iodide crystal dosimeters for use in surveys of regions of high background radiation. Science, **138**:145–46.

SOLON, L. R., W. M. LOWDER, A. SHAMBON, and H. BLATZ. 1960. Investigations of natural environmental radiation. Science, **131**:903–6.

M. EISENBUD, H. PETROW, R. T. DREW,
F. X. ROSER, G. KEGEL,
AND T. L. CULLEN

52. *Naturally Occurring Radionuclides in Foods and Waters from the Brazilian Areas of High Radioactivity*

THE LEVELS OF EXTERNAL RADIATION in the Brazilian areas of abnormally high radioactivity have been described thoroughly by Roser and his associates (Roser *et al.*, 1963, Roser and Cullen, 1963), who have shown that the levels of ambient γ radiation in the inhabited areas range as high as 1 mr/hr and that they are more than twice this value in some uninhabited regions.

Because it is known that radioactive elements present in the soil, like other trace elements, are absorbed by plants and animals and eventually find their way to man, a pilot study was undertaken to determine the extent to which the naturally occurring radionuclides are present in typical foods and waters of these regions of high radioactivity. This report summarizes the findings of these pilot studies.

It was known at the outset that the area is rich in thorium-232, and the presence of the decay products of this nuclide in foods and waters was therefore anticipated. It was also considered likely that the radioactivity in biota would be due primarily to absorption of radium-228 (mesothorium) and its daughters, rather than to the thorium isotopes,

M. EISENBUD is professor, H. PETROW is research associate, and R. T. DREW is senior research assistant, New York University Medical Center, Institute of Industrial Medicine, Environmental Radiation Laboratory, New York University, New York. F. X. ROSER is director, Institute of Physics; G. KEGEL is physicist, Institute of Physics, and director, Environmental Radiation Project; and T. L. CULLEN is assistant professor of physics, all at Catholic University, Rio de Janeiro, Brazil.

This study was supported by the Pan American Sanitary Bureau, World Health Organization, and the Division of Biology and Medicine, U.S. Atomic Energy Commission.

which have been found to be relatively unavailable to plants (United Nations, 1962). However, radiochemical studies of the minerals in these areas (Clegg and Foley, 1958) had shown traces of uranium to be present, a finding that suggested that radium-226 might also be found in the biota. This suspicion was confirmed by preliminary radiochemical analysis of the foods, and it therefore became necessary to develop procedures for differentiating radium-228 from radium-226. These procedures will be outlined later in this report.

POTENTIAL SOURCES OF EXPOSURE TO
RADIOACTIVE INTERNAL EMITTERS

Human assimilation of nuclides in the uranium and thorium chains can occur in a number of ways, including (*a*) inhalation of radon-222 and radon-220 and their daughter products; (*b*) adventitious ingestion of dust; and (*c*) ingestion of food and water.

It is likely that diffusion of radon-222 and radon-220 from the rocks and soils occurs at a rate greater than normal, but the atmospheric concentrations of these gaseous decay products in these areas have not as yet been investigated.

PLATE I

A group of children playing in an unpaved street (in Guaraparí)

PLATE II

A beach in Guaraparí. The γ-radiation levels 3 feet above the beach are greater than 1 mr/hr in some areas.

PLATE III

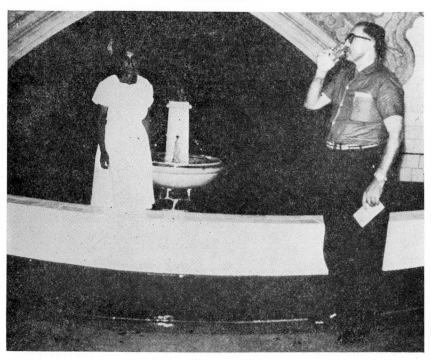

Dispensing mineral waters in Araxá

The opportunities for adventitious ingestion of radioactive dusts exist in the villages, in the country, and in the recreational areas. Plate I shows a group of children who, a short while before the picture was taken, were playing soccer on a dusty unpaved street in which the radiation levels ranged from 0.1 to 0.25 mr/hr. One of the boys can be seen with sugar cane that he chews from time to time, but when the soccer game is in progress the piece of sugar cane is laid down on the dusty street!

PLATE IV

Milk cows grazing in an area where the ambient radiation level is 0.25 mr/hr

Plate II shows one of the popular beaches at Guaraparí that is exploited for its radioactive sands. The radioactive minerals are concentrated in the darker sands, above which the γ-radiation intensities are as high as 2 mr/hr. As in many parts of the world where exceptionally high radioactivity exists, the radioactive properties of this and other radioactive areas in Brazil are exploited for their allegedly curative values, and many thousands of Brazilians take their vacations at these areas, where they come in contact with the monazite sands and drink mineral waters of high radioactive content (Pl. III). We see that there is relatively close contact with the radioactive soil of these regions, but the extent to which this results in assimilation of the naturally occurring radioactive materials has not been studied.

The most important means by which the heavy radionuclides are absorbed by man in these areas is possibly through the ingestion of food and water. Plates IV and V show milk cattle grazing and general farming activities conducted in an area where the ambient γ-radiation levels range from 0.05 to 0.25 mr/hr. Plate VI shows bananas growing

PLATE V

Farming near Araxá. The ambient γ-radiation levels are 0.25 mr/hr

in a back yard in which the radiation level is 1 mr/hr, and in Plate VII we see extensive farming in a region in which strip mining of radioactive minerals is under way. Plate VIII is a photograph of a shallow well from which both the stock animals and their owners derive drinking water.

The pilot study that will be described in this report was designed to determine the extent to which the heavy radionuclides are absorbed by the food and water in these regions of abnormally high radioactivity.

SOME PROPERTIES OF THE THORIUM
AND URANIUM SERIES

The uranium and thorium series originate in the radioactive decay of uranium-238 and thorium-232 and consist of 14 and 11 radioactive nuclides, respectively. The two chains include isotopes of 10 elements, each of which has its own metabolic properties. The 25

PLATE VI

Bananas growing in Meaipe in ambient radiation levels of 1 mr/hr

nuclides have half-lives that range from 1.4×10^{10} years for thorium-232 to 0.3 microseconds for polonium-212. From the point of view of its movement in biological systems, a radioisotope can be regarded as non-existent if the time from its formation to its decay is so short that it does not translocate from its place of formation. Thus, the biological behavior of the radionuclides is strongly influenced not only by their chemical properties but also by their half-lives.

Table 1 lists the nuclides of both the uranium-238 and the thorium-232 series, broken down into subchains having parents with half-lives of a few days or longer. This arrangement assists one in visualizing the

PLATE VII

Potato farming located adjacent to strip mining of radioactive materials near Taquarí

PLATE VIII

Shallow well near Meaipe

clusters of nuclides that are likely to be found together at various steps of the food chain. Thus, if the radium-224, having a 3.6-day half-life, should separate from the thorium-228, which has a 1.9-year half-life, then the radium-224 would be expected to be in equilibrium with 6 radioactive daughters whose lives are so short that translocation of the nuclides of this subchain is unlikely. Little is known about the kinetic behavior of the 10 elements of the uranium and thorium series, but as a first approximation one can assume that isotopes having half-lives of less than a day do not live long enough to translocate but that those with half-lives greater than a few days may translocate within the organism.

TABLE 1

NUCLIDES OF THE URANIUM-238
AND THORIUM-232 SERIES

Uranium			Thorium		
U-238	4.5×10^9 yr.	α	Th-232	1.4×10^{10} yr.	α
Th-234	24 day	β	Ra-228	5.7 yr.	β
Pa-234	1.2 min.	β	Ac-228	6.1 hr.	$\beta\gamma$
U-234	2.5×10^5 yr.	α	Th-228	1.9 yr.	α
Th-230	8.0×10^4 yr.	α	Ra-224	3.6 day	α
Ra-226	1.6×10^3 yr.	α	Rn-220	55 sec.	α
Rn-222	3.8 day	α	Po-216	0.16 sec.	α
Po-218	3 min.	α	Pb-212	10.6 hr.	$\beta\gamma$
Pb-214	27 min.	β	Bi-212	60 min.	$\alpha\beta$
Bi-214	20 min.	β	Po-212	0.3 μsec.	α
Po-214	160 μsec.	α	Tl-208	3.1 min.	$\beta\gamma$
Pb-210	20 yr.	β	Pb-208	Stable	
Bi-210	5 day	β			
Po-210	138 day	α			
Pb-206	Stable				

The relative insolubility of the long-lived thorium and uranium isotopes in the upper portions of the two chains suggests that the first isotopes to be absorbed by plants in significant amounts would be radium-226 in the case of the uranium-238 series and radium-228 in the case of the thorium-232 series. This assumption established the rationale for our analytical procedures, which were directed primarily at measurements of radium-226 and radium-228. Thorium-228 was also separated and determined.

NORMAL VALUES OF RADIUM-226 AND
RADIUM-228 IN FOOD AND WATER

Several previous investigators have analyzed foods and waters for radium-226, but there have been very few published values for radium-228 and thorium-228. Stehney and Lucas (1955) estimated the mean daily intake of radium-226 to be 1.6 pc. in the United

TABLE 2

RADIUM-228, THORIUM-228, AND RADIUM-226 CONTENT OF FOOD AND WATER

TYPE OF SAMPLE	SAMPLE No.*	RADIUM-228		THORIUM-228† (pc/kg)	RADIUM-226 (pc/kg)	γ RADIATION (mr/hr)‡
		pc/kg	pc/g Ca			
Milk.............	Ta-7	6.5	5.2	0.8	3.1	—
"	T-2	4.6	3.7	Nil	2.1	—
"	M-12	11.2	9.5	Nil	2.9	—
"	G-14	7	2.5	0.14	1	—
Cheese..........	T-3	37.9	7.9	Nil	23.2	—
Milk.............	C-4	0.32	0.28		1.1	—
Fruit:						
Banana.........	A-7	7.4	37	8		0.04
"	G-2	10.5	30	1.4	9.2	0.18
"	G-15	4	15.8	3.1	1.5	0.15
"	G-17	2.5	9.6	5.1		0.05
"	Puc-3	0.5	1.3	0.1		—
Papaya.........	M-4	2.7	5.2	1.5	0.8	0.25
"	M-5	4	7.4	1.5	2.4	0.15
"	G-16	5.5	18.2		Nil	0.08
Coconut:						
Milk.............	M-3a	<0.5§	—			1
Meat...........	M-3b	13	100	0.1		1
Husk...........	M-3c	19	51	7.2		1
Coconut:						
Milk.............	M-6a	<0.7§	—			0.125
Meat...........	M-6b		—	1		0.125
Husk...........	M-6c	3.1	17			0.125
Vegetables:						
Tomato.........	A-3	6.1	54	0.7		0.04
Cabbage........	A-5	71	35.4	21	18.3	0.04
Corncob........	Ta-1b	69	—	38.4		0.18
Lettuce.........	Ta-2	965	680	84	105	0.18
Potato..........	T-4	17	20	3.6	—	0.01
"	T-6	15	100	0.15	8.3	0.05
Corn............	M-2a	10	28	0.2	5.1	0.01
Corncob........	M-2b	61	52	7.3	8.9	0.01
Cabbage........	M-7	57	18.3	46	9.1	0.18
Bertalha........	G-3	38	188	36.8	10.1	0.09
Abobora........	G-4	4.6	25	3.6	2.1	0.20
Tomato.........	G-5	5.4	22.6	0.8	4.3	0.20
Couve..........	G-6	12	6.3	0.6	2.9	0.05
Potato..........	C-6	<1	4			—
Lettuce.........	C-7	6.5	14	1.9		—
"	Puc-2	76	84	1.3		—
Corn............	C-2	2	7.7	0.1	6	—
Spinach.........	C-3	2.5	3.5	0.3	5.2	—

* Sample nos. prefaced with A are from Araxá; Ta, from Tapira; T, from Taquarí; M, from Meaipe; G, from Guaraparí; C, from New York State.

† Thorium-228 has been corrected for ingrowth from radium-228 since time of collection in all but bone samples, assuming a 3.5-month ingrowth.

‡ Measured 3 feet from ground, using a scintillation counter.

§ Insufficient weight of ash.

TABLE 2—*Continued*

Type of Sample	Sample No.*	Radium-228		Thorium-228† (pc/kg)	Radium-226 (pc/kg)	γ Radiation (mr/hr)‡
		pc/kg	pc/g Ca			
Water:						
Spring water.....	A-1	Lost		Nil		—
Sulfur well......	A-2	1.5		Nil		—
Spring water.....	A-8	3.2		Nil	1.65	—
" "	Ta-4	<0.2		Nil		0.18
Tap water.......	Ta-12	1.5		Nil	0.78	—
Well water......	Ta-13	3.1		1.8		—
" "	T-1	<0.2		Nil	0.36	0.02
Spring water.....	T-8	<0.2		Nil		0.01
Tap water.......	M-8	1.6		Nil	0.78	—
Well water......	M-9	<0.2		Nil		1
" "	M-14	1.4		Nil		—
" "	G-7	1.4		Nil		0.04
Tap water.......	C-5	<0.2		Nil	<0.1	—
Steer Bones:						
Leg bone........	T-9	2,250	11.9	3,300	1,290	—
Jawbone........	T-10a	2,550	13.6	1,320	1,450	—
Teeth (lower)....	T-10b	1,470	4.8	2,640	835	—
Jawbone (lower)..	T-11	2,370	18.4	1,700	932	—
Leg bone........	G-8	1,566	11	1,030	262	—
Teeth...........	G-13	1,467	7.4	1,160	368	—
Leg bone........	C-1	140	1.1		152	—

States. More recently, Hallden, Fisenne, and Harley (1962) have analyzed the radium-226 content of the foods of New York, Chicago, and San Francisco and have concluded that the average daily intake in the three cities was 2.3, 2.1, and 1.7 pc. radium-226, respectively.

The radium-228 content of foods in the United Kingdom has been studied by Hill (1962) and Turner *et al.* (1958). The latter investigators made the important suggestion that the β-emitting nuclide, radium-228, is absorbed by plants to a greater degree than is its α-emitting granddaughter, thorium-228. In Table 2 are shown the radium-228 and thorium-228 content of a few low-level New York foods, together with higher level Brazilian foods.

The data of Hursh (1953) and Stehney and Lucas (1955) indicate that the normal range of the radium-226 content of public water supplies is very variable, ranging from less than 0.000X pc/l to 0.17 pc/l. The mean concentration of radium-226 in the 41 water supplies sampled by Hursh was 0.04 pc/l. Stehney and Lucas have found that water from deep sandstone wells in Illinois contain as much as 37 pc/l, which may actually exceed the maximum values currently recommended as permissible for continuous consumption.

ANALYTICAL METHODS

SAMPLE COLLECTION AND PREPARATION

The Brazilian localities from which samples were taken include Guaraparí, Meaipe, Araxá, Taquarí, Tapira, and Rio de Janeiro, the last of which was used as a control. The radiogeological features of these areas have been described by Roser *et al.* (1963).

All the samples were collected in the field, on the farm of origin. The water samples were collected in plastic jars, in which they were stored until they were received in the laboratory at New York University, where all analyses were performed. The food samples, immediately after collection, were placed in individual plastic bags, in which they remained until they reached the laboratory in Rio de Janeiro within 2–5 days after collection. Here the samples were first reduced to a char by an infrared lamp and were then ashed at 600° C. prior to being shipped to New York for analysis. In some cases the ignition of the samples was not completed until after the samples were received in New York.

The water samples were acidified and evaporated to a small volume. From this point the water samples and aliquots of the food residues were analyzed by the same procedures.

γ SPECTROMETRY

Prior to radiochemical analysis the ashed samples were examined by γ spectrometry, using a 4-inch sodium iodide well crystal in a steel cave having 6-inch-thick walls. Depending on the amount of sample available, it was counted either in the 1-inch well or in a plastic Petri dish mounted on top of the crystal.

Thorium-228 was measured quantitatively by this technique by measurement of the 2.62-mev. thallium-208 peak. The presence of radium-228 and radium-226 could be qualitatively established in some samples by the presence of well-defined peaks at 0.92 mev. in the case of radium-228 and 1.76 mev. in the case of radium-226.

Quantitative estimates of radium-228 and radium-226 were not made by γ spectrometry because of interference from the potassium-40 1.46-mev. peak and also because the outgassing factor for radon-222 was unknown.

Standards were prepared by extracting a known amount of aged (42 years) thorium-232 nitrate and all its daughters into a solution of di(2-ethylhexyl) phosphoric acid in heptane. A portion of this solution was standardized by stripping radium-224 from the solvent and α counting. The observed α count was within 1 per cent agreement with the calculated concentration of radium-224. A known portion of this solution was mixed with warm paraffin, and permanent standards were prepared for each geometrical configuration. These standards gave calibration factors of 0.039 c/m/pc in the vial and 0.019 c/m/pc in the Petri dish.

Background over thirty channels used for measuring the thallium-208 peak was 2.62 c.p.m. The minimum significant count for 95 per cent confidence is 0.69 c.p.m., corresponding to a minimum significant activity of 17.6 pc. in the vials and 36.5 pc. in the Petri dish.

RADIOCHEMICAL ANALYSIS

After examination by γ spectrometry, the ash samples were first decomposed by treatment with hydrobromic, hydrofluoric, or perchloric acids, depending on the properties of the residue being examined.

After sample dissolution, the following steps were followed in order to determine radium-226, radium-228, and thorium-228.

1. Polonium, lead, and bismuth were extracted into a quaternary amine (Aliquat 336, General Mills Company) as bromo-complexes.

2. Thorium was then extracted into an acidic-organophosphorus compound, di(2-ethylhexyl) phosphoric acid (EPHA, Union Carbide Chemical Company). This thorium-bearing solvent was then aged for a known period to allow ingrowth of radium-224 and other thorium-228 α-emitting daughters. These daughters were stripped from the solvent with dilute acid and α counted. From the observed α count and the period of α in growth, the thorium-228 content of the sample was determined through use of the Bateman equation.

3. After removal of thorium, the radium isotopes were collected on a lead sulfate precipitate. The lead sulfate was dissolved and the lead removed by solvent extraction into Aliquat 336. After aging, this radium fraction was analyzed for radium-228 by measurement of its actinium-228 daughter, and for radium-226 by the emanation technique. A low background β counter was used to detect the separated actinium-228, and radon-222 and its daughters were collected in a zinc sulfide–coated bottle and counted by an α scintillation technique.

By saving the radium and thorium fraction, these analyses can be repeated as often as desired. Furthermore, the thorium present in each sample has been preserved so that thorium-232 and thorium-230 analyses can be performed if and when they should seem desirable.

A detailed description of the method will be published elsewhere. It is derived from the procedures of Petrow et al. (Petrow et al., 1960; Petrow and Lindstrom, 1961; Petrow and Allen, 1961, 1963) and from the radon procedure of Stehney et al. (1955).

The sensitivity of these methods is of course a function of detector background and reagent blank. The reagent blank for the thorium-228 procedure was 3 counts per hour, and the background was 5 counts per hour. Since, at equilibrium, a maximum of 4 α disintegrations are obtained for each thorium-228 disintegration, as little as 0.1 pc. of thorium-228 is readily detectable. The nature of the procedure is such that the weight of sample ash should not exceed 10 grams. With the maximum sample size, the procedure is sensitive to about 3.3 pc/kg of wet bone and 0.1 pc/kg of fruit, vegetable, or dairy products.

For the actinium-228 determination, the detector background was 1.2 c.p.m. and the reagent blank 0.4 c.p.m. Thus, about 0.5 pc. of radium-228 is the minimum detectable amount. For a 10-gram ash sample, this provides a sensitivity of 0.5 pc/kg of raw vegetables or 15 pc/kg of wet bone. This limitation in sensitivity for bone samples could be a serious handicap in analyzing bones of normal radium-228 content, where a burden of 10–20 pc. of radium-228/kg is considered normal. However, the radium-228 contents of the Brazilian samples were far in excess of the limiting amount.

Finally, the sensitivity of the radium-226 procedure is approximately equal to that for thorium-228, that is, about 0.1 pc.

As regards water samples, the largest samples received were 2 liters in volume, and the minimum detectable amounts of radium-228, radium-226, and thorium-228, were 0.2 pc/l, 0.05 pc/l, and 0.05 pc/l, respectively.

CALCIUM ANALYSIS

Many of the ash samples were also analyzed for their calcium content by titration with standard EDTA solution in the presence of an indicator, Eriochrome Schwarz T.

FINDINGS

Our data for the radium-228, thorium-228, and radium-226 content of the various samples are given in Table 2. The radium-228 values are reported as pc/kg fresh material and also as pc/gm calcium. It is seen that some of the water samples are higher by a factor of about 30 than the average values reported by Hursh in this country, but the values are not nearly so high as some of those reported by Stehney in the areas of Illinois known for the high radium content of its well waters. The radium-226 and radium-228 contents of many of the food samples are considerably elevated over the normal value of about 1 pc/kg.

Table 3 shows a comparison of thorium-228 by γ spectroscopy and by radiochemistry. While, in most cases, results obtained by both techniques are in acceptable agreement, several samples give differences of a factor of 2. We believe the main source of these differences to be inhomogeneity of samples. This is particularly true of the teeth samples, which we were unable to reduce to a powder with any of the grinding equipment at our disposal. Other possible sources of error are difficulties in obtaining a reproducible sample geometry in the spectrometer and in standardizing the spectrometer. Further investigations are under way to determine the source of variability.

TABLE 3

COMPARISON OF THORIUM-228 BY RADIOCHEMICAL ANALYSIS
AND BY γ SPECTRAL ANALYSIS (PC/KG)

Sample	Type	Radio-chemistry	γ Spect.	γ Spect. / Radiochem.
A-3.........	Tomato	1.4	Nil	—
A-5.........	Cabbage	28.8	39.6	1.38
A-7.........	Banana	8.8	Nil	—
Ta-1a.......	Corn	—	Nil	—
Ta-1b.......	Cob	46	68.7	1.49
Ta-2........	Lettuce	190	163	0.86
Ta-7........	Milk	1.5	Nil	—
T-2.........	Milk	0.33	Nil	—
T-3.........	Cheese	3.1	Nil	—
T-4.........	Potato	5.5	Nil	—
T-6.........	Potato	1.8	Nil	—
T-9.........	Bone	3,300	1,830	0.55
T-10a.......	Jawbone	1,320	1,820	1.38
T-10b.......	Teeth	2,640	1,570*(P)	0.59
			2,400*(V)	0.91
T-11........	Bone	1,700	1,570	0.92
M-1........	Cow dung	—	1,350	—
M-2a.......	Corn	1.3	Nil	—
M-2b.......	Cob	13.5	Nil	—
M-3c.......	Coconut husk	9.3	10	1.08
M-12.......	Milk	0.4	Nil	—
G-2........	Banana	2.7	Nil	—
G-3........	Bertalha	41	84	2.05
G-4........	Abobora	4.1	Nil	—
G-5........	Tomato	—	Nil	—
G-6........	Couve	1.9	Nil	—
G-8........	Bone	1,030	851	0.83
G-13.......	Teeth	1,160	676	0.58
G-14.......	Milk	1.2	Nil	—
G-15.......	Banana	3.6	29.1†	8.08
G-16.......	Papaya	—	Nil	—
G-17.......	Banana	—	Nil	—

* Teeth measured both in vial and in Petri dish. No sample of teeth was homogeneous, since it was difficult to grind them.

† Gamma results gave count just above MDA. Small amount of sample may account for large difference.

DISCUSSION

A surprising finding has been the high radium content of the leafy vegetables, such as lettuce and cabbage. The lettuce from Tapira was found to have 965 pc/kg of wet weight, and Araxá cabbage was 71 pc/kg. The high value of the lettuce from Rio de Janeiro cannot be explained. It could be due to contamination in the laboratory or to the possibility that the lettuce was grown in a high-radiation area. It is interesting that the lettuce sample from New York was also somewhat higher than the other food samples.

In the United States the average radium-226 intake is about 2 pc/day (Stehney *et al.*, 1955). Since the daily calcium intake in the United States is about 1 gram, this is equivalent to about 2 pc/gm calcium, a ratio that is somewhat influenced by the relatively large fraction of the calcium contributed by dairy products in the United States. Cows tend to discriminate against radium when they produce milk. Because dairy sources are a relatively minor contributor of calcium in the diets of most Brazilians, one might expect a higher ratio of radium to calcium in their total diet.

The cow bones from the radioactive areas are elevated by a factor of 10–20. In no case is there more thorium-228 than would be expected from the decay of radium-228. It should be pointed out that thorium-228 is in transient equilibrium with radium-228, and under these conditions at equilibrium the ratio of thorium-228 to radium-228 will be 1.5.

In reviewing the foregoing data, one must bear in mind that this was a pilot study designed to approximate the extent to which the natural radioactivity contents of the foods and waters are elevated above the levels considered to be normal elsewhere in the world. The considerable scatter of these data and our lack of knowledge as to the food production of the areas and the dietary practices of the local populations would make it imprudent to attempt to assess the dosimetric implications of these observations so far as these local populations are concerned. The principal conclusion one can draw from the data given above is that additional information is badly needed.

The excellent studies of Roser and Cullen, accomplished under very difficult conditions over thousands of kilometers of poor roads, were aimed at mapping areas in which the external radiation levels were elevated and in which large numbers of people lived. These regions must now be resurveyed in order that the food-growing practices in the areas can be understood. More information is needed on the kinds and amounts of food that are grown and the manner in which the food is distributed.

ESTIMATING RADIUM-226 AND RADIUM-228 IN
THE HUMAN SKELETON

It would of course be useful to develop a program of human measurement so that the body burdens from radium-228 and radium-226 could be estimated directly. Such estimates can be made by methods of total-body counting, by radiochemical analysis of bone or teeth, by measurement of radon-222 and radon-220 in exhaled breath, and, finally, by radiochemical analysis of urine.

The use of total-body counting to estimate the body burdens of the inhabitants of these areas is contraindicated at the present time because evidence is lacking that the body burdens are elevated to a level sufficiently high to enable one to obtain useful data by methods of total-body counting. A well-designed instrument in expert hands can barely detect 10^3 pc. of radium-228 or radium-226. The required instrumentation would include a 6-inch-thick steel shield sufficiently large to accommodate the patient, a crystal at least 8 inches in diameter by 4 inches thick, and the associated electronic equipment, which would include a multichannel analyzer and readout apparatus. A shield for total-body counting could probably be improvised in the field by stacking bags of refined sugar, which in its purified form is relatively free of potassium and radium. However, the logistic problems associated with measurements of this kind would at best be quite serious because of the large geographical areas involved and the poor condition of the roads. Total-body counting should be attempted only if other methods of estimating the body burden indicate that positive data would be forthcoming. It would be wasteful to undertake an expensive program of total-body counting only to find that all the measurements are below the detectable limit of the instrument.

A more practical way to screen a population is by sampling bone or teeth. Bone samples will be difficult to obtain because of the primitive state of medical practice in these areas and the local prejudice against autopsies. However, it is possible to obtain teeth, and a program of sample collection is now under way.

A number of methods are available by which the radon-222 and radon-220 concentrations of exhaled breath can be used to estimate the body burden of radium-226 and radium-228. Of the two radioisotopes, radium-228 would be expected to be present in the bones of residents of these areas in higher amounts than would radium-226, and, for this reason, the methods developed by Evans (Evans *et al.*, 1952) for estimating the radon-220 content of the human breath have been adapted for use in the field. A few measurements have been made on residents of Guaraparí, but the results to date are equivocal and additional measurements are required. The recent development by Hursh (Hursh and Lovaas, 1962) of a new and somewhat simplified method of measuring breath radon-220 may have application to this problem, and such an instrument is currently being constructed for use in the field during the coming year.

Measurements of the urinary content of the two radium isotopes should provide a satisfactory method of screening the populations to determine whether additional work would be warranted. The urinary content of the radium isotopes would reflect both the dietary intake of radium and the amount stored in the skeleton and other tissues of the body, but it is to be expected that the former would mask the latter and that the primary purpose of urine analysis would be to estimate the daily intake of the two isotopes. Pooled samples of urine from residents of the various areas and from residents of controlled areas should be analyzed to determine whether any difference does exist.

ATMOSPHERIC RADIOACTIVITY

In view of the fact that the soils contain abnormally high amounts of the two radium isotopes, it is to be expected that their gaseous daughter products radon-222 and radon-220 may diffuse from the earth at a rate more rapid than normal. Under normal conditions of atmospheric turbulence, the additional contribution of radon-222 and radon-220 to the atmosphere of these localities would be flushed out of the region in a matter of minutes, and any significant build-up of these gases would probably be prevented. However, during periods of inversion, when the normal mixing processes of the atmosphere are greatly restricted, the radon-222 and radon-220 concentrations could conceivably rise sufficiently to represent another significant source of exposure to the local inhabitants. Whether or not this is the case can be answered only by future measurements.

OPPORTUNITIES FOR RADIOECOLOGICAL STUDIES

The foregoing discussion has been concerned primarily with uptake of the radionuclides by human foods. In addition, a number of localities seem to offer attractive opportunities for radioecological studies. In particular, the Morro do Ferro near Poços de Caldas, a hill with an area of 0.35 km.^{2}, offers the opportunity to investigate a number of interesting questions at a site that is essentially undisturbed by man. The external radiation levels at the top of the hill are more than 1 mr/hr, or about 10 r/yr. The soil of the hill is low in calcium, and preliminary measurements of the grasses indicate the radium-228 and radium-226 levels to be 10–100 times the levels observed in the foods. The skeletal dose to small mammals, the radon-222 and radon-220 exposure of underground organisms, the partitioning of radionuclides from tissue to tissue, and, finally, the radiation effects on the flora and fauna are among the numerous, fascinating, and useful questions that can be investigated at this and other sites.

REFERENCES

CLEGG, J. W., and D. C. FOLEY (ed.). 1958. Uranium Ore Processing. Reading, Mass.: Addison-Wesley Pub. Co. Pp. 436.

EVANS, R. D., J. C. AUB, L. H. HEMPELMANN, and H. S. MARTLAND. 1952. The late effects of internally-deposited radioactive materials in man. Medicine, **31**:221–329.

HALLDEN, N. A., I. M. FISENNE, and J. H. HARLEY. 1962. Radium-226 in the diet of three U.S. cities. *In* Proc. Seventh Ann. Meeting on Bio-Assay and Analytical Chemistry. Argonne National Lab. Rpt. ANL-6637, pp. 85–95.

HILL, C. R. 1962. Identification of alpha-emitters in normal biological materials. Health Physics, **8**:17–25.

HURSH, J. B. 1953. The Radium Content of Public Water Supplies. Univ. of Rochester, AEC Project Rpt. UR-257. Pp. 27.

HURSH, J. B., and A. LOVAAS. 1962. A Device for Measurement of Thoron in the Breath. Univ. Rochester, AEC Project Rpt. UR-619. Pp. 26.

PETROW, H. G., and R. J. ALLEN. 1961. Estimation of the isotopic composition of separated radium samples. Anal. Chem., **33**:1303–5.

———. 1963. Radiochemical determination of actinium in uranium process streams. *Ibid.*, Vol. **35**. (In press.)

PETROW, H. G., and R. LINDSTROM. 1961. Radiochemical determination of radium in uranium milling process samples. Anal. Chem., **33**:313–14.

PETROW, H. G., O. A. NIETZEL, and M. A. DESESA. 1960. Radiochemical determination of radium in uranium milling process samples. Anal. Chem., **32**:926–27.

ROSER, F. X., and T. L. CULLEN. 1963. External radiation levels in high-background regions of Brazil. This symposium.

ROSER, F. X., T. L. CULLEN, and G. H. KEGEL. 1963. Radiogeology of some high-background areas of Brazil. This symposium.

STEHNEY, A. F., and H. F. LUCAS, JR. 1955. Studies on the radium content of humans arising from the natural radium of their environment. Proc. 1st Internat. Conf. on Peaceful Uses of Atomic Energy, **11**:49–54.

STEHNEY, A. F., W. P. NORRIS, H. F. LUCAS, JR., and W. H. JOHNSTON. 1955. A method for measuring the rate of elimination of radon in breath. Am. J. Roentgenol., Radium Therapy & Nuclear Med., **73**:774–84.

TURNER, R. C., J. M. RADLEY, and W. V. MAYNEORD. 1958. The naturally occurring alpha-ray activity of foods. Health Physics, **1**:268–75.

UNITED NATIONS. 1962. Second Comprehensive Report of the United Nations Scientific Committee on the Effects of Atomic Radiation. General Assembly Official Records, 17th sess. Pp. 442.

F. X. ROSER, G. KEGEL, AND T. L. CULLEN

53. Radiogeology of Some High-Background Areas of Brazil

THE MAIN GEOTECTONIC FEATURES of Brazil have been investigated by a distinguished group of geologists (Oliveira and Leonardos, 1943; Guimaraes, 1951; Leinz, 1962). The central granite-gneissic plateau is of Precambrian formation. This so-called Brazilian shield or buckler is surrounded on the east by coastal plains, on the north and west by the Amazon River Basin, and on the southwest by the Paragua-Paraná River depression (Fig. 1, *a*).

The shield was formed by accretion of the remnant sial blocks of a former archeozoic continent, the Sudatlantis of the paleogeographers. When Sudatlantis broke apart, these remnants integrated, together with the Afro-Indian and the Australide-Antarctic protocontinents, the Gondwana continental complex of the Paleozoic.

Diastrophic geoclases subsequently caused a progressive fracturing of this complex during the Mesozoic era, with Brazil and Africa drifting apart, and the Atlantic trough opening up. In the Cenozoic era, the Andean mountain range was folded up on the drift front of South America, slowly tilting the continental block.

Two different geotectonic processes were brought about by these successive crustal movements.

1. Orogenetic processes produced mountain ranges and a geochemical upward migration to form mineral deposits on the surface. When the blocks were first converging to form the Gondwana complex, plasticity was still high, and the overthrusting edges resulted in arch-folded anticlines. Thus Brazil's eastern ridges (Serra do Espinhaço,

F. X. ROSER is director, G. KEGEL is geologist, and T. L. CULLEN is assistant professor of physics at the Institute of Physics, Catholic University, Rio de Janeiro, Brazil.

Study supported by the National Research Council of Brazil, the Brazilian Nuclear Energy Commission, and the U.S. Atomic Energy Commission.

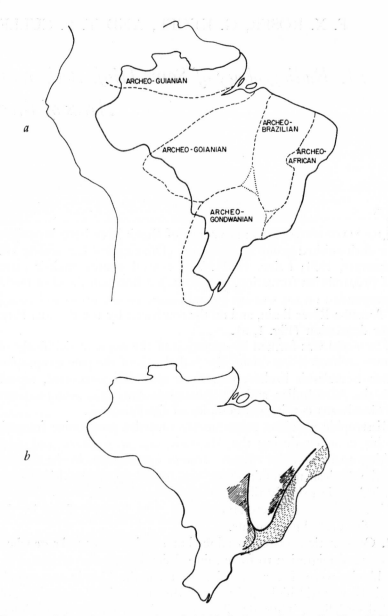

Fig. 1.—*a*, Schematic reconstruction of sial blocks integrating the Brazilian shield. *b*, Orogenic foldings and diastrophic ruptures in Brazilian shield as sites of uranium and thorium.

Serra dos Aimorés, Serra do Mar, and Serra da Mantiqueira) were formed along the line of contact between the archeo-Brazilian and archeo-African blocks. In a similar way the ridges of the west (Serra da Canastra, Serra Dourada, Serra dos Pirineus, and Serra Geral) came into being.

2. Epirogenic ruptures along the suture lines between these blocks produced later extrusions of magma and of submagmatic alkaline material. This was caused by the divergent movement of the Mesozoic era, when progressive rigidity of the shield had already fixed the definite position of the original blocks. The development of geotectonic fault lines in the triangular regions of coalescence of the archeo-Brazilian, archeo-Gondwanian, and archeo-Goiânian blocks brought forth extensive flows of basaltic magma and gave origin to a string of volcanic intrusives, extending into the state of Goiaz (Tingua, Itatiaía, Poços de Caldas, Araxá, Tapira, Serra Negra, Salitre, Mata da Corda, etc.).

These processes explain the existence of the J-shaped geotectonic belt encompassing the archeo-Brazilian block, as shown in Figure 1, *b* (Guimaraes, 1955). Numerous foci of intense mineralization are closely related to the diastrophic phases of its evolution during the ages.

Thus, abundant mineral deposits and a number of hydrothermal spas are located along the belt. In many cases radioactive minerals are associated with the lines of fracture and with the upward diffusion of metallic and metalloidic elements to the surface, resulting in local infiltration or generalized contamination of the strata. Progressive weathering of the primary formations, together with subsequent eluvial and alluvial processes, produced secondary mineral deposits that were quite frequently of higher concentration.

Five different types of radioactive ore deposits can be distinguished (Moraes, 1955):

a) In the archeozoic or proterozoic bedrocks along the upfolding edges of the sial blocks profusely permeated with quartzo-pegmatitic veins and with other infiltrations of metamorphic origin

b) In the alkaline eruptive plugs of the Triassic-Jurassic along the fault lines where intensive hydrothermal mineralization developed

c) In the phosphatic sediments associated with the calcareous stratifications of the epicontinental sea that covered the northern half of the archeo-Brazilian block during the latter part of the Paleozoic era

d) In the monazite-bearing strata of the Jurassic-Cenozoic peneplains of debris and silt that resulted from severe erosion of the mountain ranges during the Permian glaciation

e) In the Quaternary alluvial and eluvial deposits that originated from the progressive weathering of primary rocks and erosion of the sedimentary peneplains by transportation and successive concentration of monazite sand along river banks and beaches

For reasons of biological interest, our survey was confined to regions with intensity levels at least several times above normal background. Thus, only cases referred to under *b* and *e* were investigated. The monazite sand concentration along the shore line of Espírito Santo and southern Baía and the heavily contaminated alkaline extrusives along the southern border of the states of Minas Gerais and Goiaz are samples of such regions that have been studied on successive field trips (Roser and Cullen, 1958, 1962).

From a mining point of view, however, other regions of different types might prove equally interesting, since, even in cases of low concentration of radioactive elements, they are often found to be associated with gold, niobium, zirconium, and other ores and minerals of great economic value.

REGIONS OF MONAZITE SAND DEPOSITS

Among Brazil's eastern ridges, Serra do Mar and Serra dos Aimorés, paralleling the Atlantic Coast along Rio de Janeiro and Espírito Santo states, are most prominent. Their continuation to the north (Serra do Espinhaço) has almost been leveled to the ground due to its older age (Precambrian glaciation).

The ridges of today represent only the core of the original mountain foldings. Vast quantities of rock, many thousands of feet thick, have been removed through the ages until the very roots were laid bare, consisting of steeply inclined contacts and gabbrodioritic batholiths, which impart to the Brazilian landscape of today its bizarre appearance of sugar-loaf-like mountain peaks and impressive monadnocks steeply emerging from coastal waters or from surrounding alluvial plains.

The destructive influence of geological factors was very predominant during the Permo-carboniferous glaciation that covered the Gondwana complex, at that time close to the South Pole, with a widespread continental ice sheet. To judge from the thickness of its moraine deposits, it must have lasted longer—by an order of magnitude —than the Quaternary glaciations of Europe and North America.

The geosyncline between archeo-Brazil and archeo-Africa was gradually filled with layered sediments of coarse conglomerates (tillite), resulting in extensive plains of pebbles, clay, sand, and loess.

Isostatic adjustments produced a general elevation of the plains and deep climatic changes. Increasing aridity, as in the case of the

Pleistocene glaciations of the Northern Hemisphere, dominated for the rest of the period the semidesertic, steppelike plateau covered with shallow lagoons offering a hostile habitat to a scarce vegetation and to hardly any fauna.

During the Triassic, the Gondwana complex entered a new geotectonic phase of epirogenic character. Geoclases started from the south and gave origin to huge overflows of magma and to intensive volcanic activity.

In the Jurassic, a rupture line developed, rifting the sial block, between Africa and India, Australia, and Antarctica to give birth to the Indian Ocean. By the Cretaceous, the fault line had propagated around Africa to open the South Atlantic trough. During the Tertiary its further propagation eventually reached Greenland to give the North Atlantic its present form during the Quaternary.

These massive tangential dislocations and drifts modified the earth with the appearance of new mountain ranges (Andes, Himalayas, and Alps) and the diastrophic movements with all the host of phenomena connected (tectonic, magmatic, and geochemical) raised its energetic level to new periods of rejuvenescence.

The Atlantic rift left the fringe of the archeo-African block sutured to the Brazilian shield along the foldings of the mountain ranges and transformed the Mesozoic peneplains of the east to a coastal province partly submerged as a continental shelf.

This changed the drainage pattern for all the rivers that had been running north, meandering through almost level peneplains to empty into the epicontinental sea covering the equatorial part of the Brazilian shield. They turned east toward the ocean in rejuvenated activity, cutting deep canyons and ravines in their swift descent and washing great amounts of loosely bound glacial debris down to the coast, where Tertiary coastal plains of much finer texture developed. The gradual tilting of the South American continent transformed them into escarpments and cliff formations at the foot of which today's coastal plains and beaches were slowly formed. Thus the age-long, multiple-stage transformation of uranium and thorium, disseminated in the rocks into concentrates of monazite along the coast, becomes clear.

Recurrent diastrophic phenomena in the upfolding edges of the continental blocks favored the chemical upward migration of elements and permeated the rock formation with coarse-grained dikes and sheets of profusely mineralized pegmatites. They constitute the main source of Brazil's world-renowned riches of well-developed minerals and semiprecious stones. Niobium, tantalum, tungsten, beryllium, and rare earths with uranium and thorium admixtures are profusely dispersed in the pegmatites but rarely concentrate in the high-grade veins (**uraninite**).

The heavy metals occur as uranates, niobates, tantalates, and titanates in minerals of euxenite, samarskite, djalmaite, and others. The rare earths occur as phosphates, mostly of cerium ($CePO_4$), with thorium oxide impurities in minerals of monazite. Zircon ($ZrSiO_4$) (hardness 7.5, specific gravity 4.2 to 4.8), ilmenite ($FeTiO_3$), magnetite (Fe_3O_4) (hardness 5.5–6.5, specific gravity 5.2), and rutile (TiO_2) (hardness 6–6.5, specific gravity 4.3) are also present. Granite and gneiss contain only microscopic crystals of monazite, constituting about 0.1 per cent of the rocks. In pegmatite formations, however, it occurs in macroscopic crystals and in greater concentration, from 1 to 1.5 per cent. Values of 0.01 per cent of uranium oxide and of 0.02–0.07 per cent of thorium oxide are found. These exceed by a factor of 50 the normal uranium and thorium content of ordinary granite. Monazite proper consists of 70 per cent rare earths; thorium oxide, 4–6 per cent; and uranium oxide, 0.15–0.25 per cent. Usually there is 1 part of uranium oxide to 40 or 50 parts of thorium oxide.

Monazite is monoclinic but is ordinarily found in translucent yellow to brown grains with a resinous luster. Its hardness is 5–5.5, the specific gravity 4.9–5.3. It is remarkably resistant to attrition and alteration. Thus, it can be traced through more than one cycle of erosion and sedimentation.

Much controversy has revolved around the question whether thorium is an essential constituent of monazite or whether the thorium silicate is merely present in admixture. No agreement has been reached on this point, but it has been suggested that the thorium is in solid solution with the cerium phosphate (Vickery, 1953).

Weathering in geological times has resulted in the decomposition of the archeogneisses in the mountains. Disintegrated debris is separated by the double process of chemical dissolution and physical attrition. Monazite, zircon, and ilmenite are physically hard and not soluble in water. Ground to fine particles and grains, they are carried downstream by the many rivers along the coast. They are separated from other components because of their high specific gravity and concentrated in fluvial and marine placers.

Both the long stretches of Tertiary cliff formation and the churning movement of the surf between them and the rocky reefs and shoals paralleling the shore line favored the gradual build-up of long, bead-like strings of lenticular block deposits intercalated with inactive layers of ordinary sand and gave rise, besides, to a great variety of long, extended sandbanks and barriers on promontories and nearby islands (Fig. 2).

Because of the slow persistent rise of South America's Atlantic Coast line, such beach and barrier formations are frequently found at some distance from the shore, covered and even completely buried under more recent layers of soil. In other places they present a regular pattern of successive beach lines, giving the impression of huge

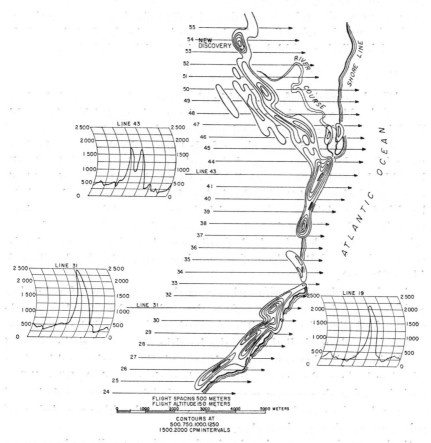

Fig. 2.—Air-borne radiometry of coastal monazite deposits. (Courtesy of Prospec S.A.)

tilled fields. Such thorium-bearing coastal strips are aligned in an almost continuous succession for over 300 miles along the shores of Espírito Santo and extend into the neighboring states to the north and south. The monazite concentration in the layered patches ranges from 5 to 20 per cent, but, even over considerable extension of ordinary sand, it may reach 0.5 per cent. An exact computation of the total available amount is still very difficult; conservative estimates based on exploratory inspection and drilling give a wide range, from 100,000 to over 1,000,000 tons of monazite. As for the amounts of fluvial sedimentation existing in many places throughout Brazil's interior, no estimates are available (Leonardos, 1955*b*, 1956*c*, 1959).

From an economic point of view, only the more concentrated layers are of interest. For the purpose of biological investigation, however, even moderately contaminated soils afford an opportunity to study the uptake of radioactive isotopes into the food chain and the internal irradiation of biological material resulting therefrom.

REGIONS OF VOLCANIC INTRUSIVES

The long series of diastrophic phenomena that spanned the later part of the Mesozoic, bringing about the continental drift, also mobilized huge masses of subcrustal magma and entailed the outflow of vast sheets of basaltic lava along fault lines and other tectonic fractures. These lava flows covered the entire southern part of the Brazilian shield roughly corresponding to the archeo-Gondwanian block remnant (Leinz, 1949). Mechanical and chemical weathering of the basaltic layer, rich in such essential elements as calcium, magnesium, potassium, phosphorus, iron, manganese, sodium, and traces of many other metals, produced today's fertile stretches of *terra roxa*, the famous red soil of Brazil's coffee-growing states (São Paulo, Paraná, Mato Grosso).

Another regional manifestation was the appearance of volcanism and the extrusion at many points along the faults of submagmatic plugs of alkaline character. The formation of such alkalic masses resulted from the assimilation by the basaltic magma of sialic material producing derivative magmas, such as nepheline-syenites, with a host of other basic rocks (pyroxenites, carbonatites, apatite). The differentiation of such magmas was accompanied by an accumulation of gases, wherefore explosive volcanism with and without emission of substratum melt came into play.

Such extrusives, appearing in the form of protruding plugs, more or less circular in shape and usually of considerable extension, are pervaded by an intricate net of lines of fracture in which subsequent intensive hydrothermal and pneumatolithic mineralization could develop. A dozen such extrusives originated in the triangular regions of coalescence of the original blocks, varying in size from 20 to less than 0.5 miles across. Another branch of diatremes propagated southward along the edge of the archeo-Gondwanian massif (Leinz, 1949).

In many places no actual extrusion took place, and only small stocks developed, filling the vents. In others, only the existence of hot springs of mineralized and radioactive waters indicates the presence of sub-crustal fissures.

Two of the most representative extrusives will be described in greater detail.

A. POÇOS DE CALDAS

The regions of Poços de Caldas (Pl. I and Fig. 3) constitute an outstanding example of such a tectonic process, in which an alkaline plug of more than 1,000 sq. km. (30 × 35 km.) was raised to a height

PLATE I

Mosaic aerial view of circular intrusive, Poços de Caldas. (Courtesy of Lasa Co.)

of 400–500 m. above the adjacent granitic substratum. Caulderon subsidence, followed by deep weathering and intense erosion, lowered the inside portion of the province, leaving intact the peripheric rim of harder rocks and producing the striking appearance of a huge volcanic "caldera" even though no truly explosive activity ever took place (Ellert, 1959; Bjoernberg, 1959).

No rocks with fluidal texture are present; all minerals are holocrystallines. Only at a few places on the outside of the rim do there

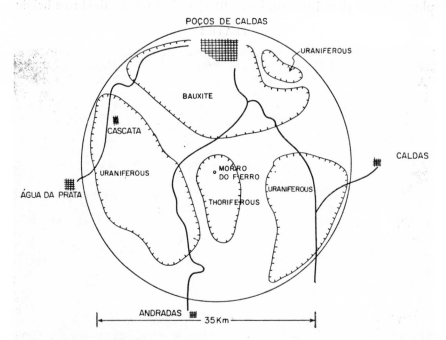

POÇOS DE CALDAS

URANIFEROUS

BAUXITE

CASCATA

URANIFEROUS

MORRO DO FIERRO

ÁGUA DA PRATA

URANIFEROUS

CALDAS

THORIFEROUS

ANDRADAS

35 Km

Fig. 3.—Schematic distribution of mineral deposits in circular intrusive, Poços de Caldas. (Courtesy of Resk Frayha.)

seem to exist small residues completely altered, which have been taken as remnants of lava.

In contrast to the metamorphosed granite of the coastal mountains, uranium and thorium are found distinctly separated in this alkaline province, uranium being associated with zirconium and thorium with iron and manganese oxides.

An almost complete and deep-reaching decomposition under subtropical conditions of the outcropping veins and of the bedding rocks, together with subsequent eluvial and/or alluvial concentration of the hard and resistant constituents, produced some 70 anomalous radioactive areas constituting 3 or 4 separate zones in which either uranium or thorium is predominant. At other sites the far advanced or complete alteration of the eruptive syenitic alkaline gave origin to the important bauxite deposits of Poços de Caldas.

The following anomalies are examples of two different types of mineralization and ore enrichment:

1. *Taquarí*. This conspicuous hill, rising 280 m. above the plateau, constitutes the most illustrative case of uranium occurrence in the whole crater area. By filling up the fissures and clefts that profusely ruptured the alkaline intrusive at this site, intensive hydrothermal action deposited gross veins and sizable bodies of "caldasite" or zirconiferous mineral, with which uranium is found to be associated.

The whole mountain is pervaded by numerous deep-lying veins and lodes of zirconiferous uranium-bearing ore, outcropping at many points near the top of the hill. Caldasite, the characteristic mineral of the region, is a mixture, in variable proportions, of zircon ($ZrSiO_4$) and baddeleyite (ZrO_2). The former contains some 60 per cent of zirconium oxide, with 0.1–0.2 per cent of uranium; the latter, 65–94 per cent of zirconium oxide, with 0.4–0.7 per cent of uranium, on the average. In many other cases (not at Taquarí) the grade is above 1 per cent, and in a remarkable outcropping of over 40 tons of rock it reaches 2.9 per cent. However, thorium is much lower (200 p.p.m. of thorium oxide). Pebbles of baddeleyite, beanlike in form and color, are very frequent in alluvial deposits and may reach considerable size (up to 50 kg.). Conservative estimates give around 100,000 tons of caldasite for the entire crater region. (Uranium oxide content averages 0.5 per cent [Leinz, 1949].)

At Taquarí mineralization took place also over the entire bedrock of the hill, enriching it with molybdenum sulfide (10 per cent) and uranium oxide (0.025 per cent) to a depth of at least 200 m. This constitutes a colossal body of some 20 million tons of rock containing about 5,000 tons of uranium oxide (Frayha, 1960; Schuhmacher, 1954). The mass effect makes this site one of the three most remarkable anomalies (for radiation intensities) of the crater, even though the still relatively low grade of specific radioactivity in the rocks would hardly arouse any interest in exploitation if it were not for the substantially higher concentration in the veins and for the molybdenum by-products to be obtained.

2. *Morro do Ferro*. Two well-developed main dikes of magnetite, with admixture of titanium and residual manganese running through the rock formation to a length of several hundred meters, are accompanied in broad lateral extension by numerous secondary fractures in which a process of intensive mineralization took place, yielding a great variety of rare-earth oxidation compounds (up to 10 per cent),

with a strong percentage of thorium oxides (0.5–1.8 per cent) and traces of uranium.

Extensive subaerial weathering of the eruptive alkaline resulted at this site in considerable secondary enrichment of the surface layers and in a more homogeneous distribution of the thorium-bearing ore. Values of 3 per cent of thorium and of 20 per cent of rare earths are not uncommon. Even plant ashes show 0.85 per cent of thorium oxide.

The mineralization zone extends over the entire slope (0.35 sq. km.) of a steeply rising hill 300 m. high. The radioactive layers are found at considerable depths (over 100 m.), as verified by drillings and proving galleries driven through the rock. The thorium content is particularly high close to the two dike formations of magnetite that, at a distance of some 100 m., follow the entire length of the slope from the bottom up to the top, splitting up all along into a great many secondary ramifications.

This constitutes one of the most important potential ore reserves of thorium in the world, not only for the colossal volume but, mainly, for the fact of its high compactness and easy removal. Except for some 50,000 tons of magnetite (not radioactive), everything else is a high-grade ore of thorium and rare earths. There are as much as 1 million tons with 1 per cent of thorium oxide; 5 million tons with 0.5 per cent. This would yield some 35,000 tons of thorium oxide and 300,000 tons of rare-earth oxides (Frayha, 1960).

Because of the thorough decomposition, primitive and remineralized rocks are intimately and evenly mixed to constitute a huge mass of heavily contaminated material; environmental radiation levels are extremely high (up to 3 mr/hr), thus making the anomaly one of the most impressive sites in the world (Pl. II).

B. ARAXÁ AND TAPIRA

Along the geotectonic fracture line to the north several other focuses of recurrent diastrophic phenomena developed, giving rise to chemical upward migration of elements that were to form important mineral deposits. Two of them, Araxá and Tapira, have attracted a great deal of professional and public attention (Schuhmacher, 1954; Guimaraes, 1947; Guimaraes and Ilchenko, 1954; Leonardos, 1955a).

Even though less markedly laid out and less impressive than the crater formation of Poços de Caldas, these two intrusives are similar to it in the sequence and results of geological events that gave rise to their somewhat smaller size (30 and 56 sq. km.). Still farther inland, several more examples of varying degrees of geomorphological evolution can be found.

1. *Araxá.* At this site the magmatic rock extrusion consists to a

PLATE II

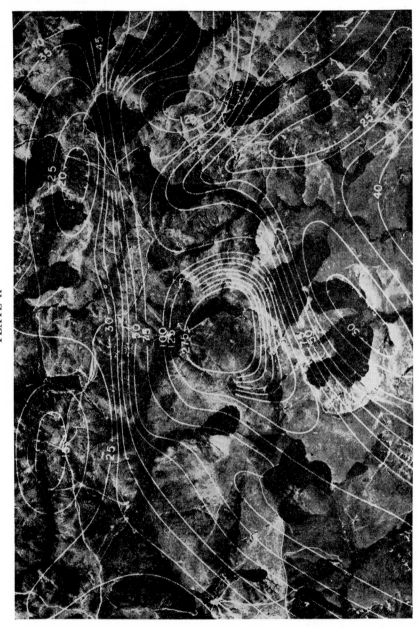

Aerial view of Morro do Ferro, showing isoradiometric lines in multiples of 0.7 μr/hr. (Courtesy of Lasa Co.)

great extent of apatite (complex fluorphosphate of calcium) as a result—it was thought—of contact interaction of the plutonic rocks with calcareous Precambrian formation existing in this region. More recently, however, synoptic studies of many similar cases in Africa, Europe, and America led to the admission of carbonatite formations as having originated directly from alkali magma when crystallizing under extreme conditions of temperature and pressure at the bottom of volcanic chimneys. Such carbonatite structures of calcite, dolomite, and—in the case of Araxá—of apatite are profusely mineralized with magnetite, perovskite, and pyrochlore (Leonardos, 1956a, b).

PLATE III

Ground view of Morro do Ferro

The radioactive material occurs almost exclusively in the form of pyrochlore or of a complex niobium-tantalo-titanate compound of rare earths with a high content of thorium and only relatively minor amounts of uranium (niobium oxide, 54 per cent; titanium oxide, 2 per cent; cerium oxide, 6 per cent; thorium oxide, 9 per cent; uranium oxide, 0.22 per cent). The average concentration of niobium of about 3 per cent (corresponding to 6 per cent of pyrochlore) throughout a considerable portion of the intrusion reaches 14 per cent in the eluvial layers of the surface, 2–5 per cent of the content being thorium. In no other place in the world have such high concentrations of pyrochlore been found.

The region of most intensive radioactive mineralization occupies an area of approximately 2 sq. km. and extends to a depth of over 50 meters, thus representing a very important occurrence of niobium and thorium. A large-scale exploitation of the remarkable intrusive would yield considerable quantities of fertilizers (apatite), niobium, and thorium.

Conservative estimates give for apatite with better than 22 per cent of phosphorus oxide, 42.6 million tons; for inferior apatite with phosphorous oxide ranging from 12 to 22 per cent, 49.2 million tons, or 91.8 million tons total. In addition, 4 million tons of niobium oxide; 130,000 tons of thorium oxide; 90,000 tons of rare-earth oxides; and 60,000 tons of uranium oxides are present (Frayha, 1960; Leonardos, 1956*a, b*).

2. *Tapira.* The evolutionary process and extent (10×6 km.) of this alkaline province were quite similar to those of Araxá. But, whereas a simple intrusion of magmatic masses took place in Araxá, in Tapira a first-stage process of intrusion followed by a phase of violent explosion (attested by copious remnants of effusive rocks) occurred, rupturing the overlying cap of surface strata. Since a considerable part of the mineralizing agent thus went off in a sudden thrust, a much smaller volume and lesser concentration of thorium-bearing mineral were obtained (one-tenth that of Araxá).

Even though the Araxá and Tapira intrusives are quite small in total extension, their mineralized areas are considerably more developed, in volume, than those of Poços de Caldas. The far advanced stage of total lateralization and deep-reaching weathering, characteristic of both alkaline provinces, left countless small crystals of chemically very resistant pyrochlore evenly distributed over large stretches of enriched residual soil. As a result, the aerometric survey sees such anomalies as extensive zones of elevated background almost coincident with the whole intrusive (Fig. 4). The anomalies of Poços de Caldas, on the contrary, exhibit extremely steep radiation gradients encircling them.

A characteristic feature of Araxá and Tapira is the very strong pattern of drainage and erosion developed in the center of the area, with egress toward an assembly of summer resorts and watering places of tepid sulfurous radioactive fountains and ponds. The temperature ($28°–31°$ C.) and high activity (0.06–25 pc/l) of these waters compare quite favorably with world-renowned places in Europe and elsewhere.

At all these sites, uranium and thorium occur only as admixtures to other minerals. True minerals of uranium and thorium, however, have been found in many other locations, notably in an extensive region of magmatic gneiss (Emboaba, state of Minas Gerais), where they occur as uranothorites, thorites, thorianites, and thorogumites.

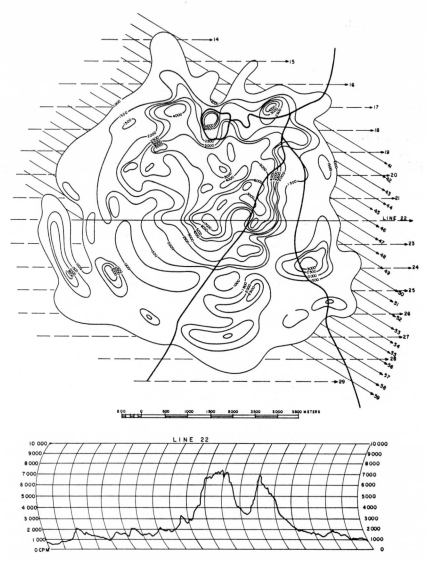

Fig. 4.—Air-borne radiometry of the alkaline intrusive with pyrochlore mineralization, Araxá. (Courtesy of Prospec S.A.)

CONCLUSION

A great deal of work remains still to be done. Brazil's western ridges and the almost limitless reaches of land beyond them still await thorough exploration. It is hard to foresee what the result of it will be.

REFERENCES

BJOERNBERG, A. J. S. 1959. Rochas clasticas do planalto de Poços de Caldas. U.S.P. Fac. Fil. Boletim 237, Geologia 18. São Paulo.

ELLERT, R. 1959. Contribuição a geologia do Maciço alcalino de Poços de Caldas. U.S.P. Fac. Fil. Boletim 237, Geologia 18. São Paulo.

FRAYHA, R. 1960. Urânio e torio no planalto de Poços de Caldas. Belo Horizonte: SICEG.

GUIMARAES, D. 1947. Origem das rochas alcalinas. I.T.I. Minas Gerais, Boletim 5. Belo Horizonte.

———. 1951. Arqui-Brasil e sua veolução geológica. Rio de Janeiro: Departmento Nacional de Produção Mineral.

———. 1955. Areas geologically favorable to occurrence of thorium and uranium in Brazil. Proc. Internat. Conf. on Peaceful Uses of Atomic Energy, 6:129–33.

GUIMARAES, D., and V. ILCHENKO. 1954. Apatita de barreiro, Araxá, Minas Gerais. Sec. Agric. Minas Gerais, Boletim 7/8, 9/10, 11/12. Belo Horizonte.

LEINZ, V. 1949. Contribuição a geologia dos derrames basálticos do Sul do Brasil. U.S.P. Fac. Fil., Geologia 5. São Paulo.

———. 1962. Geologia geral. São Paulo: Editôra Nacional.

LEONARDOS, O. H. 1955a. Araxá, a "bomba atômica" de Djalma Guimarães. Mineração e Metalurgia, Vol. 22.

———. 1955b. Monazita no Brasil. *Ibid.*

———. 1956a. Carbonatitos com apatita e pirocloro no estrangeiro e no Brasil. *Ibid.*, Vol. 23.

———. 1956b. Carbonatitos com apatita e pirocloro. Rio de Janeiro: Departmento Nacional da Produção Mineral.

———. 1956c. Sôbre a abunância de torio no Brasil. Mineração e Metalurgia, Vol. 22.

———. 1959. Disponibilidade mundial de Torio. *Ibid.*, Vol. 30.

MORAES, L. J. DE. 1955. Known occurrences of uranium and thorium in Brazil. Proc. Internat. Conf. on Peaceful Uses of Atomic Energy, 6:134–39.

OLIVEIRA, A. I., and O. H. LEONARDOS. 1943. Geologia do Brasil. Rio de Janeiro: Ministério de Agricultura.

ROSER, F. X., and T. L. CULLEN. 1958. On the intensity levels of natural radioactivity in certain selected areas of Brazil. *In* Radioactive Products in the Soil and Atmosphere. Rio de Janeiro: Instituto Brasileiro de Bibliografia e Documentação.

ROSER, I. H., and T. L. CULLEN. 1962. Radiation levels in selected regions of Brazil. Anais acad. Brasil. cienc., **34**:23–35.

SCHUMACHER, F. 1954. Relatório sôbre as jazidas uraníferas do Brasil. Rio de Janeiro (unpublished).

VICKERY, R. C. 1953. Chemistry of the Lanthanons. New York: Academic Press. Pp. 296.

R. L. BUSDIECKER AND B. W. MAXWELL

54. Environmental Radiation Surveillance in the Antarctic

Research and exploration in the Antarctic promise to yield much information about the earth that is unattainable in other parts of the world. These efforts are expensive because of the difficulty in supplying the necessities for research and living to such a remote region. The coastal region of Antarctica, such as McMurdo Station, is accessible by ship for only about 3 or 4 months during the summer season. The inland stations are reached almost exclusively by air (Fig. 1). The bulk of the supplies transported to Antarctica consists of fuel for heating and generating electrical power.

The U.S. Navy, in charge of supply at Antarctica, developed an early interest in power reactors. A nuclear power reactor was considered for installation at McMurdo Station as a source of power and heat because it would reduce greatly the bulk of supplies transported to Antarctica. The nuclear power plants that were considered were designed for complete containment of radioactivity and for storage of radioactive waste. All such wastes would be shipped back to the United States for disposal.

In August, 1960, the U.S. Atomic Energy Commission selected the proposal of the Martin Company (Martin-Marietta) of Baltimore, Maryland, to supply a packaged power plant for McMurdo Station, Antarctica. The construction of this reactor began in November, 1960, and the reactor began power production in July, 1962. McMurdo was selected as the site for the reactor because it is accessible by sea, and from there the inland stations, Byrd and South Pole, are resupplied by air. Studies have also been initiated to determine the

R. L. BUSDIECKER is chemist, and B. W. MAXWELL is geologist, at the Technical Operations Branch, Division of Radiological Health, Public Health Service, Department of Health, Education, and Welfare, Washington, D.C.

feasibility of constructing and operating a nuclear power plant at Byrd Station.

To provide conclusive data that this nuclear power plant was not releasing activity, the Public Health Service under an agreement with the U.S. Navy began environmental background radiation measurements at McMurdo in December, 1960. In February, 1962, environmental measurements were begun at Byrd Station.

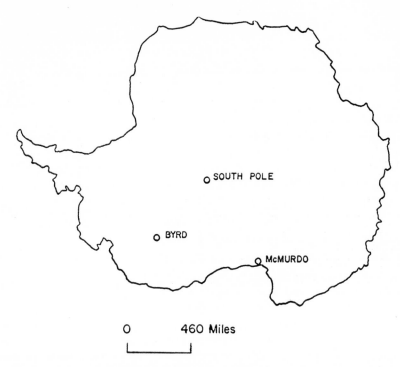

Fig. 1.—Antarctic continent

Since there was a shortage of environmental media, routine sampling was confined to air, snow, and camp water. Biota were nonexistent except for about a month during the summer, and then were found only at McMurdo. Maintenance of instruments presented some problems, since spare parts, needed during the 7–8 months of isolation during each year, had to be estimated during the previous summer. Power supply, though adequate for most needs, was often too variable for proper instrument performance. Under these circumstances the concentrations reported in the sample should not be regarded as absolute values.

PROCEDURES

A previous study of activity levels showed that the natural and fission-product activities were exceedingly low in Antarctica (Lockhart, 1960; Picciotto, 1958). Since the levels were so low, no attempt was made to do analyses for specific nuclides and no facilities were provided for radiochemical analysis in Antarctica. Estimates of activity were based only on α and β radiation. Low-level α and β counters (Nuclear Measurements Corporation PC-3A internal proportional counters for α and Sharp and Tracerlab thin-window counters for β) were used at the Byrd and McMurdo stations. No γ-detection equipment was brought in until the summer of 1961–62, just prior to the completion of the reactor at McMurdo. At that time a single-channel analyzer was set up in the McMurdo laboratory for the qualitative detection of argon-41. This activation product was the only release anticipated from the reactor site.

Procedures for the determination of activity in air and water were essentially the same as described by Setter *et al.* (Setter and Coats, 1961; Setter and Goldin, 1956). To facilitate the handling of large numbers of air samples and to simplify calculations, it was assumed that radon-222 and radon-220 daughters were in equilibrium with their respective parents, that filtering efficiency was 100 per cent, and that self-absorption losses were negligible. Particulates in the air were collected with Staplex Hi-Vol Samplers on Gelman, type E, glass-fiber filters. Water and melted-snow samples were separated into suspended and dissolved fractions by filtering through millipore membranes.

SAMPLE COLLECTION

MC MURDO

Initially, two air samplers were operated at McMurdo, one at the U.S. Public Health Service laboratory in the camp and one near the reactor site (Fig. 2). A third station at the new cosmic-ray building site, about a mile from the reactor site, was placed in operation in February, 1962. Samplers were operated for 24-hour periods before filters were replaced. The Hi-Vol Samplers were not affected appreciably by the cold, but the many oil-burning stoves and vehicles contributed large quantities of soot, which quickly plugged filters, stopped air flow, and caused sample rejection. Ice and snow particles also accumulated on filters and increased sample rejection. With transportation at a minimum, a regular schedule for collecting samples was exceedingly difficult except by walking to the station.

At McMurdo, water samples were taken from two sources, the camp water supply and a number of ponds throughout the camp area where the runoff collected only during the summer months. The camp water, which was melted snow, was treated before it was used for drinking and cooking. Samples of fresh snow were collected when available in and around the camp area. The collection of samples was hampered by the meager quantity of snow falling in the area (about 9 inches per year) and by contamination with dirt and fine dust. The dust contamination made large samples impossible to process. For comparison purposes, additional water samples were available from neighboring fresh-water lakes within helicopter flying distance. These could be sampled only during the short flying season in the summer.

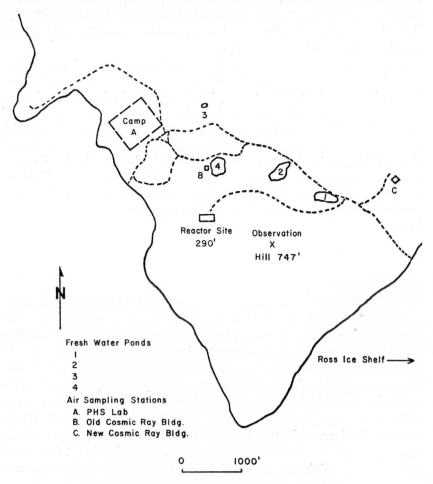

Fig. 2.—Sampling locations, McMurdo, Antarctica

TABLE 1

MONTHLY AVERAGE LEVEL OF ACTIVITY (PC/M³) IN AIR
SAMPLES FROM ALL STATIONS AT BYRD

Month (1962)	Radon-222	Radon-220	GFP*
March...............	0.5	0.03	0
April................	0.5	0.01	0
May.................	0	0	0
June................	0.2	0	0.04
July................	0.5	0	0.01
August..............	0.2	0	0.01
September..........	0.2	0	0.03
October............	0.1	0	0.03
November..........	0	0	0.11
December...........	0	0	0.05

* Gross fission products.

TABLE 2

MONTHLY AVERAGE LEVEL OF ACTIVITY (PC/M³) IN AIR
SAMPLES FROM ALL STATIONS AT McMURDO

Month	Radon-222	Radon-220	GFP*
1961			
January..........	15	0.15	0.06
February.........	10.3	0.10	0.05
March...........	7.6	0.07	0.02
April............	5.9	0.06	0.02
May.............	4.8	0.08	0.02
June.............	5.5	0.06	0.01
July.............	8.8	0.09	0.02
August...........	7.9	0.08	0.01
September........	Instrument failure		
October..........	2.6	0.06	0.02
November........	10	0.15	0.02
December.........	10.7	0.14	0.02
1962			
January..........	10.4	0.15	0.02
February.........	1.9	0.13	0.02
March...........	1.6	0.11	0.03
April............	2.4	0.16	0.01
May.............	3	0.19	0.02
June.............	1.3	0.12	0.03
July.............	3.6	0.12	0.03
August...........	2.2	0.12	0.05
September........	2.1	0.13	0.07
October..........	1.5	0.13	0.07
November........	2	0.12	0.10
December.........	2.2	0.07	0.07
1963			
January..........	1.9	0.07	0.13

The average residual a activity was less than 0.001 (trace amount).
* Gross fission products.

BYRD

Movement at Byrd was even more restricted than at McMurdo. McMurdo, being on the coast, rested on soil and rock, but Byrd Station was built on 8,000 feet of ice in the interior of the continent. Drifting snow was a serious problem; it frequently extended 10 feet above the level of the surface. Because of drifting, no buildings could

TABLE 3

Monthly Average Level of Activity in Water
and Snow Samples from McMurdo Station

Total β Activity (Dissolved plus Suspended) (pc/l)

Month	Snow	Pond	Camp Water
1961			
January.........	14	7	12
February........	8	—	12
March..........	—	—	4
April...........	—	—	—
May............	—	—	4
June...........	—	—	4
July............	—	—	3
August.........	—	—	7
September.......	—	—	—
October.........	—	—	3
November.......	—	6	6
December........	—	9	0
1962			
January.........	5	12	0
February........	—	—	35
March..........	—	—	6
April...........	—	—	5
May............	—	—	6
June...........	—	—	7
July............	—	—	—
August.........	66	—	13
September.......	130	—	7
October.........	56	—	13
November.......	249	—	8
December........	120	16	7
1963			
January.........	57	27	24

Dashes denote no sample.
Residual α activity was a trace amount.

rest on the surface, and facilities were placed in tunnels below the snow level. The roofs of the tunnels were approximately at snow level, and the ceilings were about 30 feet high.

Initially, air-sampling stations were established at Byrd in the tunnels under the snow, at the level of the snow, and on an Aurora observation tower (Fig. 3). In the tunnels the problem of air filters' becoming clogged with soot was very bad, and about 20 per cent of

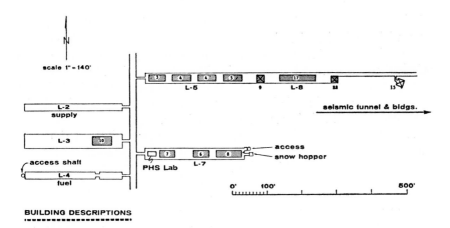

scale 1" = 140'

L-2
supply

L-3

access shaft

L-4
fuel

L-5

L-8

seismic tunnel & bldgs.

PHS Lab
L-7

access
snow hopper

0'   100'   500'

BUILDING DESCRIPTIONS

3   galley/sick bay
4   quarters
5   science bldg.
6   shop
7   communication
8   generator
9   aurora tower (outside)

10   garage
15   balloon inflation tower (outside)
17   meteorology
18   radome tower-met. (outside)
Not shown:
    radio noise −7500' N
    V.L.F.−3000' S − ionosphere,
    geomagnetics, seismo bldgs.

FIG. 3.—Byrd Station tunnels and buildings existing during 1962

TABLE 4

MONTHLY AVERAGE LEVEL OF ACTIVITY IN
WATER AND SNOW SAMPLES
FROM BYRD STATION
Total β Activity (Dissolved plus Suspended)
(pc/l)

Month (1962)	Snow	Camp Water
March.	—	8
April.	10	15
May.	22	23
June.	9	11
July.	18	16
August.	52	42
September.	41	37
October.	201	146
November.	78	58
December.	227	100

Dash denotes no sample.
Residual α activity was not observed.

all air samples were rejected because of clogged filters. The station at ground level had to be abandoned because of continual plugging of filters with snow. The sampler on the Aurora observation tower was relatively free from snow except in the worst weather. It was planned to maintain a sampler also at one of the substations, but the half-mile distance from Byrd made it too hazardous to reach the sampler during the darkness and drifting snow of the winter.

Water samples from Byrd Station were collected from the camp supply and represented untreated melted snow. Snow and water samples were collected biweekly.

Discussion

Although the air data for 1962 have not been completely processed, the snow and water data indicate increases in gross fission-product levels starting in August, 1962. No decay of the fission products was observed, over a period of a week or more, in the snow and water samples; thus these samples contained no significant increase from fresh fission products after July, 1962, when the reactor achieved power. The major source of the debris was probably the high-altitude Pacific tests. The High Altitude Sampling Program (HASP) studies have shown that stratospherically injected debris has a longer residence time than has debris from tropospheric shots (PHS, 1961). The absence of fresh debris in Antarctica indicates little tropospheric influence.

In the winter of 1961 the α-counting equipment malfunctioned. Since the problem could not be rectified until the resupply season opened in October, radon-222 and radon-220 determinations were made on the basis of β counts. The poor correlation between radon-222 and radon-220 activities, as determined separately from α and β counts, was found when defective equipment was replaced. This discrepancy tends to show that some of the assumptions, such as parent-daughter equilibrium or negligible self-absorption, were not valid. However, the same procedures were retained to indicate trends in radon-222–radon-220 levels rather than absolute values.

The air values suggest seasonal variations or trends in the level of both natural and artificial radioactivity (Fig. 4). The radon-222 and radon-220 levels are related to the amount of soil exposed. During the Antarctic winter (March–October) the ground is completely frozen and a greater amount of snow exists, but during the summer season (November–February) the ground is snow free and the top layer of the soil thaws and dries out. The peaks in fission-product activity during the summer are probably related to the seasonal mixing of the tropospheric and stratospheric air (PHS, 1961).

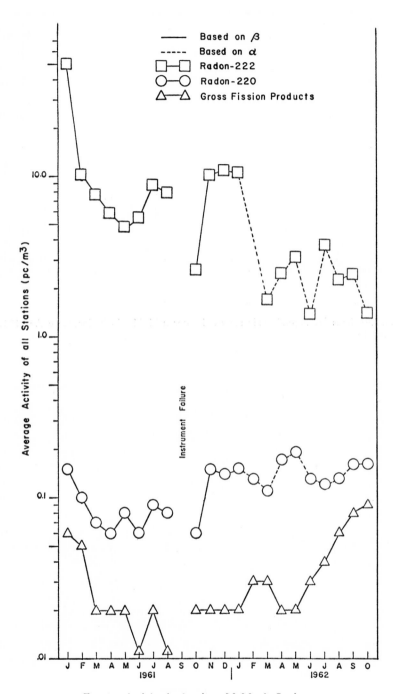

FIG. 4.—Activity in the air at McMurdo Station

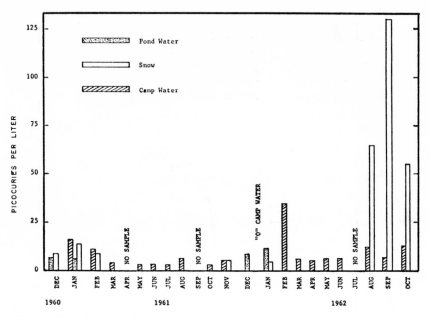

FIG. 5.—Gross fission products in snow and water at McMurdo Station, December, 1960–October, 1962.

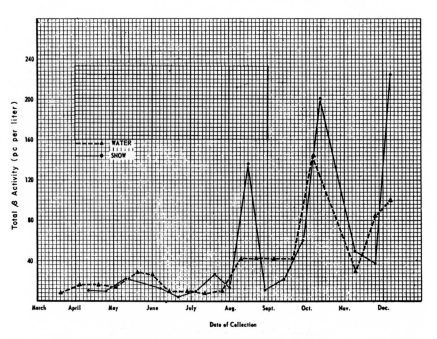

FIG. 6.—Total β activity in surface snow and camp water, Byrd Station, Antarctica, 1962

The low level of long-lived natural activity in air relative to fission-product activity makes determination of the latter very sensitive. The sensitivity of the method was limited mostly by the available counting time; the error at the 95 per cent confidence level was about 10 per cent. The sensitivity for long-lived β is about 0.003 pc/m^3.

The wide difference between snow and camp water values at McMurdo was due to two major factors (Fig. 5). Snow, used for the camp supply, was much older than the fresh surface snow. The older subsurface snow had little chance to become contaminated with recent fallout. The snow, after melting, was filtered and treated before use; the filteration undoubtedly removed still more activity.

The difference between snow and camp water values is less at Byrd than at McMurdo (Fig. 6). The melted snow was used directly at Byrd without any prior filtration or treatment. The lower levels in the water can be attributed to deposition on the storage tank and piping.

REFERENCES

LOCKHART, L. B., JR. 1960. Atmospheric Radioactivity in South America and Antarctica. Washington, D.C., Naval Res. Lab. Rpt. 5526. Pp. 10.

PICCIOTTO, E. 1958. Mésure de la radioactivité de l'air dans l'Antarctique ("Measurement of the radioactivity of the air in the Antarctic"). Nuovo Cimento, 10:190–91.

SETTER, L. R., and G. I. COATS. 1961. The determination of air-borne radioactivity. J. Am. Ind. Hyg. Assoc., 1:64–69.

SETTER, L. R., and A. S. GOLDIN. 1956. Measurement of low-level radioactivity in water. J. Am. Water Works Assoc., 48:1373–79.

U.S. PUBLIC HEALTH SERVICE. 1961. Radiological Health Data Quarterly Report, October, 2:426–29. Washington, D.C.

F. W. SPIERS, M. J. McHUGH,
AND D. B. APPLEBY

55. *Environmental γ-Ray Dose to Populations:*
Surveys Made with a Portable Meter

THESE SURVEYS were made to investigate the extent to which a representative average population dose could be determined by a limited number of on-site measurements. The results of the surveys have been published elsewhere (Court-Brown *et al.*, 1960; Spiers, 1960), and this article will be chiefly devoted to a description of the dosimeter, to a consideration of problems of calibration, and to the general method of surveying. Some results will be given to illustrate various features of the discussion.

Initially, surveys were designed for three localities in Scotland—Edinburgh, Dundee, and Aberdeen—where the local geologies and distributions of building materials were sufficiently homogeneous to give expectation of definitive results. Later, a larger area, the county of Aberdeen, was surveyed in which there were considerable variations in the radioactivity of the local rocks, in the building materials used in different localities, and in the population densities. Even in this county survey an average γ-ray dose rate, representative of the whole population of Aberdeenshire, is considered to have been obtained. It is clear, however, as will be discussed later, that the methods we have used of on-site surveying will not necessarily provide a representative dose rate in all circumstances and that there are areas where the radioactivities of rocks and building materials are so heterogeneous that other methods of measuring population dose are required.

F. W. SPIERS is professor, head of the Department of Medical Physics, M. J. McHUGH is physicist, and D. B. APPLEBY is technical officer, Medical Research Council, Environmental Radiation Research Unit, Department of Medical Physics, University of Leeds, England. M. J. McHugh is at present in the Department of Physics, Christie Hospital and Holt Radium Institute, Manchester, England.

A Portable Radiation Meter

A combination of a high-pressure ionization chamber and simple battery-operated electrometer circuit is designed to provide a portable background γ-radiation dosimeter of high sensitivity and long-term stability of response. The instrument operates by integrating the dose, and its sensitivity is such that a dose of about 1 microrad of γ radiation can be measured with a standard error of ±3 per cent. Typical background dose rates of around 10 μrad/hr can be measured satisfactorily with an observation time of only a few minutes.

CHAMBER ASSEMBLY

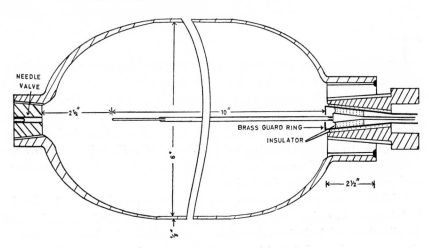

Fig. 1.—General chamber construction

The ionization chamber is a seamless steel cylinder with wall thickness, 0.125 inches; diameter, 6 inches; and over-all length, 15 inches. It has a volume of approximately 5.5 liters and is filled with nitrogen to a pressure of approximately 45 atmospheres. One end of the chamber is fitted with a standard gas inlet valve, and the other end carries a conically threaded steel nut, into which the electrode system is built. The central electrode is made of thin telescopic stainless-steel tubing, and this is brought out, together with the guard cone and separating polytetrafluoroethylene (P.T.F.E. or "teflon") insulators, through the steel nut. The whole assembly of electrode, guard cone, and insulators is coned to a half-angle of 10° to form a gas-tight plug. Mounted directly on the steel nut is an aluminum base plate carrying the aluminum canister, 6 inches in diameter, which contains the electrometer tube and auxiliary electronics. The general construction of the meter is shown in Figure 1, and the details of the electrode system and mounting of the electrometer tube in Figure 2. A photograph of the complete instrument is shown in Plate I.

For reliability, the electrometer circuit (Fig. 3) has been reduced to its simplest form and consists of a Victoreen 5800 electrometer

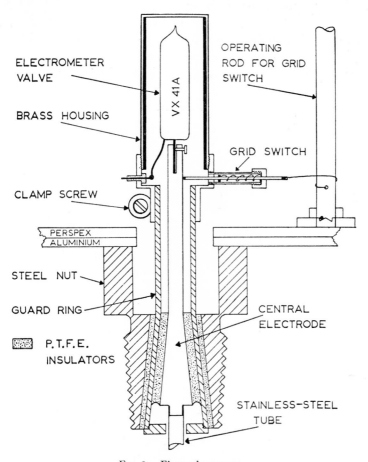

ELECTROMETER VALVE

BRASS HOUSING

CLAMP SCREW

PERSPEX
ALUMINIUM

STEEL NUT

GUARD RING

P.T.F.E. INSULATORS

VX 41A

OPERATING ROD FOR GRID SWITCH

GRID SWITCH

CENTRAL ELECTRODE

STAINLESS–STEEL TUBE

Fig. 2.—Electrode system

space-charge tetrode in a Townsend balance circuit. The chamber central electrode is connected to grid 2 of the tetrode, and both can be connected to the guard cone by a remotely operated switch in the "EARTH" position. The charge-balancing circuit is inserted in the lead from this switch to the chamber-polarizing battery. With the switch open, a measured voltage of up to 10 volts from a potentiometer circuit can be applied to the guard cone to balance the potential rise on the chamber electrode under rate-of-drift conditions. By this means, the electrometer tube is used as a null-indicator, and errors in its operation are minimized. The "standard" anode current, to which the tube is restored by operation of the balancing circuit during a measurement, is set by experiment to a value that gives minimum drift with falling battery potentials. This adjustment is made possible by the circuit design (see Fig. 3), and the optimal anode current of about 11 microamperes is also one that usually results in minimum grid current. At typical background dose rates the ionization current is of the order of 10^{-13} amperes, and the grid current is usually 10^{-15} amperes or less.

PLATE I

Ionization chamber

BASIC TESTS OF FUNCTION

Because it is not possible to test the instrument under conditions of zero radiation (except perhaps deep within or below the polar ice cap!), special care must be taken to make tests that insure that all sources of instrument drift other than background radiation are either absent or insignificant. Radioactive contamination of the ionization chamber is minimized by the use of steel, and any ionization from residual α radiation from the internal surface is reduced to negligible

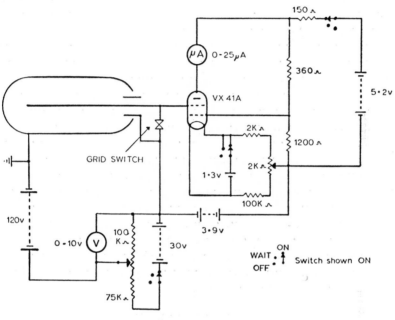

FIG. 3.—Circuit

proportions by the use of a high gas pressure and a low ion-collection field near the chamber wall. Checks are required, however, to test for (*a*) maintenance of the gas pressure, (*b*) insulator leakage, and (*c*) tube grid current.

The response of the instrument to a standard source (e.g., a few μc of radium-226) at a standard distance will test the constancy of the γ-ray sensitivity and in effect check the maintenance of the gas pressure. Because this test is made by obtaining the difference between the response to background and the response to background plus the standard source, it will not of itself give any information on the reliability of the background response as a true measure of background radiation.

The insulation of the central electrode can be tested, within the statistical accuracy of the ionization current measurement, by making observations of the background response under two conditions of rate-of-drift compensation. Measurements can be made (*a*) with continuous voltage compensation, so that the standard anode current is maintained and no potential difference is allowed to develop between the central electrode and guard cone, and (*b*) with voltage compensation only at the end of the measurement, so that an average potential difference of $V/2$ exists across the central electrode insulator, V being the final compensating voltage measured. Agreement between the two methods means that the insulator leakage current must be less than the standard error of the difference of the two measurements.

Although the grid current of the electrometer tube is measured before the instrument is assembled, it is important to be able to check the grid current under operating conditions. This can be done by temporarily uncoupling the chamber polarizing battery and connecting the chamber wall instead directly to the compensating circuit. The meter response measured under these conditions then comprises the grid current together with a very small ionization current resulting from residual stray potentials. Usually, this single observation is sufficient to establish the magnitude of the grid current for practical purposes. If, however, this "apparent" grid current is measured at three different background radiation dose rates (which are known approximately), extrapolation to a "zero radiation" condition will enable the true grid current to be estimated. The electrode assembly-grid capacity with the grid switch open is approximately 12 picofarads, and hence if, under the test conditions, the grid drifts V volts in T seconds, the apparent grid current is given by:

$$I_g = C\frac{dV}{dt} = \frac{12\,V \times 10^{-12}}{t}\ \text{amps.}$$

After an initial warming-up period, battery and circuit drifts are very small. If, on closing the grid switch at the end of a drift-rate observation, the anode current has changed slightly, a small correction can be made to the observed compensation voltage, since the relationship between a change in compensation voltage and anode current is known.

CALIBRATION OF THE DOSIMETER

Saturation curves for a number of dose rates are given in Figure 4, from which it is evident that an ion-collecting potential of about 100 volts is adequate for radiation dose rates up to 500 mrad/yr. At

125 mrad/yr and 120 volts the slope of the saturation curve is 0.06 per cent per volt. A collecting potential of 90 volts has been used in most surveys.

A polar diagram of the instrument response to point sources of radium and iodine-131 is shown in Figure 5; the reduced sensitivity for radiation incident in the directions of the filling valve and the electrometer end do not greatly affect the response to omni- or nearly omnidirectional radiation. The average response, obtained by integrating the polar curve for the radium source over 4π space, is 0.94 of

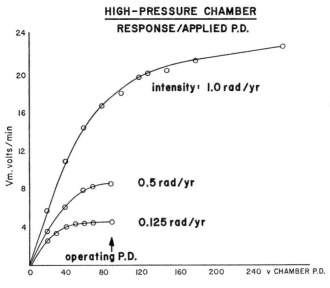

FIG. 4.—Saturation curves

that when the source is in the broadside-on position. The ratio is not significantly different if the average is taken over 2π for the lower part of the polar curve. The 4π average ratio for iodine-131 γ radiation is 0.91.

Typically, the 4π response to a dose rate of 100 mrad/yr is, for the present gas pressure, 2.19 v/min for radium γ rays and 1.83 v/min for iodine-131 γ rays. If we assume, following Vennart (1957), that the quality of background γ radiation can be represented by a mixed radiation comprising 70 per cent radium γ rays and 30 per cent iodine-131 γ rays, the effective response to 100 mrad/yr would be 2.08 v/min, and hence a drift rate of 1 v/min is equivalent to a background dose rate of 48.1 mrad/yr (5.48 μrad/hr or 5.75 μr/hr).

PERFORMANCE OF THE DOSIMETER

The long-term stability of the dosimeter can be followed by the basic tests described above. The over-all stability can also be investigated by repeated observations in a low-background steel cubicle, where the radiation level is almost entirely due to cosmic-ray events and is relatively unaffected by changes in environmental radioactivity. For example, the cosmic-ray response of the dosimeter in a steel room remained between the limits 0.490–0.515 v/min (when corrected for the effect of barometric pressure on cosmic-ray intensity) over the

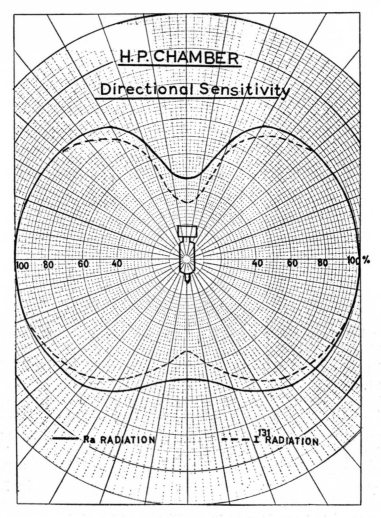

FIG. 5.—Polar diagram

period October, 1957–July, 1959, during which time the γ-radiation dose rate over undisturbed grassland in Leeds increased from 6 μrad/hr to 8 μrad/hr.

The statistical accuracy of the dosimeter measurements was investigated by taking repeated observations to establish the standard error of a single measurement of a given charge. Observations of the times to collect charges corresponding to compensation voltages of 1.67 v., 5.00 v., and 8.50 v. showed that the standard error of a single observation could be expressed as:

$$\sigma = 3.85 \, V^{-1/2} \text{ per cent} ;$$

where V is the voltage "collected" in the time of observation. A single collection of 5 v. (half the full-scale deflection of the compensating voltmeter), therefore, carries a standard error of a little over 1.5 per cent. This statistical performance was further borne out in field work, where analysis of some 120 pairs of observations showed that the difference between the two readings of a pair was less than 5 per cent in 93 per cent of the pairs observed.

DESIGN OF THE SURVEYS

In most situations the average background radiation dose received by a population depends to a great extent upon the radioactivity of the materials used in house and building construction. In open country the outdoor dose rate varies with the geology of the surface soil, but in towns and in villages it is mainly determined by the materials used for pavements and roads. The surveys were, therefore, designed to distribute the measurements in houses and outside in a manner representative of the prevalence of various building and paving materials. The numbers of houses measured in the three cities were: Edinburgh, 157; Dundee, 71; and Aberdeen, 103. In each city, measurements were also made at a similar number of outdoor sites, usually on the road or sidewalk near the houses. The Edinburgh survey occupied a total of 17 days; Dundee required 9, and Aberdeen 13 days. The numbers of house measurements followed as nearly as possible the proportions of different types of houses given by the records of the city engineer. The distribution of the 102 road measurements in Aberdeen is shown in the lower half of Figure 6; the distribution of the houses measured follows almost exactly the same pattern.

The house addresses were obtained through the auspices of a number of organizations: the staff of the city health department, the university, the city hospital, and the civil defense authorities. In each city a member of the staff of one of the local organizations undertook the preliminary task of getting the addresses of those willing to have their

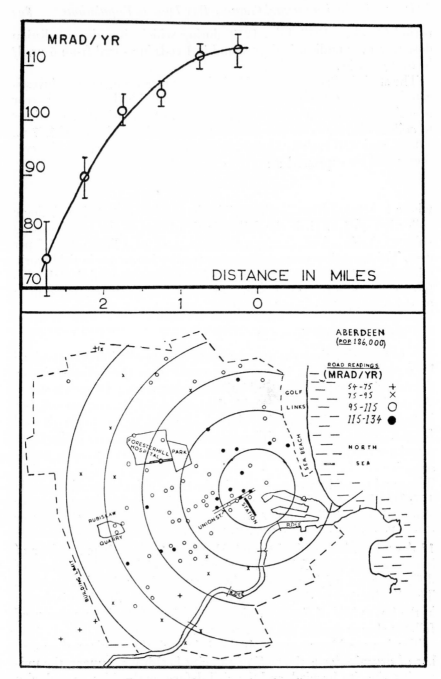

Fig. 6.—Aberdeen map and road readings

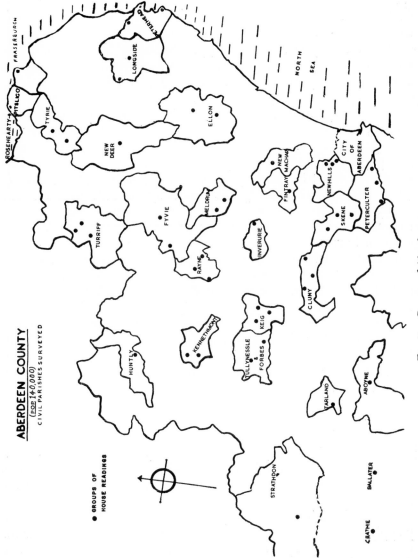

ABERDEEN COUNTY

(POP 140,000)

CIVIL PARISHES SURVEYED

● GROUPS OF
HOUSE READINGS

ROSEHEARTY
PITSLIGO
FRASERBURGH

TYRIE

NEW
DEER

LONGSIDE

PETERHEAD

ELLON

NORTH
SEA

NEW
MACHAR

FINTRAY

CITY
OF
ABERDEEN

NEWHILLS

MELDRUM

FYVIE

SKENE

PETERCULTER

TURRIFF

RAYNE

INVERURIE

CLUNY

KENNETHMONT

KEIG

TULLYNESSLE
&
FORBES

HUNTLY

TARLAND

ABOYNE

STRATHDON

CRATHIE

BALLATER

FIG. 7.—County of Aberdeen, map

houses measured and making out daily lists of appointments. This work was done during the three or four weeks preceding the measurements and contributed enormously to the efficiency of the surveys.

A more elaborate design was necessary for the survey of the county of Aberdeen. The distribution of the 172 houses in which measurements were made was arranged to take into account the variations in population density and local geology. The survey included all the 10 parishes or boroughs with populations of 3,000 or more, 4 of the 10 parishes with populations between 2,000 and 2,999, 5 of the 23 parishes with populations between 1,000 and 1,999, and 8 of the 39 parishes with populations less than 1,000. A procedure of random selection, with some stratification to insure representation of the different geological regions, was adopted for the parishes of less than 3,000 population. In each group the number of houses was fixed on the basis of about 12 houses per 10,000 inhabitants, and this number was then distributed among the randomly selected parishes. The parishes and boroughs in which measurements were made are named in Figure 7, where the black circles indicate the sites of the selected groups of houses.

In houses of two or more stories two observations of a 5-volt drift were made at each of 3 sites: in the living room, in the kitchen, and in one bedroom. The mean of all the 6 observations was taken as the dose rate in the house. If the house was a single-storied bungalow or a flat, two sites (a kitchen and one other room) were measured. Measurement at an outdoor site comprised two observations of a 5-volt drift. Many of these observations were made in a car, and a correction factor was determined that allowed for the attenuation of the γ radiation by the vehicle.

RESULTS OF THE SURVEYS

Before using the calibration factor to convert the drift rate to a dose rate in millirad per year, the cosmic-ray response of the dosimeter was subtracted from the observed value of the volts per minute. The outside cosmic-ray response was determined by measuring the response in a light wooden boat over a fresh-water lake at about half a mile from land. The cosmic-ray response inside houses was obtained by applying a correction factor to the outside observation to allow for overhead shielding by roof tiles, rafters, ceilings, floor boards, and joists. Based on data by Clay (1936), the cosmic-ray ionization is reduced to 87, 82, and 79 per cent by overhead shieldings of 13, 19, and 25 gm/cm^2, typical of situations on the first, second, and third floors, respectively, *below* a roof structure of wood and tiles.

The outdoor dose rates in Edinburgh, Dundee, and Aberdeen are shown in histograms in Figure 8, from which it is evident that the mean outdoor dose rate in Aberdeen during the survey was 104 mrad/yr compared with 48.5 mrad/yr in Edinburgh. An interesting feature of the Aberdeen dose rates is shown in the upper half of Figure 6, which gives the mean dose rates in the annular zones shown on the map in the lower part. As the survey moved in from the suburbs, with relatively wide roads and a lower density of houses, the dose rate increased from 75 mrad/yr to 113 mrad/yr in the more densely built-up

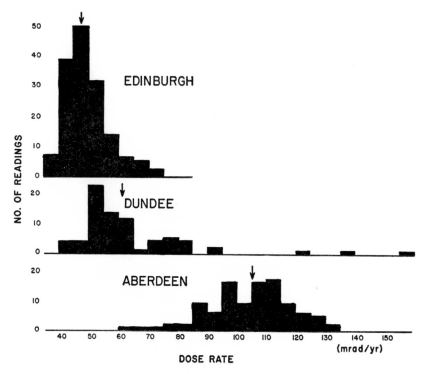

Fɪɢ. 8.—Road readings, Edinburgh, Dundee, and Aberdeen

central zone. This feature appeared only in Aberdeen, where the houses are mainly of granite and all roads are paved with granite.

The dose rates in the houses built with local stone in the three cities are shown in the histograms of Figure 9. Again, significant differences between the cities are evident. The mean dose rate in the Aberdeen granite houses, being 87 mrad/yr, is nearly twice the average in Edinburgh stone houses. The difference in dose rates between the Dundee (Old Red) and the Edinburgh (Lower Carboniferous) sandstone houses is in keeping with the higher potash content of the Dundee stone. The average dose rates in millirad per year in clay-brick houses in the three cities were: Edinburgh, 74.5 ± 1.0; Dundee, 77 ± 1.6; and Aberdeen 81 ± 4. The differences between the dose rates are not statistically significant, and the Aberdeen average is based on measurements in only 7 houses.

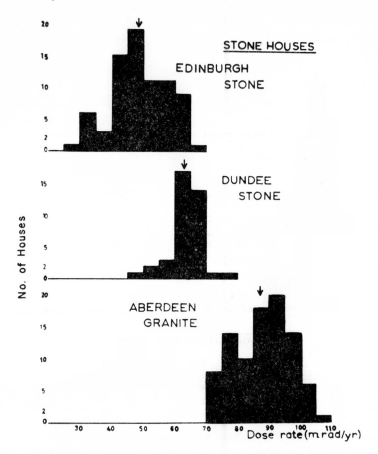

Fig. 9.—House readings, Edinburgh, Dundee, and Aberdeen

The detailed results of the surveys in Edinburgh, Dundee, and Aberdeen are given in Table 1, where for each city the numbers of different types of houses are listed together with the mean dose rates for each type. The numbers of houses are also expressed as percentages and compared with the official percentages from the city engineer's department. The mean dose rate for each city is obtained by weighting the values for each type of house by its fractional prevalence, and an estimated mean population dose rate in air is ob-

TABLE 1

RESULTS IN DETAIL FOR EDINBURGH, DUNDEE, AND ABERDEEN

EDINBURGH (POP. 465,000)

House Type	No. Surveyed	Survey (per cent)	Official (per cent)	Mean Dose Rate (mrad/yr)
Stone	76	49	63	48.5
Brick	54	35	34	74.5
Brick and stone	17	11	....	71.5
Miscellaneous	8	5	3	48

Weighted mean dose rate in houses (mrad/yr): 60 ±0.7(SE)
Mean dose rate on roads (mrad/yr): 48.5±0.6(SE)
Estimated mean population dose rate (mrad/yr in air): 57.1
Total limiting error (mrad/yr): ±2.1

DUNDEE (POP. 177,000)

House Type	No. Surveyed	Survey (per cent)	Official (per cent)	Mean Dose Rate (mrad/yr)
Stone	39	55	59	63
Brick and concrete	25	35	33	77
Brieze block	2	3	3	68
Wooden	1	1.5	2	62
Asbestos prefab	2	3	2	50
Al. prefab steel house	2	3	1.5	40

Weighted mean dose rate in houses (mrad/yr): 67.2±0.8(SE)
Mean dose rate on roads (mrad/yr): 63.0±2.3(SE)
Estimated mean population dose rate (mrad/yr in air): 66.1
Total limiting error (mrad/yr): ±2.3

ABERDEEN (POP. 186,000)

House Type	No. Surveyed	Survey (per cent)	Official (per cent)	Mean Dose Rate (mrad/yr)
Granite Type A*	55	53	....	89
Granite Type B†	22	21	72	82
Partly granite	14	14	11	88
Non-granite	12	12	17	76

Weighted mean dose rate in houses (mrad/yr): 85.3±0.8(SE)
Mean dose rate on roads (mrad/yr): 104.0±1.2(SE)
Estimated mean population dose rate (mrad/yr in air): 90.0
Total limiting error (mrad/yr): ±3.1

* Houses with two or more stories.   † One-story houses with roof bedrooms.

tained on the assumption that 18 hours per day are spent indoors and 6 hours outdoors. The total limiting error given in Table 1 is obtained by combining a limiting calibration error, estimated at ±3 per cent, with a limiting statistical error, taken as twice the standard error of the mean for the contributing dose rates.

The dose rates measured in the survey of the county of Aberdeen are shown in Figures 10 and 11. Each histogram shows a greater spread

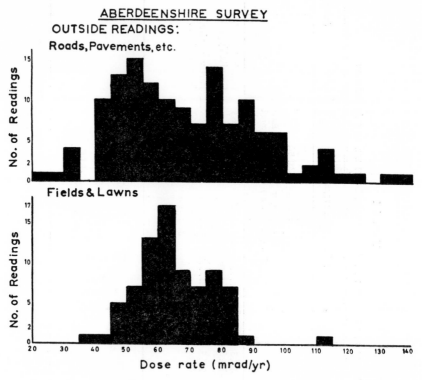

Fig. 10.—Aberdeenshire, outdoors

of dose rates than is seen in the results for the cities because considerably greater variation exists in surface geology and in the materials used for constructing roads and houses. Detailed data are given in Table 2, which includes also the mean dose rates for each area surveyed and for each population group. The local geological formations are shown in the table by code letters, and it is evident that the high outdoor dose rates correlate with the presence of granite. An average dose rate in air to the population is obtained by weighting the mean dose rate for each group by the proportion of the total population it represents.

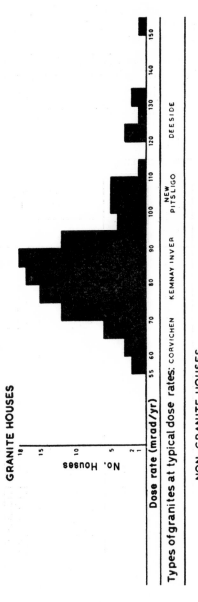

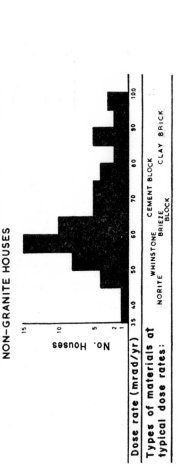

Fig. 11.—Aberdeenshire, houses

The results of the surveys were used, finally, to estimate the mean gonad and mean bone-marrow dose rates to the populations concerned. The details of the estimations are given in Table 3. On the basis of measurements by Spiers (1956) and Spiers and Overton (1962), mean attenuation factors of 0.63 and 0.65 were used to relate the dose rates to the gonads and marrow, respectively, to the dose rates measured in air in the surveys. The contributions from sources other than local γ radiation were estimated to be 49 mrem/yr to the gonads and 43 mrem/yr to the bone marrow; the cosmic-ray component in

TABLE 2

RESULTS IN DETAIL FOR ABERDEENSHIRE

PARISH	POPULATION 1951 CENSUS	No. HOUSES MEASURED	No. OUTSIDE READINGS	MEAN DOSE RATES (MRAD/YR)	
				Houses	Outside
Peterhead (G)............	15,300	18	17	83	84.5
Fraserburgh (AG).........	10,800	15	15	68	61
Petercoulter (E)..........	6,700	8	15	82.5	75
Newhills (EG)............	6,600	7	6	87	55
Inverurie (E).............	4,900	5	8	85	54
Huntley (N).............	4,800	6	8	71.5	59
Turriff (LS).............	4,400	5	5	63	59
Ellon (GE).............	3,400	5	11	67.5	58.5
New Deer (AN)...........	3,400	5	5	82	69
Fyvie (A)...............	3,000	5	5	81	63
Ballater (G).............	10 parishes	6	6	113 ⎫	103 ⎫
Longside (EG)...........	2/3,000	6	7	88 ⎪ 97	73 ⎪ 75
Newmacher (EA).........	Aggregate	7	13	84 ⎬	60 ⎬
Pitsligo (A)............	23,500	6	6	104 ⎭	65 ⎭
Aboyne (G).............	23 parishes	7	11	118.5 ⎫	93.5 ⎫
Crathie and Braemar (G)...	1/2,000	7	5	89 ⎪	90 ⎪
Meldrum (E).............	Aggregate	7	8	68 ⎬ 83.5	54 ⎬ 73.7
Skene (EG).............	34,300	7	15	87 ⎪	78 ⎪
Rosehearty (AS)..........	....	7	7	55 ⎭	53 ⎭
Cluny (G)...............	39 parishes	4	7	78.5 ⎫	71.5 ⎫
Fintray (E)..............	<1,000	4	4	84.5 ⎪	67.5 ⎪
Keig (AG)...............	Aggregate	4	5	92.5 ⎪	92.5 ⎪
Kennethmont (NL)........	23,600	4	3	63.5 ⎬ 73.5	44.5 ⎬ 65
Rayne (N)...............		4	4	51.5 ⎪	34.5 ⎪
Strathdon (HE)...........	....	4	7	70.5 ⎪	59.5 ⎪
Tarland (EG)............	....	5	8	80.5 ⎪	80.5 ⎪
Tullynessle and Forbes (A).	....	4	6	67 ⎭	68 ⎭
Aberdeenshire............	144,700	172	217	81.5	69.5

Estimated mean population dose rate (mrad/yr in air) 78.5

A = andalusite-schist; E = gneiss; G = granite; H = hornblende-schist; S = slate; N = norite; S = sandstone.

these figures was 24 mrem/yr, and the annual doses from potassium-40 were taken to be 21 millirems for the gonads and 15 millirems for bone marrow. Small additional contributions from carbon-14 and from radon were also assumed, but no contribution has been included for the cosmic-ray neutron dose. The relative biological efficiency (RBE) for cosmic-ray neutrons will depend upon the biological end-point, and, in the case of gene mutation, the relevant value is at present a matter for conjecture.

TABLE 3

TOTAL DOSE RATES TO GONADS AND BONE MARROW
(In mrem/yr)

Radiation and Significant Tissue	Edinburgh	Dundee	Aberdeen-shire	Aberdeen
Local gamma radiation (measured in air)				
Outdoors.................	48.5	63	69.5	104
In houses...............	60	67.2	81.5	85.3
24-hour average.........	57.1	66.3	78.5	90
Dose rate to gonads				
Local gamma radiation....	36	42	50	57
Other sources...........	49	49	49	49
Mean gonadal dose rate...	85	91	99	106
Dose rate to bone marrow				
Local gamma radiation....	37	43	51	58.5
Other sources...........	43	43	43	43
Mean marrow dose rate...	80	86	94	101.5

DISCUSSION

Relative values of the mean dose rates in different areas, measured with the same instrument, should be accurate to limits set by the (intrinsic) variations of the house and road dose rates. The standard errors of the mean dose rates in the city surveys lay between 1 and 3 per cent (see Table 1). In the county survey it was necessary to measure a much larger number of groups with fewer houses per group; the standard errors of the means for the group were then larger than in the city surveys, but the final averages based on 172 house and 217 road measurements are not thought to carry statistical errors much greater than the corresponding dose rates in Table 1.

Systematic errors (calibration, attenuation factors, allowance for the cosmic-ray response of the dosimeter indoors, etc.) set a limiting error of about 5 per cent on the absolute dose rates. The difference between the mean gonad or bone-marrow dose rates to the population of Aberdeen and Edinburgh is 21 mrem/yr, which, on these considerations, carries a limiting error of the order of ±1 mrem/yr. Although the four surveys were carried out over a period of one year (1958–59), during which time changes of about 5–10 per cent were observed in the γ radiation over undisturbed grassland, little change would be expected in the dose rates over streets and in houses. Observations repeated at some road sites in Aberdeen agreed with values obtained one year earlier within the limits of accuracy set on the measurements.

There are some areas in Britain, and almost certainly elsewhere, where the local γ radiation is extremely variable. In one small town that is typical of such conditions, the dose rates on the streets can vary from 50 to 300 mrad/yr because of the variable use of mine-spoil of high radioactivity or because of the outcropping of active strata. The dose rates in houses variously constructed with different building materials show similarly great variations. In these circumstances, it would be difficult, if not entirely impracticable, to design a satisfactory plan for a radiation survey in terms of on-site measurements. Here, the solution must lie in the use of personnel dosimeters if these can be designed to be small enough to be carried without inconvenience and to have the requisite stability and sensitivity.

SUMMARY

A high-pressure ionization chamber, containing approximately 6 liters of nitrogen at 45 atmospheres pressure, has been developed as a background monitor of high sensitivity and long-term stability. The ionization current, which is about 10^{-13} amperes, is measured by a miniature electrometer valve operating under rate-of-drift conditions. The full-scale deflection of the balancing voltmeter is equivalent to a dose of approximately 1 microrad, and an observation of 3 minutes' integrating time measures a typical background dose rate with a standard deviation of ± 1.5 per cent.

Methods of calibration and basic testing of the instrument are considered, and its application to the problem of determining the average dose to populations in some localities in Britain is described. Because the background dose to humans depends to a considerable extent on radiation levels inside buildings, the surveys were designed to distribute the house and the outdoor measurements in a manner representative of building materials, local geology, and population distribution. Results are given for the mean dose to gonads and to bone marrow in a number of localities.

ACKNOWLEDGMENTS

The authors are greatly indebted to Dr. W. Court-Brown of Edinburgh, Mr. H. D. Griffith of Aberdeen, and Mr. J. McKie of Dundee and to their staffs for assistance in organizing the surveys. Thanks are also due the medical officers of health, the university, hospital, and civil defense authorities for help in obtaining addresses of householders willing to have their houses included in the surveys. Acknowledgment is made particularly to Mr. Martin of the Aberdeenshire Civil Defense Authority for assistance with transportation.

The authors wish to express their thanks to Mr. G. A. Hay of the Department of Medical Physics for advice on the initial design of the radiation meter and to Mr. R. L. Corry for its mechanical construction.

REFERENCES

CLAY, J. A. VAN GEMERT, and J. T. WIERSMA. 1936. Decrease of primaries, showers and ionization of cosmic rays under layers of lead and iron. Physica, **3**:627–40.

COURT-BROWN, W. M., *et al.* 1960. Geographical variation in leukemia mortality in relation to background radiation and other factors. Brit. Med. J., **1**:1753–59.

SPIERS, F. W. 1956. The Hazards to Man of Nuclear and Allied Radiations. Report by a committee appointed by the Medical Research Council, London, H.M. Stationery Office, Appendix F, Cmd. 9780. Pp. 128.

———. 1960. The Hazards to Man of Nuclear and Allied Radiation. II. A second report to the Medical Research Council, London, H.M. Stationery Office, Appendix D, Cmd. 1225. Pp. 154.

SPIERS, F. W., and T. R. OVERTON. 1962. Attenuation factors for certain tissues when the body is exposed to nearly omni-directional gamma radiation. Phys. Med. Biol., **7**:35–43.

VENNART, J. 1957. Measurements of local gamma-ray background at Sutton, Surrey, and in London. Brit. J. Radiol., **30**:55–56.

WAYNE M. LOWDER, ASCHER SEGALL,
AND WILLIAM J. CONDON

56. *Environmental Radiation Survey in*
Northern New England

IN RECENT YEARS there has been a growing interest in the possibility that the continuous, low-level human exposure to environmental radiation may result in measurable biological effects in sufficiently large populations. The Harvard School of Public Health has been conducting studies of the incidence of leukemia, malignant neoplasms of bone, and congenital malformations in the populations of selected areas of Maine, New Hampshire, and Vermont. Since the available geological data indicate significant differences in mean bedrock radioactivity within these areas, a parallel study of the population exposure to natural radiation has been carried out. While it was considered doubtful that any significant correlation between radiation exposure and incidence of biological effect would be found, owing to the limited size of the population and the relatively small range in radiation exposure expected, it was thought that such a venture would provide considerable methodological information, as well as experience, useful in further studies of a similar nature.

Several different surveys have been undertaken to provide information on population exposure to natural radiation (Segall, 1962, 1963). These include:

1. A radiogeological survey of bedrock radioactivity in Maine, New Hampshire, and Vermont (Billings, to be published).

2. Environmental radiation measurements with portable ionization

WAYNE M. LOWDER and WILLIAM J. CONDON are physicists at the Radiation Physics Division, Health and Safety Laboratory, U.S. Atomic Energy Commission, New York, New York; ASCHER SEGALL is assistant professor of epidemiology in the Department of Epidemiology, Harvard School of Public Health, Boston, Massachusetts.

chambers, scintillation detectors, and a γ-ray spectrometer at outdoor locations and within private homes in selected areas (Lowder and Condon, to be published).

3. A personnel monitoring survey of population exposure to external radiation in these selected areas (Segall and Reed, 1964).

4. A survey of the concentration of radium-226, radium-224, and polonium-210 in teeth extracted from life- or long-time residents of these areas (Radford *et al.*, to be published).

The detailed analysis of the data will be published in the indicated reports. A summary of the findings from these surveys will be presented here, with particular reference to the relationship of bedrock radioactivity to the radiation levels to which the population is exposed. This point is of particular interest because of the occasional use of bedrock radioactivity as an indicator of levels of population exposure in the absence of direct dosimetric information.

RADIOGEOLOGY

The populated areas chosen for detailed study of environmental radiation levels are indicated in Figure 1. The northwestern Vermont regions are underlain primarily by limestone, dolomite, some shale, and sandstone. Elsewhere, the rocks, originally shale and limestone, are now regionally metamorphosed, primarily to schists and gneiss.

TABLE 1

PROPERTIES OF THE BEDROCKS

Formation	Mean eU (p.p.m.)	Main Town	Description
Dunham dolomite (Vt.).....	5	Rutland	Silicaceous buff-weathered dolomite
Beldens formation (Vt.).....	5	Middlebury	Interbedded buff to brown dolomite and white to blue-gray marble and limestone
Glacial drift (Vt.)..........	9	Bennington	Till, sand, and gravel overlying Dunham and Monkton formations
Monkton formation (Vt.)...	11	Burlington	Red quartzite with some buff and white quartzite and thick gray dolomite
Fitchburg granite (N.H.)....	23	Manchester	Heterogeneous pink granite gneiss with intrusive binary granite
Littleton formation (N.H.)..	23	Franklin	Very heterogeneous formation composed of mica schist, quartz-mica schist, and gneiss
Binary granite (N.H.)......	26	Concord	Light-gray to white granite composed of potash-feldspar, quartz, oligoclase, and some biotite and muscovite
Conway granite (N.H.).....	45	Conway	Pink biotite granite

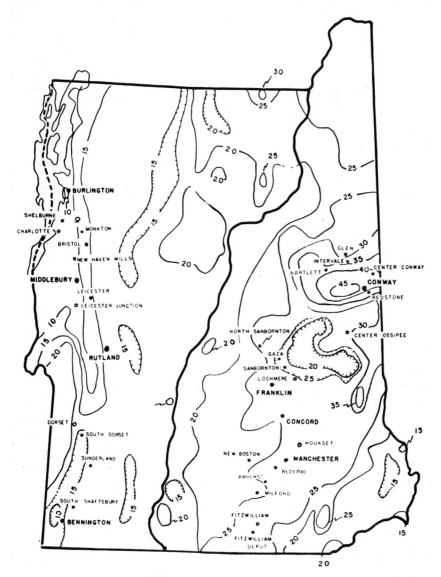

Fig. 1.—Survey areas in Vermont and New Hampshire with bedrock isorad lines (equivalent uranium p.p.m.).

Granitic rocks are quite prevalent throughout New Hampshire and include the thorium-rich Conway granite extensively investigated by Adams and his co-workers (Adams *et al.*, 1962; Richardson, 1963).

The isorad map in Figure 1 is based on determinations by Billings (to be published) of the equivalent uranium concentrations in p.p.m. of the various types of bedrock, using the best available data in the literature. The principal sources of these data are chemical analysis or α counting of rock specimens, direct radiometric measurements on outcrops during automobile traverses, calculations from the results of aerial surveys, and lithological similarity to rocks for which such data are available. The term "equivalent uranium," an imprecise but often used unit, refers to that quantity of uranium in equilibrium with its daughters that would yield the same quantity of γ radiation in roentgens as the actual radioactive contents of the particular rock. The results are summarized in Table 1. The values obtained for mean bedrock radioactivity range from 5 p.p.m. in limestone areas in Vermont to 45 p.p.m. in the Conway granite region of New Hampshire and are generally lower in Vermont.

Outdoor Radiation Survey

During July and August of 1962 a team from the Health and Safety Laboratory carried out measurements of outdoor environmental radiation in areas around the towns of Bennington, Rutland, Middlebury, and Burlington, Vermont, and Manchester, Concord, Franklin, and Conway, New Hampshire (see Fig. 1). Readings were obtained with high-pressure ionization chambers, portable scintillation detectors, and a γ spectrometer, using the techniques described in another paper at this symposium (Lowder, Condon, and Beck, 1963). The spot readings were generally made over open ground, with area surveys carried out with the portable scintillation detectors for the purpose of ascertaining the degree to which each spot reading was representative of the immediate neighborhood, including the effect of roadways and buildings.

Because of the presence of γ-emitting fallout, the spectrometer readings provided the only direct measurements of natural γ radiation. The fallout levels estimated from these readings were generally between 2 and 3 $\mu r/hr$ at the time of measurement. If this correction is applied to the approximately 100 additional readings made with the pressurized ionization chamber and portable scintillation detector, the inferred natural γ-radiation levels generally fall within a very limited range around the average values of the relatively few spectrometric determinations. If this had not been the case, a considerably larger number of readings would have been necessary before any quantitative estimates of population exposure could be justified.

The data obtained with the γ spectrometer in the eight areas are summarized in Table 2. Of particular interest is the considerable reduction in the inferred mean values of the effective equivalent uranium concentrations when the direct estimates from field spectrometric determinations over soils are used instead of the estimates based on mean bedrock geology. The spectrometric estimates were derived assuming that 2.1 p.p.m. thorium-232 in equilibrium with its daughters and 0.44 per cent potassium metal in the soil are each equivalent to

TABLE 2

RESULTS OF γ-SPECTROMETER MEASUREMENTS

REGION	BEDROCK	eU (p.p.m.)*	SPECTROMETRIC MEASUREMENTS (MEAN VALUES)					
			No.	Potassium (per cent)	Uranium (p.p.m.)	Thorium (p.p.m.)	eU (p.p.m.)	$\mu r/$ hr
Rutland, Vt.......	Dunham dolomite	5	5	1.6	1.3	5.6	7.6	5.6
Middlebury, Vt....	Beldens formation	5	5	1.9	1.1	7.2	8.8	6.6
Bennington, Vt....	Glacial drift	9	8	1.9	1.3	6.1	8.5	6.5
Burlington, Vt....	Monkton formation	11	8	1.4	1.3	5.0	6.9	5.2
Manchester, N.H..	Fitchburg granite	23	7	1.5	1.5	9.4	9.4	7.0
Franklin, N.H.....	Littleton formation	23	6	1.4	1.6	9.7	9.4	7.1
Concord, N.H.....	Binary granite	26	5	1.7	1.6	11.9	11.1	8.4
Conway, N.H.....	Conway granite	45	16	2.1	2.2	15.6	14.4	10.9

* As determined by Billings (to be published).

1 p.p.m. uranium-238 in equilibrium, in terms of air dose rates at 1 meter above the ground (see Table 3; Lowder, Condon, and Beck, 1963). It is also apparent from Table 2 that the soils over which the measurements were taken do show some relationship to the activity of the bedrock. The region of highest bedrock activity, that around Conway, New Hampshire, exhibits radiation levels significantly higher than those of any of the other regions. That the Vermont readings in general are somewhat lower than those in New Hampshire also supports this relationship.

POPULATION EXPOSURE ESTIMATE

The fact that most individuals spend a large part of their time indoors complicates the estimation of population exposure by means of spot readings. To check on the validity of the inferences made on the basis of the outdoor readings, portable scintillation-detector readings

TABLE 3

SUMMARY OF FOUR SURVEYS

| GEOLOGICAL CATEGORY | eU (p.p.m.) | RADIATION LEVELS (mr/week) | | | | | PERSONNEL MONITORING SURVEY | | | INTERNAL EMITTERS | | | | |
| | | Outdoor γ Natural | Outdoor γ Total* | Cosmic | Total Exposure† | | No. | mr/wk (1962) | No. Teeth | Radium-226 (pc/gm/ash) | Skeletal Dose Rate (mrad/yr) | Polonium-210 (pc/gm/ash) | Skeletal Dose Rate (mrad/yr) | Radium-226 in Tap Water (pc/l) |
					1962	Nat.								
Dunham dolomite	5	0.94	1.34	0.71	1.78	1.46	100	2.63	25	0.014	1.014	0.050	3.406	0.04
Beldens formation	5	1.11	1.43	0.68	1.82	1.57	100	2.52	15	0.020	1.486	0.057	3.865	0.03
Glacial drift	9	1.09	1.43	0.70	1.84	1.57	100	2.65	20	0.009	0.682	0.052	3.546	0.08
Monkton formation	11	0.87	1.27	0.63	1.65	1.33	100	2.38	20	0.010	0.744	0.047	3.153	0.02
Fitchburg granite	23	1.18	1.52	0.65	1.87	1.59	100	2.75	20	0.018	1.352	0.056	3.783	0.04
Littleton formation	23	1.19	1.63	0.66	1.96	1.60	100	2.62	20	0.016	1.184	0.061	4.136	0.04
Binary granite	26	1.41	1.78	0.64	2.06	1.77	100	2.82	20	0.014	1.008	0.050	3.397	0.02
Conway granite	45	1.83	2.27	0.68	2.50	2.14	100	3.21	20	0.025	1.814	0.059	4.025	0.01

* Includes fallout, averaged over all spectrometer locations within area.   † Obtained as indicated in text; in units of air dose rate.

were made inside approximately 170 private houses and apartments in the main towns. Several rooms were surveyed in each dwelling, usually including the living room and at least one bedroom. The vast majority of the houses in these areas are of wood-frame construction. The measured indoor levels averaged 70 per cent of the outdoor readings made at the same time. This proved to be a consistent pattern for all the towns, the indoor readings showing the same general trend as the outdoor measurements.

An estimate of mean population exposure to environmental radiation could be obtained by calculating a suitably weighted average of the indoor and outdoor readings of the portable survey instruments (Spiers *et al.*, 1963). One encouraging sign as to the validity of such an average in this case is the already referred-to pattern found in the measurements made in a particular area, observable after relatively few readings at widely scattered locations. As a first approximation, the mean air dose rate to which the population is exposed is estimated to be 80 per cent of the mean terrestrial γ levels given in Table 2 plus approximately 4 $\mu r/hr$ from cosmic radiation. Not included is the possibly quite significant dose contribution from the cosmic-ray neutron component, which would be roughly constant over the areas surveyed.

PERSONNEL DOSIMETER SURVEY

A more direct determination of population exposure can be obtained by the use of personnel dosimeters. For this purpose, a set of 200 Victoreen Model 362 condenser ionization chambers was utilized with a stable pulse-height readout system (Roesch *et al.*, 1958; Segall *et al.*, to be published). In this method, a precision stable-voltage supply is used to charge the ion chambers. After exposure, the chamber is recharged to the same voltage through a resistor. A voltage pulse is produced across the resistor proportional to the differences in voltage between the ion chamber and the charging voltage and, therefore, proportional to the measured dose. It was determined that readings of 1 ± 0.2 mr. were possible at the 95 per cent level of confidence with a single pencil (Roesch *et al.*, 1958). Mechanical and thermal stability were tested and found adequate, and corrections were made for average leakage rates (equivalent to 0.25 mr/week).

These dosimeters were distributed in pairs to 5 individuals in each of 16 areal units, 8 urban and 8 rural. The distribution was limited to 5 standard occupational categories in each area. Each individual wore the dosimeter for 1 week, after which the instruments were read and then redistributed. The experiment was conducted for 5 weeks, resulting in a total sampling of 400 individuals, 25 in each area.

Figure 2 gives the mean mr/week values obtained with the dosimeters plotted as a function of the mean equivalent uranium content of the underlying bedrock as estimated by Billings (to be published). The urban and corresponding adjacent rural data have been combined, since no statistically significant differences were observed between these categories. The best straight-line fit to the data is shown, and this trend appears to be statistically significant. The high intercept

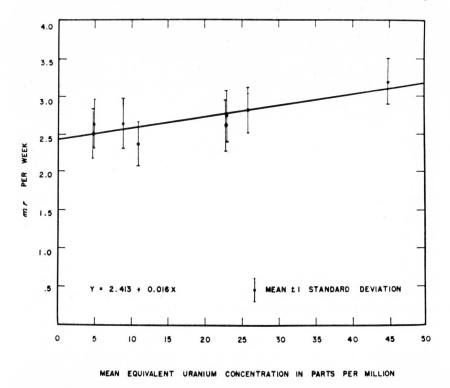

MEAN EQUIVALENT URANIUM CONCENTRATION IN PARTS PER MILLION

Fig. 2.—Personnel monitoring survey results as a function of mean bedrock radioactivity

at zero bedrock radioactivity cannot be explained on the basis of the relatively constant cosmic-ray and fallout contributions to the total dose, for these cannot amount to more than approximately 1.2 mr/week.

It is interesting to consider how Figure 2 would be changed if the equivalent uranium contents of the natural emitters in the soil, as determined by γ spectroscopy (see Table 2), were substituted for the bedrock estimates. The Conway point would then be shifted considerably to the left, and the lowest dosimeter reading would correspond to the lowest radiation level. The slope of the line would become steeper and intersect the *y*-axis at a lower value. This, of course, is expected, since the dosimeters respond to the actual radiation field. In a sense, the data presented in Figure 2 provide a qualitative indication of the influence of bedrock geology on natural radiation exposure in these particular areas. The effect is a relatively small proportion of the total exposure levels and may be practically significant only for the Conway region, where the bedrock content of natural emitters is unusually high and where the reddish sand from the original rock is ubiquitous in the populated areas. The linear trend of the data in Figure 2 may be a reflection of a fairly close relationship between the soil and bedrock in most of the areas considered.

Dose from Internal Emitters

An estimate of the skeletal dose from natural emitters within the human body in these areas has been obtained from determinations of the radium-226, radium-224, and polonium-210 contents of 160 teeth extracted from life- or long-time residents of 8 municipalities. The details of that study are being published elsewhere (Hunt *et al.*, 1964; Radford *et al.*, 1963). The mean radium-226 contents of the teeth in the various municipalities varied from 0.010 to 0.025 pc/gm ash, and the polonium-210 from 0.047 to 0.061 pc/gm ash. The values for the inferred mean skeletal dose rates, assuming equilibrium for the internal emitters, are 1.2 mrad/year from radium-226, 0.8 mrad/year from thorium-228, and 3.7 mrad/year from lead-210. No correlation with mean bedrock radioactivity was found.

Summary and Conclusions

The results of the four studies are summarized in Table 3. Error estimates are not included because the data analysis is as yet incomplete. Our expectation that the range of radiation exposure in these areas would be fairly narrow has been confirmed in all the surveys. The correlation between the estimates of population exposure obtained by spot readings and by the personnel monitoring dosimeters is quite good, although there is no adequate explanation as yet for the approximately 40 per cent systematic difference between the two sets of readings. The most obvious possibility is that the leakage rates of the dosimeters under field conditions were, on the average, substantially greater than those measured in the laboratory.

It should be mentioned at this point that the study of the incidence of cancer and congenital malformations in northern New England shows no indication of any correlation with bedrock radioactivity estimates.

The data given in Table 3 support the contention that mean bedrock geology cannot be used to make quantitative inferences as to population exposure to natural environmental radiation, although it may provide useful indications as to where regions of unusually high or low radiation levels are likely to be found. The former point can be exemplified by a comparison of the southern New Hampshire results with the Vermont data, which would indicate very little influence of mean bedrock radioactivity on the measured radiation levels. On the other hand, the central New Hampshire results show the potential value of relevant geological information in predicting where areas of elevated background levels may be found.

It is well known that most γ-radiation incident on people in their daily rounds originates in the top few inches of the ground or in man-made structures. Since most populated areas, including those considered here, are situated in a "soil" rather than in a "rock" environment, the degree to which the underlying bedrock affects the dose to the population depends strongly on the often complicated relationship between the bedrock and its overburden of soil, as well as on the prevalence of its use as building material. Both these factors can be expected to vary considerably from place to place, thus negating the possibility that a "calibration" of bedrock radioactivity in terms of radiation exposure levels obtained in one or more locales (such as Fig. 2) may be applicable to other areas a priori. There appears to be no adequate substitute for direct radiation measurements in studies in which environmental radiation exposure is a significant parameter.

It should be noted that there are many places where the nature of building materials and road surfaces is a far more significant determinant of population exposure to environmental radiation than are the soils or underlying bedrock (Spiers *et al.*, 1963). This possibility only emphasizes further the need for direct radiation measurements.

ACKNOWLEDGMENTS

The writers wish to acknowledge the able assistance of A. Spiegel in carrying out the indoor measurements and of J. Grebowsky and H. Grotch in the outdoor survey. Part of the study reported here was aided by contract SAph 73556 from the Division of Radiological Health, U.S. Public Health Service, and research grants 62-46C-6373 and RG-7615 from the National Institutes of Health.

REFERENCES

ADAMS, J. A. S., M.-C. KLINE, K. A. RICHARDSON, and J. J. W. ROGERS. 1962. The Conway granite of New Hampshire as a major low-grade thorium resource. Proc. Nat. Acad. Sci., **48**:1898–1905.

BILLINGS, M. P. Areal distribution of natural radioactive radiation from rocks in northern New England. Department of Geological Sciences, Harvard University. (To be published.)

HUNT, V., E. RADFORD, and A. SEGALL. 1964. Suitability of teeth as indicators of the natural levels of alpha-emitting isotopes in the human skeleton. Radiation Res. (In press.)

LOWDER, W. M., and W. J. CONDON. Environmental radiation survey in northern New England. Health and Safety Laboratory, USAEC, New York. (To be published.)

LOWDER, W. M., W. J. CONDON, and H. L. BECK. 1963. Field spectrometric investigations of environmental radiation in the U.S.A. This symposium.

RADFORD, E., V. HUNT, and A. SEGALL. A survey of natural radioisotopes in teeth in relation to terrestrial radioactivity in northern New England. Boston: Harvard School of Public Health. (To be published.)

RADFORD, E., V. R. HUNT, and D. SHERRY. 1963. Analysis of teeth and bones for alpha-emitting elements. Radiation Res., 19:298–315.

RICHARDSON, K. A. 1963. Thorium, uranium, and potassium in the Conway granite, New Hampshire, U.S.A. This symposium.

ROESCH, W. C., R. C. McCALL, and F. L. RISING. 1958. A pulse reading method for condenser ion chambers. Health Physics, 1:340–44.

SEGALL, A. 1962. Measurement of background radiation for epidemiologic studies. Am. J. Public Health, 52:1660–68.

———. 1963. Radiogeology and population exposure to background radiation in northern New England. Science, 140:1337–39.

SEGALL, A., and R. REED. 1964. A survey of population exposure to external background radiation in northern New England. Arch. Em. Health. (In press.)

SEGALL, A., J. SHAPIRO, and J. WORCESTER. The application of a personnel monitoring system to population dosimetry of background radiation. Boston: Harvard School of Public Health. (To be published.)

SPIERS, F. W., M. J. McHUGH, and D. B. APPLEBY. 1963. Environmental γ-ray dose to populations: Surveys made with a portable meter. This symposium.

CECIL PINKERTON, WILLIAM Y. CHEN,
R. G. HUTCHINS, AND RALPH E.
SCHROHENLOHER

57. *Background Radioactivity Monitoring of a Pilot Study Community in Washington County, Maryland*

T HE HISTORY AND DEVELOPMENT of a project leading to the investigation of environmental factors, including background radiation, in relation to the geographic distribution of cancer within a single county has been published (Lawrence and Chen, 1959).

Primarily, a health differential was sought in terms of cancer experience, based upon the human population associated with the environment of houses. There were some arguments in favor of the house environment as a suitable factor for investigation. Some physical parameters—in particular, natural radiation associated with soil and rock—relate to the immediate environment of a residence and may affect the individual through external exposure, through the air he breathes or through ingestion in homegrown foods and water from private wells.

In an over-all look by means of an air-borne survey, the town of Sharpsburg, in Washington County, Maryland, was found to be surrounded by one of the most radioactive areas of eastern Washington County, associated with well-defined differences in the underlying geology and resultant soils. It was believed that radiological measure-

CECIL PINKERTON is the health service officer, Engineering Section, Basic and Applied Sciences Branch, Division of Water Supply and Pollution Control, Robert A. Taft Sanitary Engineering Center, Cincinnati, Ohio. W. Y. CHEN is chief, Preventable and Chronic Diseases Division, Government of the District of Columbia, Washington, D.C. R. G. HUTCHINS is a soils scientist, Epidemiology Branch, Environmental Cancer Field Research Laboratory, National Cancer Institute, Hagerstown, Maryland. RALPH E. SCHROHENLOHER is on the staff of the University of Alabama Medical School, Birmingham, Alabama.

ments made within the community would be invaluable, even if not directed toward specific epidemiological objectives. The main problem seemed to lie in establishing the reliability of differences in individual-house background radiation measurements in a small area where there was a large over-all mass effect.

Various methods have been employed by different investigators to measure and relate background radioactivity to the epidemiology of certain human diseases. These methods generally fall into three categories: (*a*) direct measurements in air, that is, outdoor and indoor measurements at residences of the human-study group; (*b*) radiation counting and chemical analysis of water, vegetation, soil, rock, and air samples representing the environment; and (*c*) use of geology, soil, and other physical guides concerned with various aspects of the environment, with inference from measurements already known for such aspects. Examples of the use of direct measurements are shown by studies carried out by Court-Brown and Spiers (1960) in connection with leukemia and by Meyers (1928) and Meyers and Hess (1952) in connection with cancer. By direct sampling of air for the measurement of its radon-222 and radon-220 content, Hultqvist (1956) established the dose rate in houses of different types of construction.

As an example of the use of geological guides, Kratchman and Grahn (1959) grouped data on deaths from congenital malformations by geologic provinces of the United States. The data suggest that mortality incidence from malformation may be higher in the geologic provinces of the United States that contain major uranium-ore deposits, uraniferous waters, or helium concentrations. It is inferred that these provinces have a higher-than-average level of environmental radiation.

Wesley (1960) examined the variation of malformed deaths with world-wide background radiation as measured by the cosmic-ray energy flux, which in turn is a function of the geomagnetic latitude. Gentry (Gentry *et al.*, 1959) used geological guides to estimate background radiation of houses associated with congenital malformations in New York State residents.

Geological guides to the problem of radioactive-ore finding used by the U.S. Geological Survey have been listed by Page (1956). Specific guides have been applied in conjunction with detailed radiometric, panning, geochemical, botanical, and geophysical studies in advance of physical exploration. Increased radioactivity was believed by Stow (1955) to be associated with shear zones, faults, and geological contacts in parts of the Appalachian Valley of eastern United States. From the epidemiology standpoint, radiological measurement should not be confined to any one type with a view to looking for any single effect.

All three methods have found application for evaluating background radioactivity in the Sharpsburg community in Washington County, Maryland, where we were interested in making a clearer comparison of the different methods and their application to a small geographical area of homogeneous status. Socioeconomic conditions, sanitation, including air and water pollution, cosmic radiation, and soil-fertility were believed to be more constant in contrast to the larger geographical area comprising Washington County, Maryland. If the heterogeneous geological situation believed to prevail in the town could be shown to cause differences in natural radiation, within distances spaced by house lots, a means would be afforded to vary the radiation environment of individual houses. The contribution of soil and rock to radioactivity was first examined in a general way for an area surrounding the town. This was followed by a house-to-house hand-borne scintillation-counter survey in which larger rural areas were included for comparison. Finally, residences with an apparent background radiation differential were selected for more intensive study to determine whether there were significant variations in exposure levels that were not revealed by gross estimations, such as soil or geological guides, air-borne surveys, or hand-borne scintillation-counter surveys. When the techniques were refined, houses showing a health differential in terms of cancer experience could be studied by means of paired analysis, using index and controls of comparable person-years exposure. The information would also be useful in evaluating soils and geological and air-borne radioactivity guides to natural radiation exposure. At present, the latter are being recorded for a large percentage of all residences in Washington County in connection with cancer epidemiology studies. For each house, air-borne readings are taken from air-borne radioactivity-survey isorad contour maps, while soil type and geological formation are recorded from detailed survey maps. These form cohorts for analysis in conjunction with the human data for each residence over a 10-year period.

This account of our results is considered significant because of the necessity of correlating such measurements with other possible concomitant etiological factors.

Description of the Area

The study area, approximately $2\frac{1}{2}$ miles square, located in the southern part of Washington County, Maryland, occupies a dissected limestone valley upland with rather strongly rolling topography (Figs. 1 and 2). Maximum relief is approximately 85 feet. The town of

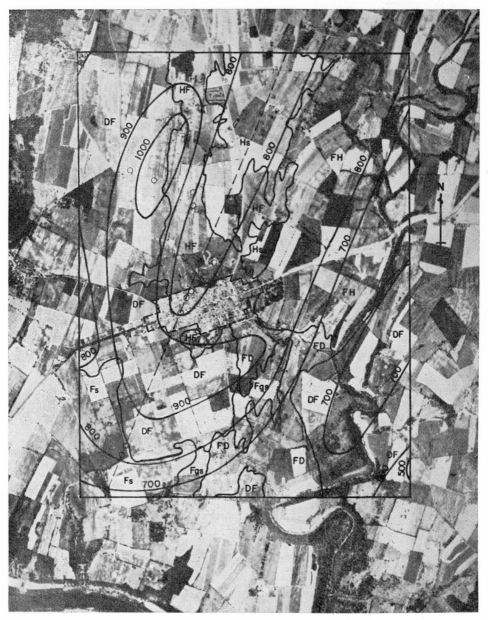

Fig. 1.—Pilot study community, showing aeroradiometric measurements in relation to major soils associations.

Sharpsburg (population 837 in 1958) lies in a depression created by the confluence of five watersheds. These waterways are expressed as areas where the Huntington silt loam, local alluvium (Hx), has been mapped (Pl. I). The streams of the waterways are intermittent to the point in the town at which a spring becomes the source of an un-named perennial stream that empties into Antietam Creek. The town of Sharpsburg is situated on a north-plunging asymmetric syncline of carbonate rocks that belong to the Elbrook limestone. Attitudes of

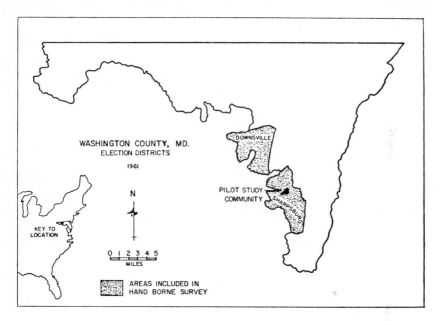

Fig. 2.—General location of study area in Washington County

the rock strata in and around the town range from almost vertical to almost horizontal. Where the dips of the beds range from 10° to 45°, a given rock stratum may act as an aquifer for one dwelling; it may outcrop in the soil substratum (at basement level) of a second house; and it may form the soil parent material of a third (Fig. 3). Rock strata of small thickness, but high in background radioactivity or trace-element content, may be of little significance where the bedding plane is practically vertical (small outcrop area) but may assume in-creased significance where the bedding plane is nearly horizontal and the stratum outcrops over an extensive area at the land surface. Con-firmation of the hypothetical cross section shown in Figure 3 must await the completion of detailed structural geology studies in Sharps-

PLATE I

Soils map of Sharpsburg town showing relation to isorad contours

burg, in progress, by the Geological Health Unit, Trace Element Census Branch, USGS.

Heterogeneity and mineralization of the upper Elbrook and lower Conococheague limestone formations have been described by Cloos (1951), King (1950), and Cannon (Cannon and Bowles, 1961). This has been confirmed in the present study by the finding of small, highly mineralized limestone sections containing copper, pyrite, and radioactive minerals.

Ridge formation near the perimeter of Sharpsburg may be geologically controlled and consequently have some effect on the tenor

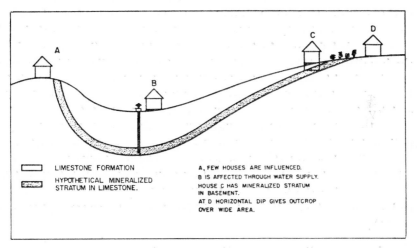

LIMESTONE FORMATION

HYPOTHETICAL MINERALIZED STRATUM IN LIMESTONE.

A, FEW HOUSES ARE INFLUENCED.
B IS AFFECTED THROUGH WATER SUPPLY.
HOUSE C HAS MINERALIZED STRATUM IN BASEMENT.
AT D HORIZONTAL DIP GIVES OUTCROP OVER WIDE AREA.

Fig. 3.—Cross section of geological syncline showing possible relation of lithology to water-bearing formation, basements, and gardens.

of radioactive minerals, including potassium. Sando (1957) and Wilson (1952) describe ridges formed by sandy or silty members of the Conococheague limestone. Whittaker (Chichilo and Whittaker, 1958, 1961) found total impurity of silica content of limestone to be a rough guide to the content of several elements, particularly aluminum, potassium, cobalt, iron, sodium, boron, vanadium, manganese, fluorine, and sulfur.

Approximately the same area was studied by E. G. Otton of USGS in a ground-water survey of the area (to be published). Certain radioactivity features are more easily investigated in the surrounding open country, where there is less man-made disturbance of soils than there is in the more congested town area. Physical features some distance from the town may influence its background radioactivity through rainfall surface runoff, erosion, underground water movement, or air flow.

General Methodology

Dose rates as established by scintillation-counter measurements of background radioactivity are comparatively easy to carry out. However, because of the fluctuating nature of external γ radioactivity, only a synoptic pattern of radioactivity can usually be discerned by a given set of measurements carried out within a short period of time. The number of houses or sites investigated must be kept small if a number of measurements are made involving different techniques or instruments. Continuous monitoring showing variation with time is more complicated, requiring a separate, continuous recording installation for all residences to be monitored. Manpower needs are increased accordingly, and private citizens may be reluctant to have their privacy invaded frequently by personnel making readings or servicing the installations. They may also fear to have complicated instruments associated with radioactivity installed in their homes. Our work has been oriented, therefore, toward establishing gross patterns of radioactivity by an air-borne radiation survey checked by appropriate ground monitoring and sampling. The steadiness of the background radioactivity has been investigated to some extent by comparison of the predicted cumulative dose, as measured by scintillation counters, with the cumulative dose measured with film badges. It is believed for the present that the background radioactivity is fairly steady. In general, our procedure follows that of Marsden *et al.* (1944) and Denson (1956).

Air-borne Background Radiation Survey

An air-borne background radiation survey of the eastern part of Washington County, Maryland, was carried out under contract for the National Cancer Institute by the U.S. Geological Survey. The relationship of the air-borne data to the geology of eastern Washington County has been shown by Moxham (n.d.)'. Flights were made on half-mile spacings so that isorad contouring in between lines represents considerable interpolation. Air-borne readings are therefore most exact for houses within a 1,200-foot-diameter area along the flight line. The relationship to the ground dose at an individual spot depends upon several factors, the most important of which appear to be the extent of rock outcrops and the contribution of bedrock to background radioactivity.

A portion of the air-borne–radiation contour map for the county was superimposed on a soils map generalized from field sheets of a soil survey of the pilot community by the USDA Soil Conservation Service, 1938–43. The map is shown in Figure 1. Air-borne radiation intensities are expressed in counts per second corrected for the cosmic component, and the isorad contour interval is 100 c.p.s. A belt of high radioactivity forms a U-shaped pattern around all except the north side of the town. This distribution is apparently due to Sharpsburg's location in the trough of a north-plunging syncline whereby the geologic formations with associated residual soils crop out in a corresponding U-shaped pattern around the town.

From a field investigation made in the area northwest of Sharpsburg, where radiation reaches its greatest intensity, Moxham concluded that potassium in shale and limestone contributes about 40 per cent of the total radioactive content of these rocks.

Relation of Air-borne Radiation Intensity to Soils

In general, the soils series mapped in the area by the USDA Soil Conservation Service bears a close relationship to the lithology of the underlying bedrock. Thus, the Frankstown soil series in Washington County is mapped on limestone and shale of the Elbrook and Conococheague geological formations. Soil types of the Duffield and Frankstown soil series, mapped as a horseshoe-shaped area surrounding the town of Sharpsburg, are associated with air-borne background radioactivity of over 900 c.p.s. The highest air-borne radioactivity (1,000 c.p.s.) of eastern Washington County was recorded for an area northwest of the town, mapped as Frankstown-Hagerstown stony silt loams. This soil association, extending into the town of Sharpsburg, represents a highly intricate complex of types and phases of varying degrees of depth, thickness of solum, surface texture, erosion phases, and rockiness. The relationship of aeroradiometric intensities to town soils is shown in Plate I. Less of the area within the town is represented by the deeper, finer-textured Frankstown soils, which may show a greater concentration of heavy minerals, giving rise to increased γ radiation. The Hagerstown soils near Sharpsburg town have extensive limestone outcrops. Such outcrops would be expected to dilute the γ radioactivity, but, in between limestone ledges, soils may exhibit more of the inherited parent-rock radioactivity in the form of unleached uranium and radium minerals. Formation of soil parent material from bedrock is the primary soil evolutionary process under these conditions. Parent material and description of the soils shown in Figure 1 are given in Table 1.

Ground Monitoring

The establishment of a gross radioactivity pattern by an air-borne survey is usually followed, in the case of mineral exploration, by efforts to find deposits of uranium and thorium. Such efforts are

TABLE 1

DESCRIPTIONS OF SOILS: SHARPSBURG TOWN AND SURROUNDINGS

Approved Map Symbol	Approved Name*	Parent Material and Brief Description
DmB2....	Duffield silt loam, 3–8 per cent slopes, moderately eroded	Limestone interbedded with thin seams of shale.
DmC2....	Duffield silt loam, 8–15 per cent slopes, moderately eroded	Deep, well-drained upland soil with yellow-colored, light silty clay loam subsoil, medium to heavy texture, and high moisture-holding capacity. Near Sharpsburg there is some shale in the subsoil, and occasional outcrops of shaly limestone occur.
FvC2.....	Frankstown very rocky silt loam, 3–15 per cent slopes, moderately eroded	Impure siliceous limestone and shale. Similar to Duffield but not as deep and always contains much more residual fragments of chert, shale, and sometimes limestone and sandstone. Outcrops of limestone containing chert and thin seams of sandstone are numerous near Sharpsburg.
FvE2.....	Frankstown very rocky silt loam, 15–45 per cent slopes, moderately eroded	Same as FvC2 except that slopes are too great for agricultural operations.
FwB2.....	Frankstown and Duffield channery, silt loams, 0–8 per cent slopes, moderately eroded	Impure siliceous limestone and shale. Moderately deep, upland soil, which exhibits the typical Frankstown series profile, similar to DmB2 but not so deep on the average, and has been developed from even more shaly or shabby limestones, which usually leave large residues of cherty gravel as well as shale and limestone fragments. Near Sharpsburg town, the Frankstown gravelly silt loam occupies the ridgelike area in the eastern section of town.
FwC2.....	Frankstown and Duffield channery silt loams, 8–15 per cent slopes, moderately eroded	Upland soil similar to FwB2 except that slopes require careful management to prevent loss of soil through erosion.
HbD2.....	Hagerstown extremely rocky silt loam, 0–25 per cent slopes, moderately eroded	Nearly pure, massive limestone. Deep, well-drained, heavy-textured upland soil with reddish-brown silty clay to clay loam subsoils. In and near Sharpsburg town the soils have profiles typical of Hagerstown series except that more residual shale and siliceous impurities are present in the soil mapped as very stony loam and the soil grades toward the Duffield and Frankstown soils in characteristics. The very irregular depth of bedrock and extremely abundant outcroppings of hard limestone (45–90 per cent of the surface) give rise to a highly intricate complex of types and phases of varying degrees of depth, thickness of solum, surface texture, erosion phases, and rockiness.
Rk.......	Rocky eroded land	Very steep slopes with vertical exposure of eroded parent material.
HeB2.....	Hagerstown silt loam, 0–8 per cent slopes, moderately eroded	Nearly pure massive limestone. Deep, well-drained, medium-textured upland soil, derived from relatively pure limestone with yellowish-red heavy silt loam or light silty clay loam in the subsoil. The surface soil is brown to dark-brown silt loam.

* To appear on published map, Washington County, Maryland, soil survey.

TABLE 1—*Continued*

Approved Map Symbol	Approved Name	Parent Material and Brief Description
HfB2	Hagerstown silty clay loam, 0–8 per cent slopes, moderately eroded	Nearly pure massive limestone. Heavy-textured upland soil with the same characteristics as Hagerstown silt loam except that the surface contains less silt and more clay.
HfC2	Hagerstown silty clay loam, 8–15 per cent slopes, moderately eroded	Upland soil similar to Hagerstown silty clay loam, 0–8 per cent slopes, except that the steeper slope presents a greater erosion hazard.
HgC2	Hagerstown very rocky silt loam, 3–15 per cent slopes, moderately eroded	Nearly pure massive limestone. Well-drained upland soil, similar to Hagerstown silt loam except that up to about 40 per cent of the surface consists of outcropping ledges and reefs of hard limestone. Variability in soil characteristics of the stony loam is discussed in the text. Near Sharpsburg town the soil grades toward the Duffield and Frankstown soils in characteristics as a result of considerable impurity in the parent rock.
HgE2	Hagerstown very rocky silt loam, 15–45 per cent slopes, moderately eroded	Nearly pure massive limestone. Upland soil similar to Hagerstown stony loam, 3–15 per cent slope, except that the steeper slope presents a greater erosion hazard.
Hx	Huntington silt loam local alluvium	Limestone material. Moderately fine-textured soils having no true subsoils but with exceptional beginning of profile development giving a substratum slightly finer in texture than the surface. Variable in depth to C2 horizon, which may be several feet down in places; colors vary somewhat from brown to reddish-brown in the surface soil. Parent material is local alluvium formed by the deposit of fine material washed directly down from surrounding upland soils. Near Sharpsburg town it occupies depressions within the limestone soil areas and around drainage heads and on foot slopes close to smaller drainage ways. It was mapped where settlement of fine material out of flood waters occurred, and here it grades in characteristics toward true flood-plain soils in some places in Sharpsburg town.
Lm	Lindside silt loam	Limestone material. Somewhat poorly drained soil of flood plain and upland depression of recent alluvium from limestone-derived soils. Moderately wet, occasionally flooded, with seasonally high water table.
FwB2	Frankstown and Duffield channery, silt loams, 3–8 per cent slope, moderately eroded	
FwC2	Frankstown and Duffield channery, silt loams, 8–15 per cent slopes, moderately eroded	Well-drained deep upland soil developed in residuum from limestone containing much shale and chert and occasionally some thin seams of limestone. Similar to Duffield soils and may be mixed with them. Southwest of Sharpsburg town, Frankstown and Duffield soils occupy plainlike area, less undulating than Sharpsburg town. The role of residual chert and shale in development of deep soils and the relationship to background radioactivity are discussed in the text.

concerned chiefly with phenomena peculiar to the unique property of radioactive disintegration. In health studies, both the external γ-ray dose and the internal dose arising from the ingestion of naturally occurring uranium, thorium, radium, and their decay daughters, together with certain radioactive rare-earth elements, are of interest. Some measurements were made, therefore, to determine the α and β radioactivity and thorium content of soil samples separated by a linear surface distance representing the contour interval of 100 c.p.s. air-borne radioactivity. The uranium content of different soils in the area shows little variation and averages approximately 3 p.p.m.

TABLE 2

RADIOACTIVITY MEASUREMENTS OF SOILS SAMPLED ALONG AERORADIOMETRIC FLIGHT LINE No. 17, $1\frac{3}{4}$ MILES SOUTHWEST OF SHARPSBURG, MD.*

SITE	DEPTH (IN.)	SOIL TYPE/ SLOPE EROSION	AIR-BORNE (C.P.S.)	GROUND (μR/HR)	THORIUM (P.P.M.)	NET C.P.M. FROM 4.5 CM.² AREA	
						α	β
Loc. No. 1, near lane.........	1–4	8/8B-2†	875	....	......	0.78 ± 0.33‡	16.0 ± 1.9
	4–7	"	......	....	......	0.17 ± 0.18	17.2 ± 1.9
	10–14	"	......	....	......	0.51 ± 0.26	19.9 ± 2.0
						Av. 0.48	17.7
Loc. No. 2, field.	0–1	8/8B-2	840	10.7	......	0.35 ± 0.21	19.8 ± 2.0
	1–4	"	"	"	10	0.47 ± 0.26	16.5 ± 1.8
	4–7	"	"	"	......	0.30 ± 0.21	18.4 ± 1.9
	10–14	"	"	"	......	0.40 ± 0.24	19.2 ± 1.9
						Av. 0.38	18.4
Loc. No. 3, woods........	4–7	8/11C-2	720	8.8	7	0.17 ± 0.19	18.7 ± 1.9
	10–14	"	"	....	......	0.37 ± 0.23	19.0 ± 1.9
						Av. 0.27	18.8

* Sampling locations are shown in Figure 1.

† Soil symbol from field sheet, Soil Conservation Service, 8/8B-2 refers to Frankstown cherty silt loam, 8 per cent slope, moderately eroded. Description of soils given in Table 1, under approved symbols FwB2 and FwC2.

‡ Statistical error.

For the sake of comparison, soil profile samples were taken at sites along an air-borne flight line southwest of Sharpsburg town. Sites 2 and 3 (Table 2) were separated by a lineal distance representing a 100-c.p.s. change (700- and 800-c.p.s. intensities) in air-borne radioactivity. Approximately the same ground distances separate the 800- and 900-c.p.s. intensity lines in Sharpsburg town along the flight line (Fig. 1). Ground dose-rate readings were made with a LaRoe Model FV-6-S Scintillation Counter calibrated against a radium source; α and β emitters were determined by radiometric counting of infinitely thick soil samples following methods of Adams et al. (1958) and

General Dynamics Corporation (n.d.); and thorium was determined chemically. The methods of determination are given in Appendix B following this article.

From the results (Table 2), a 100-c.p.s. change in the air-borne γ-ray intensity is equivalent in this instance to approximately a 1.9-μr/hr hand-borne instrument-measured ground dose, or to 1.3 p.p.m.

TABLE 3

RADIOACTIVITY OF GRID SAMPLES NORTH OF SHARPSBURG TOWN
ALONG AERORADIOMETRIC FLIGHT LINE NO. 17*

LOCATION	DEPTH (IN.)	SOIL TYPE	AIR-BORNE (C.P.S.)	NET C.P.M.—4.5 CM.^{2} a	NET C.P.M.—4.5 CM.^{2} β
Grid 64........	2–8 8–14	05/9C-2 Hagerstown very stony loam, 9 per cent slope, moderately eroded	800	0.49±0.28 0.57±0.29	22.9±2.2 22.0±2.1
				Av. 0.53	22.4
Grid 64A......	2–8 8–14	7/6B-1 Frankstown silt loam, 6 per cent slope	800	0.55±0.26 0.78±0.26	21.3±2.1 21.7±2.0
				Av. 0.66	21.5
Grid 64B......	0–3 3–8	05/5B-1 Hagerstown very stony loam, 5 per cent slope	800†	0.40±0.35 0.52±0.32	20.1±2.3 19.0±2.2
				Av. 0.46	19.5
Grid 64C......	0–3	05/9C-2 Hagerstown very stony loam, 9 per cent slope, moderately eroded	800	0.52±0.22 0.67±0.27	23.9±2.0 21.8±2.0
				Av. 0.59	22.8

SAMPLES TAKEN WITHIN 12-IN. RADIUS OF "0" POSITION OF GRID NOS. 64 AND 64A

Lab. No.		Depth	Soil Type	Air-borne	a	β
84	64-0.....	0–6	05/9C-2	800	0.48±0.23	27.0±2.1
85	64-1.....	"	"	800	0.54±0.24	23.5±2.1
86	64-2.....	"	"	800	0.49±0.23	24.5±2.1
87	64-3.....	"	"	800	0.60±0.25	24.1±2.1
88	64-4.....	"	"	800	0.87±0.23	22.2±2.1
89	64-6.....	"	"	800	0.55±0.25	22.9±2.1
90	64-7.....	"	"	800	0.53±0.23	24.1±2.1
					Av. 0.58	24.0
97	64A-0...	"	7/6B-1	800	0.41±0.24	24.9±2.3
94	64A-1...	"	"	800	0.51±0.23	22.9±2.2
95	64A-2...	"	"	800	0.52±0.22	24.3±2.2
96	64A-3...	"	"	800	0.53±0.26	22.7±2.2
					Av. 0.46	23.7

* Sampling locations are shown in Figures 1 and Plate I.
† Ground reading 7.7 μr/hr.

equivalent uranium as thorium (mineral content of soil). Radio-metrically counted for α activity, a 100-c.p.s. change in air-borne γ-ray intensity is equivalent to approximately 0.1 count per minute from an infinitely thick soil sample (4.5 cm.² area) on the ground.

North of the town, soil sampling was carried out on a grid pattern in order to establish the variability of different soil types. Measurements of soil radioactivity by radiometric counting are shown in Table 3. The grid points (64, 64A, 64B, 64C) are 500 feet apart in a spoke arrangement, with grid sample 64 at the center (Fig. 1 and Pl. I).

The average α count for Hagerstown loam is 0.56 c.p.m. compared with 0.32 for the Frankstown cherty silt loam soil type No. 8 at approximately the same air-borne intensity southwest of the town. Laboratory samples 84 through 90 were taken at about a 1-foot radius, spoke sampling arrangement, and shows the α count ranging from 0.48 to 0.87 within a 3-square-foot area on the Hagerstown loam. Samples of Duffield silt loam (Lab. Nos. 94–97) show less variability. Organic-matter content and pH varied only slightly among these samples. These results seem to confirm that, for a given air-borne γ-ray background radioactivity intensity, different soil types and geologic combinations will exhibit different α and β radioactivities. The contribution of potassium-40 to the γ-ray spectrum of other naturally occurring radioelements may result in the same air-borne γ-ray intensity through variation in potassium content of soils from place to place. Results of additional measurements along flight line No. 17 within the town of Sharpsburg (Table 8) show a difference of approximately 1.5 count of α activity for a 100-c.p.s. change in air-borne radioactivity. (Compare Lab. Nos. 22 and 126.) Other measurements made outside the town indicate the relationship of the air-borne radioactivity to the ground dose along other flight lines. (Compare Lab. No. 110 with No. 137, and No. 137 with No. 31, Table 8.) These sampling locations are shown in Figure 1 and Plate I.

HAND-BORNE MONITORING OF RESIDENCES

The pilot study community is in one of two election districts in Washington County in which a hand-borne radiometric survey was made at over 800 residences by a team of investigators during a 2-week period (October 19–November 3, 1959). At each house involved, two background radiation measurements were made with a LaRoe Model FV-6-S Scintillation Meter. All meters were calibrated against a radium source under reproducible conditions. The calibration is described in Appendix A. The method of calibration is similar to that used by Fair and Howells (1955).

The first reading at a house was taken by placing the meter against the outside of the house about 48 inches above the ground. The second measurement was obtained at a site at least 50 feet from the dwelling, with the detector held about 18 inches above the ground surface. The primary consideration of the survey involved whether or not hand-borne readings would parallel air-borne readings made within the same geographical area. Second, the contribution from different building materials used in the construction of dwellings to the expected exposure dose was of interest. Histograms of the frequency

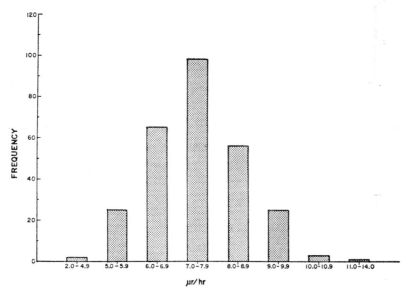

Fig. 4.—Frequencies of hand-borne radioactivity readings in Downsville district

for all hand-borne readings made in three districts are shown in Figures 4, 5, and 6. The histogram for the district with the lowest average air-borne radioactivity, Downsville (Fig. 4), has an almost normal distribution of frequencies (compared with combined frequencies for all three districts) owing to greater variation in surface geology. The hand-borne results are compared with air-borne readings in Table 4.

A histogram (Fig. 7) that combines hand-borne frequencies for the aerial reading interval, 700–900 c.p.s., from two rural districts (Downsville and Sharpsburg rural) may be compared with Figure 6, the frequency for the same air-borne interval measured over Sharpsburg town. The frequency range of hand-borne readings (for the air-borne

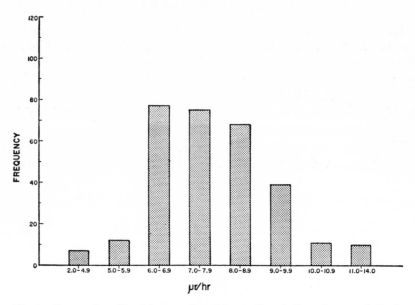

FIG. 5.—Frequencies of hand-borne radioactivity readings in Sharpsburg rural district

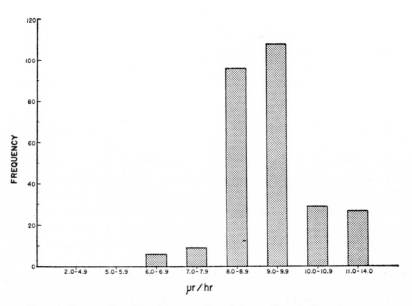

FIG. 6.—Frequencies of hand-borne radioactivity readings in Sharpsburg town

range 700–900 c.p.s.) appears to be wider in the rural areas, representing an average of many soil types and geological terrain conditions. Sharpsburg town is fairly homogeneous with respect to soils, with most of the area represented by a single soil type as mapped, although actually highly complex within itself. For the air-borne range, hand-borne reading frequencies are apparently shifted toward a narrow range of higher values.

TABLE 4

COMPARISON OF AERORADIOMETRIC DATA, 1956, WITH HAND-BORNE SURVEY RESULTS
OBTAINED AT EVERY RESIDENCE IN SHARPSBURG AND DOWNSVILLE
ELECTION DISTRICTS, WASHINGTON COUNTY, MD., 1959

SECTION I. AIR-BORNE SURVEY RESULTS

District or Area	Total Readings (No.)	Average (c.p.s.)	Median (c.p.s.)	Range (c.p.s.)	Average Range (c.p.s.)	Remarks
Sharpsburg......	576	599	800	300– 920	417– 887	Map reading*
		892	1,093	593–1,213	710–1,180	Map reading+cosmic
Downsville......	275	585	600	400– 800	491– 651	Map reading
		878	893	693–1,093	784– 944	Map reading+cosmic
Sharpsburg town.	275	818	800	700– 900	796– 861	Map reading
		1,111	1,093	993–1,193	1,089–1,154	Map reading+cosmic
Sharpsburg rural.	301	643	640	300– 920	417– 864	Map reading
		936	933	593–1,213	710–1,157	Map reading+cosmic

SECTION II. HAND-BORNE SURVEY RESULTS

District or Area	Total Readings (No.)	Average (μr/hr)	Median (μr/hr)	Range (μr/hr)	Average Range (μr/hr)	Place of Reading
Sharpsburg......	574	8.4	8.6	4.0–14.0	5.8–11.4	Yard
	570	6.5	6.1	3.1–14.0	4.2–10.8	House
Downsville	275	7.5	7.5	2.7–11.8	5.7– 9.1	Yard
	275	6.2	5.7	2.9–15.0	4.0– 9.6	House
Sharpsburg town.	275	9.2	9	6.4–12.2	7.8–11.0	Yard
	276	6.8	6.5	3.3–13.7	4.9– 9.9	House
Sharpsburg rural.	299	7.7	7.4	4.0–14.0	5.8–10.1	Yard
	294	6.3	6	3.1–14.0	4.4– 9.5	House

* Map readings show the radiation intensity in counts per second at nominal 500-foot altitude. From aeroradiometric contour maps of Moxham (n.d.). Values for dwellings between intensity lines obtained by interpolation.
   Cosmic factor: 293 c.p.s. (from aeroradiometric data).
   Average range: Average of 50 lowest readings and 50 highest readings obtained in each area described.
   Yard readings: Taken in the yard at least 50 feet away from each dwelling.
   House readings: Taken approximately 48 inches from the ground with instrument held against the exterior wall of each dwelling.

Moxham (n.d.) found a ratio between aerial reading versus dose rate on the ground that is a constant at a given air-borne reading so long as uniform conditions, presumably soil type, obtained for several hundred feet in all directions from the point of ground observation.

During the survey several readings were made at sites—that is, virgin forest areas, open meadows over homogeneous bedrock units, and bridges over river water—representing areas of slight or low human interference. In Figure 8 those readings observed are plotted against

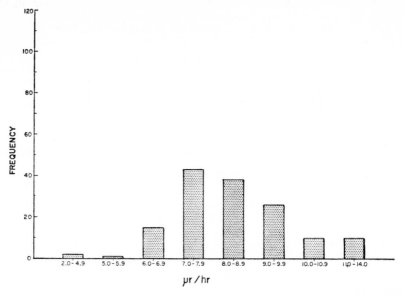

Fig. 7.—Hand-borne frequencies for air-borne interval 700–900 c.p.s. from Downsville and Sharpsburg rural district.

the air-borne intensities observed at the same (vertically plotted) locations. The curves in Figure 8, then, represent an estimate of handborne values to be expected at more homogeneous sites at that particular level of aeroradiometric intensity.

Moxham (n.d.) found a proportional relationship between handborne and air-borne readings as shown by curve *I* on Figure 8. Based on this observation of lineal gradation of hand-borne readings from air-borne lows (200 c.p.s.) to air-borne highs (1,400 c.p.s.), the straight-line curve *II* (theoretically expected) in Figure 8 is drawn through our observed hand-borne readings made at different aeroradiometric intensity areas or locations.

One might expect that instrumentation, geometry differences, and/or interpretation of the field data are possible causative factors where ground readings depart widely from a curve (*III*, Fig. 8) fitted by visual inspection through plotted points. Therefore, caution should be used in attempting to obtain absolute exposure dose values from these graphs. However, the graphs are used to show the results obtained and to illustrate how they compare with the aeroradiometric data.

The most reliable comparison between hand-borne results and air-borne results was observed in areas having near-average (872 c.p.s.) air-borne readings. Insufficient hand-borne observations within areas below the air-borne range of 700–799 c.p.s. create extreme unreliability. In comparing Figure 5 with Figure 9 (Sharpsburg rural), and

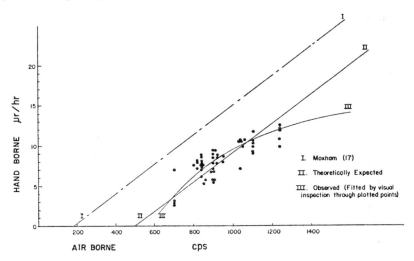

Fɪɢ. 8.—Relationship between air-borne and hand-borne readings at homogeneous sites

Figure 4 with Figure 10 (Downsville), one observes an overlap (in both areas) of the observed hand-borne and air-borne measurements. Lower than theoretically expected hand-borne readings are observed (compare Fig. 6 with Fig. 11, Sharpsburg town) above the 100-c.p.s. level noted in the apparent air-borne peak shift to the right of the plotted hand-borne results.

It was expected that, the more rocky the soil type surrounding an individual spot of interest, the lower the air-borne reading would be in relation to the hand-borne reading. Where more than one soil type was mapped near a house in this area, chances are good that a rocky type of the series was included. In Sharpsburg town more than 95 per cent of all houses have more than one type within 225 feet (Pl. I).

Differentiation of types within a soil series is made based on texture of the upper plow layer of soil and on field classification of the amount of loose stone and the degree of rockiness. Phases within a given soil type are represented by different degrees of slope, erosion, and management practices, including fertilization, and all may contribute to the tenor of radioactive minerals. Intermediate types, rockiness, or the presence of unweathered parent material, especially limestone in the solum, would seem to represent the highest degree of variability. Highly variable soil textures may be exhibited by very stony soils, depending upon the actual amount of loose stone, varying depth of bedrock, and attitude or dip of bedding. Free lime in soil may cause an increased uptake of radioactive minerals by plants and its return to the surface soil.

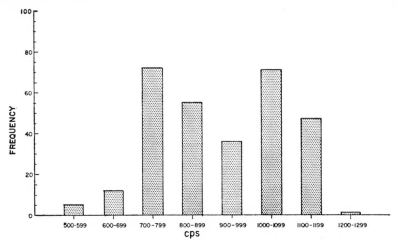

Fig. 9.—Frequency distribution of air-borne readings of houses in rural Sharpsburg

Radioactivity and slope may be related in Washington County through geological control. This is seen where nearly vertical beds of hard, resistant limestone form the crest of a ridge and softer and slightly radioactive shaly or dolomite beds on either side weather more rapidly to form the slopes. Hillsides may then exhibit higher radioactivity than do valley bottoms or hill crests.

For the reasons given above, the town of Sharpsburg, which is on Hagerstown stony loam soil on fairly steep slopes, was investigated more thoroughly with regard to background radioactivity, since it was believed that the air-borne readings represented only an average of variable ground-dose rates as measured at individual residences. Sharpsburg was also of great interest from the epidemiological standpoint, since it was important to know whether significant background radioactivity differences prevailed between paired residences

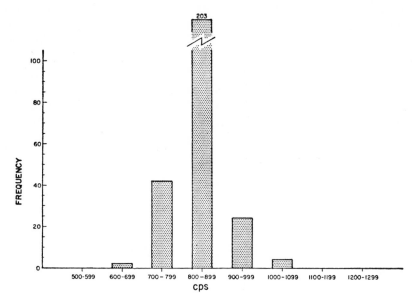

FIG. 10.—Frequency distribution of air-borne readings of houses in Downsville

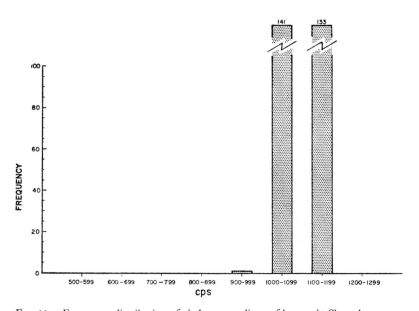

FIG. 11.—Frequency distribution of air-borne readings of houses in Sharpsburg town

that might be located close together in the area of highest house concentration. Consideration of soil and bedrock associated with the geological structure of Sharpsburg indicated that differences in actual mineral concentration in rock, soil, plants, and water might exist, but give only small differences in natural background radioactivity as measured by hand scintillation counters. Effects of urbanization, such as increase in outdoor radioactivity simply from a high density of brick buildings, might tend to equalize hand-borne readings between buildings, so that actual sampling and analysis of air, water, soil, and plants would be necessary to establish differences.

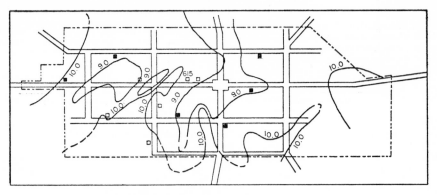

    □ GROUP I. SPECIAL STUDY HOUSES FOR MEASUREMENT
         OF RADIOACTIVITY INDOORS

    ■ GROUP II. SPECIAL STUDY HOUSES FOR MEASUREMENT
         OF RADIOACTIVITY INDOORS

      RADIATION
      (μr / hr)

FIG. 12.—Hand-borne radioactivity iso-contour map of Sharpsburg town

Since hand-borne counter readings made within the town area represented a closely spaced grid with points approximately 100 feet apart, contouring the readings could be attempted. The results (Fig. 12) support the general impression of increased radioactivity near the south edge of town. Radioactive highs are found, however, extending in the direction of the plunge axis of the geological syncline. The axis trends through the heart of town and plunges about 10° NNE. This suggested that more radioactive lithological units outcropping near the southwest perimeter of Sharpsburg plunge with the syncline and would be found at greater depth toward the center of Sharpsburg. Soil sample No. 129, Plate I, taken at a depth of about 4 feet in C_2 parent material gave the highest β activity of all samples shown in Table 8, and among the highest α activities. At 204 West Main Street, investigation of soil to a depth of 12 feet showed evidence of considerable man-made fill and washed-in alluvium above the 8-foot depth.

At about a 10-foot depth, typical subsoil materials (Lab. No. 164) of the Hagerstown soil series were countered. A black layer at a 6-foot depth indicated an old surface soil. Comparison of scintillometer readings and results of radiometric counting of soil samples were as shown in the accompanying tabulation.

| | HAND-BORNE READING | DEPTH (IN.) | NET C.P.M. IN 4.5-CM² AREA | |
LAB. NO. AND SOIL DESCRIPTION			α	β
Backyard, 204 West Main.........	0.0094	....	....	....
165 Topsoil, back yard..............	0.0094	0–12	0.67±0.29*	21.7±2.1
166 V.F. silt fill material—limestone chips abundant...............	0.010	20–30	0.66±0.24	22.0±2.1
163 Fine silt material—limestone chips abundant.....................	0.013	60–72	0.59±0.28	21.3±2.1
167 Subsoil—heavy silty clay loam....	0.015	84–96	0.69±0.23	23.8±2.2
164 Silty clay—shale particles and manganese and iron concretions abundant.........................	0.02	120	1.30±0.46	22.0±2.2

* Statistical error.

The findings here indicate some of the difficulty in investigating background radioactivity of houses where the upper layers of soil are disturbed or alluvial in nature, but it appears that near the center of Sharpsburg town the lithological units of highest radioactivity may be found at depths where shallow wells would bottom or basements would be dug. Radioactivity in basement soil and rocks is discussed later.

With such small but detectable differences in hand-borne readings (amounting to 1–3 μr/hr) from house to house, additional studies involving field measurements, soil sampling, and laboratory counting were carried out to confirm whether the differences were real or could be explained merely by variations in the ambient radiation levels. Since organic-matter content may affect the background radioactivity of surface soils and since pH is some indication of free lime in soil, these measurements are listed along with radioactivity measurements (Table 8). Sampling locations are shown in Plate I.

SUBSOIL RADIOACTIVITY OF GARDEN SOILS

Closely spaced residences on the same soil type may differ in background radioactivity as a result of soil activity background arising from a highly variable A horizon or topsoil, from building materials, or from a variable amount of background radiation arising from radon-222 in the atmosphere. It was of interest to know whether the effects due to soil at a specific residence could be determined by taking readings in a pit dug in the soil. In a hole the primary γ-ray effect might come from the subsoil, whereas effects mentioned above might be largely shielded out.

The results given in Table 5A were obtained during soil sampling in connection with growing a standard vegetable (potatoes) at 120 special-study houses in four districts and during an experiment to determine the dose rate by use of moisture-proof film badges at 12

TABLE 5A

AVERAGE DIFFERENCE OF BACKGROUND RADIOACTIVITY READINGS BETWEEN SUR-
FACE AND SUBSURFACE SOILS IN POTATO PLOTS OF RESIDENCES IN FOUR AREAS OF
WASHINGTON COUNTY, MD., COMPARED WITH AVERAGE AERORADIOMETRIC AND
YARD READINGS AT THE SAME RESIDENCES

AREA AND MAJOR SOILS ASSOCIATION	YARD		AERORADIOMETRIC		SUBSURFACE OVER SURFACE SOIL DIFFERENCE	
	No. Observations	Reading (μr/hr)	No. Observations	Reading (c.p.s)	No. Observations	Reading (μr/hr)
Sharpsburg town (Hagerstown stony loam).......	29	8.4	29	816	45*	4.2
Sharpsburg rural (Frankstown-Hagerstown stony loams)...............	15	7.1	15	630	15	3.7
Tilghmanton (Hagerstown silt loam; Hagerstown stony loam)...........	....	....	25	647	25	2.1
Downsville (Duffield-Frankstown silt loams; Murrill and Warners loams)...............	17	7	17	598	17	1.7

* Includes observations made during film-badge study.

houses in the town. Some increase in reading in the hole was expected because of changing geometry, but in the case of soil types having small variability in arrangement and thickness of soil layers, one would expect the same relative increase from place to place within a reasonable distance. Any other difference noticed is probably due to an actual increase or decrease in radioactivity of the subsoil (the subsoil representing the B horizon in some, but not all, cases measured).

Surface soil readings were first taken with the scintillation counter resting on the ground very near the excavation. A subsurface reading was then taken immediately by lowering the counter into the excavation onto the subsoil surface. Since only the difference was noted, frequent adjustment of the counter, as for the hand-borne measure-

ments shown in Table 4, was not necessary. This approach to shielding of stray radiation is similar to those used by Gibbs and McCullum
(1955), who buried a special Geiger detector in place, and Lieftinck
(1957), who placed the barrel of a scintillation counter or probe of a
Geiger-Müller counter in auger holes.

Previously it was found that a activities of soil were lower at the
4–7-inch depth (average 0.21 net cpm/4.5-cm^2 area) of the silt loam
developed on shaly or cherty limestone than at the 1–4-inch depth
(average 0.51, Table 2).

The subsoils (3–14-inch depth) of the Hagerstown stony loams, on
the other hand, showed increased a activity (average 0.58 c.p.m., as
determined by radiometric counting) compared with surface soil
(average 0.46 c.p.m., a activity, grids 64, 64B, 64C, Table 3). The
Frankstown and Duffield silt loam, somewhat similar in texture and
depth but both deeper and finer textured than the Hagerstown stony
loam, is the predominant soil in Downsville.

In Table 5A, average differences between scintillometer readings
(subsoil reading minus surface reading) are shown for three districts
and Sharpsburg town. The scintillation-counter activities for District
1 and Sharpsburg town were approximately two times higher than for
Downsville subsoils, indicating an increased subsoil activity in keeping with the radiometric counting results comparing the Hagerstown
stony loam with the Frankstown silt loam.

In Sharpsburg town, increased radioactivity in the subsoil was associated both with upland subsoils (HgC2, HbD2, DmB2, DmC2,
FwB2, FwC2, FvEw) and with alluvial subsoils (Hx and Lm). Since
these alluvial subsoils are much siltier than are subsoils of the upland
source soils (see Table 1), the results seem to indicate that particles less
than 2 microns in diameter are not of great significance in terms of γ
radioactivity. Hoogteijling and Sizoo (1948) produced evidence that
radioactive elements show a preference for the grain size 2–16 microns
of several sedimentary soils. Heavy minerals as a source of γ-ray
radioactivity in Sharpsburg soils are suggested, and the findings tend
to support Carroll's observation (1959) that upland soils contribute,
through erosion, more heavy minerals, including radioactive ones, to
alluvial soils than might be expected from an estimation of the quantity of minerals in the parent rocks.

Table 5B shows soil background readings at various depths in ex-
cavations that were made to "plant" film badges in place in 12 garden
soils of the town. The soils are divided into two groups, on the basis
of an average difference of 1.5 $\mu r/hr$ hand-borne yard readings.

The average difference between groups is 0.5 $\mu r/hr$. Large variation
is apparent within groups and also according to time (month) of

TABLE 5B

SOIL BACKGROUND READINGS IN μR/HR, YARDS OF TWELVE RESIDENCES IN SHARPSBURG TOWN, MD.

(Readings made at various depths as indicated.*)

Station No.	Soil Type 2†	Air-borne Reading (c.p.s.)	Yard Reading (μR/HR) 1959	Soil Subsurface Reading I	II	III	IV	Soil Surface Readings V	VI	VII (III–V)	VIII (IV–VI)	IX (Average, VII+VIII)
GROUP I												
532	HgC2	840	11.2	13.2	16.2	14.7	14.8	9.8	8.3	4.9	6.5	5.7
609	HgC2	800	9.8	16.6	17.8	17.2	....	10.2	....	7	....	....
615	HgC2	800	9.4	12.9	14.4	13.6	....	10.2	....	3.4	....	....
529	HgC2	800	9.2	11.2	11.2	11.1	10.8	7.4	8.3	3.8	2.5	3.1
444	HbC2	800	9.4	16.6	16.6	16.6	16.6	9.8	9.3	6.8	7.3	7
636	HgC2	820	9.0	16.6	17.8	17.2	14.8	11.6	8.4	5.6	6.2	5.9
Average.			9.7									5.4
GROUP II												
696	HgC2	840	8.7	13.2	14.4	13.8	12.8	9.8	8.3	4	4.5	4.2
664	DmC2	880	8.1	12.4	14.4	13.4	14.8	9	8.3	4.4	6.5	5.4
424	HgC2 Hx	820	8.1	14.4	16.2	15.3	12.6	11.6	8.3	3.7	4.3	4
459	HeB2	800	8.1	14	15.8	14.9	13.2	9.8	10.4	5.1	2.8	3.9
566	Hx	800	7.8	11	14.9	12.9	....	9	....	3.9	....	....
490	HgC2	800	7.8	18.4	18.4	18.4	15.4	9.8	9.6	8.6	5.8	7.2
Average.			8.2									4.9

Difference, Groups I and II 0.5

* Column heading descriptions: I, Film depth or B horizon of soil, March, 1961; II, Soil pit bottom, March, 1961; III, Average of cols. I and II; IV, Soil pit bottom, June, 1961; V, Surface soil reading, March, 1961; VI, Surface soil reading, June, 1961; VII, Increase of pit reading over soil surface reading, March, 1961; VIII, Increase of pit reading over soil surface reading, June, 1961; IX, Average difference of subsoil and surface readings, cols. VII and VIII.
† Symbol to appear on published map, Soil Survey of Washington County, Md. (to be published).

measurement. Since most houses in both groups are on the same soil type, variation is most likely accounted for by the heterogeneous nature of the parent material and possibly through the influence of lithological units of bedrock that occur at varying depths in this area, dependent upon the dip of bedding and degree of erosion of surface cover. Results of completed radiometric counting tests for α and β activity on all soils taken for the growing of potatoes may show any real difference in subsoil radioactivity according to soil type or parent material.

Radioactivity Measurements at Special-Study Houses

Scintillation-counter measurements were made at 12 specially selected houses. The houses are divided into two groups of 6 each based on the mean value for hand-borne readings made in yards of houses in Sharpsburg. This value, 9.0 μr/hr (Table 4), separates the two largest groups in the frequency range chosen (Fig. 6).

The average difference of approximately 1.5 μr/hr between the two groups of outdoor γ-ray background radioactivity measurements is approximately the limit of sensitivity of the scintillation counter; however, multiple measurements at each house at a different time established the reliability of the yard readings. Also, differences within groups in other parts of the house environment not revealed by yard readings or air-borne intensities are shown in the tables. The actual readings for the yards are given, as obtained in the survey of 1959.

Many factors contribute to the dosage indoors, compared with that of the outdoors; these factors include ventilation, shielding from external radioactivity, and contribution of the building material. The main construction material above the foundation of these houses is wood. When the foundation was sufficiently above ground, its contribution to background radioactivity was noted (3d col., Table 6). Generally, native limestone gives a lower reading than does soil in the yard.

BEDROOMS

Results of scintillation-counter measurements made in first- or second-floor bedrooms of the two groups of houses are summarized in Table 6. The average differences in readings for exterior house, yard, middle of room, and interior wall are 0.9, 1.5, 0.8, and 0.6 μr/hr, respectively, for the two groups. The outdoor measurements would of themselves, therefore, appear sufficient for comparison of dose rates indoors between paired houses of the type of construction considered here. There were no brick houses in either group.

BASEMENTS

Many of the older houses in the town have partially finished basements with earth floors and walls, and limestone outcrops are exposed where the earth has been excavated. This condition, aside from a possibly direct influence on dose rates in other rooms of the house, afforded an opportunity to compare the radioactivity of relatively unweathered soil parent material with soil from yards or gardens. Table 7 shows an average difference between Groups I and II of 0.6, —0.1, and 4.5 μr/hr, for readings made in the middle of the basement room, near overhead floor joists, and on dirt floors, re-

TABLE 6

HAND-BORNE SCINTILLATION-COUNTER READINGS: COMPARISON OF
INDOOR AND OUTDOOR READINGS, 1961, AT 12 SPECIALLY
SELECTED HOUSES IN SHARPSBURG

STATION No.	HOUSE CONSTRUCTION WHERE READING TAKEN		OUTDOOR READINGS, 1959 SURVEY (μR/HR)		INDOOR READINGS* 1961 SURVEY (μR/HR)	
	Frame	Other (Stone, etc.)	Exterior House	Yard	Middle of Room	Interior Wall
		GROUP I >9.0 μR/HR, YARD				
532*.........		X	6.5	11.2	9.1	6.2
609..........		X	5.7	9.8	4.2	3.7
615..........		X	5.7	9.4	4.2	5.4
444..........		X	6.4	9.4	3.7	....
529..........	X		6.5	9.2	3.3	3.1
636..........		X	4.5	9.0	3.7	3.1
Group I av.			5.9	9.7	4.7	4.3
		GROUP II <9.0 μR/HR, YARD				
696..........		X	5.4	8.7	4.2	3.1
664..........	X		4.9	8.1	4.2	3.1
459..........		X	5.0	8.1	1.9	4.2
424*.........		X	6.5	8.1	6.2	4.2
566..........	X		3.3	7.8	3.7	3.3
490..........	X		5.3	7.8	3.3	4.2
Group II av......			5.0	8.2	3.9	3.7
Difference, Groups I and II..			0.9	1.5	0.8	0.6

* Indoor readings obtained on first floor. (Readings obtained at stations not starred represent second-floor locations.)

spectively. Differences in basement soil between Groups I and II are much greater than the differences noted in outdoor soils. Whether the differences were due to the exposed soil and rock or merely to the geometry of the confined spaces was investigated by relative ionization measurements made on samples of fine basement soil and garden soil from Station 615. The measurements were made with a Landsverk electrometer on an infinitely thick sample of 40.4-cm.² area, with the following results.

	Div/Min
Basement soil	4.5
Garden subsoil	0.9

TABLE 7

RADIOACTIVITY MEASUREMENTS IN BASEMENTS OF TWO GROUPS
OF HOUSES DIFFERING IN OUTDOOR RADIOACTIVITY,
SHARPSBURG, MD., 1961*

STATION No.	YARD READING (μR/HR)	BASEMENT READINGS (μR/HR)		
		Middle of Room	Near Overhead Floor Joists (Film-Badge Loc.)	Other
		GROUP I >9.0 μR/HR		
532†	11.2	10	8.3	12.8 (S)
				7 (R)‡
609	9.8	4.2	3.3	. . . .
615	9.4	12.8	10	14.8 (S)
				13 (R)
444	9.4	10.3	6.2	12.8 (S)
529	9.2	8.3	5	6.2 (R)
636	9	2.9	3.1	. . . .
Group I av	9.7	8.1	6	{13.4 Soil { 8.7 Rock
		GROUP II <9.0 μR/HR		
696	8.7	4.1	5.2	8.2 (S)
				3.1 (R)
664	8.1	11	7	10 (S)
459	8.1	4.2	5	. . . .
424	8.1	10.3	5.2	9.3 (S)
566	7.8	9.1	6.2	. . . .
490	7.8	6.2	8.3	6.2 (R)
Group II av . . .	8.2	7.5	6.1	{ 8.9 Soil { 4.7 Rock
Difference, Groups I and II	1.5	0.6	− 0.1	{ 4.5 Soil { 4 Rock

* Measurement made on exposed soil.
† Cement floor in main basement. Readings made on dirt and rock in small adjacent basement.
‡ Measurement made on exposed rock.

TABLE 8

COMPARISON OF RADIOACTIVITY OF SOIL SAMPLES FROM PILOT STUDY COMMUNITY,
SHARPSBURG, MD., COMPARING AIR-BORNE, HAND-BORNE MEASUREMENTS
AND RADIOMETRIC COUNTING OF SOIL SAMPLES

LAB. No. AND LOCATION	SOIL TYPE*	pH	AIR-BORNE READINGS (C.P.S.)	HAND-BORNE READINGS (μR/HR)	ORGANIC MATTER (PER CENT)	NET C.P.M.—4.5 CM.^{2}	
						α	β
139 Boyar farm, pasture subsoil..............	HbD2	5.8	800	8.7	1.02	0.69±0.31	22.8±2.1
113 Boyar farm, pasture residual clay.........	HbD2	6.3	800	8.7	0.50	1.21±0.42	21.3±2.2
138 Boyar farm, pasture subsoil..............	HbD2	6.2	800	8.7	0.76	1.25±0.26	27.7±2.4
119† 114 W. Chapline, garden.................	Hx	7.0	800	9.4	6.28	0.39±0.20	23.8±2.2
22† 224 W. Main, lawn...	Hx	7.0	820	9.0	7.26	0.26±0.15	21.2±1.9
9† 204 W. Chapline, lawn.	HgC2	7.0	820	8.2	8.9	0.25±0.20	20.7±1.9
67† 206 S. Hall, lawn.....	HgC2	7.2	840	11.2	7.62	0.97±0.30	22.9±2.1
126† Opposite 233 W. Antietam, field...........	HbD2	6.9	860	9.0	3.6	1.77±0.51	25.9±2.3
106† Limestone quarry, west side................	HgC2	6.5	880	10.2	1.10	1.40±0.47	22.5±2.2
132† Limestone quarry, below rim..............	HgC2	7.3	880	9.6	0.57	0.79±0.31	23.1±2.2
135† Limestone quarry, south rim near rock.........	HgC2	7.2	880	9.8	0.50	1.44±0.50	26.6±2.4
137† Opposite Fred Roulette farm................	FwB2	6.0	900	10.6	1.69	0.42±0.23	20.1±2.0
31† Sheppardstown Pike, field.................	Hx	6.3	850	7.0	0.43	0.40±0.20	8.0±1.3
110† Mondell Road, field...	DmB2	7.2	1,000	12.2	2.03	0.91±0.34	27.9±2.4
130† Mondell Road, field...	FwB2	5.2	940	10.8	2.03	1.27±0.41	23.7±2.3
129 107 South Hall, road cut, subsoil 4-ft. depth	HgC2	7.7	800	10.0	0.47	1.36±0.4	31.4±2.5
29 209 W. Antietam, garden.................	HgC2	7.0	800	12.0	6.19	0.37±0.19	25.0±2.1
42 211 W. Antietam, back yard..............	HgC2	5.9	800	11.3	5.38	0.4 ±0.20	23.0±2.1
105 304 E. Main, lawn....	FwB2	7.6	800	8.3	3.38	0.59±0.30	24.1±2.2
17 205 W. Chapline, basement subsoil.........	HgC2	6.0	800	10.8	1.36	0.89±0.28	28.2±2.4
109† A. B. Deatrich farm...	DmB2	6.6	800	10.0	1.95	0.89±0.35	25.5±2.3
131† A. B. Deatrich farm, garden...............	HgC2	6.8	800	9.5	3.72	0.62±0.20	22.2±2.1
26 205 E. Main, garden..	HgC2	6.8	800	8.3	5.26	0.52±0.24	18.9±1.8
127 204 E. Chapline, garden	HgC2	6.8	800	8.7	3.67	0.52±0.23	21.6±2.1
20 104 N. Mechanic, garden.................	Hx	6.9	800	8.2	6.22	0.44±0.20	18.1±1.8
11 206 W. Antietam, garden.................	HgC2	7.1	820	11.6	2.95	0.61±0.25	20.9±2.0
70 207 S. Mechanic, garden.................	Hx	7.4	840	9.3	5.15	0.75±0.30	25.1±2.3
13 207 S. Mechanic, front lawn...............	HgC2	7.6	840	9.3	6.48	0.48±0.22	17.2±1.8
5 103 E. Antietam, garden.................	HgC2	7.0	840	9.0	5.64	0.25±0.18	17.3±1.8

* Approved soil symbol from correlation field sheet for Washington County, U.S.D.A. Soil Conservation Service. See Table 1 for description of soil.

† Location shown in Figure 1 or Plate II. All others in Plate I.

TABLE 8—*Continued*

Lab. No. and Location	Soil Type*	pH	Airborne Readings (c.p.s.)	Handborne Readings (μR/hr)	Organic Matter (per cent)	Net c.p.m.—4.5 cm.²	
						α	β
21 102 E. Antietam, garden........	HgC2	7.1	840	8.8	5.19	0.53±0.21	20.5±1.9
33 206 S. Hall, lawn.....	HgC2	5.8	840	11.2	1.98	1.24±0.44	19.8±1.9
35 301 W. Main, lawn....	Hx	7.1	840	9.0	6.69	0.73±0.30	20.6±1.9
1 205 E. Antietam, lawn.	HgC2	6.6	840	11.0	5.29	0.34±0.22	22.0±2.0
154 Lohman farm, pasture.	FvE2	6.0	860	11.0	2.41	0.28±0.20	25.1±2.3
156 Lohman farm, pasture.	DmB2	6.7	860	9.8	2.17	0.65±0.28	25.5±2.3
145 Church, High St., cemetery........	FvE2	5.5	860	11.4	2.79	0.39±0.21	20.8±2.0
128 308 W. Chapline, garden........	DmB2	7.3	880	10.4	4.22	0.46±0.27	25.2±2.3
133 311 W. Main, garden..	DmC2	7.1	880	8.1	5.60	0.29±0.21	22.4±2.1

Another sample of basement soil (Lab. No. 17, Table 8) showed increased α and high β activity compared with that of garden soils in the area. The possibility existed, of course, that daughter decay products of radon-222 gas had built up on the fine clay particles in the confined space. Measurements by C. P. Straub (1958) indicated that radon-222 levels measured at a firehouse approximately 400 feet north of Station 615 were higher than levels measured at the schoolhouse west of town. The result of further radiometric characterization of the basement soil sample is shown in Table 9. The radium-226 content is considerably higher than that of the average soil of the United States and appears to be out of equilibrium with the parent uranium-238. Hence, a concentration point for increased dosage is provided in an otherwise relatively low-background area.

The measurements reported here show the complexity of measur-

TABLE 9

RADIOMETRIC ANALYSIS OF BASEMENT SOIL AND ROCK
NO. 615, SHARPSBURG, MD.

Soil:

Radium-226...............	4 pc/gm raw soil
Thorium-232..............	4 p.p.m.
Uranium-238.............	3 p.p.m.
Total α activity..........	38.7 pc/gm raw soil
Total β activity...........	157.4 pc/gm raw soil
Potassium-40.............	28 pc/gm equivalent to 3.33 per cent K raw soil

Rock:

Uranium-238.............	2 p.p.m.

ing radiation dosage received by an individual in this type of environment and of assessing the significance of very small differences in background radiation when comparing groups of paired residences. Gross exposure doses determined by aeroradiometric measurements are of little meaning under these conditions but may be of value when large numbers of residences are considered, and they are in fact being used.

Discussion

It was not expected on the basis of geological guides that radioactive minerals in any quantity would be found associated with the carbonate sedimentary rocks of the study community. Bell (1956) states that the syngenetic uranium contents of carbonate rocks and sediments are among the lowest of all rocks of the earth's crust. It must be stressed, however, that the bedrock in the Sharpsburg area is highly complex lithologically. The finding of rather large differences of radioactivity associated with residual limestone soils of the study community may be largely attributed to two soil-formation factors—parent materials and degree of relief. Both factors vary within a short distance and locally have affected the degree of weathering. Relief has affected the rate of rainfall runoff, drainage, and erosion. When carbonate bedrock is exposed through erosion, weathering processes, as carried out through solution aided by soil and humus acids, must proceed much more slowly than they do when a thick layer of soil covers the rock. Numerous fragments of chert and shale occur in some carbonate beds and are concentrated at the surface of residual soil, where they form a protective blanket over the silt and sand below (Hack, 1957). Throughout the soil they form a skeletal framework that gives good internal drainage and allows weathering by solution to proceed rapidly. Surface soils (A horizons) from most carbonate rocks contain smaller amounts of clay, bases, iron, and certain other materials than do the subsoils (B horizons) (Carey, 1959). In contrast, quartz and other minerals that weather slowly, including zircon and potash feldspars (both sources of γ radiation), are more abundant in the surface soils than in the subsoils.

Soil development through solution of carbonate rock concentrates the resistant heavy minerals of many cubic feet of parent rock in a few inches of residual surface soil. The older and more mature the residual soil formed above any rock, the greater is the concentration of heavy minerals. Favorable sites are flat plainlike areas (Carroll, 1959). The end effect seems to be that soils with the highest level of γ-ray radioactivity due to concentration of resistates may not contain the most leachable radioactivity. The potassium content of interbedded shale and limestone soil parent material from the area is probably enhanced in the soil, since carbonate is depleted by solution and the argillaceous matter is concentrated in the residuum. Moxham's data indicate that the thorium content of soils from this locality more nearly approaches that of the average shale (12 p.p.m.) than that of the average carbonate rock and that the potassium content is about twice that of an average shale.

Small differences shown by aeroradiometric measurements and hand-borne scintillometer ground determinations near residences of Sharpsburg, Maryland, serve as guides for the further investigation of background radioactivity, especially in areas where the principal soil evolution process is the formation of soil parent material. In such areas, soil subsurface readings and readings where less weathering of bedrock is indicated, as in basements, have shown increased values, depending upon the nature of the bedrock. As in the findings of other investigators (Gibbs and McCullum, 1955; Hoogteijling and Sizoo, 1948; Marsden and Watson-Munro, 1944), the source of soil γ-ray background radioactivity seems to be the parent materials. Some limestones are without doubt very low in uranium and thorium minerals, either coprecipitated or present as detrital material. The same soil type (specifically Duffield silt loam) developed on different geological formations in Sharpsburg and Downsville communities showed considerable difference in aeroradiometric and hand-borne measurements.

SUMMARY

Aeroradiometric measurements could be related to ground dose and tenor of radioactive minerals when measurements were made over large uniform areas, as indicated by soil type, southwest of Sharpsburg town.

As shown by intensive investigation of the background radioactivity at specially selected study houses in Sharpsburg, Maryland, differences of up to a factor of 4 in γ-ray radioactivity and α-particle emanation may be indicated by sampling of basement soil at some locations having little variation in aeroradiometric or ground dose values. The aeroradiometric data will be most useful for evaluating health data in areas of Washington County, where soil material is uniform and free of rocky conditions. It would seem that enough hand-borne measurements to characterize lithological units from a detailed geological study and soil types as mapped in the typical modern soils survey offer an alternative to the use of gross aeroradiometric data. In conclusion, our principal findings are as follows.

1. There seems little doubt that the mass radiation exposure effect

in Sharpsburg is greater than it is in rural areas; this is not, however, necessarily due to γ radiation. Potential α-particle-emanating sources exist in outcropping geological strata, and, as a result, residents are living near soil or rock that exhibits a fourfold difference in α activity from place to place.

2. A somewhat uniform γ-radiation dose results from different ratios of γ-ray-emitting radionuclides in the environment of the study community from place to place.

3. A combination of geological, pedological, and radiological studies is needed to define clearly the radiation dose. Soil and rock need to be studied in profile and at depth. Detailed soils and geological guides are not subject to the usual temporal, spatial, and cultural variations that affect the radiological characterization of air, water, or vegetation.

4. Small, but detectable, differences in dose rate under otherwise uniform conditions are probably epidemiologically more significant than are larger differences under non-uniform conditions.

5. The recording of an increased dose rate in a basement with walls and floor of natural soil and rock would suggest that the evolutionary process of formation of soil parent material from carbonate rock and the relation of this process to the concentration of radionuclides should be more intensively studied to determine whether concentration points for increased dosage prevail in other, similarly unfinished basements and also in gardens, aquifers, and agricultural or quarrying areas.

Acknowledgments

The authors wish to express their appreciation to their associates at the National Cancer Institute: to Dr. V. E. Archer, under whose supervision the hand-borne scintillation-counter survey was carried out while he was medical officer in charge; to Mr. Henry Greenville for making thorium analyses; to Philip Physioc for making the pH and organic-matter determinations; to Linn Davison and K. T. Moats for their assistance in field sampling, scintillation-counter measurements, and laboratory preparation and radiometric counting of samples; and to Dr. J. C. Bryant for his encouragement. Appreciation is also due Dr. William Sando and Sam Rosenblum of the U.S. Geological Survey for technical review and for uranium analyses of soil and rock.

The authors also wish to express their appreciation to the following at the Robert A. Taft Sanitary Engineering Center: to Dr. R. L. Woodward and staff for manuscript preparation and for making possible participation at the Natural Radiation Environment International Symposium, in Houston, Texas, April 13, 1963; to Seymour Gold and Thomas Rozzell for radiometric analyses; to Mr. Kenneth Cassel, Jr., for editorial review; and to Dr. Conrad P. Straub and Mr. Lee McCabe for technical review.

REFERENCES

ADAMS, J. A. S., J. S. RICHARDSON, and C. C. TEMPLETON. 1958. Determinations of thorium and uranium in sedimentary rocks by two independent methods. Geochim. & Cosmochim. Acta, **13**:270–79.

BELL, K. B. 1956. Uranium in precipitates and evaporites. U.S. Geol. Survey Prof. Paper 300, pp. 381–86.

CANNON, H. L., and J. M. BOWLES. Distribution of trace elements in rocks, soils, and plants in relation to cancer occurrence in the eastern part of Washington County, Maryland. U.S. Geol. Survey. (In progress.)

CAREY, J. B. 1959. Soil Survey, Lancaster County, Pennsylvania. U.S. Dept. Agric., Soil Conservation Service. Pp. 121.

CARROLL, D. 1959. Sedimentary studies in the middle river drainage basin of the Shenandoah Valley of Virginia. U.S. Geol. Survey Prof. Paper 314-F, pp. 125–54.

CHICHILO, P., and C. W. WHITTAKER. 1958. Trace elements in agricultural limestones of Atlantic Coast regions, Agron. J., **50**:131–35.

———. 1961. Trace elements in agricultural limestones in the United States. *Ibid.*, **53**:139–44.

CLOOS, E. 1951. History and geography of Washington County. Stratigraphy of sedimentary rocks. Igneous rocks. Structural geology of Washington County. Mineral resources of Washington County. Ground water resources. *In* The Physical Features of Washington County, State of Maryland. Washington, D.C.: U.S. Dept. of Geol., Mines, and Water Resources. Pp. 193.

COURT-BROWN, W. M., and F. W. SPIERS. 1960. Geographical variation in leukemia mortality in relation to background radiation and other factors. Brit. Med. J. (London), **1**:1753–59.

DENSON, M. E. 1956. Geophysical-geochemical prospecting for uranium. Geol. Survey Prof. Paper 300, pp. 687–703.

FAIR, D. R. R., and H. HOWELLS. 1955. A survey of the natural gamma radioactivity in the West Cumberland area. J. Nuc. Energy, **1**:274–79.

GENERAL DYNAMICS CORPORATION. Convair Division. N.d. Technical Pub. 3-R-72R.

GENTRY, J. T., E. PARKHURST, and G. V. BULIN. 1959. An epidemiological study of congenital malformations in New York State. Am. J. Pub. Health, **49**:497–513.

GIBBS, H. S., and G. J. McCULLUM. 1955. Natural radioactivity of soils. New Zealand J. Sci. & Technol. **37B**:354–68.

HACK, J. T. 1957. Studies of longitudinal stream profiles in Virginia and Maryland. U.S. Geol. Survey Prof. Paper 294-B, pp. 85–86.

HOOGTEIJLING, P. J., and G. J. SIZOO. 1948. Radioactivity and grain size of soil. Physica, **14**:65–72.

HULTQVIST., B. 1956. Studies on naturally occurring ionizing radiation. Kungl. Svenska Vetens. Handl., Vol. **6**, Ser. 4. Pp. 125.

KING, P. B. 1950. Geology of the Elkton Area, Virginia. U.S. Geol. Survey Prof. Paper 230. Pp. 82.

KRATCHMAN, J., and D. GRAHN. 1959. Relationship between the geologic environment and mortality from congenital malformation. Div. Raw Materials, AEC, and Div. Biol. & Medicine, AEC, Rpt. TID-8204. Pp. 23.

LAWRENCE, P. A., and W. Y. CHEN. 1959. A project for studying the geographic distribution of cancer within a single county as related to environmental factors. Am. J. Pub. Health, **49**:668–74.

LIEFTINCK, J. E., JR. 1957. Radioactivity as a basis for correlation of glacial deposits in Ohio. Ohio J. Sci., **57**:375–78.

MARSDEN, E., and C. WATSON-MUNRO. 1944. Radioactivity of New Zealand soils and rocks. New Zealand J. Sci & Technol., **26** (Sec. B): 99–114.

MEYERS, J. 1928. Cancer death-rate variations in relation to combustion products of fuel, topography and population. N.Y. State J. Med., **28**: 365–72.

MEYERS, J., and V. F. HESS. 1952. Cancer death rates, topography and terrestrial radiation. N.Y. State J. Med., **52**:463–67.

MOXHAM, R. S. N.d. Aerial radiometric and geologic maps of the eastern part of Washington County, Maryland. U.S. Geol. Survey Open File.

PAGE, L. R. 1956. Geologic prospecting for uranium and thorium. U.S. Geol. Survey Prof. Paper 300, pp. 627–32.

SANDO, W. J. 1957. Beekmantown Group (Lower Ordovician) of Maryland. Geol. Soc. America Mem. 68. Pp. 161.

STOW, M. H. 1955. Report of Radiometric Reconnaissance in Virginia, North Carolina, Eastern Tennessee and parts of South Carolina, Georgia, and Alabama. U.S. Atomic Energy Commission Pub. RME-3107. Pp. 25.

STRAUB, C. P., G. R. HAGEE, B. M. BRONSON, G. J. KARCHER, and A. DOVEL. Survey for natural atmospheric radioactivity—Hagerstown, Md., 1958. Robert A. Taft Sanitary Engineering Center, Radiological Health Research, Cincinnati. (Unpublished data.)

WESLEY, J. P. 1960. Background radiation as the cause of fatal congenital malformation. Internat. J. Radiation Biol., **2**:97–112.

WILSON, J. L. 1952. Upper Cambrian stratigraphy in the central Appalachians. Geol. Soc. America Bull., **63**:275–322.

APPENDIX A

CALIBRATIONS OF LaRoE Fv-6-S SCINTILLATION COUNTERS

The LaRoe Scintillation Counter, a very sensitive γ-ray detector, has found wide acceptance for mineral prospecting and survey work. Components of the meter are carefully selected and matched for maximum sensitivity by the manufacturer. While the meter response to γ radiation was found to be linear for all four meters used in this survey, some meters of this make are more sensitive than are others and therefore require a smaller change in γ-ray flux for a given meter response. The sensitivity and stability are determined largely by characteristics of the scintillation crystal (sodium iodide, thallium activated) and the photomultiplier tube. Dark current in the photomultiplier tube may approach the same order of magnitude as that which one is attempting to measure unless discriminating circuits are used to filter out the very low-energy-level pulses coming from the photomultiplier. This may lower the sensitivity, but it eliminates the dark-current pulses produced in the photomultiplier. It is necessary to check the calibration frequently and maintain fresh batteries while in the field.

The lowest range of the LaRoe instrument is 0.0000–0.0100 mr/hr. In a low-level survey it is desirable to have nearly full-scale response at the lowest background γ radiation level anticipated. All meters used should also give the same dose rate under a fixed set of conditions, as determined from a calibration curve. In calibrating at these low levels, one must find a location where background is very low (top or basement) and then add the background to the calculated dose rate from the source. The site chosen for calibration was outdoors on the roof of a laboratory building, away from soil and other ground effects that might not be uniform or constant. Since the effect of building material was to be noted, a limestone ledge was chosen as the starting point for arbitrarily setting one meter, chosen as reference, to give a reading of 0.01, full scale. A fixed 100-μc. source of radium some 60 feet away on the roof of the laboratory building was then approached until the meter reading doubled, switching to the 0.25 scale. The γ-radiation level from the source at that distance (approximately 4.5 meters) was then calculated from the inverse-square relationship. At 1 meter, a 100-μc. radium source has a γ-radiation level of 0.084 mr/hr. At 4.5 meters the intensity is $0.084/(4.5^2)$, or 0.004 mr/hr, and is equal to background at the starting point. This value, 0.004 mr/hr, then becomes the lowest point on a calibration curve of meter readings versus actual mr/hr γ radiation obtained by taking meter readings at other points closer than 4.5 meters to the source. At each point the actual value (mr/hr) is the calculated γ radiation of the source plus 0.004. At a meter reading of 0.02, for example, the actual γ radiation is twice 0.004, or 0.008 mr/hr.

Reference to the calibration curve then gives the actual γ radiation in mr/hr for any meter reading. Because of differences in response, other meters may give a different background γ-radiation value for the starting point (limestone ledge) under the procedure described above. In practice, therefore, all meters were arbitrarily set to give a 0.0200 scale value 4.5 meters from the 100-μc. source. The meter reading at the 0.004-mr/hr point might then be somewhat more or less than 0.01 but would again give the lowest point on the meter calibration curve.

Under the conditions of this survey the radiation dose rate is defined as 0.008 mr/hr, or 8 μr/hr, when the 100-μc. radium source is placed 4.5 meters from the reference scintillometer.

A 1-μc. radium button was placed in a fixed spot on the outside of the instrument case, under the laboratory roof conditions described above, and the meter reading noted. Thereafter, the meter was adjusted to give this reading in the field with the source in place. Meters were checked at least every hour. Under these conditions any drift due to a fluctuating background condition, such as radon-222 gas increase or decrease, would be treated as instrument drift.

Appendix B

Determination of the Radioactivity of Infinitely Thick Soil Samples

The gross α and gross β activities of infinitely thick soil samples reported in this paper are the average for duplicate samples. In the event that duplicate samples did not agree at the 90 per cent confidence level, two more samples were prepared and counted.

Method

Instrumentation. An automatic internal gas-flow proportional counter (Nuclear Chicago) and a 1-inch diameter stainless-steel sample pan were employed. The efficiency of the instrument was checked frequently; a National Bureau of Standards polonium-210 source was used for the α range, and a series of ten thallium-204 samples was prepared in 1-inch-diameter stainless-steel sample pans from a thallium-204 standard solution (purchased from Nuclear Chicago Corp.) for the β range. Daily checks with secondary α and β standards were also made. The α and β efficiencies remained constant within the probably 90 per cent error. The plateaus were checked in detail once a week.

Counting. Determination of α activity was made by measuring the time required to detect 20 counts, and of β activity by measuring the time to detect 500 counts. Each sample was counted twice for each activity. In the event that the two results did not agree at the 90 per cent confidence level, the sample was counted a third time, and the best two results were used to calculate the activity of the sample. Frequent backgrounds (after every fifth counting period) were run.

Sample Preparation. Two grams of finely ground sample were placed in the sample pan. Water was added and stirred to make a thin paste. The samples were then dried to constant weight. Cracking was noted in some samples, but in most instances it was not severe and probably had little or no effect on the results.

Selection of Sample Size. To determine the quantity of soil necessary to give a sample of infinite thickness in the sample pans used and with the counter described above, a series of duplicate samples of increasing thickness was prepared from the same soil sample and counted. The results indicated that a thickness of 250 mg/cm² was sufficient for the sample tested. This would give a sample size of about 1.1 grams. A sample size of 2 grams was selected to allow for possible variation from one soil to another.

P. R. KAMATH, A. A. KHAN, S. R. RAO,
T. N. V. PILLAI, M. L. BORKAR, AND
S. GANAPATHY

58. *Environmental Natural Radioactivity Measurements at Trombay Establishment*

Background radiation measurements have been made for some years in India as a part of the preoperational environmental survey program near the atomic energy installations and the monitoring of high-radiation-background areas in the country. Vohra *et al.* (1960) have reported on the air-borne radioactivity and background γ radiation at Bombay. Gamma measurements in the monazite placer areas in Kerala have been reported by Bharatwal and Vaze (1958). Environmental surveys include sampling and analyses of soil, vegetation, water supplies, and biological materials for radioactivity content. A study of the survey results gives a good understanding of the natural occurrence of radioelements, leaching, and uptake processes taking place in the environment. Concentration of radioactivity by natural processes and through human agency has significance from considerations of human exposure. Of the natural radionuclides, radium-226 is historically known for the exposure hazard; β-emitting radium-228, a decay product of natural thorium-232, has gained considerable significance with increased knowledge of its toxicity through its α-emitting daughters. Radioactivity content of foodstuffs, fish, salt, and drinking water near areas where radiation background is above normal requires consideration.

P. R. KAMATH, A. A. KHAN, S. R. RAO, T. N. V. PILLAI, and M. L. BORKAR are scientific officers and chemists, and S. GANAPATHY is scientific assistant and chemist, at Radiation Hazards Control Section, Health Physics Division, Atomic Energy Establishment Trombay, Bombay, India.

The present paper reports on the results of systematic environmental studies carried out in the Trombay Establishment and discusses some radiochemical methods employed in these studies.

Low- and High-Background Areas

The scope of environmental studies covered regions of different radiation status (Fig. 1).

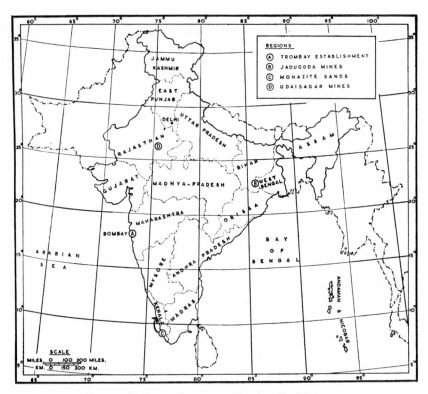

Fig. 1.—Background survey at Trombay Establishment

REGION A

The environment at Trombay is essentially a low-background area with no radioactive history in the preoperational period of the atomic energy installations. The Trombay Establishment started its activities nearly a decade ago, and they have expanded considerably since. On the site there are three reactors, processing plants for thorium and uranium production, and ancillary facilities, including fuel fabrication, radiochemical, and metallurgical laboratories. The waste effluents are discharged very much below permissible levels into the Bombay Harbor Bay, which has been the subject of considerable study for its dilution capacity (Pillai *et al.*, 1958; Kamath *et al.*, 1959, 1961). Nearly 5–10 km. away from the site, salt is produced by solar evaporation of the bay water for industrial and domestic use in Bombay and elsewhere. Within 16 km. there are areas where the general public lives in considerable numbers and where there are agricultural fields and a big center of milk production (Aarey milk colony). The main drinking-water supplies for the city of Bombay are obtained from natural rain-water reservoirs in the neighborhood of Bombay.

The environmental samples were obtained in the region up to 40 km. from the site, and effluent samples were obtained from discharge points on the site. Table 1 gives a sample survey of results for the site in the early period of operation and in recent years.

TABLE 1

GROSS ACTIVITY LEVELS 1958–62
TROMBAY ESTABLISHMENT—
REGION A

	Gross α (pc/gm)	Gross β (pc/gm)
1958–60		
Soil (top soil)........	0.4	40
Vegetation (grass)....	0.02	0.6 (wet wt.)
Fish...............	0.04–0.4	0.1–1.0
1960–62		
Soil...............	0.1 –1.0	10.0–35.0
Vegetation..........	0.01	0.3
Fish...............	0.05–0.09	2.2

REGION B

Region B is the neighborhood of the low-grade Jadugoda uranium mine in the state of Bihar. At present the mining is carried out at about 100 m. below surface. It is proposed to install ore milling and processing plants in the area. The environmental samples were obtained from a region within 3 km. from the mine. Streams from the mine site join the Suvarnarekha River, whose waters are used for drinking purposes. Machuya and Digdi, referred to in Table 8B, are small villages containing a few hutments. Mosabani is 16 km. east of the site, where the copper mines belonging to the Indian Copper Corporation, Ghatsila, are located.

The environment of the Indian Copper Corporation plant is included in this region. The mining of chalcopyrite ore is done at Mosabani about 600 m. below surface. The ore contains 0.01 per cent uranium oxide. Physical-beneficiation processes have been adopted to enrich uranium content in the tailings from the copper plant (Somasundarem and Kamath, 1958). These tailings were discharged to the river for over 30 years before beneficiation and recovery of uranium were planned. The effluents and rejects from the plant are released to the Suvarnarekha River. The samples were obtained from the mine colony, plant streams, river water, and river bed, which was dry in many places at the time of survey.

REGION C

Monazite-bearing sands and processing plants for recovery of thorium in Kerala are located in region C. The operations consist of both physical and chemical processes leading to the chemical recovery of the thorium and uranium present. Monazite contains about 9 per cent thorium oxide and 0.35 per cent uranium oxide. The samples analyzed (Vasudevamurthy, 1962) include those from the plant proper and those from the nearby river, 1.6–5 km. on either side of the discharge point.

REGION D

Region D is in the state of Rajasthan and forms the environment of the low-grade Udaisagar mine and the old Umra mine, which are about 12 miles from Udaipur. The mine is well type and about 40 m. deep. Nearby are hill tracts and some patches of agricultural production. Samples were obtained within a distance of 400 m. from the mine.

Background Build-up near Nuclear Facilities

Near ore-processing and mining regions, background radiation can build up as the result of operations carried on for years on end. From low-grade wastes dumped in the neighborhood, radioactive contaminants are picked up by soil, vegetation, and aquatic organisms (NAS, 1957; Revelle *et al.*, 1955; Dunster, 1958). These can ultimately add to a significant increase over natural background measured under the preoperational conditions. Similar addition to radiation background takes place in the environment of atomic energy installations.

Table 2 gives types and levels of activity discharged to the environment from long-term and continuous processing operations in the different regions (Somasundarem and Kamath, 1958; Vasudevamurthy, 1962; Kamath and Pillai, 1958). It would be seen that the process effluents are very much more active than background levels, but they do not contribute at present to any unsafe conditions, since adequate dilution is available and the total amount discharged is small (cf. discussion below).

As pointed out earlier, salt forms an important safety criterion in the Trombay environment. The results of analyses of salts (Pillai, 1963) are given separately in Table 3A. For the sake of comparison, results of analysis of solar-evaporated salt obtained through the courtesy of scientists abroad are incorporated in Table 3B.

TABLE 2

LONG-TERM AND CONTINUOUS DISCHARGE TO ENVIRONMENT:
ACTIVITY IN PROCESS EFFLUENTS

Region A
Thorium plant at Trombay:

Leachable activity from rare earth fluorides waste........ $\left\{\begin{array}{l}\text{Gross }\alpha, \quad 250 \text{ pc/gm} \\ \text{Gross }\beta, \quad 300 \text{ pc/gm}\end{array}\right.$

Effluent activity at discharge point.................... $\left\{\begin{array}{l}\text{Gross }\alpha, 24,000 \text{ pc/l} \\ \text{Gross }\beta, 16,000 \text{ pc/l} \\ \text{Radium }\beta, \quad 3,000 \text{ pc/l}\end{array}\right.$

Seashore mud 45 m. away from discharge point.......... Gross α, 5,800 pc/gm
Seashore mud 450 m. away from discharge point......... Gross α, 3,000 pc/gm

Region B
Ghatsila:

Tailings from plant................................. $\left\{\begin{array}{l}\text{Gross }\alpha, \quad 68 \text{ pc/gm} \\ \text{Gross }\beta, \quad 65 \text{ pc/gm}\end{array}\right.$

Stray stream from plant joining the river
Filtrate... $\left\{\begin{array}{l}\text{Gross }\alpha, \quad 60 \text{ pc/l} \\ \text{Gross }\beta, \quad 10 \text{ pc/l}\end{array}\right.$

Residue....................................... $\left\{\begin{array}{l}\text{Gross }\alpha, \quad 50 \text{ pc/gm} \\ \text{Gross }\beta, \quad 52 \text{ pc/gm}\end{array}\right.$

Vegetation from the river side....................... $\left\{\begin{array}{l}\text{Gross }\alpha, \quad 30 \text{ pc/gm} \\ \text{Gross }\beta, \quad 6 \text{ pc/gm}\end{array}\right.$

Mud from dry river bed............................ $\left\{\begin{array}{l}\text{Gross }\alpha, \quad 60 \text{ pc/gm} \\ \text{Gross }\beta, \quad 60 \text{ pc/gm}\end{array}\right.$

Region C
Alwaye plant I.R.E. Ltd.:

Effluent activity at discharge point.................... $\left\{\begin{array}{l}\text{Gross }\alpha, 25,000 \text{ pc/l} \\ \text{Gross }\beta, 10,000 \text{ pc/l} \\ \text{Radium }\alpha, \quad 9,000 \text{ pc/l} \\ \text{Radium }\beta, 10,000 \text{ pc/l}\end{array}\right.$

Sampling processes and analytical methods are briefly described in Appendix A. Appendix B gives particulars of the instruments used and counting procedures.

RESULTS

The environmental samples were analyzed for gross α, β, Sulkowitch β, potassium, radium α and β, thorium, and uranium.

The results of the survey (Borkar *et al.*, 1961; Kamath *et al.*, 1961) are summarized in the tables. There was little variation in the level of natural activity with time in most sample sites; as such, the values reported represent the general level arrived at after analyses of a large number of samples. Where individual variations are significantly different and the results tend to show a spread, the ranges of values are indicated.

In the selection of samples for background study, in regions of above-normal radiation background, care was taken to see that there was no direct radioactive contamination. In estimating process discharge and effluents release, samples were obtained at the discharge point or at known distances away from it.

TABLE 3A

COMMON SALT SAMPLES (INDIAN), REGION A

Location*	Description	Potassium (p.p.m.)	Uranium (p.p.m.)	Radium α (pc/gm)	Radium β (pc/gm)	Thorium (p.p.m.)
Rai Baynder.......	Karkus	606	0.055	0.016	0.68	
Rai Baynder.......	Kuppa	971	†	0.018	0.23	
Versova..........	White	971	†	0.03	0.27	0.016
Versova..........	White	1,119	†	0.018	†	
Bhandup..........	Karkus	861	0.009	†	0.1	
Belapur...........	Kuppa	606	†	0.021	†	
Belapur...........		1,250	†		†	
Uran.............	Karkus	348	†	0.016	†	0.02
Uran.............	Kuppa	1,001	0.008	0.008	0.26	
Kantherpara.......		†	0.022	0.007	0.28	
Sholapur..........		1,153	0.017	0.007	0.08	
Wadala salt pan....		1,422	0.05	0.003	0.11	

* All from Bombay State and produced from sea water by solar evaporation.
† Below detection limits.

TABLE 3B

COMMON SALT SAMPLES (FOREIGN)

Location	Description	Potassium (p.p.m.)	Uranium (p.p.m.)	Radium α (pc/gm)	Radium β (pc/gm)	Thorium (p.p.m.)
Foggia, S. Italy......	Sale Commune A	24	*	*	0.125	0.006
	Sale Commune B	17	*	0.01	0.05	
	Sale Raffinate	327	*	0.02	*	
	Sale Seetlo	203	*	*	*	
Vienna	Table salt	*	*	*	*	
Geneva.............	Table salt	*	0.005	0.02	*	
San Diego	Crude salt	2402	*	0.04	*	
Great Salt Lake......	Crude salt	296	0.039	0.027	*	
	Refined salt	203	*	0.019	*	
Grossmere, New Zealand......	Coarse grade	....	*	*	0.149	
	Pure grade	....	0.019	*	1.1	
Turkey.............	....	....	*	0.05	0.27	

* Below detection limits.

The results are reported for different regions in the tables for any particular type of the environmental sample. A comparative study of activity content in the environmental samples of different regions can thus be readily made.

DISCUSSION

WASTES RELEASE

Table 2 gives an account of the activity levels in wastes effluents released from processing plants in the different regions.

Region A.   The total liquid effluents activity released daily from the final thorium nitrate production does not exceed 1 mc. of α activity (Kamath and Pillai, 1958). Radium β activity is also not very high.

Region B.   The Ghatsila tailings discharged to the river have led to the contamination of the river bed in the area (Somasundarem and Kamath, 1958). However, the total activity disposed of is small and leachable activity is insignificant.

Region C.   The alkali digestion of monazite (Vasudevamurthy, 1962) releases non-thorium and non-uranium activity in the effluents. The gross activity levels were high. Gross α activity at the effluent discharge point in the plant was 25,000 pc/l, and β activity was 10,000 pc/l.

In all operations involving recovery of thorium, the activity content of radium-228 in the effluents must be carefully checked. Radium-228 is a bone seeker and reaches equilibrium with actinium-228 in 24 hours. The latter has high energy (β, 1.2 mev.; γ, 1.08 mev.) of emission. With radium-228, α-emitting radium and thorium-228 are generally present.

The radium α and β activities in the effluents were found to be 9,000 pc/l and 10,000 pc/l, respectively. These values were high compared with the following maximum permissible concentrations (168-hr. week, ICRP, 1959)—for unidentified nuclides, 100 pc/l; radium-226, 100 pc/l; and radium-228, 300 pc/l. However, no excessive build-up of activity was observed in the area from effluents released, since the total activity discharged daily was less than 2 mc. and the dilution available was adequate (Table 4).

SALT

Tables 3A and 3B show that the potassium content is high in the crude Indian salt. On an average, 15 gm. of salt is consumed daily per capita (Pillai and Ganguly, 1961). It has been reported that increased dietary potassium has protective action against the effects of increased sodium chloride (Meneely *et al.*, 1958), and the potassium to sodium

TABLE 4

DISPERSAL OF DISCHARGED EFFLUENTS AT ALWAYE, REGION C

LOCATION	GROSS ACTIVITY (pc/gm)		URANIUM (μgm/l)	THORIUM (μgm/l)	RADIUM α (pc/l)	RADIUM β (pc/l)
	α	β				
3 km. upstream from discharged point	*	*	0.07	*	0.22	*
2 km. downstream at Varapuzha	0.8	2.3	0.04	*	0.46	*
1 km. downstream at Methanam	0.24	4	0.07	*	0.58	1.1
1 km. downstream from discharged point	*	0.7	0.1	*	0.5	1.2

* Below detection limits.

TABLE 5

POTASSIUM AND SODIUM CONTENT OF SOME INDIAN FOODSTUFFS*

Foodstuffs	Botanical Name	Potassium (p.p.m.)	Sodium (p.p.m.)	Potassium-40 (pc/gm)
Rice, parboiled, home pounded	Oryza sativa	1,120	70	0.896
Rice, puffed	Oryza sativa	1,250	5,320	1
Whole-wheat flour	Triticum aestivum	3,750	130	3
Maize (dry)	Zea mays	2,750	140	2.2
Bengal gram	Cicer arietinum	7,970	1,250	6.4
Black gram (urd)	Phaseolus mungo	10,370	300	8.3
Green gram (mung)	Phaseolus aureus Roxb.	9,750	280	7.8
Lentil (masur)	Lens culinaris Medic.	5,500	280	4.4
Beans (dry)	Dolichos lablab	11,600	210	9.28
Spinach (dry)	Spinacia oleracea	35,870	41,950	28.69
Bitter gourd (dry)	Momordica charantia	29,000	1,720	23.2
Brinjal (dry)	Solanum melongerna	41,000	460	32.8
Cauliflower (dry)	Brassica oleracea botrytis	40,280	950	32.2
Lady's-finger (dry)	Abelmoschus esculentus	15,950	1,550	12.76
Potato (dry)	Solanum tuberosum	37,370	1,770	29.9
Onion (dry)	Allium cepa	7,570	5,250	6.06
Turnip (dry) (salgam)	Brassica rapa	37,750	6,020	30.2
Coconut	Cocos nucifera	6,000	980	4.8
Cashew nut	Anacardium occidentale	4,100	650	3.28
Plantain (ripe)	Musa paradisiaca	11,000	1,020	8.8
Chillies (dry)	Capsicum frutescene	19,950	980	15.96
Pepper	Piper nigrum	20,000	1,180	16
Jaggery (gur)		8,250	340	6.6
Tea extract		11,870	270	9.5
Coffee		11,750	1,160	9.4
Milk		1,000		0.8
Salt, crude (av.)		1,000	6×10^5	0.8

* Roychowdhury et al., 1962.

ratio has a physiological significance. Studies reported by Roychow-dhury *et al.* (1962) gives value for sodium and potassium content in some of the Indian foodstuffs. Their values are adopted in Table 5 to calculate potassium-40 content.

The radium α and β activity in salt appears to be well within the permissible values recommended for Group B(c) (ICRP, 1959) in India by Pillai and Ganguly (1961). (See the accompanying tabulation, in which all α and β are assumed as due to the quoted isotopes.)

	Indian Salt (pc/gm)	Foreign Salt (pc/gm)	Maximum Permissible Concentration
Radium-226 α.........	0.003–0.145	0.01–0.05	0.44
Radium-228 β.........	0.08 –0.68	0.05–1.1	1.32

Production of salt by solar evaporation is done in pans, and the crop is harvested by visual inspection. The salt produced as a result will not compare with standards of batch production. This is seen from the very differing results of different salt samples in the same region, and even from different batches in the same evaporating pan. The values should therefore be understood as being indicative of a range rather than as absolute.

The uranium contents in Indian salt samples in areas close to and those away from the Trombay site are nearly the same. This indicates that there is no build-up of uranium in the salt due to any chance release from the establishment in effluents.

SOIL, VEGETATION, WATER, AND BIOLOGICAL MATERIALS

Tables 6A–10B give the results of analyses of environmental samples of soil, vegetation, water, and biological materials.

Soil is an important subject of study for background radioactivity; its β and γ activities have been used to calculate the direct radiation dose (Lough and Solon, 1958). The average potassium content in Trombay soils is 10^{-3} gmK/gm of soil. The corresponding potassium-40 activity is 0.8 pc/gm. Lowder and Solon (1956) have reported potassium in the range of 10^{-3} to 3×10^{-2} gmK/gm soil. Fallout activity is very much lower than the total gross β activity.

Study of gross α and β activities in sea water and marine organisms indicates that concentration and uptake processes are taking place (Tables 8B, 9A, 9B, and 9C). The order of levels in sea water and fish

SOIL SAMPLES, REGION A

LOCATION	GROSS ACTIVITY (pc/gm)		SULKO-WITCH β (pc/gm)	URANIUM (p.p.m.)	THORIUM (p.p.m.)	RADIUM α (pc/gm)
	α	β				
Thana film industries...	⌈ 0.079	7–19	—	0.6	0.22	0.5
	⎜ —	—	—	0.14	0.31	0.35
	⎜ 0.083	—	—	0.73	0.11	0.34
	⌊ 0.01	—	—	0.22	0.41	0.33
Vihar Lake............	0.16	12	—	—	—	0.2
Powai Lake............	0.16	15	—	—	—	0.3
Chembur colony.......	0.45	47	—	1.6	—	0.2
Ooty (Madras)........	0.5	15–53	—	2.2	—	0.41
				0.03		
Santacruz............	0.24	12–60	0.028	—	—	0.62
Aarey milk colony.....	0.65	34	0.28	0.31	—	0.37
Sion.................	0.02	19	1.6	—	—	0.9
Vikhroli.............	0.04	6.9	0.07	—	—	—
Plant areas in Trombay region						
Uranium plant.......	52.6	21	18			
Faggots.............	0.09	320	8.4			
Gamma garden......	⌈ 0.06	5.5–70	2			
	⌊ 1.2		0.1			
Apsara..............	0.13	31–70	0.8			
CIR.................	0.3	12	0.4			
Project Phoenix......	0.2–1	2	0.7–0.9			

TABLE 6B

SOIL SAMPLES, REGIONS B, C, AND D

LOCATION	DESCRIPTION	GROSS ACTIVITY (pc/gm)		THORIUM α (pc/gm)	RADIUM α (pc/gm)
		α	β		
Region B					
Jadugoda mines ..	⌈ Soil	0.3	3.55	0.11	0.11
	⎜ Near office	1.67	7.8	1.5	1.2
	⎜ Drilling camp	0.8	1.6	0.4	1.6
	⎨ Mud from a pond	0.35	1.08	0.35	1.1
	⎜ Mud near a spring	1.1	3.7	0.46	0.34
	⎜ Suvarnarekha river bed	0.28	3.9	0.37	0.38
	⌊ Near dam site	0.31	2.4	0.24	0.24
Ghatsila.........	⌈ River-bed soil	62	60	—	—
	⌊ Water sediment	62	60	—	—
Region C (Travancore minerals)					
A-1.............	⌈ Soil opposite store	14	43	—	40
	⌊ Canteen	1.4	8.2	—	5
Region D (Udaisagar)					
Umra mines......	⌈ Near well 1	0.5	1.4	0.37	0.03
	⎜ Near well 2	0.7	1.4	0.38	0.02
	⎨ Near well 3	0.4	0.8	0.22	0.03
	⌊ Udaisagar Lake bank	0.4	1.7	0.25	0.03

TABLE 7

POTASSIUM-40 IN SOILS AND VEGETATION, REGION A

Location	Soil (pc/gm)	Vegetation (pc/gm)
Panvel Ulva bridge....................	0.88	—
Panvel Gadi bridge...................	0.38	—
Pen Goa road milestone 46..............	0.97	—
Pen Goa road milestone 52.............	0.26	—
Belapur Thana road LHS milestone 12.....	0.72	3.1
Belapur Thana road RHS milestone 12.....	1.1	1.4
Vile Parle railway crossing...............	0.78	—
Aarey milk colony unit 5................	—	3.5
Aarey milk colony unit 6................	—	2.8

TABLE 8A

WATER SAMPLES, REGION A

Location	Gross Activity (pc/l)		Sulko-witch β (pc/l)	Uranium (μgm/l)	Thorium (μgm/l)	Radium α (pc/l)
	α	β				
Thana Kalwa............	0.15–5	1.4	1.2–2	0.3	1.1	0.099
Powai Lake..............	0.32	1.6	—	0.02	2.7	0.05
Vihar Lake..............	—	2	—	0.03	2.4	0.06
Ulhas River.............	—	—	0.4–0.7	0.2	—	0.2 –0.4
Pathal Ganga............						0.2
Stream Ooty (Madras)....	1	0–13	—	—	—	0.14–0.18

VAJRESHWARI HOT SPRINGS NEAR BOMBAY

Location	Temperature ° C.	Gross Activity (pc/l)		Uranium (μgm/l)	Thorium (μgm/l)	Radium α (pc/l)	Potassium-40 (pc/l)
		α	β				
Ganeshpuri Agnikund..	55.5	0.35	4.05	—	0.74	0.4	6
Bhimeshwar kund......	50.5	0.20	—	0.20	0.53	0.25	5
Dutta kund...........	47	0.21	6.2	—	0.27	0.35	5

TABLE 8B

WATER SAMPLES, REGIONS B, C, AND D

LOCATION	GROSS ACTIVITY (pc/l)		URANIUM (μgm/l)	THORIUM α (pc/l)	RADIUM α (pc/l)
	α	β			
Region B (Jadugoda mines)					
Digdi Well water...........	0.48	*	—	0.21	0.16
Machuya pond.............	1	0.97	—	0.4	0.31
Suvarnarekha River.........	2.4	0.18	—	0.3	0.6
Matigara spring water.......	1.7	0.21	—	0.6	0.5
Mosabani drinking water....	1.8	0.6	—	0.3	0.4
ICC plant water............	0.36	*	—	0.12	0.16
Region C (I.R.E., Alwaye)					
Sample I..................	0.8	2.3	0.04	—	0.46
Sample II.................	0.24	4	0.07	—	0.58
Region D (Udaisagar)					
Udaisagar Lake.............	2.6	7.4	4.2	0.17	0.28
Fatehsagar Lake............	1.8	6.5	3.9	0.23	0.45
Mine colony...............	3.1	9.2	3.3	3.3	1.1
Mine water...............	2.8	11	36	2	1.5

* Below detection limits.

TABLE 8C

SEA WATER (SW) AND SEA MUD (SM) SAMPLES, REGION A

Location	Gross α (pc/l/sea water; pc/gm/sea mud)	Gross β (pc/l/sea water; pc/gm/sea mud)
Trombay naval jetty		
SW..................	1.1	44
SM..................	0.27	28
Peer Pau		
SW..................	0.5	18
SM..................	0.6	18
RCL jetty		
SW..................	0.98	39
SM..................	0.45	24
Mahul village		
SW..................	0.42	7.4
SM..................	0.3	15

TABLE 9A

BIOLOGICAL SAMPLES, 1958–60, REGION A

DESCRIPTION	GROSS ACTIVITY (pc/gm)	
	α	β
Fish		
Skate................	0.04	0.86
Bombay duck........	0.03	0.01
Golden anchovy......	0.4	0.01
Mackerel............	0.02	0.33
White sardine........	0.7	0.23
Sea bass............	0.93	0.38
Pomfret.............	0.006	2.2
Ribbon fish.........	0.06	2.2
Crustaceans		
Prawns.............	0.1	0.23
Blue crab...........	0.09	0.7
Shrimp.............	0.14	1.8
Mollusks		
Octopus............	0.07	2.3
Clam...............	0.01	0.7
Algae		
Green..............	0.02	0.3
Brown.............	0.15	3–7
Plankton		
Grass..............	0.5–5	2–4

TABLE 9B

BIOLOGICAL SAMPLES, 1960–62, REGION A

DESCRIPTION	GROSS ACTIVITY (pc/gm)	
	α	β
Pomfret (fish)...............	0.05	1.7
Prawns....................	0.09	2.2
Golden anchovy............	*	1.5
Prawns (Sh. Is.)............	0.02	7.2
Bombay duck..............	0.02	0.8
Black pomfret..............	0.02	1.6
Ruhu.....................	*	3
Tilapea...................	0.09	1.6

* Below detection limits.

TABLE 9C

BIOLOGICAL SAMPLES, REGIONS B, C, AND D

LOCATION	GROSS ACTIVITY (pc/gm) α	GROSS ACTIVITY (pc/gm) β	THORIUM α (pc/gm)	RADIUM α (pc/gm)
Region B				
Fish Jadugoda.............	0.012	1.56 (fresh water)	0.01	0.01
Suvarnarekha River........	0.05	2	0.06	0.03
Prawn muscle..............	0.07	2.3	—	0.05
Glyptostermum telchitta......	0.14	2.26	0.03	0.005–5
Inside of egg (hen).........	{0.003 / 0.016	0.54 / 1.2	0.02 / 0.02	0.05 / —
Inside of egg (duck)........	0.006	0.6	0.026	0.05
Region C (I.R.E., Alwaye)				
Wallage attu..............	{0.6 / 0.2	2.6 / 1.6	— / —	— / —
Carp puntius.............	{0.36 / 0.4	3.5 / 3.2	— / —	— / —
Lobster..................	0.2	—	—	—
Region D (Udaisagar Lake)				
Labeo specimen...........	0.008	0.4	Trace	0.001
Cirrhinear................	0.01	0.28	Trace	0.001

TABLE 10A

VEGETATION (GRASS), REGION A

LOCATION	GROSS ACTIVITY (pc/gm) α	GROSS ACTIVITY (pc/gm) β
Project Phoenix..............	{0.003 / 0.006 / 0.002	0.48 / 0.3 / 0.09
Aarey milk colony...........	{0.002 / 0.014	0.04 / 1.4
Kandivili...................	0.005	0.04
Peer Wadi..................	0.005	0.17
Powai area.................	0.005	1.5
Trombay Hill...............	0.02–0.06	14–66
Trombay: Vegetables		
Onions....................	0.008–0.3	0.6–0.86
Celery....................	0.05–0.7	3.2–9.6

for a large number of samples observed in region A is as shown in the accompanying tabulation. Values for vegetation and soil in a location

	Gross α (pc/gm)	Gross β (pc/gm)
Sea water.......	0.75×10^{-3}	0.27×10^{-1}
Sea fish.........	0.37×10^{-1}	0.29

in region A, where a large number of samples were analyzed, gave the representative values shown in the following tabulation. (See also Table 6A.)

	Gross α (pc/gm)	Gross β (pc/gm)
Aarey		
Soil..........	0.65	34
Grass........	0.002–0.14	1.4
Powai		
Soil..........	0.16	15
Grass........	0.005	1.7

TABLE 10B

BIOLOGICAL SAMPLES (GRASS), REGIONS B AND D

LOCATION	DESCRIPTION	GROSS ACTIVITY (pc/gm) α	GROSS ACTIVITY (pc/gm) β	URANIUM (p.p.m.)	THORIUM α (pc/gm)	RADIUM α (pc/gm)
Region B						
Murgahutta.......	Jadugoda mines	0.075	0.8	—	0.011	0.01
Rigdi............	Jadugoda mines	0.03	1.7	—	0.015	0.01
Swasspar.........	Jadugoda mines	0.05	2.4	—	0.007	0.007
Mossabani........	Jadugoda mines	0.02	0.73	—	0.026	0.045
Mati Gora........	Jadugoda mines	0.03	1.9	—	—	—
Ghatsila.........	Jadugoda mines	0.05	1.6	—	0.008	0.017
Jadugoda mines rice		0.03–6	1–2	—	0.09	0.02–0.3
Region D						
Umra mines.......	Maize	0.01	0.43	0.001	0.045	0.002
	Wheat	0.007	0.14	0.045	0.003	—
	Barley	0.007	0.89	0.001	0.046	0.002

SOIL-PLANT RELATIONSHIPS

   Interpretations of results to establish these relationships in natural systems are difficult because of the very different and heterogeneous systems involved. The soil minerals have radioactivity contents that vary greatly with the pattern of the constituents, location, and depth of the samples. The leaching characteristics of the clay minerals would be difficult to reproduce in the laboratory in order to study the mineral constituents available for plant uptake under natural conditions. The radioactivity content, as determined by leaching with different reagents, is significantly different, depending on whether the treatment restricts its action to the surface of the soil particle or possibly attacks the mineral. Since the evidence is that growing vegetation receives its nutrients and trace elements from the leachable fraction, any attempt at correlation between radioactivity content in plant and in soil must first insure a proper method for extracting leachable activity without disturbing the mineral. We propose to study these aspects in more detail.

RADIUM

A wide range of values for radium-226 have been reported in the literature (Lowder and Solon, 1956). For soils, values given are 0.09–0.8 pc/gm; for ocean sediments, the radium content is reported as 1–22 pc/gm. In the present study the radium content of soil in region A was found to be 0.3–1 pc/gm by acid leaching. For fresh-water supplies in region A the values observed were 0.05–0.2 pc/l. Hot-water springs (temp. 47°–55.5° C.) gave slightly higher values: 0.25–0.4 pc/l. In the high-background regions B, C, and D the radium content ranged from 0.4 to 1 pc/l (Table 8A).

The estimation of radium activity by radiochemical methods becomes complicated owing to the growth of daughters and the presence of several radium isotopes. In environmental samples from region C, the chemically separated radium can contain radium isotopes from the uranium and thorium decay series: radium-226 and radium-223 (from uranium) and radium-228 and radium-224 (from thorium decay). Radium-224 reaches quick equilibrium with its α-emitting daughters radon-220 and polonium-216. For measurement of β-active radium-228, the precipitate is set aside for 24 hours to reach equilibrium (Gopinath and Singh, 1960) with actinium-228. In this period there is a significant growth of lead-212 and bismuth-212, both of which are β emitters, the latter having an energy of emission of 2.25 mev.

If the precipitate containing radium-228 is set aside for a month or so for the complete decay of radium-224 (half-life of 3.65 days) and flashed to drive off radon, radium-228 can be counted without interference from other β emitters. Radium-226 is measured by counting the α activity within one hour after flashing. Contributions from uranium-235 daughters or from build-up of thorium-228 can be neglected.

For routine analyses of large numbers of samples this procedure piles up a huge backlog of work, and considerable time delays are involved. We propose to employ the α spectrometer for the study in future and correct the β activity of radium-228 measured by calculating the lead-212–bismuth-212 component activity from the polonium-216 count obtained from the spectrometer observations.

GROSS α MEASUREMENTS

An evaluation of the gross α method (Kamath and Soman, 1958) was made to determine the type of the radioactivity mostly present in the gross α precipitate. It was found that, in an environment not contaminated with transuranic elements, only thorium activity is collected by the lanthanum trifluoride carrier precipitate. This was confirmed using solutions of uranium and thorium and estimation of the

TABLE 11A

STUDY OF METHODS: COMPARISON OF GROSS α
AND THORIUM ACTIVITIES

Location	Gross α	Thorium α
Spring water, Umra........	0.31 pc/l	0.33 pc/l
Drinking water............	0.31 pc/l	0.30 pc/l
Milk, Udaisagar...........	0.29 pc/l	0.23 pc/l
Soil, Udaisagar............	0.36 pc/gm	0.26 pc/gm
Soil, Jadugoda.............	0.30 pc/gm	0.29 pc/gm
Soil, Jadugoda near road	1.1 pc/gm	1.6 pc/gm
Bottom soil, Jadugoda......	0.38 pc/gm	0.36 pc/gm

NOTE.—Gross α activity is that of thorium, and β activity of the gross α precipitate will give an idea of thorium-234 and therefore of uranium-238. In a solution of uranium, a gross α determination gave the following results:

Taken		*Found*
40	α	..
38	β	36

The method has been checked by spiking with thorium-234.

gross α activity and also with tracer experiments using thorium-234. Table 11A gives a comparative study of total thorium α activity determined in some soil samples by the gross α method and by a specific ion-exchange method for thorium (Pillai, 1963).

The gross α precipitate contains all naturally occurring thorium isotopes: thorium-232, thorium-230, thorium-234, thorium-228, thorium-227, and thorium-231. The contributions from the last two are not considered because they are low. The β activity of the precipitate is due to thorium-234 and thorium-231, which, assuming equilibrium, gives a measure of uranium-238 content. Natural thorium (thorium-232, thorium-228) content can be estimated in the precipitate by the conventional spectrophotometric method (Kamath and Soman, 1958) after conversion of the lanthanum trifluoride carrier into perchlorate and extraction of thorium with TTA in benzene. For practical purposes other thoriums do not interfere in this estimation, and a measure of natural thorium can be obtained.

A few samples of soils were analyzed for gross β activity by the Sulkowitch method during the surveys. The gross β-activity measurements using the Sulkowitch reagent indicate that these appear to be in good agreement with the values for strontium-90 activity in the sample (Table 11B). Sample measurements for fallout activity in the country are reported elsewhere (AEET, 1962).

TABLE 11B

STUDY OF METHODS: COMPARISON OF SULKOWITCH
AND STRONTIUM-90 ACTIVITIES IN SOILS

Location	Sulkowitch β (pc/gm)	Strontium-90 (pc/gm)
Vihar Lake	0.2	0.3
Pu-plant	0.3	0.24
Vikhroli	{0.07 0.09 0.15	{0.07 0.09 0.16

ACKNOWLEDGMENTS

The authors are indebted to Shri A. S. Rao, director, Electronics Group, and Dr. A. K. Ganguly, under whose guidance and encouragement this work was carried out. Thanks are due to several others who participated in these studies and standardization of procedures, in particular to Smt. Kamala Rudran, Shri K. C. Pillai, and the counting group. Thanks are also due to Shri P. S. Chhabria for assistance in preparing the paper.

REFERENCES

ATOMIC ENERGY ESTABLISHMENT TROMBAY. 1962. Measurements on the Environmental Radioactivity in India from Nuclear Weapon Tests: Data Collected during 1956–1961. AEET/A.M./26.

BHARATWAL, D. S., and G. H. VAZE. 1958. Radiation dose measurements in the monazite areas of Kerala State in India. Proc. 2d United Nations Internat. Conf. on Peaceful Uses of Atomic Energy, **23**:156–58.

BORKAR, M. D., *et al.* 1961. Environmental Survey Results 1960–61. Atomic Energy Establishment Trombay, Rpt. AEET/HP/Environ/2.

DUNSTER, H. J. 1958. The disposal of radioactive liquid wastes in coastal waters. Proc. 2d United Nations Internat. Conf. on Peaceful Uses of Atomic Energy, **18**:390–99.

GOPINATH, D. V., and H. SINGH. 1960. Activity Build-up of Naturally Occurring Radioactive Materials (Part I). Atomic Energy Establishment Trombay, Rpt. AEET/HP/Th/3.

INTERNATIONAL COMMISSION ON RADIOLOGICAL PROTECTION. 1959. Reports.

Committee 2. Permissible dose for internal radiation. New York: Pergamon Press.

KAMATH, P. R., and K. C. PILLAI. 1958. Activity of Wastes from Indian Rare Earths Factory at Trombay. Atomic Energy Establishment Trombay, Rpt. AEET/HP/RWD/1.

KAMATH, P. R., K. C. PILLAI, and A. K. GANGULY. 1959. Operation Sawdust. Atomic Energy Establishment Trombay, Rpt. AEET/HP/RWD/5.

KAMATH, P. R., and S. D. SOMAN. 1958. Bioassay Procedures at Trombay Establishment. Atomic Energy Establishment Trombay, Rpt. AEET/HP/BA/1.

KAMATH, P. R., *et. al.* 1961*a*. Tidal Movement Studies at Thana Creek. Atomic Energy Establishment Trombay, Rpt. AEET/HP/RWD/6.

———. 1961*b*. Background Radiation Surveys for Trombay Operations. Atomic Energy Establishment Trombay, Rpt. AEET/HP/Environ/1.

KUCHELA, K. S., and L. H. PESHORI. 1960. Application of bidirectional dekatron tubes in low level beta activity measurements by cancellation technique. Nuc. Instr., & Methods, 7:179–83.

LOUGH, S. A., and L. R. SOLON. 1958. The natural radiation environment. *In* Radiation Biology and Medicine (chap. 17), W. D. CLAUS, ed. Reading, Pa.: Addison-Wesley Pub. Co.

LOWDER, W. M., and L. R. SOLON. 1956. Background Radiation: A Literature Search. USAEC Rpt. NYO-4712. Pp. 43. Washington, D.C.

MENEELY, G. R., W. L. OLSOBROOK, J. M. MERRILL, O. J. BALCHUM, R. H. WEILAND, and C. O. T. BALL. 1958. Metabolism of the major mineral elements of the animal body. *In* Radiation Biology and Medicine (chap. 30), W. D. CLAUS, ed. Reading, Pa.: Addison-Wesley Pub. Co.

NATIONAL ACADEMY OF SCIENCES. 1957. Effects of Committee on Atomic Radiation on Oceanography and Fisheries. Washington, D.C., Nat. Acad. Sci., Nat. Res. Council, Pub. 551. Pp. 137.

PILLAI, K. C. Radioactive and Inactive Tracers Studies in Bombay Harbour Bay. MS thesis, Bombay University, 1963.

PILLAI, K. C., and A. K. GANGULY. 1961. Evaluation of Maximum Permissible Concentration of Radioisotopes in Sea Waters of Bombay. Atomic Energy Establishment Trombay, Rpt. AEET/HP/R/11.

PILLAI, K. C., T. SUBBARATNAM, and A. K. GANGULY. 1958. Tidal Movement and Water Renewal Rates in Bombay Harbour Bay. Atomic Energy Establishment Trombay, Rpt. AEET/HP/RWD/2.

PILLAI, T. N. V. Estimation of Ultra-trace Quantities of Heavy Radioactive Elements by Ion Exchange and Non-radioactive Toxic Compounds by Chemical Methods. M.S. thesis, Bombay University, 1963.

REVELLE, R., T. R. FOLSOM, E. D. GOLDBERG, and J. D. ISAACS. 1955. Nuclear science and oceanography. Proc. Internat. Conf. on Peaceful Uses of Atomic Energy, 13:371–80.

ROYCHOWDHURY, S. P., J. A. KHAN, and K. N. ROSE. 1962. Minor mineral contents of some foodstuffs. J. Proc. Inst. Chemists (India), 34 (Pt. 2): 89–93.

SOMASUNDAREN, S., and P. R. KAMATH. 1958. Special Survey Report. Atomic Energy Establishment Trombay, Rpt. AEET/HP/Survey/19.

VASUDEVAMURTHY, S., *et al.* 1962. Alwaye Plant of Indian Rare Earths Ltd. Atomic Energy Establishment Trombay, Rpt. AEET/HP/Survey/66.

VOHRA, K. G., P. ABRAHAM, D. N. KELKAR, and M. C. SUBBARAMU. 1960. A Preliminary Survey for Airborne Radioactivity and Background Gamma Radiation at Bombay. Atomic Energy Establishment Trombay, Rpt. AEET/AM/18. Pp. 36.

APPENDIX A

ANALYTICAL METHODS

SAMPLE PREPARATION

Soil. The soil sample is crushed to pass 100 mesh and heated at 400° C. for 3 hours.

Water. Drinking- and spring-water samples are filtered through No. 1 Whatman paper and evaporated to dryness. The residue is ashed at 400° C. Sea water is taken up after filtration.

Vegetation. Vegetation and biological samples are ashed at 400° C. If white ash is not obtained, the ash is treated with nitric acid and heated again. The ash is treated for analyses.

Salts. Salts are dissolved in water and filtered. The filtered solution is used.

GROSS β ACTIVITY

One hundred mg. of the dried material is transferred to an aluminum tray (2.54 cm.) and counted for β activity in an end-window Geiger-Müller counter.

OTHER ESTIMATIONS

The dried residues are treated as follows for other estimations. The sea-water samples are taken as such.

The residue is digested with 1:1 nitric acid for 1 hour. The supernatant is decanted. Digestion is repeated with a second addition of nitric acid. The supernates are combined.

GROSS α ACTIVITY

Take 5 gm. equivalent of the extract and adjust the acid strength to about 2 per cent. Add 100 mg. bismuth carrier and precipitate bismuth phosphate by gradual addition of 2 ml. phosphoric acid. Allow the precipitate to settle. Dissolve the precipitate in hydrochloric acid. Add purified 2.5 mg. lanthanum carrier and precipitate lanthanum trifluoride by addition of hydrofluoric acid. Wash the precipitate, transfer to stainless steel planchet, and count for α activity in a zinc sulfide scintillation counter.

SULKOWITCH β

Take 5 gm. equivalent of the extract and evaporate to about 25 ml. Add 1:1 ammonium hydroxide until the solution is neutral. Add Sulkowitch reagent (consisting of a mixture of oxalic acid, ammonium oxalate, and glacial acetic acid) until the pH is about 4–5. Allow the precipitate to settle. Centrifuge, wash twice with 2 per cent ammonium oxalate solution and twice with distilled water. Transfer the precipitate to an aluminum tray and count for β activity.

RADIUM

To 15 gm. equivalent of the extract add 3N citric acid; add lead, barium carriers, 200 mg. and 6.7 mg., respectively. Neutralize the solution with ammonia and heat to 50° C. Add sulfuric acid until a precipitate is formed. Separate by centrifuging lead and barium sulfates and dissolve them in ammoniacal EDTA. Precipitate barium sulfate with acetic acid. Wash, transfer the precipitate to stainless-steel planchet, flame, and count for α activity.

The precipitate is counted for β after 24 hours. The activity observed should be corrected for β growth from radium-226 to obtain the value for radium-228. Radium-226 contributes about 30 per cent equivalent of its α activity to total β in 24 hours.

Alternatively, the precipitate is set aside for the complete decay of radium-224 and daughters, flashed, and counted for radium α and β activity.

POTASSIUM

Take 15 gm. equivalent of the extract. Add sodium carbonate solution until all insoluble carbonates and hydroxides are precipitated. Allow to settle; filter through buchner. Acidify the filtrate with acetic acid, add 5 gm. of cobaltous nitrate, and cool the solution in ice. To the cooled solution add saturated sodium nitrate solution and stir vigorously. (If the precipitate does not appear immediately, keep the solution overnight in cooled condition.) Decant the supernatant, centrifuge, and wash the precipitate with 5 per cent acetic acid solution and finally with distilled water. Transfer the precipitate to a tared dish and weigh. Calculate potassium-40 content from the precipitate, $K_2NaCo(NO_2)_6 \cdot H_2O$, by multiplying the potassium content by 0.00012.

URANIUM

To the extract (15 gm.) in the separating funnel add aluminum nitrate in 0.5M nitric acid to get approximately 2M in aluminum nitrate. Add 90 ml. ether and gently shake to get a good mixing (ammonium nitrate may also be used for salting). When it has settled, draw off the salt phase and discard. Transfer the ether extract from the top into a beaker containing 2 ml. water and evaporate off ether over a steam bath in a hood. Take up the water extract for fluorimetric estimation of uranium.

THORIUM

1. *Spectrophotometric Method.* The lanthanum trifluoride precipitate obtained by the gross α method described earlier is taken up for thorium extraction. Decompose the anthanum trifluoride by heating with 10 ml. nitric acid and 3 ml. perchloric acid. Dissolve the residual perchlorates in 3 ml. nitric acid and quantitatively transfer to a separating funnel containing 10 ml. of 10 per cent TTA in benzene. Shake and extract thorium into the organic fraction and discard the aqueous layer. Back extract thorium in 2N nitric acid by repeated washing of the organic phase. Thorium nitrate is reconverted into the perchlorate by heating with perchloric acid. Thoronol reagent is added, and the transmittancy read on a Beckman model DU spectrophotometer at 545 μ in comparison with a standard.

2. *Ion Exchange Method.* Evaporate to dryness 5 gm. equivalent extract. Dissolve residue in 10 ml. 4N nitric acid and pass through a 2-gm. column of conditioned Dowex-50 resin. Wash the column with 10 ml. nitric acid followed with water. Elute thorium from the column with 0.5M oxalic acid (10 ml.). Evaporate the eluate to dryness with concentrated nitric acid to destroy oxalic acid. Evaporate the solution, and transfer the residue to a stainless-steel planchette, dry, and count for α activity.

APPENDIX B

SAMPLING, COUNTING, AND INSTRUMENTS

There were systematic periodic sampling programs in region A, that is, the Trombay Establishment environment. In other areas samples were obtained at random during surveys. The effluent activity data were obtained over a number of observations, and, since the processes were standardized, the samples did not vary from day to day. Studies conducted in a mine region showed that environmental activity did not vary from time to time but showed slight variations from site to site in the regions. The differences were not very significant.

For α measurements, a zinc sulfide scintillation counter was used (5 cm. in diameter), giving a background of 0.06–0.07 counts per minute. The average efficiency, as determined using a plated natural uranium source, was 25 per cent. The samples were counted for 100 α counts or for 2 hours. The sample activity was reported at 66 per cent confidence limit and 10 per cent counting error.

For β counting, a low-background Geiger-Müller counter with a background of 0.5–1 counts per minute was used, operating at a power level of 1,200–1,250 volts and employing the cancellation technique (Kuchela and Peshori, 1960). The distance between the sample and window was 3 mm., and the thickness of the mica window was 2.5 mg/cm². The average efficiency of the counter determined using a potassium source was 15 per cent. The samples were counted for a minimum of 400 counts or 2 hours and reported at 66 per cent confidence and 5 per cent error of counting.

M. GEYH AND S. LORCH

59. Determining the Contribution of γ Rays to the Natural Environmental Radiation at Ground Level

THIS REPORT deals with the results of measurements carried out over a period of several years concerning the contribution of γ rays to the natural environmental radiation in the Federal Republic of Germany. These measurements were performed at the request of the former Bundesministerium für Atomkernenergie (Federal Atomic Energy Ministry) by the Department of Geophysics in the Bundesanstalt für Bodenforschung (Federal Geologic Research Office), Hannover. Apart from purely scientific investigations, these measurements were also intended to determine the maximum natural radiation dose that man is exposed to.

For the purpose of getting a general view, these measurements were made not only in the open terrain but also in and around buildings. Figure 1 shows the region of investigation, consisting of an area of approximately 250,000 sq. km. Most of the measurements here were made along roads. In order to survey less accessible terrain, a jeep was employed to assist in the work.

NATURAL ENVIRONMENTAL RADIATION

In order to find the detector most suitable for these measurements of the natural radiation, about 25 samples were selected from the area under observation; they were examined with a single-channel γ spectrometer using a sodium iodide well-type crystal. By this means, it was shown that the radiation was predominantly from potassium-40, radium-226 and daughter elements, and thorium-232 and daughter elements.

M. GEYH is physicist and S. LORCH is doctor at Niedersächsisches Landesamt für Bodenforschung, Hannover, Germany.

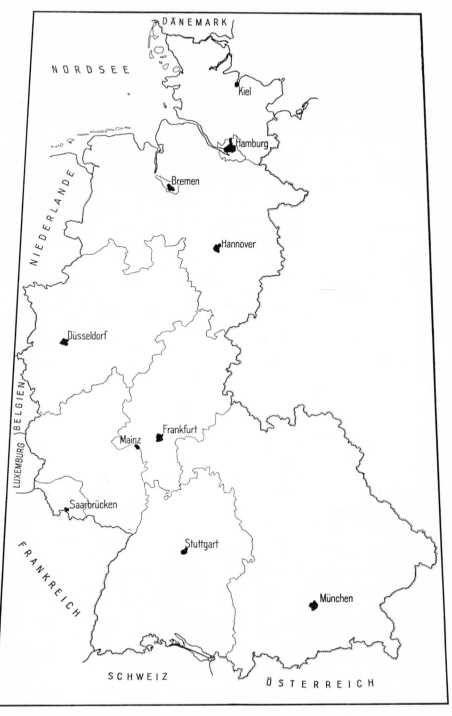

F𝐈𝐆. 1.—The area of the Federal Republic of Germany

Our results correspond, within the limits of error, to the statements of Jeffreys (1959), who found:

In basalt
  Radium 0.86×10^{-12} gm/gm; thorium 6.34 p.p.m.
In granite
  Radium 3.26×10^{-12} gm/gm; thorium 30.5 p.p.m.

More accurate quantitative statements could not be obtained with our measuring apparatus.

Since the energies of γ quanta emitted by the above-mentioned radionuclides vary between 0.4 and 2.6 mev., a scintillation counter was considered the best detector.

DESCRIPTION OF THE APPARATUS AND CALIBRATION

A suitable scintillation counter for road measurements was developed and constructed in our laboratory; it was firmly mounted outside the jeep at a height of 90 cm. above the ground. Figure 2 shows the layout on the jeep.

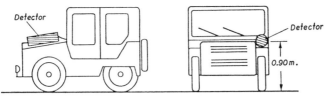

FIG. 2.—View of the car complete with car-borne counter

This detector is fitted with a sodium iodide crystal 7.5 cm. in height and 7 cm. in diameter. The recording of the signal was done graphically by means of an automatic Varian recorder whose paper transport, connected to the wheels of the vehicle, was kept in proportion to the driving speed. When stations were entered and left, and also at other marked points, location references were noted on the recording paper. The time constant of the apparatus was about 2.1 sec. The driving speed was restricted to 30 m.p.h. for the purpose of avoiding any suppression of anomalies. In addition, we constructed, for measurements in buildings, a portable counter with a rate meter. The sodium iodide crystal used had a height of 5 cm. and a diameter of 2.5 cm. The time constant in the normal measuring range of 0–10 μr/hr was about 2.1 sec.

The relative variations of the environmental radiation were such that scientific investigations could be made with values given in c.p.m. The additional problem, however, was to find out the dose of the environmental radiation. This necessitated a special calibration of the detector. Since, as is known, there exists a complex mixture of radio-nuclides of variable composition and since, moreover, the recording of the γ quanta is a function of energy for the scintillation counter (Fig. 3), it was at first assumed that the dose efficiency constant is variable for geologic bodies of variable composition. To check this assumption, measurements on the dependence of the dose efficiency on the counting rate of our probes were carried out at places where the three major components occurred separately. The appropriate places for such measurements were the talus of potash mines, of pitchblende, and of thorium mines. The work was done in co-operation with the Radiologische Institut Freiburg (RIF) and the Max-Planck-Institut in Frankfurt (MPF). The comparison measurements

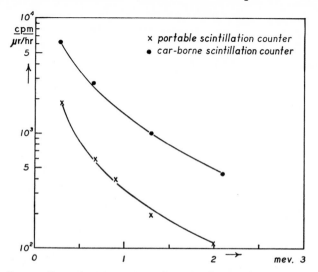

Fig. 3.—Dependence on energy of γ rays of the detector (NaI[Tl])

were made relative to an absolutely calibrated ionization chamber. The results revealed a linear dependence of the dose efficiency on the counting rate (Fig. 4). The standard deviation was calculated at 30 per cent, which was considered adequate for the solution of the problem. The error includes variations in the relative amounts of radio-nuclides (different rocks) of the objects under investigation, statistical errors, drift in the electronics, and other factors.

For the car-borne counter we obtained a conversion factor of

350 c.p.m. = 1 μr/hr for the portable counter;

1,620 c.p.m. = 1 μr/hr for the counter on the jeep.

Furthermore, the directional characteristic of the car-borne detector was measured by following the γ count rate from a cobalt-60 source as a function of angle (see Fig. 5). Thus it became possible to compute as a first approximation, on a road 8 meters wide with adjacent flat terrain, that 70 per cent of the detected radiation emanates from the road, assuming a uniform distribution of radioactivity. This value may vary considerably if there are elevations or depressions in the neighborhood of the road. More exact statements will be given in a paper to be published in the near future. It was established that the terrain within a circumference of 20 meters produces about 95 per cent of the radiation detected by the instrument (Fig. 6).

The "background" for the portable counter was 270 c.p.m., which

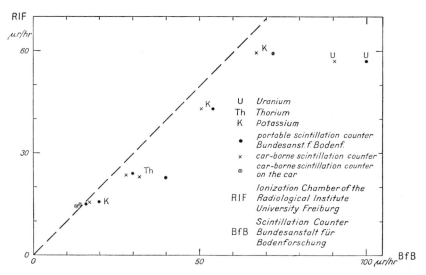

Fig. 4.—Comparison of calibration values between ionization chamber and scintillation counter.

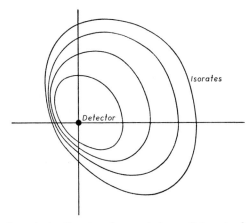

Fig. 5.—Section through the directive characteristic parallel to the face of the counter

corresponds to 0.76 μr/hr. The "background" measurements were carried out over the Bodensee. Since the "background" in most measurements is less than 10 per cent of the total counting rates, it can be said that the measurements reflect essentially the terrestrial contribution. This seems justified regarding the error of ±30 per cent. The considerable discrepancy between our measured contribution from cosmic rays and the value given in the literature (3.2 μr/hr) for middle

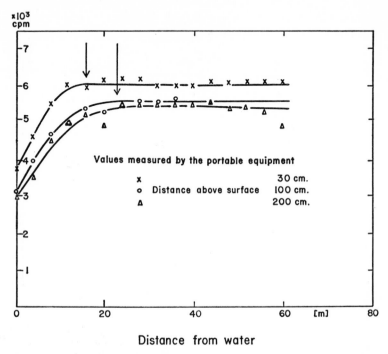

Fig. 6.—Ascertaining the comprehended zone during measuring with a portable scintillation counter.

latitude at sea level is explained by the energy dependence of the scintillation-counter efficiency.

All measured values were corrected for the absorption of the vehicle. The effect of dirt particles or contamination in the vehicle could not be determined. For the purpose of correcting drift in the rate meter, frequent calibrations were made with a radium-226 source.

RESULTS

It has frequently been stated that in car-borne measurements the radiation of the closer environs or of the terrain beyond the road cannot be determined; this seemed obvious from the knowledge of the directional characteristic of the detector and from consideration of the geometric factor.

To check this assumption, we measured the counting rate at more than 400 points with our portable counter exactly in the center of the road and also in the field about 25–30 meters distant from both road edges, and then we took the average. However, in comparing the

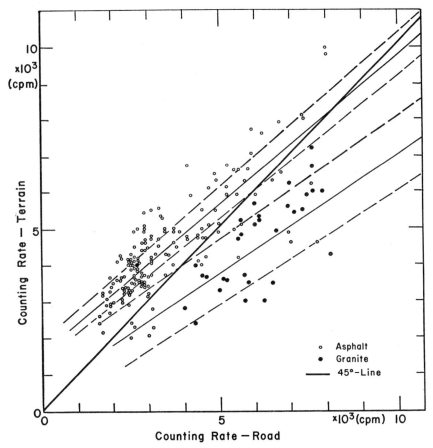

FIG. 7.—Comparison between measuring values on the road and in the terrain

values obtained on the road with those on the terrain, the relationship between the two cannot be denied (Fig. 7). In the attempt to clear up this phenomenon, the assumption was expressed that such relation arises through the use of local material in road construction. The assumption is justified if, as a first approximation, a road surface is assumed to be an absorber with a specific characteristic activity, to be neglected in the case of asphalt but with considerable influence in the case of a granite and slagstone surface. Thereby the specific activity appears as a parameter, as can be seen from Figure 7. A formulation of such a model takes the form:

$$N = \frac{N_f - N_r}{\eta_r + \eta_f e^{-\mu d}},$$

where $N_f =$ the measured value in the infinitely extended field, $N_r =$ the measured value on the infinitely extended road surface, $d =$ the thickness of the road surface, $\mu =$ the specific absorption coefficient, η_f, $\eta_r =$ the fraction of the total counting rate originating from the field or from the road, respectively, and $N =$ the counting rate measured on a road crossing the terrain.

This expression indicates that the intersection points of the curve sets always lie upon the negative ordinate axis or at zero. Furthermore, it is evident that the slopes of curves could never be less than 45°. In the interpretation, we obtained—assuming a linear regression—slopes for asphalt of 0.81 and for granite of 0.69, whereas the corresponding one for field paths oscillated around the 45° line, as was expected. The intersection points of the ordinates near 1,800 c.p.m. and 500 c.p.m. point to an error in the statement of the model. If, however, the characteristic activity of the road scintillator is taken as a function of terrain values (local or autochthonous material), the solution of this discrepancy is then reached. A regression of asphalt values bit by bit from low to higher values shows that the slope of the asphalt curve gradually decreases from 1.65 to 0.81, which means that the curve is not simply linear.

The difference between the granite and the asphalt curves is statistically significant (about 1,300 c.p.m.). The mean error of the single measurement was found to be about 25 per cent. The results show that, for considerable variances, the different morphology can be made responsible to only a small extent; they are, rather, due to considerable activity variations of soils with a homogeneous outer appearance.

GEOLOGY

The differences in measuring values are essentially due—apart from the geometric factor—to the alterations in the composition of the soil material, namely, the geology. The following standard values were established: for sandstone, 5 μr/hr; for effusive rock in the Kaiserstuhl, 15 μr/hr; and for alluvium on the banks of the Rhine River, 7 μr/hr. In addition, there is a variety of geologic material, each of which covers a relatively large specific activity interval.

Generally, such regions overlap each other. Figure 8 illustrates the variations for a section taken up in the Hegau (Germany), related to the measuring values of peat. As indicated above, a clear correlation of the dose efficiency to the different geological bodies is not possible.

However, experience has shown that in most cases contacts of different geological layers can be located very well, which is a substantial help in the reconnaissance survey. Furthermore, our measurements showed that the maximum dose internationally accepted for civil populations (60 μr/hr) is exceeded in some regions by an unusually high activity of the rocks. We found 10 μr/hr as a mean value for Germany. The normal activity interval in Germany varies between 2 and 24 μr/hr, except for extreme cases.

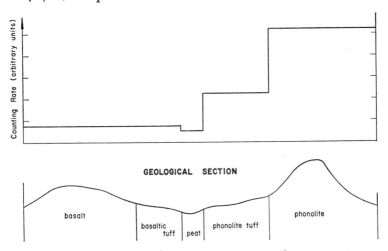

Fig. 8.—Hegau section of activity (counting rate referring to peat)

FURTHER OBSERVATIONS

Finally, the measurements also gave evidence that the values obtained on streets in populated places are often higher than those obtained on roads in the open terrain. The additional radiation from buildings is a minor effect only. The greater effect is to be assigned to the frequently greater radioactivity of road surfaces, owing to the use of granite or basalt. The phenomenon quoted in literature, that is, that counting rates on overgrown lands are higher than on those not overgrown, was confirmed by our measurements. Generally, the increase amounts to 7 per cent. At first it was assumed that this phenomenon could be explained by the growth of plants. This assumption was then contradicted by measurements on plain fields (without growth), which did not reveal any alteration in the values either before or after fertilization with potash (now plowed in). On the other hand, according to a personal communication from Frank, it can be assumed that the plants contain all solid constituents of the fallout and thus are partly responsible for the observed increase of measured values. A clear solution has not been reached as yet.

The opinion that rain and snow are of no consequence on the surface dose was not confirmed by our measurements. On the contrary, we attained differences up to 25 per cent in values when recordings were made at one and the same place, with and without snow covering. Reworking or weathering processes cannot be made responsible for this feature because, after the melting of the snow, the values had their original quantity again.

Measurements in Buildings

Measurements made in dwellings gave counting rates that were slightly smaller in rooms of new buildings than were those in old ones (prewar buildings). The cause is assumed to be the variable composition of building material and its distribution in construction. In general, the values averaged from 1,150 measurements for buildings in Bavaria were about 7.7 ± 0.2 $\mu r/hr$, and hence they do not differ essentially from the average dose in Germany. The average obtained in streets of 98 cities in Bavaria was 6.5 ± 0.2 $\mu r/hr$; for roads 8,770 km. in length, 5.3 $\mu r/hr$.

Conclusions

In summary, it can be stated:

a) The portable scintillation counter represents a useful auxiliary means for geological mapping, and it will gain in importance in the near future.

b) It is also proved that an approximate measurement of the activity in the terrain can be attained from the road. The accuracy in measuring lies near 30 per cent.

c) The measurements have proved that, for a quantitative expression of the natural radiation in a large district, a closely spaced survey can be dispensed with because the results obtained with considerably fewer points were sufficiently reliable.

References

Hultqvist, B. 1956. Studies on naturally occurring ionizing radiation. Kungl. Svenska Vetenskapsakad. Handl., Vol. **6**, Ser. 4, No. 3. Pp. 125.

Jeffreys, H. 1959. The Earth: Its Origin, History and Physical Constitution, pp. 295–96. 4th ed. Cambridge: Cambridge University Press.

Lauterbach, R. 1962. Gamma-Spektrogramme von Basalten. Geophys. & Geol., 4:100–101.

Solon, L. R., W. M. Lowder, A. Shambon, and H. Blatz. 1960. Investigations of natural environmental radiation. Science, **131**:903–6.

United Nations. 1958. Report of the United Nations Scientific Committee on the Effects of the Atomic Radiation. New York: General Assembly Official Records, 13th sess., Supplement 17. Pp. 228.

EDUARDO RAMOS AND MARGARITA CELMA

60. Environmental Radioactivity Surveys in Spain

THE ENVIRONMENTAL RADIOACTIVITY STUDIES started by the Board of Nuclear Energy in Spain in 1958 were intended to cover two fundamental objectives: (a) the monitoring of the contamination around the reactor installed at La Moncloa and around the uranium treatment plant in Andújar (Jaen) and (b) the surveillance of the levels of exposure of the general population caused by environmental radiation from all sources (fallout, natural radioactivity, etc.). Some of the measurements carried out are common to both objectives, even though they differ in each case as to their frequency and technique of measurement.

The accompanying diagram (Fig. 1) presents the plan of the environmental radioactivity survey established by the Board of Nuclear Energy, whose goal is to obtain a view of the whole situation within the national boundaries. The distribution of the sampling points in the Iberian Peninsula is given in Figure 2 and Table 1.

ATMOSPHERIC ACTIVITY

The measurements of atmospheric radioactivity were begun with continuous recording apparatus that have been maintained in the surveys of the nuclear establishments. They are gradually being replaced by methods of discontinuous measurements of successive periods of 24 hours in those places where the measurement of the activity in the air has as its object the detection of radiation levels dangerous to the general population. The continuous recording devices were obtained from the firms of Landis and Gyr (Switzerland) and Frieseke and Höpfner (Germany). The discontinuous measuring devices have been

EDUARDO RAMOS is head and MARGARITA CELMA is senior chemist, Medicine and Protection Section, Nuclear Energy Board, Ministry of Industry, University City, Madrid, Spain.

constructed mainly with local components (some still use Landis pumps). The paper used to sample dust is Whatman 41 (68 mm. in diameter). The total quantity of air inspired is some 500 m³/hr.

Sampling is carried out by changing the filter papers each 24 hours at the different stations and sending them to the central laboratory at La Moncloa, where they are measured without any pretreatment of the sample. These samples are counted on a β scintillation counter whose efficiency, determined with strontium-90–yttrium-90 standard, is 27.2 per cent. The background count with 5 cm. of lead shielding is 140 c.p.m. The results of the measurements are expressed in pc/m³.

Figure 3 gives the average monthly values for the station in Madrid. The existing situation in the atmosphere at different times is clearly reflected. The β activity, which showed a high value at the end of

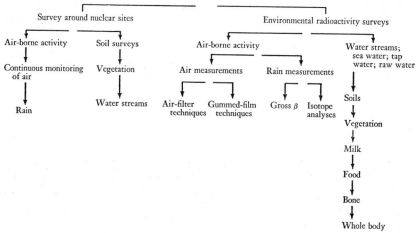

Fig. 1.—Environmental radioactivity survey plan

Fig. 2.—Sampling points in the Iberian Peninsula

TABLE 1

DISTRIBUTION OF SAMPLING POINTS IN THE IBERIAN PENINSULA

Sample	Measurement	Sampling Location	Frequency
AIR — Continuous	Gross β	Madrid, Las Palmas, Valencia Barcelona, Andújar, Bilbao	Continuous
AIR — Discontinuous (24 hr.)	Gross β	Madrid, Barcelona, Valencia, S. Fernando	Daily
AIR — Gummed film	Gross β; γ spectroscopy	Madrid, Valencia, Las Palmas Barcelona, S. Fernando	Weekly
RAIN WATER	Gross β; γ spectroscopy	Madrid, Barcelona, Valencia Las Palmas, S. Fernando	Each precipitation
RAIN WATER	Isotope analyses (Sr-90, Sr-89, Cs-137, Ce-144)	Madrid, S. Fernando	One bulk sample, monthly
DRINKING WATER	Gross β; Ra-226; Sr-90	Madrid (3 points) Barcelona (6 points)	Monthly
SOILS, VEGETATION; MILK	Sr-90, Cs-137	Madrid (16 points—soils)	Two times/year
SOILS, VEGETATION; MILK	I-131 (milk)	Madrid (4 points—veget., milk)	Two times/year
SOILS, VEGETATION; MILK	I-131 (milk)	Madrid (4 points—soil)	Monthly
SOILS, VEGETATION; MILK	I-131 (milk)	Barcelona (6 points)	Two times/year
SOILS, VEGETATION; MILK	I-131 (milk)	La Coruña	Two times/year
SOILS, VEGETATION; MILK	I-131 (milk)	Vitoria	Two times/year
SOILS, VEGETATION; MILK	I-131 (milk)	Oviedo	Two times/year
SOILS, VEGETATION; MILK	I-131 (milk)	Burjasot (Valencia)	Two times/year
SOILS, VEGETATION; MILK	I-131 (milk)	Badajoz	Two times/year
SOILS, VEGETATION; MILK	I-131 (milk)	Córdoba	Two times/year
SOILS, VEGETATION; MILK	I-131 (milk)	Murcia	Two times/year
SOILS, VEGETATION; MILK	I-131 (milk)	Tenerife	Two times/year
SOILS, VEGETATION; MILK	I-131 (milk)	Valladolid	Two times/year
SOILS, VEGETATION; MILK	I-131 (milk)	Jerez dela Frontera	Two times/year

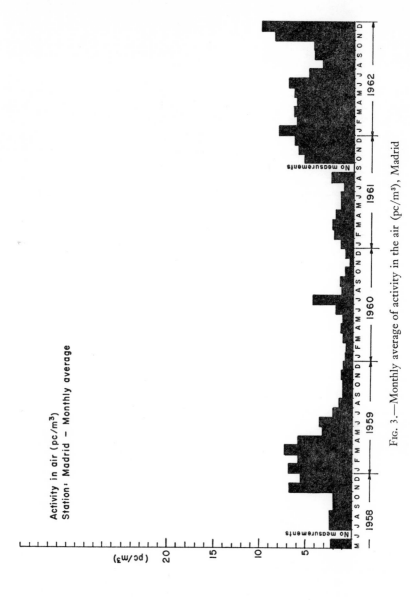

Activity in air (pc/m³)
Station: Madrid — Monthly average

Fɪɢ. 3.—Monthly average of activity in the air (pc/m³), Madrid

1958 that was maintained into the spring of 1959, began to decrease at the beginning of the summer of 1959. During all 1960 it remained low, with the exception of the sharp spring rise during the months of June and July. The spring rise was almost imperceptible in 1961. After the autumn of that year, the renewal of the nuclear tests and the accelerated rate with which they were conducted raised the values of activity to those existing in 1958. At the beginning of 1962, the values surpassed the previous highs in 1958 and 1961, descended slightly in the summer and autumn, and rose again at the end of 1962. The average corresponding to December of 1962 showed the highest value of activity in the atmosphere for the entire period studied.

The values obtained at the other sampling stations follow, in general, the same variations; thus, their presentation is omitted here.

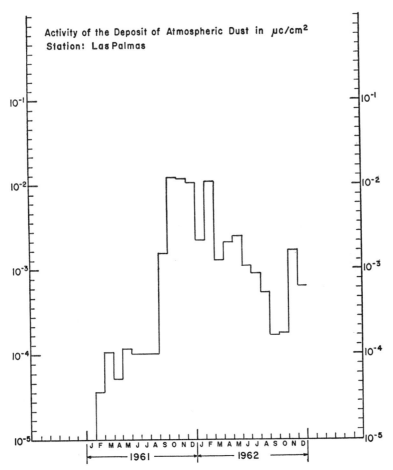

FIG. 4, *A*.—Activity of the deposit of atmospheric dust in $\mu c/cm^2$, Las Palmas

The determinations of the total β activity of deposited atmospheric dust by the gummed-film method were started in 1961. At all stations there is a parallel variation with time in the values determined, and there is a sharp increase in the values at the beginning of September, 1961 (Fig. 4).

With this method, the papers are exposed during one week and are sent to the central laboratory at La Moncloa, where they are ashed at 450–500° C. and counted on conventional equipment that has been calibrated with potassium-40. The decay of these samples takes place during a period of at least three months. The results are expressed in $\mu c/m^2$.

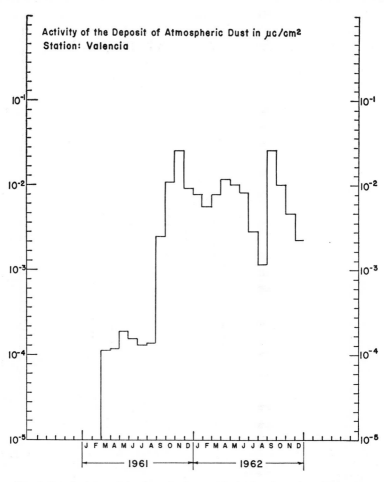

Fig. 4, *B.*—Activity of the deposit of atmospheric dust in $\mu c/cm^2$, Valencia

Those samples showing high specific activity were studied in the multichannel pulse-height analyzer (RCL; 256 channels) to identify their γ emitters. In all these samples, two preponderant fission products, ruthenium-106–rhodium-106 and zirconium-95–niobium-95, are observed. Other γ emitters, such as ruthenium-103, cerium-144–praseodymium-144, and barium-140–lanthanum-140, appear occasionally in some of the samples studied.

ACTIVITY IN RAIN WATER

The measurement of total β activity in rain water is carried out for each individual precipitation. The water is filtered, an adequate aliquot portion is concentrated to dryness, and the activity corre-

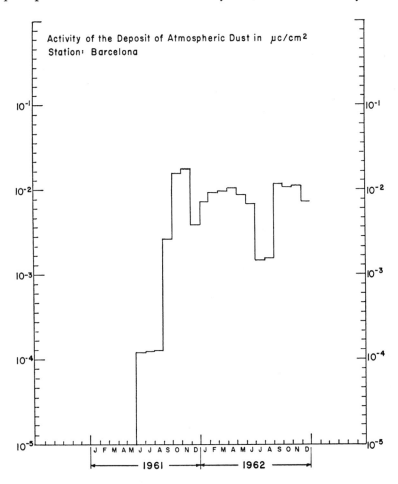

FIG. 4, *C.*—Activity of the deposit of atmospheric dust in $\mu c/cm^2$, Barcelona

sponding to liquids and solids in suspension is measured separately. Then the total activity is given in $\mu c/l$ or $\mu c/m^2$. The equipment used has been calibrated with respect to strontium-90–yttrium-90. According to the activity in the sample, different equipment is used.

For samples of high activity, conventional apparatus, such as the Geiger-Müller counter with a window of 3.7 mg/cm², is used. The apparatus has efficiencies varying between 18 and 23 per cent for strontium-90–yttrium-90 and backgrounds of 14–19 c.p.m.

Samples of lower activity are measured in anticoincidence equipment with a flow counter or in a sealed Geiger-Müller counter with an automatic sample changer. The background of this equipment is

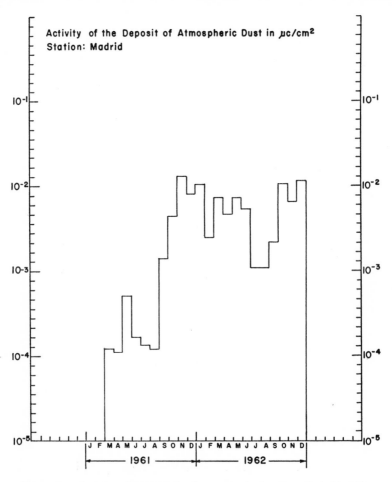

Fɪɢ. 4, *D.*—Activity of the deposit of atmospheric dust in $\mu c/cm^2$, Madrid

2.5 c.p.m., and its efficiency for strontium-90 is 56 per cent when the flow counter is used and 30.4 per cent when the Geiger-Müller counter is used.

Other anticoincidence equipment, of the type described by LeVine *et al.* (1959), is composed of three β counters of the "pancake" type, with a double semicircle of guard counters, and is provided with mercury shielding around the β counters and a total shield of 10 cm. of lead. It has a background of 0.5 c.p.m. It is generally reserved for the measurements of activity due to specific isotopes.

In monthly samples composed of rain water, determinations of specific isotopes are carried out by radiochemical methods, such as determinaton of γ emitters by spectrometry. The spectra obtained

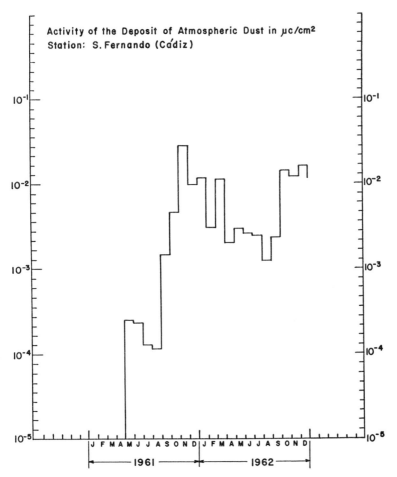

Fig. 4, *E*.—Activity of the deposit of atmospheric dust in μc/cm², San Fernando (Cádiz)

TABLE 2

ISOTOPIC ANALYSES

Date	β Total (μc/l)	Strontium-90 (pc/l)	Strontium-89 (pc/l)	Cesium-137 (pc/l)	Cerium-144 (pc/l)
RAIN WATER: MADRID					
12/60–1/61.......	2.4×10^{-4}	1.6	....	Trace	....
2/61–4/61.........	6×10^{-5}	5.2	....	....	15.6
8/4/61–29/4/61....	2.1×10^{-5}	2	Trace	....	3.4
4/61–6/61.........	6.3×10^{-5}	3.3	0.18	....	4.3
6/61–7/61.........	....	4.3	2.9	....	22.4
7/61–9/61.........	1.3×10^{-4}	2.1	3.8	....	1.7
28, 29/9/61.......	2.08×10^{-3}	....	....	....	15.5
19/9/61...........	2.1×10^{-4}	3.2	Trace	....	52
9/61–10/61.......	4.7×10^{-4}	4.4	25.9	....	8.8
					8.27
10/61–11/61......	3×10^{-4}	1.2	87.1	....	15.1 Ce[141]
					14.7 Zr[95]
11/61–12/61......	9×10^{-4}	4.3	268.6	....	48
12/61–1/62........	9.6×10^{-4}	0.6	6.5	....	48.5
1/62–3/62.........	6.3×10^{-4}	10	156	3.1	65
8–3—21–3/62.......	8.8×10^{-4}	12	403.5	2.6	6.3
3/62–4/62.........	3.6×10^{-4}	15.2	110.3	0.8	2.8
4/62–5/62.........	4.6×10^{-4}	4.3	57.1	0.2	29.9
5/62–6/62.........	3.1×10^{-3}	14.8	56.5	6.7	140
13/6—16-6/62.....	6.2×10^{-4}	22.4	146.2	2.7	3.7
6/62–7/62.........	....	....	328.1	34.7	....
7/62–9/62.........	....	106.3	....	3.7	17.4
6/62.............	....	19.7	....	35.8	....
10/62............	....	7.3	....	4.3	....
11/62............	....	....	....	23.7	....
SNOW: MADRID					
2/11/61...........	5.6×10^{-4}	2.55	141	Trace	0.16
15/11/61..........	5.7×10^{-4}	0.26	2.4	2.6	0.32
1/3/62............	1.5×10^{-4}	0.41	1.8	1.6	0.3
9/4/62............	2.98×10^{-4}	4.1	82.2	1.9	2.5
RAIN WATER: SAN FERNANDO					
10/61–11/61.......	8.3×10^{-5}	1.9	108	0.73	17.8
11/61............	5.2×10^{-4}	2.5	159	....	45.8
12/61............	1.03×10^{-4}	6.1	74.1	1.33	4.6
12/61............	2.1×10^{-4}	3	56.6	0.75	6.8
12/61............	....	5.4	216	....	53.2
1/62.............	2.11×10^{-4}	15.7	21.7	0.72	0.54
1/62–2/62.........	4.09×10^{-4}	2.6	1,688	0.8	6.5
3/62.............	3.93×10^{-4}	3.1	209	0.56	8.3
3/62.............	1.5×10^{-4}	1.5	250	3.2	2
RAIN WATER: BARCELONA					
6/62.............	....	25	....	....	....
7/62.............	....	6.6	....	2.4	....
8/62.............	....	203	....	3.8	....
9/62.............	....	9.9	....	7.3	....
10/62............	....	8	....	4.4	....
11/62............	....	7.2	....	4.4	....

from rain water do not differ essentially from those obtained from atmospheric dust. There are observed two photopeaks at energies of 0.513 mev. and 0.765 mev., corresponding to the γ emission of ruthenium-106–rhodium-106 and zirconium-95–niobium-95, respectively. Occasionally, the following have been observed to be present: cerium-144–praseodymium-144, ruthenium-103, yttrium-90, and barium-140–lanthanum-140. The spectra obtained from the liquid fraction do not differ from those obtained from the solid residue of the precipitation, except, as is to be expected, in the case where the presence of iodine-131 is detected more frequently in the residue, which was not submitted to any treatment.

The long-lived isotopes determined by radiochemical procedures are strontium-90, cesium-137, strontium-89, and cerium-144. All the strontium analyses are based on the known separation of the strontium from the group of the alkaline-earths by means of 75 per cent nitric acid, purification of the contaminating activities, and the formation of yttrium-90 that arises from the purified strontium. The method used for the cerium analyses is based on the bromate ion oxidizing the cerous ion to ceric ion, followed by the precipitations of the latter by iodate. For the determinations of cesium, its nearly insoluble chloroplatinate, which permits separation from other alkalies, is utilized. The *Manual of Standard Procedures* (HASL, 1962) describes all these methods; therefore they are not given in detail here. Table 2 shows some values obtained from these analyses.

ACTIVITY IN POTABLE WATER

Just at the renewal of the nuclear tests in 1961 and as a survey of the contamination introduced in the potable water, measurements of total β activity were started on the waters that are provided to the populations of Madrid and Barcelona. The values obtained vary around 10^{-5} μc/l, as may be seen in Figure 5, which reflects the measurements for the Madrid station. The same occur in Barcelona.

STRONTIUM-90 ACTIVITY IN SOILS,
VEGETATION, MILK, AND BONES

As a contribution to the evaluation of possible contamination of the human organism by the long-lived radionuclides, strontium-90 determinations were started on the materials mentioned. At the beginning of the surveys, sampling was limited to a zone around Madrid having a radius of 25 kilometers. Later sampling was extended to include all the stations listed on the map (Fig. 2). These sampling points were selected in the directions of cardinal compass points and their

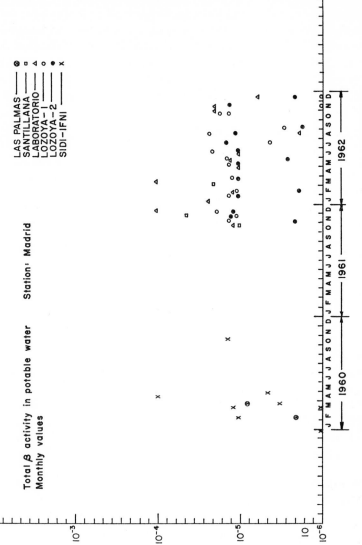

Fig. 5.—Monthly values of total β activity in potable water, Madrid

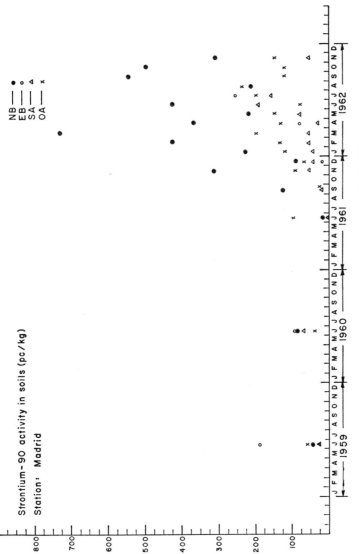

Fig. 6.—Strontium-90 activity in soils (pc/kg), Madrid

TABLE 3

STRONTIUM-90 ACTIVITY
IN SOILS

Station	Date	gm Ca/kg	pc Sr^{90}/kg	pc Sr^{90}/gm Ca	No. Samples
Madrid					
NA........	June, 1959	1.5241	80.4	52.62	2
NB........	"	1.1240	47.36	41.90	3
NEA......	"	13.230	109.0	15.78	2
NEB......	"	2.6891	67.76	25.19	3
EB........	"	4.8510	184.76	27.77	3
SEA......	"	1.1180	64.60	57.78	3
SA........	"	15.1868	30.00	1.97	2
SOA......	"	3.5412	35.58	10.01	2
SOB......	"	0.6706	62.79	93.61	2
OA........	"	1.3062	60.30	46.16	3
NOB......	"	2.1902	31.40	14.40	2
NB........	June, 1960	1.2040	87.0	72.2	3
NEA......	"	20.500	66.2	3.22	3
NEB......	"	1.400	82.3	58.8	3
EB(X).....	"	17.5286	86.8	5.05	3
SEA......	"	0.8600	73.2	85.1	2
SEB......	"	27.470	Trace	Trace	2
SA........	"	9.300	74.5	8.01	3
SB........	"	23.470	Trace	Trace	2
SOA......	"	5.502	212.4	38.6	3
SOB......	"	0.460	69.4	148.6	3
OA........	"	0.730	38.08	52.1	3
OB........	"	0.860	20.0	23.3	3
NOA......	"	0.470	129.5	275.5	3
NOB......	"	2.270	47.2	20.8	3
NB........	June, 1961	1.090	12.0	11.0	2
NEA......	"	21.890	....	....	.....
NEB......	"	1.600	212.4	132.0	1
SEA......	"	1.120	4.4	6.6	1
SEB......	"	26.060	6.2	0.32	2
SA........	"	13.650	3.0	0.22	2
SB........	"	24.300	19.1	0.75	3
SOA......	"	4.600	74.2	16.1	2
SOB......	"	0.654	98.0	149.8	3
OA........	"	0.8580	96.5	113.6	3
OB........	"	0.8600	21.6	25.1	3
NOA......	"	0.5006	Trace	....	2
NOB......	"	1.7533	225.6	128.6	2
NB........	Sept., 1961	1.210	125.5	103.6	3
NEA......	"	18.070	38.5	2.03	3
NEB......	"	1.723	235.0	202.8	3
SEA......	"	0.910	13.4	14.7	2
SEB......	"	21.700	20.8	0.9	2
SA........	"	10.610	18.4	1.73	1
SB........	"	23.300	....	....	.....
SOA......	"	5.389	132.0	24.0	3
SOB......	"	0.543	112.0	206.9	2
OA........	"	0.947	21.7	22.8	3
OB........	"	0.891	119.4	133.9	3
NOA......	"	0.510	....	....	.....
NOB......	"	2.560	129.8	50.7	3

TABLE 3—*Continued*

Station	Date	gm Ca/kg	pc Sr⁹⁰/kg	pc Sr⁹⁰/gm Ca	No. Samples
Madrid—*Continued*					
NB........	Nov., 1961	1.305	314.9	247.0	3
SA........	"	9.800	54.4	5.7	3
OA........	"	0.753	85.3	113.3	2
NB........	Dec., 1961	0.866	85.2	98.3	1
EB........	"	2.935	13.8	4.8	2
SA........	"	10.55	40.5	3.9	3
OA........	"	1.161	61.1	52.6	....
NB........	Jan., 1962	1.442	225.4	157.2	3
SA........	"	7.646	44.8	5.8	2
OA........	"	1.560	115.9	74.2	3
NB........	Feb., 1962	1.114	423.6	360.2	3
SA........	"	10.312	53.2	5.1	2
OA........	"	2.204	131.0	59.4	2
NB........	Mar., 1962	1.200	730.8	609.0	2
SA........	"	8.895	54.4	6.6	3
OA........	"	0.818	198.0	242.0	3
NB........	Apr., 1962	1.220	369.0	302.0	2
NEA......	"	17.541	89.8	5.1	2
NEB......	"	1.800	88.4	49.2	2
EB........	"	2.735	77.6	28.3	3
SEA......	"	0.922	Trace	Trace	1
SA........	"	9.788	31.6	3.22	3
SOB......	"	0.614	76.6	124.7	2
OA........	"	0.951	126.3	133.0	2
OB........	"	0.862	203.1	235.6	3
NOA......	"	0.600	257.3	429.3	3
NOB......	"	2.411	310.2	128.6	2
NB........	May, 1962	1.198	217.4	181.8	3
SA........	"	9.650	79.6	8.2	2
OA........	"	0.871	145.2	166.7	....
NB........	June, 1962	1.202	....	....	....
SA........	"	9.542	188.0	19.7	2
OA........	"	0.889	75.4	84.8	3
NB........	July, 1962	1.203	452.0	367.8	2
EB........	"	2.735	256.0	93.5	2
SA........	"	9.580	163.6	17.0	3
OA........	"	0.872	195.0	223.0	3
NB........	Aug., 1962	1.291	209.3	162.1	3
SA........	"	9.488	....	....	....
OA........	"	0.890	239.8	269.4	3
NB........	Sept., 1962	1.112	540.0	486.0	2
SA........	"	9.792	....	....	....
OA........	"	0.865	120.4	139.2	3
NB........	Oct., 1962	1.112	495.0	445.0	3
SA........	"	9.500	....	....	....
OA........	"	0.890	119.2	133.8	3

TABLE 3—*Continued*

Station	Date	gm Ca/kg	pc Sr^{90}/kg	pc Sr^{90}/gm Ca	No. Samples
Madrid—*Continued*					
NB........	Nov., 1962	1.229	311.0	253.2	3
SA........	"	9.792	57.6	5.8	2
OA........	"	0.865	145.0	167.6	3
Córdoba......	Oct., 1962	....	494.0	....	3
Tenerife......	Nov., 1962	....	173.2	....	3
Las Palmas....	16/12/59	29.670	10.58	0.35	3
	"	26.040	52.60	2.01	3
	"	10.102	21.54	2.132	3
	"	11.250	117.18	10.40	3
	3/3/60	5.784	12.26	2.25	2
	"	8.942	138.80	15.90	2
	28/3/60	10.150	94.0	9.60	2
	15/4/60	2.666	11.85	4.48	3
	"	8.604	68.80	8.00	2
	20/5/60	5.134	127.3	24.8	3
Sidi Ifni......	14/12/59	3.560	25.36	7.12	2
	"	4.666	25.60	5.5	3
	"	15.968	210.60	13.19	2
	8/1/60	3.049	23.45	7.5	3
	8/2/60	7.895	18.88	2.39	2
	27/2/60	5.930	39.60	6.66	3
	"	11.880	28.82	2.41	3
	10/3/60	2.944	43.60	14.8	2
	"	8.360	52.10	6.23	3
	25/3/60	6.352	83.02	13.08	3
	"	1.490	3.05	2.05	2
	14/4/60	3.662	74.80	20.6	2
	"	2.096	6.80	3.2	2
	1/5/60	3.310	61.80	18.7	2
	"	5.374	22.57	4.2	3
	15/5/60	2.315	68.80	30.4	2
	"	7.337	29.78	3.94	2
	"	2.950	86.40	27.8	2
	30/7/60	8.477	167.6	20.6	2
	"	6.089	89.5	14.6	3
	15/8/60	4.833	106.0	21.9	3
	21/9/60	2.631	63.5	24.2	3
	13/10/60	2.990	84.8	29.8	2

TABLE 3—*Continued*

IN MILK

Station	Date	pc Sr^{90}/l	pc Sr^{90}/gm Ca		No. Samples
Madrid.......	5/5/59	3.53	2.88		3
	"	4.89	6.1		3
	"	1.065	....		2
	30/9/61	9.2	10.5		2
	"	7.0	8.0		3
	6/11/61	18.0	....		2
	11/11/61	9.9	11.5		2
	11/1/61	12.2	....		2
	28/1/63	21.0	....		2
Barcelona.....	5/10/61	2.76	4.0		3
	"	0.96	1.1		3
	"	1.93	2.6		3
	"	2.20	3.5		3
	12/11/62	12.7	....		2
	"	9.2	....		2
	"	12.4	....		2
Las Palmas....	16/12/59	8.29	10.3		2
	12/2/60	2.34	2.9		2
	3/3/60	0.91	1.1		2
	24/3/60	1.45	1.8		2
	19/5/60	0.63	0.8		1
Sidi Ifni......	12/2/60	1.15	1.4		1
	1/3/60	1.10	1.4		3
	19/5/60	0.91	1.4		3

IN VEGETATION

Station	Date	gm Ca/gm Ashes	pc Sr^{90}/gm Ashes	pc Sr^{90}/gm Ca	No. Samples
Madrid.......	2/4/60	....	0.772	....	3
	"	....	1.190	....	3
Sidi Ifni......	15/12/59	....	....	2.42	3
	"	....	....	0.77	3
	11/3/60	....	....	1.33	3
Barcelona.....	15/11/61	0.0283	5.57	196.6	3
	"	0.0302	1.01	33.4	3
	"	0.0168	3.6	214.0	3
	"	0.0268	5.1	192.5	3
	"	0.0301	5.1	169.0	3
	"	0.0265	2.6	98.0	3
	"	0.0333	3.0	90.0	3

IN ANIMAL BONE

Station	Date			pc Sr^{90}/gm Ca	No. Samples
Madrid.......	5/5/60			9.73	2
Las Palmas....	16/12/59			1.00	2
	13/4/60			1.68	2

bisectrices. The points lie on two concentric circles having radii of 15 and 25 kilometers. Thus there are 16 points. These points are sampled in the spring and fall for strontium-90 in soil, bone, vegetation, and milk. Attempts are made to have the dates of sampling for all substances coincide.

Since the renewal of nuclear-weapons testing in the fall of 1961, 4 points, 1 each in the directions of the cardinal compass points, were chosen for monthly sampling in order to follow the evolution of cumulative strontium-90 deposition in the soils. For example, in the milk samples, the method for determination of strontium-90 proceeds from the ashes of milk. Simultaneously there is carried out the method of non-incineration, which utilizes the precipitation of the proteins of milk by TCA (trichloroacetic acid) and the separation of the same. In the liquid part, the yttrium is extracted with TBP (tributylphosphate). This method seems on some occasions to be advisable for its rapidity. The results obtained by both procedures agree to within about 10 per cent.

The measurements are performed as described above in the anticoincidence equipment, which has background of 0.5 c.p.m. Some of the values obtained for activity of strontium-90 in soils for the Madrid Station, expressed in pc/kg, are given in Figure 6. Table 3 presents the values obtained at some stations for activity due to strontium-90 in soils, milk, vegetation, and animal bone.

The Board of Nuclear Energy is continuing and extending the surveys described to put into effect the whole of the planned national network. In this brief note we give only information related to those measurements of routine character that have already been made.

REFERENCES

ALEXANDER, L. T., E. P. HARDY, and L. MACHTA. 1961. Strontium-90 on the Earth's Surface: Summary and Interpretation of a World-wide Soil Sampling Program. U.S. Atomic Energy Commission Rpt. TID-6567. Pp. 28.

BRYANT, F. J., A. C. CHAMBERLAIN, A. MORGAN, and G. S. SPICER. 1956. Radiostrontium Fallout in Biological Materials in Britain. Great Britain, Atomic Energy Research Establishment Rpt. AERE HP/R-2056. Pp. 49.

BRYANT, F. J., A. MORGAN, and G. S. SPICER. 1958. Radiostrontium in Soils, Herbage, Animal Bone and Milk Samples from the United Kingdom. Great Britain, Atomic Energy Authority Research Group, Atomic Energy Research Establishment, Rpt. HP/R-2730. Pp. 23.

CAMPO, R. G. DEL. 1962. Técnica de contaje de bajo fondo, utilizando plásticos de centelleo. Junta de Energía Nuclear Rpt. S.M.P./N.2, pp. 1–9.

CHAMBERLAIN, A. C., R. J. GARNER, and D. WILLIAMS. 1961. Environmental monitoring after accidental deposition of radioactivity. Reactor Sci. & Technol., 14:155–67.

DUNSTER, H. J., H. HOWELLS, and W. L. TEMPLETON. 1958. District surveys following the Windscale Incident, October, 1957. Proc. 2d United Nations Conf. on Peaceful Uses of Atomic Energy, 18:296–308.

HEALTH AND SAFETY LABORATORY. 1962. Manual of Standard Procedures, with Revisions and Modifications to August, 1959. N.Y. Operations Office, USAEC Rpt. NYO-4700.

HINZPETER, A. 1961. Routinenverfahren zur Sr-90 Bestimmung mit Ionenaustauschern ("Routine methods for Sr-90 determination with ion exchangers"). *In* Radiostrontium, Strahlenschutz No. 18. Pp. 34–42. Munich: Gersbach & Sohn Verlag.

KIEFER, H., and R. MAUSHART. 1961. Choice of equipment for monitoring emergency situations. Seminar on Agricultural and Public Health Aspects of Radioactive Contamination in Normal and Emergency Situations. Scheveningen, Netherlands.

KULP, J. L., and A. R. SCHULERT. 1962. Strontium-90 in Man, Vol. 1. Columbia University, Lamont Geological Observatory, Geochemistry Laboratory, Summary.

KULP, J. L., and R. SLAKTER. 1958. Current strontium-90 level in diet in United States. Science, 128:85–86.

LANGHAM, W. H., and E. C. ANDERSON. 1958a. Strontium-90 and its uptake in man. Bull. Suiss Acad. Med. Sci., Vol. 14. *Cited in:* Environmental Contamination from Weapon Tests, pp. 282–307. Health and Safety Lab., USAEC, Rpt. HASL-42.

———. 1958b. Entry of radioactive fallout into the biosphere and man. Bull. Schweiz. Akad. med. Wiss., 14:434–78.

LEVINE, H. D., L. CHARLTON, and R. T. GRAVESON. 1959. Low Background Nuclear Counting Equipment. New York Operations Office, Health and Safety Lab. (AEC) Rpt. HASL-60. Pp. 30.

LOCKHART, L. B. 1959. Radiochemical Analyses of Air-Filter Samples Collected during 1958. Washington, D.C., Naval Res. Lab. Rpt. NRL-5390. Pp. 25.

Methods of Radiochemical Analysis. 1959. Report of a Joint WHO/FAO Expert Committee. Atomic Energy Series, No. 1. Pp. 116.

ORGANIZATION FOR EUROPEAN ECONOMIC CO-OPERATION (OEEC). 1962. Matières radioactives en suspension dans l'air. Rpt. 40770, pp. 6–31.

PALMER, G. H. 1958. Environmental Survey around an Atomic Energy Site. United Kingdom Atomic Energy Authority, Research Group, Atomic Energy Research Establishment, England, Rpt. AERE-HP/R-2742. Pp. 33.

Schriftenreihe des Bundenministers für Atomkernenergie und Wasserwirtschaft. 1961. Symposium des Sonderausschuss Radioaktivität. Strahlenschutz No. 18. Pp. 15–340.

(Numbers in parentheses refer to chapter authored by discussant.)

Lockhart (15, 18, 23) asked Foote (45, 46) about the possibility of looking at the distribution of the radium daughters in the atmosphere by facing his detector upward and perhaps correlating these results with the activity missing at the ground surface. Foote replied that this would not be feasible without a very large iron or lead shield to remove the ground radiation, since the contribution of the aerial radioactivity is only a small percentage of the total readings.

Lucas (17) noted that Eisenbud (52) was using urine analyses as a measure of the intake of radioactive elements and questioned the fraction of intake that might be excreted in urine. Eisenbud (52) replied that he had only mentioned the possibility of making the urine analyses; some urine specimens have been collected but not analyzed. He suggested that the concentration of radioactivity in the urine could not be used as a measure of either intake or body burden; however, urine analyses might be useful to detect anomalously high daily intakes by local populations. Eisenbud (52) estimated that only a small percentage of the intake might be excreted in the urine.

Adams (30, 34) emphasized the need for large samples in regional studies of the Brazilian type. Although thorium-rich monazite dominates the external radiation environment in the Brazilian studies, the important amounts of uranium daughters found internally can arise from very small percentages of xenotime and other uranium-rich minerals associated with monazite. Only large samples of some kilograms or the use of portable spectrometers (34) could give representative data about the environment. Eisenbud (52) discussed the suggestion that a whole-body counter with a shadow shield might prove useful in Brazil, as it had in the cesium-137 studies in Finland and Alaska. He concluded that the radium-226 body burdens may not be much above the 0.002 microcuries that can be detected by such instruments in expert hands. Hill (24) noted that Eisenbud's (52) data indicated twice as much thorium-228 as radium-228 in steer bones. Thus the thorium-228 must be entering the bones independently of radium-228, and he asked Eisenbud (52) to comment on this observation. Eisenbud (52) replied that part of this difference, about 10 per cent, may arise from a failure to correct for thorium-228 ingrowth. There is also the possibility that thorium-228 is retained more effectively than is radium-228. Per gram of calcium, the bones contained 40 times as much radium-228 as the milk from the same herd.

Lucas (17) noted that his group had shown that in a human constant calcium intake the thorium-228 generated from radium-228 was retained quantitatively and that the concentrations of both nuclides were equal. Lucas (17) suggested that in the steer bones described by Eisenbud (52) the problem was not one of the correction for ingrowth of thorium-228, but one of variation in diet and that this would give rise to a transient equilibrium rather than a secondary equilibrium.

Adams (30, 34) asked Cullen (51, 53) whether he had any information as to why Brazil nuts contained so much α activity. Cullen (51, 53) replied that Brazil nuts grew in northern Brazil, outside his areas of study. These northern areas of Brazil have not been considered particularly radioactive, and it is not clear how the Brazil nuts acquire such unusually high concentrations of α emitters.

R. LOWRY DOBSON

61. *Radiation and Other Environmental Factors in Human Biology*

MAN AND HIS ENVIRONMENT are a continuum. Indeed, life and the inanimate are parts of the same thing. The universe contains atoms, thermal energy, protoplasm, ionizing radiation, and galaxies. At the molecular level the living and the non-living blend; it is difficult to draw a dividing line.

The surface of the earth is covered and crawling with life: trees, furry animals, worms, brush, insects, grasslands. The soil is teeming with bacteria and fungi; a drop of ocean water under the microscope is an organic spectacle. Man is part of this mass of life. He shares the planet with myriads of living things, large and small, advanced and primitive, all existing experimentally, living in this or that relationship to one another, exchanging energy and highly-ordered molecular configurations.

Man is an animal, reacting with other animals and with himself; with diseases, war, floods, and volcanoes; with food and the lack of it; with chemicals, heat, and radiation—with his total environment. He exists today because he has been successful in the past.

Mankind evolved gradually from prehuman ancestors probably about a million years ago. He has continued to evolve and is evolving today. His contemporary evolution deserves particular attention because modern man is tinkering with his environment and with himself to an extent that he has not done before. And, since evolution results from interplay between changing environment and changeability in living organisms, it behooves man, if he has any real interest in his destiny as a species, to look carefully at his present biological situation, to study and understand it.

This address was given at the Symposium banquet.

R. LOWRY DOBSON, M.D., Ph.D., is chief medical officer, Radiation and Isotopes Section, World Health Organization, Geneva, Switzerland.

Like his other animal friends, man is an elaborate complex of chemical reactions taking place in an intricate spatial system. His body is a vast array of reaction vessels, supply channels, communication networks, and facilities for data collection, storage, and analysis. The reactions are his mechanism. With their energetics they are the biochemical and biophysical basis of his life process—of metabolism, emotions, intellect, and human aspirations.

Data-handling, particularly storage, retrieval, and analysis, is the biological specialty of *Homo sapiens*. Otherwise, man resembles other animals in the most charming ways. His primitive, slow-moving body with its four appendages and fingers and toes is so much like that of a salamander that only a blind man could fail to see the relationship. Livers, pancreases, nerves, thyroids, and kidneys man shares with a vast number of other species. Moreover, just as humans are social animals, working for and depending upon their fellows, the individual is himself a colony, a community of cells living together in collaboration. These cells show great diversity, specialization, and division of labor. Some are the telephone systems, some the gendarmerie; some run the power stations, and others are middlemen. Labor problems are not unknown. Foreign invasions are frequent, and they call forth emergency mechanisms. But the colony is sure to be defeated in the end, or to run down: all men die. Biological continuity is maintained in two ways: one through the germ cells, if the individual leaves progeny; the other through what he does, the cultural mark he leaves on his species.

Man's social structure and behavior, which are as much a part of human biology as his cellular chemistry and tissue organization, took many thousands of years to develop. His biochemical and physiological mechanisms have a history of many millions of years, for man inherited them from prehuman parents; they in turn obtained them from more primitive ancestors, and so on back through geological time to the origin of life.

What is life, of which man is a part? What differentiates it from the inanimate? Of course, we think of such attributes as movement, irritability, metabolism, reproduction. But, in terms of first principles, what is it? Does it respect the second law of thermodynamics?

Life is persistent. It is characterized by a high state of order and a remarkable stability—a durability of complex organization. Arising on earth some time probably between one-half and two and one-half billion years ago, it has remained, has grown, has become progressively more highly ordered, and has now reached the stage at which man—an almost incredible phenomenon—looks at himself, his world, his universe, and thinks about such questions.

Living stuff and non-living are made of the same materials, the same chemical elements. They share each other's molecules. But living stuff has a special type of survival. A mountain is different; it has no such persistence. It is born out of geological upheaval and spends the rest of its time being worn down and degraded. Why does not life also decay into thermodynamic equilibrium and death? Surely we do not escape the second law. We do not want to invent new physical laws to explain life unless we have to, yet we must take care of the entropy riddle.

Living stuff escapes thermodynamic degradation by, as Erwin Schrödinger has expressed it, feeding upon negative entropy. It must compensate the inevitable entropy increase associated with its living, and it does this by "continually sucking orderliness from its environment." Man does it by nourishing himself with matter that is in a very highly organized state, complicated organic compounds, which he takes in his food and returns to the environment in a degraded form.

Entropy is a measure of molecular disorder and is always increasing. Life is characterized by molecular order, which it must always maintain. Ionizing radiation impinging upon the organized system of a living cell tends to produce random disturbances in its chemical balance, and therefore runs against the grain of life.

Both environment and living things change. We see changes during a lifetime; on the geologic time scale they are dramatic. Why is life not destroyed by environmental alterations? It is. In vast amounts. But, so long as there is the thread of continuity and enough varied talent in the living system to allow success under new, changed conditions, life will continue. Living organisms are forever experimenting, testing, scanning, generation after generation. But why do we have generations? And what is the biological significance of sexual reproduction, from which we get generations?

Here I should like to divert for a moment, and at this symposium in Houston, Texas, in 1963, refer to another symposium, which was held in Athens some two thousand three hundred and fifty years ago. The participants were a remarkable group and included such persons as Socrates, Phaedrus, and Aristophanes. The proceedings of this earlier symposium were written up by Plato. I should like to quote a bit. Aristophanes is speaking about love and human nature. He proposes the hypothesis that the sexes were not as they are now; they were originally three in number, man, woman, and a combination—an androgynous type. And he says:

". . . the primeval man was round, his back and sides forming a circle; and he had four hands and four feet, one head with two faces, looking opposite ways, set on a round neck and precisely alike; also four ears, two privy members, and the remainder to correspond. He could walk upright as men now do, backwards or forwards as he pleased, and he could also roll over and over at a great pace, turning on his four hands and four feet, eight in all, like tumblers going over and over with their legs in the air; this was when he wanted to run fast. Now the sexes were three, and such as I have described them; because the sun, moon, and earth are three; and the man was originally the child of the sun, the woman of the earth, and the man-woman of the moon, which is made up of sun and earth, and they were all round and moved round and round like their parents."

He goes on to recount how great was their might and strength; they were a threat to the gods, so Zeus devised a plan to "humble their pride and improve their manners." He decided to cut each one in half so that they would have to walk upright on two legs, and he made the reservation that if they continued to be insolent they would be split again and hop about on one leg, and

"as he cut them one after another, he bade Apollo give the face and the half of the neck a turn in order that the man might contemplate the section of himself: he would thus learn a lesson of humility. Apollo was also bidden to heal their wounds and compose their forms. So he gave a turn to the face and pulled the skin from the sides all over that which in our language is called the belly, like the purses which draw in, and he made one mouth at the center, which he fastened in a knot (the same which is called the navel); he also moulded the breast and took out most of the wrinkles, much as a shoemaker might smooth leather upon a last; he left a few, however, in the region of the belly and navel, as a memorial of the primeval state. After the division the two parts of man, each desiring his other half, came together, and throwing their arms about one another, entwined in mutual embraces, longing to grow into one, they were on the point of dying from hunger and self-neglect, because they did not like to do anything apart; and when one of the halves died and the other survived, the survivor sought another mate, man or woman as we call them—being the sections of entire men and women—and clung to that. They were being destroyed, when Zeus in pity of them invented a new plan: he turned the parts of generation round to the front, for this had not been always their position, and they sowed the seed no longer as hitherto like grasshoppers in the ground, but in one another; and after the transposition the male generated in the female in order that by mutual embraces of man and woman they might breed, and the race might continue."

This is a beautiful and poetic description, but as an adequate explanation it will hardly bear the scrutiny of modern scientific inquiry. However, Aristophanes was dealing not only with something of interest in the sociological sense but with something of profound significance biologically. The basic point is that sexual reproduction is a magnificent scheme for producing genetic variability. It does this by gene segregation and recombination.

A genetically homogeneous population of living organisms might easily be exterminated by some environmental change—say, a geologically induced alteration in salt concentration or temperature or what not—which proved too severe to be compatible with the life processes of the organisms. On the other hand, a population having a suitable amount of variation would be expected, on the basis of probability, to be more likely to have some of its members adequately equipped to cope with the new conditions. Most of the original population might die, but a new one could grow up from the survivors.

This, then, in the long time perspective, is the way life survives and develops. The environment mercilessly prunes away individuals unable to make the grade. And the species must be able not only to make up its lost numbers but also to have enough variability to meet new conditions. This is natural selection operating upon genetic variation. Man is subject to it just as is any other organism. And this is what was meant when it was said earlier that life is forever experimenting, testing, scanning, generation after generation.

Genetic variation comes about not only through segregation and recombination, that is, through separating out and making new combinations of existing genes, but through the appearance of *changes* in genes. These mutations may be caused by ionizing radiation, chemical mutagens, heat, and who knows how many other factors we do not understand.

However, too much genetic variation may be as bad as too little. And it might be expected that any species would have an optimum mutation rate for given environmental conditions. We might illustrate this in the following way. In a static environment, a well adjusted species could not afford too high a mutation rate because, on the basis of probability, a random mutation in an evolved animal would be overwhelmingly more likely to be detrimental than beneficial; while, on the other hand, if the environment suffers a change, members of the previously adjusted species would now find themselves unable to keep up and would face biological failure unless they could adjust to the change. One is tempted by the idea that an optimum mutation rate for a species would be a function of the rate of change of the environment.

There is reason to believe that rates of evolution have been accelerated during times of unusual changes or increased rates of change in the environment, such as occurred in certain periods of geological unrest. At such times, while many animal forms disappeared, others seem to have made their appearance.

Now, if we might be allowed to speculate a bit, we might suggest the possibility that rates of evolution at certain times were accelerated, at least in part, by increased mutation rates associated with elevations in radiation exposure. Let us look at two possibilities; we can reject them if they are unsatisfactory.

Our distant ancestors had many adventures millions of years ago. They were knocked to and fro by the changing environment to which they were ever having to adjust. And this uncertainty in their fate, combined with genetic variation and natural selection, molded them and moved them through their long, never ending evolutionary path. Probabilistic phenomena guided them. Radiation was always bombarding them and shaking them up. Other things were, too. They could never rest; they were nervous and seeking; otherwise they would die.

They had two tremendous adventures. Both were slow motion and lasted a very long time; they were separated by millions of years. One was when they left the blissful stage of living as single independent cells and moved through simple colonial forms to truly multicellular organisms; the other, when they left the protected thermostatic and cushioned life in mother ocean to venture forth onto the dry land. Imagine what was involved! But, since we are particularly interested at the moment in radiation, let us focus, in the first case, on naturally radioactive potassium-40 and the kidney (an organ of special interest to theoretical biology) and, in the second, on cosmic and terrestrial radiations and their differences in intensity on land as compared to those in the sea.

We keep in mind that radiation produces gene mutations, that gene mutations are raw materials for genetic variability, and that it is out of genetic variation that new living forms are devised and grow, particularly if living conditions are strenuous or unfamiliar. Our hypothesis is that an elevated mutation rate—produced by an increased radiation exposure—in the presence of an intensified selection pressure from a changing environment, could facilitate and accelerate evolution.

This mechanism could have provided assistance in both situations, giving to our ancestors an amplification—a feedback intensification—of evolutionary change during these two biological transitions when it was most needed. This is a rather pretty picture. How might it have been?

Well, cellular metabolism produces chemical waste products (metabolites), which must be got rid of for osmotic and other reasons. An organism—in phylogenetically leaving the one-cell stage, where it could easily rid itself of wastes by simple diffusion into the aqueous environment, and becoming an organized multicellular animal—finds it necessary to have a mechanism for transporting away its chemical refuse, which by this time is coming from large numbers of contiguous cells and would otherwise build up dangerous concentrations and threaten the living chemical machinery. The mechanism that animals did in fact develop is the kidney, simple at first in primitive animals, more sophisticated in later models.

As the kidney pumped away unwanted metabolites, however, it had to pump away some of the animal's extracellular fluid as a vehicle. This had to be replaced. The source of replacement was of course the surrounding sea water, which diffused into the animal to make up the water deficit. But the ocean was saline, and ions went in also. And here is the point: the hydrated potassium ion, being more mobile than those of sodium, magnesium, and calcium (other prominent cations of the Archean sea), entered the organism preferentially. With it went potassium-40, and the animal's genes were exposed, by virtue of this chain of events, to an elevated level of chronic irradiation. This, according to our scheme, increased the rate of mutation and thus aided an increased rate of organic evolution at a time when our ancestor was challenged and restive.

Let us jump ahead now a few million years. An ancestor is either curious (like Columbus) or is being driven by some hostile monarch out of his comfortable ocean home. He explores various possibilities, including dry land, and finally makes the grade. How? For one thing, by taking his ocean environment with him. He must have done this, for each of us today carries about a remnant of our ancestral marine home for his own vast community of cells. Our blood and tissue fluid, and that of other animals—marine, fresh-water, or terrestrial—are remarkably similar in relative ionic composition to the ancient sea whence we ventured in the latter part of the Paleozoic era some 200–300 million years ago. Our kidneys became even more critically important, and we have kept the high concentration of potassium. We are getting today approximately one-sixth of our natural radiation exposure from the potassium-40 in our tissues.

Two other things happened when our ancestor climbed out onto the shore. His radiation dose from cosmic rays increased as he left the shielding of his aquatic environment, and he was also exposed to terrestrial radiation from radioactive materials in the earth's solid crust.

Now that he was on land, he needed a higher genetic variability, for environmental conditions there were much more unpredictable than they had been in the sea. Terrestrial evolution has very likely gone much faster than marine evolution.

But enough of this fanciful ancient history. It has been dwelt upon here simply to stress the dynamic aspects of changing life and to put mankind and human biology into time perspective.

What about present-day radiation exposures and contemporary human evolution? We gain a degree of orientation immediately from knowing that no more than a small fraction of man's natural so-called "spontaneous" mutation rate can be accounted for by natural radiation (United Nations, 1962). What produces the rest of these mutations is not well understood. It is an extremely interesting and fundamental question. Research is needed.

We do know, however, that, just as radiation is mutagenic, so also are many chemical substances. How many are active in this way, we do not know, because so few have been tested. Ethyl alcohol has been shown to cause chromosome breaks, and caffeine is mutagenic in bacteria. While we have always lived in a chemical world, we are now adding new features, synthesizing new molecules that life has not been faced with before, using some as drugs and some as insecticides and putting others into our urban atmospheres as pollutants. Nitrogen mustard, a war gas, was the first recognized chemical mutagen. We have a growing list now; and it may be that the more we look, the more we shall find. This certainly warrants investigation (World Health Organization, 1962).

While we spoke earlier about the importance of mutations in the evolutionary progress of our forebears, we should not leave the subject without reminding ourselves that in our present state the gene pool of human populations is extremely diversified. Any increase in mutation is not likely to add new ones that we have not experienced before. On the other hand, as we saw earlier, any mutation, whether it be "spontaneous," radiation-induced, or a product of chemical mutagenesis, has an overwhelming probability of being detrimental. The statistical basis of this is exemplified by the watch analogy: a fine timepiece is a product of a long series of improvements made to an originally careful design. If one thrusts a screwdriver randomly into its works, the chance of improving it is vanishingly small; the chance of damaging it is very great.

We are in urgent need of sound data—obtained from human beings wherever possible—on the genetic effects of radiation (particularly at relatively low-level exposure), of chemicals, and of other agents. Areas of the world where there are human populations living in the presence of high natural radiation offer possibilities for such research (World Health Organization, 1957).

The World Health Organization has for a number of years been studying such possibilities (World Health Organization, 1959), in India with our Indian colleagues, in Brazil with our Brazilian colleagues, and, more recently—in fact in the earlier part of this year—in Ceylon.

Any effects that might be found would be expected to be very small indeed. The studies should be large scale, long term, and detailed. Meticulous care in research design and data collection is essential to make these investigations worthwhile. They are extremely difficult. Their importance lies in the fact that other opportunities for obtaining the necessary human data do not present themselves.

In summary, then, we see man as a product of millions of years of life's trial, error, and trial again. His history did not begin when he took his present form a million years ago; it is a history of hundreds of millions of years. In a sense, his inheritance goes back beyond the origin of life to the autocatalytic reactions and chemical evolution that preceded it. But man—humble though he may be—is a cosmological phenomenon of a very special sort. In some ways his biological evolution has not reached as high a level as, for example, that of the insects, but man has transcended bounds by evolving one special organ: his brain. It is a physical-chemical-biological mechanism of astounding significance. On it he depends. With it he is at work today rapidly changing his environment physically, chemically, and biologically. One hopes that, on the average, these changes will be for the better. But there is no demonstrable law of nature that says it must be so. Species have arisen, blossomed, and disappeared before.

Clearly, we must look very carefully at the changing circumstances of human life, the external and the internal influences. Radiation exposure is one. There is a host of others. We have become responsible for ourselves as no other species has ever been. More than that—we are now responsible as well for the other species, with whom we share the earth, the solar system, and the universe.

REFERENCES

UNITED NATIONS. 1962. Report of the United Nations Scientific Committee on the Effects of Atomic Radiation. General Assembly Official Records, 17th Sess., Suppl. 16 (A/5216). New York: United Nations.
WORLD HEALTH ORGANIZATION. 1957. Effect of Radiation on Human Heredity. Report of the Study Group. Geneva: World Health Organization.
———. 1959. Effect of Radiation on Human Heredity: Investigations of Areas of High Natural Radiation. First Report of the Expert Committee on Radiation, World Health Organization Tech. Rept. Ser., No. 166. Geneva: World Health Organization.
———. 1962. Radiation Hazards in Perspective. Third Report of the Expert Committee on Radiation, World Health Organization Tech. Rept. Ser., No. 248. Geneva: World Health Organization.

Appendixes

APPENDIX 1

Concluding Motion

Sir ernest marsden moved and the members of the symposium adopted a motion of thanks for the support and sponsorship of the Special Projects Branch of the United States Atomic Energy Commission, the Division of Radiological Health of the United States Public Health Service, and the William Marsh Rice University. Thanks and appreciation were also voted to the organizers, John A. S. Adams and Wayne M. Lowder.

The hope was expressed that a second symposium might be held to continue and extend the fruitful and useful exchanges about the natural radiation environment.

Editor's Postscript

THE OPINION has been expressed earlier in these proceedings that it would be most useful if more interdisciplinary symposia on the general subject of radiation and its sources in the environment could be held at suitable intervals. But there is also a need for meetings on a much smaller scale and on more restricted subjects. At least one such meeting has occurred recently that was directly stimulated by the Rice symposium and in particular by the intercomparison of instrumentation that took place on the first day (see Appendix 3). This meeting will be briefly reported here because of its relevance to many of the problems discussed at the symposium.

The meeting took place at Argonne National Laboratory (ANL), Argonne, Illinois, U.S.A., on February 24–25, 1964, and was devoted to the subject of measuring soil and rock radioactivity. In addition to those from the host laboratory, representatives of working groups at the University of California's Lawrence Radiation Laboratory (UCLRL), Rice University, the University of Illinois, Massachusetts Institute of Technology, and the U.S. Atomic Energy Commission's Health and Safety Laboratory (HASL) attended and made informal presentations. Two main areas of concern were the problem of detailed intercalibration of the various laboratory systems for the analysis of rock and soil radioactivity and the question of the effect of radon-222 migration *in situ* and its loss or buildup in samples on the inferred radium-226 or uranium-238 contents.

During September, 1963, sets of soil samples collected by HASL in northern New England and by a team from Rice University in North and South Carolina and at Rice were sent to three laboratories (Rice, ANL, and UCLRL) for analysis of their potassium-40 and equivalent uranium-238 and thorium-232 contents (assuming radioactive equilibrium in each decay series). Field measurements were taken at the sampling sites by both Rice and HASL. In general, the laboratory determinations of potassium and thorium soil contents fell within a range of ±10 per cent about the mean. The variation was considerably greater for the uranium determinations. The Rice and HASL field measurements of thorium were in very close agreement, generally indicating a somewhat lower value than did the laboratory soil analyses. This same effect showed up in the HASL inferred potassium and uranium values, the potassium figures generally being 10–20 per cent below the mean of the various laboratory determinations of soil potassium content. For uranium, the effect was considerably larger in some cases, the deficit in the field as compared to the laboratory measurements ranging from 10 to 50 per cent. The explanation for these systematic differences between laboratory and field measurements lies, partly, in the fact that the field measurements refer by definition to *in situ* earth material, while the laboratory determinations are usually in terms of weight of dry soil. A water content of 10–20 per cent by weight in the soil is not unusual and would produce some of the observed effects. The larger observed differences in the uranium figures are also related to variations in the degree of equilibrium of radon-222 with its parents, the *in situ* situation quite probably being different (generally with more enhanced disequilibrium) from that of the soil sample. Finally, there is doubt in many cases that the laboratory sample is fully representative of the system measured *in situ*.

During the discussion of these data it gradually became clear that, if a precise intercalibration of the various laboratory and field techniques was desired, it would have to be by means of truly homogeneous samples with the uranium series in equilibrium. It was decided that a small rock sample would be prepared by Rice University and sent to all the various laboratories in turn. The University of California representative also suggested the possibility of carrying out a field intercalibration over an outcrop of homogeneous bedrock in the Sierras. This possibility will be investigated.

The second half of the meeting was devoted to consideration of the general problem of how radon-222 and its properties affect measurements of natural radiation and of soil and rock radioactivity. Only relatively recently has it been realized that the movement of free radon out of the upper layers of the soil and into the atmosphere generally *reduced* the dose-rate contribution of uranium series at ground level. It also appears that much more consideration is going to have to be given to radon in terms of obtaining reproducible and comparable results in soil sampling procedures. In addition, the relating of laboratory soil analyses for uranium-238 or radium-226 to field spectrometric determinations of the same radioisotopes must take into account the fact that a significant fraction of the radon-222 and its daughters is generally missing from the upper layers of the soil *in situ,* but not necessarily so from soil samples, particularly if they have been sealed for some time.

One interesting result of the radon discussion was the discovery that three different types of analyses had been carried out on the soil at the Rice campus site, providing the following three estimates of the radium-226 contents of the soil:

1. Rice, ANL, UCLRL ..... $(0.68\pm0.08) \times 10^{-12}$ gm/gm soil
2. Illinois .... $(0.46\pm0.08) \times 10^{-12}$ gm/gm soil
3. HASL .... $(0.3 \pm0.1) \times 10^{-12}$ gm/gm soil

The first result is the mean of the laboratory soil analyses carried out at the three laboratories. The second is an estimate of the radium-equivalent of the free radon-222 completely removed from a soil sample. The third is in effect (see Chapter 35) an estimate, obtained by means of field spectrometry, of the radium-equivalent of the radon-222 retained by the soil material. Since the soil sample analyses were presumably carried out under conditions of near-equilibrium, the second and third results should add up to the first if the various assumptions involved in interpreting the measurements are reasonably correct. This is in fact the case—within the experimental uncertainties. These results indicate an emanation coefficient for the Rice soil of 60–70 per cent, a high but not unreasonable value that goes far toward explaining the low uranium series dose rate observed at that location (see Appendix 3). This is one more example of the kind of information that comes out of interdisciplinary and intergroup co-operation in these areas.

In view of the lack of more formal channels of communication, the editors would be interested in hearing informally from research groups interested in participating in future co-operative studies, intercalibration experiments, or conferences in relevant areas. Any such projects would certainly benefit from the more widespread participation that would follow from such contacts.

Intercalibration Experiment

O<small>N</small> A<small>PRIL</small> 10, 1963, the day preceding the opening of the formal sessions of the symposium, a number of the participants gathered to conduct an informal intercalibration of some of the instrumentation used for environmental radiation measurements. Three locations were chosen for the measurements:

1. Rice University campus, Houston, Texas—on a broad grassy area between the Keith-Wiess Geological Laboratories and the stadium.
2. Beach at Galveston, Texas—centered on the sand at least 150 feet from the water and from the salt grass or storm line.
3. Zircon sand pile at the Wah Chang Tin Smelter, Texas City, Texas— measurements made on or near the large sand pile. The location quoted in Table 1 was on the floor of the building approximately 20 feet from the sand.

It was anticipated that location 1 would be a representative area and that locations 2 and 3 would be low- and high-background areas, respectively.

Six laboratories have submitted the results of their measurements for a detailed comparison. In each case, the instruments used and the methods of calibration have been described in papers presented at the symposium. The participating groups and the instruments used are listed below:

1. Argonne National Laboratory, Argonne, Illinois, U.S.A. (ANL)— a muscle-equivalent ionization chamber as described by J. Kastner et al. (chap. 39) and a laboratory γ spectrometer (for soil-sample analysis) as described by P. Gustafson and S. Brar (chap. 31).
2. Edgerton, Germeshausen, and Grier, Inc., Santa Barbara, California, U.S.A. (ARMS)—aerial survey equipment as described by J. E. Hand (chap. 41).
3. Health and Safety Laboratory, U.S. Atomic Energy Commission, New York, N.Y., U.S.A. (HASL)—a high-pressure argon-filled

ionization chamber and γ spectrometer for field use, as described by W. M. Lowder *et al.* (chap. 35).

4. Lawrence Radiation Laboratory, University of California, Berkeley, California, U.S.A. (UCLRL)—a portable scintillation detector and laboratory γ spectrometer (for soil-sample analysis) as described by H. A. Wollenberg and A. R. Smith (chap. 32).

TABLE 1

LOCATION	LABORATORY	INSTRUMENT	DOSE RATES (μR/HR)						
			Cosmic	Natural γ			Fall-out	Total γ	Total
				K	U	Th			
Rice Univ......	ANL	Ion chamber	3.8					9.1	*12.9*
		Spectrometer		1.0	1.8	3.4	4.7	10.9	
	HASL	Ion chamber	3.4					9.4	12.8
		Spectrometer	(1) 3.4	*0.9*	*0.6*	*3.0*	5.6	10.1	13.5
			(2) 3.4	*0.9*	*0.6*	*3.0*	4.8	*9.3*	12.7
	UCLRL	Scintillation det.	3.4					9.6	13.0
		Spectrometer (lab.)		0.9	2.0	3.4	(high)		
	Leeds	Ion chamber	3.4					9.8	13.2
	Rice	Spectrometer		1.3	1.6	3.4			
Galveston......	ANL	Ion chamber	3.8					4.1	*7.9*
		Spectrometer		0.8	1.0	0.9	1.7	4.4	
	HASL	Ion chamber	3.4					*3.4*	6.8
		Spectrometer	(1) 3.4	*1.2*	*0.7*	*0.9*	0.5	3.3	6.7
			(2) 3.4	*1.2*	*0.7*	*0.9*	0.5	3.3	6.7
	UCLRL	Scintillation det.	3.4					3.3	6.7
		Spectrometer (lab.)		1.2	1.6	1.9			
	Leeds	Ion chamber	3.4					*3.3*	6.7
	Rice	Spectrometer		1.2	0.9	1.2			
Texas City, near zircon sand........	HASL	Ion chamber						*38.9*	
	UCLRL	Scintillation det.						*38.6*	
	Leeds	Ion chamber						*38*	

5. University of Leeds, Leeds, U.K.—a high-pressure nitrogen-filled ionization chamber as described by F. W. Spiers *et al.* (chap. 55).
6. Rice University, Houston, Texas, U.S.A.—a portable γ spectrometer as described by J. A. S. Adams and G. E. Fryer (chap. 34).

The results of the various measurements are tabulated in Table 1, with all data reduced to units of μr/hr in air. The outdoor sea-level cosmic-ray intensity was taken to be that quoted by Shamos (chap. 37), in close agreement with that of Burch (Proc. Phys. Soc. A., **67**:421, 1954). The only exception is the value given for the ANL measurements, experimentally determined on Lake Michigan by Kastner (chap. 39). The dose-rate values for the individual components of the natural radiation field were taken from spectrometric measurements of soil content of the various radioisotopes by applying the following conversion factors:

$$1\,\mu r / hr = 0.58 \text{ per cent potassium,}$$

$$= 1.32 \text{ p.p.m. uranium-238 in equilibrium,}$$

$$= 2.78 \text{ p.p.m. thorium-232 in equilibrium.}$$

These values were calculated from the data given in Table 3, chapter 35.

Two sets of figures are given for the HASL spectrometer readings, the first being the results of the total absorption peak analyses, and the second giving the same figures for the natural components, a fallout estimate from the difference between the natural levels and the total spectrometer reading, and the total reading as obtained from the calculation of total energy absorbed by the crystal, considering 0.15–3-mev. events (see chap. 35).

TABLE 2

Altitude (ft.)	γ c.p.s.	Ground Dose Rate* ($\mu r/hr$)	Conversion Factor (c.p.s./$\mu r/hr$)
100	1,500	9.5	158
300	900	9.5	95
500	530	9.5	56

* Inferred from ground measurements cited above.

The values in italic in Table 1 are the dose rates directly inferred from field readings of the instruments by means of calibration factors. The remaining figures were determined indirectly, for example, by subtractive techniques, or by utilizing data from other sources.

The γ readings obtained by the ARMS-II system at various altitudes over the Rice campus location are given in Table 2. The γ count rate data are not directly interpretable in terms of ground dose rate at a point in the absence of ground measurements because of the energy dependence of γ attenuation in air and because of the large size ($> 1,500$ feet in diameter) of the effective infinite ground source area at 500 feet above the ground.

Four samples of the zircon sand from the Texas City location were analyzed at three laboratories, with the results shown in Table 3. It should be noted that the uranium is contained largely in the zircon itself, whereas the thorium is contained largely in a small (< 0.5 per cent) but highly variable impurity of monazite.

In general, the agreement between the various measurements is quite satisfactory, especially considering the somewhat difficult conditions under which the experiment was conducted. Approximately thirty people were in fairly close proximity to the instruments during the measurements, which were necessarily taken with these instruments separated by 10–20 feet. In addition, none of the locations proved to be an entirely homogeneous radiation source. Scintillation detector surveys carried out by UCLRL indicated significant changes in the total γ readings from point to point at all locations, particularly at Galveston beach on a line perpendicular to the shore line where the observed count rate rose from 75 c.p.s. at the water's edge to 200 c.p.s. on the back beach beyond 200 feet from the water. Under these conditions, small differences in total intensity readings between the

TABLE 3

Laboratory	Potassium (per cent)	Uranium (p.p.m.)	Thorium (p.p.m.)
ANL............	0.2 ± 1.0	288 ± 10	575 ± 15
UCLRL			
(1)............	0.6	226	486
(2)............	1	220	498
Rice			
Laboratory......	—	228	679
With field γ spectrometer......	—	226	571

various instruments and perhaps larger differences in the component dose rates inferred from the relatively small soil samples might be expected. Aerial data obtained in such a situation would be extremely difficult to interpret, and the ARMS-II data obtained at Galveston have not been included here. It should be noted that the Texas City zircon sand source produced a strongly non-uniform radiation field, but in this case considerable care was taken to obtain readings with the various instruments at the same location.

While the results of the experiment are not conclusive in any quantitative sense, the data are of considerable interest and certainly show that the results published by a number of widely separated groups are reasonably compatible. Another such experiment, on a larger scale and under more controlled conditions, would certainly be most useful, particularly if more widespread participation could be encouraged.

Properties of the Uranium and
Thorium Series

THE NATURALLY OCCURRING RADIOISOTOPES uranium-238 (99.3 per cent of natural uranium) and thorium-232 (effectively 100 per cent of natural thorium) are both long-lived α emitters and parents of radioactive decay chains that contribute, along with potassium-40, practically all the natural environmental radiation of terrestrial origin. In Tables 1 and 2 are listed, in order, the members of these two radio-active series and their more important properties. The "historical" names listed are still found in the literature, although they have not been used in the symposium papers and generally appear to be going out of usage. The figures given in parentheses in the last column are percentage yields of radiation of the given energy in units of particles (or photons) per 100 disintegrations of the particular isotope. Under conditions of radioactive (secular) equilibrium, this unit may refer alternatively to 100 α disintegrations of uranium-238 or thorium-232, since the activities of all members of the chain are identical. The only exceptions to this rule are polonium-212 and thallium-208 in the thorium series, which are on parallel branches in the chain. For polonium-212 the given percentage yields should be divided by 1.5, and for thallium-208 by 3, to obtain the yields in terms of thorium-232 disintegrations, since roughly one-third of the bismuth-212 decays are α decays leading to thallium-208, and two-thirds are β decays leading to polonium-212. The less abundant members of the other parallel branches in the two series have not been included in the tables, since the decay is almost entirely by the main branch (> 99.9 per cent) in each case. These omissions follow the general rule adopted in the tables of neglecting components with yields of less than 1 per cent.

The data presented in the tables have been derived from several recent references, listed below, which summarize the known properties of these nuclides. The available information is often quite incon-

TABLE 1

URANIUM (RADIUM) SERIES

Isotope	Symbol	Historical Name	Half-life	Radiation	Energy (mev.)
Uranium-238..	$_{92}U^{238}$	Uranium I	4.5×10^9 yr.	α	4.18(77), 4.13(23)
Thorium-234..	$_{90}Th^{234}$	Uranium X_1	24.1 day	β	0.19(65), 0.10(35)
				γ	0.09(15), 0.06(7), 0.03(7)
Protactinium-234.......	$_{91}Pa^{234}$	Uranium X_2	1.18 min.	β	2.31(93), 1.45(6), 0.55(1)
				γ	1.01(2), 0.77(1), 0.04(3)
Uranium-234..	$_{92}U^{234}$	Uranium II	2.50×10^5 yr.	α	4.77(72), 4.72(28)
				γ	0.05(28)
Thorium-230..	$_{90}Th^{230}$	Ionium	8.0×10^4 yr.	α	4.68 (76), 4.62(24)
Radium-226...	$_{88}Ra^{226}$	Radium	1622 yr.	α	4.78(94), 4.59(6)
				γ	0.19(4)
Radon-222....	$_{86}Rn^{222}$	Radon	3.82 day	α	5.48(100)
Polonium-218..	$_{74}Po^{218}$	Radium A	3.05 min.	α	6.00(100)
Lead-214......	$_{82}Pb^{214}$	Radium B	26.8 min.	β	1.03(6), 0.66(40), 0.46(50), 0.40(4)
				γ	0.35(44), 0.29(24), 0.24(11), 0.05(2)
Bismuth-214..	$_{83}Bi^{214}$	Radium C	19.7 min.	β	3.18(15), 2.56(4), 1.79(8), 1.33(33), 1.03(22), 0.74(20)
				γ	2.43(2), 2.20(6), 2.12(1), 1.85(3), 1.76(19), 1.73(2), 1.51(3), 1.42(4), 1.38(7), 1.28(2), 1.24(7), 1.16(2), 1.12(20), 0.94(5), 0.81(2), 0.77(7), 0.61(45)
Polonium-214..	$_{84}Po^{214}$	Radium C'	160×10^{-6} sec.	α	7.68(100)
Lead-210......	$_{82}Pb^{210}$	Radium D	19.4 yr.	β	0.06(17), 0.02(83)
				γ	0.05(4)
Bismuth-210..	$_{83}Bi^{210}$	Radium E	5.0 day	β	1.16(100)
Polonium-210..	$_{84}Po^{210}$	Radium F	138.4 day	α	5.30(100)
Lead-206......	$_{82}Pb^{206}$	Radium G	Stable		

TABLE 2

THORIUM SERIES

Isotope	Symbol	Historical Name	Half-life	Radia-tion	Energy (mev.)
Thorium-232..	$_{90}Th^{232}$	Thorium	1.41×10^{10} yr.	α	4.01(76), 3.95(24)
				γ	0.06(24)
Radium-228...	$_{88}Ra^{228}$	Mesothorium I	6.7 yr.	β	0.05(100)
Actinium-228..	$_{89}Ac^{228}$	Mesothorium II	6.13 hr.	β	2.18(10), 1.85(9), 1.72(7), 1.13(53), 0.64(8), 0.45(13)
				γ	1.64(13), 1.59(12), 1.10, 1.04, 0.97(18), 0.91(25), 0.46(3), 0.41(2), 0.34(11), 0.23, 0.18(3), 0.13(6), 0.11, 0.10, 0.08
Thorium-228..	$_{90}Th^{228}$	Radiothorium	1.91 yr.	α	5.42(72), 5.34(28)
				γ	0.08(2)
Radium-224...	$_{88}Ra^{224}$	Thorium X	3.64 day	α	5.68(95), 5.45(5)
				γ	0.24(5)
Radon-220....	$_{86}Rn^{220}$	Thoron	54.5 sec.	α	6.28(99+)
Polonium-216..	$_{84}Po^{216}$	Thorium A	0.158 sec.	α	6.78(100)
Lead-212......	$_{82}Pb^{212}$	Thorium B	10.64 hr.	β	0.58(14), 0.34(80), 0.16(6)
				γ	0.30(5), 0.24(82), 0.18(1), 0.12(2)
Bismuth-212..	$_{83}Bi^{212}$	Thorium C	60.5 min.	α	6.09(10), 6.04(25)
				β	2.25(56), 1.52(4), 0.74(1), 0.63(2)
				γ	0.04(1), with α 2.20(2), 1.81(1), 1.61(3), 1.34(2), 1.04(2), 0.83(8), 0.73(10), with β
Polonium-212..	$_{84}Po^{212}$	Thorium C'	0.30×10^{-6} sec.	α	8.78(100)
Thallium-208..	$_{81}Ti^{208}$	Thorium C''	3.1 min.	β	2.37(2), 1.79(47), 1.52, 1.25
				γ	2.62(100), 0.86(14), 0.76(2), 0.58(83), 0.51(25), 0.28(9), 0.25(2)
Lead-208......	$_{82}Pb^{208}$	Thorium D	Stable		

sistent, and the figures given in the tables, particularly those for the yields, are rough averages only. No attempt has been made to determine a "best" value based on a critical evaluation of the literature.

Information on the other natural radionuclides, including the actinium and neptunium series and the singly occurring isotopes, can be obtained from several of the references (Belousova and Shtukkenberg, 1961; Lowder and Solon, 1956; United Nations, 1962).

REFERENCES

BELOUSOVA, I. M., and YU. M. SHTUKKENBERG. 1961. Natural Radioactivity, chap. 5. Moscow: State Publishing House of Medical Literature.

COFIELD, R. E. 1959. Radioactivity of Thorium and Feasibility of in Vivo Thorium Measurements. Union Carbide Nuclear Co. (Oak Ridge, Tenn.), Report Y-1280.

DZELEPOW, B. S., N. N. ZHUKOVSKY, S. A. SHESTOPALOVA, and I. F. UCHEVATKIN. 1958. Gamma-ray spectrum of radium in equilibrium with its decay products. Nuc. Phys., **8**:250.

EMERY, G. T., and W. R. KANE. 1960. Gamma-ray intensities in the thorium active deposits. Phys. Rev., **118**:755.

GRIFFIOEN, R. D., and J. O. RASMUSSEN. 1961. Analysis of long-range alpha-emission data. Phys. Rev., **121**:1774.

HULTQVIST, B. 1956. Studies of Naturally Occurring Ionizing Radiation, with Special Reference to Radiation Doses in Swedish Houses of Various Types. Kungl. Svenska Vetenskapsakad. Handl. 6, Ser. 4. Pp. 125.

LOWDER, W. M., and L. R. SOLON. 1956. Background Radiation: A Literature Search. Health and Safety Laboratory (USAEC) Report NYO-4712, pp. 13–18.

SISIGINA, T. I. 1957. Spectral composition of γ radiation of elements of the uranium and thorium series. Izv. Akad. Nauk. SSSR, ser. geofiz., No. 12, p. 1484. (Trans. from the Russian in Bull. Acad. Sci. U.S.S.R., Geophys. Ser., No. 12, p. 65.)

STEHN, J. F. 1960. Table of radioactive nuclides. Nucleonics, **18**:186.

STROMINGER, D., J. M. HOLLANDER, and G. T. SEABORG. 1958. Table of isotopes. Revs. Mod. Phys., Vol. **30**, Part II.

UNITED NATIONS GENERAL ASSEMBLY. 1962. Report of the scientific committee on the effects of atomic radiation. Annex E, p. 218. New York: United Nations.

Abbreviations

bev., billion electron volts $= 10^9$

c., curie(s) or count(s)

c/cm³, counts per cubic centimeter

cc. or cm.³, cubic centimeter(s)

c.f.m., cubic feet per minute

c/gm, curies or counts per gram

c/m/mg, counts per minute per milligram

c/m/pc, counts per minute per picocurie

c/min, counts per minute

c/min/gm, counts per minute per gram

c/sec, counts per second

cm²/gm, square centimeters per gram

cm²/sec, square centimeters per second

c.p.h., counts per hour

c.p.m., counts per minute

cpm/pc, counts per minute per picocurie

c.p.s., counts per second

cwt., hundredweight

d/m/gm, disintegration per minute per gram

d.p.m., disintegration per minute

ev., electron volt(s)

ft., foot (feet)

gev., giga electron volt(s) $= 10^9$ electron volts

gm/cc or gm/cm³, grams per cubic centimeter

gm/cm², grams per square centimeter

gm/gm, grams per gram

gm/l, grams per liter

kev., thousand electron volts

l/m, liters per minute

lb/gal, pounds per gallon

m/day, meters per day

m/yr, meters per year

mc., millicurie(s)

mc?mi², millicuries per square mile

mev., million electron volts

mg/cm², milligrams per square meter

ml., milliliter(s)

m.p.h., miles per hour

mr., milliroentgen(s)

mr/hr, milliroentgens per hour

mr/yr, milliroentgens per year

mrad/yr, millirads per year

mrem/yr, millirems per year

mv., millivolt(s)

mv/sec, millivolts per second

n/cm²-sec, neutrons per square centimeter per second

nc/ft², nanocuries per square

foot $= 10^{-9}$ curies per square foot

pc., picocurie(s) $= 10^{-12}$ curies

pc/gm, picocuries per gram $= 10^{-12}$ curies per gram

pc/kg, picocuries per kilogram $= 10^{-12}$ curies per kilogram

pc/l, picocuries per liter $= 10^{-12}$ curies per liter

pc/m³, picocuries per cubic meter $= 10^{-12}$ curies per cubic meter

pf., picofarad(s) $= 10^{-12}$ farads

pgm/gm, picograms per gram $= 10^{-12}$ grams per gram

p.p.m., parts per million

p.s.i., pounds per square inch

r/hr, roentgens per hour

r/yr, roentgens per year

r.p.m., revolutions per minute

v., volt(s)

v/min, volts per minute

μ, micro(n) $= 10^{-6}$ meters

μc., microcurie(s)

μc/l, microcuries per liter $= 10^{-6}$ curies per liter

μc/m², microcuries per square meter $= 10^{-6}$ curies per square meter

μgm., microgram(s) $= 10^{-6}$ grams

μgm/cm², micrograms per square centimeter $= 10^{-6}$ grams per square centimeter

μr., microroentgen(s) $= 10^{-6}$ roentgens

μr/hr, microroentgens per hour $= 10^{-6}$ roentgens per hour

μrad/hr, microrads per hour $= 10^{-6}$ rads per hour

μsec., microsecond(s) $= 10^{-6}$ seconds

Indexes

Author Index

Subject Index

Abbreviations, 1037, 1038
Accelerators and background count rate, 529
Actinium, 129, 155, 163, 818, 819
Actinium-227, 154, 397, 455, 456, 457; daughters, 398
Actinium-228, 844, 848, 963, 972, 1035; reagent blank, 849; sensitivity, 849
Activities, Antarctica, 875; snow, 878; water, 878
Activity: average, in air, 992; average, in fish, 959; of dolomite, 908; of glacial drift, 908; of granite, 908; gross, in soil, 959; gross, in vegetation, 959; in potable water, 999, 1000; in thorium plant, 961; in uranium mine, 961
Aerial gamma, aircraft position subsystem, 692
Aerial radiological instrumentation, 707, 708
Aerial Radiological Measuring Survey (ARMS), 90, 703, 705–21, 1031; coverage, 706
Aerial radiological surveys: altitude, 708, 709, 723, 739; area of, 709, 720, 723; carbonate rocks, 717; cross calibration, 716; data points, 711, 712; flight lines, 707, 710, 711, 712, 714, 725, 739; gamma spectral data, 719; spacing of flight lines, 709, 723; volcanic rocks, 717
Aerial radiometry, 87–90
Aerial surveying, 90, 910; *see also* Aerial Radiological Surveys
Aerial surveys: altimeter, 697; altitude compensation, 703; calibration, 696, 713, 797; cosmic count rate, 700; count rate, 696; equipment, 1029; evaluation of, 698, 725; lower limit of sensitivity, 701; radiation calibration, 697; radiation level uncertainties, 702; survey plane, 694; survey speed, 695; system uncertainties, 701, 703
Aeroradiation intensity, calculated, 742
Aeroradioactivity, 723–35; areal geology, 727, 728; compilation of data, 709, 710–

12; correlation between areal data, 716, 718; data, 713, 716, 718, 720, 727, 728, 733; Nevada, 717
Aeroradioactivity detection equipment: altitude, 725; area of response, 725; compensation, 725; sensitivity of, 725
Aeroradioactivity surveying, areas surveyed, 723
Aeroradioactivity surveys: chalk, 733; crystalline schist, 728; diabase, 728; gabbro, 728; glauconite, 733; granite intrusives, 728; mafic intrusives, 728; phyllite, 728; quartzite schist, 728; sand, 733; serpentine, 728; shale, 728, 733
Aeroradioactivity units, 712, 713, 717, 718; diabase, 718; geologic structure, 718; granite, 718; limestone, 718; quartzite, 718; shale, 718
Aeroradiometric measurements: of major soils, 922; related to ground dose, 951
Age determinations, absolute, 129
Aggregates, low-radioactivity, 549, 553, 557
Air-borne readings, 926, 936, 937, 938, 939; relationships to hand-borne, 934, 936, 937
Air-borne surveys, results of, 935
Air-dose rates, 911; calculated and observed, 507, 508
Air filtration, 279, 832, 833
Air-monitoring program, 280, 284, 287
Air samplers, Antactica, 875, 880
Air sampling unit, automatic, 673, 674
Alabama: beaches, 92; dunes, 108
Alaska, 332, 333, 334, 340, 341, 342
Allanite, 46, 129, 137, 142, 145; alpha activity of, 138; uranium in, 47, 139
Alpha: gross, in India, 973; homogeneously distributed, 408; net, in lead, 574
Alpha activity: in air particulates, 378; of allanite, 138; of aluminum, 569, 571; of biotite, 135; of bismuth, 569; of bones, 809; of Brazil nuts, 1010; of cadmium, 569; concentration of, 413; of copper, 569, 571; of detector materials, 419; of

1049